Materials Engineering and Sciences Division

Presentations at the 2008 AIChE Annual Meeting

Philadelphia, Pennsylvania
16 - 21 November 2008

Volume 1 of 2

ISBN: 978-1-61567-223-3

Printed from e-media with permission by:

Curran Associates, Inc.
57 Morehouse Lane
Red Hook, NY 12571

Printed by Curran Associates, Inc. (2009)

For permission requests, please contact AIChE
at the address below.

AIChE
3 Park Avenue
New York, NY 10016-5991

Phone: (203) 702-7660
Fax: (203) 775-5177

www.aiche.org

Additional copies of this publication are available from:

Curran Associates, Inc.
57 Morehouse Lane
Red Hook, NY 12571 USA
Phone: 845-758-0400
Fax: 845-758-2634
Email: curran@proceedings.com
Web: www.proceedings.com

TABLE OF CONTENTS

VOLUME 1

VOLUME 2

15a Shaped Films of Ionotropic Hydrogels Fabricated Using Templates of Patterned Paper

Malancha Gupta, Paul J. Bracher, and George M. Whitesides. Department of Chemistry and Chemical Biology, Harvard University, 12 Oxford Street, Cambridge, MA 02138

Ionotropic hydrogels are an important class of biomaterials with applications in the fields of controlled drug delivery, cellular immobilization, and biomimicry. This talk will describe the use of paper templates to fabricate shaped films of ionotropic hydrogels. Solutions of polymers such as alginic acid, carrageenan, and carboxymethyl cellulose form films with defined shapes when brought into contact with patterned templates of paper wetted with aqueous solutions of multivalent cations. This method allows the production of complex 2D and 3D shapes, such as interlocking rings and Möbius strips. The shaped films can be made magnetically responsive by using paramagnetic ions like holmium as the cross-linking ions or by suspending ferrite microparticles in the hydrogels. Heterogeneous films of ionotropic hydrogels can be fabricated through the use of more complex templates. These heterogeneous structures include single films where a pattern of one hydrogel polymer is surrounded by another polymer ("gel-in-gel" structures), multilayered hydrogels, and hydrogels that contain a gradient in the concentration of cross-linking agent.

15b Development of a Hydrolytically Degradable PEG Hydrogel with Tunable Degradability and Solute Release

Silviya L. Petrova and Jennie B. Leach. Chemical and Biochemical Engineering, University of Maryland Baltimore County, 1000 Hilltop Circle, Baltimore, MD 21250

A degradable poly(ethylene glycol) (PEG) hydrogel with tunable physical properties is developed. A 3D scaffold is made by covalently cross-linking 4-arm PEG vinyl sulfone and PEG-dithiol ester. The hydrogel is then characterized in terms of degradation time and diffusion profiles and aimed towards solute delivery applications. This is one of the few degradable PEG-based scaffolds that does not utilize copolymers (e.g. PEG-polylactic acid), rely on UV exposure for polymerization (PEG-diacrylate), or have a non-specific cross-linking chemistry (PEG-amine). The hydrogel is biocompatible, inert, and has tunable mechanical properties. Protein diffusion from hydrogels is greatly influenced by the degradability of the polymer, the mesh size of the network, and the solute size. Therefore, the focus of our characterization is on these three parameters.

The degradability of the hydrogel is controlled by three distinct strategies: by varying the molecular weight of the cross-linker, by varying polymer density, and by using cross-linkers with different numbers of methylene groups between the ester and thiol groups. It should be noted that degradation gradually increases the mesh size and decreases stiffness until the network is highly disrupted and the degradation complete. Thus the change of mesh size is monitored continuously. The change in mechanical properties (i.e., shear modulus) and the solute release are also tested as a function of degradation time. Furthermore, three solutes of different molecular weight are used to test the dependence of solute release on the size of the solute.

The developed degradable hydrogel has a variety of applications in the area of drug delivery and tissue engineering due to its unique properties, tunability, and ability to form 3D matrices under physiological conditions.

15c Synthesis and Characterization of Magnetic Hydrogel Nanocomposites for Hyperthermia Applications

Samantha A. Meenach, J. Zach Hilt, and Kimberly W. Anderson. Chemical & Materials Engineering, University of Kentucky, 177 F. Paul Anderson Tower, Lexington, KY 40506

Hydrogel nanocomposites have been investigated with increasing interest over the last several years due their ability to improve or enhance the properties of conventional hydrogels. These improvements may include increased mechanical strength or the ability for the gels to be remotely-controlled via external stimuli such as light or magnetic fields. These systems can be utilized in a wide variety of biomedical applications such as drug delivery, tissue engineering and hyperthermia treatment for cancer. In this work, the nanocomposites studied involved a temperature-responsive, poly(N-isopropylacrylamide)-based system and a stealth, poly(ethylene glycol)-based systems, both with and without iron oxide magnetic nanoparticles incorporated into the hydrogel matrices. The addition of iron oxide nanoparticles allows for the nanocomposites to be heated upon exposure to an alternating magnetic field. The heating of these magnetic nanocomposites can allow them to be used for hyperthermia-based cancer treatment by damaging cancer cells while leaving normal healthy cells and tissues unharmed due to the higher sensitivity of the cancer cells. This remote-controlled hyperthermia may also lead to new multi-modality cancer treatments with combined treatment options such as chemotherapy and/or radiotherapy. Heating analysis of the gels was completed to show the heating capability of the gels in their dry and swollen states. The nanocomposites were also analyzed for the swelling characteristics and show the potential to also be used for drug delivery. Cytocompatibility analysis performed on the nanocomposites and nanoparticulates to show that they can be safely used in implant applications.

15d Non-Synthetic Polymer Biomodification Using Gold Nanoparticles

Craig D. Buckley[1], Kathleen A. Vermeersch[1], J. T. Westerfield[1], and Jessica O. Winter[2]. (1) Chemical and Biomolecular Engineering, The Ohio State University, Columbus, OH 43210, (2) Department of Chemical and Biomolecular Engineering, The Ohio State University, Columbus, OH 43210

Nanoparticles, due to their unique size-dependent properties, provide an opportunity for modification of many biomaterials. Gold nanoparticles are ideal for many of these applications because of their optical characteristics, ability to be synthesized in a variety of sizes, simple conjugation chemistry, and biocompatibility, especially compared to other nanoparticles such as quantum dots. Applications of gold nanoparticle-biomaterial composites include the creation of materials with increased biocompatibility and with the ability to detect the presence of specific biomolecules.

Whereas polymeric biomaterials display excellent mechanical characteristics and compatibility to native tissue, they do not readily support cell adhesion. Unfortunately, modification of these materials can be difficult. For example, agarose and poly (ethylene glycol) (PEG) hydrogel neural tissue mimetics only weakly support cell growth, and cell adhesion molecules must be added to improve the cell-material interface. Methods to chemically modify agarose and PEG hydrogels have been developed, but these methods tend to be difficult and time consuming. A new technique for modification, using gold nanoparticles embedded within a hydrogel matrix, offers a solution to these problems. The particles serve as attachment points for cell adhesion peptides to facilitate bioconjugation. These methods can be applied to many types of hydrogels with different pore sizes simply by changing the nanoparticle size, as opposed to developing novel synthetic chemistry. Several sizes of gold nanoparticles have been synthesized, entrained in agarose hydrogels, and tested to show that the bulk of particles remain in the gel for a substantial length of time. This method does not affect the gel's chemical structure, so changes in mechanical properties should be minimized, allowing the gel to best match the corresponding native tissue. A cell-binding peptide has successfully been conjugated to gold nanoparticles, demonstrating feasibility. The extent of peptide attachment is currently being quantified, and the effect of the peptide on cell growth and adhesion is being studied. Due to its flexibility, this technology is not limited to a single biomaterial, but can be applied to all areas of tissue engineering, providing novel methods of non-synthetic bioconjugation.

In addition to biomodification, these materials offer the opportunity for integrated sensing, due to the well recognized optical properties of gold nanoparticles. Biosensor detection is based on the absorbance shift resulting from surface plasmon resonance (SPR) experienced by aggregated AuNPs. For example, two bound gold nanoparticles experience a SPR-induced absorbance shift as a result of proximity. When the particles are separated, the absorbance returns to its original value. In our proof-of-concept device, particle aggregation is achieved using a cell binding peptide (CGGGRGDSGGGC), whereas cleavage is produced by an enzyme that promotes cell detachment (trypsin), returning particles to their initial unaggregated state. Particles are also modified with tri(ethylene glycol) mono-11-mercaptoundecyl ether, a stabilizing agent that protects the AuNPs from unwanted aggregation. Although our proof-of-concept system examines cell adhesion using the RGD peptide/trypsin protease system, this biosensor could be customized to almost any enzyme-substrate combination. Any substrate with thiol ends (which can be added through cysteine termination) has the ability to bind the AuNPs together, and any substrate specific enzyme can cleave the peptide bond activating the sensor. Thus, analyte sensing can be directly built into the modified hydrogel.

This modification method has numerous advantages. Both the increased biocompatibility and sensing applications of gold nanoparticle-biomaterial composites are improvements over systems based on just hydrogels and polymers or just nanoparticles alone. The combined system provides the hydrogel biomaterials with increased functionality without the requirement of complicated syntheses. In addition, the nanoparticles are provided with a supportive framework. Some of the most promising biosensor models employ aqueous nanoparticles, which are not inherently portable and operate only in the liquid phase. A hydrogel support permits the development of portable devices with potential for gas phase operation. The methods described here are also very flexible as a result of the ability to functionalize the gold nanoparticles with a wide array of biomolecules, providing a composite system with a variety of features.

15e In Vivo Evaluation of Amorphous Nano Tricalcium Phosphate in Flexible Composites and Injectable Bone Cements

Oliver D. Schneider[1], Tobias J. Brunner[1], Stefan Loher[1], Franz E. Weber[2], Brigitte von Rechenberg[3], and Wendelin J. Stark[4]. (1) Department of Chemistry and Applied Biosciences, ETH Zurich, Wolfgang-Pauli-Strasse 10, Zurich, 8093, Switzerland, (2) Department for Cranio-Maxillofacial Surgery, University Hospital Zurich, Frauenklinikstrasse 24, Zurich, 8091, Switzerland, (3) Musculoskeletal Research Unit, Equine Clinic, University of Zurich, Winterthurerstrasse 260, Zurich, 8057, Switzerland, (4) Chemistry and Applied Biosciences, ETH Zurich, Wolfgang-Pauli-Str. 10, ETH Hönggerberg, Zurich, 8093, Switzerland

Bone repair and regeneration of defects with bioresorbable implant materials are in great demand in reconstructive surgery. Many investigated biomaterials consist of calcium phosphates which excel in their biocompatibility, bioactivity and osteoconductivity. However current commercially available products are limited for clinical applications because of their brittleness, incompressibility and difficulty to shape.

The present contribution show two *in vivo* studies testing flexible bone implants for their advantageous clinical applications. We have prepared an open structured and highly porous network of amorphous calcium phosphate nanoparticles (ATCP) with biodegradable poly(lactide-co-glycolide) (PLGA) fibers [1]. *In vitro* bioactivity tests in simulated body fluid at 37 °C showed the deposition of a nano-sized hydroxyapatite layer on the fiber surface (Figure) within on single day. *In vivo* bioactivity tests were performed by creating circular defects in the cranium of rabbits and filled with cotton wool like PLGA/ATCP fibers. Another promising application for ATCP is the application as an injectable bone cement (Figure). A comparison to micron-sized α-TCP [2] showed that our amorphous TCP hardened within 7 – 15 minutes. Most promising results (resorption rate and new bone formation, histology, etc.) are given in an *in vivo* study in the femur of sheep.

Figure: Scanning electron microscopy images of cotton wool like appearance of TCP containing poly-(lactide-co-glycolide) fibers having a high *in vitro* bioactivity (left). Injectable bone cement consisting of amorphous TCP / α-TCP (right).

References

[1] O.D. Schneider, S. Loher, T.J. Brunner, L. Uebersax, M. Simonet, R.N. Grass, H.P. Merkle and W.J. Stark, J. Biomed Mater Res B, 2007, 84B(2), 350-62.

[2] T.J. Brunner, M. Bohner, C. Dora, C. Gerber, W.J. Stark, J. Biomed. Mater. Res., 2007, 83B(2), 400-7.

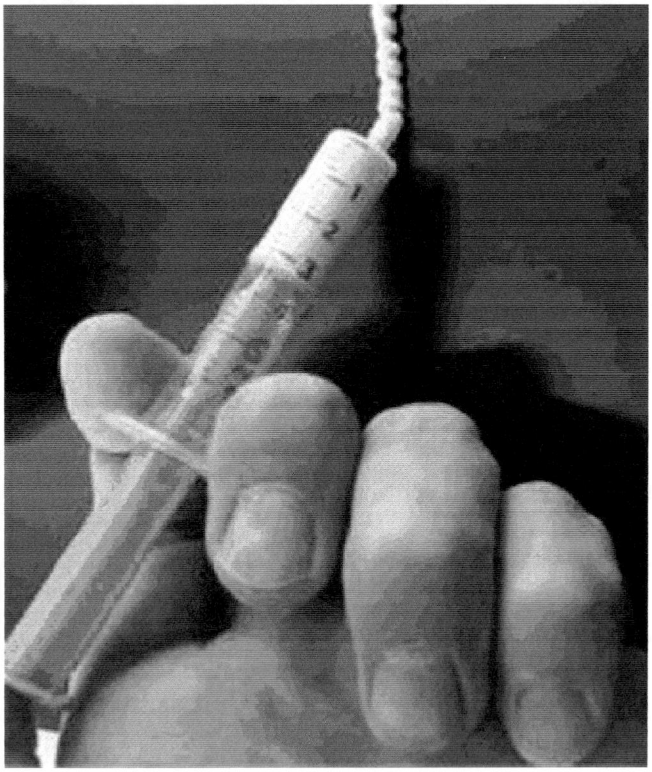

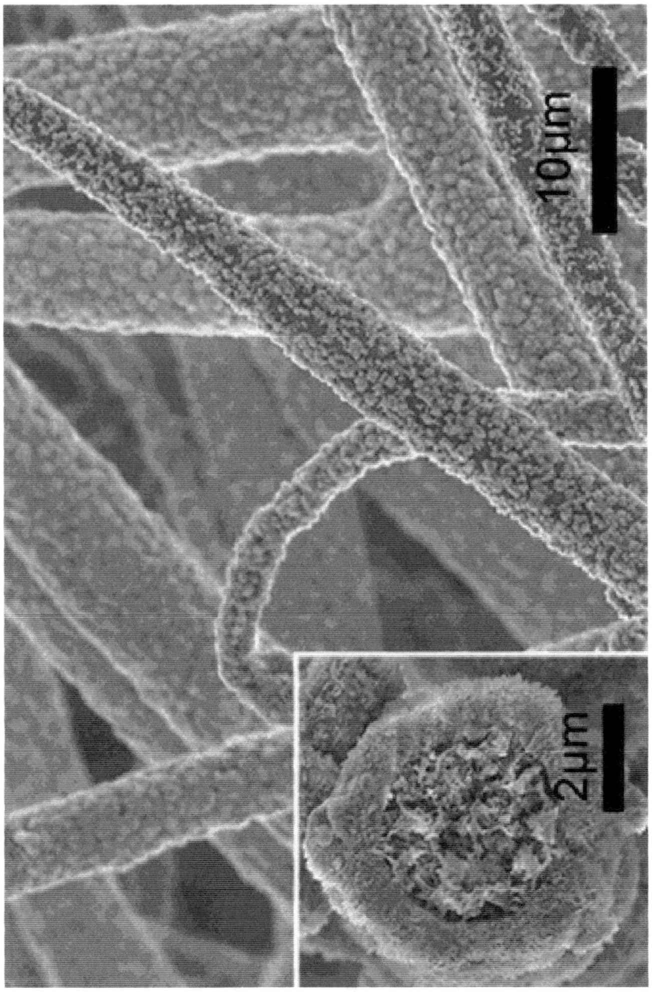
10μm
2μm

15f Nanotextured Titanium for An Improved Bone-Implant Interface

Joshua R. Bush[1], Barada K. Nayak[2], Lakshmi S. Nair[1], Mool C. Gupta[2], and Cato T. Laurencin[1]. (1) Department of Orthopaedic Surgery, University of Virginia, Charlottesville, VA 22902, (2) Charles L. Brown Department of Electrical and Computer Engineering, University of Virginia, Charlottesville, VA 22902

Introduction:

Titanium and its alloys are frequently used as joint replacement materials. However, these materials eventually fail as the implant loosens due to a poor bone-implant interface. As the average life span increases, the need for better and longer lasting joint implants increases as well. The most successful metal implant materials currently have relatively smooth surfaces on the micron size scale. They are held in place with cements that have been largely successful, with most failures not occurring until after ten years [1]. However, to move beyond this limiting time scale researchers have been working on methods to modify the metal surface to enhance integration of the implant directly with surrounding bone.

A varied surface topography of both micron and submicron (>100 nm) sized features have been found to be of the largest benefit to osteoblast response and bone growth. Both size scales seem to be required for optimal cell attachment and differentiation into mature bone forming cells [2]. The ability to control the lateral spacing of these structures has also been shown to be important [3]. Certain textures have also been able to control cell growth, being more preferable to osteoblast adhesion than to other cell types [4].

A flexible, single step process has been developed that results in a surface texture that exhibits micron scale features with superimposed submicron scale ridges. The process uses femtosecond pulses of a laser to create a complex surface topography of micron scale peaks and troughs with submicron scale ridges across the entire surface. By varying the number of pulses and the overlap of the laser as it scans the surface, both the micron scale roughness and lateral spacing of features can be varied for optimization. This surface has been shown to have great promise as a biomaterial with future application in arthroplasty procedures.

Materials and Methods:

Titanium sheets, 50x50x25 mm, were purchased from Goodfellow (USA). Rabbit osteoblasts were isolated from the bone marrow of New Zealand White rabbits and human mesenchymal stem cells were purchased from Cambrex (USA). Promega MTS assay kits were purchased from Fisher Scientific (USA). All other cell culture materials were purchased from the Tissue Culture Core Facility at the University of Virginia.

Laser Texturing of Titanium: The titanium samples were placed on a stage inside a vacuum chamber (base pressure ~1 mTorr) and mounted on a high precision computer controlled X-Y stage. The samples were exposed to laser pulses of 800 nm wavelength and 130 fs pulse duration at a repletion rate of 1 kHz from a regeneratively amplified Spectra Physics Ti-sapphire laser system. The laser beam was focused along the normal onto the sample surface by a 0.5 m focal length coated lens and the laser fluence was adjusted by using a Glan calcite laser polarizer. The X-Y stage translation speed was adjusted to vary the average number of laser pulses impinging on a given surface area of the sample. The samples were then washed in acetone and methanol to remove metal debris and sterilized.

Contact Angle Analysis: The contact angle of deionized water and cell culture media was determined on the textured titanium surface and compared to the contact angle on control titanium samples with no surface modification. The analysis was performed with a Model 200 Standard Goniometer from ramé-hart instrument co. The contact angle was measured by image analysis using DROPimage software. Before contact angle analysis, the samples were sterlized by several different means to ensure consistent behavior across different protocols.

Cell Cultures and Proliferation Assay: Both textured and control samples of titanium were placed in the bottom of the wells in 24 well plates where they covered the majority of available surface area. The titanium was irradiated with UV light on each side for sterilization. Rabbit osteoblasts and human mesenchymal stem cells were seeded on the titanium samples at 10,000 cells/cm2 in 24 well plates and cultured for 3 days. The cell proliferation was analyzed at two time points, 24 hours and 72 hours. At these time points the titanium samples were washed with phosphtate buffered saline and transferred to new well plates. The MTS assay was performed at each time point to determine the number of viable cells on the titanium.

Results and Discussion:

The laser texturing process creates a hierarchical surface on metals. After treatment, the surface is textured with submicron scale ridges that measure approximately 250 nanometers in width superimposed on micron scale pillars and troughs. By changing the laser texturing conditions the pillars can be made with different height to diameter aspect ratios and cover the surface in varying density per unit area. This flexibility is key to creating an optimized surface in the single step process. The samples used in the discussed cell studies were textured with pillars that measured 10 microns in height and 15 microns at the base, tapering to approximately 1 micron at the peak.

The textured titanium surfaces remained very hydrophilic with no special treatment during the texturing process or during transfer from texturing chamber to cell culture. While carbon contamination can cause some chemically etched surfaces to become extremely hydrophobic [2], the titanium samples textured by this method remained hydrophilic with no measurable contact angle even after several weeks in normal atmosphere. The samples were sterilized by several methods, including steam autoclaving, UV irradiation, and liquid CO_2 treatment. The same hydrophilic behavior was retained across all sterilization treatment methods.

Rabbit osteoblasts and human mesenchymal stem cells adhered well on the textured surface as evidenced by SEM images and the MTS assay. SEM qualitatively indicated an increase in cell number from day 1 to day 3 over that observed on an untreated control titanium surface. At the later time point, the cells spread well on the micron scale pillars and started forming bridges between neighboring pillars. This is corroborated by the MTS assay which indicated a larger increase in the number of viable cells from day 1 to day 3 on the textured surfaces than control titanium with no laser treatment.

Conclusions:

A novel method to create textured metals with a combination of micron and superimposed submicron scale features has been shown to enhance cell proliferation at early time points on titanium. This one step method of surface texturing has several advantages when compared to chemical treatment methods, including the retention of hydrophilic behavior without special processing and the ability to control lateral spacing without additional lithography. Both osteoblasts and mesenchymal stem cells respond favorably to the hierarchical structure of this unique surface with micron and submicron scale features. The synthesis method is flexible enough to continue optimization of surface features for further improvements in cell attachment and differentiation.

References:

1. Gioe TJ, Novak C, Sinner P, Ma W, Mehle S. Knee arthroplasty in the young patient. Clinical Orthopaedics and Related Research. 2007;464:83-87.

2. Zhao G, Raines AL, Wieland W, Schwartz Z, Boyan BD. Requirement for both micron- and submicron scale structure for synergistic responses of osteoblasts to substrate surface energy and topography. Biomaterials. 2007;28:2821-2839.

3. Khang D, Lu J, Yao C, Haberstroh KM, Webster TJ. The role of nanometer and sub-micron features on vascular and bone cell adhesion on titanium. Biomaterials. 2008;29:970-983.

4. Richert L, Vetrone F, Yi JH, Zalzal SF, Wuest JD, Rosei F, Nanci A. Surface nanopatterning to control cell growth. Advanced Materials. 2008;20:1488-1492.

23a Electrospun Mats Modified by Layer-by-Layer Assembly for Use as Proton Exchange Membranes

J. Nathan Ashcraft and Paula T. Hammond. Chemical Engineering, Massachusetts Institute of Technology, 77 Massachusetts Avenue, 66-525, Cambridge, MA 02139

Recently, we have developed very promising layer-by-layer (LBL) assembled polymer films for use as the proton exchange membrane (PEM) in direct methanol fuel cells (DMFCs). These materials have the highest ionic conductivity values (**35 mS/cm** at 25°C and 98% RH) ever obtained from an LBL system. [1] Additionally, these multilayer systems exhibit low liquid methanol permeability which allow for their incorporation into DMFCs. We have demonstrated that simply coating traditional fuel cell membranes with 3 to 5 bilayers of these LBL films improves the power output of DMFCs by up to 50%.

In an effort to develop PEMs where the LBL materials serve as the only ion conductor, we have utilized electrospun mats as the substrate. We anticipate these composite membranes will be more mechanically robust as compared to the as-assembled LBL films. Because these composite material systems can be greatly modified at the molecular level to alter localized mobility, mechanics, and stability simply by changing the relative compositions of each adsorbed nanolayer of polymer or by altering the nature or composition of the underlying electrospun network, the systems are infinitely tunable and the architectures of the films can be modified across several length scales to achieve highly optimized and readily processable PEMs. Preliminary results indicate these composite PEMs maintain the high ionic conductivity of the original LBL materials.

[1] Argun, A. A.; Ashcraft, J. N.; Hammond, P. T., *Highly Conductive, Methanol Resistant Polyelectrolyte Multilayers.* Advanced Materials **2008**, 20, (8), 1539-1543.

23b Dramatic Changes In Glass Transition Temperature and Physical Aging Via Covalent Attachment or Attractive Interactions Between Polymer and Nanofiller

Perla Rittigstein, Department of Chemical and Biological Engineering, Northwestern University, 2145 Sheridan Road E136, Evanston, IL 60208 and **John M. Torkelson**, Departments of Chemical and Biological Engineering and of Materials Science and Engineering, Northwestern University, 2145 Sheridan Rd, Tech E136, Evanston, IL 60208.

Our group has previously developed simple fluorescence-based methods to characterize how polymer-substrate interfaces, free surfaces (polymer-air interfaces), and nanoscale confinement itself can alter the glass transition temperature (Tg) and physical aging (relaxation toward equilibrium in the non-equilibrium glassy state) of ultrathin polymeric films from their bulk response. Here, we extend these studies to cases in which there is either covalent attachment between polymer and nanofiller or attractive, hydrogen-bonding interactions between polymer and nanofiller. Using a 'grafting to' approach involving simple condensation-type reactions, we produce 99/1 wt% poly(methyl methacrylate) (PMMA)/carbon nanotube (CNT) nanocomposites in which there is effective covalent bonding between PMMA and the CNT surface. Depending on the size and flexibility of the unit employed in attaching the ester side group of PMMA to the CNT, the PMMA Tg can be enhanced by 8-32 K relative to neat PMMA. Accompanying these increases in Tg are major reductions in physical aging. Based on short-term physical aging monitored via fluorescence, we find that the physical aging (defined on a rate of change relative to logarithmic rather than linear time) can be reduced by up to an order of magnitude or more by the covalent attachment of polymer to the CNT. This dramatic reduction in physical aging rate is further supported by long-term physical aging studies conducted via enthalpy relaxation measurements. For example, in the 99/1 wt% PMMA-CNT sample with the longest, most flexible linking unit between the polymer and CNT and thereby exhibiting the least effect on Tg and physical aging rate, we observed the same amount of enthalpy relaxation in the nanocomposite after 100 days of physical aging at room temperature as we observe after only 4 days of physical aging at room temperature with neat PMMA. This suggests that a potentially important application of polymer nanocomposites can be in producing glassy polymeric materials which exhibit little or no physical aging during in-use application.

We have obtained related Tg and physical aging data in polymer nanocomposites in which polystyrene is covalently attached at low levels to silica nanofiller and in cases in which silica or alumina nanofillers undergo attractive hydrogen bonding interactions with the polymer (PMMA or poly(2-vinyl pyridine). In contrast, polymer nanocomposites that have wetted interfaces with only van der Waals interactions between polymer and nanofiller (e.g. PS/silica nanocomposites with no covalent bonding) exhibit Tg values and physical aging rates that are within error identical to that of the neat polymer. The underlying causes of the effects of nanoconfinement and interactions on Tg and physical aging will be discussed.

23c Influence of Interfacial Reactivity and Viscoelastic Response on Adhesion

E. Jason Robinette, Andres Bujanda, Robert Jensen, and Steven McKnight. Materials Division, Army Research Laboratory, Attn: AMSRD-ARL--WM-MA, Building 4600, Aberdeen Proving Ground, MD 21005

The strength of an adhesive bond strongly depends on the viscoelastic properties of the adhesive and its chemical interaction with the substrate. In this work, we present initial attempts to correlate interfacial bonding and adhesive viscoelastic parameters on the adhesion of urethane-acrylate adhesives to glass substrates. Interfacial bonding was controlled through silane chemistry by treating substrates with various ratios of reactive and non-reactive silanes. Dynamic mechanical analysis (DMA) was utilized to obtain WLF viscoelastic constants for the adhesive. A 90° peel test correlated peel strength and surface reactivity for various strain rates under ambient conditions. It was found that strain rate influenced the effect of surface reactivity on peel strength and the mode of failure (adhesive vs. cohesive). Future work will investigate several peel temperatures and use the viscoelastic constants to generate master curves isolating the effects of strain-temperature and interfacial bonding of the adhesive with the surface.

Developing Composites of Polymer Microgels and Titania Nanoparticles for Photocatalytic Degradation

Cecil A. Coutinho and Vinay K. Gupta*

Department of Chemical & Biomedical Engineering
University of South Florida, Tampa, FL – 33620

ABSTRACT

Microcomposites composed of titanium dioxide nanoparticles embedded within cross-linked, thermally responsive microgels of poly(*N*-isopropylacrylamide) were prepared. These microcomposites showed rapid sedimentation, which is useful for gravity separation of the titania nanoparticles in applications such as environmental remediation. To investigate the degradation kinetics using these microcomposites in aqueous suspensions, methyl orange was employed as a model contaminant. The decline in the methyl orange concentration was monitored using UV-Vis spectroscopy. Degradation of methyl orange was also measured using only nanoparticles TiO_2 (Degussa™ P25) for comparison with the microcomposites. Experiments were performed at different pH conditions that spanned acidic, neutral, and basic conditions to gain insight into the interplay of TiO_2 surface charge, ionization of the polyelctrolyte chains in the microcomposites, and ionization of the methyl orange.

INTRODUCTION

Titanium dioxide is a common and widely studied photocatalyst due to its appealing attributes such as non-toxicity, chemical inertness and high photocatalytic activity[1-3]. The large band gap of TiO_2 (3.2eV) permits it to absorb photons in the UV region, which results in production of electron-hole pairs that participate in redox reactions known to degrade simple organic species[4]. In recent years, there has been increased interest in use of nanosized titania powders due to enhancements in photocatalytic activity[5-7]. Because separation of suspended titania nanoparticles from water has been a major obstacle, use of very fine particles of titania in applications such as waste-water treatment have been limited. In a recent report[8], we demonstrated the synthesis of novel microcomposites in aqueous media comprising of polymer gels on micron length scales that were loaded with Degussa™ P25 TiO_2 nanoparticles. Microcomposites with high mass fractions of titania (50-75wt%) were prepared that showed rapid sedimentation (~minutes), which is a useful characteristic for gravity separations. In this study we build on our previous body of work by investigating the photodegradation of a model organic dye, methyl orange, using the novel polymer-titania microcomposites. Kinetics of the photodegradation are evaluated by monitoring the changes in the methyl orange concentration using UV-Vis spectroscopy. The influence of pH of the solution, which influences the interactions between the poly(acrylic acid) (PAAc) in the polymer microgels, the titania surface and the methyl orange adsorbate is studied. Degradation of methyl orange using freely suspended titania is also conducted for comparison with the microcomposites.

EXPERIMENTAL: PHOTOCATALYTIC STUDY

Degradation studies were done using aqueous solutions containing 5 ppm of methyl orange (MO) and suspended microcomposites or free Degussa™ P25 particles such that the overall concentration of TiO_2 was 50, 100, or 200 ppm. The pH was adjusted using 0.1M HCl or 0.1M NaOH and degradation kinetics are reported here at a pH of 2±0.1 and 6.5±0.2. Photocatalysis was performed under illumination using two commercially available 15W Philips F15T8 black-light fluorescent bulbs

* Author to who correspondence should be addressed: vkgupta@eng.usf.edu

(model 392233) that have spectral energy distribution centered at 352 nm (UVB radiation). The intensity of the radiation reaching the solution surface ($3.5mJ/cm^2$) was detected using a Chromaline UV Minder radiometer (UVM226) connected to a remote probe (UVM226S). Control experiments were performed under UV irradiation without the addition of any catalyst or microcomposites. Negligible decolorization (<1%) of the MO was observed confirming that the degradation of MO predominately occurs by photocatalysis using titania rather than photolysis. Dark adsorption was conducted for at least three hours before irradiation for adsorption of MO onto the TiO_2 surface. Aliquots of 1.5ml of the suspension were collected at regular intervals during the degradation experiments, centrifuged (10000g, 30mins) to completely remove any particles, and the peak absorbance was analyzed using a V-530 UV-Vis spectrophotometer (Jasco, MD).

RESULTS AND DISCUSSION

The microcomposites used in this study consist of porous, cross-linked PNIPAM microspheres with interpenetrating chains of PAAc that allow easy and efficient loading of titania nanoparticles within the IP-microgels[8]. The degradation of MO by both the microcomposite particles and freely suspended titania (Figure 1) is well described by a mono-exponential curve, suggesting that a pseudo-first-order reaction model can be used for describing the kinetic behavior of the photocatalysis. The apparent rate constant was obtained directly using a regression analysis of the experimentally observed decline in the peak height of MO absorbance as a function of time.

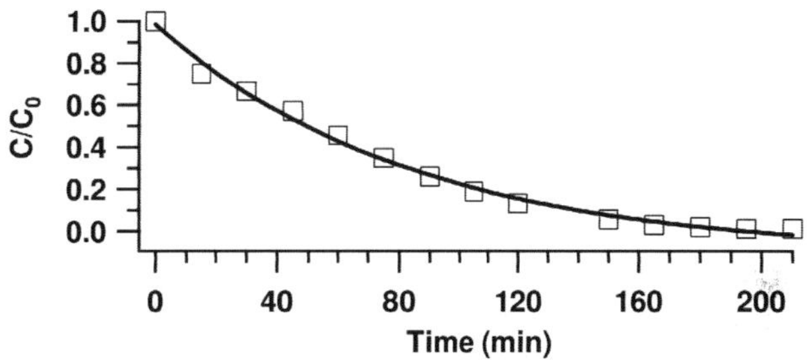

Figure 1: Photocatalytic degradation at pH 2 using the microcomposites (square). Line represents 1st order fit.

Table 1: Degradation rate constants

Concentration (PPM)	Freely Suspended TiO_2		Microcomposite	
	k (min^{-1}) pH 2	k (min^{-1}) pH 6.5	k (min^{-1}) pH 2	k (min^{-1}) pH 6.5
200	0.0130	0.0189	0.0136	0.0046
100	0.0093	0.0154	0.0084	0.0053
50	0.0036	0.0086	0.0033	0.0036

Table 1 shows that the reaction rate constant for the photodegradation of MO by the freely suspended TiO_2 at the acidic pH (~2) is 57% higher than the value obtained when the pH is near neutral conditions (~6.5). This effect has been reported in literature[9, 10] and can be attributed to the increased active sites available for MO degradation. Regarding the pH variation our results correlate well with that of Kansal and co-workers[11], who also reported faster degradation kinetics under neutral (or basic) conditions using freely suspended TiO_2. At a basic pH, both the surface of the titania (ISP ~ 6.5) and the MO are negatively charged, which impacts surface adsorption of MO. However, the presence of large quantities of hydroxyl ions on the particle surface as well as in the reaction medium favors the formation of OH• radicals that are responsible for the photocatalytic degradation of MO. Even though the concentration of OH- ions in bulk solution is reduced at a neutral pH, the electrostatic repulsion between the titania surface, which is now relatively uncharged, and the MO is reduced favoring adsorption of the dye and its oxidation. In an acidic medium, although the adsorption of the MO is improved on the positively charged TiO_2 surface, the reduction in the concentration of the OH- ions leads to a decrease in the rate constant.

Table 1 shows that a number of significant observations can be made when comparing the performance of the freely suspended titania with the microcomposites. At acidic conditions, the photodegradation by the microcomposites and the free titania show similar rate constants and variation with the concentration of titania. In contrast, near a neutral pH the photocatalytic degradation of MO using the microcomposites has much smaller rate constant and remains largely unaffected even when the concentration of titania is increased. A closer look at the structure of the microcomposite can give some insight into these differences. The PAAc chains that are present as interpenetrating chains[8] in the microcomposite contain carboxylic acid groups that are mostly deprotonated above pH 4 and known to functionalize inorganic oxide surfaces[12-14]. Near a neutral pH, photocatalytic degradation of MO using the microcomposites is less because the negatively charged carboxylate groups interact with the oxide surface and can disrupt the adsorption of negatively charged hydroxyl species onto the titania surface due to electrostatic repulsion. Since the number of oxidative species is diminished, the lower photocatalytic oxidation reaction is experimentally manifested as a lower reaction rate constant. However, at pH 2 the PAAc is protonated and the titania surface within the microcomposite remains primarily unhindered. As a result, the photocatalytic performance of both the microcomposites and the freely suspended titania are comparable under acidic conditions.

Increasing the overall TiO_2 concentration in the solution from 50 to 200 ppm requires increasing the weight fraction of microcomposites in the solution. The resulting increase in polymer fraction has little influence at acidic conditions and the primary effect from increased titania surface sites dominates leading to faster photocatalysis. In contrast, at neutral pH conditions, the increase in the PAAc fraction balances the increase in titania and little change in the reaction rate constant can be observed as TiO_2 content changes from 50 ppm to 200 ppm.

CONCLUSIONS

The photocatalytic behavior of novel polymer-titania microcomposites that show rapid sedimentation in aqueous dispersions was studied. Using the photodegradation of a model organic dye, methyl orange, the photocatalytic behavior of the microcomposites was compared with freely suspended titania over a range of pH values. Under acidic conditions, the reaction rates were found to be identical while at a pH of 6.5, the freely suspended titania showed faster rate kinetics. However, at both pH values the microcomposites showed rapid sedimentation with settling velocities nearly 100 times faster than the freely suspended titania, which makes them promising candidates for applications such as wastewater remediation where the use of nanoparticles of TiO_2 is advantageous for photocatalysis but separation of the nanoparticles is difficult and time consuming.

ACKNOWLEDGEMENT

Financial support in the form of a graduate teaching assistantship from a NSF grant on Curriculum Reform (EEC-0530444) to CAC is gratefully acknowledged. The authors would also like to thank Bijith D. Mankidy for help with photocatalytic experimentation.

REFERENCES

1. Earle, M. D., The electrical conductivity of titanium dioxide. *Physical Review* **1942,** 61, 56-62.
2. Fujishima, A.; Rao, T. N.; Tryk, D. A., Titanium dioxide photocatalysis. *Journal of Photochemistry and Photobiology, C: Photochemistry Reviews* **2000,** 1, (1), 1-21.
3. Serpone, N.; Khairutdinov, R. F., Application of nanoparticles in the photocatalytic degradation of water pollutants. *Studies in Surface Science and Catalysis* **1997,** 103, 417-444.

4. Hoffmann, M. R.; Martin, S. T.; Choi, W.; Bahnemann, D. W., Environmental Applications of Semiconductor Photocatalysis. *Chemical Reviews* **1995,** 95, (1), 69-96.
5. Chun, H.; Yizhong, W.; Hongxiao, T.,: Preparation and characterization of surface bond-conjugated TiO2/SiO2 and photocatalysis for azo dyes. *Applied Catalysis, B: Environmental* **2001,** 30, (3,4), 277-285.
6. Lee, S. A.; Choo, K. H.; Lee, C. H.; Lee, H. I.; Hyeon, T.; Choi, W.; Kwon, H. H., Use of Ultrafiltration Membranes for the Separation of TiO2 Photocatalysts in Drinking Water Treatment. *Industrial & Engineering Chemistry Research* **2001,** 40, (7), 1712-1719.
7. Pozzo, R. L.; Giombi, J. L.; Baltanas, M. A.; Cassano, A. E., The performance in a fluidized bed reactor of photocatalysts immobilized onto inert supports. *Catalysis Today* **2000,** 62, (2-3), 175-187.
8. Coutinho, C. A.; Harrinauth, R. K.; Gupta, V. K., Settling characteristics of composites of PNIPAM microgels and TiO2 nanoparticles. *Colloids and Surfaces, A: Physicochemical and Engineering Aspects* **2008,** 318, (1-3), 111-121.
9. Bacsa, R. R.; Kiwi, J., Effect of rutile phase on the photocatalytic properties of nanocrystalline titania during the degradation of p-coumaric acid. *Applied Catalysis, B: Environmental* **1998,** 16, (1), 19-29.
10. Gupta, V. K.; Jain, R.; Mittal, A.; Mathur, M.; Sikarwar, S., Photochemical degradation of the hazardous dye Safranin-T using TiO2 catalyst. *Journal of colloid and interface science* **2007,** 309, (2), 464-9.
11. Kansal, S. K.; Singh, M.; Sud, D., Studies on photodegradation of two commercial dyes in aqueous phase using different photocatalysts. *Journal of hazardous materials* **2007,** 141, (3), 581-590.
12. Zhu, F.; Zhang, J.; Yang, Z.; Guo, Y.; Li, H.; Zhang, Y., The dispersion study of TiO2 nanoparticles surface modified through plasma polymerization. *Physica E: Low-Dimensional Systems & Nanostructures* **2005,** 27, (4), 457-461.
13. Kurihara, M.; Matsuyama, H. Organic polymer-modified colloidal silica and its transparent antisoiling aqueous emulsion coatings. JP Patent 2004067740, 2004.
14. Gupta, V.; Kumar, A.; Coutinho, C.; Mudhivarthi, S. Polymeric microgels for chemical mechanical planarization processing. WO 2008052216, 2008.

23e Highly Magnetic Hydrogel Composites for Magnetic Actuators or Artificial Muscle Applications

Fabian M. Koehler, Roland Fuhrer, Evagelos K. Athanassiou, and Wendelin J. Stark. Chemistry and Applied Biosciences, ETH Zurich, Wolfgang-Pauli-Str. 10, ETH Hönggerberg, Zurich, 8093, Switzerland

Magnetic and electrically conducting polymer composites offer significant advantages over their metallic or ceramic counterparts. Therefore they offer alternatives in a wide range of potential applications such as magnetic actuators, electronic devices, or even artificial muscles.

This contribution reports on the preparation of highly magnetic hydrogel polymer composites. Similar to previous studies on magnetic polymers [1] carbon coated cobalt nanoparticles [2-3] were directly incorporated into the polymerization of 2-hydroxyethyl methacrylate (HEMA). To overcome the characteristic problem of the homogeneous dispersion of the nanoparticles in the polymer, the carbon shell of the particles was covalently functionalized [2] with an organic component similar to the monomer. Later favored the in-situ polymerization of the particles with the monomer and resulted in the formation of a highly magnetic and mechanical stable polymer. Regarding the used mold, different polymers (30 wt% nanoparticles with respect to monomer) could be produced, which could offer an interesting alternative for magnetic actuators or artificial muscles.

Figure 1: Photo of the as prepared homogeneous magnetic hydrogel (left). Upon the application of an external magnetic field the polymer can contract or elongate making it appropriate in application such as for magnetic actuator or even artificial muscles (right).

23f Metal-Support Interactions In High-Temperature Stabilized Nanocomposites

Rahul D. Solunke[1], Tengfei Liu[2], and Götz Veser[2]. (1) Chemical Engineering Department, University of Pittsburgh, 1249 Benedum Hall, Pittsburgh, PA 15261, (2) Department of Chemical Engineering, University of Pittsburgh, Pittsburgh, PA 15261

Metal-oxide nanocomposite materials have gained much attention in recent years as a possible configuration for stabilization of metal nanoparticles. In particular, high-temperature reactive environments, such as in high-temperature catalysis, pose a severe challenge for the utilization of nano-sized and/or nanostructured materials. However, such environments are of great importance, for example, for new and emerging energy technologies, ranging from hydrogen generation to clean combustion.

Chemical looping combustion (CLC) is an emerging technology for clean energy-production from fossil and renewable fuels. In CLC, an oxygen carrier (typically a metal) is first oxidized with air. The hot metal oxide is then reduced in contact with a fuel in a second reactor, thus combusting the fuel. Finally, the reduced metal is transferred back to the oxidizer, closing the materials "loop". CLC is an emerging 'green' combustion technology which allows for flame-less, NOx-free combustion without requiring expensive air separation, and produces sequestration-ready CO2-streams without significant energy penalty. However, the very demanding high-temperature, cyclic redox conditions impose a significant constraint on the selection of suitable oxygen carriers for CLC.

We have previously demonstrated that the embedding of metal nanoparticles into a ceramic matrix can result in unusually active and sinter-resistant nanocomposite materials which combine the high reactivity of metals with the high-temperature (~1000°C) stability of ceramics. Here, we extended our previous work onto a range of metal-oxide combinations to test the flexibility of our synthesis approach. In particular, the role of metal-support interactions in these nanocomposites and their effect on the stability of the resulting materials in multiple high-temperature redox cycles was investigated.

A range of different supports (hexaaluminates, MgO and SiO2) and metals (Ni, Cu and Fe) were synthesized via a reverse-microemulsion template sol-gel approach. The nanocomposites were characterized via TEM, XRD, EDX, BET, and chemisorption, and then analyzed with regard to their stability in reduction-oxidation cycles at temperatures between 700 and 900oC, mimicking the CLC environment. We find a wide range of different behaviors, depending on the specific metal-support combination. The formation of solid solutions results in complete deactivation of some nanocomposites by making the metal unavailable for reduction/oxidation, while for other metal-support combinations the formation of solid solutions appears to be reversible, keeping the metal available for the redox process. Furthermore, the melting point of the metal component has a significant impact on the stability of the resulting nanocomposite.

Synthesis, characterization, and results from the high-temperature redox tests will be discussed in detail in the presentation.

26a Diffusion, Mass Uptake, and Free Volume Behavior of Polymer Thin and Ultra-Thin Films

Richard Lawson, Peter J. Ludovice, and Clifford L. Henderson. School of Chemical & Biomolecular Engineering, Georgia Institute of Technology, 311 Ferst Drive NW, Atlanta, GA 30332-0100

Diffusion of penetrants through polymer thin films is important in a variety of applications ranging from microelectronics to gas separations. Previous studies of diffusion behavior in ultra-thin films have shown a significant decrease in diffusion coefficient as the film thickness decreases. [1-3] These studies only investigate films up to a thickness of around 200 nm; at this thickness, the diffusion coefficient is still several orders of magnitude lower than that of the bulk. Our recent experiments have shown that the decrease in diffusion coefficient occurs over a much larger range of thicknesses than would be expected based on an interfacial effect argument that has been previously used to explain the behavior. Quartz crystal microbalance experiments have been carried out in a variety of polymers to determine the diffusion coefficient and mass uptake behavior for a number of different film thicknesses and film aging times. For example, the diffusion coefficient for water uptake into a poly(methyl methacrylate) film increases by over four orders of magnitude as film thickness is increased approximately 20 nm to approximately 2 microns in thickness where the film behavior becomes bulk like. Over this range, the diffusion coefficient scales with thickness squared. A discussion of the effect of molecular weight and aging time on diffusion coefficient and mass uptake in polymer thin films will also be presented along with a discussion of the physical origin of this scaling relationship of diffusion coefficient. Finally, positron annihilation lifetime spectroscopy studies will be presented and analyzed to elucidate the free volume behavior in such thin film materials and the connections between this free volume and the observed diffusion behavior.

[1] C. L. Soles, R. L. Jones, J. L. Lenhart, V. M. Prabhu, W.L. Wu, E. K. Lin, D. L. Goldfarb, M. Angelopoulos. Proc. SPIE 5039 (2003) 366.

[2] B. D. Vogt, C. L. Soles, H. J. Lee, E. K. Lin, W. L. Wu. Polymer 46 (2005) 1635.

[3] B. D. Vogt, C. L. Soles, H. J. Lee, E. K. Lin, W. L. Wu. Langmuir 20, (2004), 1453.

26b Modeling Surfactant and Drug Transport from P-Hema Hydrogels

Yash Kapoor and Anuj Chauhan. Chemical Engineering, University of Florida, Bldg 723, Room 411, Department of Chemical Engineering, University of Florida, Gainesville, FL 32611

Surfactants are commonly incorporated into hydrogels to increase solute loading and attenuate the drug release rates. Drug transport in these systems is strongly coupled to the surfactant transport. We have investigated the microstructure, and transport of drug and surfactant in poly-hydroxy ethyl methacrylate (p-HEMA) gels loaded with Brij surfactants and an immunosuppressant drug, Cyclosporine A (CyA). This talk will focus on measurements and modeling of the drug and the surfactant transport, along with measurements of swelling, mechanical and optical properties. The SEM imaging clearly shows presence of about 50 nm aggregates inside the gel. The aggregates are distributed heterogeneously as patches that are a few microns in size and the patches are separated by a few microns. Results also show that the transport of CyA through the surfactant-laden gels is controlled by diffusion through the gel but the transport rates are reduced due to partitioning of the drug in the surfactant aggregates that form inside the gel. The transport rates for drug are coupled to the transport rates of surfactant, which are also diffusion controlled. The models developed are validated by experiments conducted with gels of different thicknesses and surfactant loadings. The models for the surfactant and drug release are fitted to the experimental data to obtain partition coefficient between drug concentration inside the micelle and that in the gel. These models can be very helpful in tuning the drug release rates from hydrogels by controlling the surfactant concentration. The results also show that surfactant laden p-HEMA gels exhibit an extended release of CyA and so contact lenses made with this material can be used for extended ocular delivery of CyA which is currently used to treat dry eye disease.

26c Diffusion and Interfacial Behavior of PDMS-Water Systems

Ahmed E. Ismail, Performance Assessment and Decisions Analysis Department, Sandia National Laboratories, 4100 National Parks Hwy, Carlsbad, NM 88220, Gary S. Grest, Surface and Interface Sciences, Sandia National Laboratories, PO Box 5800, MS 1395, Albuquerque, NM 87185-1395, David R. Heine, Science and Technology Division, Corning, Inc., SP-TD-01-1, Corning, NY 14831, Mark J. Stevens, Sandia National Laboratories, MS 1411, Albuquerque, NM 87185-1411, and Mesfin Tsige, Dept. of Physics, Southern Illinois University at Carbondale, Mail Code 4401, Carbondale, IL 62901.

Polydimethylsiloxane (PDMS) is a main constituent of silicone adhesives, which have a wide use as adhesives. Often these adhesives are used as sealants. The interaction between water and PDMS is of fundamental importance. To improve our understanding at the molecular level, we have performed molecular dynamics (MD) simulations of PDMS in the presence of water, with the long-term goal of studying how water molecules effect debonding at the surface. Knowledge of the basic interfacial properties of a multicomponent system, such as the surface tension, contact angle, and diffusion constant, are essential to obtain the proper dynamic behavior in a molecular simulation of adhesion and wetting processes. Explicit-atom simulations have been used to study the structure and dynamics of the PDMS-water interface and of water diffusing through bulk PDMS. In particular, we focus on the effects of temperature and concentration dependence.

26d Surface Infusion of Colorants and Nanoparticles into Processed Thermoplastics

Ronald C. Hedden[1], **Daniel M. Lentz**[1], and Robert A. Pyles[2]. (1) Materials Science & Engineering, Penn State University, University Park, PA 16802, (2) Future Business, Bayer MaterialScience, 100 Bayer Road, Pittsburgh, PA 15205-9741

Colorants, UV stabilizers, antioxidants, and nano-fillers are examples of additives that are traditionally blended with thermoplastic polymer resins prior to conducting melt-state shaping operations such as die forming or injection molding. This presentation concerns a novel approach to plastics modification whereby colorants or other compounds are infused into a molded, cast, or extruded plastic article after melt-state processing using an aqueous plasticizer solution (AURA® Color Infusion Technology, Bayer MaterialScience). Infusion of colorants or other compounds into the plastics is achieved at a temperature well below the melt-processing temperature by immersing the articles in a mixture of water and organic surfactants, which plasticizes a thin surface layer and allows entry of the infusant. The infusant is able to penetrate 30 micrometers or more into the surface of the article, achieving coloration or other surface modifications within a few minutes, usually without degradation of surface gloss. Because the infused compounds are not subjected to melt-state processing conditions, the infusion technique provides a novel approach to introducing additives that ordinarily present problems with thermal degradation, microphase separation, or agglomeration. In addition, because the infusion affects a thin surface layer, it is especially applicable to processes in which the desired outcome is modification of surface characteristics with minimal change to bulk properties.

Earlier work successfully demonstrated rapid infusion of standard organic dyes into commercially important plastics including polycarbonates, thermoplastic polyurethanes, acrylics, and others. Recent work has explored infusion of metal-containing compounds and infusion of metal nanoparticles using a two-step redox process. To infuse metal nanoparticles, the plastic article is first immersed in a solution of a transition metal salt in the plasticizing solvents, which allows the salt to enter the plastic surface. The article is then treated with a second solution of an oxidizing or reducing agent, generally in water. The resulting reaction-diffusion process produces metal or metal oxide particles within the topmost layer of the plastic. Precipitation of the nanoparticles is arrested by the high viscosity of the polymer matrix, producing a stable dispersion of particles (< 100 nm typical diameter). Thus, it is possible to create polymer-metal nanocomposites that could not be processed in the melt state without precipitation of the nanoparticles. We demonstrate the effectiveness of the technique by examining infusion of gold (Au0) nanoparticles into thermoplastic polyurethanes and a polyester-coated acrylic sheet. Manipulation of processing conditions permits infusion of particles having controlled size and good dispersion in these materials, though poor solubility of the metal precursor compounds is found to limit the infusion in other plastics.

26e Investigation of the Dissolution Behavior of Waste Polymers In Biodiesel

Ying Zhang, Surya K. Mallapragada, and Balaji Narasimhan. Chemical & Biological Engineering, Iowa State University, Ames, IA 50011

The dissolution of polymers is an important phenomenon in polymer science and engineering. It has found significant applications in a variety of areas such as drug delivery, microlithography, and membrane science. Polymer dissolution also plays a major role in the treatment of unsorted plastics for recycling. In the present study, investigation of the dissolution behavior of common waste plastics in biodiesel was carried out. The polymers of interest are polystyrene (PS) and low-density polyethylene (LDPE), which are commonly found in waste. The solvent of interest in this work was biodiesel (i.e., composed of methyl esters in this study). Biodiesel is a non-petroleum-based diesel fuel consisting of short chain alkyl (methyl or ethyl) esters, typically made by transesterification of vegetable oils, recycled cooking greases or oils, or animal fats, which can be used (alone, or blended with conventional petroleum-based diesel) in unmodified diesel-engine vehicles. Since biodiesel is produced from domestic, renewable resources, it can act as a clean burning alternative fuel. The objective of our research is to provide experimental data for dissolution of waste plastics in biodiesel, which combined with engine performance testing, will help obtain optimal conditions for use of the polymer solutions for energy applications.

Our experiments demonstrate that PS is completely soluble in methyl esters over the range of molecular weights studied (80,000-980,000 Da). LDPE is also soluble in biodiesel, but only at elevated temperatures and at low polymer molecular weight (below 6500 Da). The dissolution kinetics of PS and LDPE in methyl esters was monitored by gravimetric methods and FTIR imaging. Dissolution rates of the two polymers as a function of polymer molecular weight and temperature were extracted from the experiments. From the data, parameters such as activation energies and scaling indices (of dissolution rate vs. polymer MW) were estimated. Other properties of the PS (LDPE)/biodiesel solution such as shear viscosity and particle size in solution were also characterized. Together, the dissolution and the shear viscosity experiments provided conditions in which these solutions could be used as fuel in engines. To complement the experimental investigations, molecular dynamics simulations were employed to construct models of methyl esters, and parameters such as cohesive energy densities and solubility parameters were calculated. These properties were used to predict thermodynamic polymer/ solvent interaction parameters and based on these calculations, phase diagrams were constructed. The modeling work also provides a predictive toolbox to design biodiesels with different compositions (e.g., mixed solvents, and additivies) that will dissolve other common plastics.

26f Atomistic Simulation of the Diffusion of Small Gas Molecules in Polyisobutylene

Kavitha Chelakara Satyanarayana[1], Jens Abildskov[1], Rafiqul Gani[2], Georgia Tsolou[3], and Vlasis Mavrantzas[4]. (1) Department of Chemical Engineering, Denmark Technical University, Room # 232, Bldg. 227, Søltoft Plads, Kgs. Lyngby, 2800, Denmark, (2) CAPEC, Department of Chemical and Biochemical Engineering, Technical University of Denmark, Søltofts Plads, Building 229, 2800 Kgs. Lyngby, Denmark, (3) Institute of Chemical Engineering and High temperature Chemical Processes (ICE-HT/FORTH), Senior Researcher, Patras, 26504, Greece, (4) Chemical Engineering, University of Patras, Patras, 26504, Greece

Abstract

Computer simulation approaches to explain the markedly low permeability properties of polyisobutelene (PIB) to small-molecule penetrants have led in the past to predictions that, unfortunately, deviate significantly from the experimental values [1-3]. We have recently introduced a new simulation strategy [4] based on a more accurate and detailed united-atom (UA) model for PIB, which has opened up the way towards understanding the molecular origin of the superior barrier properties of this polymer. For example, by borrowing well-equilibrated representative PIB structures obtained with the new UA forcefield and using them as input to explicit, all-atom sorption simulations with the COMPASS force-field has led to solubility predictions for small molecules such as He, Ar, N2 and O2 in PIB that compare extremely favourably with available experimental data. In this work, we will present additional simulation results for the diffusion coefficient of these penetrants in PIB, obtained by detailed, brute-force molecular dynamics simulations in the NPT statistical ensemble at small enough loadings. The simulations cover a wide range of temperatures (from 500K down to room temperature) and, in addition to the values of the diffusion coefficients of the small penetrants, provide extremely useful information for the local packing in PIB and the free volume and its distribution, all as a function of temperature. A comparison against other elastomers with similar chemical structure (like polyethylene PE) will also be presented. How the new molecular-based permeability results can be integrated into grand hierarchical design methodologies for quantifying structure-property relationships in polymer science will also be analyzed and discussed.

References

[1] A.A. Gusev, U.W. Suter, J. Chem. Phys. 1993, 99, 2228.

[2] F. Müller- Plathe, S.C. Rogers, W.F. Vangunsteren, J. Chem. Phys. 1993, 98, 9895..

[3] G.J. van Amerongen, J. Polym. Sci. 1950, 5, 307.

[4] G. Tsolou et al., Macromolecules, 2008, submitted.

46a Thermoplastic Elastomers Containing Crystalline and Glassy Components from Single-Phase Melts

John P. Bishop and Richard A. Register. Chemical Engineering, Princeton University, Princeton, NJ 08544

Introduction

Thermoplastic elastomers (TPEs) are typically symmetric ABA triblock copolymers made up of a majority component rubbery midblock (B) and minority component glassy endblocks (A) [1]. In this system, interblock repulsion between the hard glassy segments and the rubbery midblocks causes microphase separation, creating a physical crosslinking of the rubbery midblock chains, thus producing elastomeric behavior. In these types of TPEs where both the hard and soft blocks are amorphous, it is desirable to have a strong degree of incompatibility between the blocks so microphase separation can occur. However, the resulting microphase separated melts have an adverse effect on the processability of the material as the segregated melts have a high viscosity. If instead the hard segments were crystalline, microphase separation can be driven by crystallization itself from a lower viscosity single-phase melt and therefore strong interblock repulsion is not required. The use of crystalline endblocks in TPEs can also confer properties such as solvent resistance to the material [2]. Indeed, there has been a considerable amount of research on TPEs from single-phase melts containing crystalline endblocks, primarily using hydrogenated high-1,4 polybutadiene as the crystalline component [2-4]. Unfortunately, these crystalline TPEs were plagued by poor tensile strengths and high hysteresis relative to their amorphous counterparts.

Here, we report the synthesis of a TPE formed from a single-phase melt containing both crystalline and glassy hard segments. The idea here was to combine the superior tensile strength of a glassy block with the solvent resistance of a crystalline block, and allow microphase separation to be driven solely by crystallization from a single-phase melt. In order to accomplish this, we synthesized a symmetric pentablock copolymer with the architecture crystalline-glassy-rubbery-glassy-crystalline. With this architecture and appropriate selection of block lengths, crystallization from a single-phase melt will cause a glassy block-rich layer to form around the crystalline lamellae, forming an effective hard block with both crystalline and glassy components.

Experimental

We use living ring-opening metathesis polymerization (ROMP) and subsequent hydrogenation to synthesize our TPEs. This synthesis method provides access to model crystalline, glassy, and rubbery blocks suitable for incorporation into TPEs. These include hydrogenated polynorbornene (hPN), a highly crystalline polymer with an equilibrium melting temperature of 156°C [5]; hydrogenated poly(5-hexylnorbornene) (hPHN), a rubbery amorphous polymer with T_g = -22°C [6]; and hydrogenated polymethyltetracyclododecene (hPMTD), a glassy polymer with T_g = 163°C. ROMP polymerizations were conducted in a nitrogen atmosphere and involved the addition of a weighed amount of a Schrock-type initiator to toluene (previously purified by vacuum distillation from sodium benzophenone ketyl), followed by the addition of monomer charges (previously purified by vacuum distillation from metallic sodium). For the synthesis of triblock copolymers, the polymers were synthesized by sequential addition of monomer charges, and the reaction was terminated with a hundred-fold molar excess of benzaldehyde. For the synthesis of the pentablock, half of the polymer was synthesized by sequential monomer addition, which was followed by coupling the living ends together by terminating with a stochiometric amount of isophthalaldehyde to form the desired symmetric pentablock [7]. Hydrogenations were carried out in a 2 L Parr reactor at 500 psig H_2 and 100°C using $Pd^0/CaCO_3$ as the catalyst.

Gel permeation chromatography (GPC) in THF was used to characterize each polymer's molecular weight distribution. Differential scanning calorimetry (DSC) data were acquired with a Perkin-Elmer DSC-7, calibrated on heating with indium and tin, with a scan rate of 10°C/min. Small-angle X-ray scattering (SAXS) patterns were acquired with an Anton-Paar compact Kratky camera equipped with a hotstage and an M. Braun OED-50M position sensitive detector. Uniaxial tensile stress-strain curves were obtained with either an Instron Model 1122 or 5865 on ASTM D1708 dogbones stamped from molded sheets at room temperature and a crosshead speed of 2 in/min.

Results and Discussion

An hPN-hPMTD-hPHN-hPMTD-hPN pentablock with both glassy and crystalline hard blocks was synthesized with target molecular weights of 7.5-7.5-120-7.5-7.5 kg/mol. The target molecular weights were chosen so that the polymer would form a homogeneous melt in order for microphase separation to be driven by crystallization of the hPN block. For comparison to our pentablock TPE with crystalline and glassy hard segments, we also synthesized a triblock TPE with just crystalline endblocks formed from a single-phase melt (verified by SAXS and the observation that the polymer flows under its own weight when heated above its melting temperature [8]), hPN-hPHN-hPN with a target molecular weight of 10-80-10 kg/mol, and a triblock TPE with only glassy endblocks, hPMTD-hPHN-hPMTD with a target molecular weight of 10-80-10 kg/mol. For this TPE, interblock repulsion must drive the microphase separation necessary for physical crosslinking.

SAXS and DSC experiments were performed to probe the microphase behavior of the pentablock and confirm that microphase separation in the polymer is driven by crystallization. The room temperature SAXS pattern of the pentablock shows a broad first-order peak consistent with a crystallization-induced microstructure. The microstructure persists until ~117°C, which coincides with the crystalline melting peak of the polymer observed by DSC. At 117°C and above, the SAXS pattern is featureless indicating a single-phase melt. The pentablock was also observed to flow under its own weight when heated above its melting temperature, further indicating that the polymer forms a single-phase melt [8]. The crystalline melting peak of the pentablock is very small (~1 J/g compared to 65 J/g for hPN homopolymer [5]), perhaps due to the formation of an hPMTD-rich layer surrounding the crystalline lamellae vitrifying during the early stages of crystallization, thus impeding the growth of the crystalline lamellae. The low level of crystallinity results in poor solvent resistance as the polymer is easily dissolved at room temperature in cyclohexane.

Uniaxial tensile testing of the pentablock shows desirable strain hardening behavior typical of all-amorphous TPEs and, more importantly, does not show any yielding. This is in stark contrast to the hPN-hPHN-hPN TPE, which shows plastic deformation of the crystals at even moderate strains, yielding a polymer with poor elastomeric properties. The pentablock's tensile strength of 7.4 MPa is lower than the 13 MPa of the amorphous hPMTD-hPHN-hPMTD triblock, perhaps due to the low levels of crystallinity in the pentablock discussed previously. We are currently working on obtaining greater levels of crystallinity by reducing the T_g (and delaying vitrification) of the hPMTD-rich layer by reducing the molecular weight of the hPMTD block and by reducing the relative amount of hPMTD in the pentablock (thereby increasing the relative amount of crystalline hPN).

References

1. Legge, N.R. *Rubber Chem. Technol.* **1987**, *60*, G83.

2. Morton, M. *Rubber Chem. Technol.* **1983**, *56*, 1096.

3. Morton, M.; Lee, N.-C.; Terrill, E.R. In *Elastomers and Rubber Elasticity;* Mark J.E., Lal, J., Eds.; American Chemical Society: New York, 1982; p. 101.

4. Seguela, R.; Prud'homme, J. *Polymer* **1989**, *30*, 1446.

5. Lee, L.-B.W.; Register, R.A. *Macromolecules* **2005**, *38*, 1216.

6. Hatjopoulos, J.D.; Register, R.A. *Macromolecules* **2005**, *38*, 10320.

7. Myers, S.B., Ph.D. Thesis, Princeton University, **2008**.

8. Bates, F.S.; Bair, H.E.; Hartney, M.A. *Macromolecules* **1984**, *17*, 1987.

46b Thin Film Morphologies of Rod-Coil Block Copolymer

Manas R. Shah and Venkat Ganesan. Department of Chemical Engineering, The University of Texas at Austin, CPE 3.402, 1 University Station, Code C0400, Austin, TX 78712

The interplay of microphase separation and liquid crystalline ordering in rod-coil block copolymers leads to formation of complex morphologies distinct from that of conventional flexible block copolymer phases. In order to be used for many electronic and semi-conducting applications, rod-coil block copolymers must be patterned into thin films. The final morphology and the nature of orientation of rod units would now depend (in addition to the constituent interactions) on the interactions of the blocks with the confining surfaces. We combine the self-consistent field theory models of rod-coil block copolymers in a thin film framework to understand the effect of confinement on the morphology and the nature of orientation of rod-units. Also for nearly symmetric rod-coil copolymers, we analyze the parallel – perpendicular lamellae transitions using a free energy framework.

46c Symmetric Diblock Copolymers In Nanopores: Self-Consistent Field Calculations

Dong Meng, Chemical & Biological Engineering, Colorado State University, Fort Collins, CO 80521 and Qiang Wang, Chemical Engineering, Coloraldo State University, 905 W Laurel St, Apt 319, Fort Collins, CO 80521.

We have performed parallel self-consistent field (SCF) calculations in continuum to study the self-assembled morphology of symmetric diblock copolymers in nanopores. The SCF equations are solved with high accuracy in real space, without a priori knowledge of the possible morphologies. The surface preference and pore diameter are varied systematically, and various morphologies are found as stable phases. Our calculations reveal the formation mechanisms of these morphologies, which can be understood from their counterparts in thin films confined between two homogeneous surfaces with different preference. We have also compared our SCF calculations with available experiments, Monte Carlo simulations, and a strong-stretching theory.

46d Tunable Nanophase Segregation of Gradient Copolymers: Ordering In Novel Materials with Sinusoidal Composition Profiles across Lamellar Nanodomains

Michelle Mok, Department of Materials Science and Engineering, Northwestern University, Evanston, IL 60208, Christopher J. Ellison, Dept. of Chemical Engineering and Materials Science, University of Minnesota, 421 Washington Ave SE, Minneapolis, MN 55455, Wesley Burghardt, Department of Chemical and Biological Engineering, Northwestern University, 2145 Sheridan Road, Room E136, Evanston, IL 60208, and John M. Torkelson, Chemical and Biological Engineering and Materials Science and Engineering, Northwestern University, 2145 Sheridan Rd, TECH E136, Evanston, IL 60208.

Gradient copolymers are a class of polymers that exhibit a gradual change in composition along the entire chain from mostly A-monomer to mostly B-monomer. Theoretical work has predicted that gradient copolymers organize into sinusoidal composition profiles rather than the step-like profiles seen for block copolymers. Their unique combination of composition profiles and morphology in the segregated state raises questions regarding the extent to which gradient copolymers with composition gradients along the whole copolymer backbone will exhibit the signatures of block copolymer nanophase segregation such as long-range ordering and shear alignment. Here, small-angle x-ray scattering and low amplitude oscillatory shear studies were performed to investigate the impact of gradient design and comonomer choice on their unique ordering. Samples showed a variety of temperature-dependent, non-terminal behaviors consistent with their chain architecture relative to block copolymers, indicating highly tunable nanophase segregation. Scattering results were consistent with the degree of nanophase segregation observed via rheology. In addition, it was demonstrated for the first time that application of high amplitude oscillatory shear induced domain shear alignment in a manner similar to block copolymers, even though gradient copolymers do not possess distinct domain boundaries.

46e Novel Characterization of Critical Micelle Concentrations of Block and Gradient Copolymers In Homopolymer: Effects of Sequence Distribution, Composition and Molecular Weight

Robert W. Sandoval[1], Daniel Williams[2], Christopher Wong[3], Jungki Kim[4], Connie B. Roth[4], and John M. Torkelson[5]. (1) Department of Chemical and Biological Engineering, Northwestern University, Evanston, IL 60208, (2) Chemical & Biological Engineering, Northwestern University, Evanston, IL 60208, (3) Materials Science and Engineering, Northwestern University, 2145 Sheridan Rd, Evanston, IL 60208, (4) Chemical and Biological Engineering, Northwestern University, 2145 Sheridan Rd, Evanston, IL 60208, (5) Chemical and Biological Engineering and Materials Science & Engineering, Northwestern University, 2220 Campus Drive, Cook Hall Room 2036, Evanston, IL 60208-3108

Over the past thirty years, many theoretical studies have examined the effects of block copolymer composition and molecular weight as well as homopolymer molecular weight on the critical micelle concentration (CMC) of block copolymer in homopolymer. However, because of a lack of appropriate experimental methods, very few experimental studies have been performed to test the theoretical results. Here we describe a simple, novel, fluorescence-based method for characterizing the CMC values of styrene-containing block copolymers in homopolymer and extend this approach to a novel class of materials known as gradient copolymers, in which the sequence distribution along the length of the copolymer chain gradually varies from one comonomer to another.

At concentrations below the CMC, the fluorescence spectrum of the styrene units in styrene/methyl methacrylate block or gradient copolymers in blends with poly(methyl methacrylate) exhibit a strong peak centered at 280-285 nm, which is associated with fluorescence from a single excited-state phenyl ring and is called monomer fluorescence. At concentrations above the CMC, a second peak centered near 330 nm becomes evident. This lower energy fluorescence arises from excited-state dimers which form from two phenyl rings arranged in a sandwich-like conformational state and is called excimer fluorescence. This is easily explained because when micelles are formed, there is a high local concentration of styrene units in the micelle core, which accommodates excimer formation and fluorescence. A plot of the ratio of excimer fluorescence intensity to monomer fluorescence intensity as a function of weight percent copolymer increases abruptly at the CMC, allowing for simple, quantitative determination of CMC values. Our studies reveal that the CMCs of block copolymer in homopolymer are approximately an order of magnitude lower than those of gradient copolymers of similar molecular weight and composition. Studies to test particular theoretical pictures of micelle formation of block copolymers in homopolymer are underway, and our investigation is being extended to examine a broad range of styrene-containing copolymer and homopolymer systems.

46f Well Ordered Polymer Melts from Low Molar Mass Surfactants upon Blending with Selectively Associating Additives

Vikram Daga[1], Vijay R. Tirumala[2], Curran Chandler[3], Alvin H. Romang[1], Eric Anderson[3], Eric K. Lin[2], and James J. Watkins[4]. (1) Department of Chemical Engineering, University of Massachusetts-Amherst, 159 Goessman lab, 686 N. Pleasant st, Amherst, MA 01003, (2) Polymers Division, National Institute of Standards and Technology, 100 Bureau Dr Stop 8541, Gaithersburg, MD 20899, (3) Polymer Science and Engineering, University of Massachusetts - Amherst, Amherst, MA 01003, (4) Department of Polymer Science and Engineering, University of Massachusetts, 120 Governors Dr., Amherst, MA 01003

Electronic materials have benefitted from a continual decrease in feature sizes over the last few decades resulting in reduction in size of the electronic components in parallel to higher computational speed and reduction in power consumption. The use of block copolymer melts as nanostructured templates with sub-10 nm domains in applications is limited by their low segregation strength (chi*N) and thus impractical (very low) order-to-disorder transition temperatures. Since increasing molar mass also increases the interdomain spacing, nanostructured periodic templates with sub-10 nm domains spacing can only be obtained by increasing the Flory-Huggins segment-segment interaction parameter, chi. Through an enhancement in the apparent chi, we have shown that disordered block copolymer surfactant melts with a molar mass less than 15 kg/mol can be induced to undergo disorder-to-order transition at practical processing temperatures yielding well-ordered morphologies with features having domains as small as 5 nm when blended with a selectively associating homopolymer. Examples of the block copolymer surfactants include commercially available copolymers such as poly (oxyethylene-oxypropylene-oxyethylene) (Pluronics) and poly(ethylene-oxyethylene) (Brij). Examples of homopolymers included poly (acrylic acid), poly (4-vinyl phenol) and poly (styrene sulfonate). We have studied the effect of homopolymer molar mass on the segregation of a disordered Pluronics resulting in lamellar, cylindrical and spherical morphologies in the ordered phase. The ability to generate well defined morphologies using additives having various chemical and physical functionalities further enable subsequent modification of the templates and thus pathways to a variety of applications.

47a Crystalline Order In Nanoparticulate Thin Films by Continuous Convective Assembly

J. Alex Lee, Chemical Engineering and Materials Science Dept., University of Minnesota, Department of Chemical Engineering and Materials Science, 421 Washington Avenue SE, Minneapolis, MN 55455-0132 and Michael Tsapatsis, Department of Chemical Engineering and Materials Science, University of Minnesota, 421 Washington Ave. SE, Minneapolis, MN 55455-0132.

Ordered films assembled from discrete nanoparticles have been a topic of interest in the recent literature. Convective assembly as a general phenomenon, and extensions to it such as continuous substrate withdrawal and controlled solvent evaporation, seems to be a promising route to cheap and scalable nanoparticulate film fabrication. While there are some recent works that branch out to explore the application of convective assembly to non-spherical particles, we find that even spherical nanoparticles (5 to 25 nm silica) in clean aqueous sols prove difficult to order into colloidal crystal films via convective assemby. That is despite the fact that settling and drying these sols in a non-convective assembly results in monoliths with an apparently high degree of crystallinity. This presentation will discuss some experimental explorations of processing conditions during convective assembly on the quality, i.e. structural order and uniformity, of the final nanoparticulate film.

47b Rapid Fabrication of Metal Organic Framework Thin Films Using Microwave-Induced Thermal Deposition

Yeonshick Yoo and Hae-Kwon Jeong. Chemical Engineering, Texas A&M University, University Dr., TAMU 3122, College Station, TX 77843

Nanoporous metal-organic frameworks (MOFs) have attracted considerable research interest due to their potential applications for hydrogen storage1, selective gas adsorption2,3 catalysis and enantioselective separation4. Metal-organic framework materials (MOF) generally consist of metal-oxygen polyhedera containing divalent (Zn^{2+}, Cu^{2+})5 or trivalent (Al^{3+}, Cr^{3+})6 metal cations interconnected with a variety of organic linker molecules, resulting in tailored nanoporous materials. With a judicious choice of organic linker groups, it is possible to fine-tune size, shape, and chemical functionality of the cavities and the internal surfaces2. This unique structural feature offers unprecedented opportunities in small-molecule separations as well as chiral separations and catalysis.7,8,9

While most of the previous works have focused on the discovery of new MOFs and their applications in powder form, there are only few studies on the preparation of MOF thin films on substrates.10,11 The highly inter-grown thin MOF films will find technologically important applications as gas separation membranes, chemical sensors, and nonlinear optical devices. However, it is yet to be shown that highly inter-grown continuous MOF films can be prepared.

In this work, we will present a novel method, microwave-induced thermal deposition (hereafter, MITD), to rapidly prepare MOF-5 seed layers on substrates under microwaves and to subsequently grow the seed layers into highly inter-grown continuous films.12 MITD is fundamentally different from the conventional microwave-assisted method for film preparation. MITD takes advantages of electron-phonon scattering (Joule heating) to rapidly increase kinetics of heterogeneous nucleation of MOF crystals on porous substrates coated with thin conductive layers. Therefore, our method does not depend on the properties of the substrates as long as a thin conductive layer can be deposited on the substrates. We will talk about the effect of microwave-induced Joule heating on the formation of MOF-5 seed crystal layers on the substrates coated with various conductive materials. The mechanical strength of MOF-5 thin film on the substrate will also be discussed. We will then present the selectively deposited MOF-5 patterns and the multi-layers of MOF-5 and silicalite-1 films on porous substrate. Finally, we will discuss a novel strategy to grow the MOF seed layers into well-intergrown continuous membranes and their performance for gas separations.

1 N. L. Rosi, J. Eckert, M. Eddaoudi et al., SCIENCE 300 (5622), 1127 (2003).

2 M. Eddaoudi, J. Kim, N. Rosi et al., SCIENCE 295 (5554), 469 (2002).

3 A. R. Millward and O. M. Yaghi, Journal of the American Chemical Society 127 (51), 17998 (2005).

4 J. S. Seo, D. Whang, H. Lee et al., Nature 404 (6781), 982 (2000).

5 H. Li, M. Eddaoudi, M. O'Keeffe et al., Nature 402 (6759), 276 (1999).

6 G. Ferey, C. Mellot-Draznieks, C. Serre et al., Science 309 (5743), 2040 (2005).

7 S. Kitagawa, S. Noro, and T. Nakamura, Chemical Communications (7), 701 (2006).

8 D. Maspoch, D. Ruiz-Molina, and J. Veciana, Chemical Society Reviews 36 (5), 770 (2007).

9 J. L. C. Rowsell and O. M. Yaghi, Microporous and Mesoporous Materials 73 (1-2), 3 (2004).

10 S. Hermes, D. Zacher, A. Baunemann et al., Chemistry of Materials 19 (9), 2168 (2007).

11 S. Hermes, F. Schroder, R. Chelmowski et al., Journal of the American Chemical Society 127 (40), 13744 (2005).

12 Y. Yoo and H.-K. Jeong, Chemical Communications (DOI:10.1039/B800061A) (2008).

47c Tuning the Wall Thickness, Pore Size, and Crystallinity of Mesoporous Titania Thin Films

Qingliu Wu, Department of Chemical & Materials Engineering, University Of Kentucky, 177 Anderson Hall, Lexington, KY 40506-0046 and Stephen E. Rankin, Chemical and Materials Engineering, University of Kentucky, 177 Anderson Hall, Lexington, KY 40506.

Abstract: Mesoporous titania films are of intense and growing interest as components for photovoltaic and photoelectrochemical power conversion. Our group has introduced a method to prepare titania films with vertically oriented, 2D hexagonally close packed nanopores. However, for widespread commercial use of these materials, the geometry of the films must be controlled, including the pore size, the wall thickness, and crystallinity of the walls. Mesoporous titania thin film are synthesized by sol-gel chemistry in the presence of tri-block copolymer P123 ($(EO)_{20}$ $(PO)_{70}$ $(EO)_{20}$) template. The films are deposited on glass slides modified with cross-linked P123 to orient the pores normal to the substrate, and then the fresh films are either exposed to the air during aging, or contacted with an identical modified slide to sandwich the films and give complete orientation. The effects of the molar ratio of template P123 to titanium (R) and the micelle swelling agent butanol are investigated. The results indicate that wall thickness and pore size of the mesoporous titania thin films can be tuned without damaging the well ordered mesostructure by varying R and the amount of butanol. Furthermore, when the films are sandwiched, mesoporous titania thin films with channels perpendicular to the substrates can still be obtained even as R decreases and the amount of butanol increases. The only limitation on this seems to be when the system moves into a different region of the phase diagram (e.g. when close-packed globular micelle phases form). To further improve the properties of the mesoporous titania thin films, a post synthesis treatment approach is explored in which the films are heated slowly to a temperature below 400 °C for a prolonged time, followed by a short treatment at high temperature (> 400 °C). This approach is hypothesized to preserve mesopore order while giving larger crystallites, especially when thicker walls are produced by decreasing R, and we will present data to confirm the status of the films during the post-synthesis treatment.

Key words: mesoporous, titania, thin films, crystallization, energy

47d Synthesis of Highly C-Oriented AlPO4-5 Membranes

Enping Hu, Yi Li Winnie Huang, and Zhiping Lai. Division of Chemical and Biomolecular Engineering, Nanyang Technological University, 62 Nanyang Drive, Singapore, Singapore

AlPO4-5 is one of the most important molecular sieves. In powder form it has been used as shape-selective catalysts for alkylation of many important hydrocarbons. Its unique large unidimensional and straight channels have been used as hosts for supramolecular organization. The ability of controlling molecules in the molecular level has leaded to novel applications such as microlaser, which is achieved by aligning dye molecules within the straight channels. However, the small crystal size has limited their practical use. The length scale of alignment can be greatly extended if a compact thin film consisting of well oriented AlPO4-5 single crystals or grains can be formed. Such a thin film will also greatly extend the application of AlPO4-5 into many novel applications such as separations, membrane reactors and chemical sensors, etc.

In this presentation we will demonstrate a modified seeded growth technique to make highly c-oriented AlPO4-5 film. This modified seeded growth technique has been used to prepare the first functional b-oriented MFI membrane. The basic idea of this technique is to convert an oriented seed layer into an oriented continuous thin film by secondary growth. Similarly in this work the synthesis starts from a c-oriented AlPO4-5 seed layer followed by secondary growth to convert this seed layer into a continuous AlPO4-5 film. The condition of the seondary growth was controlled so that the orientation of seed layer was maintained untill a continuous and highly c-oriented AlPO4-5 film was formed.

47e Preparation and Magnetic, Microwave Absorption Properties of Rare-Earth Doped Nano-Barium Ferrite Films

Ying Huang, Yuqing Li, Shuhua Qi, and Yongfeng Yang. Applied chemistry, Northwestern Polytechnical University, Xi'an, 710072, China

Abstract

Rare-earth doped Nano-barium ferrite films combining characteristics of nano-materials and thin films have good prospects in magnetic recording and Electromagnetic wave absorption material. We succeed preparing Rare-earth doped Nano-barium films by Sol-gel technique, and discussing magenetic properties and electromagnetic microwave absorption properties of the films prepared by changing the rare-earth doped rate and the rare earth categories (La^{3+},Ce^{3+}). Corresponding to other element doped Barium ferrite, the maximum loss tangent values at the frequency of 9.6 GHz of La doped and Ce doped Barium ferrite films increased about 110% and 196% respectively.

The phase composition and microstructure of the $BaLn_xFe_{12-x}O_{19}$ (Ln included La and Ce, x=0, 0.2…8) films were characterized by X-ray Diffraction Analysis (XRD), Field Emission Scanning Electron Microscope (FESEM) and Atomic Force Microscopy (AFM). The quantitative compositional analysis of the films was measured by X-ray Energy Dispersive Analysis (EDS). The complex permittivity and complex permeability which characterized microwave absorption properties of the films were measured by Microwave Vector Network Analyzer. The magnetic properties were measured by Vibrating Sample Magnetometer (VSM).

Results showed that: (1) according to the non-stoichiometric amount of Fe:Ba ratio, we obtained good electromagnetic properties of the single phase composition, and at Fe:Ba=9, we got the single phase composition of Rare-earth doped Nano-barium ferrite films; (2) the grain size of barium ferrite films doped by La was 40nm-80nm, length of it was 100nm, and the grain size of barium ferrite films doped by Ce was 60nm-100nm, length of it was 150nm; (3) the maximum coercivity (Hc) of La doped Barium ferrite films ($BaLa_xFe_{12-x}O_{19}$) was 366.33 kA/m at x=0.2, and the maximum saturation magnetization (Ms) of them was 49.06 A•m2/kg at x=0.6. In magnetic characteristics research of Ce doped Barium ferrite films($BaCe_xFe_{12-x}O_{19}$), the maximum coercivity (Hc) was 326.93 kA/m at x=0.6, the maximum saturation magnetization (Ms) was 26.70 A•m2/kg at x=0.4; (4) Both La doped and Ce doped Barium ferrite films had good microwave absorption properties in the range of 9.5GHz ~ 10.5GHz.

Keywords: Sol-Gel; Barium ferrite films; Doped; Magnetic properties; Microwave absorption properties

47f Preparation of Rare-Earth Doped Nano-Strontium Ferrite Films

Yuqing Li[1], **Ying Huang**[2], and Yongfeng Yang[2]. (1) Northwestern Polytechnical University, Xi'an, 710072, China, (2) Applied chemistry, Northwestern Polytechnical University, Xi'an, 710072, China

Abstract

Nano-magetic thin films are intensively researched in the field of ultrahigh-density magnetic or magnetooptical recordings and microwave absorption material. Hexagonal ferrite thin films are of great interest as potential media for magnetic recording material and microwave absorbers in the field of aeronautics and astronautics. They possess high magnetic loss tangent, coercive force and saturation magnetization, good mechanical rigidity and chemical stability, and huge uniaxial magnetic anisotropy. The present study is an attempt to demonstrate the feasibility of producing high quality Rare-earth doped Sr-ferrite films by sol-gel method and to explore the relation between crystal quality, magnetic properties and microwave absorption properties. Finally we succeed increasing magnetic properties of Sr-ferrite films by doping La ,Ce or Nd ,and improving the films microwave absorption properties by doping La or Ce.

Ln-doped nano-strontium ferrite films, $SrLn_xFe_{12-x}O_{19}$(Ln included La, Ce and Nd) (x=0,0.2…1.0), were prepared firstly from metal nitrates and citric by the citrate sol-gel method for the precursor, secondly depositing on SiO_2 substrate. The phase composition and microstructure of the $SrLn_xFe_{12-x}O_{19}$ films were characterized by X-ray Diffraction Analysis (XRD), Field Emission Scanning Electron Microscope (FESEM) and Atomic Force Microscopy (AFM). The quantitative compositional analysis of the films was measured by X-ray energy dispersive analysis (EDS). The complex permittivity and complex permeability which characterized microwave absorption properties of the films were measured by Microwave Vector Network Analyzer. The magnetic properties were measured by Vibrating Sample Magnetometer (VSM).In the research the three doped strontium ferrite films were compared and analyzed in the magnetic and microwave absorption properties.

Results showed that: (1) the rare-earth ions completely entered into the strontium ferrite lattice, and the doped element did not change the strontium ferrite appearance. The grain size of rare-earth doped strontium ferrite films was approximately 40nm-80nm, length of it was 100nm; (2) The magnetoplumbite structure of strontium ferrite was proved through testing a concertina form of the crystal grain by AFM analyzer; (3) the maximum coercivity (Hc) of $SrLn_xFe_{12-x}O_{19}$ films (Ln were La, Ce and Nd respectively) was 476.59 kA/m, 494.07 kA/m and 505.73 kA/m at x=0.2; (4) the La doped films possessed a wider microwave absorption frequency range with better gross loss angle tangent from 9GHz to 10.5GHz, and the gross loss angle tangent of $SrLa_{0.2}Fe_{11.8}O_{19}$ arrived the maximum value 0.2709; otherwise the Ce doped films had a more excellent microwave absorption property than the La doped in the frequency range of 11.5GHz-12GHz.

Keywords: Nano-Strontium films ; Rare-earth doped; Microwave absorption properties; Magnetic properties

56a Past and Future of Modeling and Simulation for Plasma Processing

Demetre J. Economou, University of Houston, 4800 Calhoun Rd. Chem. Eng. Dept. S239, Houston, TX 77204

PAST AND FUTURE OF MODELING AND SIMULATION FOR PLASMA PROCESSING

A pervasive feature of plasma processing is the great disparity of length and time scales that must be treated. Length scales range from the reactor (10s of cm), to the sheath (100s of microns), to the surface features (nm), to the lattice spacing (Angstroms). Time scales range from events happening on the surface right after ion bombardment (ps), to the plasma frequency (ns), to the gas residence time (10s of ms), to surface profile evolution (min). Judicious approximations must be made to "separate" these disparate processes. At the same time, intimate "two-way" coupling among some of these processes can make the problem extremely complicated.

Over the past 25 years, plasma reactor modeling and simulation has evolved from low-dimensional (0-D and 1-D) models with almost no chemistry, to full 3-D simulations including electromagnetics and fairly realistic chemistries. Similarly, feature profile evolution simulations have progressed from the rather primitive "string model" to full 3-D Monte Carlo-based simulations. At the same time, molecular simulations (primarily Molecular Dynamics) have provided a wealth of information on surface processes influenced by energetic ion bombardment.

This paper will review the progress made in modeling and simulation of plasma processes over the years and discuss the challenges that are facing the field moving forward.

56b Atomistic Simulations of Feature Scale Etch Profile Evolution

David B. Graves, UC Berkeley, Department of Chemical Engineering, Berkeley, CA 94720 and Joseph J. Vegh, Chemical Engineering, UC Berkeley, Berkeley, CA 94720.

Plasma etch is currently used for virtually all critical pattern transfer steps in semiconductor device manufacturing. The advantages of plasma-etch subtractive processing to form patterns are considerable, not the least of which is the existence of decades of expertise and knowledge of plasma etch tool design and control. However, the relentless shrinking of device critical dimensions (CD) and the associated stringent demands on reducing CD variability prompts the question of ultimate etch limits: how small a feature can be created by plasma etch? Not only is etch itself becoming more challenging, but the etch step is now used to minimize lithographic limitations. Further, novel etch masks at nanometer length scales are emerging, including di-block and tri-block co-polymer self assembled structures and various nano-imprint methods. The limitations and characteristics of plasma etch with these masks has hardly begun to be explored and much remains to be learned about controlling nano-structural features during etching with these novel masks.

Our strategy for studying the science underlying nano-scale plasma etch is to develop and apply atomistic simulation methods. In this talk, I will present recent results using molecular dynamics simulations of etching features (holes and trenches) on the order of several nanometers, both with and without explicit masks. One aspect that is obvious even from the simplest simulations is the role of ion bombardment in creating sidewall 'passivation' layers, even without depositing neutral species. Other important questions that challenge current MD simulations of feature shape evolution include etch mask treatment, slow neutral transport and reaction dynamics within the feature, and feature charging and ion deflection.

56c Overcoming Patterning Challenges at the 32 Nm Nodes

Arpan Mahorowala, IBM, Yorktown Heights, NY 10504

Overcoming Patterning Challenges at the 32 nm Nodes

56d From the Sawin Lab to NASA: Uses of Very Long Vacuum Tubes

Brett A. Cruden, NASA Ames University Affiliated Research Center, NASA Ames Research Center, Mail Stop 223-2, Moffett Field, CA 94035

Professor Sawin and his group have pioneered plasma-materials interactions for over 25 years. Work in the Sawin lab has included plasma characterization and spectroscopy, among other things. The cornerstone of the Sawin lab since the mid-90s has been a long Vacuum Sample Transfer (VAST) Tube, that was used to mimic cluster processing, allowing treatment and analysis of samples without ambient exposure.

This talk will review current and recent work at NASA Ames in the Electric Arc Shock Tube (EAST) facility being applied to the design of NASA's new crew exploration vehicle, drawing some parallels to experiences and knowledge gained in the Sawin lab. The EAST facility was built over 40 years ago for the Apollo program and has been used since to characterize radiative conditions of shock waves, similar to the conditions encountered during re-entry into the earth and other planetary atmospheres. Shock wave plasmas are produced in the EAST by dumping up to 1.2 MJ of energy at 40 kV via a high pressure (100 psi) arc into a volume of 1.3 L. The arc creates a pressure shock that travels down the tube, ionizing and dissociating a driven gas, and can be measured spectroscopically. While this presents a very different plasma regime from low temperature materials processing plasmas, the basic principles of plasma physics and spectroscopic characterization still apply. Present tests in the EAST facility are focused on tests simulating earth re-entry for the new Orion spacecraft. These tests are of high importance to the new vehicle heat shield design because models predict that radiative heating by the shockwave will constitute a significant fraction (i.e. more than half) of the overall heat load seen by the spacecraft. The current test condition operates at velocities near 10 km/s with air pressures of 0.1-1.0 torr. The spectra obtained are primarily comprised of atomic O and N emission in the near IR and N_2 and CN impurity emissions in the UV/visible. Of particular emphasis in the most recent studies is in the vacuum ultraviolet. Simulations have predicted VUV emission to account for approximately half of the total energy flux. Preliminary measurement and model validation in this spectral region will be presented.

56e Highlights on Plasma-Surface Interactions – Honoring the Distinguished Career of Herbert H. Sawin

Jane P. Chang, Department of Chemical and Biomolecular Engineering, University of California at Los Angeles, Los Angeles, CA 90095

This presentation will highlight the advances in understanding plasma-surface interactions, to honor the many contributions by Professor Herbert Sawin in plasma science, as he retires from Massachusetts Institute of Technology. It will feature tributes from industrial and academic leaders in the field, including the early days of plasma science, the experimental and theoretical advances in the last few decades, and its application beyond silicon processing and information technology.

70a Targeted, Endosomolytic Polymer Delivery Vehicles for siRNA

Danielle S.W. Benoit, Craig L. Duvall, Anthony J. Convertine, Allan S. Hoffman, and Patrick S. Stayton. Bioengineering, University of Washington, 1705 NE Pacific St., Foege Hall, Seattle, WA 98105

siRNA has been used to silence many genes in a variety of cell types with high specificity and is of great interest as a cancer therapeutic. siRNA, similar to other macromolecular drugs, once endocytosed, is predominantly trafficked to the lysosome and degraded enzymatically. In addition, serum proteases are capable of rapid siRNA degradation. Therefore, an optimal therapeutic siRNA delivery approach would combine serum-stability, long circulation times, and tissue-specific targeting, with efficient endosomal uptake and escape prior to lysosomal trafficking. Previously, we developed a carrier that addressed serum-stability and endosomal uptake and escape. This carrier, polymerized via reversible addition fragmentation chain transfer polymerization, combines positively-charged dimethylaminoethyl methacrylate (DMAEMA) as a siRNA-condensing block with a pH-responsive block to enable endosomal escape. To further develop our therapeutic approach, we have utilized a novel chain-extension polymerization technique to include either folate or lactose targeting moieties to (1) increase in vitro uptake by target cells and subsequent knockdown efficiency and (2) reduce in vivo dosing. Therefore, by introducing specific targeting moieties enabling receptor-mediated versus non-specific endocytosis, siRNA effects can be augmented.

70b Novel Graft Copolymers Enhance Cationic Lipid Mediated Delivery of Oligonucleotides

Lavanya Peddada[1], Nicole Harris[2], David Devore[2], and Charles M. Roth[1]. (1) Biomedical Engineering, Rutgers University, 599 Taylor Road, Piscataway, NJ 08854, (2) NJ Center for Biomaterials, Bevier Road, Piscataway, NJ 08854

Antisense oligodeoxynucleotides (ODNs) represent a useful tool in experimental biology and hold therapeutic promise for the treatment of major disease classes such as cancer, autoimmune, and cardiovascular. The successful delivery of ODNs to the intracellular site of action requires the passage of many barriers, such as interaction with extracellular serum proteins, efficient uptake by endocytosis, and escape from endosomes before lysosomal degradation [Sundaram and Roth].

The system that we have used to deliver ODNs utilizes a cationic liposome, dioleoyl-3-trimethylammonium-propane (DOTAP), with the incorporation of a pH-sensitive polymer, poly (propylacrylic acid), PPAA. Previously, we found efficient antisense delivery mediated by the ternary DOTAP/PPAA/ODN system under serum-free media conditions [Lee and Roth]. However, the effectiveness of this system is greatly reduced by the presence of serum for some cell types. We decided to improve the performance of our current system by modifying the backbone of PPAA by grafting hydrophilic polymer poly (ethylene oxide), PEO, or an analogue, (poly (oxyalkylene amines)), Jeffamine.

We found that, in the presence of serum-containing media conditions, the addition of graft copolymer PPAA-g-Jeffamine to the DOTAP/ODN system enhances ODN delivery and gene silencing effect, while the addition of PPAA-g-PEO hinders ODN delivery. Interestingly, we also observe cell type related differences in the ability of parent polymer PPAA and graft copolymer PPAA-g-Jeffamine to mediate an antisense effect. Therefore, we are investigating the mechanisms of endocytosis by which these various DOTAP based delivery systems are taken up by cells and how this influences antisense activity. In summary, our work 1) introduces a novel excipient for ODN delivery and, 2) aims to understand the influence of carrier chemistry and cell specific characteristics on efficiency of ODN delivery and antisense activity.

70c Local Gene Delivery from Ecm Coated Poly (lactide co glycolide) Multiple Channel Bridges after Spinal Cord Injury

Laura De Laporte[1], Anna Yan[2], and Lonnie D. Shea[1]. (1) Chemical and Biological Engineering, Northwestern University, 2145 Sheridan Road, Tech Building E136, Evanston, IL 60208-3120, (2) Chemistry, Northwestern University, 2145 Sheridan Road, Tech Building E136, Evanston, IL 60208-3120

Spinal cord regeneration is limited by multiple barriers, for which gene delivery provides a versatile approach to target the various cellular processes. However, the ability to influence cellular responses is limited by the delivery efficiency. Biomaterial scaffolds have been used to deliver DNA locally in order to maintain an elevated concentration in the microenvironment. An approach successfully used in vitro is surface-mediated delivery, in which DNA is immobilized to the biomaterial surface. Previous studies with encapsulation and release of 800 μg plasmid into a spinal cord lateral hemisection demonstrated localized transgene expression for two weeks, with decreasing levels further away from the injury site. Relative to controlled release strategies, the surface delivery mechanism generally employs DNA complexed with a transfection reagent, which both facilitates immobilization and cellular trafficking. Additionally, this approach has the potential to reduce the quantity of DNA required.

In this report, we investigate delivery of plasmid and lipoplexes in the injured spinal cord by immobilization to a multiple channel bridge. Initial studies characterized binding, release, and transfection in vitro. The bridges were made of poly (lactide co glycolide) (PLG), coated with different extracellular matrix (ECM) components, and DNA was adsorbed to the surface comparing three different methods: i) incubation with the bridge, ii) drying DNA onto ECM coated bridges, and iii) DNA mixed with ECM and subsequent drying onto the bridge. DNA was complexed with the lipid Transfast in a 0.5:1 weight ratio. In vitro experiments indicated that fibronectin coated PLG surfaces enhanced DNA adsorption (80% efficiency) compared with collagen I, laminin, and no ECM. The two drying methods using laminin resulted in similar adsorption efficiencies as fibronectin. Importantly, transgene expression was greatest on fibronectin for all methods, with the incubation technique producing greater expression than the two drying methods. The incubation method also produced a homogeneous distribution of rhodamine-labeled DNA on the surface and had the greatest number of transfected cells. Increasing the amount of DNA incubated over 3 ug did not enhance the levels of transgene expression. Fibronectin coated bridges incubated with DNA resulted in a slightly slower release for adsorbed lipoplexes relative to plasmid (98.7% versus 74.5% after 1 day, and 99.4% and 88.4 % after 21 days).

For in vivo studies, bridges were incubated with 16 μg of DNA for 16-20 hrs and implanted in a rat spinal cord lateral hemisection model. Interestingly, adsorption of lipoplexes resulted in 4 times higher levels of transgene expression at the injury site compared with plasmid adsorption after 1 week. Expression levels 0.5 cm caudal of the injury site were increased 650-fold, whereas segments 0.5 cm rostral and 1 cm caudal had expression increased 170 fold. These results suggest that lipoplexes are more efficient in the spinal cord than plasmid, allowing for transfection of cells adjacent to the implant site. Compared with the encapsulation method of 800 μg plasmid, surface-mediated delivery of lipoplexes enhanced total transfection in the spinal cord 11.5 times. Longer timepoints are currently under investigation to assess whether the lipoplexes allow for a longer duration of transgene expression, exceeding 2 weeks. A system for efficient gene delivery into the spinal cord can be used to identify therapeutic targets and may ultimately enable regenerative therapies.

70d Target Cell Controlled and Spatially Arranged Gene Delivery from Fibrin Hydrogels

Pedro Lei[1], Roshan Padmashali[1], and Stelios T. Andreadis[2]. (1) SUNY at Buffalo, 916 Furnas Hall, Amherst, NY 14260, (2) Chemical and Biological Engineering, SUNY-Buffalo, 908 Furnas Hall, Buffalo, NY 14260-4200

Biomaterial-mediated gene transfer has several advantages including close proximity of vectors and target cells, low cytotoxicity, and preservation of the gene delivery vehicle. Herein, we used fibrin to immobilize DNA/lipofectamine polyplexes and facilitate gene transfer in two different settings: entrapment of DNA in fibrin gels followed by addition of target cells or co-localization of DNA with cells inside the fibrin matrix. We found that co-localization of DNA with target cells yielded the highest transfection efficiency. The optimum level of gene transfer was observed at intermediate DNA concentrations (2-5 μ g). Interestingly, the effect of fibrinogen concentration on transfection efficiency was strongly dependent on the mode of gene transfer. When cells were added to DNA-containing fibrin gels, transfection decreased significantly as the fibrinogen concentration increased. By contrast, cells co-localized with polyplexes were affected by fibrinogen to a lesser extent. Fibrinolytic inhibitors such as e-aminocaproic acid or aprotinin reduced gene transfer, indicating that fibrin degradation is necessary for cells to find the polyplexes and initiate gene transfer. In addition, fibrin-mediated gene delivery significantly decreased the cytotoxic effect of polyplexes. Finally, immobilized DNA polyplexes did not diffuse out of the matrix resulting in precisely localized gene transfer on defined micropatterns. Our study suggested that fibrin gels can be used to contain DNA vectors to achieve localized gene delivery in vivo as well as for design of gene arrays for high throughput gene delivery in vitro.

70e DNA Delivery from Enzymatically Degradable Synthetic Hydrogels to Invading Cells Results In Sustained Transgene Expression

Yuguo Lei and Tatiana Segura. Department of Chemical and Biomolecular Engineering, University of California, Los angeles, 420 Westwood Plaza, Los Angeles, CA 90095

We are interested in engineering artificial extracellular matrixes (aECM) that are capable of delivering nucleic acid-based bioactive signals to promote tissue regeneration when implanted at diseased or injured sites. The general idea is to design a scaffold that allows and supports cellular infiltration while also delivers genes, encoding for bioactive signals, to the infiltrating cells. Here the delivery of DNA/ polyethylene imine (PEI) polyplexes from matrix metalloproteinase (MMP) degradable polyethylene glycol (PEG) hydrogels to the invading cells is demonstrated. Michael addition chemistry of PEG-vinyl sulfone (PEG-VS) with protease degradable bis-cystine consisting peptides was used for crosslinking. To allow cellular attachment, PEG-VS was modified with Arg-Gly-Asp (RGD) prior to gelation. Although the DNA/PEI polyplexes were physically encapsulated into the hydrogel scaffold during hydrogel gelation, the DNA release profiles showed limited diffusion (< 15%) of the polyplexes out of the gel in PBS and mouse cloned mesenchymal stem cell (D1) conditioned media, indicating that the majority of the polyplexes remain entrapped in the gel and are available for internalization by the invading cells. No changes in mechanical properties were observed due to the encapsulation of polyplexes. Cells were seeded inside the hydrogels using two approaches, homogeneously throughout the hydrogel or inside a fibrin cluster. The fibrin cluster cell plating approach was designed to mimic cells infiltrating the hydrogel post implantation in vivo. Low toxicity to cells seeded inside the polyplex containing hydrogels was observed using the LIVE/DEAD assay. The reporter gene expression pattern of invading cells was found to be different from that of cells homogenously suspended inside the hydrogel. Invading cells showed multiple surges in reporter gene expression throughout the 21-day incubation, which resulted in a sustained cumulative transgene expression and suggest that the cells are transfected several times as they infiltrate the scaffold. In contrast, cells seeded homogeneously only showed one surge in transgene expression with cumulative transgene expression leveling off after 7-days, indicating that the cells were transfected only once during the 21-day incubation. Importantly, we found that using green fluorescent protein as a reporter gene, which is widely used by the other research groups to study gene transfer from hydrogel scaffolds, often generates false positives, with cells that were not transfected or transfected by a different reporter plasmid showing green fluorescence. The ability to repeatedly transfect cells invading DNA containing aECMs maybe result in prolonged expression of bioactive factors in vivo which may have implications in tissue regeneration applications.

70f Targeting of Polymer Gene Carriers Increases Uptake in Cells but Reduces DNA Release and Overall Expression Efficiency

Yen Cu[1], Michael J. Caplan[2], and W. Mark Saltzman[1]. (1) Biomedical Engineering, Yale University, 55 Prospect St, New Haven, CT 06511, (2) Cellular and Molecular Physiology, Yale University, 55 Prospect St, New Haven, CT 06511

DNA delivery to cells can be improved by using particulate carriers made from biodegradable polymers such as poly(lactic-glycolic)acid. Modifications to the particle surface can be incorporated to enhance cellular uptake and gene expression. Taking advantage of known receptor/ligand interactions in intestinal and renal epithelial cells, we formulated PLGA particles coated with BSA. The modified particles exhibited a marked increase in uptake by cells expressing the albumin receptor, megalin, and a higher stability following prolonged incubation at physiological and low pH compared to uncoated particles. However, the slower, sustained release of DNA from particles due to incorporation of BSA resulted in significantly lowered gene expression in cells. Adding features to the design of 'smart' particles for DNA delivery, thus, can introduce unfavorable characteristics that reduces treatment efficacy. Our study highlights the importance of considering these factors in engineering polymer drug carriers, and the need to understand better the mechanisms that govern the processing of particles in cells.

70g Intracellular Trafficking of Pentablock Copolymer/ DNA Polyplexes In Cultured Human Cancer and Normal Cells for Gene Delivery

Bingqi Zhang and Surya K. Mallapragada. Chemical & Biological Engineering, Iowa State University, Ames, IA 50011

The ideal transgene vetors are expected to have no or low effect on normal cells but high transfection efficiency on carcinoma cells. In this study, the novel poly(diethylaminoethylmethacrylate)(PDEAEM)/ Pluronic® F127 pentablock copolymers were found to be able to mediate high-efficiency transfection of human epithelial ovarian carcinoma (SKOV3) cells while showing significantly lower efficiency of human epithelial retina (ARPE-19) cells an another animal normal cell line (swiss 3T3). To further examine the intracellular routes of polyplexes in SKOV3 and ARPE-19 cells, pentablock copolymers and plasmid DNA were covalently attached with Alexa Fluor® 647 and Ethidium monoazide (EMA), respectively. Confocal microscopic images showed that plenty of polyplexes translocated into nuclei within 3h of transfection in SKOV3 cells, yet only few polyplexes were observed in APRE-cells. This difference in the amount of polyplexes in nuclei was also found in the cells at 10h and 24h posttransfection, indicating less nuclear entry may result in the lower efficiency of transfection. Since Brdu staining showed that SKOV3 proliferated much faster than ARPE-19 did, the nuclear entry of polyplexes was assumed to be correlated with the proliferation rate, and we hypothesized our pentablock copolymer could mediate gene delivery selectively in fast growing cells. Furthermore, in APRE-19 cells free DNA showed a weak signal or was localized around the cell membrane area, which implied that the uncomplexed DNA may have been degraded by the nuclease in the cytoplasm, or might have been exported out of the cell via exocytosis or other mechanisms. Although the validity of the hypothesis that our pentablock copolymer could selectively transfect hyperproliferative cells needs further examination, this present work provides a new perspective to design targeting vectors based on different characteristics among specific cells.

70h Novel pH-Triggered Nanoparticles for Gene Delivery Have High Transfection and Low Cytotoxicity

Jin-Oh You and Debra T. Auguste. School of Engineering and Applied Sciences, Harvard University, 40 Oxford St. ESL 224, Cambridge, MA 02138

Nonviral gene delivery strategies may show reduced immunogenicity and oncogenicity in comparison to viral vectors. However, this has generally been at the cost of endosomal release and transfection efficiency. In this study, we used poly (N,N-dimethylaminoethyl methacrylate (DMAEMA) / 2-hydroxyethyl methacrylate (HEMA)) nanoparticles as gene carriers due to their biocompatibility and ability to swell at low pH. Once internalized via the endocytosis pathway, the nanoparticles undergo osmotic swelling facilitating endosomal escape of DNA to the cytoplasm. Monodispersed nanoparticles were synthesized by O/W emulsion followed by polymerization. Particles having 200 nm average diameters were characterized by transmission electron microscopy, dynamic light scattering, electrophoresis, and cytotoxicity. For intracellular tracking, particles were conjugated with quantum dots and examined by confocal microscopy at different time points. To examine the efficacy of DMAEMA/HEMA nanoparticles in facilitating endosomal escape of pDNA encoded EGFP, HeLa cells were incubated with DNA-loaded nanoparticles for 24 and 48 h. We demonstrate that high swelling ratios are achieved at low pH, low crosslinking density, and high content of DMAEMA. After 48 h incubation, the transfection efficiency obtained by using DNA-loaded DMAEMA/HEMA nanoparticles was 45 ± 4.8%, in comparison to 40 ± 3.1% for PEI/DNA complexes with a reduction in cytotoxicity of 93 ± 6.1%. Our results show that DMAEMA/HEMA nanoparticles lead to high gene expression which is comparable to the efficacy using polyethyleneimine (PEI)/DNA complexes without the drawback of cell death.

80a Environmentally-Benign Polymer/Silicate Nanocomposites for Biomedical and Packaging Applications

Ioannis Zuburtikudis[1], Kyriaki Tornikidou[2], Sotirios, I. Marras[1], Elpiniki Panayiotidou[1], and Georgia Christofidou[1]. (1) Industrial Design Engineering, TEI of Western Macedonia, Kozani, Greece, (2) Industrial Design Engineering & Private Dental Practice, TEI of Western Macedonia, Kozani, Greece

Polymer/layered silicate nanocomposites are a new class of materials, obtained by the dispersion of the very high-aspect-ratio clay particles into a polymer matrix [1]. These materials have distinct advantages over their macro- and micro-composite counterparts including higher modulus, outstanding diffusion barrier properties and flame retardancy. Recently, significant attention has been given to biocompatible and/or biodegradable polymers, both from biomedical and ecological perspectives. In case these polymers come from renewable sources, they are even more exciting. If they are mixed with the silicate particles via a solvent-free technique, they can offer an integrative route for producing such advanced hybrid materials in accordance with the principles of green materials design and production.

In the present study, the results from an initiative undertaken to produce such environmentally-benign nanocomposites with potential applications in the field of biomedical products (high-added value products) and in packaging (low-added value products) are presented. More specifically, nanocomposites made of three different polymers, namely poly(ε-caprolactone) (PCL), poly(3-hydroxybutyrate) (PHB) and high-amylose starch, with various amounts of organically modified montmorillonite were produced in a ThermoHaake MiniLab• micro-exruder/-compounder. The morphology of the prepared hybrids was investigated by X-ray diffraction, while their thermal characteristics were studied by differential scanning calorimetry. Thermal degradation studies by thermogravimetric analysis revealed that the nanocomposites display increased thermal stability, while their mechanical testing showed improved properties over the virgin polymers.

Acknowledgements

This work was supported by the Region of Western Macedonia (Greece) through the PEP grant with MIS 105545 to I. Z.

References

1. Marras, S.I., Zuburtikudis, I., Panayiotou, C., Eur. Polym. J., 43 (2007) 2191.

2. Ray, S.S. and Bousmina, M. Prog. Polym. Sci., 50 (2005) 962.

80b Composite Resins for Military Applications

Ian M. McAninch, Steven E. Boyd, and John J. La Scala. U.S. Army Research Laboratory, AMSRD-ARL--WM-MC, Building 4600, Aberdeen Proving Ground, MD 21005

The choice of resin in composite applications is sometimes dictated by legacy choices or by a lack of time and resources to properly choose the best resin for a given application. Composite resins with a wide range of mechanical properties: elongations ranging from 5-300%, strengths ranging from 1-110 MPa, and moduli ranging from 10-3000 MPa, were evaluated for their potential in future composite applications. The neat resins were characterized by dynamic mechanical analysis (DMA) and by flexural, fracture, and impact testing. Fracture morphology was also examined using optical and electron microscopy. The glass transition temperatures of the evaluated resins, as measured by DMA, varied from -20 to 160°C. As neat resin properties do not always directly correlate to composite properties, composites were made using S-2 glass fibers. In one such comparison, a high toughness neat epoxy showed lower composite impact resistance than a brittle neat epoxy. Composites were evaluated by flexural, short beam shear, and impact testing. The resulting neat resin/composite data base of properties will enable quicker and better resin selection in future composite applications.

80c Polymer Nanocomposite Fabrication for G-M-R Sensor Application

Zhanhu Guo[1], Amar B. Karki[2], Hongfei Lin[3], H. Thomas Hahn[4], David P. Young[2], Kimberly M. McGrath[5], Doug Carpenter[5], and Kevin D. Maloney[5]. (1) Chemical Engineering Department, Lamar University, Beaumont, TX 77710, (2) Department of Physics and Astronomy, Louisiana State University, Baton Rouge, LA 70803, (3) Chemical Engineering Department, University of California, Santa Barbara, Santa Barbara, CA, (4) Mechanical and Aerospace Engineering, University of California Los Angeles, 420 Westwood Plaza,, Eng IV, 48-121, Los Angeles, CA 90095, (5) QuantumSphere, Inc, 2905 Tech Center Drive, Santa Ana, CA 92705

Polymeric composites reinforced with inorganic fillers have attracted much interest due to their reduced weight, high homogeneity, cost-effective processability and tunable physical such as mechanical, magnetic, optical, electric and electronic properties. The applications have extended into the marine (Naval submarine) and airplane (Boeing 787) industries. Furthermore, nanoparticles (NPs) within the polymeric matrix render the nanocomposite potential electronic device applications such as fuel cells, photovoltaic (solar) cells, batteries and magnetic data storage. On the other side, the functional groups of the polymer surrounding the nanoparticles enable these polymer nanocomposites suitable for variable applications such as site-specific molecule targeting application in the biomedical areas.

Particle dispersion together with the interaction between fillers and polymer matrix are major challenges in the polymer composite manufacturing. The particle agglomerates and voids resulting from the poor bondage will serve as defects, which will definitely give deleterious physical properties such as lower tensile strength for structural material application and poorer electron transport path for integrated polymer composite electric/electronic device applications.

We have demonstrated strengthened polymer nanocomposite fabrication by surface engineering the particles.[1] However, functionalization is an extra cost for production with high particle loading. Consequently, several simple and low-cost methods (surface-initiated-polymerization and monomer stabilization method)[2-3] were developed for high-quality nanocomposite fabrication.

Polymer nanocomposites were developed into a granular giant magnetoresistance (GMR) sensor [4-5] with the highest signal among these systems. Compared with metallic matrix GMR, the polymer matrix could be facile fabrication, low-cost usage without any packaging requirement and suitable for harsh environmental applications, ready to be used in specific biomedical areas.

In this presentation, the composite fabrication methodologies and the application in the granular GMR sensor will be discussed.

Ref:

[1]Z. Guo; T. Pereira; O. Choi; Y. Wang; H. T. Hahn; Journal of Materials Chemistry, 16, 2800-2808

(2006)

[2]Z. Guo; S. Park; S. Wei; T. Pereira; M. Moldovan; A. B. Karki; D. P. Young; H. T. Hahn, Nanotechnology, 18, 335704 (2007).

[3]"Facile Monomer Stabilization Approach to Fabricate Iron/Vinyl Ester Resin Nanocomposites," Composites Science and Technology, accepted.

[4] Z. Guo; S. Park; H. T. Hahn; S. Wei; M. Moldovan, A. B. Karki; D. P. Young, Applied Physics Letter, 90, 053111 (2007).

[5]"Magnetic and Magnetoresistance Behaviors of Particulate Iron/Vinyl Ester Resin Nanocomposites," Z. Guo; H. T. Hahn;H. Lin; A. B. Karki; D. P. Young, Journal of Applied Physics, accepted.

Web Page: www.seas.ucla.edu/~zhanhu

80d Layer-by-Layer (LBL) Assembled Highly Conductive, Transparent and Robust Thin Carbon Nanotube Films for Optoelectronics

Jian Zhu[1], Bong Sup Shim, and Nicholas A. Kotov[2]. (1) Department of Chemical Engineering, University of Michigan, Ann Arbor, MI 48105, (2) Chemical Engineering Department, Materials Science Department, Biomedical Engineering Department, University of Michigan, 2300 Hayward, Ann Arbor, MI 48109

Thin Conductive transparent films (TCFs) play an essential role in many optoelectonic devices, including Touch Screens, Liquid Crystal Displays (LCDs), Organic Light-Emitting Diodes (OLEDs), and Photovoltaics (PV) as electrodes. Indium Tin Oxide (ITO) was selected as an appropriate candidate for this application long ago in industry, due to its low resistivites[1] at the magnitude of 10^{-4} ohms-cm with high transmittance in the visible spectrum. However, it falls short in several aspects, such as increasing price due to limit of availability, high patterning cost, brittleness, and low transmittance to the infrared region of solar spectrum[2], challenging the research and marketing in the field of flexible OLEDs and PVs. Two alternative materials, conductive polymers and composites with conductive fillers, are proposed to fulfill the challenge. After a decade of research and development, conductive polymers like Poly(3,4-ethylenedioxythiophene) (PEDOT) has comparable properties in conductivity and transmittance to ITO (according to product information on Baytron• PH500[3]), however, they suffer from thermal and environmental stability problems[4] and are usually colored, which prevent their widespread use in commercial applications. Currently, more interest is drew upon making highly transparent, conductive thin composite electrode using highly conductive fillers, like Single Walled Carbon Nanotubes (SWNTs). Networks of SWNTs combine many advantageous properties such as mechanical flexibility, optical transparency, high electrical conductivity, simple preparation.

Layer-by-layer (LBL) Assembly, well known for its potential to build highly tuned, functional thin films with nanometer-level control of film composition and structure, has demonstrated its significance in building ultrastrong nanocomposites[5, 6], biomaterials[7] and other materials like species-selective thin films[8] and electrochromic thin films[9], etc. We propose here that this technology can be applied to make thin SWNTs electrodes with properties equivalent to ITO, based on the understanding that it can enable the homogenous distribution of SWNTs, which is fundamental to extend nanoscale properties, like high strength and high conductivities of individual SWNTs, to the bulky composite materials.

Reference:

[1] C. G. Granqvist, Solar Energy Materials and Solar Cells 2007, 91, 1529.

[2] Z. C. Wu, Z. H. Chen, X. Du, J. M. Logan, J. Sippel, M. Nikolou, K. Kamaras, J. R. Reynolds, D. B. Tanner, A. F. Hebard, A. G. Rinzler, Science 2004, 305, 1273.

[3] www.baytron.com.

[4] G. P. Crawford, Flexible Flat Panel Displays, John Wiely and Sons, 2005.

[5] A. A. Mamedov, N. A. Kotov, M. Prato, D. M. Guldi, J. P. Wicksted, A. Hirsch, Nature Materials 2002, 1, 190.

[6] P. Podsiadlo, A. K. Kaushik, E. M. Arruda, A. M. Waas, B. S. Shim, J. D. Xu, H. Nandivada, B. G. Pumplin, J. Lahann, A. Ramamoorthy, N. A. Kotov, Science 2007, 318, 80.

[7] E. Jan, N. A. Kotov, Nano Letters 2007, 7, 1123.

[8] D. M. Sullivan, M. L. Bruening, Chemistry of Materials 2003, 15, 281.

[9] C. A. Cutler, M. Bouguettaya, J. R. Reynolds, Advanced Materials 2002, 14, 684.

80e Enhanced Melt Processing of Conductive Polymer Composites for Use In Multilayer Coextrusion

Randy A. Mrozek[1], Phillip J. Cole[2], and Joseph L. Lenhart[1]. (1) Sandia National Laboratories, PO Box 5800 MS 1411, Albuquerque, NM 87185, (2) NNSA Satellite Programs, Sandia National Laboratories, P.O. Box 5800, MS-1411, Albuquerque, NM 87122

Multilayer coextrusion combines multiple polymers in a layered structure to produce properties that are not found in a single polymer. For many potential applications one or both of the layers needs to include large volumes of filler. Amongst the desired properties, one of the largest challenges is to obtain conductivity within an individual layer. Traditionally, conductive polymer materials have been difficult to melt process. Inherently conductive composites will often degrade before melting or have a high viscosity as a result of a rigid backbone. Conductivity as the result of added filler requires high loadings that can significantly increase the viscosity. In addition, as the number of layers increases the layer thickness will decrease to below a micron. At reduced layer thicknesses the filler must be on the nanometer size scale that will have a more pronounced impact on the viscosity relative to larger fillers. High viscosities in coextrusion are problematic because a viscosity mismatch between layers can produce layer breakup. As a result, we have developed several strategies to produce conductive materials that have a reduced impact on the viscosity to aid in the production of filled multilayer structures by coextrusion.

80f Shear Rheology and Microstructure of a Concentrated Short Glass Fiber-Filled Polybutylene Terephthalate (PBT)

Aaron P. R. Eberle, Chemical Engineering, Virginia Polytechnic Institute, 154 Randolf Hall, Blacksburg, VA 24061, Gregorio M. Velez, Macromolecular Science and Engineering Department, Virginia Polytechnic Institute, 154 Randolf Hall, Blacksburg, VA 24061, Dr. Donald G. Baird, Chemical Engineering, Virginia Tech, 154 Randolph Hall, Blacksburg, VA 24061, and Peter Wapperom, Mathematics Department, Virginia Polytechnic Institute, 154 Randolf Hall, Blacksburg, VA 24061.

Short glass fiber composites represent an interesting class of materials where the properties can be highly anisotropic. This is a direct consequence of the anisotropic fiber orientation generated during processing and makes simulation of the fiber orientation and stresses imperative to part stability and functionality. Since most parts manufactured from short glass fiber composites are processed in the melt state the rheological behavior and associated fiber microstructure of the composite fluid is of technical relevance to constitutive relation model development and flow simulation.

The purpose of this work is to give a quantitative experimental description of the transient shear rheological behavior as it relates to the microstructure of a concentrated short glass fiber commercially available composite fluid. A series of stress growth measurements in start-up of flow and flow reversal tests are performed at various shear rates. Rheological measurements were performed with a Rheometrics Mechanical Spectrometer (RMS-800) using a novel sample geometry which minimizes excessive fiber boundary interaction while maintaining a homogeneous shear field. The fiber orientation is characterized using confocal laser microscopy at points of interest on a stress vs. strain plot. Experimental results are compared to predictions based on the generalized Jeffery equation with the addition of a non-affine motion function that slows the rate of fiber reorientation. Model predictions show good agreement with experimental results.

81b Membranes for Olefin Paraffin Separations

Mita Das[1], William J. Koros[1], Stephen Wilson[2], Douglas Galloway[2], Lisa Knight[2], and Chunqing Liu[2]. (1) Chemical & Biomolecular Engineering, Georgia Institute of Technology, 311 Ferst Drive, Atlanta, GA 30332, (2) UOP LLC, Des Plaines, IL 60017

Propylene is among the most important chemical feedstocks of the petrochemical industry. A significant amount of this olefin is found in petrochemical product streams and must be separated. Already much work has been done with pure polymer, to increase the permeability and selectivity of propylene/ propane separations.

This research has primarily two goals. 1) Keep the efficiency (selectivity) high at higher pressure by suppressing plasticization. 2) Increase the selectivity and permeability with the use of Mixed Matrix membranes for commercial use.

Burns and Koros have identified the so-called "propylene/propane upper bound curve" in 2001. We have identified and successfully synthesized 6FDA-6FpDA polyimide for Propylene/Propane separation. Our pure polymer selectivity is 19 at 35ºC. Propylene has shown a trend of plasticization above 2 atm. pressure with pure polymer. We have been able to suppress the plasticization occurrence by performing the experiment at 70ºC and synthesizing very high molecular weight and narrow PI polymer. Under these conditions, plasticization does not occur till 4.35 atm.. High molecular weight appears to reduce segmental free volume and mobility in the polymer.

Our next step is to disperse molecular sieves in a polymer to create a hybrid material with superior properties. We are able to see permeability and selectivity enhancements with more than one dense film pure gas permeations. Currently we are trying to verify our positive results with available mathematical models (Cussler and Maxwell). Once successful, we will focus on high temperature and pressure permeations to approximate industry application.

81c Water Permeability and Water/Salt Selectivity Tradeoff in Polymers for Desalination

Ho-Bum Park[1], Alyson C. Sagle[2], James E. McGrath[3], and **Benny D. Freeman**[2]. (1) Chemical Engineering , Hanyang University, Seoul, 680-749, South Korea, (2) Chemical Engineering, The University of Texas at Austin, 10100 Burnet Rd., Bldg 133, Center for Energy and Environmental Resources, Austin, TX 78758, (3) Macromolecules and Interfaces Institute, Virginia Tech, Blacksburg, VA 24061

Polymer membrane-based desalination technologies (e.g., reverse osmosis (RO) and nanofiltration (NF)) have been extensively developed since the 1960s and are well-established processes. The separation performance of polymeric membranes for desalination is usually described in terms of water flux (or permeance) and salt rejection. Based on a survey of available data, water permeance and NaCl rejection are often inversely correlated, and there may be an upper bound, similar to that observed in gas separation membranes, beyond which there are very few data points. However, water permeance and salt rejection are not intrinsic material properties since they are influenced by sample size (i.e., membrane thickness in the case of permeance) and measurement variables (e.g., pressure and salt concentration in the case of salt rejection). Use of water permeability, rather than water flux or permeance, and water/salt permeability selectivity, rather than rejection, in a tradeoff analysis provides a clearer comparison of properties that depend only on the fundamental transport characteristics of the materials under study. When water and salt transport data are presented on a log-log plot of water permeability versus water/NaCl permeability selectivity, a clear tradeoff relation and upper bound emerges. Further analysis suggests that both water/NaCl solubility and diffusivity selectivity contribute to the high water/NaCl permeability selectivity, but diffusivity selectivity is the dominant factor. Both solubility selectivity and diffusivity selectivity exhibit tradeoff and upper bound features when plotted as a function of water solubility and water diffusivity, respectively.

81d An Examination of Structural Parameters Affecting the Dynamics of Penetrant Transport In Glassy Polymers

Adam K. Ekenseair[1], Richard A. Ketcham[2], and Nicholas A. Peppas[1]. (1) Department of Chemical Engineering, The University of Texas at Austin, Austin, TX 78712, (2) Department of Geological Sciences, The University of Texas at Austin, Austin, TX 78712

Glassy polymers are increasingly used in the construction and fabrication of high-tech devices. These devices are often used in applications, such as microfluidics, membrane separations, food packaging, artificial tissues, bones, and organs, and controlled drug delivery, where the polymers come into contact with fluids capable of penetrating the polymer network. To examine the behavior of such systems both on the macroscopic and molecular level, it is important to study the penetrant diffusion process. In such transport, the macromolecular chains rearrange toward new conformations, where the rate of relaxation depends on the penetrant concentration. The relative rates of penetrant diffusion and macromolecular chain relaxation determine the nature of the transport process and lead to a wide variety of penetrant transport phenomena, such as Fickian, anomalous (non-Fickian), and Case II sorption behavior.

With trends leading to smaller device scales and increasingly complex polymer structures, there is a need for both a qualitative and quantitative understanding of the manner in which a polymer's network structure alters not only the rate, but also the mode of penetrant transport. To this end, poly(methyl methacrylate) (PMMA) was synthesized in an iniferter-mediated radical polymerization process to produce a variety of well-characterized polymer network structures. The effects of basic network parameters, including the degree of crosslinking, polymer mesh size, and the crosslink size, on the transport process were studied. The effects of sub-Tg annealing/aging, temperature, and the presence of un-reacted monomer were also investigated.

Utilizing both gravimetric sorption data and in-situ ultra-high-resolution X-ray computed tomography studies, the transport of methanol in PMMA discs was investigated. Changing the polymer network structure is shown to not only influence the rate of sorption but also the nature of the transport process.

81e Non-Fickian Diffusion of Water In Nafion

Daniel T. Hallinan Jr.[1], Maria Grazia De Angelis[2], Marco Giacinti Baschetti[2], Giulio C. Sarti[2], and Yossef A. Elabd[1]. (1) Chemical and Biological Engineering, Drexel University, 3141 Chestnut St., Philadelphia, PA 19104, (2) Chemical Engineering, Mining and Environmental technology, University of Bologna, Via Terracini 28, Bologna, Italy

Polymer electrolyte membranes (PEMs), such as Nafion, have been extensively explored in applications including fuel cells, actuators, and membrane-based separations. Fuel cells in particular require high PEM proton conductivities, which are strong functions of water content. Maintaining sufficient conductivity for fuel cell performance requires high water content. Therefore, a fundamental understanding of water sorption and dynamics in Nafion is critical. Although water sorption isotherms for Nafion have been reproduced in many laboratories, reported diffusion coefficients of water in Nafion vary by four orders of magnitude.

In this study, the diffusion of water vapor in Nafion was measured using time-resolved Fourier-transform infrared, attenuated total reflectance (FTIR-ATR) spectroscopy. At moderate vapor activities, Fickian behavior was observed. However, at both low and high vapor activities, two different forms of non-Fickian behavior were observed. Models have been developed for both non-Fickian regimes, where a reaction between water and immobile sulfonic acid sites within the polymer has been considered at low activities and polymer relaxation has been considered at high activities. Both diffusion-reaction and diffusion-relaxation models regress well to the experimental data. In addition to modeling non-Fickian behavior at low and high vapor activities in order to extract true diffusion coefficients, the determination of accurate diffusion coefficients over the entire activity range will be explained in terms of an appropriate assessment of the vapor-polymer boundary conditions. More specifically, water sorption dynamics were a strong function of the vapor flow rate. Inaccurately accounting for this boundary condition will result in diffusion coefficients an order of magnitude lower than the true values.

81f Enhancement of Barrier Properties of Poly(ethylene terephthalate) by Antiplasticization

Jong S. Lee[1], Robert Kriegel[2], and William J. Koros[1]. (1) Chemical & Biomolecular Engineering, Georgia Institute of Technology, 311 Ferst Dr. NW,, Atlanta, GA 30332, (2) Coca Cola, The Coca-Cola Company TEC 243A 1 Coca-Cola Plaza, Atlanta, GA 30313

An advanced barrier material was produced by using antiplasticization with poly(ethylene terephthalate) (PET). The incorporation of low molecular weight additives into PET led to antiplasticization characterized by reduction in both excess free volume and glass transition temperature. It also reduced O2 permeability. Solid state 13C-NMR experiment demonstrated a reduction in local chain mobility of carbonyl carbon. The Flory interaction parameter, estimated by Materials Studio® produced better estimation of interaction forces between PET and each diluent than those estimated by solubility parameters. The permeability of CO2 in PET and each modified PET was also studied. Dynamic Mechanical analysis (DMA) was employed to determine the mechanical property changes in modified PET.

Web Page: www.chbe.gatech.edu/koros/index.php?do=grpmbr&mbr=jong

100a Morphological Control of Structures Self-Assembled from Strongly Interacting Nanoparticles and Colloids

Eric Jankowski, Chemical Engineering, University of Michigan, Ann Arbor, MI 48109 and Sharon C. Glotzer, Chemical Engineering and Materials Science and Engineering, University of Michigan, 2300 Hayward Street, Ann Arbor, MI 48109.

Identifying particles that will self-assemble into a desired structure is a difficult task due to the enormous parameter space that includes inter-particle interactions, particle shape, and thermodynamic parameters such as density and temperature. Computer simulation plays a valuable role in exploring this design space, but the increasing complexity of available building blocks can render traditional simulation methods useless. In this work, we demonstrate the use of a new computational tool we have developed to study the assembly possibilities of a family of similar strongly interacting particles. By efficiently generating energy-minimizing configurations of these particles, we identify those that have the potential to assemble into interesting and desirable structures. Some of these structures include patterns with multiple levels of periodicity, discrete and continuous domains, and open structures with uniform pore sizes. In addition to identifying energy-minimizing structures for a family of particles, our algorithm suggests possible assembly strategies for generating these configurations.

100b Modelling Surfactant Dynamics In a System Containig Coallescent Particles

James E. Magee and **Flor R. Siperstein**. School of Chemical Engineering and Analytical Science, University of Manchester, Po Box 88, Sackville Street, Manchester, M60 1QD, United Kingdom

The use of surfactant self-assemblies and block copolymers as templates in the synthesis of structured materials is widely spread. Although the first materials synthesized with this approach were silica based, the technique has been extended to other metal oxides and inorganic components [1]. The prediction of the final structure and properties of the templated materials is difficult, specially when self-assembly and inorganic condensation occur simultaneously [2], but even when a preformed liquid crystal is used as a template, as in the true liquid crystal templating approach [3], condensation of the inorganic framework around the soft template can lead to its deformation.

Previous work has indicated that equilibrium phase diagrams obtained from Monte Carlo simulations are useful to understand general trends in the behaviour of surfactant-inorganic systems, where phase separation between a surfactant rich and surfactant poor phases is a result of strong interactions between the surfactant head and the inorganic components [4-5]. Nevertheless, in some cases, real systems may not reach equilibrium and the dynamics of the surfactant self-assemblies as inorganic condensation takes place becomes important for the description of the actual material.

We have developed a simple potential (similar to a DLVO potential) that can correctly describe reaction limited aggregation using dissipative particle dynamics (DPD). We have used this potential to model silica nanocasting using a symmetric surfactant liquid crystal as a model template. We will present the evolution of systems starting in different configurations, where the silica is originally a neutral solvent, and therefore dispersed throughout the system, and where the silica initially shows preference for the surfactant heads. The ability of the system to accommodate a silica framework without leading to phase separation is analyzed using different strengths of aggregation.

References:

[1] Y. Wan, Y. Shi, and D. Zhao, Chemical Communications, 2007, 9, 897–926.

[2] G. S. Attard, J. C. Glyde, and C. G. Goltner, Nature 1995, 378, 366–368.

[3] A. Monnier, F. Schuth, Q. Huo, D. Kumar, D. Margolese, R. S. Maxwell, G. D. Stucky, M. Krishnamurty, P. Petroff, A. Firouzi, M. Janicke, B. F. Chmelka Science 1993, 261, 1299–1303.

[4] F. R. Siperstein and K. E. Gubbins, Langmuir, 2003 19, 2049-2057.

[5] A. Patti, A. D. Mackie, and F. R. Siperstein, Langmuir 2007, 23, 6771–6780.

100c Molecular Simulation of the Formation of Ultrathin Silica Films from Teos (tetraethoxysilane) Using Kinetic Monte Carlo

J. M. Don MacElroy, **Taslima Akter**, Thomas C. McDermott, and Damian A. Mooney. Chemical and Bioprocess Engineering, University College Dublin, Belfield, Dublin, Ireland

Ultra-thin, nanoporous silica membranes have the potential to be used in high temperature gas separation processes (notably combustion gases at temperatures above 750K). Amorphous silica layers formed via chemical vapor deposition (CVD) have been shown to exhibit high permselectivities at elevated temperatures, but the influence of the process conditions on the microscopic details of the membrane formed needs to be better understood.

In this work we investigate the thermally induced deposition of silica layers from TEOS (tetraethoxysilane) on realistic amorphous nonporous silica substrates via chemical vapour deposition techniques. To model this process we use a hybrid kinetic Monte-Carlo (KMC) algorithm. Lattice KMC is used for the elementary reactions and an off-lattice method is employed for silica network relaxation and bond switching moves.

This kinetic model involves 12 different surface reactions which are considered to play a primary role in the build-up the SiOx layer on the substrate surface. Lattice KMC is used for these elementary reactions which include direct TEOS deposition and a variety of dissociation reactions which to take place within the layers as they form. All of the reactions have been carried out using reported reaction rate constants for this system.

The characterization of the deposited structure (density, pore size distribution and composition) and effects of the deposition temperature, initial concentration of surface reactive sites and water vapour pressure will be presented.

100d One-Pot Synthesis of Functional Mesoporous Materials

Alessandro Patti, Department of Physics and Astronomy, Utrecht University, Princetonplein 5, Utrecht, 3584 CC, Netherlands, Allan D. Mackie, Departament d'Enginyeria Química, Universitat Rovira i Virgili, Av dels Paisos Catalans, 26, Tarragona, 43007, Spain, and **Flor R. Siperstein**, School of Chemical Engineering and Analytical Science, University of Manchester, Po Box 88, Sackville Street, Manchester, M60 1QD, United Kingdom.

Mesoporous silicas can be obtained from the self-assembly of amphiphilic building blocks in the presence of pure silica or organosilica precursors. Solvent-extraction of the soft organic template leads to a periodic ordered mesoporous material. Such materials find application in several industrial fields, and are especially important in adsorption and catalysis. Functional groups can be added to the porous framework through a post synthesis treatment, or included in the framework at the time of the synthesis itself. The former has the advantage that the structure of the framework is known prior to adding the functional groups, while the latter has the advantage of a reduced number of steps in the synthesis procedure. One of the disadvantages of the one-pot synthesis method is that high quality materials cannot be obtained using pure organosilica precursors, and often mixtures of pure silica and organosilica precursors are used.

In this work, lattice Monte Carlo simulations in the NVT ensemble have been performed to study the formation of mesoporous materials containing functional organic groups into their pore walls. Four components systems, composed by a surfactant, a pure silica precursor, an organosilica precursor, and a simple solvent, have been studied. Different organosilica precursors have been modeled in order to establish to which extent they affect or transform the original surfactant mesophases. In particular, we observe that small functional groups can lead to the destruction of hexagonally ordered phases obtained with pure silica precursors, while sufficiently large solvophobic functional groups can transform hexagonal phases into lamellar phases.

100e Stability Analysis of H2 Clathrate Hydrates by Ab Initio Calculations

Jae W. Lee[1], Sangyong Lee[2], and Prasad Yedlapalli[1]. (1) Department of Chemical Engineering, The City College of New York, 140th street and Convent Ave, New York, NY 10031, (2) Texas A&M University-Kingsville, MSC 193, 700 Univrersity Blvd, Kingsville, TX 78363-8202

Gas hydrates are crystalline compounds, also known as clathrates. The guest (gas) molecules are entrapped in a cage-like structure of hydrogen bonded water molecules. Their general formula is Mn (H2O)p where M: gas, n: number of gas molecules, p: number of water molecules. Hydrogen clathrates are a potential storage medium for safe storage of hydrogen with a storage density of 5wt%. Hydrogen form Structre II (sII) clathrate hydrates and pure H2 hydrates form at 200 MPa and 250 K3. Recently, adding THF (tetrahydrofuran) dramatically reduces hydrogen hydrate formation pressure (5 MPa) but the H2 storage capacity of binary THF+H2 clathrate hydrates (1-4 wt %) is less than that of pure H2 hydrates (5.03 wt%). In binary THF+H2 clathrate hydrates, THF enters only the large cavities and H2 enters both the small and large cavities. The occupancy of H2 in large cavity is 4 and in small cavity it is either 1 or 2.

sII clathrate hydrates of pure hydrogen and binary hydrates of THF + H2 are studied using ab initio calculations to determine the stable occupancies of small cavities. We have considered one dodecahedron (small cavity) and one hexakaidecahedron (large cavity). These two cavities are attached to each other as in Str. II hydrate to form a double cavity. One or two H2 molecules are placed in the small cavity and one THF (or 4H2 molecules) molecule is placed in the large cavity. We have then determined the binding energies of the singly occupied and the double occupied double cavities at MP2 level using various basis sets (3-21G, 3-21G(2p), 3-21++G(2p), 6-31G, 6-31G(2p), and 6-31++G(2p)). Different basis sets yielded different stable occupancies. The results from the highest basis set (6-31++G (2p) with ZPE correction) indicate that single occupancy is as favorable as double occupancy in both the cases of pure H2 hydrates and THF+H2 double hydrates.

References:

1. W. L. Mao, H. K. Mao, A.F. Goncharov, V. V. Struzhkin, Q. Guo, J. Hu, J. Shu, R. J. Hemley, M. Somayazulu, and Y. Zhao, "Hydrogen Clusters in Clathrate Hydrate", Science, 297, 2247-2249 (2002).

2. L. J. Florusse, C. J. Peters, J. Schoonman, K. C. Hester, C. A. Koh, S. F. Dec, K. N. Marsh, and E. D. Sloan, "Stable Low-Pressure Hydrogen Clusters Stored in a Binary Clathrate Hydrate", Science, 306, 469-471 (2004).

3. H. Lee, J-W Lee, D. Y. Kim, J. Park, Y-T Seo, H. I. Zeng, L. Moudrakovski, C. I. Ratcliffe, and J. A. Ripmeester, "Tuning Clathrate Hydrates for Hydrogen Storage", Nature, 434, 743-746 (2005).

Synthesis and 2-Dimensional Modeling of Self-Propagating High-Temperature Synthesis of $La_{0.6}Sr_{0.4}MnO_3$

Jiri Selig
Sidney Lin
Lauren Griffin

Chemical Engineering Department
Lamar University

Solid Oxide Fuel Cell

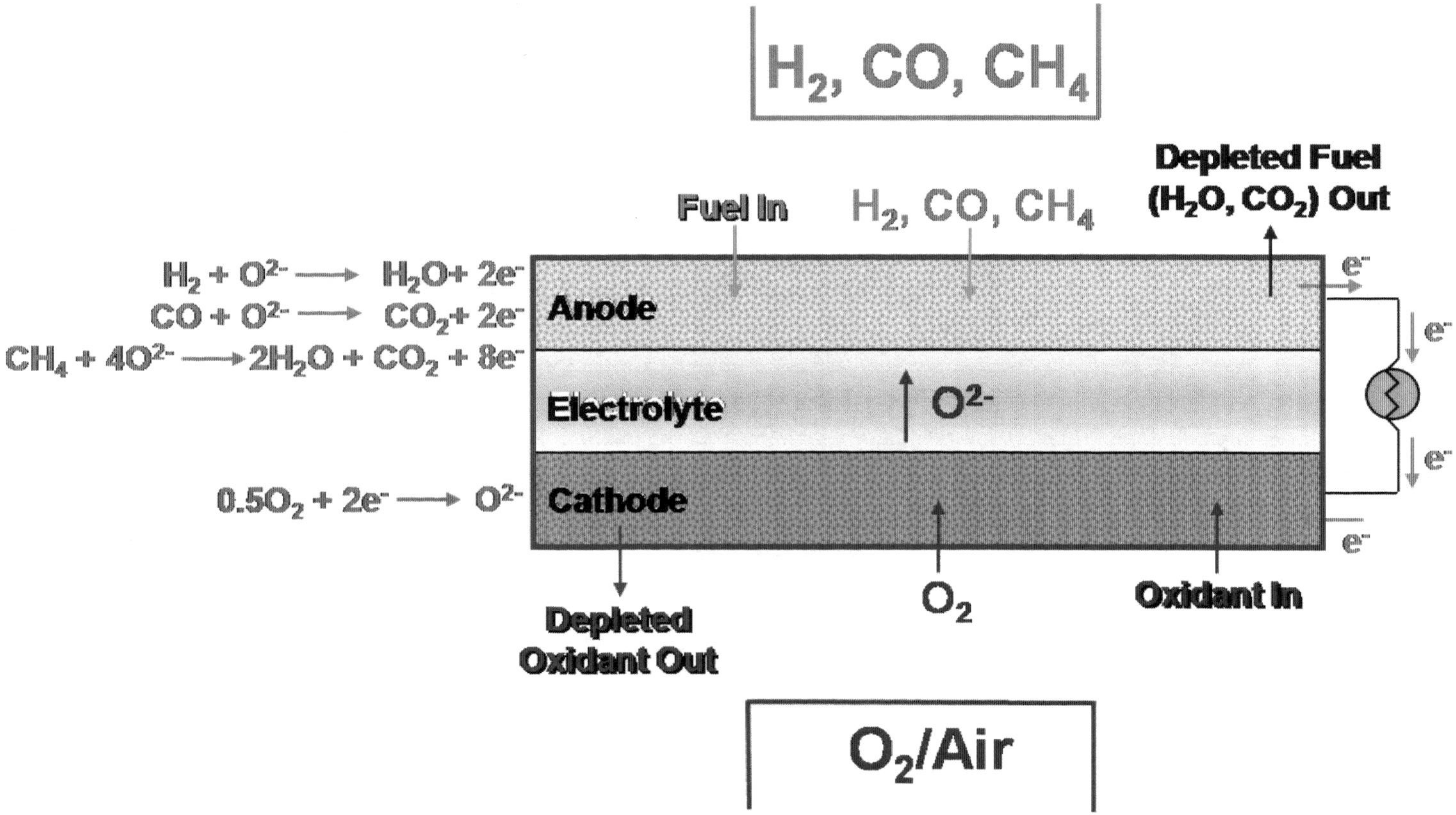

SHS - Fast and Cheap Synthesis

- Low cost (low energy requirements)
    - Cheaper then sintering
- Fast reaction (a few seconds)
- Highly exothermic reaction
- Fine powders
- High purity

Self-propagating High-temperature Synthesis of SOFC Cathode Materials

System Description

$0.3\ La_2O_3 + 0.4\ SrO_2 + 0.75\ Mn + 0.125\ Mn_2O_3 + 0.4625\ O_2 \rightarrow La_{0.6}Sr_{0.4}MnO_3$

- Adiabatic temperature: 3,200K
- Heat of reaction: -436 kJ/mol
- Green mixture is oxygen deficient
- O_2 by diffusion
- Slow but steady propagation

Sample Preparation

- Mixed in stoichiometric ratios
- Mixed in ball mill (with no balls) for 4 hrs
- Pressed into pellets in a hydraulic press
  - 0.75 metric ton, 3 cm long, 7/8" diameter
- Diluted with previously reacted product
  - SHS + 8 hours in furnace at 1,200 °C

Experimental Setup

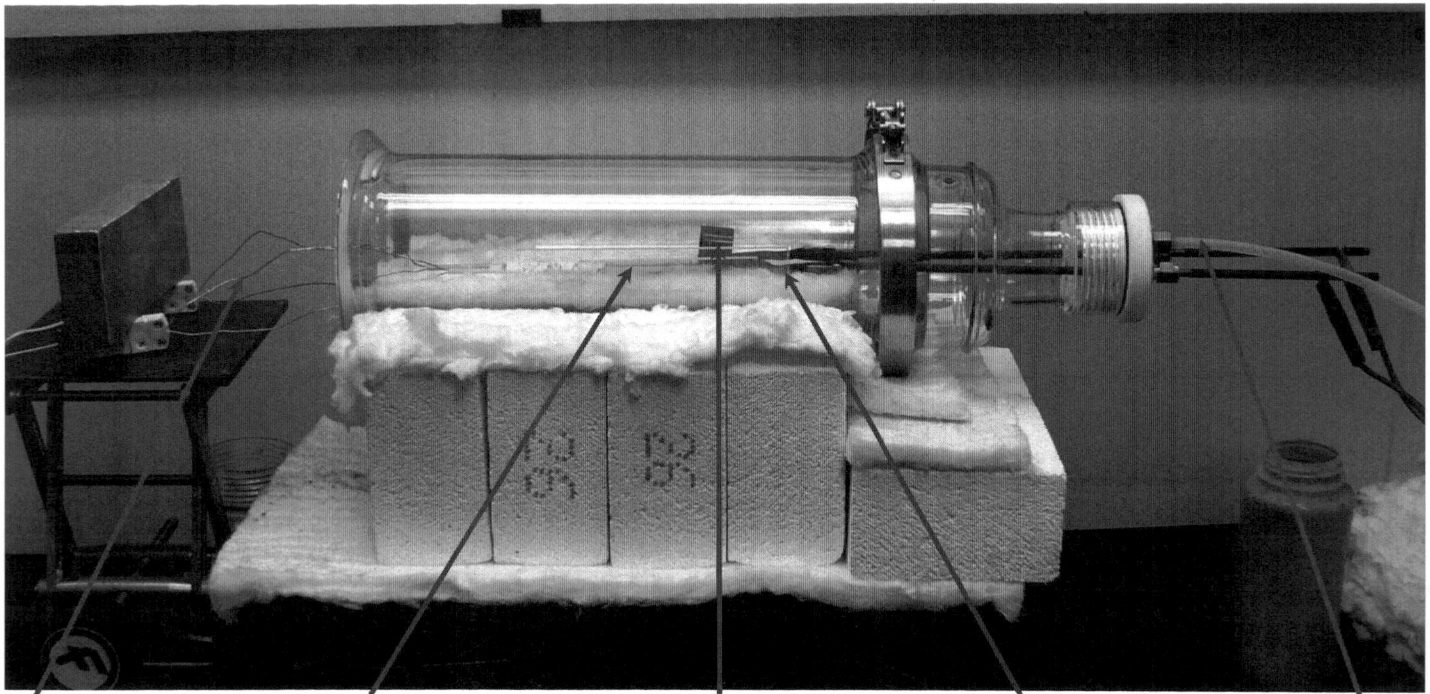

Thermocouples

Quartz Pellet Holder

Assumptions

- Constant oxygen flow (independent of temperature)
- Constant diffusivity
- Constant densities (LSM and reactants)
- All heat is released during Mn oxidation

Geometry

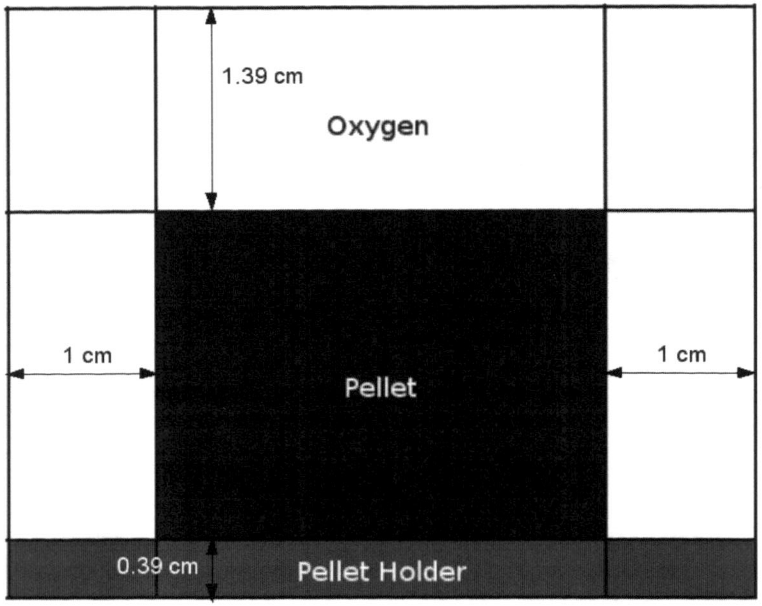

- Pellet: 3 cm x 7/8 inch, 0.39 cm quartz pellet holder
- 1 cm of oxygen on both sides and 1.39 cm above

Model Description

Energy Balance

$$\rho C_p\left(\frac{\partial T}{\partial t}+\mathbf{u}\cdot\nabla T\right)=\nabla\cdot(k\nabla T)+Q$$

$$\mathbf{n}\cdot\left[\left(k_1T_1-\rho_1C_{p1}T_1\mathbf{u}_1\right)-\left(k_2T_2-\rho_2C_{p2}T_2\mathbf{u}_2\right)\right]=q_0$$

$$q_0=q_r+q_{nc}+q_{iq}$$

$$q_r=\varepsilon\sigma\left(T^4-T_\infty^4\right)$$

$$q_{nc}=h_{nc}\left(T_f-T_\infty\right)$$

$$q_{iq}=3\times10^6(W/m^3);\ (t>1.5\ s)$$

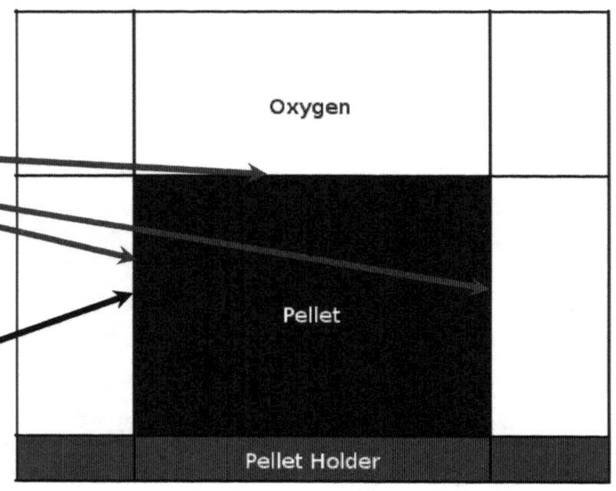

Model Description

Mass Balance

$$\frac{\partial C_i}{\partial t} + \nabla \cdot \left(-D_i \nabla c_i\right) = R_i - \mathbf{u}\nabla \cdot c_i$$

$$\frac{\partial C_{SrO_2}}{\partial t} = -k_{SrO_2} C_{SrO_2}$$

$$\frac{\partial C_{O_2}}{\partial t} = k_{SrO_2} C_{SrO_2} - \frac{\partial \eta}{\partial t} 0.75 \eta C_{Mn_0}$$

$$D_s = D\frac{\varepsilon}{\tau}$$

Kinetics (Mn Balance)

$$-\Delta H_r = \frac{\partial \eta_r}{\partial t} \rho \int C_p \, dT$$

$$\frac{\partial \eta_r}{\partial t} = \left(1 - \eta_r\right)^n k_r e^{\frac{-E_0}{RT}}$$

Temperature Data

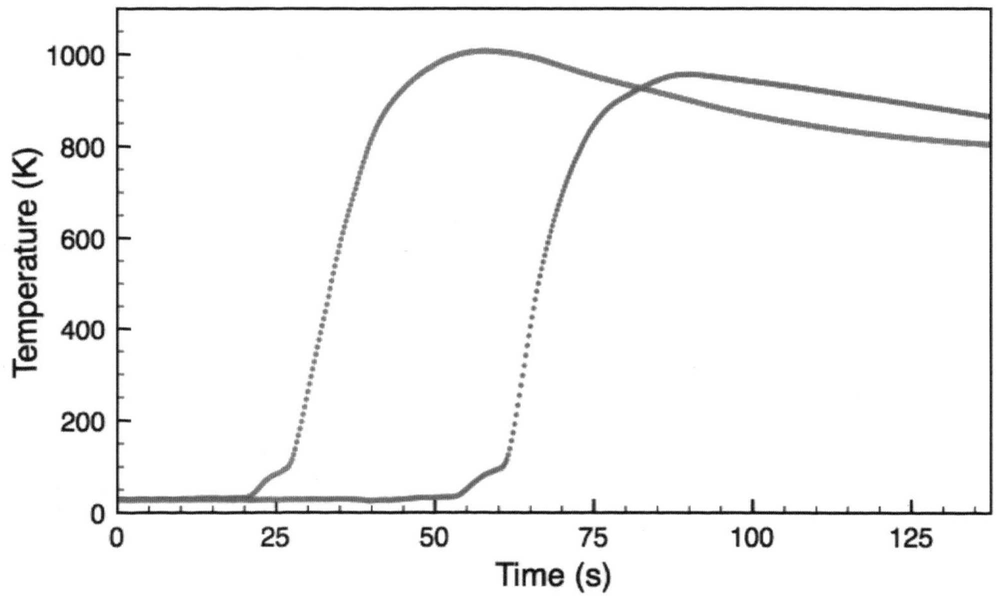

Sample recorded temperature data (5% wt dilution)

- Data points were collected at 4/s
- Used to estimate velocity (difference between the two peaks)
- Combustion front movement velocity = 0.44 mm/s

Activation Energy

Calculated from experimental data: 35kJ/mol

$$u_p = e^{\left(\frac{-E}{2RT_m}\right)}$$

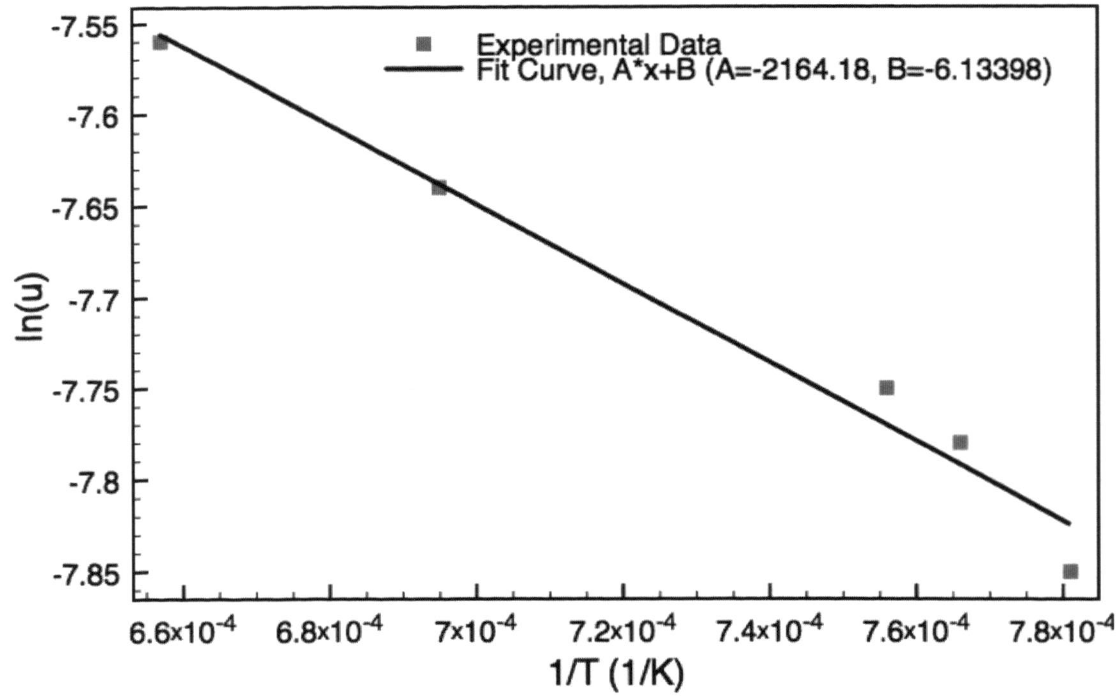

Flow Profile

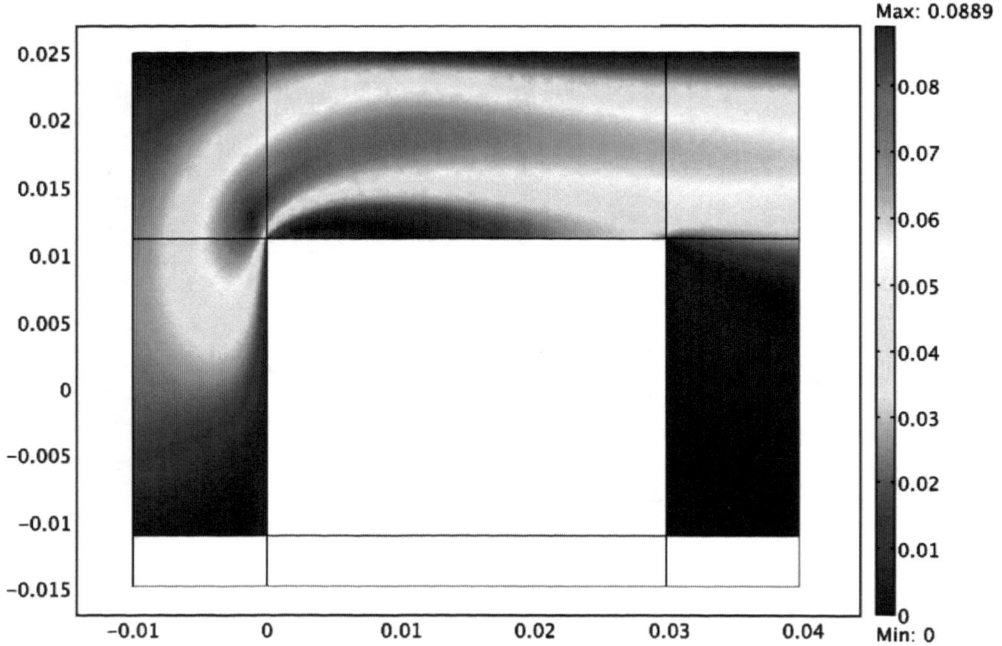

Flow profile in the reactor

Parabolic profile at inlet (mean x-velocity is 0.02 m/s)
Maximum velocity is at the upper left corner (0.0889 m/s)

Temperature Profile

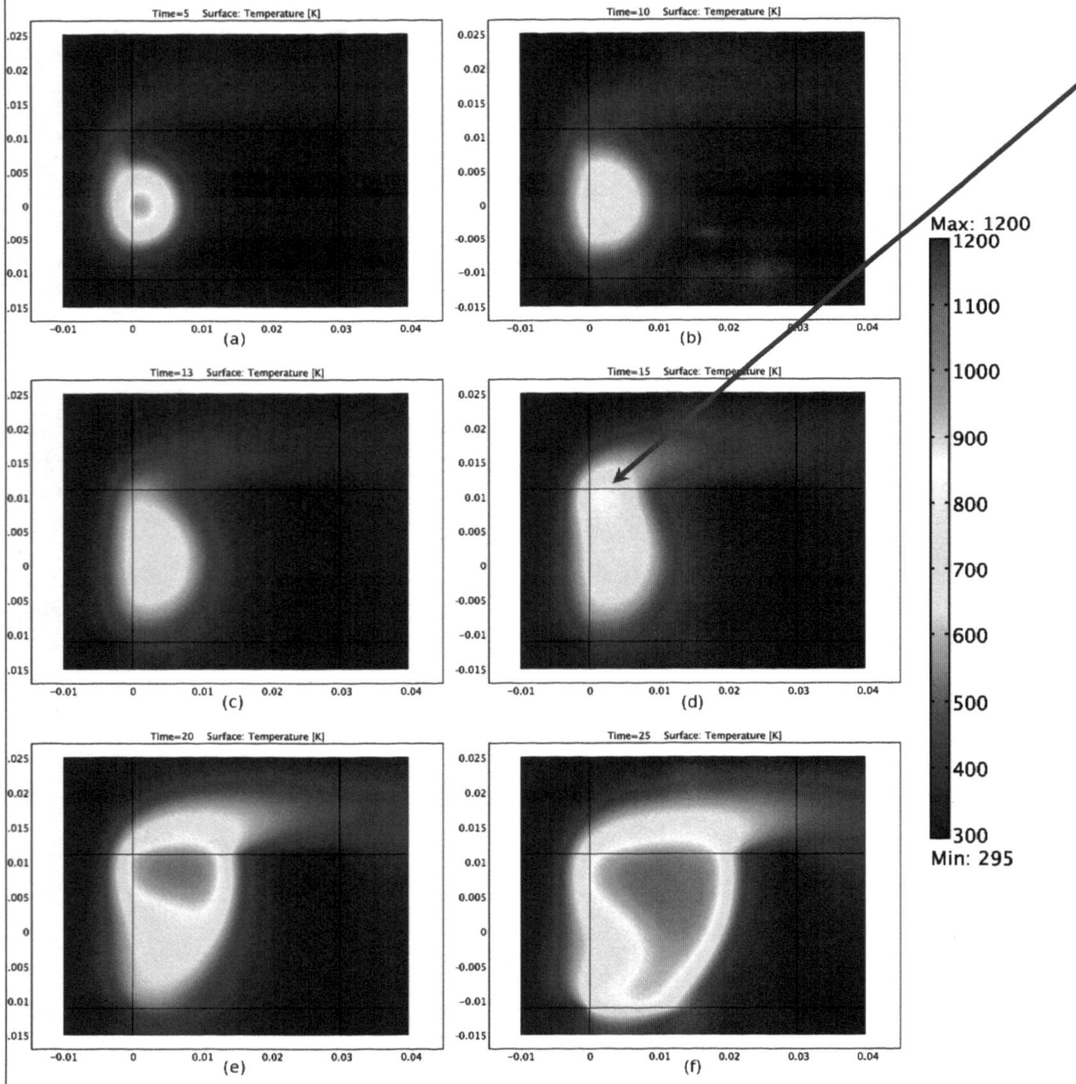

Hotspot

- Unsteady propagation
- Initial heating - ignition
- Hotspot - heat accumulation
- Lower T near pellet holder
- Heat loss to the holder
- Close to experimental data

Reaction Rate

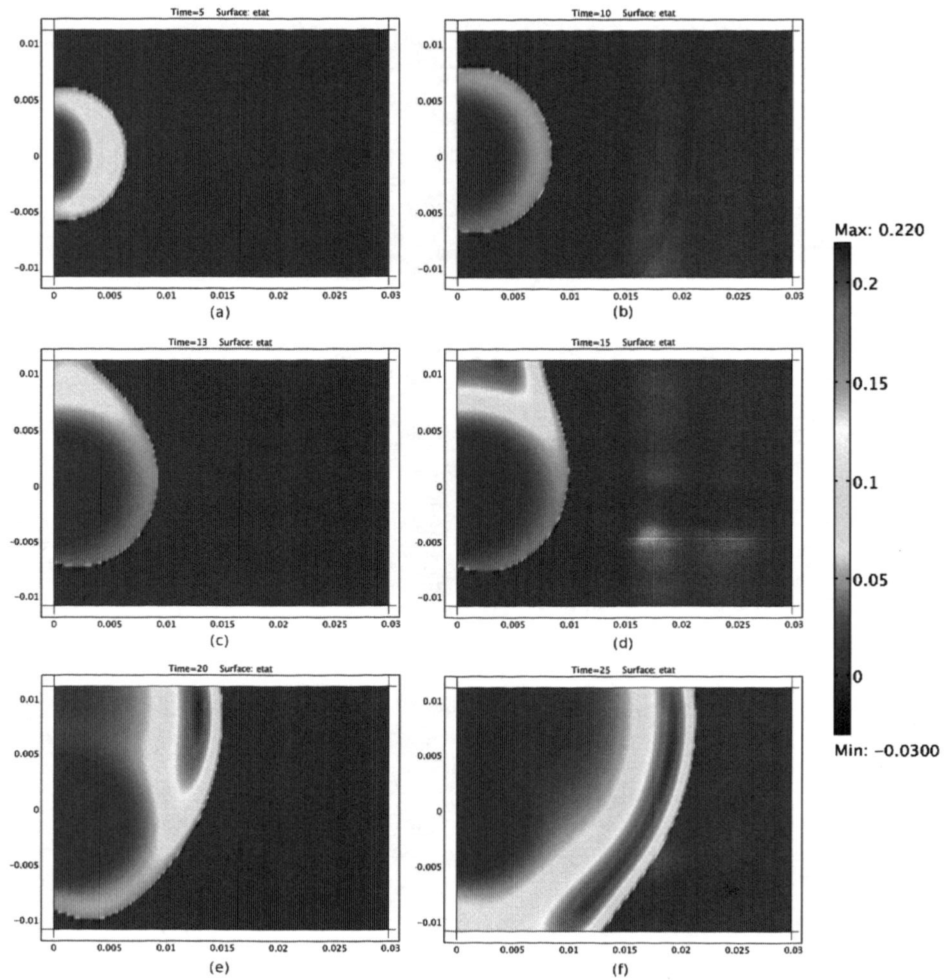

- Unsteady reaction
- Starts slowly then increases
- Reaction is slower near the bottom
- Constant maximum rate

Conversion

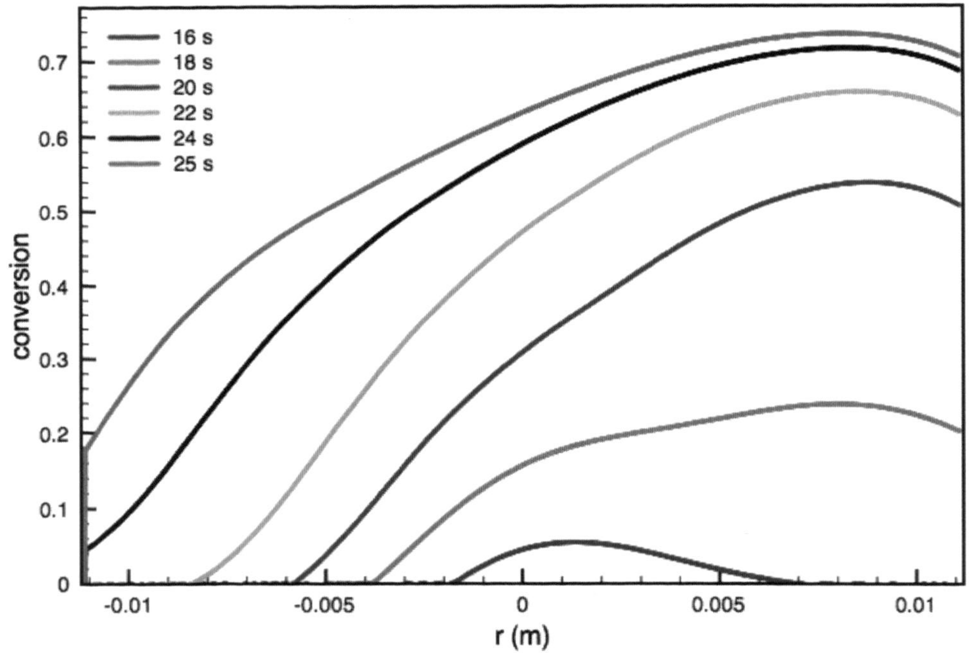

Conversion at 1cm as a function of radial position

- Upper half (r > 0) reacts first due to higher reaction rate
- Eventually levels off, only regions near the pellet remain unreacted

Reaction Rate

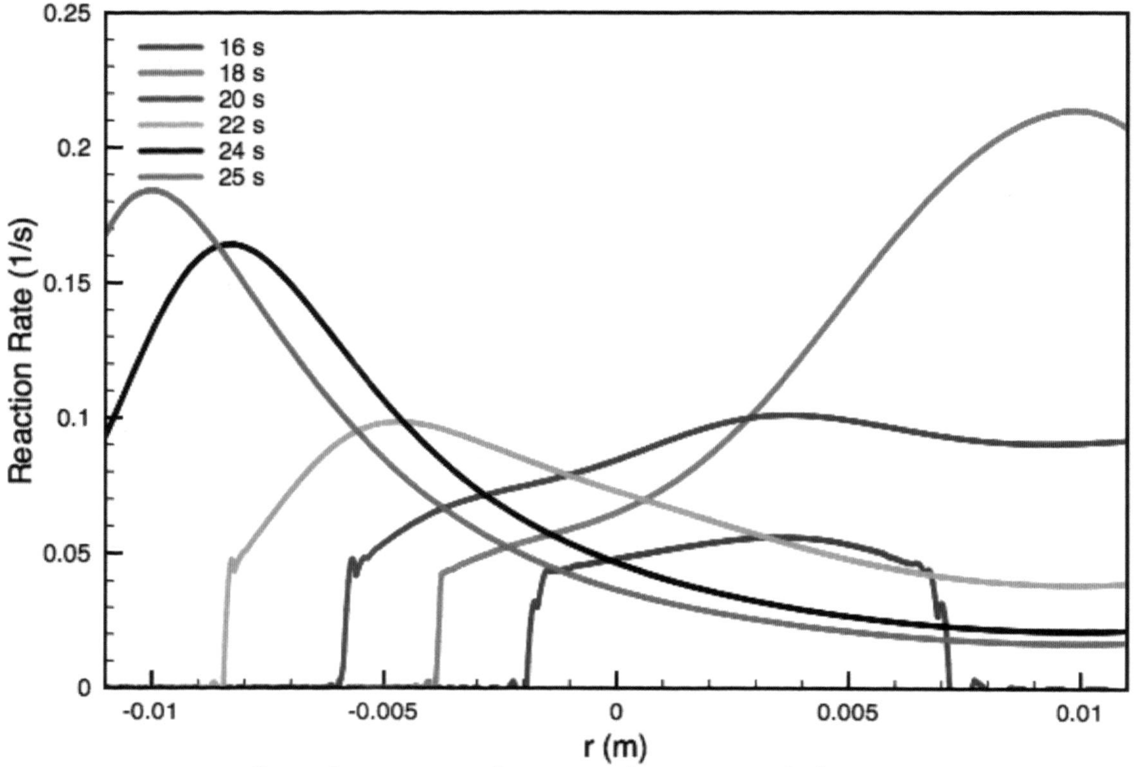

Reaction rate at 1cm as a function radial position

- Reaction propagates first in the upper half
- Reaction shifts from top to bottom

Center and Surface Temperatures

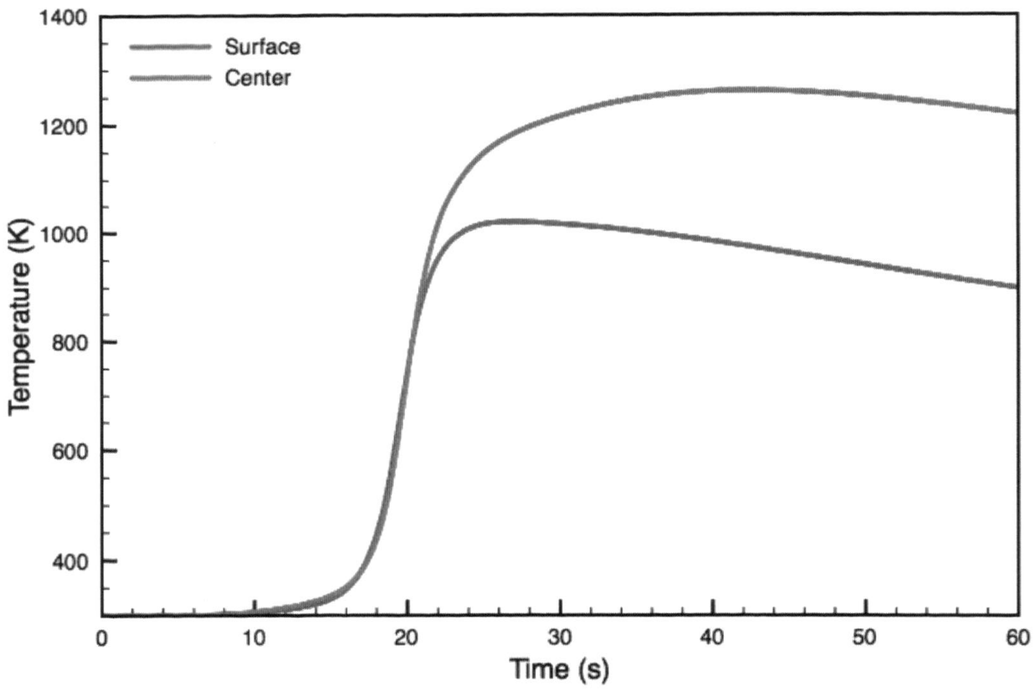

- Surface temperature significantly lower
- Surface cools at faster rate, center stays at maximum temperature longer

Comparison of Experimental Data with Model

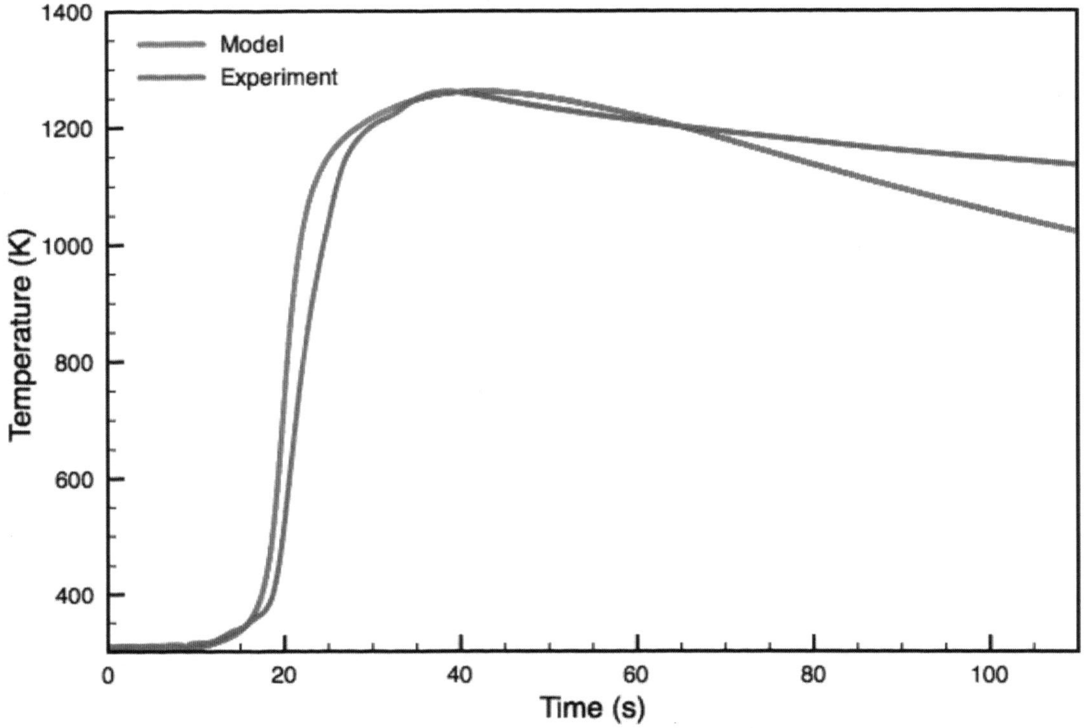

- Similar heating trend
- Model cools down at a faster rate

Acknowledgement

This work is supported by:

Lamar University Research Enhancement Grant,

Texas Air Research Center, and

Texas Hazardous Waste Research Center

103a Novel, Nanostructured, Immiscible a/b Polymer Blends Made by Conventional Melt Processing: Addition of a/c Gradient Copolymer as An Interfacial Compatibilizer

Robert W. Sandoval, Department of Chemical and Biological Engineering, Northwestern University, Evanston, IL 60208, Jungki Kim, Chemical and Biological Engineering, Northwestern University, 2145 Sheridan Rd, Evanston, IL 60208, and **John M. Torkelson**, Chemical and Biological Engineering and Materials Science and Engineering, Northwestern University, 2145 Sheridan Rd, TECH E136, Evanston, IL 60208.

Novel, compatibilized, nanostructured polymer blends have been made using a single melt-mixing step in which a 90/10 wt% A/B blend is mixed with a small amount of an A-C gradient copolymer as compatibilizing agent and in which there is an attractive dipole-dipole interaction between the B and C units. This is the first time that nanostructured blends have been made with such as a large level of the minor blend component without the aid of reactive processing or in situ polymerization of one component in the presence of the other polymer. In particular, when a 90/10 wt% polystyrene (PS)/poly (ethylene oxide) (PEO) mixture is melt blended with 7 wt% styrene (S)-methyl methacrylate (MMA) gradient copolymer, we obtain an average PEO dispersed-phase diameter of 105-110 nm. In contrast, without the addition of the gradient copolymer and using the melt processing conditions, the average PEO dispersed-phase diameter is 520-525 nm. The 90/10 wt% PS/PEO blend with added gradient copolymer is shown to be compatibilized by the virtual absence of coarsening of the dispersed-phase diameter during long-term, high-temperature static annealing (at 443 K for 90 min). However, using the same annealing protocol, the 90/10 wt% PS/PEO blend without added gradient copolymer undergoes significant coarsening, yielding an average PEO dispersed-phase diameter of 735 nm. When compared to random and block copolymer addition, the blends made with gradient copolymer show superior compatibilization efficiency, due to random copolymers unable to prevent coalescence and only very limited amounts of the block copolymer being able to reach the blend interfaces. This is apparent when observing the non-isothermal crystallization behavior. In blends with added gradient copolymer, PEO is confined to nanoscale domains, and the PEO crystallization occurs almost exclusively by homogeneous nucleation (at ~ 258 K) rather than heterogeneous nucleation at temperatures above room temperature. Thus, nanostructured PS/PEO blends that are maintained at temperatures near or above room temperature have a glassy matrix with nanoscale rubbery domains dispersed in the matrix while normal PS/PEO blends have a glassy matrix with semi-crystalline domains dispersed in the matrix. This study illustrates in a simple manner that nanostructured blends can yield major property differences from conventional polymer blends and that gradient copolymers, with a composition gradient along most or all of the copolymer chain length, can be useful in producing nanostructured blends using a very simple, commercially attractive processing method.

103b On the Equivalence of Glass Transition Temperatures In Polymer Nanocomposites and Polymer Thin Films

Venkat Ganesan[1], **Jamie Kropka**[1], and Victor Pryamitsyn[2]. (1) Department of Chemical Engineering, The University of Texas at Austin, 1 University Station, Code C0400, Austin, TX 78712, (2) Chemical Engineering, University of Texas at Austin, 1 University Station, Code C0400, Austin, TX 78712

Motivated by recent experiments, we examine within a percolation model whether there is a quantitative equivalence in the glass transition temperatures (T_g) of polymer thin films and polymer nanocomposites (PNCs). Our results indicate that while the qualitative behaviors of these systems are similar, a quantitative equivalence cannot be established in general. However, we propose a phenomenological scaling collapse of our results which suggests a simple framework by which the results of the thin films may be used to quantitatively predict the properties of PNCs.

103c High-Toughness, Hierarchically Structured Polyurethane Films from Exponential Lbl Assembly

Paul Podsiadlo[1], Eugene Kheng[2], Amit K. Kaushik[2], Harish Iyer[2], Hyoung-Sug Kim[3], Si-Tae Noh[4], Anthony M. Waas[5], Ellen M. Arruda[2], and Nicholas Kotov[1]. (1) Chemical Engineering, University of Michigan, 2300 Hayward St., 3074 H. H. Dow Bldg., Ann Arbor, MI 48109-2136, (2) Mechanical Engineering, University of Michigan, 2350 Hayward St., Ann Arbor, MI 48109, (3) R & D Center, Hepce Chem Co., Ltd., Ansan 425-836, Kyounggi-Do, South Korea, (4) Chemical Engineering, Hanyang University, Ansan 425-791, Kyounggi-Do, South Korea, (5) Aerospace Engineering, University of Michigan, 1320 Beal Ave., Ann Arbor, MI 48109

Nanostructured thin films prepared with the layer-by-layer (LBL) assembly process have shown remarkable versatility in transferring the unique properties of nanomaterials into functional, macroscale structures. The daunting problem of the LBL method is the slow speed of film growth, which is especially important for the preparation of ultra-strong nanocomposites. A recently discovered mode of the LBL assembly, the "exponential LBL" (e-LBL), offers a solution to this problem by greatly increasing the growth speed. Here we show the first preparation of high-toughness multilayered polymer films from water-born polyurethanes by the e-LBL process. We further describe a simple and general method for consolidating the thin films into hierarchical laminates which brings about a change in the paradigm for thin LBL films. The laminated structures show great improvement of mechanical properties: 4x in modulus, 4x in yield strength, 2.5x in ultimate strength, 3x in and toughness when compared to the base polyurethane. Further ability to incorporate functional materials into individual layers of the laminate allows for preparation of multifunctional polymeric materials with multi-level hierarchical organization.

103d Polythiophene-Gold Nanoparticle Hybrid Systems: Langmuir-Blodgett Assembly of Nano-Structured Films

Sundaramurthy Jayaraman[1], Ting Yu Liew[2], Prashun Gorai[2], and MP Srinivasan[2]. (1) Chem and Biomol Engg, NUS, 4 Engg Drive 4, Singapore, Singapore, (2) ChBE, NUS, 4 Engg Drive 4, Singapore, Singapore

Langmuir-Blodgett (LB) assembly technique is one of the most powerful tools to control molecular orientation, organization and for building uniform multilayer assemblies, and is a logical method to construct nano-structured thin films for micro- and nano-scale devices. Conducting polymers such as polythiophenes with excellent electronic and fluorescent properties can be deposited using the LB technique. Polythiophenes are one of the important classes of conducting polymers and the thin films of polythiophene finds application in field effect transistors, biosensors, organic photovoltaic cells and many more. Attempts have been made to increase the conductivity of ultra thin films of polythiophene either by doping, functionalization or incorporation of conducting nanoparticles.

In this work, we have fabricated nano-scale hybrid systems from polythiophene LB ultra thin film assembly with incorporated Au nanoparticles (AuNP). Investigation of electronic properties such as oxidation potential, conductivity and the electron mobility showed increase in measured parameters with increase in number of layers. Blends of an amphiphilic regioregular, ether-terminated polythiophene with a poly alkyl thiophene were also employed. Interaction studies showed that a net repulsive force exists between the two thiophene species – making the film assemblies potential candidates for patterned incorporation of AuNP. A number of interesting phenomena in terms of affinity of AuNP for thiophenes were observed by the various microscopic and spectroscopic characterization techniques. Electronic properties were investigated using Hall and surface resistivity measurements. A significant enhancement was observed in the conducting properties upon incorporation of AuNP.

103e Synthesis of Single Wall Carbon Nanotube-Polymer Composites by a Novel Two Steps Approach

Pu Zhang, Chemical & Biological Engineering, University of Missouri-Rolla, 143 Schrenk Hall, Rolla, MO 65401 and David Henthorn, Chemical Engineering, University of Missouri-Rolla, 143 Schrenk Hall, Rolla, MO 65401.

One of the most promising applications for carbon nanotubes is as a support material for different applications. In this work, we successfully developed a new "graft from" strategy to functionalize the surface of carbon nanotubes. In the first step, the carbon nanotubes were functionalized by initiator, providing active sites for subsequent polymer attachment. Depending on the application, polymer chains can be grown from the nanotubes surface either by ultraviolet light initiated polymerization or slow thermal reactions. For bio-oriented application, an ultraviolet light initiated "graft from" polymerization method was used to fabricate single wall carbon nanotube-polymer composites with pendant chains of various functionalities. Grafted polymer molecules with pendent PEG side chains were used to boost aqueous dispersibility while glycidyl methacrylate (GMA) monomer provided reactive epoxy functional groups for protein immobilization. A model enzyme, alkaline phosphatase, was used to study the loading efficiency as well as the retention of enzymatic activity. Samples with various ratios of the two monomers poly(ethylene glycol) methyl ether methacrylate (PEGMA) and GMA (PEGMA: GMA=4:1, 1:1, 1:4, 1:10) were fabricated to optimize their use in aqueous environments, and an optimum formula (1:1) of the composition was developed.

Polymer nanotubes composites rich in the GMA monomer was dispersed in epoxy resin and cured. An epoxy composite containing 0.5 wt% functionalized multiwall carbon nanotube (MWNT) was produced. With the addition of 0.5 wt% of functionalized MWNT, the reinforced epoxy exhibited a 48% enhancement in Young's modulus and 33% enhancement in ultimate strength.

Nanoscale ordering of polymer adsorbed on nanotubes

Simcha Srebnik*, Inna Gurevitz, and Stanislav Levchenko, Technion – Israel Institute of Technology, Haifa, Israel 32000

Abstract

We study the interaction of a dilute solution of semiflexible polymers with a weakly attractive infinitely long cylinder (i.e., nanotube) using Monte Carlo simulation. Apart for bending stiffness of the polymer chains, the only interactions considered in our model are weakly attractive short-ranged Lennard-Jones interactions between the monomers and with the surface. These nonspecific interactions are found to result in stable helical and multi-helical adsorbed conformations for semiflexible chains. Adsorption of these chains is found to occur in a sequential manner through tight wrapping of the polymer around the nanotube. Adsorption occurs quickly and is characterized by a sharp peak in the heat capacity. A second transition follows whereby opening and reorganization of the adsorbed chains into nearly perfect helices and multiple helices. Extension of the model to block and triblock copolymers reveals rich conformational behavior. These results are discussed on physical grounds and implications towards polymer-carbon nanotube composites are offered.

Keywords: Monte Carlo, polymers, adhesion, carbon nanotubes, block copolymers

* Email: simchas@technion.ac.il. Current address: Fischell Department of Bioengineering, University of Maryland, College Park, MD 20742

It is well known that carbon nanotubes (CNTs) possess extraordinary properties, such as regular electronic structure and high mechanical strength. However, since in their pristine state CNTs exist in a disordered agglomerated mass due to van der Waals (vdW) interactions (on the order of 0.5-1 eV per nm of nanotube-nanotube contact [1, 2]), the theoretical potential of their extraordinary properties falls far short of the actual performance [3]. The first challenge is, therefore, to obtain a homogeneous dispersion of CNTs for practical applications.

Regular helical wrapping of nanotubes, displayed by a number of polymers [4-8], may be used advantageously in the dispersion and control of properties of CNT/polymer composites, for example, through the helical pitch which directly controls the linear charge density of the complex and is predicted to be an important parameter controlling separation in ion-exchange chromatography [9, 10]. Recently, using coarse-grained MC simulations, in which only short-ranged weakly attractive nonbonded interactions were considered, we have shown that semiflexible chains will adsorb in nearly perfect helices on CNTs [11-13]. When multiple chains are allowed to adsorb on CNTs, simulations predict the formation of multiple helices [12, 13].

Block copolymers offer an additional parameter over homopolymers that can be used to control the distance between NTs, and thus may be used to tailor the structure and performance of polymer/NT composites.[14] In a recent study,[13] we showed that a sufficiently long hydrophobic block (H) will adsorb in a helical manner, while the polar block (P) extends to the bulk. Triblock (HPH) copolymers may provide additional advantage, in principle, by allowing for the formation of loops, anchored to the CNT surface by the wrapped hydrophobic blocks. The polar loops extending to the bulk can in principle entangle with free polymer, leading to strong adhesion between the CNTs and bulk polymer.

In this work, we utilize a minimalist Monte Carlo model in order to concentrate on the physical effects that drive long polymer chains to associate with CNTs and adopt ordered adsorbed conformations. The monomers are modeled as soft spheres, with Lennard-Jones (LJ) interaction potential between non-bonded monomers,

$$\frac{U_{LJ}}{kT} = 4\varepsilon_{ij}\left[\left(\frac{\sigma_{ij}}{r_{ij}}\right)^{12} - \left(\frac{\sigma_{ij}}{r_{ij}}\right)^{6}\right] \quad (1)$$

where k is the Boltzmann constant, T is the temperature, ε_{ij} is the average 2-body LJ interaction parameter between monomers i and j ($\varepsilon_{ij} = \left(\varepsilon_{ii}\varepsilon_{jj}\right)^{1/2}$), σ_{ij} is the average effective size of monomers i and j ($\sigma_{ij} = \left(\sigma_i + \sigma_j\right)/2$), and r_{ij} is the distance between the two interacting nonbonded monomers. The well potentials used for the hydrophobic and polar interactions are $\varepsilon_H = 1$ and $\varepsilon_P = 0.01$, respectively. A cut-off radius of $2^{1/6}\sigma$ for the monomer-monomer interactions is used to allow for the effect of short-range attractive forces at a low computational cost. We note that for the monomer-monomer LJ interactions, the potential cut-off has a minor affect on chain conformational behavior when compared with the bending potential (Eq. 2, below) at the temperature, chain length, and concentration studied, especially for the stiffer chains. Thus, the attractive well of the potential really only comes into play when the chains are adsorbed on the cylinder.

Chain stiffness is modeled by the following harmonic expression for the energy required to bend the angle θ formed between consecutive bond vectors,

$$U_b/kT = \kappa(\cos\theta - 1)^2 \quad (2)$$

where κ characterizes the resistance of the bond angle θ_j about monomer j to bending. is defined by $\cos\theta_j = \mathbf{b}_{j-1} \cdot \mathbf{b}_j / (|\mathbf{b}_{j-1}||\mathbf{b}_j|)$, where $\mathbf{b}_j = r_{j+1} - r_j$, and r_j is the absolute position of monomer j. The bending stiffness parameter κ appearing in Eq. 2 can be easily related to the persistence length, l_p, through a bond-angle correlation function which decays exponentially with l_p, thereby obtaining a power law dependence of l_p on κ [15-18]. Note that using $\cos\theta$ instead of θ in Eq. 2 simply provides a different correlation between κ and the persistence length of the polymer, but simplifies the computational calculation by avoiding the use of the inverse cosine function. These simulation parameters correspond to good solvent conditions.

Since CNTs exhibit an even charge distribution [19], electrostatic interaction between nanotubes and polymers is relatively weak, and therefore in the absence of chemical functionalization, the polymer-CNT interaction is governed by van der Waals forces [20, 21]. Therefore, we model the CNT as a smooth infinitely long and impenetrable cylindrical shell with a continuous LJ potential field averaged over the tube length. The polymer-tube interactions are obtained by integrating the dimensionless LJ potential over the length of the tube [17, 22], to obtain a potential that depends on tube dimensions (inner radius ρ_i and outer radius ρ_o) and the perpendicular distance of a monomer from the surface of the tube, D. Integration of Eq. 1 over the vertical z-axis of the tube yields and an analytical equation, whose integration over the radial and azymuthal axes needs to be carried out numberically. The resulting potential was adjusted such that the potential minimum of tubes of different radii correspond to the potential minimum of a tube with ρ_o=2σ in order to concentrate on the effect of surface curvature.

Simulations were carried out on chains consisting of N_m monomer units consisting of b blocks, where each block is made up of N_H hydrophobic (H) segments or N_P polar (P) segments. Bond length was taken to be 1σ in all our simulations. All segments were taken to be of equal size, σ. A cubic simulation box with periodic boundary conditions was considered with the nanotube placed in the center of the box, aligned along the z-axis. The simulation is initialized by placing N_c identical chains randomly within the box. We use N_c=3 for the homopolymer and N_c=1 for the block copolymer (unless otherwise stated). Equilibration is achieved using a combination of reptation and kink-jump moves [23], tried with approximately 0.1 and 0.9 probability, respectively, during the production stage of the simulation. The move probability is adjusted at the course of equilibration to achieve approximately 50% Metropolis acceptance rate. To ensure equilibration, 10^6 MC steps are performed prior to calculating average properties. Averages are calculated over an additional 10^6 - 10^7 MC steps, sampled every 10^2 MC steps. Although we studied nanotubes of various radii, we concentrated on ρ = 2σ, which corresponds to the dimension of single-walled CNTs [12].

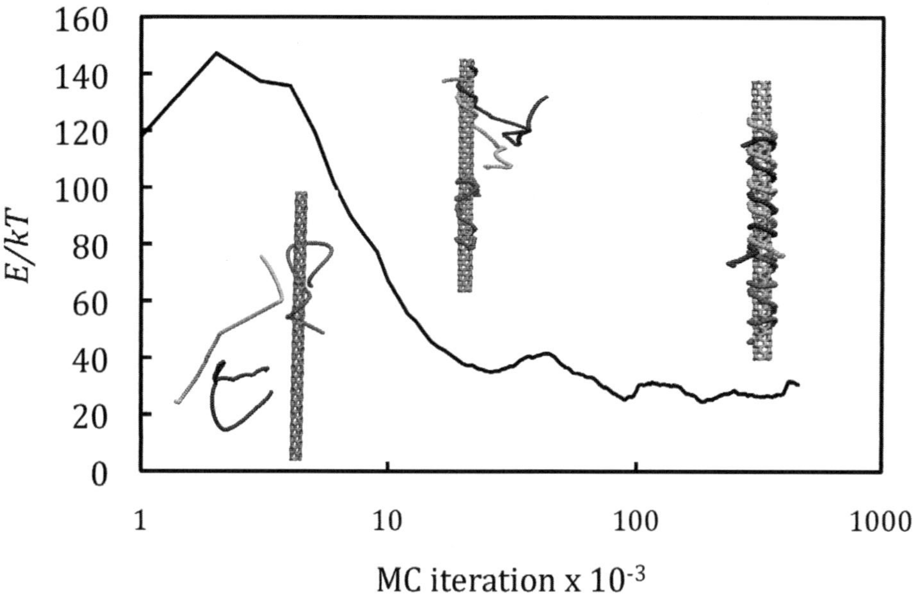

Figure 1. Energy as a function of simulation time; $\kappa = 40$, $\rho = 2\sigma$, $N_m = 100$, $kT = 1$. Snapshots, from left to right, correspond to typical conformations at these conditions after 10^3 iterations, 10^4 iterations, and 10^5 iterations, respectively.

Recently, we have shown that there exists a range of parameters for which polymers will adsorb on CNTs in an ordered helical manner [11, 13, 17]. As seen in **Figure 1**, the energy reaches its equilibrium value quite quickly, whereby nearly perfect helical conformations are observed. Adsorption is reversible with temperature (**Figure 2**), indicating that the observed helical conformations are a thermodynamically stable state, also at finite temperatures. Desorption of the polymers with increasing temperature has also been reported experimentally [5].

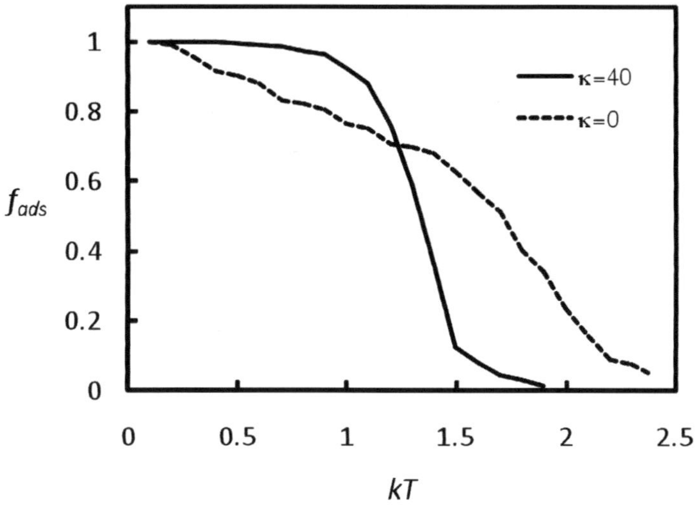

Figure 2. Adsorbed fraction as a function of temperature; $\rho = 2\sigma$, $N_m = 100$.

During the course of the simulation, the polymers spiral up and down the surface of the tube. This behavior suggests that the polymers adopt a conformation with an optimal pitch. The helical pitch is predicted to be an important parameter controlling, e.g., separation in ion-exchange chromatography since it directly controls the linear charge density of the polymer-CNT surface, and thus strongly influences the electrostatic field near the surface [19, 24]. Thus, nonspecific association that is responsible fore helical wrapping may also result in an optimal pitch for a particular polymer-CNT configuration. That an optimal helical pitch may be determined by nonspecific structure and interactions is important since all known preparations of CNTs give mixtures of nanotube chiralities [25].

The dependence of the adsorbed fraction and of α on chain length obtained from the simulation is shown in **Figure 3**. The dashed line represents the adsorption transition marking the approximate location of the peak in the fluctuations C_a of the adsorbed fraction, i.e., $C_a = \langle f_{ads}^2 \rangle - \langle f_{ads} \rangle^2$. **Figure 3**a suggests that short chains remain largely in the bulk, while longer chains adsorb readily. Chain length then plays an important factor in determining the adhesion between the polymers and CNT, with the transition between oligomeric behavior to polymer behavior in this study occurring at $N_m \approx 15$ (dashed line). Polymer chains are therefore expected to be much more efficient at solubilizing CNTs then their oligomeric counterparts [6]. We also learn from **Figure 3**b that oligomeric chains of all flexibilities as well as long flexible polymers adsorb in random orientations ($\alpha = \pi/2$), while the longer stiff polymers adsorb in a conformation characterized by $\alpha < \pi/2$. The random adsorbed orientation of the short stiff oligomers results from their relatively low entropic state (compared with a long polymer chain), allowing oligomers to readily fluctuate between and adsorbed and desorbed states.

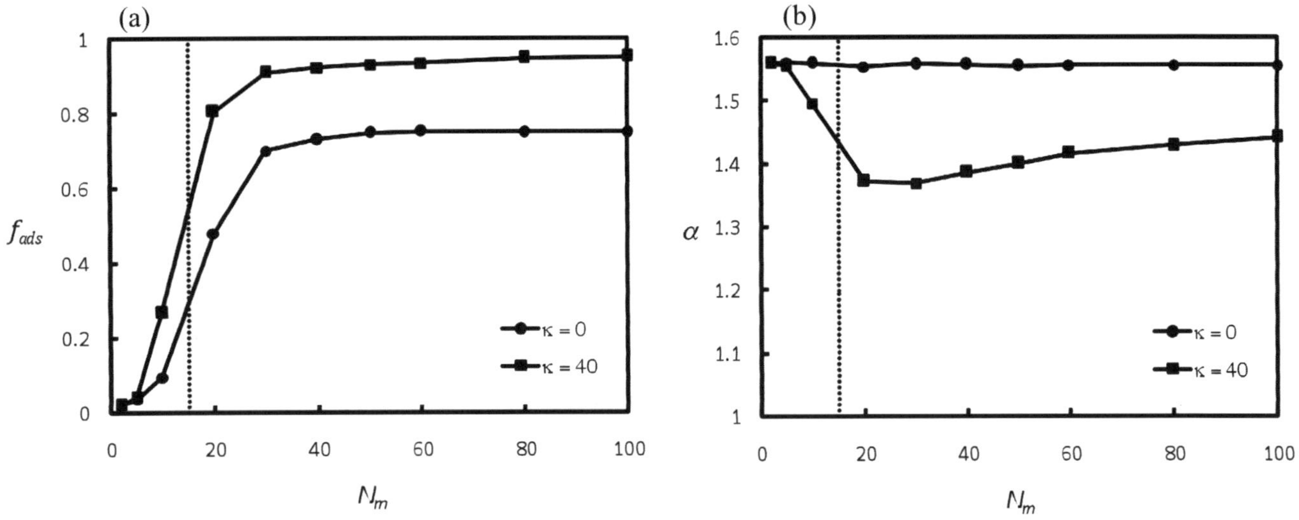

Figure 3. (a) Adsorbed fraction and (b) helical pitch as function of chain length; $\rho = 2\sigma$, kT=1. The dotted lines represents the adsorption transition obtained from the peak in the fluctuations of the adsorbed fraction.

The average pitch angle α of the adsorbed polymers is found to be nearly independent of chain length for sufficiently long chains (**Figure 3**). α can be approximated from the Kratky-Porod [26] (KP) wormlike chain assumption that the angle θ formed by bond vectors separated by x segments decays exponentially, $\langle \cos\theta_x \rangle \sim \exp(-xl/l_p)$, where l_p is the persistence length of the chain. Assuming that the optimal conformation occurs when the average bond angle between two consecutive segments (x = 1) corresponds to the average bond angle θ of the adsorbed polymer. The relation between the pitch angle α and bond angle θ can then be obtained from geometric considerations, and may be shown to be independent of chain length [27]. The prediction of the helical pitch thus obtained as a function of chain stiffness is shown by the dashed curve of **Figure 4**. The symbols show the calculated pitch angle from simulations, using the relation $l_p \approx \kappa^{0.6}$ [12]. A better agreement is obtained when θ is corrected by 2° (solid curve). That is, the adsorbed chains have a somewhat smaller bending angle (they are more flexible) then predicted by the unconfined KP model. The probable cause is the interaction with the CNT surface, which competes with the bending interactions and thereby reduces the effective size of the chain. The transition to extended conformations ($\alpha < \pi/2$) in **Figure 4** occurs when chain stiffness, determined from the average bending angle, $\theta = \cos^{-1}\left(-e^{-l/l_p}\right)$, approximately equals the minimum radius of curvature of the cylinder, $\theta_c = 2\cos^{-1}(l/2\rho)$. Polymers with l_p beyond this critical value will adsorb in extended helical conformation to reduce the effect of surface curvature.

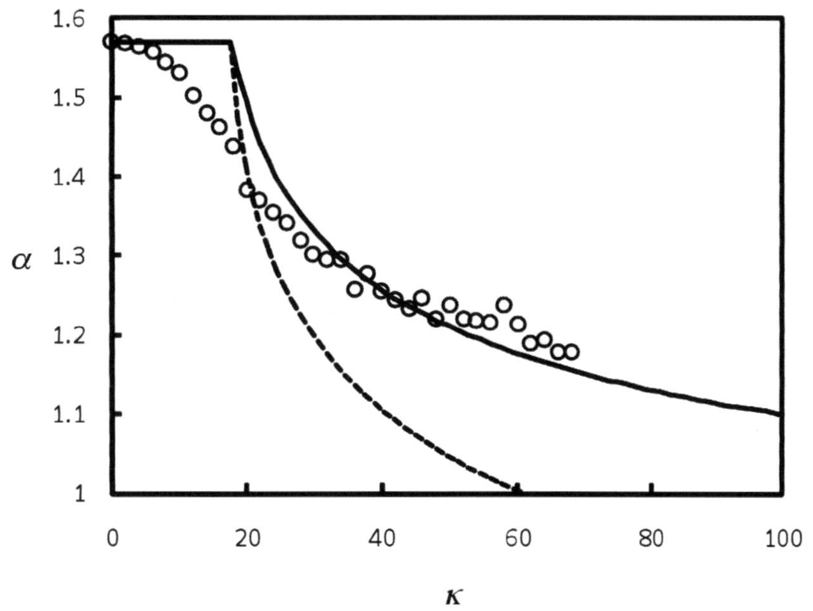

Figure 4. Helical pitch as a function of κ. Symbols show simulation results ($\rho = 2\sigma$, kT=1), dashed curve shows model predictions for the same conditions, and solid curve shows model predictions with a 2° correction to α.

Our previous simulations indicate that semiflexible homopolymers under the conditions studied adsorb in a monolayer on the surface of the CNT, for the range of concentrations studied [13]. Such conformation may not be ideal for polymer reinforcement, where strong adhesion between the polymer matrix and CNT is required. For such applications, block copolymers (BCPs) consisting of monomers with different solvent selectivity may provide a stronger interface between the CNT and polymer [28]. In such a scenario, one block is anchored to the CNT surface through strong physical association, while the other block extends to solution. This configuration not only provides steric repulsion between CNTs but also increases adhesion between the CNT-polymer interface [14] by allowing for entanglement between the adsorbed polymers and matrix polymers.

We studied the adsorption and wrapping mechanism of diblock and triblock copolymers on CNTs consisting of hydrophobic (H) blocks that favorably interact with the CNT ($\varepsilon_H = 1$) and 'polar' (P) blocks that interact with the CNT through weakly attractive interactions ($\varepsilon_P = 0.01$). Such a choice of interactions represents adsorption in a hydrophobic solvent, and results in strong microphase separation of the hydrophobic and polar segments of the polymer in absence of the CNT [29]. In general, adsorbed BCPs show significant loss of order when compared to homopolymers under the same conditions, though helical conformations of the H blocks may also be observed (Figure 5). However, at the relatively short simulation times considered, helical conformations appear to occur fortuitously if H segments near the interface between the two blocks adsorb first, allowing for sequential wrapping of the other H segments. Thus, a mixture of helical and ordered configurations might be expected, as has also been reported experimentally for alternating copolymers [30]. Similar to the homopolymer, the randomly adsorbed state of the copolymer is likely to be a metastable state. However, for the copolymer, rearrangement to the helical state requires significantly longer (simulation) times due to the competing interactions of the polar block. Nonetheless, the possibility of formation of loops with the triblock copolymers, shown in Figure 5, might offer particular advantage towards interface reinforcement, by addressing both adhesive and cohesive failure in polymer/CNT composites.

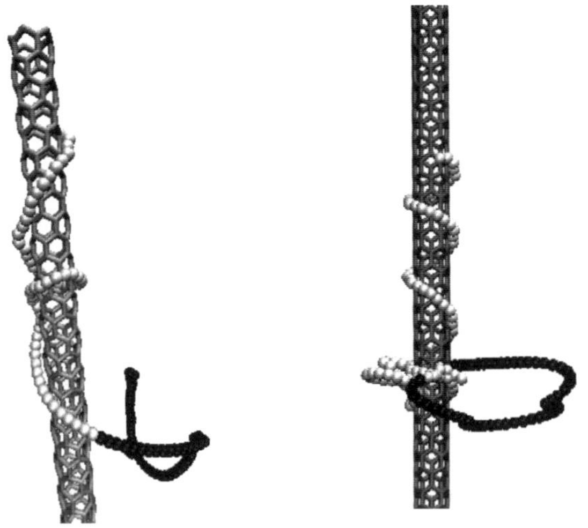

Figure 5. Simulation snapshots of diblock (left) and triblock (right) copolymer-CNT systems; $\kappa_H = \kappa_P = 40$, $\rho = 2\sigma$, $N_m = 60$ per block.

References

1. Dyke, C.A. and J.M. Tour, Journal of Physical Chemistry A, 2004. **108**(51): p. 11151-11159.
2. Nap, R. and I. Szleifer, Langmuir, 2005. **21**(26): p. 12072-12075.
3. Bokobza, L., Polymer, 2007. **48**(17): p. 4907-4920.
4. Kim, O.K., et al., Journal of the American Chemical Society, 2003. **125**(15): p. 4426-4427.
5. Tang, B.Z. and H.Y. Xu, Macromolecules, 1999. **32**(8): p. 2569-2576.
6. Yuan, W.Z., et al., Macromolecules, 2008. **41**(3): p. 701-707.
7. Fu, C., et al., Macromolecular Rapid Communications, 2007. **28**(22): p. 2180-2184.
8. Dieckmann, G.R., et al., Journal of the American Chemical Society, 2003. **125**(7): p. 1770-1777.
9. Lustig, S.R., et al., Journal of Physical Chemistry B, 2005. **109**(7): p. 2559-2566.
10. Manohar, S., T. Tang, and A. Jagota, Journal of Physical Chemistry C, 2007. **111**(48): p. 17835-17845.
11. Kusner, I. and S. Srebnik, Chemical Physics Letters, 2006. **430**: p. 84-88.
12. Gurevitch, I. and S. Srebnik, Chemical Physics Letters, 2007. **444**(1-3): p. 96-100.
13. Gurevitch, I. and S. Srebnik, The Journal of Chemical Physics, 2008. **128**(14): p. 144901.
14. Shvartzman-Cohen, R., et al., Langmuir, 2004. **20**(15): p. 6085-6088.
15. Cerda, J.J., T. Sintes, and A. Chakrabarti, Macromolecules, 2005. **38**(4): p. 1469-1477.
16. Micka, U. and K. Kremer, Journal of Physics - Condensed Matter, 1996. **8**(47): p. 9463-9470.
17. Gurevitch, I. and S. Srebnik, Chemical Physics Letters, 2007. **444**: p. 96-100.
18. Kusner, I. and S. Srebnik, Macromolecules, 2007. **40**(17): p. 6432-6438.
19. Wall, A. and M.S. Ferreira, Physical Review B, 2006. **74**(23).
20. Yang, M.J., V. Koutsos, and M. Zaiser, Journal of Physical Chemistry B, 2005. **109**(20): p. 10009-10014.
21. Xie, Y.H. and A.K. Soh, Materials Letters, 2005. **59**(8-9): p. 971-975.
22. Milchev, A. and K. Binder, Journal of Chemical Physics, 2002. **117**(14): p. 6852-6862.
23. Baschnagel, J., J.P. Wittmer, and H. Meyer, *Monte Carlo Simulation of Polymers: Coarse-Grained Models*, in *Computational Soft Matter: From Synthetic Polymers to Proteins*, N. Attig, et al., Editors. 2004, John von Neumann Institute for Computing: Julich. p. 83-140.
24. Wall, A. and M.S. Ferreira, Journal of Physics-Condensed Matter, 2007. **19**(40).
25. Moniruzzaman, M. and K.I. Winey, Macromolecules, 2006. **39**(16): p. 5194-5205.
26. Kratky, O. and G. Porod, Recl. Trav. Chim. Pays-Bas, 1949. **68**: p. 1106.
27. Srebnik, S., Journal of Polymer Science Part B-Polymer Physics, submitted.
28. Vaisman, L., H.D. Wagner, and G. Marom, Advances in Colloid and Interface Science, 2006. **128**: p. 37-46.
29. Marrink, S.J., A.H. de Vries, and A.E. Mark, Journal of Physical Chemistry B, 2004. **108**(2): p. 750-760.
30. McCarthy, B., et al., Synthetic Metals, 2001. **121**(1-3): p. 1225-1226.
31. Strano, M.S., et al., Journal of Nanoscience and Nanotechnology, 2003. **3**(1-2): p. 81-86.

127a Plasma-Induced Structural Transitions In Materials

Eray S. Aydil and Michael Behr. Chemical Engineering and Materials Science, University of Minnesota, 421 Washington Avenue SE, Minneapolis, MN 55455

Atomic hydrogen plays a key role in plasma assisted deposition of group IV films. Indeed, it is well known that plasma deposition with silane, methane and germane gas heavily diluted in hydrogen results in microcrystalline silicon, diamond and germanium, respectively. In fact, thin films of hydrogenated amorphous silicon undergo chemically-induced disorder-to-order structural transformations and crystallize upon exposure to H atoms created through plasma. Remarkably, this transition occurs at temperatures (~25-300 ^{o}C) much lower than those required for thermal crystallization. While the critical role of atomic H created through plasma dissociation of H_2 has been established experimentally, the mechanism of how H enables the deposition of microcrystalline group IV films had remained a mystery. Recently, we have shown that hydrogen induced crystallization of amorphous silicon is mediated by the insertion into strained Si-Si bonds of H atoms as they diffuse through the film and by the concerted rearrangements of the bonds in the vicinity of the insertion reaction. Taking advantage of these reactions we have even deposited microcrystalline silicon at room temperature.

Based on the hypothesis that this H-induced crystallization mechanism is general, and may be operative in other covalently-bonded group IV solids, we have exposed multiwall carbon nanotubes (MWCNTs) to H atoms produced by electron impact dissociation of H_2 molecules in an inductively coupled radio-frequency (rf) plasma. We hypothesized that reactions with H atoms like those observed in silicon may help bind carbon atoms in neighbouring concentric CNTs and transform them into diamond nanocrystals. Indeed, we observed hydrogen-induced transformation of MWCNTs to various crystalline carbon structures, even at room temperature. Specifically, exposure of multiwall carbon nanotubes (MWCNTs) or multiwall carbon nanofibres to atomic hydrogen transformed them into other carbon allotropes including cubic diamond and hexagonal diamond (lonsdaleite). Nanometer-size crystals of diamond appear gradually in webs of long strings of crystallites, ~ 5-50 nm in diameter, where nanotubes once laid. This H-induced transformation of MWCNTs to diamond is observed even at room temperature. Moreover, high-resolution transmission electron microscopy of the H-exposed material reveals the presence of carbon nanocrystals whose electron diffraction patterns and lattice spacings could not be accounted for by known crystalline phases of carbon, such as diamond and lonsdaleite, or by contaminants. In addition to cubic and hexagonal diamond, interactions of H atoms with the concentric graphene layers of the MWCNTs produce new crystalline carbon phases that have not been observed previously. Specifically, we show, unambiguously, the existence of a face-centered cubic carbon phase with lattice constant a=0.426 nm.

Herb Sawin and DuPont: 25 Years of Cooperative Research and Development

Michael T. Mocella, DuPont Electronic Technologies, Wilmington, DE

In the early 1980's, Herb Sawin and DuPont began a joint research program which, although never involving more than single-year commitments, nevertheless endured on a continuous basis for twenty-five years. There were both many scientific discoveries during the course of this work, as well as insights into the nature and workings of industrial-academic collaborations.

In this presentation, I will blend these two program components, by proposing five guidelines for effective collaboration, and using illustrations from the program's results to show their application. These guidelines are:

1. *Start from where you are.* While there may or may not be tightly specified goals at the beginning of such a collaboration, each team brings specific areas of expertise and insight to the relationship. By carefully examining the partners' skills, new and unexpected areas for collaboration may develop.
2. *Lead the science in emerging areas.* Early results can draw external contacts and input on new areas for fruitful work. The parties' standing can be enhanced by using these insights to produce results of particular interest to the industrial and academic communities exploring the technology.
3. *Understand and manage differing organizational priorities.* Corporations like to make money, and universities like to publish! While these orientations may lead to short-term disconnects, proper management can ensure that they will be harmonized within the overall program goals.
4. *On occasion, stretch towards the improbable.* High risk R&D offers the potential for high reward for the industrial partner, and unique science for the academic one. So, along with work expected with good confidence to lead to useful outcomes, some part of the program should be directed towards lower confidence but potentially game-changing directions.
5. *Put important empirical processes on a sound scientific footing.* It is not uncommon in the semiconductor industry for important processes to be highly optimized, but little understood. By developing good science around the process knowledge, new insights of both scientific and commercial value may result.

127c Modeling Plasma-Surface Interactions and Their Role in Inducing Structural Transitions

Dimitrios Maroudas, Department of Chemical Engineering, University of Massachusetts, 686 N Pleasant Street, Amherst, MA 01003

Modeling plasma-surface interactions and their role in inducing structural transitions

127d Reaction Mechanisms in Plasma Etching of High-K Materials

Jane P. Chang, Department of Chemical and Biomolecular Engineering, University of California at Los Angeles, Los Angeles, CA 90095

The continuous down-scaling of the microelectronic integrated circuits dictates the implementation of high dielectric constant materials and the corresponding design of plasma chemistry to achieve the etching specificity, selectivity, and anisotropy in the definition of these high-k materials. In this talk, I will discuss the reaction mechanisms in etching hafnium based high-k dielectrics in chlorine-based chemistries and present a generalized etch rate model that accounts for the transition between physical- and ion-enhanced etching in Cl_2 plasmas and the transition between deposition and etching in BCl_3 plasmas, as the ion energy increased.

127e Reception to AIChE Area 8e Special Symposium

Jane P. Chang, Department of Chemical and Biomolecular Engineering, Chemical Engineering Department, Cambridge, MA 02139

Area 8e will present a one-day invited symposium that highlights the advances in understanding plasma-surface interactions and thin film processing, to honor the many contributions by Professor Herbert Sawin in plasma science, as he retires from Massachusetts Institute of Technology. This invited symposium, entitled "Topics In Plasma Science and Thin Film Applications - In Honor of Herbert H. Sawin" will feature industrial and academic leaders around the world, amongst them many of Herb Sawin's former students. They are Eray Aydil (U. Minnesota), Demetre Economou (U. Houston), David Graves (UC Berkeley), Dimitrios Maroudas (U. Massachusetts), Michael Mocella (Du Pont), Jane Chang (UCLA), Brett Cruden (NASA Ames Research), Evangelos Gogolides (IMEL, Greece), Scott Lawing (Rohm and Haas), Arpan Mahorowala (IBM), Anthony Muscat (U. Arizona), Steve Vitale (MIT Lincoln Lab) and Colin Wolden (Colorado School of Mines).

The reception will start at 2:45pm, in between the two technical sessions in the afternoon.

135a A Mechanistic Modeling Approach to the Design and Evaluation of Polymeric Drug Delivery Systems

Ashlee N. Ford, Chemical & Biomolecular Engineering, University of Illinois, 600 South Mathews Avenue, 201 Roger Adams Laboratory, Box C-3, Urbana, IL 61801-3602, Daniel W. Pack, Chemical & Biomolecular Engineering, University of Illinois Urbana-Champaign, 600 S. Mathews, Box C-3 MC 712, Urbana, IL 61801, and Richard D. Braatz, Department of Chemical and Biomolecular Engineering, University of Illinois at Urbana-Champaign, 600 South Mathews Avenue, Box C-3, 293 Roger Adams Laboratory, Urbana, IL 61801-3602.

Controlled-release drug delivery systems are being developed as an alternative to conventional medical drug therapy regimens which require frequent administrations due to short pharmaceutical in vivo half-life and poor oral bioavailability. Controlled-release systems have the potential to provide better control of drug concentrations, reduce side effects, and improve compliance as compared to conventional regimens. The model-based design of controlled-release devices, such as biodegradable poly(lactic-co-glycolic acid) (PLGA) polymer microspheres, is challenging because of incomplete understanding of the mechanisms that regulate the release of drug molecules. This research focuses on modeling the autocatalytic polymer degradation and release of dispersed drug molecules from PLGA microspheres to capture size-dependent heterogeneous degradation behavior observed experimentally but not accounted for by existing models. Recently, other researchers have suggested that the autocatalytic polymer degradation is the primary mechanism by which the diffusive drug release is accelerated, and this process should depend strongly on particle size. The hypothesis of the present work is that simultaneously modeling the mathematics of diffusion, autocatalytic chemical reactions, chemical equilibria, and pore formation, the phenomena which are considered to contribute to the degradation of polymer particles, rather than independently modeling any of the phenomena in a purely sequential manner will accurately mimic the actual overall release process. The developed model tracks acid concentration as a function of space and time for determination of intraparticle pH while modeling degradation kinetics, molecular weight distribution variation, and drug transport with varying diffusivity coupled to the concentrations of other reacting species—all of which influence drug release from the polymer microspheres. Acid is considered from two sources—acid found in the external media and the acid produced from dissociation of the carboxylic acid end groups of PLGA. The chemical reaction mechanism including autocatalytic effects is coupled to a simplified diffusion model and pore formation model to incorporate spatial variations in degradation rate for all species within the microspheres. The inclusion of the spatial variation of autocatalytic effects is a unique contribution of this modeling work.

The presentation will also discuss the application of the model for the design and evaluation of different hydrolytically degradable polymeric materials which decompose by the bulk-eroding mechanism in drug delivery systems as PLGA does. Material design variables include the molecular weight distribution, the pore size distribution, and the chemical structure of the polymer. These variables can be optimized simultaneously with the particle size to provide for a desired controlled release profile.

135b Cell Uptake of Morphologically Distinct Polymer Micelles

Takamasa Harada, Karthikan Rajagopal, and Dennis E. Discher. Chemical and Biomolecular Engineering, University of Pennsylvania, Room 129 Towne Building, 220 South 33rd Street, Philadelphia, PA 19104-6393

Polymer based micelles show considerable promise as drug delivery systems, with some already in clinical trials, but all of them possess spherical morphology. Cylindrical worm-like micelles seem to possess several advantages over spherical micelles, particularly longer blood circulation and higher drug loading capacity (1,2). Although in vitro and in vivo studies of worm-like micelle have demonstrated efficiency in the delivery of anti-cancer drug, little is known as to how cells internalize worm-like micelles. Detailed understanding of the cell uptake mechanism and subsequent intracellular trafficking of worm-like micelles could be helpful for development of worm-like micelles; for example, a single microns-long micelle carries enough drug to kill a single cell whereas hundreds of micelles would be required to deliver an equivalent dose. In the present work, both worm-like and spherical micelles are prepared from amphiphilic diblock copolymer poly(ethylene oxide)-b-poly(ε-caprolactone) (OCL), a bio-compatible copolymer. The volume fraction of the poly(ethylene oxide) segment determines the preferred morphology upon assembly in aqueous media (3), although sonication and other disruptive treatments can generate kinetically stable sphere micelles regardless of preferred morphology. OCL-based micelles are visualized by integration of fluorescent dyes and fluorescence microscopy used to visualize intracellular localization as a function of time in culture with human lung carcinoma cell A549. Current studies suggest that worm-like micelles are scissioned as they are internalized by cells, highlighting the strength of the fission machinery in internalization.

[Reference]

1. Geng et al. Nature Nanotechnology 2, 249-255 (2007)

2. Cai et al. Pharmaceutical Research 24, 2099-2109 (2007)

3. Discher et al. Annu, Rev. Biomed. Eng. 8, 323-341 (2006)

135c Synthesis and Evaluation of Peptide-Functionalized Polymersomes for Targeted Drug Delivery to Colon Cancer Cells

Todd Pangburn, Frank S. Bates, and Efrosini Kokkoli. Department of Chemical Engineering and Materials Science, University of Minnesota, 421 Washington Ave. SE, Minneapolis, MN 55455

Recently, polymersomes (polymeric vesicles) have gained increased attention as systemic drug delivery vehicles due to their exceptional mechanical stability and increased stealth properties compared to pegylated liposomes. Self-assembled from amphiphilic block copolymers, polymersomes contain a hydrophobic core (e.g., polybutadiene) and a hydrophilic corona (usually poly(ethylene oxide) or PEO). This combination offers a host of advantages. Our goal is to bring biofunctionality to polymersomes in order to facilitate their potential use as a drug delivery agent. To improve the selectivity of the polymersome for different therapies, a novel peptide synthesized in our lab, called PR_b, has been introduced to the polymeric vesicle structure in order to provide selectivity for the alpha(5)beta(1) integrin, an adhesion molecule over-expressed on the cell surface of colon cancer cells. Following self assembly of amphiphilic block copolymers in water to form the polymersomes, the vesicle surface is covalently decorated with the PR_b peptide utilizing "click" chemistry. The polymers and peptide-functionalized polymeric nanoparticles have been characterized with a variety of experimental techniques such as SEC, NMR, MALDI, TEM, and DLS. The PR_b functionalized polymersomes have been further characterized with in vitro studies and our flow cytometry and confocal imaging data will be discussed. Our studies demonstrate that the peptide-functionalized polymersomes show superior targeting and internalization to CT26 colon cancer cells compared to non-functionalized polymersomes, thus making them a promising agent for targeted drug delivery.

135d Fully Bioresorbable Polymeric Vesicles for Therapeutic Applications

Dalia H. Levine[1], P. Peter Ghoroghchian[2], Jaclyn Freudenberg[3], Geng Zhang[3], Guizhi Li[4], Kevin P. Davis[5], Frank Bates[5], Michael J. Therien[4], Ramachandran Murali[3], and Daniel A. Hammer[6]. (1) Chemical and Biomolecular Engineering, University of Pennsylvania, 210 S. 33rd St., Skirkanich Hall, Rm. 335S, Philadelphia, PA 19104, (2) Bioengineering, University of Pennsylvania, 210 S. 33rd St., Skirkanich Hall, Rm. 335S, Philadelpia, PA 19104, (3) Department of Pathology & Lab. Medicine, University of Pennsylvania, 243 John Morgan Building, Philadelphia, PA 19104, (4) Chemistry, University of Pennsylvania, 231 S. 34 Street, Philadelphia, PA 19104, (5) Chemical Engineering and Materials Science, University of Minnesota, 151 Amundson Hall, 421 Washington Avenue SE, Minneapolis, MN 55455, (6) Chemical and Biomolecular Engineering/Bioengineering, University of Pennsylvania, 210 S. 33rd St., Skirkanich Hall, Rm. 335S, Philadelphia, PA 19104

Polymersomes (polymer vesicles) have been shown to possess a number of attractive biomaterial properties compared to liposomes (phospholipid vesicles), including prolonged circulation times, increased mechanical stability, as well as the unique ability to incorporate numerous large hydrophobic molecules within their thick lamellar membranes and hydrophilic molecules within their core. We have previously shown the ability to generate two types of self-assembled nano-sized vesicles ranging in size from 100's of nanometers to 10's of microns; one type comprised of a biocompatible diblock copolymer consisting of polyethyleneoxide (PEO) and polybutadine (PBD) and a second fully-bioresorbable vesicle consisting of two FDA-approved building blocks: polyethyleneoxide (PEO) and polycaprolactone (PCL). In addition, we have successfully loaded imaging agents, such as porphyrin-based near infrared (NIR) fluorophores, and therapeutics such as doxorubicin, an anti-neoplastic agent, into these polymersomes and tracked their release in situ and in vivo.

NIR-emissive polymersomes, loaded with porphyrin, can be used for biodistribution studies, to track the location of the polymersomes, and potentially for diagnostic studies. Here, we utilize NIR-emissive polymersomes to determine polymersome biodistribution in tumor bearing mice using a noninvasive small animal optical imaging instrument which detects the NIR fluorescence signal. Passive accumulation of NIR-emissive polymersomes in tumor tissues of mice, as well as other organs, is evidenced. Using porphyrin polymersomes for biodistribution studies will greatly decrease the number of animals required since the location of the polymersomes can be determined without sacrificing animals at multiple time points to perform histology on the excised organs.

Doxorubicin, an anti-neoplastic agent, was encapsulated to serve as a model system for the release of a physiologically relevant compound from the PEO-b-PCL polymersomes. While the kinetics of the release varied at pH 7.4 and pH 5, an initial burst release phase was observed for both pH's followed by a more controlled pH dependent release. The therapeutic potential of doxorubicin loaded polymersomes is shown; drug loaded bioresorbable polymersomes were administered in vivo and their capability to retard tumor growth was assessed using such metrics as tumor size and body weight. Doxorubicin loaded polymersomes were able to retard tumor growth in a live animal on a par with the commercially available DOXIL, liposomal doxorubicin. Furthermore, mouse weights remained within +/-1.5g, for all treatment groups throughout the study. Therefore, this study highlights the enormous promise for polymersomes as in vivo imaging and drug delivery agents.

135e A Novel Linear Dendritic Amphiphilic Block Copolymer as a Viable Drug Delivery Carrier

Zhiyong Poon[1], Paula T. Hammond[2], and Shujun Chen[2]. (1) Mit, Cambridge, MA 02139, (2) Chemical Engineering, Massachusetts Institute of Technology, 77 Massachusetts Avenue, 66-525, Cambridge, MA 02139

A linear dendritic amphiphilic block copolymer (LDP) that self assembles into micelles in aqueous solution was synthesized and developed as a potential drug delivery carrier. In these micelles, the hydrophobic linear block forms the micellar core, which can encapsulate hydrophobic drugs, while the hydrophilic dendron forms the outer shell. The polymer forms stable micelles in water with a low CMC of 10-8 M. It is expected that such stability will allow the LDP micelles to withstand dilution in blood. Dynamic light scattering data showed that these micelles are about 70 nm in diameter, which is in the appropriate size window for a drug delivery carrier in the body. These micelles exhibited good circulation times (~20 h) in mice and are non-toxic to mice up to 12 mg/ml.

135f Polymer Nanoparticles That Overcome the Mucus Barrier and Their Efficacy In a Mouse Model of Human Lung Cancer

Benjamin C. Tang[1], Michelle R. Dawson[1], Samuel K. Lai[1], Ying-Ying Wang[2], Ming Yang[2], Jie Fu[1], D. Neil Watkins[3], and Justin Hanes[1]. (1) Department of Chemical and Biomolecular Engineering, The Johns Hopkins University, 221 Maryland Hall, 3400 N. Charles Street, Baltimore, MD 21218, (2) Department of Biomedical Engineering, The Johns Hopkins University, 221 Maryland Hall, 3400 N. Charles Street, Baltimore, MD 21218, (3) Cancer Biology, The Johns Hopkins University, 246 Clayton Rd, Baltimore, MD 3168

Inhalation of polymeric nanoparticles holds promise for the localized delivery of diagnostic and therapeutic agents within the lungs. However, the highly viscoelastic and adhesive mucus layer that lines the airways is a critical barrier for conventional nanoparticles. Mucus layers rapidly immobilize standard nanoparticles, which are then removed from the lungs within minutes to hours via mucociliary clearance. As a result, controlled drug release in the lung airways for more than a few hours has been impossible to date, which strongly reduces the effectiveness of local drug therapies for diseases like lung cancer.

We have previously developed diblock ether-anhydride copolymers, composed of polyethylene glycol (PEG) and poly(sebacic anhydride) (PSA), and shown that these polymers are capable of: (i) efficient aerosolization into model lungs; (ii) biodegradation at rates that can be controlled by polymer chemistry, and (iii) sustained delivery of a wide range of drugs for several days. We have also recently shown that coating non-degradable latex nanoparticles with a dense surface layer of low MW PEG allows surprisingly large nanoparticles (up to 500 nm) to rapidly move through human mucus barriers.

Here we report that nanoparticles made using PEG-PSA block copolymers are capable of rapid transport in human mucus, the first such demonstration for a biodegradable polymer. Particles were incubated in native human cervicovaginal mucus and 20 second videos were captured using a high speed camera. Mean-square displacements and effective diffusivities were obtained from particle trajectories obtained through particle tracking software. PEG-PSA particles rapidly diffused through human mucus at rates >100-fold faster than similar-sized uncoated latex nanoparticles, biodegradable PSA nanoparticles, and PLGA nanoparticles.

Next, we show that the PEG-PSA polymer is capable of efficient encapsulation of etoposide, the front line drug for small cell lung cancer. In a single step process, drug was encapsulated up to 40% w/w into PEG-PSA particles. Etoposide was released from PEG-PSA particles in a sustained manner for up to 6 days in vitro, and retained cytotoxic activity in vitro against NCI-H82, a human small cell lung cancer.

Finally, etoposide-loaded particles were administered to nude mice bearing subcutaneous H82 tumor xenografts and tumor volume was measured over time. A single localized injection of etoposide in PEG-PSA particles suppressed tumor growth for over four weeks. In comparison, an equivalent dose of free etoposide injected locally or systemically over 3 days had no effect on tumor growth compared to non-treated control mice. Median survival of mice, as assessed by tumors reaching a volume of 1 cm^3, increased from 15 days for local or systemic delivery of free etoposide, to 47 days for mice treated with a single dose of etoposide in PEG-PSA nanoparticles. Studies are currently underway to test the efficacy of this system following inhalation in an orthotopic lung tumor mouse model.

135g Dynamics of NOVEL Multifunctional, Intelligent Mucoadhesive Copolymers for Oral Protein Delivery

F. Michael Marks III and Anthony M. Lowman. Department of Chemical Engineering, Drexel University, 3141 Chestnut Street, Philadelphia, PA 19104

Since the discovery that insulin can be used for treatment of diabetic patients in 1922, there have been copious amounts of research with little headway in administration options for insulin. To this day, the main course of administering insulin to diabetic patients is through injection. This is a painful and tedious experience that patients must endure. The shots that are administered into the peripheral circulation of the blood stream through subcutaneous methods result in problems mimicking insulin released into portal circulation that is seen in normal, healthy individuals. These problems can severely limit patient compliance which can ultimately cause severe medical problems such as blindness and kidney failure due to the lack of insulin. Even with compliance other problems can arise with a loss of the direct effect on hepatic metabolic processes due to the bypassing of the portal vein.

The lack of compliance with insulin is an area of great concern due to the large amount of people affected. There are 14.6 million people diagnosed with diabetes and an estimated 6.2 million that are yet to be diagnosed in the United States alone. These numbers increase every year due to increased obesity problems. With these numbers growing annually, there is an obvious need for a change in administration methods in order to increase patient compliance. With this and the problems mentioned previously the development of a novel oral delivery system could be the solution.

Graft copolymer networks of poly (methacrylic acid) (PMAA) containing well characterized tethers of high molecular weight poly (ethylene glycol) (PEG) containing different adhesive moieties with tetraethylene glycol dimethacrylate (TEGDMA) being used as a crosslinking agent were prepared using free radical solution UV-polymerization. These new tethers will be characterized to determine the interactions between the mucus layer and the polymer, and to better understand the interactions between the functional groups on the PEG tethers and the cell ligands along the epithelial cell layer of the small intestine. The copolymers were prepared in different ratios of PEG and PEG with adhesive moieties. The high molecular weight PEG was incorporated as an adhesion promoter to improve the already mucoadhesive PMAA through interpenetration into the mucous layer that covers the epithelial cells of the small intestine. Accomplishing this, should result in an improved residence time of the carrier and higher absorption at the drug delivery site.

The copolymers were characterized through dynamic swelling and adhesion testing. Swelling studies were carried out using dimethylglutaric acid (DMGA) buffer. Ten buffers were made each with a different pH ranging from 3.2 to 7.6. The initial or dry weight was recorded initially and then again after swelling to get the swollen weight. These two weights were used to calculate the weight swelling ratio of the polymer disks and plotted against pH to determine any pH sensitivity. Adhesion testing was carried out through the use of an Instron tensile testing machine. A polymer disk was placed on the upper clamp of the Instron through the application of cyanoacrylate medical adhesive. Porcine mucin was mixed with DI water in several concentrations and placed on the bottom clamp. These two clamps were brought into contact and then gradually pulled apart until there was no longer any contact between the mucin solution and the polymer disk. Data was gathered and plotted to gain a detachment force (N) versus displacement (m) curve, of which the area underneath was calculated and found to be the work of adhesion (*f*ÝJ). This procedure was also done for polymer-polymer adhesion where another polymer disk was substituted for the mucin solution on the lower clamp.

The swelling results indicate that there is definite pH sensitivity within all ratios.

Initial adhesion results suggest that the addition of the high molecular weight PEG tethers with adhesive moieties causes an increase in the adhesion to the mucus layer.

Further characterization will be done on these adhesive moieties some of which are present in several small intestines attacking.

142a Interfacial Shear Strength Studies of Plasma-Treated Ultra-High Molecular Weight Polyethylene Fibers

Jacqueline H. Yim, Chemical and Biological Engineering, Drexel University, 3141 Chestnut Street CAT 288, Philadelphia, PA 19104, Daphne Pappas, U.S. Army Research Laboratory, Aberdeen Proving Ground, MD 21005, Denis Kissounko, Center for Composite Materials, University of Delaware, Newark, DE 19716, Alexander Fridman, Department of Mechanical Engineering and Mechanics, Drexel University, Philadelphia, PA 19104, and Giuseppe R. Palmese, Department of Chemical and Biological Engineering, Drexel University, 3141 Chestnut St., CAT 283, Philadelphia, PA 19104.

One particular aspect of research surrounding fiber reinforced polymer composites has been the engineering of the interphase that exists between the fiber and polymer matrix. Surface treatment of fibers is a commonly used route to improve the adhesion of the fibers to the polymer matrix. Based on prior work, atmospheric plasma treatments have shown to successfully functionalize the surfaces of ultra-high molecular weight polyethylene (UHMW-PE) fibers with polar groups such as hydroxyl, carboxylic, and amines groups. These functional groups serve as reactive sites where covalent bonds can form between the fibers and polymer matrix to improve upon the interfacial strength. In this investigation, different plasmas – N2, O2, and N2+H2 gases and various treatment conditions were utilized to surface treat UHMW-PE fibers. The interfacial shear strength (IFSS) of these plasma-treated fibers was assessed using a single-fiber Microdroplet shear test. X-ray Photoelectron Spectroscopy (XPS) was used to correlate the interfacial strength to the type of chemical functional group.

142b One-Pot Synthesis of Silver Nanoparticles Embedded In Siloxane Rubber Matrix

Anubha Goyal[1], Ashavani Kumar[1], Shaily Mahendra[2], Pedro Jose Alvarez[2], and P.M. Ajayan[1]. (1) Mechanical Engineering and Materials Science, Rice University, 6100 Main, Houston, TX 77005, (2) Civil and Environmental Engineering, Rice University, 6100 Main, Houston, TX 77005

Poly(dimethylsiloxane) PDMS is a chemically and biologically inert material which has been used extensively for micro-fluidic channels, tissue culture and biomedical applications. This polymer is also transparent, non-conductive and flexible. Embedding nanoparticles in PDMS matrix would lead to new advanced materials where collective properties of both nanoparticles and polymer can be utilized for various diverse applications such as gas separation membrane, optoelectronic devices etc. The general approach of dispersing as prepared nanoparticles suffers from poor dispersibility in polymer matrix and requires multiple steps. We have developed simple one step synthesis of silver nanoparticles embedded PDMS films which involves mixing of silver salt in the siloxane precursor followed by curing. The curing agent not only cross links the siloxane precursor but also reduces the silver benzoate to silver nanoparticles. The in-situ generated nanoparticles are uniformly distributed in the polymer matrix and were characterized by UV- visible spectroscopy, Transmission electron microscopy and X-ray photoemission spectroscopy. This nanocomposite material shows good anti-microbial properties making it superior to PDMS film for biomedical application and it also has better tolerance to wide range of organic solvents for use in petrochemical industry.

142c Influence of Nanoparticles on the Local Mechanical Properties of Polymeric Materials

Robert A. Riggleman, Department of Chemical and Biological Engineering, University of Wisconsin, Madison, 1415 Engineering Drive, Madison, WI 53706, Gregory N. Toepperwein, Department of Chemical and Biological Engienering, University of Wisconsin, Madison, 1415 Engineering Drive, Madison, WI 53706, and Juan J. de Pablo, Department of Chemical Engineering, University of Wisconsin - Madison, 1415 Engineering Drive, Madison, WI 53706.

Nanocomposite materials are encountered in a number of high-performance applications, but the source of the improved structural and mechanical attributes of nanocomposites has remained elusive. In particular, little is known about the ageing of nanocomposite glasses, or the change of their mechanical attributes over extended periods of time. Fundamental studies of nanocomposite polymeric materials have often been challenged by the proclivity of nanoparticles to phase separate or aggregate, raising questions about the equilibrium nature (or lack thereof) of existing reports. Development of improved nanocomposite materials would benefit considerably from a molecular-level understanding of the structure and dynamics of nanocomposite polymeric glasses as a function of nanoparticle shape, size, and interaction parameters. In this work, we have performed molecular simulations of an entangled polymer and an entangled polymer nanocomposite, and we have studied the effects of ageing and deformation on the mechanical properties of each system. Recent simulation studies have shown that the mechanical properties of amorphous solids are heterogeneous, and such materials can even exhibit local domains with a negative modulus. We find that, as both the pure polymer and the nanocomposite age in the glassy state, they become more mechanically homogeneous; that is, their respective distributions of local mechanical properties becomes more narrow, and the nanocomposite is shown to age more quickly. Upon deformation, the mechanical properties become more heterogeneous, analogous to the way in which mechanical deformation can "rejuvenate" a glassy material. Additionally, we determine the effect of the nanoparticles on the plateau modulus of the polymer by reducing our polymeric systems to their entanglement network. We find that nanoparticles increase the density of entanglements near their surface, and the nanoparticles themselves behave as entanglement traps. As the material is deformed, it is found that the primitive paths of the polymer begin to condense onto the nanoparticles, further deforming the entanglement network. Finally, we discuss how heterogeneity in the shape of the nanoparticles affects the local mechanical properties and the entanglement network.

PHYSICAL AND CHEMICAL PROPERTIES OF POLYMER IMPREGNATED CONCRETE ON THE PREPARATION CONDITIONS

Won Mook Lee[*], J Priya Nair, Chul Woo Lee, Du Hyun Ku, Jeong Soon Park, Hun Young Park[**]

Department of Chemical Engineering, Hanbat National University, Daejeon 305-719, South Korea

[**] Kyung Dong Engineering Co. Limited, Trading Center Building, Anyang city, Kyunggee-Do , South Korea

• Email: wmlee@hanbat.ac.kr

Introduction

Polymers have been employed as concrete admixtures to improve durability of the concrete structures that have degraded due to exposure to weather and polluted environments. Based on the mode of their addition polymer cement concrete is classified as – Polymer Modified Cement Concrete or Mortars (PMC/PMM), Polymer Concrete or Mortars (PC/PM), Polymer Impregnated Concrete or Mortars (PIC/PIM) (1). The latter is one the oldest polymer cement composites and is prepared by impregnating the monomer into the concrete or mortar structures and then by polymerizing the monomer. PICs possess superior strength, chemical resistance and low water absorption compared to other polymer cement composites. Some of the applications where these composites find use are reactors, bridges, pipes, storage bunkers for chemicals and structures in marine environments (2,3).

Conventionally PICs are prepared by immersing the precast concrete structure in a mixture of a monomer and an initiator for a few hours at room temperature and polymerizing the monomer by thermal methods (4). This results in monomer loss during polymerization at high temperature thus affecting the properties of the PIC. In this work we describe a method to increase the monomer impregnation thereby reducing such losses. The monomer is introduced into the voids and pores of the precast concrete by subjecting the experimental setup to ultrasonic vibrations thereby enhancing the chances of monomer impregnation to a greater degree than the conventional procedure. The impregnated specimens are dried, vacuum packed to prevent monomer loss and subsequently polymerized using microwaves. When compared with composites prepared by conventional methods and ordinary precast concrete, it was observed that this procedure increased the degree of polymerization of the monomer in the concrete structures thereby improving both the mechanical and chemical resistant properties of the composite.

Experimental program

The precast cement mortar specimens used were first impregnated with methyl methacrylate and then polymerized using two different procedures. The mechanical and chemical resistant properties of these PICs after exposure to 5M sulphuric acid for different periods were evaluated and their microstructures analyzed.

Materials

Compressive strength was evaluated for precast cement mortars cubical specimens of dimensions 5cm X 5cm X 5cm weighing around 250g and flexural strength for specimens of dimensions 4cm X 4cm X 16cm weighing 530g approximately. Ordinary Portland cement and sand conforming to Korean standard KS L 5100 were used in the preparation of the specimens. The water to cement ratio was 0.48 while the cement, sand and water were mixed in the proportion 1:2.45:0.48.

Impregnation and Polymerization

The precast mortar specimens were dried in a hot air oven (Samwoo Science make) at 80 C for 12 hours. These were annealed to room temperature and weighed before impregnation. The monomer, Methyl methacrylate (MMA) was mixed with 1% by weight of 2, 2'-Azobisisobutyronitrile **(AIBN)** as initiator using a magnetic stirrer. The mortar samples were immersed into this mixture and impregnated by placing the setup in a water bath inside an ultrasound vibration system for 4.5 hours at room temperature. The samples were dried, weighed and vacuum packed in PET packets to prevent loss of monomer during polymerization.

The samples were thermally polymerized by two different procedures viz. using conventional methods and microwaves. In the conventional method the packed impregnated mortar samples were immersed in hot water at 80 C to achieve uniform heating for three hours. Polymerization of the specimens in the microwave reactor was done at a frequency of 2450MHz (400W) at 80C for two hours. After polymerization the samples were removed from the PET packets, cooled to room temperature and weighed.

Performance

The mechanical properties evaluated for the PIC specimens were compressive and flexural strengths. The durability of these composites when exposed to various chemical environments was evaluated by calculating the weight loss.

Compressive strength is a basic property used to assess the performance of hardened concrete made for a particular application. It is calculated by dividing the maximum load applied when the cubical specimen fails by its cross sectional area resisting the load. In this work the cubical specimens of 5cm X 5cm X 5cm were employed and the evaluation of these properties was performed in accordance to KS L 5105 standards (5). The maximum load needed to break the conventional cement mortar and PIC samples was determined using a Servo UTM US-200.

In concrete, cracks can propagate very easily in tension and excessive cracking of concrete causes serviceability and durability problems. In this work, the flexural strengths of conventional cement mortar and PIC specimens of dimensions 4cm X 4cm X 16cm were evaluated according to ASTM C293 - 08. A Digital Flexure Tensile Tester HJ-1171 was employed to calculate the flexural strengths of the conventional cement mortar and the PIC specimens.

The decrease in durability of concrete structures due to increasing environmental pollution is a matter of great concern. Many admixtures including polymers are employed to increase this durability. In this work, two series of specimens viz. conventional cement mortar and PIC, were weighed and immersed in 5M sulphuric acid for different periods of exposure time – 3,5,7 and 14 days to assess the extent of damage caused due to the interaction of the samples with the external chemicals. The cumulative weight losses of the PIC specimens on exposure to the acid were compared to those obtained for ordinary Portland cement mortar samples.

Microstructural studies

The degree of polymerization and molecular weight distribution of the polymer plays a vital role in determining the properties of the polymer impregnated concrete (R). In this study the above parameters were calculated using Gel Permeation Chromatography (Model Waters) and refractive index detector. A few grams of the PIC prepared from conventional thermal as well as microwave methods were mixed with tetrahydrofuran (THF) so as to extract and dissolve the PMMA. The solution was then analyzed for calculation of the above.

Porosity and pore size distribution are inversely proportional to the compressive strength of conventional cement mortars since the effective area to resist the applied load on them decreases as the voids and pores increase. A Micromeritics make automated mercury Porosimeter 9520, was employed to determine the pore characteristics in both the conventional cement mortar and PIC specimens.

Investigations using scanning electron micrographs revealed porosity and micro cracks in the conventional cement mortars before and after impregnation and subsequent polymerization of the monomers. A Scanning electron microscope (Jeol Make JSM-6390 Model) was used to carry out these studies. Samples of 1-2 mg weight were subjected to gold sputtering to make them conducting.

Results and discussion

Effect of modes of polymerization on the strength of the polymer Impregnated mortar

In this work two procedures were employed to polymerize the monomer that has been impregnated into the cement mortar – the conventional thermal method and the current microwave method

As inferred in Fig.1 (a) the strength of composites prepared by microwaves was also found to be far superior to those prepared from conventional methods.

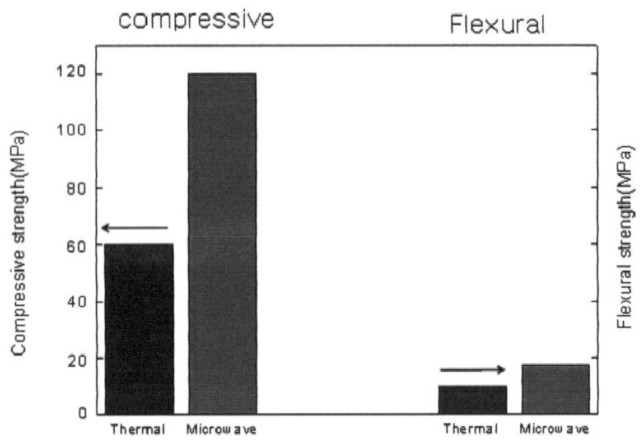

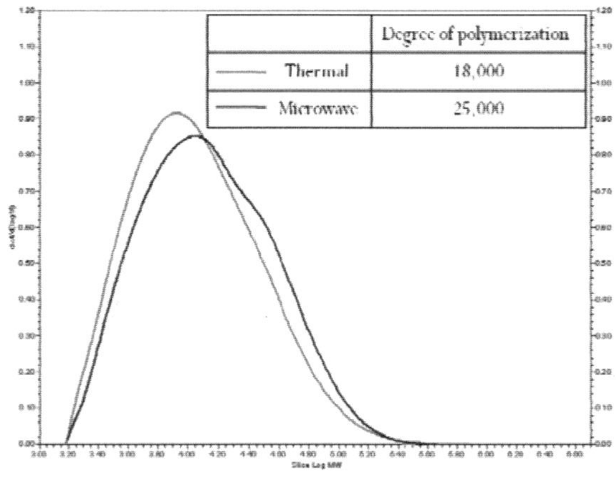

(a) (b)

Fig.1 (a) Comparison of the strengths of PIC specimens prepared from thermal and microwave methods

(b) Degree of polymerization of monomer in thermal and microwave methods

The compressive and flexural strengths of the PICs prepared using microwave heating were found to be around two and three times that of the ones prepared using the conventional heating methods respectively. This could be attributed to the higher degree of polymerization of the monomers within the pores of the precast cement mortars as observed in Fig.1(b). It is seen that the degrees of polymerization of the monomer in the conventional and the microwave heating methods were estimated to be 18000 and 25000 respectively. When the monomer is converted to the polymer within the pores, it occupies only 80% of the original volume of the pores (2). It may therefore be inferred that the effective area to resist the applied load on the specimen is directly proportional to the degree of polymerization. It was also noted that the time taken for heating in the latter method was far less resulting in a substantial reduction of energy consumed.

Durability

Fig.2 and Table 1 depicts the chemical resistant properties of the PICs on exposure to 5M sulphuric acid were compared with those obtained for conventional cement mortar by calculating the weight losses for different periods of exposure time.

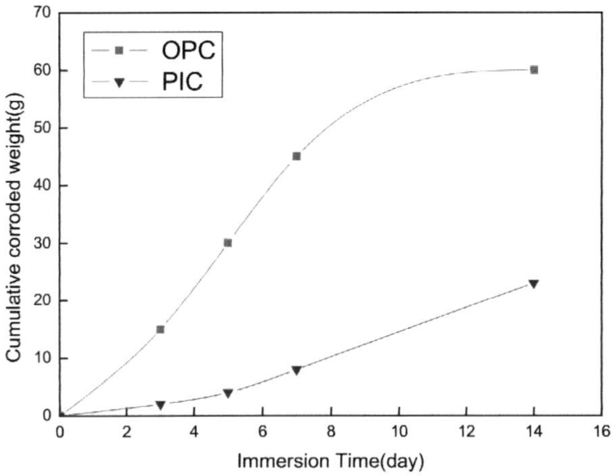

Fig.2 Effect of preparation conditions on chemical resistance (After dipping in 5M – sulphuric acid solution on dipping time, checking the corroded weight of the sample)

Immersion Time (day)	**Cumulative corroded weight(g) - OPC**	**Cumulative corroded weight(g) - PIC**
0	0	0
3	15	2
5	30	4
7	40	8
14	60	23

Table 1 Loss of weight of the samples when exposed to sulphuric acid

Investigations revealed that the cumulative corroded weight was greater for conventional cement mortars than PICs. The sulphuric acid underwent interactions with conventional OPC thereby dissolving a part of the set cement. This results in loss of weight and weakening of the material progressively by removal of the cementing constituents (8). The presence of the polymer in a PIC not only envelopes the cement mortar but also seals the voids formed during the cement hydration. This prevents cement-acid interactions that would otherwise result in loss of weight of the composite.

Microstructural studies

As part of the microstructural studies the porosity and morphology of the polymer cement matrix in the PIC was studied.

From Fig.3 the porosities of conventional cement mortars and the PICs were found to be 17.3% and 9.8% respectively.

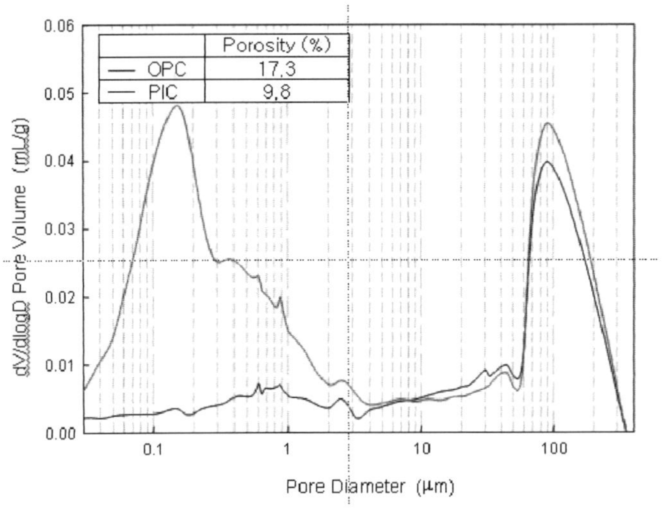

Fig.3 Comparison of the Pore diameter in OPC and PIC specimens

The decrease in porosity in the latter case could be attributed to the sealing of the voids and micro-cracks by the polymer in the precast cement mortar thereby increasing durability and strength of the cement structure as seen in the previous section.

Further, the morphological studies conducted using the scanning electron microscope (Fig.4 (a) and (b)) revealed the presence of large pores and voids in the conventional cement mortars that were not impregnated by the polymer.

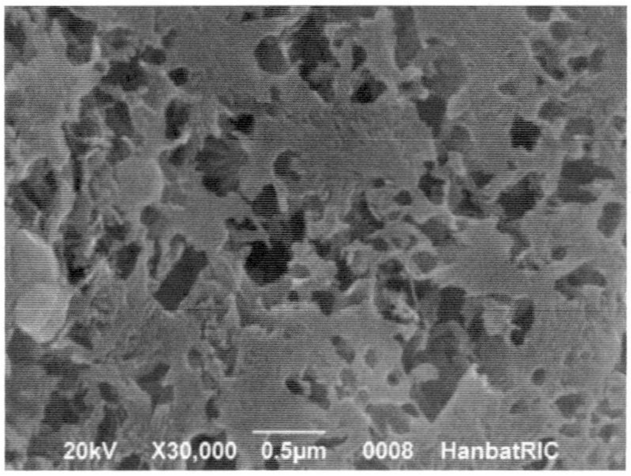

(a) OPC

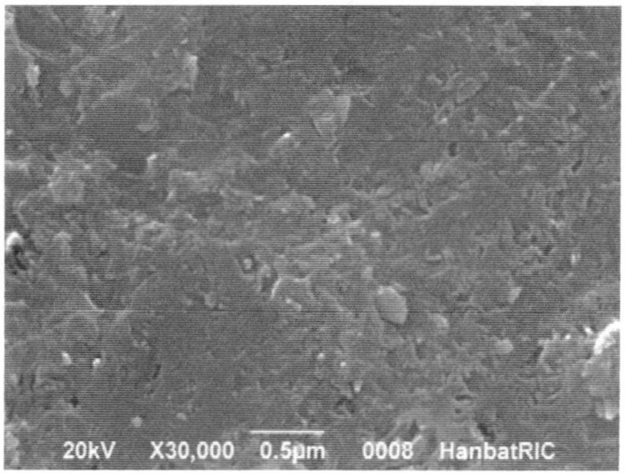

(b) PIC

Fig.4 Scanning electron micrographs of (a) OPC and (b) PIC

The microstructure of PICs shows a more compact and uniform structure with smaller voids due to the presence of polymer in the cracks and voids.

Conclusions

From the above studies the following conclusions were inferred:

1. The degree of polymerization of monomer is greater in case of PIC specimen prepared by microwaves than ones prepared from conventional thermal methods.

2. The mechanical and chemical resistant properties of PIC composites are superior to the conventional cement mortar.

3. Porosity of the conventional cement mortar is greatly reduced when it is impregnated with polymers thereby increasing its durability when it is exposed to chemically polluted environments.

4. Polymers give a more compact structure to the cement matrix and seal the cracks in cement mortar matrix.

REFERENCES

1. Ohama, Y.; Cement and concrete composites, Vol. 20, pp 189-212, 1998.

2. A.Auskern and W.Horn, J.Am. Ceram. Soc., 54(6), 282, 1971.

3. Ahmed M. Gadalla, Mohamed E. El-Derini, Journal of Applied Polymer Science, Volume 29 Issue 10, Pages 3027 – 3040, 1984.

4. Ohama, Y., Chandra, S.; Polymers in concrete, 5th chapter, Noyes Publications, 1994.

5. KS L 5105: Testing Method for compressive strength of Hydraulic cement Mortar

6. KS L ISO 679: Methods of testing cement – Determination of strength

7. S. Saccubai , M. Sarojadevi, Aravamudan Raghavan, Journal of Applied Polymer Science, Volume 61, Issue 3, Pages 577 – 584, 1998.

8. Ismail M.R. et al, Journal of applied Polymer Science, Vol.73, pp 685-693, 1999.

142e Long Glass Fiber Orientation In Thermoplastic Composites Using a Model That Accounts for the Flexibility of the Fibers

Kevin C. Ortman, Chemical Engineering, Virginia Tech, 154 Randolph Hall, Blacksburg, VA 24061, Gregorio M. Velez, Macromolecular Science and Engineering Department, Virginia Polytechnic Institute, 154 Randolf Hall, Blacksburg, VA 24061, Aaron P. R. Eberle, Chemical Engineering, Virginia Polytechnic Institute, 154 Randolf Hall, Blacksburg, VA 24061, Don Baird, Virginia Polytechnic Institute, 133 Randolph Hall, Blacksburg, VA 24061, and Peter Wapperom, Mathematics Department, Virginia Polytechnic Institute, 154 Randolf Hall, Blacksburg, VA 24061.

Mechanical properties of long glass fiber composites, used in various industrial applications, are dependant upon the fiber orientation within the part. To date, however, simulations with the ability to predict fiber orientation as a function of mold design are not available. In this study, a several options are explored to predict the orientation of long glass fibers in the concentrated regime which take the flexible nature of these fibers into account. Flow through a center gated disk geometry is simulated numerically for a high concentration of long glass fiber in a polypropylene (PP) matrix. For this, a 2D finite element (FEM) analysis was performed using the traditional Galerkin method for the balance equations and the discontinuous Galerkin method for the constitutive equations. Numerical results based on the Hele-Shaw approximation are compared with experiment to verify the models.

142f Single Walled Carbon Nanotube Fluorescence Modulation In Response to Hydrogel Swelling

Paul W. Barone, Chemical and Biomolecular Engineering, Univeristy of Illinois at Urbana/Champaign, 104 RAL, 600 S. Mathews Ave., Urbana, IL 61802, René Ortiz, Technical University of Munich, Munich, Germany, Jingqing Zhang, Chemical Engineering, Massachusetts Institute of Technology, Cambridge, MA 02139, and Michael S. Strano, 66-566 Department of Chemical Engineering, Massachusetts Institute of Technology, 77 Massachusetts Avenue, Cambridge, MA 02139.

Single walled carbon nanotubes (SWNT) fluoresce is still a new and exciting area of research. Modulation of SWNT fluorescence has previously been demonstrated by adsorption of electron withdrawing molecules[1,2], DNA rearrangement due to cation adsorption[3] and via DNA hybridization on the SWNT surface[4]; however, new methods to modulate SWNT fluorescence will allow SWNT use in new applications. We demonstrate that SWNT immobilized in a poly(vinyl) alcohol hydrogel show shifts in fluorescence emission maxima during hydrogel swelling / deswelling. The mechanism of SWNT fluorescence modulation and potential applications of SWNT based hydrogels will be discussed.

1. Barone, P.W.; et al., Nature Materials (2005), 4, 86-92.

2. Satishkumar, B.C.; et al. Nature Nanotechnology (2007), 2, 560-564.

3. Heller, D.A.; et al. Science (2006), 311, 508-511.

4. Jeng, E.S.; et al. Nano Letters (2006), 6, 371-375.

Effect of Block Copolymer Coated Nanoclay on Polystyrene Foams under supercritical Carbon Dioxide

Weibin Zha, Bin Zhu, Jintao Yang and L. James Lee
Department of Chemical and Biomolecular Engineering
The Ohio State University

Polystrene-*b*-polymethyl methacrylate (PS-*b*-PMMA) was end-tethered on Montmorillonite (MMT) by atom transfer radical polymerization (ATRP) to make exfoliated nanoclay composites. These nanocomposites were mixed with polystyrene (PS) by melt blending. The mixtures were then batch formed with supercritical CO_2. It was found that the cell density of foams based on these blends with 1 wt% modified MMT is higher than that based on neat PS and PS compounded with neat PMMA or PS-*b*-PMMA at the same foaming conditions. The presence of nanoclay dramatically increases the viscosity of the mixture. The heterogeneous nucleation effect of the nanoclay and nanocomposites will be discussed.

Introduction

The traditional physical blowing agents for PS foams, chlorofluorocarbon (CFC), have been banned all around the world due to its effect of ozone-depletion and negative impact on global warming. Although the replacement, hydrochlorofluorocarbons HCFCs (such as HCFC-142b (CF2ClCH3) and HCFC-22 (CHF2Cl), which are used in PS foaming process.) have lower ozone-depletion potential (ODP), they will also be banned by 2010 in the United States. Carbon dioxide (CO_2) is a good substitution for these hazardous blowing agents because it is cheap, nontoxic, nonflammable, and environmentally benign. Yet CO_2 has relatively low solubility and high diffusivity in polystyrene, which makes it difficult to obtain low foam density and small cell size.

The well dispersed nanoparticles in polymer matrix are found to be good heterogeneous nucleation agents for bubble generation [1-4]. Montmorillonite (MMT) is the most widely used one in nanocomposites synthesis. The large surface area combined with its fine dimensions may greatly influence cell nucleation and growth behaviors. But the incompatibility between the MMT surface and polymer matrix largely reduced the capacity for clay to intimately contact with polymer matrix. So MMT surface modification is necessary to prepare well dispersed nanocomposites especially if the polymer matrix has no favorable interaction with the clay surface.

Blending with the polymer which has high CO_2 solubility is another way to increase cell density [5]. However, to obtain uniform morphology from two immiscible polymer blends like PS and PMMA is difficult.

In this study, the CO_2-philic PMMA was grafted on MMT surface by ATRP. PS block was subsequently copolymerized to PMMA to increase the compatibility between the nanocomposites and PS matrix. This copolymerized nanocomposite was melt blended with PS and the mixture was then batch foamed with supercritical CO_2.

Experimental

The initiator of ATRP was synthesized according to the literature [6]. Scheme 1 describes the material used and the synthesis procedure.

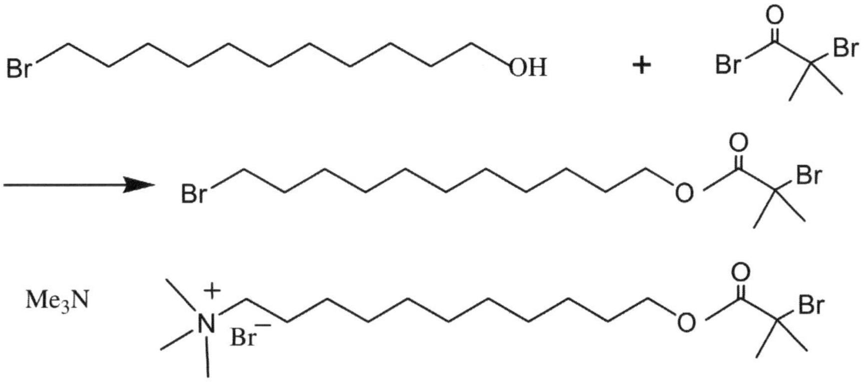

Scheme 1

The synthesized cationic initiator was dissolved in deionized water and drop-wisely added to MMT-water dispersion under vigorous stirring. The exchanged clay was filtered and washed with water and methanol respectively. It was then dried in the vacuum oven at 60°C for 24 hours. The modified clay was ground into fine powders (300 meshes) before use. The amount of immobilized initiator was determined by thermal gravimetric analysis (TGA) as shown in Figure 1.

Graft reactions were carried out in Schlenk flasks equipped with a magnetic stirring bar. Predetermined amount of modified clay, CuBr, 1,1,4,7,10,10-hexamethyltriethylene tetramine (HMTETA) and methyl methacrylate (MMA) were transferred to a Schlenk flask under nitrogen gas. The polymerization was stirred at 90°C for 3h. The Schlenk flask was then subjected to a vacuum to remove unreacted MMA monomer. Styrene monomer was then added to the flask and the polymer was dissolved in styrene monomer. The flask was then heated to 120°C for 6h. The weight percentage of the block copolymer in the nanocomposites can be adjusted by varying the monomer amount. Figure 1 gives an example of 10 wt % nanoclay coated with blockcopolymer measured by TGA. A similar procedure was used to synthesize pure PS-b-PMMA block copolymer except that the initiator was *p*-toluenesulfonyl chloride.

The synthesized nanocomposites or block copolymer was melt blended with PS using DATA mini-twin-screw mixer at 200°C for 5 min. The blends were then compression molded into 1 mm thick, 15 mm diameter disks. For batch foaming, the sample disk was first saturated with supercritical CO_2 at 120°C, 2000 psi for 24h. Then the pressure was quickly released and the sample was put into ice water for 10 min.

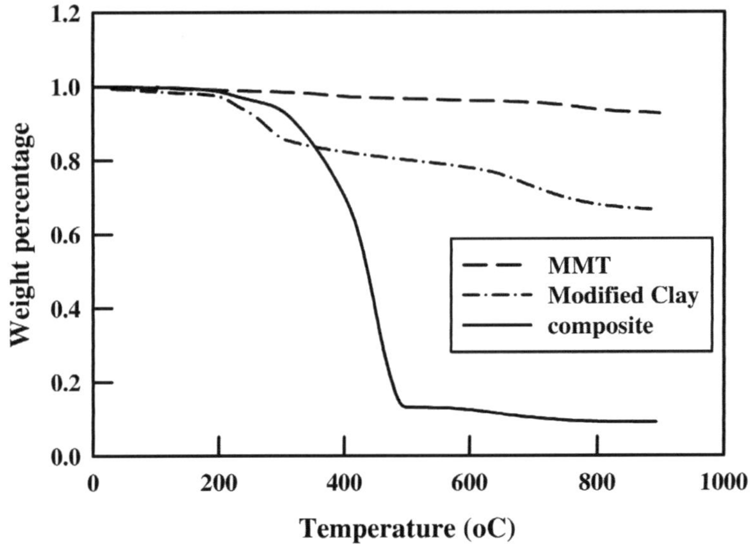

Figure 1. TGA spectrum of MMT, modified clay and PS-*b*-PMMA grafted clay nanocomposite.

Results and Discussion

X-ray diffraction patterns of the MMT, modified clay, and PS-b-PMMA grafted nanocomposite containing 10 wt% of modified clay are shown in Fig. 2. By using the Bragg equation $\lambda=2d\sin\theta$, the d-spacing values of the samples were calculated and shown in Figure 2. The interlayer spacing of MMT increases from d_{001}=1.16 nm to d_{001}=1.88 nm when MMT was modified by the initiator we synthesized. The TGA result from Figure 1 reveals that the initiator content of 25 wt% which corresponds to 0.67 mmol immobilized initiator per g modified clay.

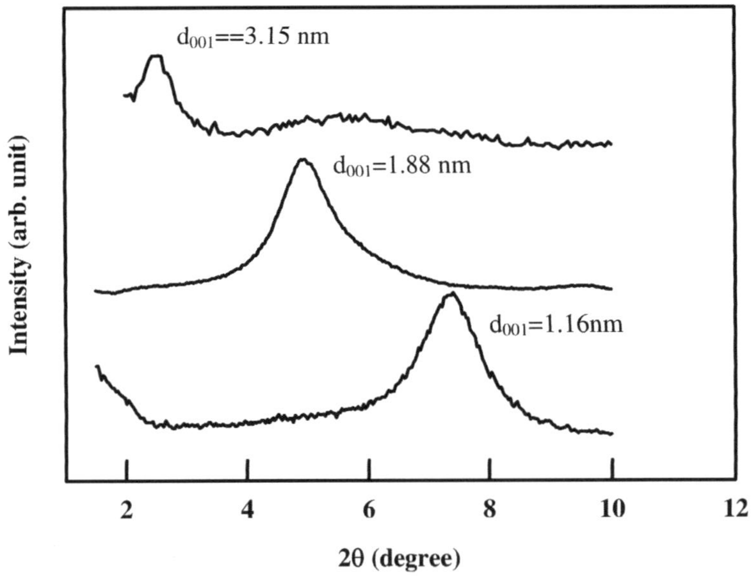

Figure 2. XRD diffraction patterns for (1) MMT; (2) Modified MMT; (3) PS-*b*-PMMA grafted nanocomposites.

After the "living" ATRP of PMMA and PS, the peak around 2θ=5.1 degree become broader and weaker and a new peak at 2θ=2.8 appears, which indicating the further increase of the interlayer distance of the modified clay. TEM image Figure 3a confirmed X-ray's result. In Figure 3a, it is worth noting that clear phase separation of PS-*b*-PMMA exists in some area. The white area indicates PMMA block and the dark area indicates PS block. According to the sequence of polymerization, all the clay should stay in PMMA phase. Yet we found in Figure 3a that there's PMMA phase without any clay in it. Although some researchers used very similar polymerization method to make block copolymer nanocomposites,[7-9] no literature has ever published such result. We speculated that during the polymerization, some initiator was detached from the clay surface. The broadened X-ray diffraction peak in Figure 2(3) indicates that the interlayer distance of some of the modified clay becoming smaller after the polymerization. The polymer propagated from detached initiator or the polymer detached from surface becomes free polymer and they may have phase separation depending upon the molecular weight. Apparently, the molecular weight of the polymer we synthesized is large enough to induce the phase separation.

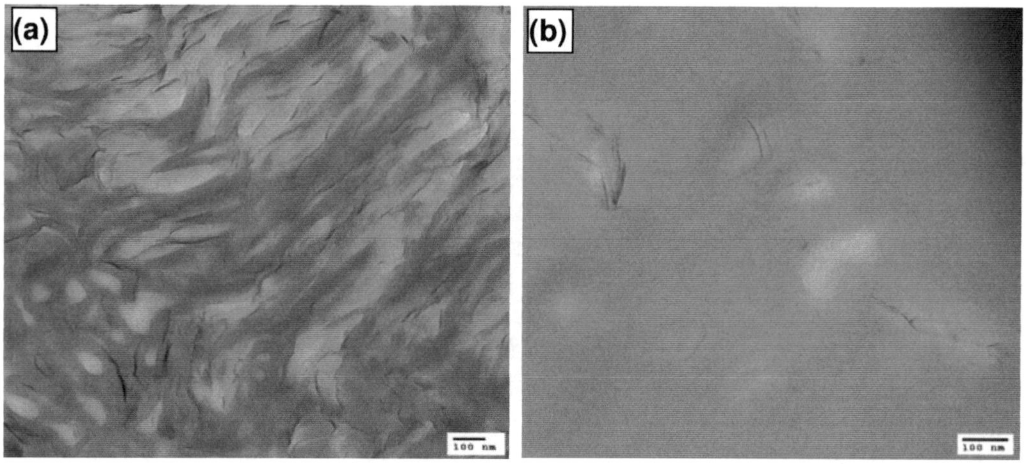

Figure 3. TEM images of (a) PS-b-PMMA grafted nanocomposite and (b) PS/(PS-*b*-PMMA grafted nanocomposite)=9/1 melt blended.

Figure 3b is the TEM image of one melt blended sample of PS and block copolymer nanocomposite described above. It is noticeable that most of the clay stay in the PMMA microdomain and the domain size is around 50-300 nm range. The interface between PS and PMMA is not so clear as the interface between PS and PMMA blends published in the literature.[5] This small microdomain and weak interface separation are attributed to the existence of PS block in nanocomposites. The presence of PS block makes the block copolymer nanocomposite be well dispersed in PS matrix.

Figure 4 shows the complex shear viscosity $|\eta^*|$ versus frequency (ω) for pure PS, PS/PMMA blend, PS/(PS-*b*-PMMA) blend and PS/(PS-*b*-PMMA nanocomposite) blend at 180 °C. The PS/(PS-*b*-PMMA nanocomposite) blend shows the highest shear viscosity among all the blends. This is reasonable because well dispersed clay has huge surface area. It is interesting that the PS/(PS-*b*-PMMA) blend shows the lowest shear viscosity according to Figure 4. Whether it has some rational behind this or it is just an experimental mistake is still under investigation.

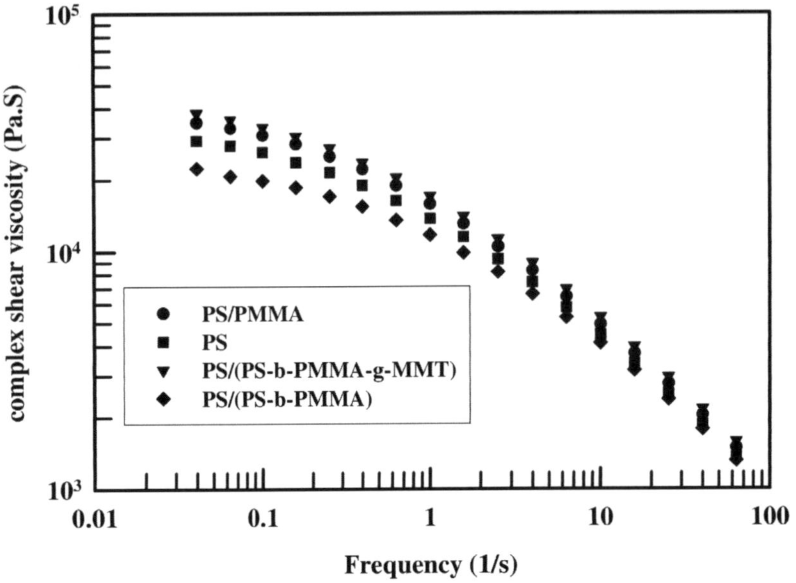

Figure 4. Complex shear viscosity of PS, PS/PMMA blend, PS/(PS-b-PMMA) blend and PS/(PS-*b*-PMMA nanocomposite) blend at 180 °C.

The pure PS and other three blends described above were all batch foamed in one batch at 120 °C, 2000 psi. The same batch condition assures that all the materials are foamed at the same saturation condition, same temperature and same pressure release speed. This is important because any condition difference will lead to different foam structure.

Figure 5 shows the foam morphology we obtained from pure PS three blends. In Figure 5 we found that the blend of PS/nanocomposite has the smallest cell size and highest cell density. This observation proves that well dispersed clay which is surrounded by CO_2-philic PMMA is very good heterogeneous nucleation agent. The PS/PMMA blend also has smaller cell size than pure PS. CO_2 has higher solubility in PMMA than in PS which makes the concentration of CO_2 is higher in PMMA under the same saturation condition. High CO_2 concentration means more homogeneous nucleation in PMMA so as to smaller cell size. Han etc. has already showed us that the cell size decreases with the decrease of PMMA domain size in PS/PMMA blends.[5]

One thing abnormal in Figure 5 is that the cell size of PS/(PS-*b*-PMMA) is even larger than pure PS. But this large cell size is consistent with the low viscosity we obtained in Figure 4. It is well known that the higher viscosity, the smaller cell size. Associating the cell size in Figure 5 with the viscosity in Figure 4, we can clearly observe this conclusion. Yet we still need to explain the low viscosity of the PS/(PS-*b*-PMMA) blend.

Conclusion

Well dispersed nanocomposite was prepared by grafting PS-b-PMMA on MMT surface by ATRP. The nanocomposite was then melt blended with PS and then batch foamed with supercritical CO_2. It was found that the cell size of the foam based on this block copolymer nanocomposite is smaller than pure PS or PS/PMMA blends. The nucleation effect of the nanoparticles surrounded by PMMA is well displayed.

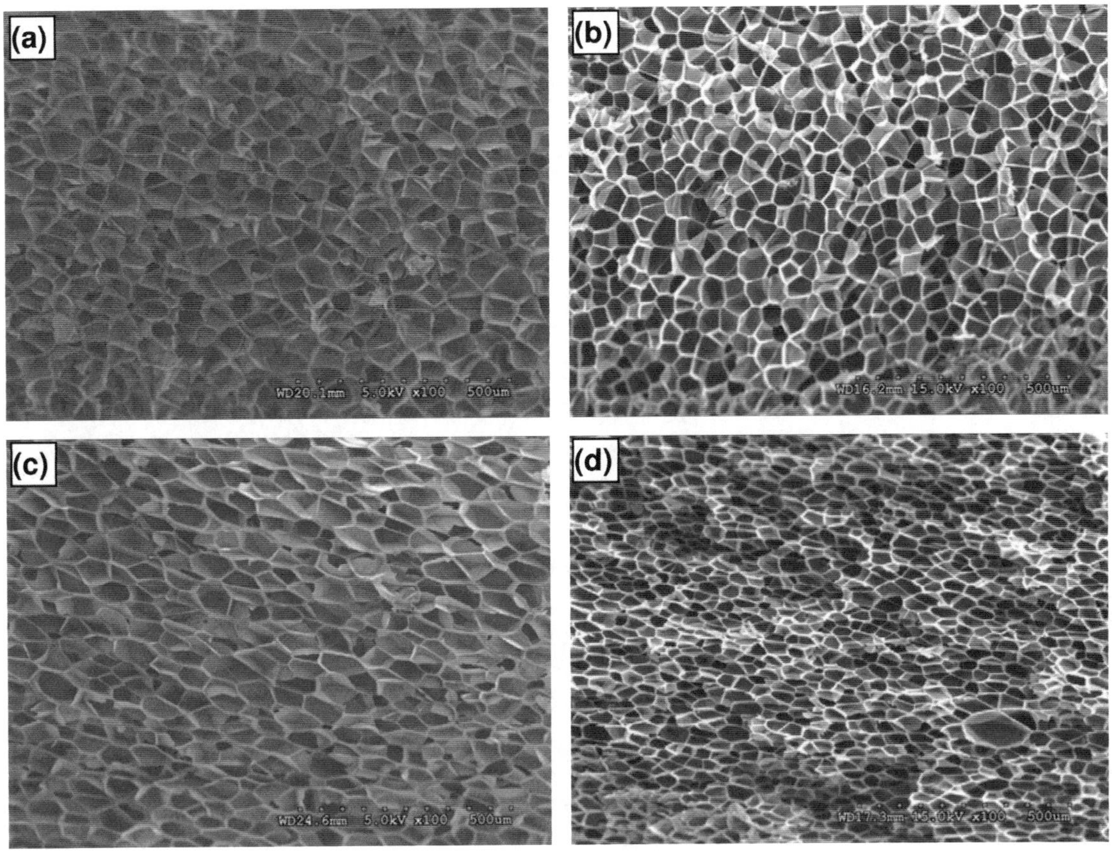

Figure 5. Foam structure at a foaming temperature of 120°C and a saturation pressure of 2000 psi: (a) pure PS (cell size, 54.8 μm; cell density, 1.5×10^7 cells/cc), (b) PS/PMMA(9/1) (cell size 45.9 μm, 17.2 mm; cell density, 1.8×10^7 cells/cc), (c) PS/(PS-*b*-PMMA)(9/1) (cell size 60.4μm, cell density, 1.2×10^7 cell/cc) and (d) PS/(PS-*b*-PMMA nanocomposite) (9/1) (cell size 29.1 μm, cell density, 3.1×10^7 cell/cc)

Literatures

1. P.H. Nam, P. Maiti, M. Okamoto, T. Kotaka, T. Nakayama, M. Takada, M. Ohshima, A. Usuki, N. Hasegawa, and H. Okamoto, *Polym. Eng. Sci.*, 2002, 42,1907.

2. P.H. Nam, M. Okamoto, P. Maiti, T. Kotaka, T. Nakayama, M. Takada, M. Ohshima, A. Usuki, N. Hasegawa, and H. Okamoto, *Nano Lett.*, **2001**, 1, 503.

3. C. Zeng, X. Han, L. J. Lee, K. W. Koeling and D. L. Tomasko, *Advanced Materilas*, **2003**, 15, 1743.

4. X. Han, C. Zeng, L.J. Lee, K.W. Koelling, and D.L. Tomasko, *Polym. Eng. Sci.*, **2003**, 43, 1261.

5. X. Han, J. Shen, H. Huang, D. L. Tomasko, L. J. Lee, *Polym. Eng. Sci.*, **2007**, 47, 103.

6. H. Bo¨ttcher, M. L. Hallensleben, S. Nuß, H. Wurm, J. Bauerb and P. Behrensb, *Journal of materials chemistry*, **2002**, 12, 1351.

7. H. Zhao and D. A. Shipp, *Chem. Mater.* **2003**, 15, 2693.

8. J. Di and D. Y. Sogah, *Macromolecules*, **2006,** 39, 5052.

9. H. Zhao, B. P. Farrell, D. A. Shipp, *Polymer*, **2004**, 45, 4473.

145a Elastic Niches and Capillaries Tuned for 3D Matrix Control of Mesenchymal Stem Cell Differentiation

Matthew Raab[1], **Karthikan Rajagopal**[2], Florian Rehfeldt[3], Manu Tewari[2], Chi Wang[4], and Dennis E. Discher[2]. (1) Bioengineering, University of Pennsylvania, Philadelphia, PA 19104, (2) Chemical and Biomolecular Engineering, University of Pennsylvania, Philadelphia, PA 19104, (3) Physics and Astronomy, University of Pennsylvania, Philadelphia, PA 19104, (4) Chemical Engineering, National Cheng Kung University, Tainan, Taiwan

While most cells are studied on 2-dimensional tissue culture plastic or glass that is appropriate for epithelium and endothelium, most mesenchymal cells generally function within a 3-dimensional solid tissue environment in the body. And while collagen gels are often used for "3D" studies of cells, the fibrous nature of such matrices means these systems are essentially 1D with ill-characterized elasticities. Interest has also rapidly grown in defining and creating niches for mesenchymal stem cells because it is increasingly believed that the microenvironment contributes or controls self-renewal and differentiation. Our past work has focused on characterizing and controlling the cell-scale micro-stiffness of various hydrogels by altering the concentrations of polymer and crosslinker, and then measuring the gel's elasticity with AFM. The elasticity of these hydrogels has proven to be a key insoluble cue in 2D for mesenchymal stem cells (MSCs), profoundly affecting cytoskeletal organization and eventually differentiation. The results suggest that the elasticity of scaffolds used in tissue engineering might also need to be fine tuned for cell engraftment and for suitable restoration of tissue function. Toward this end as well as a more basic understanding of elasticity effects in 2D versus 3D, we have engineered hydrogels by various methods, one of which involves creating a network of collagen-caoted microchannels for cells to crawl into; the cells within are thus exposed to a 3D matrix of tunable stiffness. MSCs not only crawl into the microchannels and seem to fill the lumen, but they also clearly adhere and appear to assemble cytoskeletal stress fibers. Engineering 3D microenvironments that can be tuned for stiffness should thus provide more appropriate means for studying cells because the mesenchymal cells can engage receptors over their entire surface rather than being polarized top-to-bottom like epithelial cells.

145b Combinatorial Biomaterials as An Engineered Niche Platform for Inhibition of Stem Cell Transformation and Promotion of Stem Cell-Based Regeneration

Er Liu[1], Hiral Patel[1], Hak-Joon Sung[2], Joachim Kohn[2], and Prabhas Moghe[1]. (1) Biomedical Engineering, Rutgers University, 599 Taylor Rd, Piscataway, NJ 08854, (2) New Jersey Center for Biomaterials, Rutgers University, 599 Taylor Rd, Piscataway, NJ 08854

This study investigates the possibility of engineering a stem cell niche for inhibition of stem cell transformation, while promoting stem cell growth necessary for regenerative therapies against cancer. Cancer therapy necessitates surgical resection of malignant tissues with consequent tissue regeneration, which may be accomplished by implanting biomaterial scaffolds containing or attracting multipotent stem cells such as mesenchymal stem cells (MSCs). However, this approach requires further development, as recent studies demonstrate that MSCs can become malignant due to abnormal signaling from interactions with cancerous microenvironments. Indeed, cancer reoccurrence is shown to derive from a subpopulation of stem cells. It is therefore desirable to design biomaterials that prevent abnormal MSC transformation while preserving favorable performance. There is also increasing need for quantitative methods probing abnormal cell transformation. In this study, we developed an in vitro carcinogenic model inducing aberrant MSC transformation by nickel sulfate, and applied multi-photon imaging-based quantitative analysis of nuclear protein (NuMA) distribution to evaluate relationships between nuclear structure and MSC transformation. Tyrosine-derived polycarbonates, copolymerized in various molar ratios with (i) lipophilic monomers, (ii) hydrophilic monomers, poly(ethylene glycol) (PEG), and (iii) negatively charged monomers (DT), were used to identify bioactive components influencing MSC transformation. We found DT, as opposed to PEG, to both promote MSC growth and reduce abnormal transformation. Our results suggest a new concept for biomaterial design in regenerative cancer therapy.

145c Tendon Tissue Engineering Using Mechanical Stimulation of Mesenchymal Stem Cells

Rita Abousleiman[1], Peter S. McFetridge[2], and **Vassilios I. Sikavitsas**[2]. (1) Bioengineering Department, University of Oklahoma, 100 E Boyd, T-335, Norman, OK 73019, (2) Chemical, Biological, and Materials Engineering, University of Oklahoma, 100 E. Boyd, Rm T-335, Norman, OK 73019

Tendon injuries require prolonged durations of rest and immobility. When surgery becomes inevitable four material options are available: autografts, allografts, xenografts, and prosthetic devices. Autografts have limited availability, and result in a double incision, while allografts and xenografts often result in an immune response. Prosthetic devices, on the other hand, are a temporary solution since their integrity is compromised with increased usage. To augment the limitations in the current treatments of tendon defects and injuries, researchers have attempted to engineer tendon-like tissue in-vitro that is biocompatible and possesses comparable mechanical properties to the innate tissue.

The goal of this study was to investigate the effect of cyclic mechanical stimulation on mesenchymal stem cells (MSC) seeded within human umbilical veins (HUVs), and to determine the potential of the engineered constructs to function as tendon tissue replacement models. Decelulerized HUVs were seeded with MSC's and cultured for periods of 1 and 2 weeks. A novel bioreactor was designed to house and mechanically stimulate the constructs where controls were left un-tensioned. Mechanical stimulation resulted in 8 fold higher proliferation rates and significantly stronger (156%) and stiffer (109%) constructs compared to un-tensioned samples. Microscopically, cyclically tensioned samples showed parallel orientation of collagen fibers and spindle-shaped cell nuclei mimicking the morphology of native tendons. This study further investigated the effect of different daily loading durations on constructs. 1 hour of mechano-stimulation per day resulted in higher proliferation rates and stronger constructs than static cultures. However, extended 24 hour cyclic stretching resulted in cell lyses and significant weakening of constructs.

Directed Differentiation of Embryonic Stem Cells to Cardiomyocytes in a Bioreactor

Abhirath Parikh[1], *Dong H. Jing*[1], *Daniel E. Kehoe*[1] *and Emmanuel (Manolis) S. Tzanakakis*[1], (1)Chemical and Biological Engineering, State University of New York at Buffalo, Buffalo, NY 14260.

Introduction

Heart diseases are top ranked as causes of morbidity and mortality in the US and most developed countries[1-3 4 5]. Most heart maladies are associated with the death of significant portion of cardiomyocytes leading to myocardial infarction/heart attacks and permanently impaired heart function.

Presently, the only therapeutic modality that effectively reconstitutes the heart function is heart transplantation. However, its widespread utility is hindered due to the scarcity of donor organs and complications associated with immune suppressive regimens. Hence, the search for alternative sources of heart cells including embryonic stem cells (ESCs) has intensified in recent years. ESCs are pluripotent, i.e. they are able to differentiate into all cell types of the body. More importantly, ESCs have an unlimited proliferation capacity if cultured under certain conditions and hence can serve as inexhaustible source of cardiomyocytes. Transplantation studies have shown that cardiomyocytes from differentiated ESCs have the potential to integrate with the damaged heart[6-10]. A major challenge however has been to reproducibly and efficiently generate a pure population of cardiomyocytes from ESCs.

We are developing methodologies for more efficient cardiac differentiation of embryonic stem cells, and subsequently scalable systems for their production in large quantities as required for clinical applications. We have used mouse ESCs (mESCs) as a model system and translated important findings to human ESCs (hESCs) because of the physiological similarities between the two species. Mouse ESCs were selected as a model system because (1) of their relatively high proliferation rate which enables more experiments to be done in a shorter time, and (2) of their relatively cheaper culture cost. We have efforts to better understand customary cardiac differentiation methods and to identify vital culture parameters to improve the outcome of the differentiation. Subsequently, we determined conditions that optimize the differentiation. The current differentiation methods have major limitations such as highly varied outcome[11, 12] and low yield of cardiomyocytes[6, 8, 11, 13, 14]. The differentiation methods rely on serum for nutrition and differentiation cues. The exact composition of serum is undefined but it appears to provide a mixture of proteins, hormones and other factors that support the growth of cells. It is speculated that the some of these growth factors promote cardiac differentiation, however serum-based differentiation generates a highly heterogeneous differentiated cell population with low percentage of cardiomyocytes. It is has also been difficult to separate the cardiomyocytes from the rest of the population. The differentiation outcome is also greatly variable with each batch of serum due to unknown differences in growth factor content. But studies for isolating factors that can yield a pure cardiomyocyte population have been hindered because of dependence on serum for nutrition. The signals originating from the serum interfere with agonists we attempt to study by inhibiting, suppressing or synergizing their effect. Therefore, there have been extensive efforts to develop a defined serum-free growth medium (DSFM) which can propagate cell culture without affecting differentiation. This can allow for studies to understand processes that govern ESC differentiation and subsequently to develop agonists and conditions that can control these processes to obtain the desired differentiation outcome, i.e. a pure population of cardiomyocytes. Another drawback of serum-based culture is that these cells are not fit for transplantation. Cells cultured in serum can pass on pathogens from its animal source, can trigger foreign body reactions and can cause rejection of transplanted cells due presence of animal proteins. Using a DSFM can also reduce such a risk.

We utilized a DSFM prepared with Knockout Serum Replacer (KSR, Invitrogen, Carlsbad, CA) because it supports well ESC culture and differentiation[15]. Apart from the currently used non-specific cardiomyocyte stimulants such as DMSO and Retinoic Acid (RA)[16, 17],we investigated physiologically relevant growth factors to direct cardiac differentiation. Studies on embryonic heart development have shown that bone morphogenetic protein (BMP) signaling has a direct role in cardiac induction and also terminal differentiation of cardiac tissue[18, 19]. We treated mESCs exogenously with BMP-4 to initiate this signaling. Our results indicate that BMP-4 successfully directs cardiac differentiation of mESCs and the stimulant RA was ineffective under DSFM conditions.

Conventional ESC culture is carried out on dishes which are not optimal for cell production in a clinically relevant scale. Stirred suspension bioreactors promote three dimensional non-adherent cell culture and therefore have potential for maximal scale up of cell production. We developed a method for directing cardiac differentiation of mESCs with BMP-4 in bioreactors without serum.

Our current efforts are focused on translating these findings to hESC differentiation aiming at the development of a scalable system for the generation of cardiomyocytes for cell therapies. Concurrently experiments with mESCs are yielding data to further our knowledge of ESC specification for designing differentiation protocols characterized by high efficiency.

Results:

BMP-4 directed differentiation in mESCs

Under defined serum-free conditions (DSFM), mESCs if untreated with stimulants tend to differentiate towards neural-like cells. Onset of such ectodermal differentiation is characteristic of invitro ESC differentiation in the absence of any influencial signals. Treating mESCs in DSFM with BMP-4 did not reveal any neural cells indicating that ectodermal and mesodermal differentiation are separate routes that ESCs can commit to in vitro. Additionally we observed beating cells which demonstrates that BMP-4 initiates signaling that leads to cardiomyocytes. This is in accordance with previous studies that show that BMPs have a direct role in cardiac induction and differentiation during embryonic cardiogenesis [18, 19].

We characterized the beating cell clusters obtained by BMP-4 treatment in DSFM and compared the results with serum-based differentiation. Unlike the serum-based differentiation, the curve representing the percentage of beating clusters with time of mESCs treated with BMP-4 was reproducible. The first appearance of beating was day 10 and the curve reached a maximum by days 14-15 consistently. The beating percentage was also consistently higher – approximately 90% of EBs were beating (Figure 1)- compared to the serum-treated counterparts.

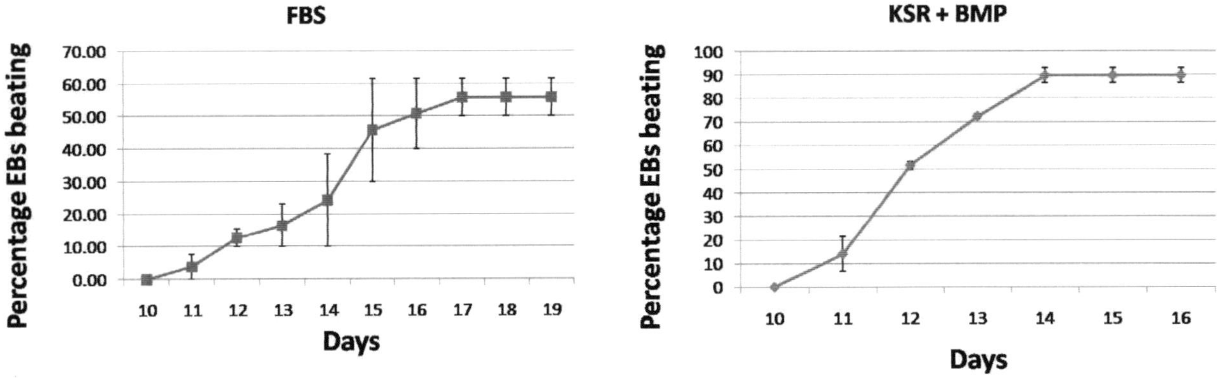

Figure 1: Comparison of beating curves between FBS-based and BMP-based differentiation.

Reverse transcription-PCR (RT-PCR) shows that the BMP treatment caused upregulation of cardiomyocyte-specific genes such as α-myosin heavy chain (α-*MHC*), *GATA-4*, *Nkx2.5*, and *ANF* compared to the BMP untreated control. The expression of α-MHC was much higher in the BMP- than the serum-based differentiation. This is expected because of the higher percentage of beating clusters in the BMP differentiation. Likewise, the BMP untreated KSR controls expressed minimal α-MHC (Figure 2).

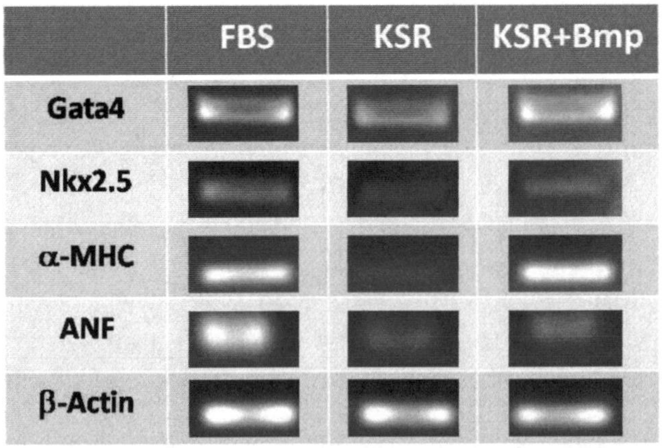

Figure 2: RT-PCR data for cardiomyocyte genes for serum, DSFM (KSR) and BMP + DSFM conditions

Addition of the BMP antagonist Noggin inhibited the ESC differentiation into cardiomyocytes. Noggin-treated cells did not show any beating despite BMP treatment. Quantitative PCR (qPCR) showed a down-regulation of cardiomyocyte genes *MHC* and *Nkx2.5* in noggin-treated cells when compared to BMP-treated cells and similar to BMP untreated control. These results confirm that BMP signaling induces cardiac differentiation. Inhibition of BMP signaling causes non-specific differentiation with only a small percentage of cells becoming cardiac cells.

Scale up of BMP-directed differentiation in a bioreactor

We explored the feasibility of scaling up BMP-based directed differentiation of mESCs cultured in dishes to continuously stirred bioreactors. First, we asked if the ESC differentiation is affected by the hydrodynamic conditions in the bioreactor. Mechanical cues stemming from different environment can affect ESC differentiation and could also alter the action of BMPs. Therefore, we compared differentiated cells obtained from dish and bioreactor cultures. Beating cells obtained from the bioreactor upregulated cardiomyocyte markers in a similar fashion as assessed by qPCR and immunostaining.

A reproducible curve depicting the percentage of beating cell clusters vs. time for bioreactor cultures was obtained and was similar to that for static cultures. In both cases, the first beating clusters appeared on day 10 and the maximum beating percentage was reached by day 14. The maximum beating percentage was also similar - approximately 90% EBs beating.

Next, we plan to compare the cell proliferation and cardiomyocyte yield between dish and bioreactor cultures.

Studies on human embryonic stem cells (hESCs)

The understanding of mESC differentiation to cardiomyocytes under defined culture conditions provides clues for the differentiation of hESCs aiming to generate cardiomyocytes for cell therapies. To successfully translate mESCs differentiation methodologies to hESCs differentiation, we first attempted to differentiate hESCs in FBS-containing medium (similar to the mESC protocol).

hESCs differentiation to cardiomyocytes was initiated by forming EB in non adherent (Petri) dishes (Figure 3b). Beating cell clusters were first identified at differentiation day 10 and the percentage of beating cells was usually increased with time.

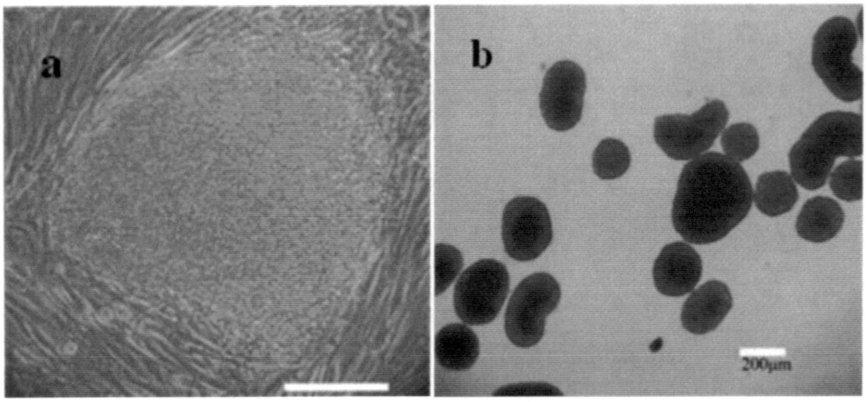

Figure 3: Differentiation of hESCs to cardiomyocytes. (a) The morphology of hESCs maintained on MEF (Bar=250μm). (b)EBs in suspension culture (Bar=200μm).

Quantitative PCR (Figure 4 Upper Panel) revealed an increased expression of heart cell-specific genes in differentiating hESCs compared to undifferentiated. These genes include the atrial natriuretic factor (ANF), human α-myosin heavy chain (hα-MHC) and human β-Myosin Heavy Chain (hβ-MHC). The expression of cardiac markers NKX2.5 and α-actinin were also identified by immunofluorescence (Figure 4, lower panel).

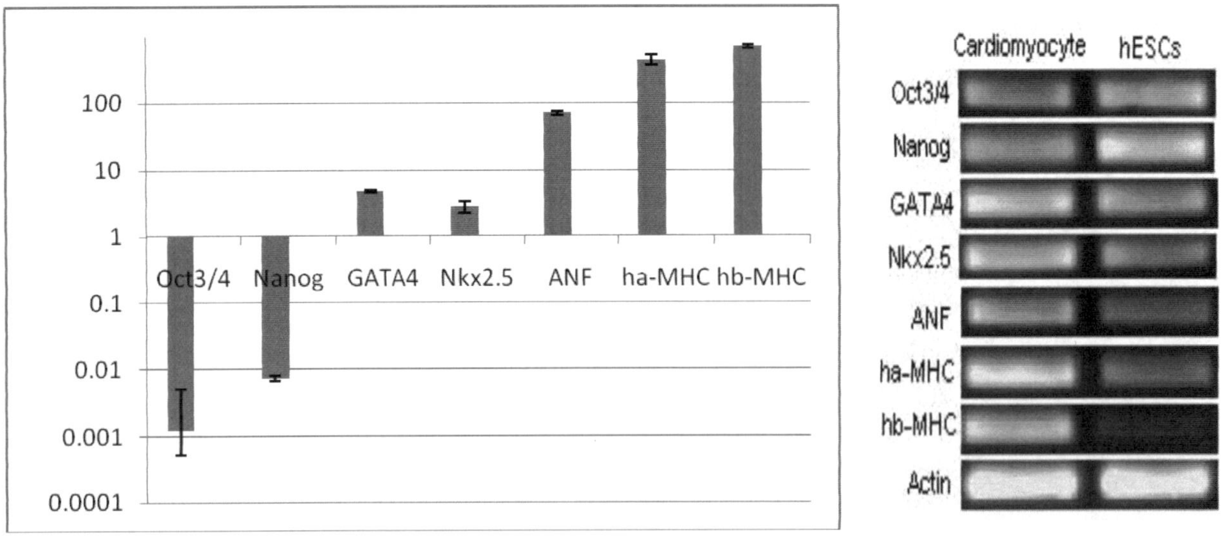

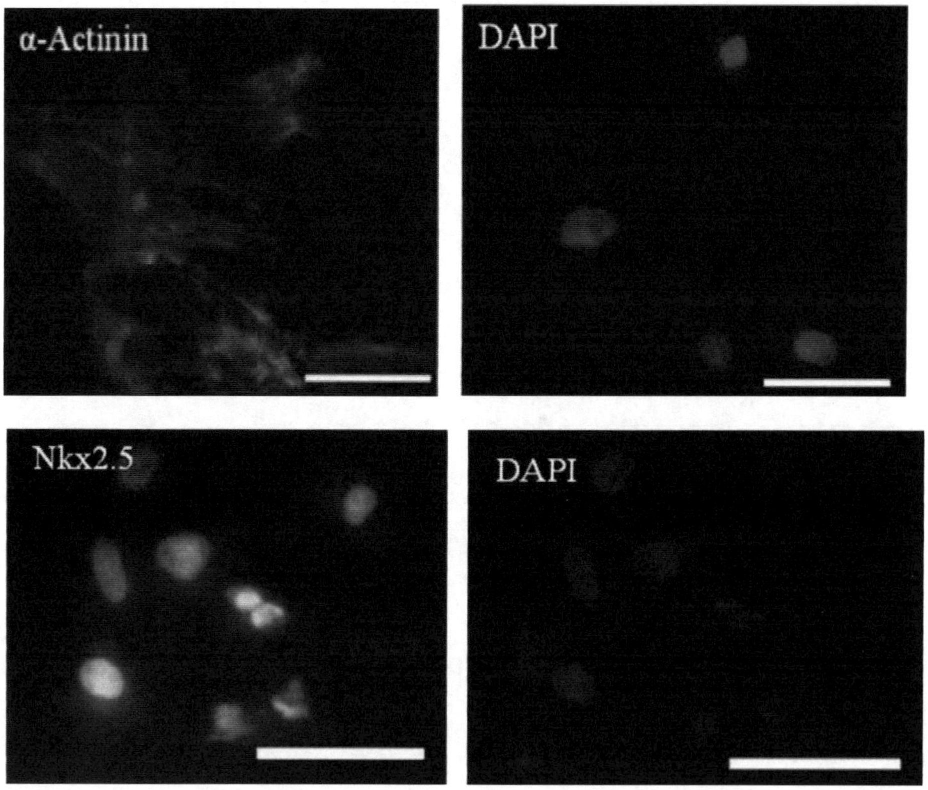

Figure 4: Cardiac marker analysis of cardiomyocytes. Upper Panel: RT-PCR analysis of Oct4, Nanog, GATA4, Nkx2.5, ANF, hα-MHC and hβ-MHC in undifferentiated and differentiated cells. Lower Panel: After seeing cells contracting, they were dissociated as single cells to culture for an additional 3 days and assessed for α-Actinin and Nkx2.5 immunoreactivity. DAPI was used for staining nuclei. Bar=50 μm.

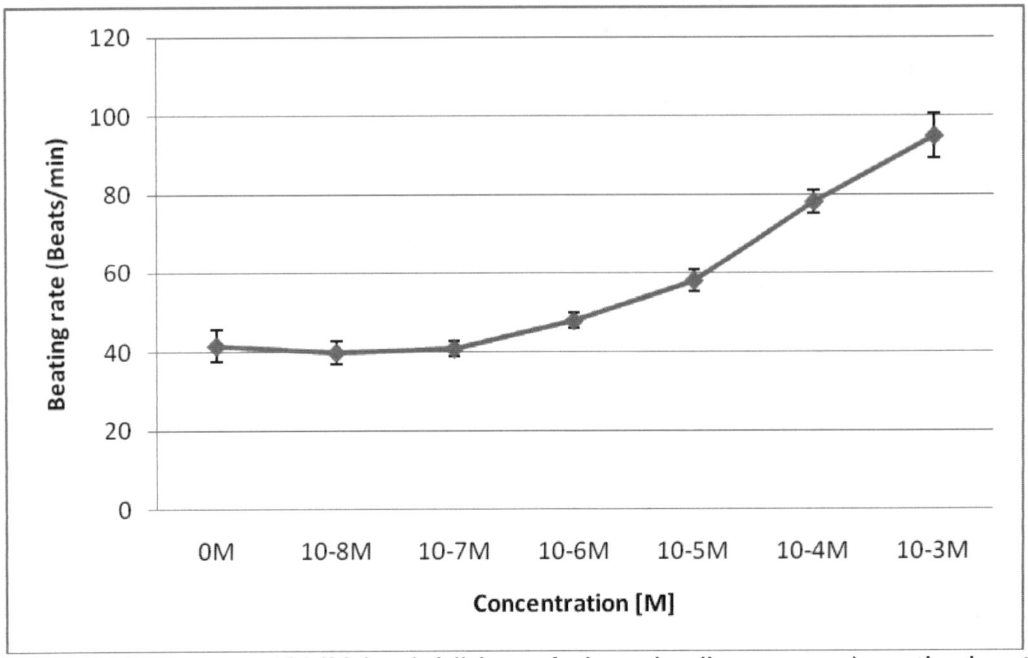

Figure 5: Effect of IBMX (an inhibitor of phosphodiesterases) on the beating rate of hESCs derived cardiomyocytes.

Furthermore, we assessed the functional characteristics of cardiomyocytes compared to native heart muscle cells. Stimulation by isobutyl methylxanthine (IBMX, an inhibitor of phosphodiesterase) resulted in a dose-dependent increased contraction rate (Figure 5). These results indicate the presence of a cAMP-dependent mechanism that mediated the contracting of cells.

Although our hESCs-derived cardiomyocytes exhibit characteristics of heart muscle cells, the current efficiency of differentiation is very low. Thus, our efforts are focused on developing a protocol for differentiation of hESCs to heart muscle cells by growth factors (BMP, Activin) in a directed way.

References

1. Secretary, O.o.t.P. *American Heart Month.* A Proclamation By the President of the United States of America. Washington D.C., 2007.
2. AM Miniño, M.H., SL Murphy, KD Kochanek. *Deaths: Final Data for 2004* National Vital Statistics Reports. United States Center for Disease Control., 2007.
3. UK, N.S. *Heart disease leading cause of death in England & Wales.* Health Statistics Quarterly, 2006.
4. *Heart Disease and Stroke Stastistics.* American heart association. Dallas, Texas, 2008.
5. *Cardiovascular disease: prevention and control.* World Health Organization: Global Strategy on Diet, Physical Activity and Health, 2003.
6. Xu, C., Police, S., Hassanipour, M., and Gold, J.D. Cardiac bodies: a novel culture method for enrichment of cardiomyocytes derived from human embryonic stem cells. *Stem cells and development* 15, 631-9, 2006.
7. Laflamme, M.A., Chen, K.Y., Naumova, A.V., Muskheli, V., Fugate, J.A., Dupras, S.K., Reinecke, H., Xu, C., Hassanipour, M., Police, S., O'Sullivan, C., Collins, L., Chen, Y., Minami, E., Gill, E.A., Ueno, S., Yuan, C., Gold, J., and Murry, C.E. Cardiomyocytes derived from human embryonic stem cells in pro-survival factors enhance function of infarcted rat hearts. *Nature biotechnology* 25, 1015-24, 2007.
8. Klug, M.G., Soonpaa, M.H., Koh, G.Y., and Field, L.J. Genetically selected cardiomyocytes from differentiating embronic stem cells form stable intracardiac grafts. *The Journal of clinical investigation* 98, 216-24, 1996.
9. Kehat, I., Khimovich, L., Caspi, O., Gepstein, A., Shofti, R., Arbel, G., Huber, I., Satin, J., Itskovitz-Eldor, J., and Gepstein, L. Electromechanical integration of cardiomyocytes derived from human embryonic stem cells. *Nature biotechnology* 22, 1282-9, 2004.
10. Dai, W., Field, L.J., Rubart, M., Reuter, S., Hale, S.L., Zweigerdt, R., Graichen, R.E., Kay, G.L., Jyrala, A.J., Colman, A., Davidson, B.P., Pera, M., and Kloner, R.A. Survival and maturation of human embryonic stem cell-derived cardiomyocytes in rat hearts. *Journal of molecular and cellular cardiology* 43, 504-16, 2007.
11. Xu, C., Police, S., Rao, N., and Carpenter, M.K. Characterization and enrichment of cardiomyocytes derived from human embryonic stem cells. *Circulation research* 91, 501-8, 2002.
12. Kehat, I., Kenyagin-Karsenti, D., Snir, M., Segev, H., Amit, M., Gepstein, A., Livne, E., Binah, O., Itskovitz-Eldor, J., and Gepstein, L. Human embryonic stem cells can differentiate into myocytes with structural and functional properties of cardiomyocytes. *Journal of Clinical Investigation* 108, 407-14, 2001.
13. Zweigerdt, R., Burg, M., Willbold, E., Abts, H., and Ruediger, M. Generation of confluent cardiomyocyte monolayers derived from embryonic stem cells in suspension: a cell source for new therapies and screening strategies. *Cytotherapy* 5, 399-413, 2003.
14. Zandstra, P.W., Bauwens, C., Yin, T., Liu, Q., Schiller, H., Zweigerdt, R., Pasumarthi, K.B., and Field, L.J. Scalable production of embryonic stem cell-derived cardiomyocytes. *Tissue engineering* 9, 767-78, 2003.
15. Kehoe, D.E., Lock, L.T., Parikh, A., Tzanakakis, E.S. Propagation of Embryonic Stem Cells in Stirred Suspension without Serum. *Biotechnol. Prog.* 24, 2008.
16. Edwards, M.K., Harris, J.F., and McBurney, M.W. Induced muscle differentiation in an embryonal carcinoma cell line. *Molecular and cellular biology* 3, 2280-6, 1983.

17. Wobus, A.M., Kaomei, G., Shan, J., Wellner, M.C., Rohwedel, J., Ji, G., Fleischmann, B., Katus, H.A., Hescheler, J., and Franz, W.M. Retinoic acid accelerates embryonic stem cell-derived cardiac differentiation and enhances development of ventricular cardiomyocytes. *Journal of molecular and cellular cardiology* 29, 1525-39, 1997.

18. Zaffran, S., and Frasch, M. Early signals in cardiac development. *Circulation research* 91, 457-69, 2002.

19. Abdelwahid, E., Rice, D., Pelliniemi, L.J., and Jokinen, E. Overlapping and differential localization of Bmp-2, Bmp-4, Msx-2 and apoptosis in the endocardial cushion and adjacent tissues of the developing mouse heart. *Cell and tissue research* 305, 67-78, 2001.

145e Integrated Expansion and Differentiation of Embryonoic Stem Cells In Fibrous Bed Bioreactors

Ning Liu, Chemical and Biomolecular Engineering, The Ohio State University, 125A Koffolt Lab, 140 w. 19th Ave., Columbus, OH 43210 and Shang-Tian Yang, Chemical and Biomolecular Engineering Department, The Ohio State University, 140 w. 19th Ave, Columbus, OH 43210.

Cell therapy using embryonic stem cells (ESCs) requires reproducible, well-defined culture systems for producing large numbers of well-characterized cells. Current ESC differentiation process involving three steps, expansion of ESCs, formation of embryoid bodies and induction of lineage-specific differentiation, is laborious and difficult to scale-up. In this study, a simple process using fibrous bed bioreactors was developed for integrated ESC expansion and differentiation. When cultured in a PET fibrous matrix with a growth medium for 30 days, ESCs were expanded 125-fold to reach a high cell density of 1x10^8 cells/ml matrix without significant differentiation. Following the expansion of ESCs in a fibrous bed bioreactor, continually refreshing the bioreactor with a differentiation medium induced ESC differentiation. Multi-lineage differentiation was confirmed using immunocytochemistry and FACS analyses. This simplified, integrated ESC expansion and differentiation process can meet the stringent requirements for developing large-scale cultures for producing ESCs for potential clinical applications.

145f Effect of Organismal Aging on Bone Marrow Derived Smooth Muscle Progenitor Cells

Juhee Han[1], Jin Yu Liu[1], Daniel D. Swartz[2], and Stelios T. Andreadis[1]. (1) Bioengineering Laboratory, 908 Furnas Hall, Department of Chemical and Biological Engineering, SUNY at Buffalo, Amherst, NY 14260, (2) Women and Children's Hospital of Buffalo, SUNY-Buffalo, Buffalo, NY 14214

Recently we demonstrated that functional smooth muscle progenitor cells can be derived from bone marrow (BM-SMPCs) and have high potential as an autologous cell source for vascular tissue engineering. However, normal cell function and turnover frequency decline with aging so that understanding the effect of aging on cell phenotype and function is very important. To address this question, we used a smooth muscle alpha actin promoter (SMαA) driving expression of EGFP to isolate BM-SMPCs from neonatal (n=3) or adult sheep (n=3) and examined the effect of animal aging on the properties of stem cells and tissue engineering vascular constructs. Our results showed that the proliferation capacity and clonogenic ability of neonatal cells were significantly higher than adult BM-SMPC. In addition, tissue engineered blood vessels (TEVs) constructed from neonatal cells demonstrated significantly higher level of receptor and non-receptor mediated vascular contractility as well as mechanical strength compared to TEVs from adult BM-SMPCs. Differential gene expression profiling using quantitative real-time PCR showed that adult BM-SMPCs expressed higher levels of matrix metalloproteases but neonatal cells expressed significantly higher levels of extracellular matrix molecules e.g. tropoelastin. These results are currently extended to obtain a comprehensive gene expression profile as a function of aging and determine potential strategies to reverse these effects in adult BM-SMPCs. Ultimately these efforts may enhance the potential of bone marrow mesenchymal stem cells for tissue regeneration.

The profound effects of ECM-associated biochemical signals on cell behavior have long been recognized. However, the impact of these biochemical signals on stem cell or progenitor cell differentiation have proven difficult to explore in a controlled manner, since most synthetic and natural materials adsorb a range of plasma proteins from the cell culture media. These adsorbed proteins are major determinants of cell behavior, in addition to any molecule deliberately added by the researcher. For controlled investigation of the effects of specific biochemical stimuli on cellular response, we therefore need a scaffold material that is essentially "non-biofouling". Poly(ethylene glycol) diacrylate (PEGDA) is a "non-biofouling" scaffold, and thus, in the absence of biochemical modification, presents a biochemical "blank slate" to cells. The photoactivity of PEGDA permits desired biochemical moieties to be readily conjugated into the PEGDA hydrogels at defined levels. In the present work, we investigated the efficacy of defined biochemical stimuli on driving mouse embryonic progenitor (10T½) cells toward a mature smooth muscle cell (SMC) phenotype. To explore these effects, mouse embryonic progenitor cells were incorporated into PEGDA hydrogels into which defined levels of collagen type IV, fibronectin, laminin, fibrinogen, or heparin had been conjugated. Each of these molecules have been implicated in SMC differentiation or in maintenance of a mature SMC phenotype. 10T½ mouse embryonic progenitor cells encapsulated in hydrogels containing heparin and fibrinogen showed significantly higher levels of SMC markers SM-a-actin, calponin h1, and SM-g-actin than gels containing laminin or fibronectin. Similarly, collagen and elastin levels were higher in the fibrinogen and heparin hydrogels than in laminin and fibronectin gels. Focused microarray analyses indicate that variable modulation of the Wnt pathway may underlie the differing responses observed in the present study.

165a Covalent Molecular Assembly of Conducting Polymers on Silicon

Sundaramurthy Jayaraman, Chem and Biomol Engg, NUS, 4 Engg Drive 4, Singapore, Singapore and MP Srinivasan, ChBE, NUS, 4 Engineering Drive 4, Singapore, Singapore.

Organic functionalization of semiconducting surfaces is emerging as an important area in the development of new semiconductor based materials and devices. Polythiophenes are one of the most important classes of conjugated polymers with a wide range of electronic properties and applications as conducting films, electrochromics, chemical sensing, memory, light emitting devices and field effect transistors. By self assembling the polythiophene film on silicon interfaces, these characteristic properties of polythiophenes can be incorporated into silicon-based electronic devices.

In this work, we report formation of nano-structured, regio-regular conducting polythiophene films by covalently binding the thiophene to silicon. The covalently bound nanostructured films possessed uniformity, good thermal and chemical stabilities. Mobility and conductivity characterizations carried out showed increase in conductivity with doping.

We further investigated the deposition of multilayers of covalently bonded polythiophene films on silicon substrates using an intermediate linker as well as by modifying the functional end group of the polythiophene. The electronic properties of the polythiophene films can be significantly improved by multilayer assemblies. The covalent bonding of polythiophene was confirmed using the spectroscopic techniques such as X-ray photoelectron spectroscopy (XPS), Fourier transform - infrared spectroscopy (FTIR), and UV-Vis spectroscopy.

By covalent bonding with the substrate, conducting polymers can be incorporated as constituents micro- and nano scale devices such as chemical sensors and Field effect transistors.

165b Novel Nanocomposites Made from Polymer and as-Received, Unmodified Graphite: Effects of Graphite Level and Dispersion on Mechanical and Electrical Properties, Crystallization, and Thermal Stability

Katsuyuki Wakabayashi[1], Philip Brunner[2], Cynthia Pierre[3], Sheldon Hewlett[3], and **John M. Torkelson**[4]. (1) Department of Chemical Engineering, Bucknell University, Lewisburg, PA 17837, (2) Chemical and Biological Engineering, Northwestern University, Evanston, IL 60208, (3) Materials Science & Engineering, Northwestern University, 2220 Campus Drive, Cook Hall Room 2036, Evanston, IL 60208-3108, (4) Chemical and Biological Engineering and Materials Science and Engineering, Northwestern University, 2145 Sheridan Rd, TECH E136, Evanston, IL 60208

Many hundreds of experimental studies have been published related to polymer-carbon nanotube nanocomposites or hybrids. In contrast, only a relatively small number of experimental studies have been published related to polymer-graphite nanocomposites or hybrids, with almost all of those focused on chemically modified or thermally pre-treated graphite, e.g., graphite oxide, expanded graphite, or thermally exfoliated graphite oxide. This is because effective dispersion or exfoliation of as-received, unmodified graphite is practically impossible with conventional melt processing. Even with pre-treatment, nanocomposite production by conventional processing is challenging due to thermodynamic and/or kinetic limitations, sometimes leading to limited property enhancement. However, relative to carbon nanotubes and modifed graphite, there are significant economic and environmental advantages associated with using the natural, as-received, unmodified graphite as a nanofiller.

To overcome the challenges associated with using graphite as a nanofiller, we have employed solid-state shear pulverization (SSSP) to produce polymer-graphite nanocomposites that are not subject to the thermodynamic/kinetic limitations associated with conventional processes. We have demonstrated that the continuous, scalable SSSP process can result in polypropylene (PP) nanocomposites containing well-dispersed unmodified, as-received graphite at levels up to 2.5 - 3.0 wt% graphite. When PP/graphite (2.5 wt%) nanocomposites are made by SSSP, the resulting materials exhibit a 100% increase in Young's modulus and a ~60% increase in yield strength in comparison with neat PP. Even at extremely low levels of graphite (0.3 wt%), the resulting PP/graphite nanocomposites made by SSSP can exhibit extroardinary enhancements in mechanical properties relative to neat PP, including a 50% increase in Young's modulus and 45% increase in yield strength with little to no reduction in impact strength or elongation at break. However, we have found that increasing the graphite to levels well in excess of 2.5 - 3.0 wt% can lead to poor exfoliation of the graphite via SSSP, with only improvements in Young's modulus and yield strength being no better than with 0.3 wt% graphite and with substantial embrittlement of the nanocomposite.

Measurements of the shear storage moduli of our resulting PP/graphite nanocomposites reveal a solid-like rheological response at low frequency with graphite levels of 2.5 wt% or greater and a liquid-like rheological response with graphite levels of 0.3 - 1.0 wt%. These results indicate the presence of a network-like percolated structure at graphite levels of 2.5 wt% graphite and greater, which is accordance with our measurements of electrical conductivity which indicate a percolation threshold between 1.0 wt % and 2.5 wt% graphite.

Besides mechanical properties and electrical conductivity, we will also describe the effects of graphite level and dispersion on the crystallization behavior and thermal degradation of PP.

165c Synthesis and Characterization of Cubic Co3O4 Nanocomposites

Dwayne Vickers, Tuskegee University, Chemical Engineering Department, Tuskegee, AL 36088, **Tamara Floyd-Smith**, Chemical Engineering Department, Tuskegee University, 522A Luther Foster Hall, Tuskegee, AL 36088, and Lynden Archer, Cornell University, 120 Olin Hall, Cornell University, Ithaca, NY 14853.

A procedure for synthesizing narrow size distribution cube-shaped Co_3O_4 nanoparticles ranging from 40 to 60 nm is reported. The rheological properties of suspensions of cubes in low molecular weight polyethylene glycol were studied over a wide range of particle volume fractions. At low volume fractions these suspensions display Newtonian fluid behavior, while at high volume fractions they manifest complex rheological characteristics, including shear thinning, shear thickening and a yield stress. Pronounced shear thinning was observed at particle volume fractions Φ as low as 0.14, and shear thickening observed at $\Phi \approx 0.36$. For particle volume fractions ranging from 0.36-0.62, the onset of shear thickening is found to occur at increasingly lower shear rates, reaching a minimum of $40s^{-1}$ at Φ = 0.62. These findings are discussed through comparisons to suspensions of spherical particles and models designed to predict the relative viscosity and shear thickening transition in colloidal suspensions.

Functionalization Multi-walled carbon nanotube and Its Bismaleimide Composites of Properties

Changwen Song, Hongxia Yan, Yi Zhang, Yusheng Tang

Department of Applied Chemistry, School of Natural of Applied Science,

Northwestern Polytechnical University, Xi'an 710072, P R China

Abstract：Carbon nanotubes (CNTs) were one of nanoparticles with unique structure and significant advantages. But the application of CNTs was limited because of the inert surface of CNTs. In this paper，fluorinating Multi-walled carbon （F-MWNTs）was synthesized by solid-phase method which taken an organic fluoride compound as the main raw material, make function carbon nanotubes. As a comparison, raw MWNTs and F- MWNTs have been characterized by using FTIR、XRD、XPS. The results showed that C–F bonds were created on the surface of carbon nanotubes, successfully synthesis fluoride carbon nanotubes. The mole contents of fluorin atom can be reached 3.59 mol%. The demonstrated C-F functionalization provides a new synthetic method and the F-MWNTs can be used as precursors for the preparation ofbismaleimide-MWNTs polymer materials.The bismaleimide (BMI)/MWNTs composites were prepared by in situ polymerization. The results showed that BMI with MWNTs fillers has extraordinary significance for the development of advanced materials with enhanced mechanical properties.

Key Words：MWNTs，Functionalization，Bismaleimide，Properties

1. Introduction

Carbon Nanotubes(CNTs) are unique nanostructured materials with remarkable physical, mechanical, chemical and electronic properties. Theses properties make them attractive for applications in many scientific and technological fields such as electronic structures, polymer composites, and biological systems[1-5]. CNTs/polymer nanocomposites have attracted considerable attention by scientists in resent years[6-9]. However, the limited ability to process and disperse in organic solvent CNTs hinders the realization of their full potential. To overcome this obstacle, surface modification of CNTs was very important in application of modified polymer [10].

In our present work, at first fluorinating multi-walled carbon nanotubes (F-MWNTs) was synthesized by solid-phase method which was a new synthetic method being differently from gas-phase method[11,12].Then the F-MWNTs can be used as precursors for the preparation of bismaleimide-MWNTs polymer materials by in situ polymerization.

2. Experimental

2.1 Materials

MWCNTs used in the paper were provided by Zhejiang University which were prepared by catalytic chemical vapor deposition method. The powder of polytetrafluoroethylene (PTEF) was supplied by HuBei Fengguang Chemicals. Bismaleimide prepolymer (BMI) was prepared by group of Rongchang Ning in Northwestern Polytechnical University (China).

2.2 Fluorinated MWNTs

Fluorinating multi-walled carbon nanotubes (F-MWNTs) was synthesized by solid-phase method which an organic fluoride (PTEF) and multi-carbon nanotubes (MWNTs) were heated to 480-500°C for 4h in sealed vessel. The proportion of MWNTs and PTEF is 1:10. As a comparison, raw MWNTs and F-MWNTs have been characterized by using FTIR, XRD and XPS.

2.3 Properties measurements

2.3.1 Sample preparation:

At first , bismaleimide prepolymer (BMI) were melt by heating. Then F-WMNT were dispersed in acetone with ultrasonically stirring for 40 minutes at room temperature. Then the suspension was added into melted BMI at 120°C and kept that temperature at stirring for 40mins. Then the mixture was poured into warm-up mould with silicone as a release layer, and the mould with the mixture was kept in a vacuum chamber and deaerated for 30 min. Following, curing:150°C/1h+180°C/2h +200°C/2h; post-curing: 220°C/4h.

2.3.2 Mechanical properties and analysis:

Composites castings were cut using a grinding wheel cutter into samples for mechanical testing in term of GB2571-1981, GB2570-1995 for determining flexural and impact strength properties, respectively. Five specimens were tested for each test and average values are reported.

The micro morphology of the cured resin and the composites were investigated by scanning electron microscopy (SEM) on HITACHIS-570 SEM. The fracture surface of was coated with gold by vapor deposition.

3. Results and Discussion

3.1 Functionalization MWNT

The nature of the surface groups of the F-MWNTs and the raw MWNT were investigated using FTIR spectroscopy (Fig.1). Both samples have two large peaks, one at 1570 cm^{-1}

and another at 2250cm^{-1}, which are assigned to the CNTs skeletal motions [10, 11]. Furthermore, the second-order curve displays a broad peak centered at 1050 cm^{-1} in Fig.1(b), due to the existence of C−F bonds [11]. The results of FTIR spectra confirmed the formation of new chemical bonds on the raw MWNT after treatment.

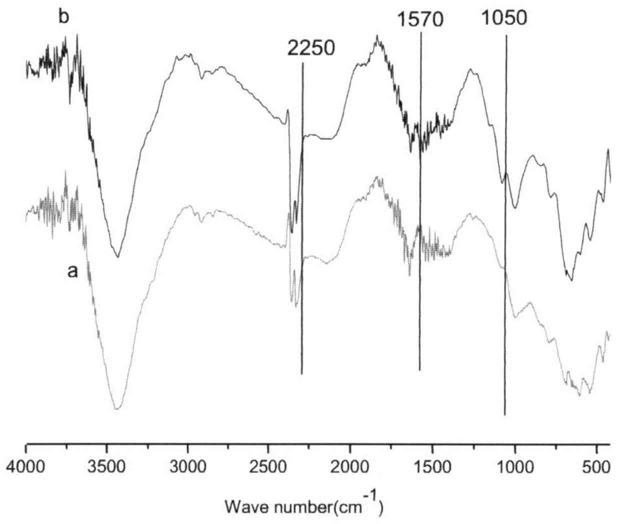

a: the raw MWNT ; b: F-MWNT

Fig. 1 FTIR spectra of the raw MWNT and F-MWNT.

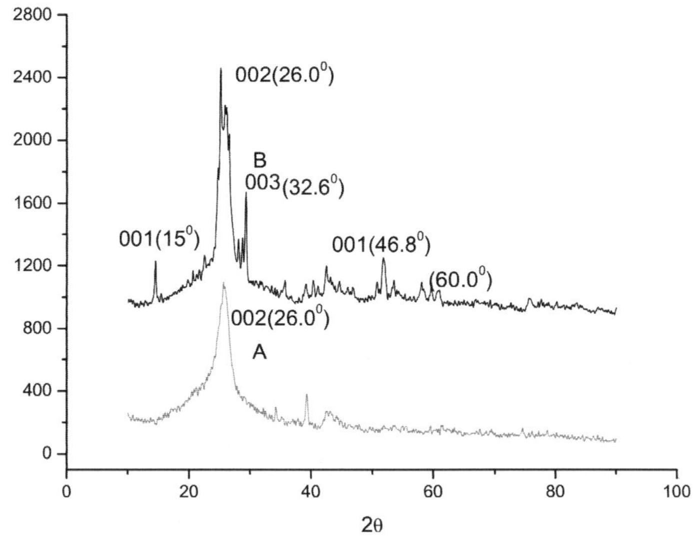

a: the raw MWNT ; b: F-MWNT

Fig. 2 XRD spectra of the raw MWNT and F-MWNT

As shown in Fig. 2, the XRD results revealed the characteristic microstructure of MWNT. the XRD patterns indicate the persistence of the main reflection of the original CNT, at $2\theta \approx 26^o$ (Fig. 2(a)). Same results were reported in other studies [12-14]. Therefore, the structure of cylindrical concentric carbon layers is preserved. In addition, there were two characteristic peaks at $2\theta \approx 15^o$ and $2\theta \approx 32.6^o$ on the second-order curve in Fig. 2(b). Some new layered patterns (46.8^o, 60°) in Fig. 2(b) could also be seen.

A survey XPS spectrum of a F-MWNTs sample, and for comparison, of non-fluorinated MWNTs, both are presented in Fig. 3, Fig. 4. Amount of carbon, fluorine and oxygen in the surface layer is presented in Table 1. As shown in Fig. 4 there is a new minor peak component at the binding energy of 687.8 eV corresponding to F 1*s* of the functional reaction on MWNT, which does not exist in the Fig. 3. As shown in Table 1, the carbon

C 1*s* *a*reas observed at 71125 in the raw sample is increased by 76395 to a higher binding energy after treatment, partly due to the appearance of electrophilic groups [15].

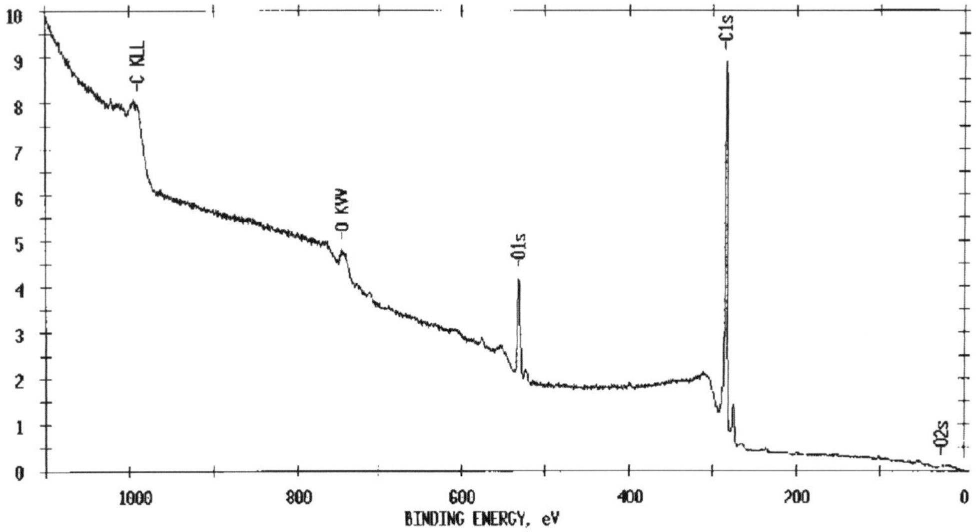

Fig. 3 XPS spectra of raw MWNT

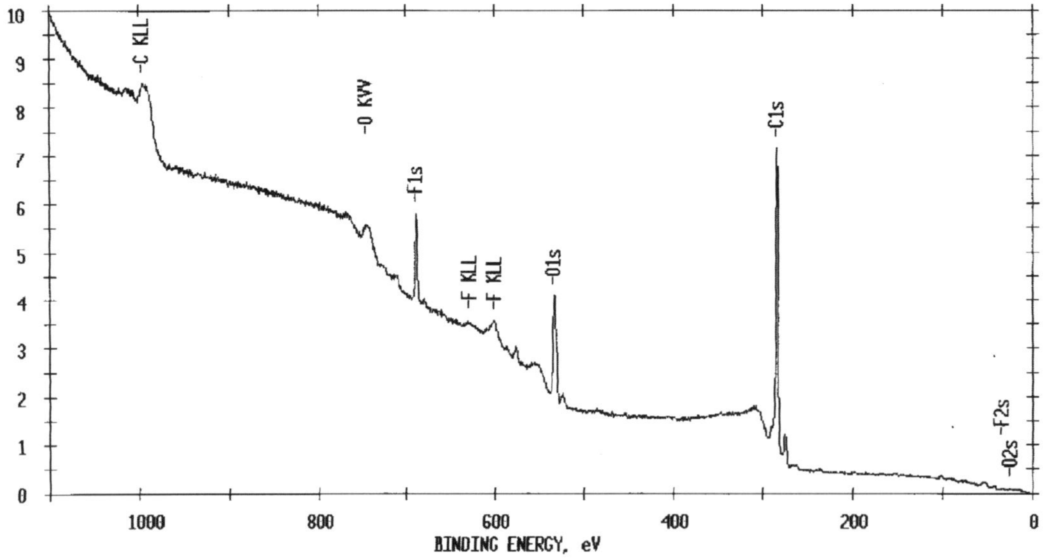

Fig. 4 XPS spectra of F-MWNT

Table 1 The result of analyzing CNTs and F-CNTs by XPS

Samples	Element	Area /cts-eV/s	Sensitivity Factor	Concentration /%
Raw MWNT	C1s	71125	0.296	88.78
	O1s	21582	0.711	11.22
	F1s	—	0	—
F-MWNT	C1s	76395	0.296	82.32
	O1s	31417	0.711	14.09
	F1s	11245	1.000	3.59

3.2 Mechanical properties

Table 2 shows the relationship between impact strength and the content of F-MWNT. From Table 2 we can see that the impact strength of BMI was improved by adding F-MWNT. First with the increase of the content of F-MWNT the impact strength of the system was enhanced and reach peak when the content is 0.6%. The impact strength was 14.68 kJ/m^2, which is much higher than the impact strength of pure BMI that was 9.48 kJ/m^2.

Tabel 2 The relationship between impact strength and the content of F-MWNT

F-MWNT (wt%)	Impact Strength （KJ/m^2）
0.0	9.48
0.2	11.98
0.4	12.79
0.6	14.68
0.8	11.15
1.0	10.82

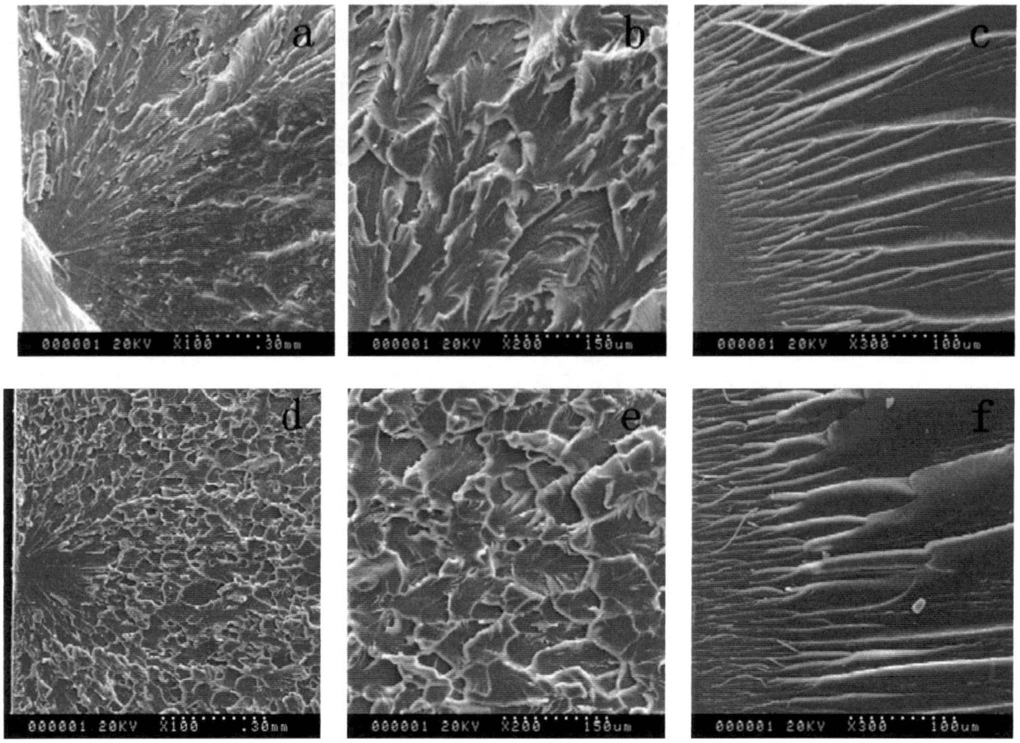

Fig. 6 SEM of fracture surfaces in the impact test of composites

a , b , c : pure BMI ; d , e , f : BMI/F-MWNT (F-MWNT:0.6%)

Fig. 6 is SEM of fracture surfaces in the impact test of samples. Photographs a, b, c are pure BMI. In these photographs, the fracture surfaces were comparatively slippery and the fracture headstream is wide. Photographs d, e, f were about BMI/F-MWNT (the content of F-MWNT was 0.6%). In contrast to pure BMI the fracture headstream of BMI / F-MWNT was small and the fine cracks were dumpy and in particular the number of toughness eddy, which was obvious bigger, is more in unit area.

4. Conclusions

A solid-phase method fluorinated MWNTs was different from the gas-phase method which usually taken strong poisonous fluorine gas as the main raw material was reported. The results in FTIR, XPS and XRD spectrums confirmed that the surface of MWNT was successfully functionalized by PTEF. Furthermore, F-MWNT was well dispersed in polar solvents. The mechanical properties of bismaleimide resin were improved by adding F-MWNT. When the content of F-MWNT was 0.6wt%, the impact strength of BMI/F-MWNT composite could be increased 54.8%.

References

1 Chen W X, Tu J P, Wang LY, et al. Tribological application of carbon nanotubes in a metalbased composite coating and composites. Carbon, 2003, 41(2):215~22.

2 Abu Barkar Sulong, Joohyuk Park. Wear behavior of functionalized multi-walled carbon nanotube reinforced epoxy matrix. Composite Mater 2006, 40(21):1947 ~ 1960.

3 Yang Jie, Shen Zeng min, Xiong Tao. The preparation and properties of carbon nanotubes in-situ encapsulated by polyaniline. New Carbon Materials, 2003, 18(2):95 ~ 99.

4 Paiva M C, Zhou B, Fernando K A S, et al. Mechanical and morphological characterization of polymer-carbon nanocomposites from functionalized carbon nanotubes. Carbon, 2004, 42(14): 2849 ~ 285

5 D Q. Yang, J F. Rochette, E Sacher. Functionalization of multiwalled carbon nanotubes by mild aqueous sonication. Physical Chemistry B, 2005,109(16) 7788 ~ 7794

6 Mohamed Abdalla, Derrick Dean, Pamela Robinson, et al.Cure behavior of epoxy/MWCNT nanocomposites: The effect of nanotube surface modification. Polymer 2008, 49, 3310 ~ 3317.

7 F Chamssedine, D Claves. A kinetic, morphological and mechanistic approach of the fluorination of multiwall carbon nanotubes. Chemical Physics Letters, 2008, 454, 252 ~ 256.

8 Susumu Arai, Akihiro Fujimori, Masami Murai, et al. Excellent solid lubrication of electrodeposited nickel-multiwalled carbon nanotube composite films. Materials Letters, 2008, 62, 3545 ~ 3548

9 Jin shan Lu. Effect of surface modifications on the decoration of multi-walled carbon nanotubes with ruthenium nanoparticles. Carbon , 2007, 45 , 1599 ~ 1605.

10 M S P Shaffer, X Fan, A H Windle. Dispersion and packing of carbon nanotunanotubes, Carbon, 1998, 36(11), 1603 ~ 1612

11 Fadel Chamssedine, Daniel Claves. Three different modes of fluorine chemisorption at the surface of single wall carbon nanotubes. Chemical Physics Letters, 2007, 443 102 ~ 106

12 Z Gu, H Peng, R H Hauge, et al. Cutting single-wall carbon nanotubes through fluorination, Nano Letter. 2002,(2), 1009 ~ 1013.

13 W Zhao, C Song, B Zheng, J Liu, et al, Thermal recovery behavior of fluorinated single-walled carbon nanotubes. Physical Chemistry B, 2002, 106(2), 293 ~ 296.

14 S J Park, H J Jeong, C Nah. A study of oxyfluorination of multi-walled carbon nanotubes on mechanical interfacial properties of epoxymatrix nanocomposites. Materials Science and Engineering A, 2004, 385(1-2),13 ~ 16.

15 Giraudet J, Dubois M, Claves D, et al. Modifying the electronic properties of multi-wall carbon nanotubes via charge transfer by chemical doping with some inorganic fluorides.Chemical Physics Letters, 2003, 381(3-4): 306 ~ 314.

16 Okotrub A V, Yudanov N F, Chuvilin A L. Fluorinated cage multiwall carbon nanoparticles. Chemical Physics Letters, 2000, 322(3-4): 231 ~ 236.

17 Y M Shulga, TaChang Tien , ChiChen Huang, et al. XPS study of fluorinated carbon multi-walled nanotubes. Electron Spectroscopy and Related Phenomena 2007,160, 22 ~ 28.

165e Effect of Nanoclays on the Network Formation and Phase Separation in Epoxy/thermoplastic Hybrid Nanocomposites

Juan A. Martínez-Sánchez, Fabiola Sánchez-Cervantes, Rocio Ortega-Pérez, Wendy P. Mas-Ku, Alejandro Gonzalez-Alvarez, and **Martin Arellano**. Department of Chemical Engineering, University of Guadalajara, Blvd. Gral. Marcelino García Barragan # 1451, Guadalajara, Jal., 44430, Mexico

Toughened thermosetting polymers are basis for many commercially important structural adhesives and composites systems. Several approaches have been employed to improve the fracture behavior of epoxies including the use of elastomers, thermoplastics and nanoclays. The mechanical properties of the hybrid nanocomposites depends on the morphology, crosslinking density and exfoliation degree. In this work, we proposed to study the effect of the addition of different organically treated montmorillonita clays on the network formations and phase separation of an epoxy-polysulfone (PSF) blend. The epoxy system was based on diglycidyl ether of bisphenol A and diamino diphenyl sulfone. The thermoplastic modifier was polysulfone of Mn=26,000 g/mol, varying the PSF concentration (5, 10 and 15 phr PSF). The clays used were Nanomer I.30E (NI30E), Nanomer I.28E (NI28E), Cloisite 15A (C15A) and Cloisite 10A (C10A). The curing reaction was followed by differential scanning calorimetry. The hybrid nanocomposites were characterized by dynamical mechanical analysis and scanning electron microscopy. The presence of NI30E accelerates the curing reaction, decreases the glass transition temperature (Tg) and increases the PSF domain size. On the other hand, there is a negligible change on the curing reaction and blend morphology for NI28E, C10A and C15A modified systems.

Preparation of SiO_2/CE/BMI Composite Using Microwave Curing

Jingfei Zhao [a, b], Hongxia Yan [a], Hao Wu [a], Rongchang Ning [a]

([a] Department of Applied Chemistry, School of Natural and Applied Science, [b] School of Materials Science and Engineering, Northwestern Polytechnical University, Xi'an 710072, P R China)

INTRODUCTION

Microwave is a type of electromagnetic waves, of which the frequency ranges from 300 MHz to 300 GHz. The microwave frequencies used in the industry are 915 MHz, 2.45 GHz, 5.8 GHz, 24.124 GHz, etc. 2.45 GHz is used most frequently[1]. Presently, thermal curing is always used to cure the polymers. The mechanism of conventional thermal heating is that the thermal energy is conducted to a material through such means as conduction, convection and radiation, so that the material is heated from the outer part to the inner part[2]. The mechanism of microwave heating is that the microwave energy which is absorbed by the material is transformed into the kinetic energy and potential energy of the molecules of the material, and the outer and inner parts of the material are heated simultaneously[2]. As a new curing method, microwave curing has a lot of advantages such as high heating rate, uniform curing, high utilization rate of thermal energy, no overheating at the surface and convenient process control. It has been widely used in the processing of polymers and polymer matrix composites[3-5].

Cyanate Ester (CE) is one of the high performance resins, having a lot of advantages such as good heat resistance, low dielectric constant, very low dielectric loss, good mechanical properties and good fire resistance property. But CE also has the disadvantage of lacking toughness. If CE is modified by bismaleimide with active diluent agent (BMI), the toughness of CE can be increased and the other advantages can be kept. Furthermore, the modified system has the low viscosity, and it is beneficial to the dispersion of the filler[6,7].

In our present work, a new type of materials used for integrate circuit board—SiO_2/CE/BMI composite—was prepared using microwave assisted curing and conventional thermal curing. The fillers used are nano-sized and micro-sized SiO_2. And the dielectric and impact properties of the cured samples were measured and compared in our research work.

EXPERIMENTAL STUDIES

1. Materials

Bisphenol A Cyanate Ester (CE), industrial quality; Bismaleimide with active diluent agent (BMI), provided by Professor Ning Rongchang and used as received; Nano-sized and micro-sized SiO_2, industrial quality; etc.

2. Preparation of SiO_2/CE/BMI Composite

Before the preparation process, the nano-sized and micro-sized SiO_2 were dried at 150 ^{o}C for 1-2 h and then were modified by γ-(2, 3-epoxypropoxy) propyl trimethoxysilane (KH-560) and dispersed using ultrasound for 30 minutes. CE was dried at 70 ^{o}C and vacuum conditions for 1-2 h.

BMI (half the mass content of CE) was heated up to about 110 ^{o}C in a flask, and then CE was added into it. The mixture was stirred for 10-15 min to form the prepolymer. Then the modified SiO_2 was added into the prepolymer, and the new mixture was stirred for another 10-15 min. Then the mixture was poured into a PTFE mould, and the mould with the mixture was put into a vacuum drying chamber for 1-2 h to degas, and then was put into a microwave oven to cure using microwave radiation. To prevent the mixture from reacting tempestuously at the beginning of the curing reaction, intermittent curing method was adopted and the radiation power of the microwave oven was relatively low. As the curing reaction progresses, the radiation power could be higher and the time for continuous radiation could be longer. After curing in the microwave oven for about half an hour, the sample was postcured using conventional thermal curing in an oven for 2 h at 150 °C in order to cure it thoroughly. In order to compare the microwave assisted curing with conventional thermal curing, the sample of the same formulation was also prepared by conventional thermal curing method. The thermal curing procedure was 150 °C/2 h + 180 °C/2 h.

3. DSC analysis of SiO_2/CE/BMI Composite

Similar to the preparation process, after the solution of SiO_2 was added into the prepolymer of CE and BMI, the new mixture was dispersed again using ultrasound and stirred. Then put the mixture into the vacuum drying chamber to degas for 4-8 h, make sure that the acetone volatilized thoroughly and yellow liquid was formed. Differential Scanning Calorimeter was tested from ambient temperature to 300 °C with nitrogen atmosphere using about 4 mg of the liquid. The heating rates were 5 °C/min, 10 °C/min and 15 °C/min, respectively.

4. Impact and dielectric testing

The specification of the samples for impact testing is 15mm×10mm×4mm. And the specification of the samples for dielectric testing is 27mm×27mm×4mm. The samples were at first incised to the specification and then were tested on the relevant equipments.

RESULTS AND DISSCUSSION

1. Microwave curing technology

Research work shows that the composite can be cured well by using power-changing and intermittent curing method. Comparing to thermal curing, a lot of time can be saved by using microwave curing. During the microwave curing process, the output power of the microwave oven is very important. To prevent the mixture from reacting tempestuously at the beginning of the curing reaction, intermittent curing method was adopted and the radiation power of the microwave oven was relatively low. As the curing reaction progresses, the radiation power could be higher and the time for continuous radiation could be longer.

2. The curing kinetics of SiO_2/CE/BMI composite

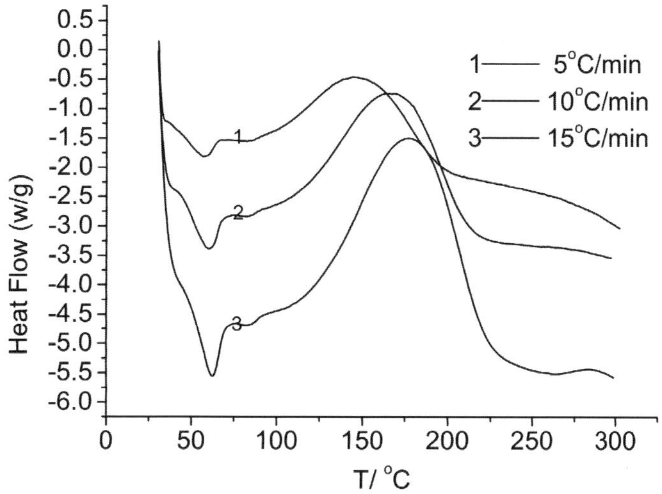

Fig.1 DSC curves of the composite at different heating rates

Fig.1 shows the DSC curves of SiO_2/CE/BMI composite, in which the mass of nano-sized and micro-sized SiO_2 were 5.0% and 30.0% of the mass content of CE, at the heating rates of 5 °C/min, 10 °C/min and 15 °C/min, respectively. From Fig. 1, we can see that there is an endothermic peak at about 58 °C at the three different heating rates. But the position of the exothermic peaks is different. With the increase of the heating rate, the exothermic peak shifts to higher temperature. The proposal reason is, polymer is not a good conductor of heat, the inner temperature of the samples may differ from the environment temperature, this leads to the phenomena of heat lagging. The higher the heating rate, the more serious the phenomena.

Tab. 1 The peak temperatures of the DSC curves at different heating rates

Heating rate °C/min	5	10	15
Onset Temperature /°C	96.95	108.60	117.50
Peak Temperature(T_m) /°C	150.77	169.81	180.54
Ending Temperature /°C	218.46	233.62	248.36

Tab. 1 is the peak temperatures at different heating rates got from the DSC curves. According to the data listed in Tab 1, linear regression was used to get the curing technology parameters of the SiO_2/CE/BMI composite, they are: gel temperature 87.13 °C, cure temperature 137.27 °C and post-treated temperature 203.58 °C.

Thereinafter, Kissinger and Ozawa method[8] were employed to calculate the curing kinetics parameters of SiO_2/CE/BMI composite system.

(1) Kissinger method

According to the relationship between heating rates and the variety of peak temperature of the DSC curves, the curing kinetics parameters were calculated, including the calculation of activation energy and the reaction order.

The activation energy can be calculated by Kissinger equation, shown as Eq. 1:

$$\frac{d\left[\ln\left(\beta / T_m{}^2\right)\right]}{d\left(1/T_m\right)} = -\Delta E / R \qquad \text{(Eq.1)}$$

Where β is the heating rate (K/min), T_m is peak temperature (K), ΔE is the activation energy (J/mol), R is 8.314 J/(mol·K). From Eq. 1 we know that -ln$\left(\beta/T_m^2\right)$ is linear to $1/T_m$ and Eq. 2 is the regression equation:

$$-\ln(\beta/T_m^2) = 4.019818 + 665.167357(1/T_m) \qquad \text{(Eq. 2)}$$

From the slope of the line, ΔE was calculated to be 5.530 kJ/mol.

Kissinger equation was further deduced by Crane, and then Crane equation was obtained:

$$\frac{d\ln\beta}{d\left(1/T_m\right)} = -\frac{\Delta E}{nR} - 2T_m \qquad \text{(Eq. 3)}$$

Where n is the reaction order. When $\Delta E/nR >> 2T_m$, $2T_m$ can be neglected, so lnβ is linear to $1/T_m$, Eq. 4 is the regression equation:

$$\ln \beta = 8.18576 - 993.174\left(\frac{1}{T_m}\right) \qquad \text{(Eq. 4)}$$

From the slope of the line and ΔE, the reaction order n can be calculated to be 1.493.

(2) Ozawa method

The equation of Ozawa method is shown as Eq. 5:

$$\Delta E = \frac{-R}{1.052} \cdot \frac{\Delta \ln \beta}{\Delta \left(1/T_m\right)} \qquad \text{(Eq. 5)}$$

According to Eq. 4, the activation energy of Ozawa method was calculated to be 7.854 kJ/mol.

Consider Kissinger and Ozawa method together, we get the average activation energy is 6.692 kJ/mol, the reaction order is 1.493.

3. The impact strength of SiO_2/CE/BMI composite

Fig. 2 is the comparison of microwave cured samples and thermal cured samples with different addition of SiO_2. From Fig. 2 we can see that the impact strength of microwave cured samples are higher than the impact strength of thermal cured samples, this may be due to the different heating mechanism of microwave and conventional heating[2]. In addition, from Fig. 2 we can see that, for microwave and thermal cured samples, from sample 1,2,3 and sample 4,5 we can conclude that when the content of nano-sized SiO_2 is constant, the impact strength of the sample increase with the augment of the micro-sized SiO_2; From sample 2, 4 and sample 3, 5 we can conclude that when content of micro-sized SiO_2 is constant, the impact strength of the sample increase with the augment of the nano-sized SiO_2, but the variety of thermal cured samples are not as obvious as microwave samples. Furthermore, from Fig. 2 we know that the sample with the biggest impact strength in our experiment is n5.0%+m40.0% (represented that the content of nano-sized SiO_2 and micro-sized SiO_2 are 5.0% and 40.0% of the mass content of CE respectively, the same below) sample, and its value is 13.8383 kJ/m^2.

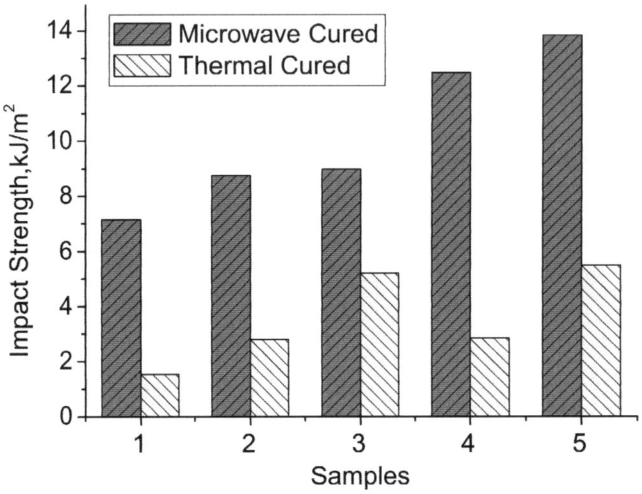

Fig. 2 The impact strength of microwave cured and thermal cured samples

(sample 1: n3.0%+m20.0%, 2 : n3.0%+m30.0%, 3 : n3.0%+m40.0%,

4 : n5.0%+m30.0%, 5 : n5.0%+m40.0%.)

4. The dielectric properties of SiO_2/CE/BMI composite

Fig. 3 and Fig. 4 are the dielectric constant and dielectric loss tangent of the composite as a function of frequency respectively. From Fig 3 and Fig 4, we can conclude that: (1)For the composite with the same component, the dielectric constant and dielectric loss tangent of the microwave cured samples are lower than the dielectric constant and dielectric loss tangent of the thermal cured samples in a large range of frequency in our experiment; (2)For microwave cured samples, when the content of nano-sized SiO_2 is 3.0% of the mass content of CE, with the augment of micro-sized SiO_2, the dielectric constant and dielectric loss tangent of the composite decrease at first and then increase. In the range of our experiment, when the content of micro-sized SiO_2 is 30.0% of the mass content of CE, the composite has the lowest dielectric constant and dielectric loss tangent in a large range of frequency. But when the content of nano-sized and micro-sized SiO_2 are 5.0% and 30.0% of the mass content of CE respectively, only in high frequency does the composite has lower dielectric constant; When the mass of micro-sized SiO_2 is 20.0% and 40.0% of the mass content of CE, the dielectric constant and dielectric loss tangent of the composite with 5.0% nano-sized SiO_2 is lower than the composite with 3.0% nano-sized SiO_2, but when the content of micro-sized SiO_2 is 30.0% of the mass content of CE, the case is the opposite. We can deduce that there is interaction between the content of nano-sized and micro-sized SiO_2 on the effect of the dielectric properties of SiO_2/CE/BMI composite; (3) In our experiment, microwave cured sample with 3.0% nano-sized SiO_2 and 30.0% micro-sized SiO_2 has the lowest dielectric constant and dielectric loss tangent in a large range of frequency.

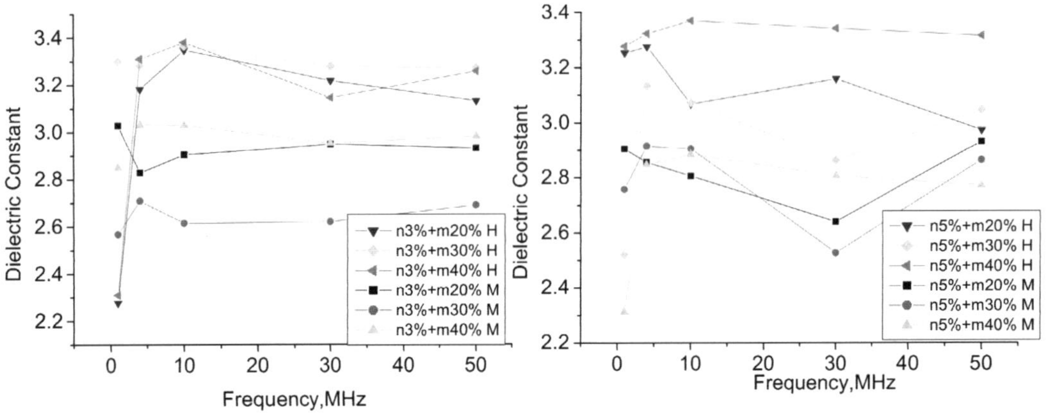

Fig.3 The dielectric constant of composite with different component as a function of frequency (M refers to microwave cured sample and H refers to thermal cured)

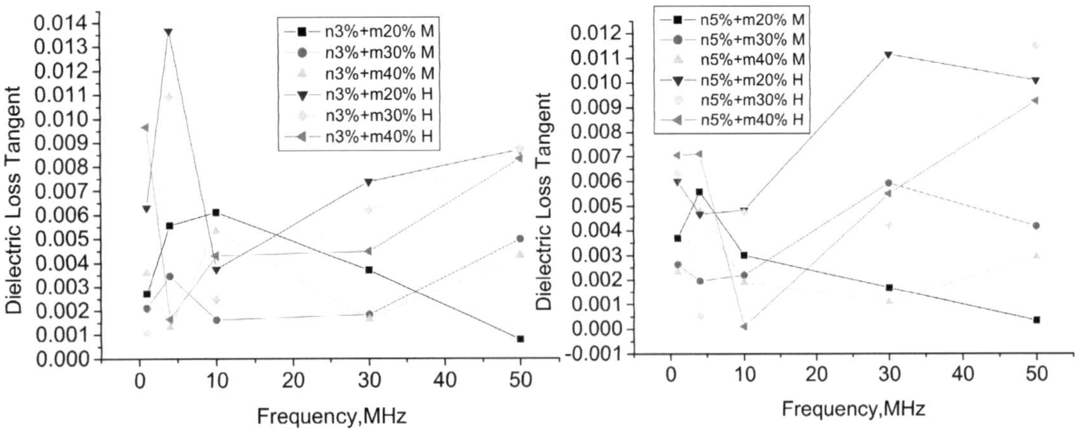

Fig. 4 The dielectric loss tangent of composite with different component as a function of frequency (M refers to microwave cured sample and H refers to thermal cured)

CONCLUSIONS

1. The composite can be cured well by using power-changing and intermittent curing method. And a lot of time can be saved by using microwave curing.
2. The curing kinetics parameters of SiO_2/CE/BMI were calculated by Kissinger and Ozawa method: the average activation energy is 6.692 kJ/mol, and the reaction order is 1.493.
3. For the composite with the same component, the impact strength of the microwave cured samples are higher than the impact strength of the thermal cured samples. The materials cured by microwave have good toughness. In the range of our experiment, the impact strength increase with the content of the filler.
4. For the composite with the same component, the dielectric constant and dielectric loss tangent of the microwave cured samples are lower than the dielectric constant and dielectric loss tangent of the thermal cured samples in a large range of frequency. In the range of our experiment, the variety of dielectric constant and dielectric loss tangent with the content of the filler does not have uniform rule, there may be interaction between the content of nano-sized and micro-sized SiO_2.

REFERENCES

[1]Yarlagadda Prasad K D V, Cheok E C. Study on the microwave curing of adhesive joints using a temperature-controlled feedback system. Journal of Materials Processing Technology, 1999, 91 (1-3): 128–149

[2]Zhou Jian, Shi Chun, Mei Bingchu, et al. Research on the technology and the mechanical properties of the microwave processing of polymer. Journal of Materials Processing Technology, 2003,137 (1-3): 156–158

[3]Boey F Y C, Yap B H. Microwave curing of an epoxy–amine system: effect of curing agent on the glass-transition temperature. Polymer Testing, 2001, 20 (8): 837–845

[4]Olofinjana Ayodele, Yarlagadda Prasad K D V, Oloyede Adekunle. Microwave processing of adhesive joints using a temperature controlled feedback system. International Journal of Machine Tools & Manufacture, 2001, 41 (2): 209–225

[5]Yarlagadda Prasad K D V, Hsu Shu-Hau. Experimental studies on comparison of microwave curing and thermal curing of epoxy resins used for alternative mould materials. Journal of Materials Processing Technology, 2004, 155–156 (11): 1532–1538

[6]Gu Aijuan. High performance bismaleimide / cyanate ester hybrid polymer networks with excellent dielectric properties[J]. Composites Science and Technology, 2006, 66 (11-12): 1749-1755

[7]Hamerton I, Herman H, Mudhar A K, et al. Multivariate analysis of spectra of cyanate ester / bismaleimide blends and correlations with properties[J]. Polymer, 2002, 43 (11): 3381-3386

[8] Jagadeesh K S, Siddaramaiah. Kinetics of curing epoxy formulations with diaminodiphenyl ether [J]. Polymer International, 2007, 26(2): 129-132

169a Novel Catalysts for Selective Oxidation of Lower Alkanes

Prakash Biswas, Jungwon Woo, and Vadim V. Guliants. Department of Chemical and Materials Engineering, University of Cincinnati, Cincinnati, OH 45221-0012

Olefins are industrially important starting molecules for the production of many oxygenated compounds and nitriles, obtained respectively by the oxidation and ammoxidation reactions. However, the high cost of such feedstocks have stimulated research on catalytic oxidation processes based on C2-C4 alkanes. Among different catalysts investigated, Mo-V-Te-Nb-Ox mixed metal oxide catalysts are very promising for oxidation, ammoxidation and oxidative dehydrogenation of lower alkanes. Recently, the Mo-V-Te-Nb-Ox catalysts with orthorhombic M1 phase structure received much attention as catalysts for (amm)oxidation of propane to acrylonitrile. However, the current M1 phase catalysts suffer from low thermal stability due to Te volatility and limited selectivity to acrylonitrile.

In this study we investigated the use of Fe, Ru and Au as promoters of thermal stability and selectivity in (amm)oxidation of C2-C4 alkanes over the M1 phase catalysts. Trivalent Fe is well known as an active component of catalysts employed for partial oxidation for ethane and propane, whereas small Ru and Au nanoparticles (<10 nm) exhibit unexpectedly high catalytic activities towards many types of catalytic reactions, such as selective oxidation, steam reforming and water-gas shift reaction. Selective (amm) oxidation of lower alkanes were investigated over hydrothermally synthesized Mo-V-Nb-Te-Ox, Mo-V-Nb-Fe-Ox as well as Au- and Ru-doped Mo-V-Nb-Te-Ox M1 phase catalysts. Au ad Ru dopants were introduced to the Mo-V-Nb-Te-Ox M1 phase catalysts were by the incipient wetness impregnation. The structural characteristics of multi-component Mo-V-Nb-Te-Ox, Mo-V-Nb-Fe-Ox, Ru- and Au-doped Mo-V-Nb-Te-Ox catalysts were investigated by X-ray diffraction (XRD), scanning electron microscopy (SEM), transmission electron microscopy (TEM) and N2 adsorption at 77 K. Selective (amm)oxidation of lower alkanes were carried out in a tubular reactor at 300-4300C. The total flow rate as well as the He: C2-C4: O2: NH3 molar ratios were varied to optimize the conversion of feed, selectivity and yield of the desired products. The experimental results showed that the Mo-V-Nb-Fe-Ox catalysts was least active as compared to all other catalysts tested, which may be due to the presence of a different phase other than the M1 phase. The best results were obtained over Au-doped Mo-V-Nb-Te-Ox catalysts in terms of alkane conversion, selectivity and yield of desired products. The enhanced catalytic activity of the Au-doped Mo-V-Nb-Te-Ox M1 phase catalysts may be attributed to the presence of the supported Au nanoparticles.

169b Tailoring CO Bond Energies Via Compositional Tailoring in Bimetallic Nanocatalysts

Anmin Cao, Chemical and Petroleum Engineering, University of Pittsburgh, 1249 Benedum Hall, 3700 O' Hara street, Pittsburgh, PA 15261 and Götz Veser, Department of Chemical Engineering, University of Pittsburgh, Pittsburgh, PA 15261.

The past two decades have seen an explosive development in the synthesis of "nanocatalysts", i.e. catalytic materials which are synthesized with well-defined control over the structure of the materials from the nano-scale up. A growing number of studies have shown that "nano-sizing" the active component in a catalyst can lead to dramatic increases in catalytic activity, and even – as in the well-known case of "nano"-gold – to entirely new catalytic properties. At the same time, it has been well-known for many decades that the addition of a second metal can alter the catalytic properties (activity and, more significantly, selectivity) of a metal-based catalyst. However, relatively few studies have to-date looked at the combined effect of tailoring both size and composition in catalysis.

We are reporting on the synthesis of bimetallic nanocomposite catalysts, in which size and composition of the metal component as well as the morphology of the support structure are well-controlled. Bimetallic nanoparticles are embedded and confined into the pores of a nanostructured ceramic oxide via a microemulsion-templated sol-gel synthesis approach. These bimetallic nanocomposites are synthesized, characterized, and then evaluated with regard to their CO adsorption/desorption properties. The CO bond energy is chosen as a characteristic parameter since it plays a crucial role in a wide range of energy-related catalytic reactions (ranging from water-gas shift, to preferential CO oxidation and alcohol synthesis from syngas), and at the same time is a fairly easily characterized quantity that allows direct comparison between different catalyst materials. Based on a wide range of studies in the published literature, it is known that both particle size and the addition of promoters have a pronounced effect on CO bond energies, and it can hence be expected that a well-controlled tailoring of size and composition of the active component should allow significant control over this important catalytic parameter.

In the present contribution, we report on the synthesis of Pt-based bimetallic nanocomposites. PtM-BHA (M= Fe, Sn, Cu, Au; BHA= barium hexa-aluminate) are synthesized in a two-step process: First, Pt-BHA is prepared through a 'one-pot' synthesis of metal-ceramic nanocomposites, a route which has been developed in our laboratory in recent years for monometallic nanocomposite. This (monometallic) nanocomposite is then transformed into a bimetallic material via solution-based conversion strategies, either through a polyol-mediated process or a galvanic replacement route. A range of Pt-based nanocomposites were synthesized and then characterized by SEM, TEM(EDAX), XRD, BET/BJH, and CO TPD. Formation of true bimetallic nanoparticles is confirmed via XRD and electron diffraction.

We find that the 'conversion' process does not affect the morphology of the original nanocomposite significantly. The bimetallic particles are well dispersed in the ceramic matrix and the nanocomposites retain high specific surface areas of >300 m2/g. However, introduction of the second metal fundamentally changes the CO chemisorption property of PtM-BHA in comparison to pure Pt-BHA. Significant peak shifts are observed in CO-TPD both for PtSn/BHA and PtFe/BHA catalysts, while PtCu/BHA appears to show a superposition of the (monometallic) Pt/BHA and Cu/BHA CO-TPD patterns. Most significantly, the CO desorption energy can be gradually adjusted via adjusting the Pt:M ratio, as demonstrated for PtSn/BHA catalysts: With increasing Sn content, desorption peaks gradually shift towards lower temperatures, i.e. lower CO bond energies.

Synthesis, characterization, and reactive test of these bimetallic nanocomposites will be discussed in detail in the presentation.

169c Surfactant-Templated Synthesis and Catalytic Properties of Ordered Nanoporous Titania Supports Loaded with Platinum Nanoparticles

Arijit Bose, Department of Chemical Engineering, University of Rhode Island, 16 Greenhouse Road, Room No: 217 Crawford Hall, Department of Chemical Engineering, Kingston, RI 02881, Jayashri Sarkar, Chemical Engineering, University of Rhode Island, 16 Greenhouse Rd, Kingston, RI 02881, Christopher J. Brooks, Honda Research Institute, 1381 Kinnear Road, Columbus, OH 43212, Vijay T. John, Chemical and Biomolecular Engineering, Tulane University, 300 Lindy Boggs Bldg., New Orleans, LA 70118, and G. Ramanath, Materials Science and Engineering, Rensselaer Polytechnic Institute, Troy, NY 12180.

The highly organized nanoscale structures in the AOT/lecithin/water/IO system have been used as robust templates for a variety of reactions in the aqueous and organic nanochannels to produce highly ordered composite materials. In one example, TiO2/Pt nanocomposite structures, formed by a low temperature synthesis route, has been used as a catalyst for the oxidation of carbon monoxide. The facetted Pt nanoparticles, their homogeneous distribution throughout the support pores, and the highly ordered interconnected structure created by the surfactant template dramatically increase the performance of the catalyst compared to commercial catalysts. The general principles used in the synthesis can be transferred to a large range of nanocomposite systems, and additional examples will be provided.

A Comparison of Mesopore Templating Methods for Ceria-Zirconia-Yttria Catalysts

Prince Anyaba[1], *Paul T. Fanson*[2], *Monica Javornik*[1], *and* ***David Bruce***[3]

(1) Clemson University, 127 Earle Hall, Clemson, SC 29634-0909, (2) Catalyst Materials, Toyota Motor Engineering & Manufacturing North America, Inc., 2350 Green Rd., Ann Arbor, MI 48105, (3) Department of Chemical and Biomolecular Engineering, Clemson University, 127 Earle Hall, Clemson, SC 29634-0909

1.0 Introduction

High surface area mesoporous ceria and zirconia based oxides continue to be in great demand for applications that include emissions abatement, solid oxide fuel cells, electronics, sensors and catalysis. To date, the most cost effective methods for preparing mesoporous materials are via techniques that employ templates or structure directing agents. These templates can be divided into two groups: endo-templates (i.e., soft templates, such as surfactants, dendrimers, and block copolymers) and exo-templates (i.e., hard templates, such as porous carbons and resins).[1-10] This work has critically evaluated the effectiveness of each of these methods to produce highly stable, mesoporous ceria-zirconia-yttria (CZY) mixed oxides with high area surface areas. Specifically, CZY materials were prepared from sol-gel mixtures of the appropriate metal salts using a rage of endo- and exo-templates, and the resulting oxides were evaluated using nitrogen physisorption (BET surface area and BJH Pore Volume), TEM, SEM, TGA, DSC, elemental analysis, and catalytic testing.

2.0 Soft Template Techniques

The general synthesis strategy for all of the materials prepared using soft templates are listed in Figure 1. Three basic procedures are involved: synthesis, drying, and template removal. Four groups of reagents are required for the synthesis step: the metal oxide precursors, solvent(s), hydrolyzing agent or catalyst, and the template. Often, metal alkoxides are used as the metal oxide precursors for sol-gel systems. However, the cost of the CZY alkoxide precursors were higher than desired, so the corresponding inorganic metal salts (e.g., nitrates, carbonates, or oxalates), which are lower in cost, were used instead. Our preference was the nitrate salts because the anion is readily converted to gaseous byproducts upon calcination. Specifically, the list of oxide precursors includes: cerium (III) nitrate hexahydrate, ammonium cerium (IV) nitrate, zirconyl nitrate hexahydrate, and yttrium nitrate hexahydrate. The sol-gel solvent is typically deionized water, but alcohol cosolvents are frequently used, especially with metal alkoxide precursors. Moreover, certain templates are not readily water soluble; for example, the block copolymers and amine functionalized dendrimers have low water solubility. Thus, they are dissolved in water-alcohol mixtures before being added to the other synthesis components. In this work, we used 1-propanol or 1-octanol as the alcohol source. The general synthesis procedure is depicted by the process flow diagram shown in Figure 1.

The two solutions containing either the metal precursors or the soft template (e.g. CTAB) and hydrolyzing agent are prepared separately. The hydrolyzing agent used is ammonium hydroxide, urea, or sodium hydroxide. Both solutions are continuously stirred at room temperature until all solids are dissolved. The precursor salt solution is then mixed with the hydrolyzing agent/CTAB solution and stirred continuously for a 1-4 hr. The synthesis is carried out at pH > 10. Aging of the resulting sol-gel is carried out in a sealed PFA Teflon reaction vessel that is maintained at a constant temperature. The aging temperature and duration have been found to be very significant factors affecting final oxide properties. The major advantage of this method is the relatively low synthesis temperature condition of the oven. The usual range has been between 70-130 °C. The aging time also varies from 1-5 days. The samples are removed from the oven and then washed thoroughly with water and centrifuged at least two times. The samples are dried at 110 °C for 24 hrs after which the samples were calcined in air at low heating rate to 600 °C for 4h.

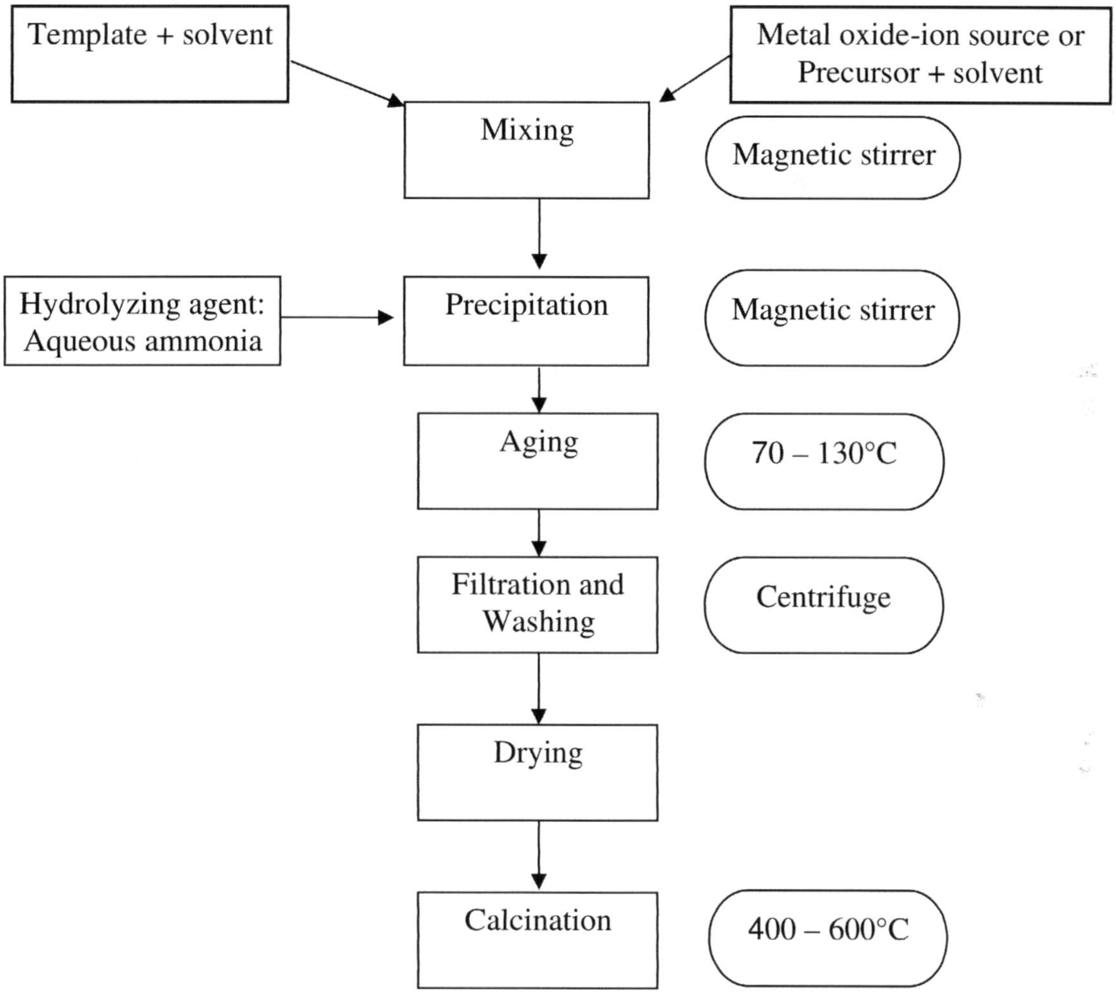

Figure 1: Schematic representation of the synthesis procedures for non-siliceous mesoporous materials prepared using soft template materials, such as surfactants.

Of the reagents required during the synthesis steps for soft template techniques, the most important is the template source. Three types of soft templates were studied: surfactants, block copolymers and dendrimers.

2.1 Surfactant Template Technique

The most frequently used surfactant for the synthesis of mesoporous materials has been cetyltrimethylammonium bromide (CTAB), a cationic surfactant. Just recently, the first successful synthesis of mesoporous ceria-zirconia-yttria mixed oxide (CZY) using CTAB as template was reported by Feng[11] *et al.* The synthesis procedure was quite similar to that shown in Figure 1. The final CZY material was calcined at a heating rate of 1 ºC/min from room temperature to 550 ºC and held for 8hr at this temperature. After calcination, the sample was characterized and the specific surface area for CZY was measured to be 139 m^2/g, and the pore diameter was approximately 6.8 nm. Hung et al. also reported the synthesis of yttria stabilized zirconia (YSZ) using only CTAB as the template. Thermal degradation studies of the YSZ sample revealed that the specific surface area significantly reduced from 137 m^2/g to 3.3 m^2/g upon heating from 600 ºC to 1200 ºC. Feng[11] *et al.* also reported synthesizing zirconia-yttria mixed oxides (ZYO) with a specific surface area of 200 m^2/g. Kapoor[12,13] *et al.* , using a different synthesis procedure, reported synthesizing mesoporous CZY. In their procedure no hydrolyzing agent was used in the synthesis. However, it is important to note that most transition metal oxide precursor solutions are frequently strongly acidic; thus, they implicitly used the acidic nature of the precursor solution as the hydrolyzing agent. The CZY synthesized had BET surface areas that were between 365 m^2/g and 229 m^2/g depending on the CZY compositions.

Many researchers have used co-solvent or cosurfactant in combination with the surfactant to enlarge the size of the micellar aggregates during the synthesis of silica based mesoporous material[4,14,15]. Cosurfactants are mostly medium chain length alcohols ranging from propanol to 1-octanol[4]. Non-siliceous mesoporous oxide based materials like titanium oxophosphate have also been reportedly synthesized using co-solvents[4]. We used 2-propanol as the cosurfactant.

Another way frequently used to enlarge the size of the micellar aggregates during synthesis of mesoporous materials has been with the addition of micellar swelling agents. Swelling agents are large molecular weight organic compounds that are hydrophobic in nature and tend to be readily soluble in organic solvents. Examples of swelling agents are benzene and mesitylene. The core of the supramolecular aggregates of the surfactant can solubilize these swelling agents. Thus, the swelling agents find their way to the core of the supramolecular aggregates and thereby cause the supramolecular aggregates to swell and experience a slight increase in their diameter. The theoretical basis for this phenomena is well established[4,16]. We found that the physical properties of the CZY samples synthesized using CTAB only, CTAB/1-propanol, and CTAB/1-propanol/benzene as the SDA showed some differences in the BET surface area, BJH pore size distribution and average pore diameter.

2.2 Block Copolymer Template Technique

The block copolymer templated synthesis of mesoporous oxides has been reported extensively in the literature.[17-26] Further, the walls of the mesoporous materials synthesized using block copolymers are known to be thicker than those made from surfactants.[5] The triblock copolymers poly (ethylene oxide)-ploy (propylene oxide)-poly (ethylene oxide) (PEO-PPO-PEO) have been by far the most widely studied. A number of publications on the use of block copolymers as SDA in the synthesis of mesoporous oxides of ceria, zirconia, and yttria have been reported.[27-31] In our lab, we have extended the technique to synthesize CZY. Pluronic P123 triblock copolymer is the block copolymer used during all syntheses. The Pluronic P123 structure directing agent was dissolved in a water-propan-2-ol mixture to synthesize CZY as outlined in Figure 1. The BET surface area and the average pore diameter are 96.0 m^2/g and 3.8 nm, respectively.

2.3 Dendrimer Templated Technique

Dendrimer templated mesoporous materials abound in the literature.[32-38] Unlike supramolecular aggregates of surfactants and block copolymers that are capable of assuming various forms and shapes; for example, the hexagonal structure of MCM-41, the bilayer structure of MCM-48, and the cubic structure of MCM-50[5], the 4th and 5th generation polypropyleneimine, which are also known as DAB-Am-32 and DAB-Am-64, respectively, are unique in that they do not self-associate like the surfactants and block copolymers. They have a preformed pseudo-spherical structure. The use of DAB-Am-32 and DAB-Am-64 as SDA for the synthesis of silica based mesoporous materials was first reported by Larsen et al.[38] The reason for the use of these particular dendrimers was that Larsen et al. found that their pseudo-rigid structure enabled them to maintain their overall structure better than the more flexible and more common PAMAM dendrimers. Polypropyleneimine dendrimers are currently quite expensive, so their use for the large scale synthesis of porous catalysts is unlikely but their unique structure affords the possibility of examining another type of templating platform. CZY material were synthesized via the same synthesis procedure outlined in Figure 1, with the polypropyleneimine SDA dissolved in 2-propanol.

3.0 Hard Template Techniques

A variety of hard templating synthesis techniques have been examined for mostly siliceous materials, but some non-silica compositions have also been studied.[3,39-44] In general, the synthesis procedure for mesoporous oxides employing hard templates is quite different from that used with soft templates. Figure 2 shows a schematic representation of the synthesis process used when rigid porous materials, which may or may not be functionalized with hydrophilic or hydrophobic groups, are used as structure directing agents.

The precursor solution containing all of the metal precursor salts is mildly heated and stirred continuously. When the precursor solution becomes homogeneous, it is then slowly combined with the hard template by a wet impregnation technique. The solid-solution is properly mixed to ensure complete filling of the interstices and that the solid is laden with the solution. The filtration step can be skipped in most cases but to be sure only the interstices are covered with the precursor solution the solid-solution mixture can be vacuumed filtered. The drying step is occasionally useful if increased loading of the precursor solution are desired, otherwise this step may be omitted. Calcination of the sample is carried out by heating at the rate of 1 °C/min from room temperature to 600°C and maintaining this temperature for 4h.

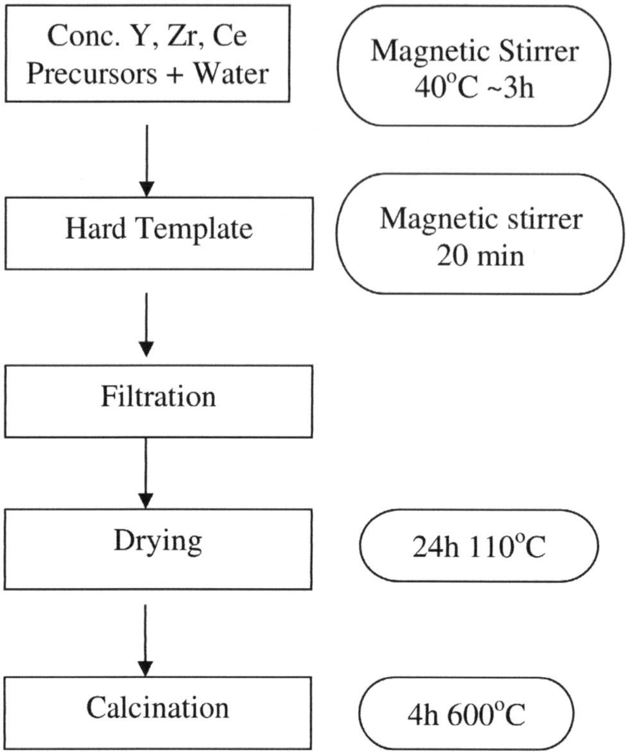

Figure 2: Schematic representation of the synthesis procedure for non-siliceous mesoporous materials prepared using hard templates.

3.1 Activated carbon Template Technique

The earliest report of the use of activated carbon as SDA for the synthesis of mesoporous materials was reported by researchers at the Toyota Central R&D lab in Japan.[44] More recently, the technique has been widely used in the synthesis of a number of metal oxides, including pure single component oxides and mixed oxides containing two or more active metals.[3,39-42] It is important that the ash content of the activated carbon template used be as low as possible and that it should be highly porous and with high surface area. The general preparation method involves dissolving the metal oxide precursors, in this case the nitrate salts of cerium, zirconium, and yttrium, in water. The activated carbon powder is loaded with the precursor solution by the incipient wetness method. The paste is kneaded together properly before it is dried at 110 °C for at least 12 hr. The dried solid is further calcined in air at a heating rate of 1 °C/min to 600 °C and then held for 4h at this temperature. The BET surface area and BJH pore size distribution are measured to be 118m^2/g and 9.0nm, respectively. The TEM images of the resulting calcined oxide show that the CZY particles as relatively agglomerated and crystalline in nature, with the average particle size of the primary particles being approximately 5.0 nm in diameter.

3.2 Polymeric Resin Template Technique

The use of polymeric resins as templates for the synthesis of mixed oxides has been reported in the literature.[45-48] There have been excellent articles written about similar techniques, called opal techniques. This technique is fascinating because it confers a negative replica of its structure unto the porous metal oxide that is formed. The method of preparation is very similar to that laid out for the activated carbon template. The precursor salts are dissolved in water and the resulting solution is mixed with the polymeric resin. The polymeric resin-precursor solution is then dried in at 110 °C for at least 12h before being calcined in air at a heating of 1 °C/min to 600 °C and then held for 4h at this temperature. The BET surface area and BJH pore size distribution for the CZY sample is 98.0 m^2/g and 3.4 nm, respectively. The TEM images revealed of the resulting oxide particles showed that they are agglomerated and crystalline in nature. The powder X-ray diffraction pattern revealed that the CZY contained nanocrystalline particles.

4.0 Results and Discussion

The CZY sample composition typically used in this study were 45 mol% Ce : 50 mol % Zr : 5 mol% Y. The challenge in the synthesis of ordered mesoporous non-siliceous oxides is two-fold when soft templates are required. First is the control of the kinetics of crystallization of the metal oxide precursors in relation to the self-assembly of the organic template during synthesis. Unlike ordered mesoporous silica based oxides for which the crystallization of the silica precursor and self-assembly of the soft template occur in complete complimentary fashion[49]; the same cannot be said of non-siliceous oxide precursors, which have a tendency to crystallize faster than it would take the soft templates to self-assemble. Second is the rapid collapse of the mesostructure of the non-siliceous oxides during template removal. The goal of this research was to study ways in which we could mitigate or eradicate completely the effects of the above mentioned challenges. Tables 1-3 below show a summary of a fraction of the results obtained during our experimental efforts.

Table 1. Effect of Synthesis Temperature on the Physical Properties of CZYs Made from Different Combination of Structure Directing Agent and Calcined at 600°C for 4h

	BET Surface Area (m^2/g)				Avg. Pore Diameter (nm)			
	Hydrothermal Treatment Temperature (°C)				Hydrothermal Treatment Temperature (°C)			
Sample	70	90	110	130	70	90	110	130
CZY/no Template	102	93	101	84	3.7	3.7	4.7	3.6
CZY/CTAB	80	94	94	107	4.7	4.7	3.3	5.4
CZY/CTAB/1-Propanol	85	87	89	121	6.3	5.4	4.6	2.2
ZY/CTAB/1-propanol/benzene	81	98	92	95	4.1	4.8	5.3	3.3

From Table 1, it is observed that the effects of hydrothermal treatment temperatures on the physical properties of the CZY are quite moderate to say the least. A significant impart though of the hydrothermal treatment temperature is on the aging time. At high hydrothermal treatment temperatures, crystallization of the CZY occurred much faster as expected.

On the other hand, the use of hydrolysis retarding agents like acetylacetone (ACAC) or triethanol amine (TEA) on the physical properties of the CZY synthesized using CTAB and P123 soft template was significant as shown in Table 2. While ACAC led to an increase in the BET surface area from 82 to 106 m^2/g and from 58 to 78 m^2/g for CZY synthesized via CTAB and P123, respectively; however, TEA led to a decrease in the BET surface area of the CZY samples from 82 to 14 m^2/g and from 58 to 44 m^2/g for CZY synthesized via CTAB and P123 SDA's.

Table 2. Effect of Hydrolysis Retarding Agents on the Physical Properties of CZY's made from Different Structure Directing Agent and calcined @ 600°C for 4h

	BET Surface Area (m^2/g)	Avg. Pore Diameter (nm)
CZY Template	20°C/min Heating Rate During Calcination	20°C/min Heating Rate During Calcination
CTAB	82	6.2
CTAB/ACAC	106	5.3
CTAB/TEA	14	3.8
DAB-AM-32	97	3.0
DAB-AM-64	40	2.8
P123	58	2.8
P123/ACAC	78	3.6
P123/TEA	44	3.6

Table 3 suggests that carrying out the calcination of the templated CZY samples at higher heating rates lowered BET surface area of the CZY sample than when it was calcined at lower heating rate. When heating rates was increased from 1 to 20°C/min, the BET surface area of the resin and activated carbon templated CZY decreased from 98 to 78m^2/g and 118 and 103m^2/g, respectively. For P123 templated CZY, the BET surface area similarly dropped from 96 to 78m2/g and 50 to 44m2/g for the ACAC and TEA treated samples, respectively.

Table 3. Effect of Heating Rate on the Physical Properties of CZY made from Different Structure Directing Agent and calcined @ 600°C for 4h

	BET Surface Area (m^2/g)		Avg. Pore Diameter (nm)	
	Heating rate (°C/min)		Heating rate (°C/min)	
CZY Template	1	20	1	20
CTAB	78	82	6.2	6.2
CTAB+acac	106	106	5.4	5.3
CTAB+TEA	9	14	3.8	3.8
P123+acac	96	78	3.8	3.6
P123+TEA	50	44	3.8	3.6
Resin	98	78	3.4	3.4
Activated Carbon	118	103	9.0	9.0

5.0 Conclusion

The results shown above are our attempt at providing the following answers: does hydrothermal treatment temperature help in slowing down the rates of non-siliceous oxide precursor crystallization during synthesis; which is a better hydrolysis retarding agent between acetylacetone and triethanol amine; and what heating rates best preserve the mesostructure of the synthesized CZY during calcination? In conclusion, we summarize our findings as follows: at high temperatures the crystallization rates increases; acetylacetone is a much better hydrolysis retarding agents than triethanol amine; and adopting a calcination protocol that uses a low heating rate helps to minimizes the heat evolution rates, thus, preventing hot spots during calcination. The CZY synthesized using the various templates all showed that we successfully made mesoporous oxides even though there is little evidence they are highly ordered. It is hoped that attempts in the future would be focused on making highly ordered mesoporous non-siliceous oxides with appropriate combination of hydrolysis and condensation retarding agents, and suitable calcination protocol.

6.0 References

(1) Ciesla, U.; Froba, M.; Stucky, G.; Schuth, F. *Chemistry of Materials* **1999**, *11*, 227-234.
(2) Ciesla, U.; Schuth, F. *Microporous and Mesoporous Materials* **1999**, *27*, 131-149.
(3) Crocker, M.; Graham, U. M.; Gonzalez, R.; Jacobs, G.; Morris, E.; Rubel, A. M.; Andrews, R. *Journal of Materials Science* **2007**, *42*, 3454-3464.
(4) Czuryszkiewicz, T.; Kleitz, F.; Schuth, F.; Linden, M. *Chemistry of Materials* **2003**, *15*, 3704-3709.
(5) Di Renzor, F.; Galameau, A.; Trens, P.; Fajula, F. In *Handbook of Porous Solids*; Schüth, F., Sing, K. S. W., Weitkamp, J., Eds.; Wiley-VCH Verlag GmbH & Co. KGaA: Weinheim, Germany, 2002; Vol. 3, p 1311-1395.
(6) El Haskouri, J.; Cabrera, S.; Caldes, M.; Alamo, J.; Beltran-Porter, A.; Marcos, M. D.; Amoros, P.; Beltran-Porter, D. *International Journal of Inorganic Materials* **2001**, *3*, 1157-1163.
(7) Fajula, F.; Galarneau, A.; Di Renzo, F. *Microporous and Mesoporous Materials* **2005**, *82*, 227-239.
(8) Schuth, F. *Chemistry of Materials* **2001**, *13*, 3184-3195.
(9) Schuth, F. *Angewandte Chemie-International Edition* **2003**, *42*, 3604-3622.
(10) Schuth, F. In *Mesoporous Crystals and Related Nano-Structured Materials* 2004; Vol. 148, p 1-13.
(11) Feng, R. M.; Yang, X. J.; Ji, W. J.; Au, C. T. *Materials Chemistry and Physics* **2008**, *107*, 132-136.
(12) Kapoor, M. P.; Ichihashi, Y.; Kuraoka, K.; Shen, W.-J.; Matsumura, Y. *Catalysis Letters* **2003**, *88*, 83-87.
(13) Kapoor, M. P.; Raj, A.; Matsumura, Y. *Microporous and Mesoporous Materials* **2001**, *44*, 565-572.
(14) Fang, H.; Zhang, M.; Shi, W. H.; Wan, T. L. *Journal of Non-Crystalline Solids* **2006**, *352*, 2279-2283.
(15) Kleitz, F.; Blanchard, J.; Zibrowius, B.; Schuth, F.; Agren, P.; Linden, M. *Langmuir* **2002**, *18*, 4963-4971.
(16) Luechinger, M.; Pirngruber, G. D.; Lindlar, B.; Laggner, P.; Prins, R. *Microporous and Mesoporous Materials* **2005**, *79*, 41-52.
(17) Bhattacharyya, S.; Lelong, G.; Saboungi, M. L. *Journal of Experimental Nanoscience* **2006**, *1*, 375-395.
(18) Kramer, E.; Forster, S.; Goltner, C.; Antonietti, M. *Langmuir* **1998**, *14*, 2027-2031.
(19) Ma, Y. R.; Qi, L. M.; Ma, J. M.; Wu, Y. Q.; Liu, O.; Cheng, H. M. *Colloids and Surfaces a-Physicochemical and Engineering Aspects* **2003**, *229*, 1-8.
(20) Matos, J. R.; Kruk, M.; Mercuri, L. P.; Jaroniec, M.; Zhao, L.; Kamiyama, T.; Terasaki, O.; Pinnavaia, T. J.; Liu, Y. *J. Am. Chem. Soc.* **2003**, *125*, 821-829.
(21) S. Horiuchi, M. I. S. Y. N. *Advanced Materials* **2000**, *12*, 1507-1511.
(22) Schmidt-Winkel, P.; Glinka, C. J.; Stucky, G. D. *Langmuir* **2000**, *16*, 356-361.
(23) Yang, P. D.; Zhao, D. Y.; Margolese, D. I.; Chmelka, B. F.; Stucky, G. D. *Chemistry of Materials* **1999**, *11*, 2813-2826.
(24) Yu, C.; Tian, B.; Fan, J.; Stucky, G. D.; Zhao, D. *J. Am. Chem. Soc.* **2002**, *124*, 4556-4557.
(25) Yu, S. Q.; Zhao, R. Y.; Liu, C. G.; Guan, C. *Abstracts of Papers of the American Chemical Society* **2003**, *226*, U263-U264.

(26) Yue, Y. H.; Ma, Z.; Hua, W. M.; Gao, Z. *Acta Chimica Sinica* **2000**, *58*, 777-780.
(27) Brezesinski, T.; Antonietti, M.; Groenewolt, M.; Pinna, N.; Smarsly, B. *New Journal of Chemistry* **2005**, *29*, 237-242.
(28) Hung, I. M.; Wang, H. P.; Lai, W. H.; Fung, K. Z.; Hon, M. H. *Electrochimica Acta* **2004**, *50*, 745-748.
(29) Suzuki, K.; Sinha, A. K. *Journal of Materials Chemistry* **2007**, *17*, 2547-2551.
(30) Hung, I. M.; Hung, D. T.; Fung, K. Z.; Hon, M. H. *Materials Letters* **2008**, *62*, 1147-1150.
(31) Rezaei, M.; Alavi, S. M.; Sahebdelfar, S.; Yan, Z. F. *Journal of Porous Materials* **2008**, *15*, 171-179.
(32) Larsen, G.; Lotero, E.; Marquez, M. *J. Phys. Chem. B* **2000**, *104*, 4840-4843.
(33) Hukkamaki, J.; Pakkanen, T. T. *Microporous and Mesoporous Materials* **2003**, *65*, 189-196.
(34) Volden, S.; Glomm, W. R.; Magnusson, H.; Oslash; ye, G.; Sj; ouml; blom, J. *Journal of Dispersion Science and Technology* **2006**, *27*, 893 - 897.
(35) Mitra, A.; Bhaumik, A.; Imae, T. *Journal of Nanoscience and Nanotechnology* **2004**, *4*, 1052-1055.
(36) Cho, B. K.; Jain, A.; Nieberle, J.; Mahajan, S.; Wiesner, U.; Gruner, S. M.; Turk, S.; Rader, H. J. *Macromolecules* **2004**, *37*, 4227-4234.
(37) Larsen, G.; Spretz, R.; Lotero, E. *Chem. Mater.* **2001**, *13*, 4077-4082.
(38) Larsen, G.; Lotero, E.; Marquez, M. *Chemistry of Materials* **2000**, *12*, 1513-+.
(39) Schwickardi, M.; Spliethoff, B.; Schmidt, W.; Schuth, F. *Zeitschrift Fur Physikalische Chemie-International Journal of Research in Physical Chemistry & Chemical Physics* **2005**, *219*, 939-948.
(40) Schwickardi, M.; Johann, T.; Schmidt, W.; Busch, O.; Schuth, F. In *Scientific Bases for the Preparation of Heterogeneous Catalysts* 2002; Vol. 143, p 93-100.
(41) Schwickardi, M.; Johann, T.; Schmidt, W.; Schuth, F. *Chemistry of Materials* **2002**, *14*, 3913-3919.
(42) Blanco, J.; Petre, A. L.; Yates, M.; Martin, M. P.; Suarez, S.; Martin, J. A. *Advanced Materials* **2006**, *18*, 1162-1165.
(43) Xu, Q.; Fan, H. J.; Guo, Y. Q.; Cao, Y. X. *Materials Science and Engineering a-Structural Materials Properties Microstructure and Processing* **2006**, *435*, 158-162.
(44) Ozawa, M.; Kimura, M. *Journal of Materials Science Letters* **1990**, *9*, 446-448.
(45) Deshpande, A. S.; Niederberger, M. *Microporous and Mesoporous Materials* **2007**, *101*, 413-418.
(46) Shchukin, D. G.; Caruso, R. A. *Chemistry of Materials* **2004**, *16*, 2287-2292.
(47) Shi, Z.-G.; Xu, L.-Y.; Feng, Y.-Q. *Journal of Non-Crystalline Solids* **2006**, *352*, 4003-4007.
(48) Yan, Y. J.; Huang, Z. R.; Dong, S. M.; Jiang, D. L. *Journal of the American Ceramic Society* **2006**, *89*, 3585-3588.
(49) Beck, J. S.; Vartuli, J. C.; Roth, W. J.; Leonowicz, M. E.; Kresge, C. T.; Schmitt, K. D.; Chu, C. T. W.; Olson, D. H.; Sheppard, E. W.; McCullen, S. B.; Higgins, J. B.; Schlenker, J. L. *Journal of the American Chemical Society* **1992**, *114*, 10834-10843.

169e Synthesis and Characterization of Novel Hierarchically Structured Aluminosilicate Catalysts

Sikander H. Hakim, Chemical and Biological Engineering, Iowa State University, 2114 Sweeney Hall, Dept. of Chemical and Biological Engineering, Ames, IA 50011 and Brent H. Shanks, Department of Chemical and Biological Engineering, Iowa State University, 2119 Sweeney Hall, Ames, IA 50011.

Hierarchical materials with pores at two different length scales (meso and macro) integrated in the same particle are of interest in the design of novel structured catalysts. These materials with the combined advantages of high surface area and reduced diffusion resistance have the potential to be highly efficient catalysts. Previously, our group reported the first one-pot synthesis of hierarchically structured meso-macroporous alumina. Beginning from alkoxide precursors, these unique morphologies are obtained in solution by spontaneous self-assembly of metal oxide nanoparticles as mesoporous walls around macroporous channels. This relatively simple synthesis approach does not require the presence of any templating agent. After the discovery of hierarchical alumina, hierarchical materials from other metal oxides such as titania and zirconia were successfully synthesized for comparative studies. The work to be presented here discusses unprecedented synthesis routes to novel hierarchically structured aluminosilicates that were recently developed based on our previous work with individual alkoxides. The synthesized materials were characterized for their structural morphology and textural properties as well as catalytic properties through multiple characterization techniques. The synthetic routes were evaluated both on the basis of the creation of acidity and the quality of multi-scale porous structure, since both are crucial for the catalyst performance. Presented will be results demonstrating the synthesis of hierarchical meso-macroporous aluminosilicate materials having the presence of both Brönsted and Lewis acidity. These materials were catalytically active and were investigated further for their performance in the acid-catalyzed esterification of free fatty acids with methanol to produce methyl esters.

169f Synthesis of Hierarchical Pore Zsm-5 on Silica Gel and Its Catalytic Performance for Fcc Reaction

Hui Feng[1], Xiaoyin Chen[2], and Johannes W. Schwank[2]. (1) Chemical Engineering, University of Michigan, USA; China University of Petroleum, 2300 Hayward, H.H.Dow 3074, Ann Arbor, MI 48109, (2) Chemical Engineering, Transportation Energy Center, University of Michigan, 2300 Hayward, H.H. Dow 3074, Ann Arbor, MI 48109-2136

ZSM-5 zeolite is a preferred active component for low-carbon alkene production during fluidized catalytic cracking (FCC). However, during industrial catalyst preparation, ZSM-5 is inevitably mixed with binder, thereby reducing the contact with the feedstock and lowering the efficiency of ZSM-5. It is interesting to grow a layer of ZSM-5 zeolite on the mother-support. Here we report a method for the direct in-situ crystallization of ZSM-5 on the surface of hollow silica gel microspheres. The objective of using such composite materials is to improve mass transfer and the FCC product distribution. In this study, ZSM-5 zeolite has been synthesized hydrothermally using n-butylamine on silica gel microspheres as the template. The composite materials were characterized by X-ray diffraction (XRD), Scanning Electronic Microscope (SEM), and BET surface area via liquid N2 adsorption. The effects of initial gel composition (Na+/SiO2, OH-/SiO2, H2O/ SiO2, BuNH2/SiO2), crystallization time, and crystallization temperature on crystallinity, morphology and pore structures were systematically investigated. Different morphologies and crystal sizes of ZSM-5 were obtained. Hierarchical ZSM-5 on the surface of silica gel microsphere will be synthesized by various methods, such as surfactants addition and steam treatment. Composite materials ZSM-5/SiO2 with different pore distributions will be obtained. Finally, the effects of crystal morphology, pore distribution, and acidity on the selectivity/yield for FCC products will be investigated.

169g Preparation of Magnetic Hollow Zsm-5/Ni Composite Spheres and Fibers

Lixiong Zhang[1], Jianfeng Yao[1], and **Xiaohua Lu**[2]. (1) Sate Key Laboratory of Materials-oriented Chemical Engineering, Nanjing University of Technology, Xinmofan Road 5#, Nanjing, China, (2) State Key Laboratory of Materials-oriented Chemical Engineering, Nanjing University of Technology, Xinmofan Road 5#, Nanjing, China

A simple template method was developed for the preparation of magnetic hollow ZSM-5/Ni composite spheres and fibers. This was fulfilled by first coating Ni layers on polystyrene (PS) microspheres, poly (vinylidene fluoride) (PVDF) hollow fibers microfiltration membranes, and cellulose acetate fibers from cigarette filters, respectively, by electroless plating, followed by growing ZSM-5 layers on the resulting Ni/template composites. Magnetic hollow ZSM-5/Ni spheres and fibers were subsequently prepared by calcination and reduction (if necessary) of the ZSM-5/Ni/ template composites. The samples were characterized by SEM, XRD, TG, EDX, temperature-programmed reduction (TPR) and a vibrating sample magnetometer (VSM). The results indicated that the hollow ZSM-5/Ni composites inherited magnetic property from the nickel layers. Their magnetization values could be adjusted by regulating the thickness of the nickel layer of the Ni/template composites, which was implemented by controlling the time of electroless plating. By controlling the hydrothermal treatment time and temperature, the thickness of the zeolite layers can be regulated. For the preparation of the hollow spheres, a continuous ZSM-5 layer with a thickness of 400 nm was formed on the surface of Ni/PS spheres after one cycle of hydrothermal treatment at 80 ^{o}C for 36 h. The saturation magnetization (*Ms*) and the coercivity (*Hc*) values of the hollow ZSM-5/Ni spheres with the Ni/PS spheres prepared by ultrasonic electroless plating for 4 min were about 14.08 emu/g and 88.95 Oe. When PVDF hollow fibers were used as the template, the resulting magnetic hollow ZSM-5 hollow fibers showed a round structure in cross-section with inner and outer diameters of ca. 600 and 800 μm, respectively; while the magnetic hollow ZSM-5 hollow fibers showed a trefoil structure in cross-section with an inner size of ca. 20 μm when the cellulose acetate fibers were used. The saturation magnetization value and the BET specific surface area of the magnetic hollow ZSM-5/Ni fibers were over 20 emu/g and 100 m^2/g, respectively.

Keywords: hollow sphere, hollow fiber, magnetic, zeolite

Acknowledgement: This work is supported by Program for Changjiang Scholars and Innovative Research Team in University (PCSIRT 0732) and National Natural Science Foundation of China (No. 20676059)

183a Plasma Etching for Metal Gate Electrode Fabrication in Conventional and Sub-Threshold Transistor Integration

Steven A. Vitale, Jakub Kedzierski, Nisha Checka, and Craig L. Keast. Advanced Silicon Technology, MIT Lincoln Laboratory, 244 Wood Street, Lexington, MA 02420

The 45nm technology node is witnessing the introduction of high-k dielectric and metal gate materials into CMOS logic production, after about a decade of materials research and process development. Coincidentally, the emerging field of ultra-low power electronics is investigating the integration of metal gate materials with a completely different set of performance needs than those of high-performance CMOS logic. A discussion of the physical and electrical requirements of the gate materials for these two technologies will be presented, along with an introduction to the "gate first" vs. "gate last" integration approaches.

A key challenge of integrating high-k / metal gates is the plasma etching of the gate stack. Results from etching of dual work function band-edge CMOS logic gates will be presented, demonstrating the feasibility of "gate first" integration. Plasma etching results on blanket and patterned mid-gap metal gate stacks will also be presented, illustrating how the fundamentals of plasma chemistry and ion-enhanced etching can be used to develop new plasma processes enabling integration of novel gate stack materials.

*This work was sponsored by the Air Force under contract #FA8721-05-C-0002. Opinions, interpretations, conclusions, and recommendations are those of the author and are not necessarily endorsed by the United States Government.

183b Interface Preparation for High Mobility Substrates

Anthony J. Muscat, Chemical and Environmental Engineering, University of Arizona, Harshbarger Bldg., Room 134, 1133 E. James Rogers Way, Tucson, AZ 85721

Significantly higher electron mobilities compared to silicon make III-V compound semiconductors the materials of choice for high speed, low power devices. The poor quality of thermal oxide layers grown on these surfaces, however, has limited the use of III-V materials in microelectronics. Recent advances in depositing high-*k* films on silicon using atomic layer deposition (ALD) could expand the use of III-V materials in high volume manufacturing. ALD offers precise control of film thickness, low processing temperatures, and excellent conformality on high aspect ratio structures. ALD relies on self-terminating chemisorption of a reactant on a surface producing up to a monolayer coverage. Recent results show improved gate stack quality in the absence of interfacial layers that are formed typically by reoxidation after cleaning or by reaction during high-*k* deposition. Controlled removal of the native oxide and passivation/activation of the surface prior to depositing high dielectric constant materials are important process to develop and understand. Etching techniques to remove native oxides on III-V compounds such as liquid phase chemical etching, thermal desorption, and ion bombardment face issues of partial oxide removal, roughened surfaces, and selective etching, which can produce changes in surface stoichiometry. Gas phase processing provides an alternative etching technology that not only addresses the issues of process repeatability and uniformity, but also reduces environmental impact and thermal budget by minimizing the use of ultra pure water and reducing the necessity for recleaning. This study investigated gas phase etching of In-based III-V materials using mixtures of anhydrous HF and water vapor at a total pressure of 100 Torr and a temperature of 29ûC. Native oxide removal and surface termination of InAs(100) and InSb(100) using liquid and gas phase HF chemistries were studied using x-ray photoelectron spectroscopy. Aqueous HF etching removed the native oxides on InAs and produced elemental As, which reoxidized when exposed to air. On InSb the native oxides were not completely removed due to F-termination, which passivated the surface. Gas phase HF etching of InSb native oxide completely removed Sb_2O_5 producing a stoichiometric semiconductor surface terminated by F atoms on primarily In sur- face sites. On InAs gas phase HF completely removed As_2O_3 producing two surface stoichiometries. For the majority of HF to water molar ratios studied, a stoichiometric bulk metal and an As-rich overlayer was produced. For a lean HF composition, an As-rich bulk metal and In-rich overlayer was produced. Deposition of Al_2O_3 by atomic layer deposition (ALD) at 170°C directly onto F-terminated InSb produced a chemically sharp Al_2O_3/InSb interface. ALD of Al_2O_3 on an In-rich overlayer on InAs resulted in an interfacial layer containing As-oxide.

183c Applying Lessons Learned from An Etching Lab: The Role of O Atoms In PECVD

Colin A. Wolden, Colorado School of Mines, 1500 Illinois, Golden, CO 80401

I was fortunate to spend the final 18 months of my PhD working in the Sawin laboratory. In this talk I will briefly recount that experience and illustrate the vital impact it had on the evolution of my own research in developing plasma-assisted technologies for the deposition of electronic oxides and thin film photovoltaics. My experience in the Sawin laboratory led directly to our development of high vacuum plasma-assisted chemical vapor deposition, which was used for the synthesis of semi-conducting zinc oxide. A modification of this approach was used for the plasma-assisted co-evaporation copper chalcopyrite photovoltaic absorbers. In studies of metal oxide synthesis by standard PECVD our work has focused on the dominant role of O atoms in controlling the gas-phase chemistry. This work lead to the development of low frequency pulsed PECVD as a high rate alternative to ALD for self-limiting deposition (~ 1 Å/pulse) of metal oxides. We describe the deposition process and highlight critical aspects of reaction design and operation, using examples from several material systems. With cycle times < 1 s, net rates > 30 nm/min were obtained by pulsed PECVD while retaining high quality and digital control.

183d Polymer Nano-Texturing and Stochastic Nano-Patterning Using Plasma Processing

Evangelos Gogolides, Angeliki Tserepi, Nikos Vourdas, Maria Elena Vlachopoulou, Katerina Tsougeni, and Dimitrios Kontziampasis. Institute of Microelectronics IMEL, NCSR 'Demokritos', Terma (End of) Patriarhou Gregoriou Street, Aghia Paraskevi, Attiki, 15310, Greece

Plasma processing induces roughening of polymers at the nanoscale, and forces one to abandon the idea of a flat surface, after processing, commonly used at the macro and micro scales. Plasma induced roughening of polymers may have detrimental effects when fabricating an electronic nano-device, or may have beneficial effects when it imparts anti-reflective or superhydrophobic properties on surfaces to mention only three examples. Thus, roughness should be reduced in some cases, while it should be maximized or controlled in others. Plasma induced polymer nano-roughness control may thus lead to new nano-manufacturing processes and products. This paper summarizes the recent finding of our group on plasma processing of both organic and inorganic (Silicon-containing) polymers.

A Helicon Plasma reactor is used for polymer plasma processing. Fractal theory based nanometrology protocol has been developed for the analysis of AFM images of the etched surfaces. (1+1) and (2+1) Monte Carlo Simulations are used for the explanation of observed behaviour.

First, plasma nanotexturing has been used to control the wetting and reflectivity of polymer surfaces. Wettability and its control is an important feature of surface science and technology finding also new micro- and nano- scale applications such as droplet frictionless motion in microfluidics, biocompatibility tuning, protein and cell adsorption and others. Both very hydrophilic (VH) and super hydrophobic (SH) (self-cleaning) surfaces may be desirable. In addition to wetting control, other parameters such as optical properties should be also exactly controlled. For polymers the production of SH, optically transparent, and antireflective surfaces is a very attractive objective. Wettability depends on both surface chemistry and topography.

We developed a generic, fast, all-dry process for VH and SH surface production for organic and silicon- containing polymers. The organic polymer such as PMMA, PEEK, PS, PET sheet is loaded in a low-pressure reactor equipped with a Helicon source to generate cold plasma. First oxygen-based plasma is applied to etch the surface and create nano-scale surface roughness (nanotexturing). The time of the process may differ from 1 min to several min depending on the roughness amplitude and on the degree of transparency desired. Surfaces treated in this way become superhydrophilic with 0 contact angle. The longer the processing time the longer the hydrophilicity resists ageing. Surfaces treated for 20min age in weeks back to the initial contact angle, while surfaces treated for 1-5 min age within hours-days. After this first step, the gas chemistry may be altered into a fluorocarbon one which leads to a fluorocarbon film (FC) deposition, thus altering the surface chemistry. CA and CA hysteresis measurements reveal the SH character of the treated surfaces.

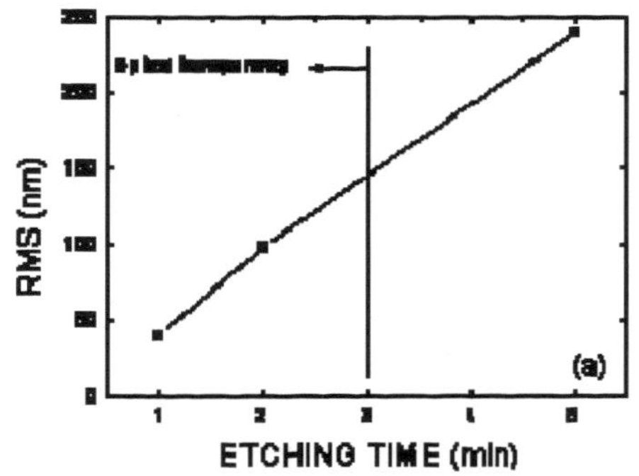

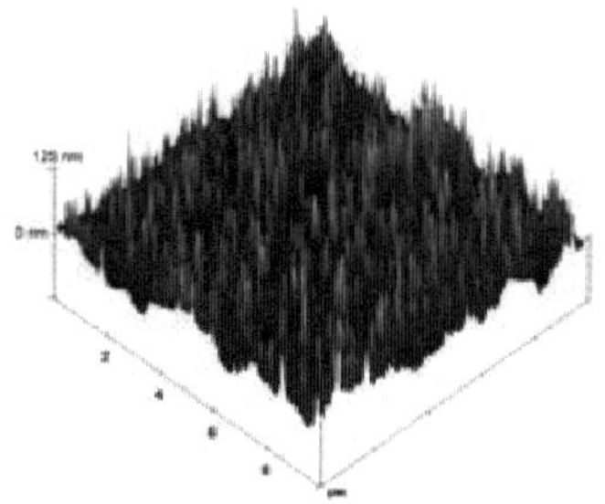

Figure 1 PMMA nanotexturing in O2 plasmas, and Supehydrophobic (SH) Surface production . From Left to Right:

Nanoroughness (RMS) versus etch time. Contact angle and contact angle hysteresis after plasma deposition of fluorocarbon (FC) thin layer. Morphology of the surface after 1min etching time. Conditions: 2000 W source power, 0.75 Pa, 100 sccm, -100 V bias, 10 °C

When an organosilicon polymer such as PDMS is etched in SF6 plasma (instead of Oxygen) high aspect ratio nanostructures are also created. After plasma deposition of a thin fluorocarbon layer, these surfaces become SH as well.

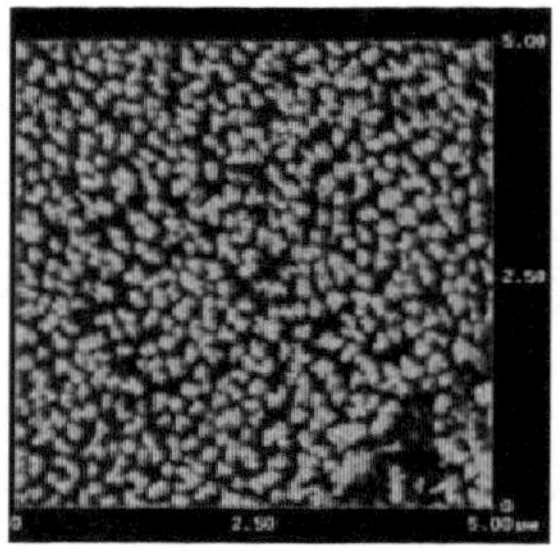

Figure 2 PDMS nanotexturing in SF6 plasmas, and Supehydrophobic (SH) Surface production . From Left to Right:

SEM image of a PDMS elastomer (Sylgard 184) surface after a 6 min SF_6 plasma treatment. 1.45 ìm-high nano-columns are shown.

AFM top view image of the same elastomer surface after 2 min treatment in SF_6 plasma (roughness rms ~ 133 nm and periodicity ~ 240 nm). Image of a water droplet during its rolling off the surface in (a) after being conformally coated with a 20 nm-thick FC film.

Inorganic polymers treated with SF_6 plasmas gave high Aspect Ratio Nanopillar like morphology. However, for the same polymers ripple-like and periodic morphology was produced on O_2 plasma etched PDMS (see top left figure 3), leading to sticky – hydrophobic surfaces most of the times after fluorocarbon deposition. We have also demonstrated 2-10 fold increase of protein adsorption on such nanotextured surfaces.

In all the above applications nanotexture but not nanopattern has been achieved. Could we do nanopatterning with plasmas? The answer to the question is yes, and we show an example below. A plasma-based methodology to fabricate oriented spontaneously formed nanostructures on Poly (dimethylsiloxane) PDMS films and stamps was discovered. Oxygen plasma treatment of PDMS produces spontaneously-formed wavy structures with high nano-scale amplitude and with periodicity of a few 100's nm as revealed by Atomic Force Microscopy (AFM). These nanostructures may be produced by two methods. a) In the first method, a stencil mask was placed on top of a thin PDMS film and O_2 plasma treatment followed. b) In the second method, an elastomeric stamp was prepared by casting Sylgard 184 elastomer (Dow Corning) over lithographically created masters and then treated in O_2 plasma. The dimensions of the elastomeric stamp determine the wavelength of the periodic nanostructures formed (see fig. 3)

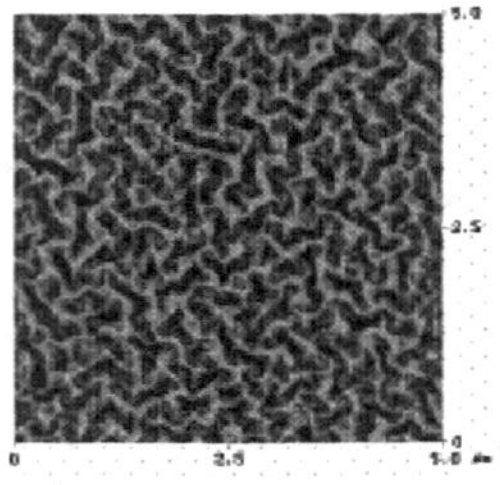

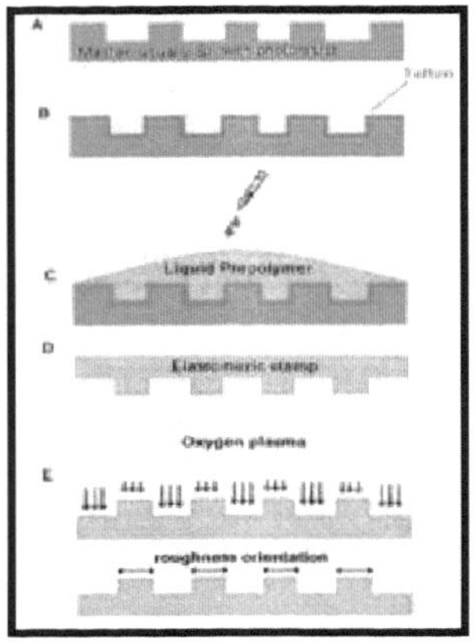

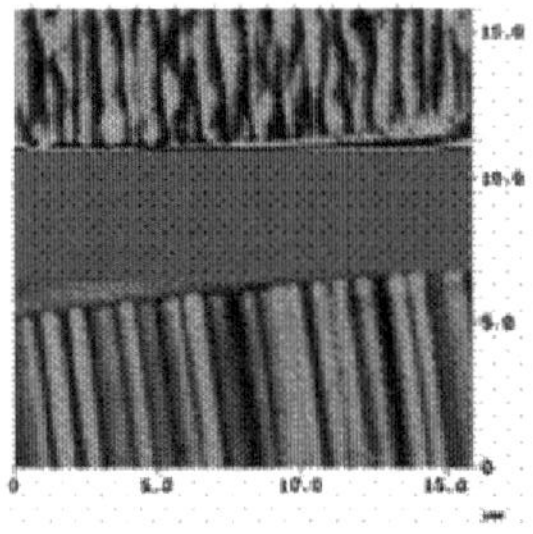

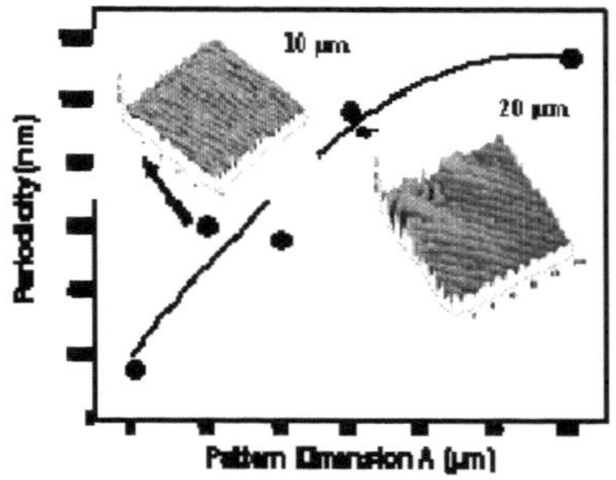

Figure 3. Plasma Based Nanopatterning of polymers. PDMS nanopatterning with Oxygen plasmas. From Left to Right:

AFM image of a PDMS surface after 7 min treatment in O_2 plasma (roughness rms =48 nm, periodicity ~ 302 nm, surface area ratio ~1.95.

Process for fabrication of periodic nanostructures on PDMS using oxygen plasma processing.

AFM 2D image of a PDMS stamp after exposure to O_2 plasma for 10 min at the transition from the protruding parts A to the recessed part B of the stamp.

Periodicity of an elastomeric stamp versus lithographic pattern dimension

This simple oxygen plasma treatment to orient spontaneously formed nanostructures, with wavelength in the sub-micrometer range, may be used to print, for example, fine patterns of biologically active molecules onto surfaces with high spatial frequency.

We will also demonstrate plasma-based nanodot formation on polymers, and propose plasma driven self-assembly processes.

In conclusion, plasma processing proves to be an attractive path to nanotexture surfaces of polymers for various applications, and for nanopattern formation in a stochastic periodic manner.

References

[1] Katerina Tsougeni, George Boulousis, Evangelos Gogolides and Angeliki Tserepi Oriented spontaneously formed nanostructures on Poly (dimethylsiloxane) films and stamps treated in O_2 plasmas In Press, available on Line, Microelectronic Egineering 2008, doi:10.1016/j.mee.2008.01.007

[2] Tsougeni K, Tserepi A, Boulousis G, V. Constantoudis, E. Gogolides. "Tunable poly(dimethylsiloxane) topography in O2 or Ar plasmas for controlling surface wetting properties and their ageing", (2007) JPN J Applied Physics 1 46 (2): 744-750.

[3] Tsougeni K., Tserepi A., Boulousis G., Constantoudis V., Gogolides E. "Control of nanotexture and wetting properties of polydimethylsiloxane from very hydrophobic to super-hydrophobic by plasma processing", (2007) Plasma Processes and Polymers, 4 (4), pp. 398-405.

[4] Vlachopoulou M.-E., Tserepi A., Beltsios K., Boulousis G., Gogolides E. "Nanostructuring of PDMS surfaces: Dependence on casting solvents", (2007) Microelectronic Engineering, 84 (5-8), pp. 1476-1479.

[5] Vourdas N., Tserepi A., Gogolides E. "Nanotextured super-hydrophobic transparent poly(methyl methacrylate) surfaces using high-density plasma processing", (2007) Nanotechnology, 18 (12), art. no. 125304.

[6] Tserepi A.D., Vlachopoulou M.-E., Gogolides E. "Nanotexturing of poly(dimethylsiloxane) in plasmas for creating robust superhydrophobic surfaces"(2006) Nanotechnology, 17 (15), art. no. 062, pp. 3977-3983.

[7] Gogolides E., Constantoudis V., Patsis G.P., Tserepi A. "A review of line edge roughness and surface nanotexture resulting from patterning processes", (2006) Microelectronic Engineering, 83 (4-9 SPEC. ISS.), pp. 1067-1072.

[8] Tserepi A., Gogolides E., Misiakos K., Vlachopoulou M.-E., Vourdas N., Greek Patent application number 20050100473; PCT application number GR 2006 / 000011.

Web Page: www.imel.demokritos.gr/plasmagroup

183e CMP Pad Surface Analysis and the Components of Pad Surface Texture

Andrew S. Lawing, Rohm and Haas Electronic Materials CMP Technologies, 451 Bellevue Road, Bldg 6, Newark, DE 19711

Pad surface analysis is a powerful tool to help understand the impact of pad texture on CMP process performance as well as the implications of pad texture on process and consumable design. Pad surface data can be obtained from any number of surface analytical techniques. The most effective are non-contact optical methods such as confocal microscopy or vertical scanning interferometry. These optical techniques can provide accurate 3-dimensional representations of the pad surface. Three dimensional pad surface textural data can be interpreted in the form of pad height probability distribution functions, which provide a statistical representation of how the pad surface is distributed in space. The pad height probability distribution function can be deconvolved into three components arising from: 1) the native porosity of the pad material, 2) the surface texturing process applied to the pad material and 3) polishing process effects on the pad near surface. These data can be used to help drive pad design, pad manufacturing process improvements and CMP process improvements and consumable design. In this talk we will discuss examples of the respective textural components including the characterization of two different pad types illustrating the effect of conditioning on pad materials with differing native porosity and an example illustrating the competing effects of textural restoration from pad conditioning, and pad asperity wear.

199a Benign Silica and Germania Synthesis: From Monodisperse Nanoparticles and Nanocrystals to Porous Inorganic Hollow Shells and Thin Films

Zhuopeng Wang, State Key Laboratory of Inorganic Synthesis and Preparative Chemistry, Jilin University, College of Chemistry, Changchun 130012, China, **Mark A. Snyder**, Department of Chemical Engineering, Lehigh University, 111 Research Drive B-322, Iacocca Hall, Bethlehem, PA 18015, Tracy M. Davis, Chevron (Current), 100 Chevron Way, Bldg 50, Room 1134A, Richmond, CA 94802, and Michael Tsapatsis, Department of Chemical Engineering and Materials Science, University of Minnesota, 421 Washington Ave. SE, Minneapolis, MN 55455.

Drawing motivation from ubiquitous biomineralized silica structures in nature, we recently identified the formation of stable, highly monodisperse silica nanoparticles in aqueous amino acid solutions.[1,2] The novelty of the synthesis derives from its simplicity and benign nature (i.e., more neutral pH), where hydrolysis of a silica source is accomplished in an aqueous solution of L-Lysine and a range of handles exist for tuning particle size (ca. 5-25 nm). Spurred by an interest to explore the influence of biomolecules on other oxide systems, we have extended the amino acid-mediated synthesis techniques to that of germania,[3] an oxide bearing interesting reactive and optical properties. Here, we will first discuss the facile synthesis of nanometer-sized germania crystals and amorphous germania nanoparticles (ca. 1 nm) through hydrolysis of germanium tetraethoxide and subsequent condensation of germania in both pure water and aqueous lysine solutions. We will analyze similarities between the lysine-mediated germania and silica synthesis as related to the transition from germinate species to nanoparticles in solution. In addition, we will highlight the remarkable room temperature aqueous crystallization of germania, and will discuss the influence of lysine, nanoparticles, and germania concentration on the onset of crystallization and the resulting crystal size, morphology, and monodispersity.

In general, the room temperature and near-neutral conditions for nanoparticle formation and crystallization, especially in the presence of the basic amino acid lysine, make both the silica and germania systems attractive for benign materials applications. As such, we will describe techniques for successful assembly of the nanoparticles into nanoparticle-crystals and ordered multi- and monolayer coatings, where micro and mesoporosity imparted by the interstitial particle spacing can be engineered by fine control of particle size. We will also describe how materials properties of both metal oxide systems (i.e., charge, stability, solubility) can be leveraged individually and simultaneously for novel and benign (i.e., temperature, pH) synthesis of metal oxide core-shell structures, highly monodisperse hollow porous shells, and novel nanoparticulate and structured thin films.

References

1. Davis, Snyder, Krohn, Tsapatsis, *Chemistry of Materials*, 2006, **18**(25), 5814-5816.

2. Snyder, Lee, Davis, Scriven, Tsapatsis, *Langmuir*, 2007, **23**(20), 9924-9928.

3. Davis, Snyder, Tsapatsis, *Langmuir*, 2007, **23**(25), 12469-12472.

199b Nanoparticle Precursors and Phase Selectivity In Hydrothermal Synthesis of Zeolite Beta

Nathan D. Hould, Center for Catalytic Science and Technology, Department of Chemical Engineering, University of Delaware, 150 Academy St., Newark, DE 19716 and Raul F. Lobo, Center for Catalytic Science and Technology, Department of Chemical Engineering, Univesity of Delaware, 150 Academy St., Newark, DE 19716.

The crystallization mechanism of zeolite beta in solutions composed of 1 SiO_2 : 80 H_2O : 0.5 TEAOH : 0.1 NaOH : Y $NaAlO_2$: Z $B(OH)_3$ where Y is 0 and 0.02 and Z is 0 and 0.05 was investigated using small angle x-ray scattering, dynamic light scattering, XRD, and electron microscopy. In this system there is a critical aggregation concentration (cac) for silica above which silica monomers and oligomers aggregate into nanoparticles and there is a negative correlation between the cac and the boric acid concentration. Initially primary nanoparticles (< 3 nm) are formed above the cac. When these solutions are heated some of the primary particles transform into larger secondary particles (6 - 50 nm). In solutions without aluminum the secondary particles remain stable for extended periods of time. In solutions with aluminum the secondary particles aggregate into tertiary particles (zeolite beta, > 200 nm). Initially the tertiary particles have poor long range order and an oblate morphology. They evolve into zeolite beta crystals with longer range order and smooth square bipyramidal morphology by solution mediated monomer dissolution and reattachment.

199c The Spectroscopic Signature Nitrogen-Substituted Zeolites

Karl D. Hammond[1], Fulya Dogan[2], Geoffrey A. Tompsett[1], Murad Gharibeh[1], Vishal Agarwal[1], W. Curtis Conner[1], Clare P. Grey[2], and Scott M. Auerbach[3]. (1) Chemical Engineering, University of Massachusetts, 159 Goessmann Laboratory, Amherst, MA 01003, (2) Chemistry, SUNY Stony Brook, State University of New York at Stony Brook, Stony Brook, NY 11794, (3) Chemistry, University of Massachusetts, 701 Lederle Graduate Research Tower, Amherst, MA 01003

Nanoporous acid catalysts in the form of zeolites are the foundation of catalytic technologies in petroleum refining. Recent interest in fuel production from biomass, however, has created demand for similarly shape-selective base catalysts because of the high oxygen content of biomass-derived feeds. Strongly basic zeolites, however, are notoriously difficult to make due to the strongly acidic nature of aluminosilicates. Much research has been conducted into the production of strongly basic zeolites by treating ordinary zeolites with amines, but to date there is little direct evidence that nitrogen is actually incorporated into the zeolite framework during such treatments. We detail synthesis, NMR spectroscopy, and quantum chemical calculations showing that nitrogen likely adds to both surface and interior sites, preserving the framework structure. The resulting zeolites show unique catalytic activity. This finding is of critical importance to the rational design of biomass refining catalysts, and allows over fifty years of zeolite science to be applied to the catalytic synthesis of biofuels.

199d Facile Synthesis of Zeolite Mcm-22 Via a Two-Step Sol-Gel Route

Jun Wang, Yajing Wu, and **Xiaohua Lu**. State Key Laboratory of Materials-oriented Chemical Engineering, Nanjing University of Technology, Xinmofan Road 5#, Nanjing, China

Zeolite MCM-22 was first synthesized by Mobil researchers in 1990. It presents a peculiar structure with two independent pore systems, both accessible through 10-membered rings. In recent years, the synthesis of zeolite MCM-22 has become one of the increasing interests due to its important applications in a variety of catalytic processes. In general, MCM-22 is synthesized in a hydrothermal system containing an alumina source, a silica source and a template. Very often the product obtained is dependent on the silica source selected. Although TEOS was the most preferred organic silica source for the synthesis of zeolite because of its monomeric nature and moderate rate of hydrolysis, no attempt to the synthesis of MCM-22 using TEOS as silica source appears so far. On the other hand, the physicochemical properties of zeolites are strongly influenced by the framework composition. Zeolite MCM-22 can be readily crystallized when the framework SiO2/Al2O3 molar ratio is 20-50. Actually, intensive efforts targeting at the expansion of the SiO2/Al2O3 ratio out of this range have been tried, however, a substantial increase of the SiO2/Al2O3 ratio directly by the conventional synthesis is difficult. Therefore, the direct hydrothermal synthesis of high-silica MCM-22 zeolite is still a rich field of research.

In this work, pure MCM-22 zeolite with high crystallinity was facile synthesized via a novel two-step sol-gel route using TEOS as silica source. This involved the first step of the hydrolysis of TEOS catalyzed in a strong acid media with a low pH value of less than 2, followed by the second step of the hydrothermal reactions of the hydrolyzed precursor with template and alumina source in a base media with a high pH value of 11-12. This synthesis strategy is very different from traditional synthetic methods, where the simultaneous hydrolysis and crystallization reactions occur in a multi-component system under the same basic conditions. Various parameters that significantly influenced the synthesis results during both the hydrolysis reaction of TEOS in acid media and the hydrothermal reaction in base media were investigated in detail. The produced samples were characterization by XRD, ICP, and SEM techniques. It is demonstrated that the this new two-step synthesis route can lead to the well crystallized zeolite MCM-22 in a broad window of reaction conditions. The increase of temperature and pH value for the hydrolysis of TEOS favors the shortening of crystallization time and the decrease of particle size of MCM-22 product. Moreover, by adjusting the hydrothermal reaction parameters, such as increasing the crystallization temperature, the alkalinity of the synthesis gel mixture and decreasing the SiO2/Al2O3 ratio of the synthesis gel mixture, the crystallization is accelerated substantially. Also, rose-like or lamellar-like morphologies of MCM-22 particles with different sizes can be obtained by changing these reaction parameters. The distinguished advantages of this synthesis strategy are the shortness of crystallization time, low cost of template, and the possibility of directly synthesis of very high silica zeolite MCM-22 with a SiO2/Al2O3 ratio up to 100.

Keywords: hydrothermal synthesis, zeolite MCM-22, tetraethyl orthosilicate

Acknowledgements: This work is supported by Program for Changjiang Scholars and Innovative Research Team in University (PCSIRT 0732)

199e Zeolite Crystallization in Crosslinked Chitosan Hydrogels: Crystal Size Control and Chitosan Removal

Dan Li and Huanting Wang. Chemical Engineering, Monash University, Wellington Road, Clayton, Melbourne, 3800, Australia

For the purpose of controlling zeolite crystal size, crystallization of zeolite NaA and NaY in glutaraldehyde -crosslinked chitosan (GA-CS) hydrogels was studied in this work. Zeolite crystals were obtained by penetration of Na_2O-Al_2O_3-H_2O alkaline solution into GA-CS hydrogels containing colloidal silica, followed by hydrothermal treatment and removal of GA-CS hydrogels. The synthesis parameters including the amounts of silica, chitosan, and glutaraldehyde, aging and cooking times were systematically investigated. A hydrogen peroxide treatment method was introduced as an effective way for removing GA-CS hydrogels, and the conventional calcination can be avoided. X-ray diffraction (XRD), light scattering, scanning electron microscopy (SEM), thermogravimetric analysis (TG), and N_2 sorption were used to characterize the zeolite samples. This work showed that GA-CS could be a good alternative space-confinement medium for the synthesis of zeolite nanocrystals with tunable crystal sizes and excellent dispersibility in water.

199f Metal Oxide/Zeolite Core-Shell Nanostructures: Synthesis and Characterizations

Easir A. Khan, Enping Hu, and Zhiping Lai. Division of Chemical and Biomolecular Engineering, Nanyang Technological University, 62 Nanyang Drive, Singapore, Singapore

In this presentation we will report the synthesis of metal oxide/zeolite core-shell nanostructures by adsorption of zeolite nanoparticles on metal oxide particles followed by secondary hydrothermal growth. The metal oxide surface was first modified by layer by layer self-assembly of polyelectrolyte and then attached with zeolite nanoparticles by coulombic forces. By monitoring the zeta potential vs pH, this method can be applied for different metal oxides, such as ZnO and TiO2, with versatile surface properties. The secondary growth was promoted to convert zeolite nanoparticles into a continuous, well-intergrown shell. Although the secondary growth condition is highly alkaline and corrosive along with significantly different from the coating conditions, the growth of the zeolite nanoparticles is fast enough to form a closed layer before the coating falls apart as well as avoid damage to the metal oxide core particles. Such a metal oxide/zeolite core-shell microstructure can be used as membrane microreactor and may find potential applications in the field of metal oxide based heterogeneous catalysis.

199g Porous Glass Beads with a Core-Shell Structure: Preparation and Applications

Yiwen Sun, Yujun Wang, Yangcheng Lu, and Guangsheng Luo. Department of Chemical Engineering, The State Key Lab of Chemical Engineering, Tsinghua University, Tsinghua University, Beijing, 100084, China, Beijing, China

A simple and environment friendly method, subcritical water treatment, was proposed to prepare porous glass beads with a core-shell structure from ordinary soda-lime glass beads in one step. In this method, reactive subcritical water rather than any other chemical additive was utilized to selectively corrode the glass and mainly leach the alkali ions from the glass. The core-shell structure has been characterized by SEM observation. The treated alkali-lime-silicate glass beads have mesopores narrowly distributed around 4nm, and a layer of uniform nanoflake on the surface. The shell structure can be easily tailored by changing the treatment temperature and water flow rate. A possible mechanism of corrosion-ion immigrating-recondensation pattern was hypothesized to explain the formation of the core-shell porous structure. The product has high adsorption capacity and fast mass transfer properties due to their structure, and can serve as inorganic adsorbent. Accordingly, its applications in fast separation of protein mixture as well as ion-exchange of heavy metal ions were studied.

Key words: Porous glass, Core-shell structure, Subcritical water treatment, Fast separation

204a Molecular Layer Deposition of Nanoscale Organic Films

Paul W. Loscutoff and Stacey F. Bent. Chemical Engineering, Stanford University, 381 North South Mall, Stanford University, Stanford, CA 94305

Atomic layer deposition (ALD) has gained attention by allowing for nanoscale control of inorganic films via a sequence of self-limiting surface reactions. In an analogous technique, nanoscale organic thin films are deposited by molecular layer deposition (MLD), utilizing a series of self-limiting reactions of organic molecules. In this study, 1,4-phenylene diisocyanate (PDIC) and ethylenediamine (ED) are employed to deposit polyurea films on a silicon substrate. 3-aminopropyltriethoxysilane is vapor deposited on the hydroxylated SiO_2 surface to yield an amine-terminated surface, as confirmed by ellipsometry and XPS. Following amine termination, PDIC and ED are dosed in a binary cycle, and this cycling is repeated to yield the desired thickness of organic film. Ellipsometry and SEM indicate a linear growth rate of 6.7 Å/cycle, with each of the reactants demonstrating saturation behavior with increasing exposure time. The urea coupling moiety is confirmed by FT-IR, and films are shown to have stoichiometric composition by XPS. Temperature dependent measurements show that the films have good thermal stability. Density function theory is used to elucidate the reaction pathway and predicts an estimated growth rate of 6.4 Å/cycle. Experimental and theoretical studies of related MLD systems, including combinations of diisocyanates and diisothiocyanates with diols, dithiols, and diamines, will also be presented.

204b Ald Copper-Palladium Thin Films for Molecular Electronics

Irene Hsu, University of Delaware, Newark, DE 19716 and Brian G. Willis, Chemical Engineering, University of Delaware, Newark, DE 19716.

Atomic layer deposition (ALD) is a powerful tool for nanoelectronics. In addition to its role for growing ultrathin films in microelectronics, ALD can be used for controlling critical dimensions on the nanometer scale. In this work we present a unique technique for growing monolithic molecular electronic tunnel junctions using ALD. In molecular tunnel junctions, the spacing between the metal electrodes must be small enough to adsorb individual or small groups of molecules, on the order of 1-2 nm. This requirement makes ALD an ideal process to fabricate nanoelectronic devices because it provides subnanometer thickness control. Compared to other molecular electronics devices, ALD grown junctions have a number of advantages including the ability to measure and control the electrode chemical composition and atomic structure. Electrical transport measurements through molecules have established that contact effects are significant, so the ability to tune structure and composition is a considerable advantage. Transmission electron microscopy (TEM), x-ray photoelectron spectroscopy (XPS), and glancing incidence x-ray diffraction (XRD) were used to investigate the chemical composition and atomic structure of ALD-grown copper films on palladium seed layers after various growth conditions. XPS showed that palladium is present on the surface of the film, regardless of how much copper is deposited, and a pure copper surface is not obtained. This is explained by the miscibility of the two metals and the formation of alloys under growth conditions. It is proposed that the presence of palladium on the surface may promote the growth of copper, and this characteristic may be useful to control copper growth and surface layer composition. In addition, XRD and electron diffraction have been used to characterize the electrode structure. It will be demonstrated that ALD grown molecular tunnel junctions are a promising approach to understanding and engineering molecular electronic devices.

204c Lanthanum Stabilization of Ald-Grown Hafnia

John G. Ekerdt and Tuo Wang. Chemical Engineering, University of Texas at Austin, University of Texas at Austin, Austin, TX 78712

La doped HfO_2 films have been deposited on Si(100) by atomic layer deposition (ALD) to investigate the film phase transformation after high temperature annealing. The incorporation of La was achieved by depositing HfO_2 and La_2O_3 alternatively, using tetrakis (ethylmethylamino) hafnium, tris[N, N-bis (trimethylsilyl)-amide] lanthanum, and H_2O as precursors. X-ray photoelectron spectroscopy compositional analysis shows that the Hf and La atomic percentage ratio is well controlled by varying the Hf and La precursor dosing cycle ratio. 10nm layered structures were grown from multiple cycles of HfO_2 and a single cycle of La_2O_3 ALD to examine the effects of composition and composition profile on amorphous stabilization. We used X-ray diffraction, spectroscopic ellipsometry and cross section transmission electron microscopy to study the microstructure of the films and found that the introduction of La causes the film crystallization temperature to increase from 500 °C for a pure 10nm HfO_2 film to more than 1000 °C for doped films of the same thickness and a La concentration higher than 20%. 10% La produced stable films at 800 °C. This result indicates that ALD doping La is a potential method to solve the crystallization problem for HfO_2 high-k dielectric gate materials.

204d Atomic Layer Deposited Y_2O_3 Thin Films Using Novel Cyclopentadienyl-Type Yttrium Precursor

Christos G. Takoudis, Chemical Engineering and BioEngineering, University of Illinois at Chicago, 810 S Clinton St, Chicago, IL 60644

The increased miniaturization of complementary metal oxide semiconductor transistors and memory storage capacitors is driving the need for ultrathin nanometer layered film structures of high dielectric constant (κ) metal oxides in order to replace SiO_2 in gate oxide layer of transistors and charge storage media of capacitors. High-κ metal oxides including HfO_2, ZrO_2 and Y_2O_3, serve as potential candidates to replace SiO_2 and enhance scaling limits of device manufacturing. Y_2O_3 has high dielectric constant (~15), good thermal stability, relatively large conduction band offset (2.3 eV), and valence band offset (2.2 eV). For these reasons, Y-containing films may likely help serve the needs for high dielectric insulating films in nanometer dimensioned transistors and memories.

In this study, thin films of Y_2O_3 are deposited on silicon using tris (ethylcyclopentadienyl) yttrium as the metal precursor and water as the oxidizer. The film growth kinetics of these films is examined as function of several deposition parameters. The deposition rate of Y_2O_3 films is 1.7 Å/cycle at 250 °C with self limiting growth characteristics. These films are subsequently analyzed using X-ray photoelectron spectroscopy (XPS), X-ray diffraction (XRD), and atomic force microscopy. The as-deposited Y_2O_3 films are also found to be stoichiometric with carbon content below detection limit as probed by XPS and amorphous in nature as indicated by XRD measurements. In addition, the properties of Y_2O_3-doped HfO_2 films are investigated in the context of crystallization and interfacial behavior due to high temperature annealing. The $HfYO_x$ films of different compositions are grown by altering the deposition cycle ratio of the two metal precursors. The resulting films are then characterized after annealing at different temperatures using XRD and XPS. These results will be discussed in the context of high-κ materials for gate dielectric and capacitive memory applications.

204e Surface Reaction Mechanism during the Atomic Layer Deposition of Titanium Dioxide from Titanium Tetraisopropoxide and Ozone

Sumit Agarwal and Vikrant R. Rai. Department of Chemical Engineering, Colorado School of Mines, 443 Alderson Hall, 1500 Illinois Street, Golden, CO 80401

In this presentation, the authors will discuss the surface reaction mechanism during the atomic layer deposition (ALD) of TiO_2 using titanium tetraisopropoxide (TTIP) as the metal precursor, and O_3 and H_2O as the oxidizers. The film growth mechanism was investigated using a combination of in situ, real-time attenuated total reflection Fourier-transform infrared (ATR-FTIR) spectroscopy, quartz crystal microbalance (QCM), and quadrupole mass spectrometry (QMS). The ATR-FTIR spectroscopy setup is ideal for studying ligand-exchange surface reactions during ALD due to its sub-monolayer sensitivity for surface adsorbates. In our experiments, deposition was done on ZnSe internal reflection crystals. The refractive indices of ZnSe and amorphous TiO_2 are closely matched (~2.2) and, therefore, internal reflection on the top surface occurs at the vacuum-film interface. In our geometry we have 25 reflections on each surface and, hence, the IR beam passes through the film 50 times, greatly enhancing the signal-to-noise ratio. Figure 1 shows IR spectra recorded during exposure of the surface to TTIP, using an O_3-exposed surface as the reference. Sub-monolayer surface absorbates were easily detected in real time and were identified in these spectra. To quantify the surface coverage, and to determine the growth per cycle (GPC), the mass uptake during each half-cycle was determined using the QCM and the surface reaction products were detected with a differentially pumped QMS. Finally, the total thickness and the refractive index of the films were characterized using *ex situ* spectroscopic ellipsometry. Using O_3 as the oxidizer, relatively contaminant-free TiO_2 films were deposited at 150 °C: this is nearly a 100 °C lower than the widely-accepted minimum temperature for the TTIP-H_2O ALD window. Our data shows that the reactive sites for the adsorption of TTIP after the O_3 cycle are surface carbonate species, which have symmetric and anti-symmetric absorptions in the 1400 – 1700 cm^{-1} region, shown in Fig. 1. These surface carbonates can be present as mono-, bi-, or poly-dentates. Since all these forms of surface carbonates are known to be stable on a TiO_2 surface at the growth temperature, and their absorption bands overlap, it was not possible to determine their relative surface concentration. Presence of surface carbonates was further confirmed by detecting CO_2 as a reaction product during both the TTIP and O_3 cycles. During O_3 exposure, CO_2 was formed due to the combustion of isopropoxy ligands. The surface carbonates were then formed due to the back reaction of some of this CO_2 with a hydroxyl-terminated TiO_2 surface. When the carbonate terminated surface is exposed to TTIP, CO_2 is eliminated from the surface. This is in contrast to the commonly-studied metal alkoxide-H_2O ALD, where the alkoxides react directly with surface hydroxyl groups that are formed after H2O exposure. During our experiments, we also observe non-ideal ALD behavior. First, we observe that the surface coverage, measured through the integrated absorbance change due to the CH_x stretching modes in the ~3000 cm^{-1} region, is greater for a higher TTIP flux for an otherwise identical precursor dose. Second, the surface carbonates that are formed during the following O_3 cycle have different absorption bands, and a lower integrated absorbance for a lower flux. We explain this behavior using a simple surface site balance, and show that each TTIP molecule is able to exchange more ligands with the surface under low flux conditions, resulting in a low apparent coverage in the IR. Finally, under high flux conditions the GPC was measured over the temperature range of 150-275 °C using a QCM and spectroscopic ellipsometry. The GPC was almost constant over this temperature range at an average value of 0.52 Å/cycle, in good agreement with the literature.

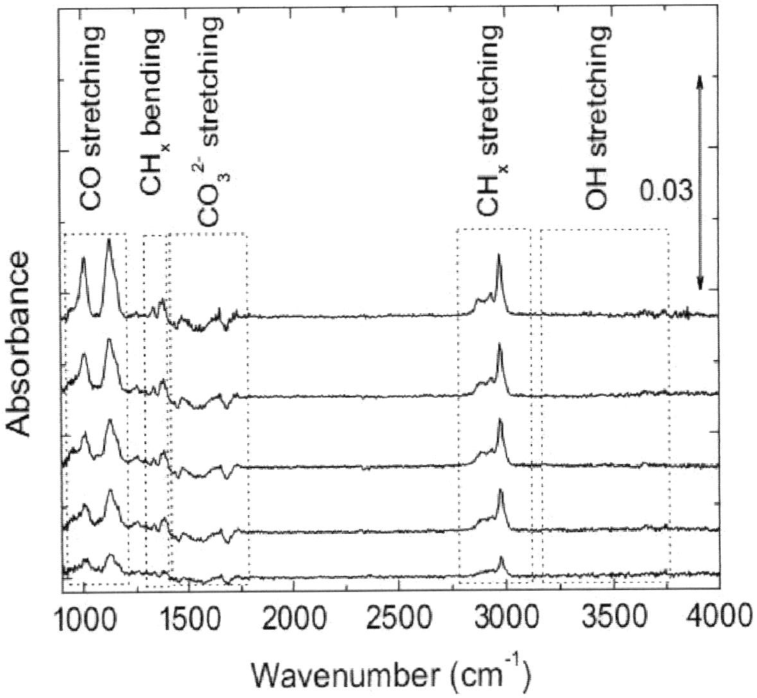

Fig. 1 Temporal evolution of IR spectra recorded during TTIP exposure of a TiO_2 surface at a substrate temperature of 150 °C. The background was collected after O_3 exposure. Stretching vibrations of hydrogen-bonded and free –OH groups are absent in the 3200-3600 cm^{-1} region.

204f Effect of Yb^{3+} Co-Doping on the Luminescent Properties of $Er^{3+}:Y_2O_3$ Thin Films

John Hoang, Chemical and Biomolecular Engineering Department, UCLA, Los Angeles, CA 90095 and Jane P. Chang, Department of Chemical and Biomolecular Engineering, University of California at Los Angeles, Los Angeles, CA 90095.

Radical enhanced atomic layer deposition (ALD) has been previously shown to enable the control of the Er^{3+} spatial distribution in Y_2O_3 thin films, thereby achieving an enhanced direct absorption cross section at 1540 nm and much improved photoluminescence with well-resolved Stark features.[1] Given the larger index of refraction of Y_2O_3 compared to that of SiO_2, this suggests that $Er:Y_2O_3$ thin films has the potential to achieve compact optical amplification. In this work, we report the utilization of radical enhanced atomic layer deposition to control the concentrations and spatial distances of Yb and Er in Yb^{3+} co-doped $Er:Y_2O_3$ thin films and the corresponding improvement in their optical characteristics.

The electronic energy level and a large absorption cross section of Yb^{3+} make it an effective sensitizer for Er^{3+}. Thin films of approximately 10 nm are synthesized by sequential radical-enhanced ALD of Y_2O_3, Er_2O_3, and Yb_2O_3 at 350°C. The composition, microstructure, cation distribution, local chemical bonding and optical properties of the as-synthesized thin films were determined by x-ray spectroscopy, electron microscopy and photoluminescent measurements. To optimize the effect of the sensitizer and minimize the concentration quenching, the concentration of Yb^{3+} and Er^{3+} were controlled by changing the global deposition cycle sequence. Extended x-ray absorption fine structure analysis verified the spatial control of Yb^{3+} and Er^{3+} in the Y_2O_3 host. Much improved effective absorption cross sections were estimated using the photoluminescence yield as a function of the pump power, as compared with measurements from thin films without Yb^{3+} sensitizers. From these measurements, an optimum concentration and spatial arrangement is chosen for the design of spiral ridge waveguide devices. These waveguides can achieve potential gains of two orders of magnitude over $Er:SiO_2$ and an order of magnitude over ion implanted Al_2O_3.

[1] Hoang, J., T. T. Van, et al. Journal of Applied Physics. 101(12), 123116 (2007).

205a Molecular Relaxation Behavior of Fatty Acid Based Vinyl Ester Resins

Steven E. Boyd, Army Research Laboratory, AMSRD-WMRD-WM-MC, Apg, MD 21005

The experimental characterization of the time-dependent properties of fatty acid based vinyl ester resins with reduced styrene content and emissions was conducted and compared to that of various commercial vinyl ester resins. Constant heating rate and isothermal, multi-frequency sweep experiments were conducted over a wide temperature range using dynamic mechanical analysis. Storage and loss modulus master-curves were formed using time – temperature superposition (TTSP) and analyzed to quantify the molecular relaxation behavior using accepted techniques and theories. Special attention was focused on determining the effect of reducing styrene weight percent on the derived viscoelastic properties. The fatty acid based vinyl ester resins were found to have similar or slightly inferior thermo-mechanical properties and a more pronounced viscoelastic response compared to the commercial resins. However, the research definitively demonstrates that the evaluated fatty acid vinyl ester resins are a viable replacement to commercial resins in certain applications with concomitant attractive environmental benefits.

205b Modification and Functionalization of Soybean Oil Resins for Dielectric Applications

Mingjiang Zhan and Richard P. Wool. Chemical Engineering, University of Delaware, 150 Academy Street, Colburn Lab, Newark, DE 19716

The most widely used and successful dielectric materials for printed circuit boards (PCB) are FR-4 and are primarily epoxy based. The epoxies have good combination of dielectric, mechanical, and thermal properties which are very important for printed circuits. However, the epoxy dielectrics are petroleum based and non-sustainable. Furthermore, the petroleum based materials are not cost effective, which becomes obvious in the recent years.

We presented a new soybean oil based resin to replace the petroleum based epoxies for PCBs, which can avoid sustainability issues. Unsaturated soybean oil has been functionalized with different reactive groups and used in different applications. Unfortunately, the thermal and mechanical properties such as glass transition temperature, modulus, etc. are not good enough for PCBs.

Improving the properties of soybean oil resin in a framework of sustainability and cost effectiveness is desirable. In this work, we proposed different chemical routines to modify and functionalize the soybean oil resins for dielectric applications. Multifunctional agent was used to increase the crosslink density of the final polymer. Maximum utilization of functional groups on the molecules was achieved. The rigidity of the soybean oil was increased to improve the mechanical properties. The resin system was also designed to minimize its dielectric constants in order to increase the operation speed of the printed circuits.

Finally, natural fibers such as chicken feather fibers were incorporated as reinforcement to produce biobased composites for dielectric devices such as PCBs.

This project was supported by the National Research Initiative of the USDA Cooperative State Research, Education and Extension Service, grant number 2005-35504-16137.

205c Bio-Based Composite Repair Resins Containing No Hazardous Air Pollutants

Kevin Andrews[1], John La Scala[2], Scott Bingham[3], James M. Sands[1], and Giuseppe R. Palmese[4]. (1) Materials Division, Army Research Laboratory, Building 4600, Aberdeen Proving Ground, MD 21005, (2) U.S. Army Research Laboratory, Attn: AMSRD-ARL--WM-MC, Building 4600, Aberdeen Proving Ground, MD 21005, (3) Colorado School of Mines, CO, (4) Drexel University, Dept of Chemical and Biological Engineering, 3141 Chestnut St, Philadelphia, PA 19104

Unsaturated polyester-based repair resins, such as Bondo™, are widely used for automotive repair, marine repair, sporting equipment repair, and household repair of metal, composites, plastics, and wood. Commercial repair resins usually contain 10-40 wt% styrene, which is a hazardous air pollutant (HAP). The reduction of HAP usage in the Army has become a high materials need because of EPA regulations, including the Defense Land Systems for Miscellaneous Equipment (DLSME) National Emissions Standard for Hazardous Air Pollutants (NESHAP) and the Reinforced Plastic Composites NESHAP. Non-volatile fatty acid monomers can be used to replace the styrene in these repair resins. Inorganic fillers, such as talc, calcium carbonate, dolomitic limestone, sodium metaborate, glass microspheres, and milled fiberglass, were added to a vinyl ester – fatty acid binder at various loading levels to elucidate the effects of individual fillers on the thermal, mechanical, rheological properties of the resin, as well as the effects on storage stability. Inorganic additives had little effect on the glass transition temperature of the cured resin, with the exception of sodium metaborate, which increased the temperature of the loss modulus peak by roughtly 12 ^{o}C at 40 wt%. The inorganic particulate fillers increased the storage modulus of the cured resin from roughly 750 MPa (zero filler) to approximately 2 GPa (40 wt% inorganic filler) at 25 ^{0}C. However, hollow glass microspheres had a smaller influence on the modulus and milled fiberglass increased the modulus much more than all of the other fillers. In general, increasing the amount of inorganic particulate fillers in the resin increased the viscosity exponentially. The degree of viscosity increase and shear- and time-dependent effects were influenced by particle size, morphology, and particle-matrix interactions.

205d Rheological Analysis of Cellulose Nanocrystal Aqueous Suspensions and Phase Behavior

Esteban E. Ureña-Benavides and Christopher L. Kitchens. Chemical & Biomolecular Engineering, Clemson University, 127 Earle Hall, Clemson, SC 29634

Cellulose nanocrystals (CNCs) have potential to become a renewable and cost effective alternative for biocompatible polymer nanocomposites with a high strength to weight ratio. However, these particles are usually very polydispersed and non-compatible with hydrophobic polymers. Aqueous suspensions of rod shaped CNCs were prepared by sulfuric acid hydrolysis of kraft wood pulp from a local paper mill and ashless cotton cellulose powder. Atomic force microscopy was used to image the particles from both sources. The cotton crystals are on average 104 nm long, and 13.9 nm wide; while the wood pulp crystals have an average length of 107 nm and a width of 12.1 nm. Aqueous CNCs suspensions phase separate into a chiral nematic – isotropic equilibrium which is dependent on the particle size, cellulose concentration and solvent properties. Rheological analyses have been used to characterize the fluid behavior of the nanocrystal suspensions for different cellulose concentrations, average lengths, and crystal ordering; including the monophasic and biphasic regions. The relative viscosity, G', and G'' were all measured. Samples of different crystal lengths were obtained by fractionation using the size dependent phase partitioning. Nanocomposites filled with CNCs have usually shown little improvement on their properties, presumably because of the poor compatibility with the polymeric matrix. Approaches to increase the dispersibility of CNCs within polymer matrices for new sustainable nanocomposites are currently being explored.

205e Thermal and Mechanical Properties of Polylactide-Nanoporous Zeolite Composites

Isinay E. Yuzay, Rafael Auras, and Susan Selke. School of Packaging, Michigan State University, 130 School of Packaging, East Lansing, MI 48824

Polylactide (PLA) and its composites have gained significant attention due to their bio-based origin. This study investigates the inclusion of nanoporous zeolite particles in PLA with the purpose of obtaining multifunctional polylactide composite materials. PLA (94 % L-lactide) containing 0, 1, 3, and 5 wt. % zeolites (Type 4A) was successfully fabricated using a mini twin screw extruder and injection molder. The effects of the amount of zeolite on morphology, mechanical, and thermal properties of PLA composites were investigated. Scanning electron microscopy (SEM) and transmission electron microscopy (TEM) studies revealed a homogeneous distribution of zeolites in the PLA matrix and good interfacial adhesion between the zeolites and the PLA matrix. The addition of zeolites was shown to increase the crystallinity of PLA. The storage modulus measured by dynamic mechanical analysis (DMA) increased with increasing zeolite content. A similar trend was also observed for tan delta values. The tan delta peaks were much higher and broader for composites of PLA with zeolite compared with neat PLA, indicating improved damping properties. Thermal degradation behavior of PLA and PLA/zeolite composites was investigated using thermogravimetric analysis (TGA). The activation energies for thermal degradation (Ea) were estimated at different weight loss values of 2.5 – 90 % using Flynn-Wall-Ozawa Methods. The activation energies of PLA and PLA with 5 wt. % zeolite composites were found to be in the range of 110-119 kJ/mol and 71-101 kJ/mol, respectively.

205f Surface Modification of Wood Flour by Acid Chloride Esterification: Characterization and Performance of High Density Polyethylene/ Wood Flour Composites

Yongcheng Zhang[1], **Hossein Toghiani**[2], Charles U. Pittman Jr.[1], and Yibin Anna Xue[3]. (1) Department of Chemistry, Mississippi State University, Mississippi State, MS 39762, (2) Dave C. Swalm School of Chemical Engineering, Mississippi State University, Mississippi State University, MS 39762, (3) Center for Advanced Vehicular Systems, Mississippi State University, Starkville, MS 39759

Modification of wood fiber/flour (WF) surfaces has the potential to improve its compatibility with hydrophobic plastic matrices and to reduce water uptake in WF/Plastic composites. Wood flour was modified by esterification with octanoyl chloride and with palmitoyl chloride. The modified WF was analyzed by FT-IR. A higher degree of esterification was achieved in highly polar dimethylformamide (DMF) compared to the less polar methyl tert-butyl ether (MTBE). DMF penetrates into the fiber far more than MTBE, thereby increasing the number of –OH groups available for esterification. Increasing the acid chloride chain length from C8 to C16 decreased the extent of esterification. Longer chains cover surface –OH groups, retarding reactions with nearby hydroxyls after an esterification event. Longer chain acid chlorides also have lower reactivity and penetrate into the hydrophilic wood fiber more slowly. Modified wood flour surfaces were covered by a hydrophobic waxy layer of surface ester groups (SEM). The surfaces of modified wood flour and the fracture surfaces of WF/HDPE composites were studied by SEM. C8-modified wood flour (60 wt%)/HDPE composites exhibited a large drop in equilibrium water absorption after a 24 hr immersion compared to unmodified WF (60 wt%)/HDPE composites. However, esterified WF/HDPE composites exhibited lower flexural strengths and moduli. In contrast to octanoyl chloride modification, the addition of maleated polypropylene (MAPP) to WF/ HDPE composites improved composite mechanical performance.

206a Polylysine Modified Hydrogel Coatings to Enhance the Neuro-Electrode Interface

Shreyas Rao, Chemical and Biomolecular Engineering, the Ohio State University, 140 W 19th Ave, Columbus, OH 43210, Michael Owens, Chemical and Biomolecular Engineering, Ohio State University, 140 W 19th Ave, Columbus, OH 43210, and **Jessica O. Winter**, Chemical and Biomolecular Engineering/Biomedical Engineering, Ohio State Unversity, 140 W 19th Ave, Columbus, OH 43210.

Neural prostheses have significantly advanced treatment options for patients with traumatic injury or neurodegenerative disease. Prostheses typically consist of a signaling unit, which creates an electrical response to an external stimulus (e.g., light, sound, input from sensory nerves). This signal is then passed to an electrode array that replaces the function of missing or damaged nerves by providing the signal to remaining neural circuitry. One of the major limitations of these devices has been the inability to integrate electrode arrays with target neural tissue. Poor interfacing can result in incompatibilities of the electrode material, anatomical changes resulting from disease or injury progression, and the trauma of prosthetic implantation.

Our research is focused on improving electrode-host integration by applying tissue engineering strategies to prosthetic devices. To this end, we have developed electrode coatings consisting of hydrogels, cross-linked, hydrophilic, swollen polymers. Hydrogels make excellent brain mimetics because of their structural and mechanical similarity to glycosaminoglycans, which make up brain connective tissue. However, because of their synthetic nature, polymeric hydrogels often do not support cell adhesion. Our previous research efforts identified polylysine as a potential biomolecule to enhance neuron adhesion, extension, and regeneration at the biomaterial interface. Here we describe our efforts to incorporate polylysine into polymeric poly (ethylene glycol) – poly (caprolactone) (PEGPCL) hydrogels to enhance neurite extension.

PEGPCL hydrogels were synthesized as described previously [1,2] with modifications to enhance yield. Polylysine was conjugated to NHS-PEG-Acrylate through standard NHS bioconjugation methods producing a polylysine-PEG-acrylate compound, which could be incorporated into hydrogels through UV photopolymerization. The yield of polylysine conjugation was characterized using FITC-conjugated polylysine and quantified with a fluorescent plate reader. We are currently assessing the cell response to these materials using PC12 cells and primary rat cortical neurons. In particular, we are evaluating neuronal adhesion (cells/micron^2) and neurite extension produced for different levels of polylysine conjugation versus unconjugated control hydrogels. The coating described here could be used to enhance the electrode-host interface, possibly lowering electrical requirements for stimulation which will improve prosthesis resolution and size.

1. Sawhney AS, Pathak CP, Hubbell JA. Bioerodible Hydrogels Based on Photopolymerized Poly (Ethylene Glycol)-Co-Poly(Alpha-Hydroxy Acid) Diacrylate Macromers. Macromolecules 1993;26 (4):581-587.

2. Winter JO, Cogan SF, Rizzo III JF. Neurotrophin-Eluting Hydrogel Coatings for Neural Stimulating Electrodes. J Biomed Mater Res B 2007;81B(2):551-563.

206b Matrix Dimensionality Alters Integrin Signaling and Neurite Outgrowth

Andreia S. Ribeiro[1], Erin Voss[1], Elizabeth M. Powell[2], and Jennie B. Leach[1]. (1) Chemical and Biochemical Engineering, University of Maryland Baltimore County, 1000 Hilltop Circle, Baltimore, MD 21250, (2) Anatomy and Neurobiology, University of Maryland School of Medicine, Baltimore, MD 21201

Translating information from two-dimensional (2D) culture into three-dimensional (3D) systems has been a major hurdle in the use of tissue engineered biopolymers for neuronal repair applications. The ultimate goal of peripheral nerve tissue engineering is to provide rationally designed alternatives to grafted tissue through a deeper understanding of the dynamic 3D interactions between the cellular and extracellular environment. Neurons sense ligands in the surrounding extracellular matrix (ECM) and respond through integrin signaling pathways which influence cell fate and process outgrowth. Neural signaling response controls neurite outgrowth and is thought to be different in 2D and 3D systems, with in vitro 3D environments being a better representation of in vivo systems. While 2D substrates have been incredibly valuable in revealing intricacies of cell biology, previous studies with non-neuronal cells demonstrate that signaling pathways are dramatically altered when cells are placed in a 3D matrix. For example, integrin expression in migrating cells in 3D substrates more closely resembles the in vivo pattern. We hypothesize that 3D culture imposes changes in matrix ligand organization and alters neuronal behavior by modulating β1-integrin cytoskeletal signaling. We test our model using dorsal root ganglion (DRG) neurons isolated from E13.5 mouse embryos cultured in Type I collagen, a major component of the ECM and ligand for β1-integrin. We examine the expression patterns and phosphorylation of downstream molecules including Rho GTPases and focal adhesion kinase (FAK). Preliminary results show that 3D culture demonstrates more diffuse integrin mediated signaling as neurons are embedded in collagen gels when compared to the punctate signaling complexes formed in 2D culture in the location of point contacts. Tissue engineering principles are also being implemented to develop a new in vitro model system based on a synthetic polyethylene-glycol (PEG) crosslinked hydrogel that allows for independent tuning of the mechanical properties of the material (e.g. stiffness) and introduction of specific biological ligands. Culturing neurons in 3D in tunable materials offers vast potential for deciphering both the morphogenic and molecular response to the in vivo ECM. These results provide a foundation to design optimal biomaterials for controlled nerve growth which is critical for the development of therapeutics for nerve repair.

206c Reduction of Chondroitin Sulfate Proteoglycans to Increase Axonal Outgrowth In a Neuronal Co-Culture Model

Hannah M. Tuinstra[1], Amy C. Sebeson[2], and Lonnie D. Shea[1]. (1) Chemical and Biological Engineering, Northwestern University, 2145 Sheridan Rd, Tech E136, Evanston, IL 60208, (2) Department of Biochemistry, Molecular Biology and Cell Biology, Northwestern University, Evanston, IL 60208

Regeneration after spinal cord injury is limited in part by an environment that inhibits axonal outgrowth, and overcoming this barrier is a necessary step. In particular, deposition of an inhibitory molecule, chondroitin sulfate proteoglycan (CSPG), is increased through the activation of astrocytes at the injury site. Our laboratory is developing an *in vitro* neuronal-astrocyte co-culture model to investigate methods to degrade and prevent the production of CSPGs after spinal cord injury. Neu7, the astrocyte cell line used in this model was derived from rat cerebral cortex astrocytes, produces CSPGs and inhibits axon growth. Several approaches have been investigated to decrease CSPGs associated with Neu7 cells, including enzyme degradation, non-viral delivery of shRNAs, and siRNA. These approaches can be implemented by either encapsulating factors within poly(lactide-co-glycolide) bridges for localized release or by immobilization to the culture surface. Initial studies targeted degradation of CSPGs by treatment with the enzyme chondroitinase ABC. CS-56 staining for CSPGs revealed a 20% decrease in average fluorescence intensity relative to untreated cells. Subsequent experiments attempted to limit the production of CSPGs using shRNA. Neu7s with shRNA vectors against enzymes involved in assembling CSPGs (such as chondroitin polymerizing factor) could lead to a reduction in CSPGs and allow for greater axonal outgrowth from the co-cultured neurons. Neu7 cells were transfected with Transfast-pEGFPLuc lipoplexes resulting in luciferase transgene expression of 300 RLU per μ g protein, with 5% of the cells transfected. This low percentage of transfected cells was expected to have minimal effect on axonal outgrowth. The final approach was to employ siRNA to target enzymes involved in CSPG assembly, as siRNA is active within the cytoplasm and was expected to have greater efficacy. Preliminary results with siRNA transfection indicate that transcript levels can be reduced by 85%. Results will be presented for axonal outgrowth in the neuronal-astrocyte co-culture model. These studies will identify the most effective ways to overcome the CSPG inhibition of axonal outgrowth.

206d Microparticle-Mediated Delivery of Interleukin-10 Plasmid DNA for the Treatment of Neuropathic Pain

Ryan Soderquist[1], Evan Sloane[2], Scott Johnson[1], Linda Watkins[2], Erin Milligan[3], and Melissa Mahoney[1]. (1) Chemical and Biological Engineering, University of Colorado at Boulder, ECCH Rm. 111, 424 UCB, Boulder, CO 80309, (2) Department of Psychology and Center for Neuroscience, University of Colorado at Boulder, UCB 345, Boulder, CO 80309, (3) Department of Neurosciences, University of New Mexico, MSC08 4740, Albuquerque, NM 87131-0001

Neuropathic pain caused by inflammation and/or damage to peripheral nerves afflicts millions of people worldwide. Non-viral plasmid DNA (pDNA) encoding for the anti-inflammatory cytokine interleukin-10 (IL-10) can alleviate symptoms of neuropathic pain in animal models for more than 40 days when administered to the spinal cord. This approach, however, requires two high-dose injections over the course of three days which limits its clinical utility. The focus of this study was to improve the clinical feasibility of pDNA-IL-10 therapy by delivering pDNA-IL-10 via degradable PLGA microparticles. We hypothesized that microparticles with sufficient pDNA loading could achieve long-term therapeutic efficacy following a single injection with reduced pDNA dosage requirements by simultaneously eliciting phagocytic immune cell recruitment and providing prolonged pDNA-IL-10 exposure. To test the hypothesis, the extent to which phagocytic cell populations are stimulated by microparticles in vivo was examined by immunohistochemical analysis of cell populations isolated from cerebrospinal fluid and sectioned spinal cord tissue 24 hours after intrathecal delivery of fluorescently labeled microparticles. An increased presence of classically and alternatively activated macrophages (ED1 positive and ED2 positive cells) was observed in the cerebrospinal fluid after microparticle administration. In tissue sections, microparticles were distributed throughout the meningeal tissue surrounding the spinal cord and elevated levels of phagocytic immune cells were identified (MHCII and OX-42 positive) in areas of high microparticle accumulation. Overall, microparticles recruited and activated phagocytic cells in the intrathecal space, which is a previously demonstrated requirement for successful therapy using multiple doses of unencapsulated pDNA. When administered as a single in vivo intrathecal injection, microparticles loaded with a 14-fold lower dose of pDNA alleviated neuropathic pain in animal models for a prolonged time period (> 60 days), greatly enhancing the potential clinical utility of pDNA-IL-10 therapy for neuropathic pain.

Support: NIH grant DA018156

207a Development of Super-Low Fouling and Bacterial Resistant Surface Coatings

Matthew Bernards, Gang Cheng, and Shaoyi Jiang. Chemical Engineering, University of Washington, Benson Hall, Box 351750, Seattle, WA 98195

Surface coatings that resist non-specific protein adsorption and bacterial adhesion are important for a number of applications including biosensors, biomaterials, and marine coatings. Recent efforts have shown that surface coatings composed of a mixture of charged groups with an overall charge neutrality result in surfaces that are highly resistant to non-specific protein adsorption. Examples of polymers that show these super-low fouling characteristics include zwitterionic materials such as poly(sulfobetaine methacrylate) (pSBMA) and co-polymers formed from a mixture of charged monomers. It has been shown that low-fouling surface coatings composed of poly(oligo(ethylene glycol) (pOEG) reduce bacterial adhesion as well. In this work we show that these mixed-charge, super-low fouling polymer brush coatings present an excellent approach for preventing bacterial adhesion and biofilm formation.

In this study polymer brushes were grafted to gold coated glass slides using atom transfer radical polymerization (ATRP). The composition of the polymer brushes were characterized using electron spectroscopy for chemical analysis (ESCA), to determine reaction conditions that resulted in a statistical copolymer. The thickness of the polymer brush was characterized by atomic force microscopy (AFM) and the brushes were shown to be 25-35 nm in thickness. The non-specific adsorption of proteins to polymer brush coated surfaces was quantified using a surface plasmon resonance (SPR) biosensor. Solutions of 1 mg/mL of fibrinogen (FBG), lysozyme (LYZ), and bovine serum albumin (BSA) were used to confirm the super-low fouling characteristics of the polymer brushes. It was seen that there is no non-specific protein binding to these polymer brush surfaces following 10 minutes of adsorption. Bacterial adhesion of both *S. epidermidis* and *P. aeruginosa* was quantified after both 2 hours and 24 hours with fluorescent microscopy using a live/dead stain. In this study, it was seen that the pSBMA surface coating significantly decreased the bacterial adhesion as compared to the control glass surface. Additionally, it was shown that the pSBMA brushes reduced the bacterial adhesion to the same extent as pOEG brushes. Finally, the biofilm formation was characterized on the polymer brush coated surfaces after 24 hours. It was seen that the mixed charge surface coatings greatly reduced the biofilm formation as compared to the control surfaces. Additionally, the biofilm formation was reduced to a similar level as that seen for pOEG surface coatings. This indicates that mixed charge polymer brushes composed of either zwitterionic or mixed charge co-polymers represent a very promising surface coating for the next generation of non-fouling materials.

This work has shown that super-low fouling polymer brush coatings can be prepared from both zwitterionic and mixed charge co-polymers. These surface coatings greatly reduce non-specific protein binding as well as the adhesion and accumulation of bacteria. Due to the susceptibility of pOEG to oxidation damage, these mixed charge materials represent a promising new approach for the development of bacteria resistant coatings.

207b Chemical Cross-Linking of Select Polyelectrolyte Nanofilms to Control Mechanical Properties for Use In Cell Contacting Applications

Jennifer A. Phelps and Paul R. Van Tassel. Chemical Engineering, Yale University, New Haven, CT 06520

Polyelectrolyte nanofilms, formed by the layer-by-layer (LbL) assembly of oppositely charged polymers, are promising biomaterials due to their highly tunable film properties (i.e. viscosity, elasticity, thickness, degree of hydration). While film properties may be controlled through solution variables (pH and ionic strength) and choice of polymers, the controlled formation of chemical cross-links between functional groups on the polymers offers perhaps the greatest degree of property manipulation. We investigate here the influence on polymer film mechanical and cell-contacting properties brought about by standard chemical cross-linking employing EDC-NHS chemistry, where carboxylate groups on polyanion species within the film are activated and spontaneously form cross-links with amine groups on polycation species. We also investigate the possibility to confine cross-linking to the outer region of the film by adding a few layers of activated polyanion to the top of a film. (This type of inhomogeneous cross-linking is motivated by a desire to achieve films that are simultaneously rigid and bioactive.) We characterize film mechanical properties in situ using quartz crystal microgravimetry with dissipation (QCM-D) and optical waveguide lightmode spectroscopy (OWLS), and investigate cell-contacting properties using human hepatocellular carcinoma (HepG2) cells.

207c Resorbable Polyurethane/bone Composites for Bone Tissue Engineering

Jerald Dumas and Scott A. Guelcher. Chemical Engineering, Vanderbilt University, 107 Olin Hall, 2400 Highland Avenue, Nashville, TN 37212

There is a recognized need for improved biomaterials that promote orthopedic healing. Traditional care includes metallic rods and bioceramics, which are typically bioinert. These options provide a satisfactory level of clinical outcome for typical civilian applications. However, some cases such as those in the battlefield environment introduce an elevated level of fracture management with challenges such as infection. We are addressing elevated concerns and aim to improve orthopedic fracture management with bone/polymer composite biomaterials that will facilitate fracture reduction and enhance biological healing and biomechanical function.

Composites synthesized from bone and thermoplastic polymers have been shown to remodel in vivo.[1] However, the polymers must be heated above the glass transition or melting temperature and compression-molded to achieve high mechanical properties. We are pursuing a reactive liquid molding approach wherein bone/polyurethane composites are mixed and compression-molded at ambient temperature, thus rendering them useful for injectable applications as well. Furthermore, by enhancing the reactivity of the surface of the bone particles, the interfacial binding, and therefore the mechanical properties, can be improved.

We have synthesized bone/polyurethane (PUR) composites from lysine diisocyanate (LDI), polycaprolactone (molecular weight 300) (PCL 300), and mineralized bone powder (100-500 microns) (MBP). Polymers synthesized from LDI and polyester polyols have been reported to support cell attachment and biodegrade to non-cytotoxic degradation products.[2-4] We have synthesized composites incorporating 75 wt% MBP over a range of isocyanate indices (index = 100 x NCO:OH equivalent ratio). Mechanical properties were measured using compression and 3-point bending. DMA frequency sweeps have shown that surface demineralization enhances storage modulus by up to 1 GPa. Composites with compression modulus up to 2.5 GPa and compression strengths up to 80 MPa have been prepared. Our results show that index has a large effect on mechanical properties, with a maximum in the modulus occurring at an index of 140. The glass transition temperature was observed to be 65 ¢ªC using DMA, while it was observed to be 51 ¢ªC using DSC. Degradation experiments show that the composites retain greater than 90% of their original mass over a 12 week period, and the rate can be changed by manipulating the backbone of the polyol. In vitro studies have shown that the MBP/PUR composites support MC3T3 osteoprogenitor cell attachment. A 3 rabbit in vivo study was completed in January 2008 in which MBP/PUR composite plugs were implanted into 6 mm femoral defects. The rabbit femurs were extracted, and there were no signs of infection. Histology results are currently being processed to confirm bone resorption. These preliminary studies suggest that MBP/PUR composites are promising biomaterials for bone tissue engineering.

1. Boyce TM, Winterbottom JM, Lee S, Kaes DR, Belaney RM, Shimp LA, Knaack D. Cellular Penetration And Bone Formation Depends Upon Allograft Bone Fraction In A Loadbearing Composite Implant. 2005. p 133.

2. Guelcher SA, Patel V, Gallagher K, Connolly S, Didier JE, Doctor J, Hollinger JO. Synthesis and biocompatibility of polyurethane foam scaffolds from lysine diisocyanate and polyester polyols. Tissue Eng 2006;12(5):1247-1259.

3. Guelcher SA, Gallagher KM, Srinivasan A, McBride SB, Didier JE, Doctor JS, Hollinger JO. Synthesis, in vitro biocompatibility and biodegradation, and mechanical properties of two-component polyurethane scaffolds: effects of water and polyol composition. Tissue Eng In Press.

4. Guelcher SA, Didier JE, Srinivasan A, Hollinger JO. Synthesis, mechanical properties, biocompatibility, and biodegradation of cast poly(ester urethane)s from lysine-based polyisocyanates and polyester triols. Manuscript in preparation.

Polysaccharide-Based Tissue Adhesives for Closure of Surgical Wounds

Sujata K. Bhatia, **Samuel D. Arthur**, **H. Keith Chenault**, **Garret D. Figuly**, **Sharon L. Haynie** and **George K. Kodokian**, Materials Science and Engineering, DuPont Central Research & Development, Experimental Station, E328, P.O. Box 80328, Wilmington, DE 19880-0328

Reliable closure of surgical wounds remains one of the great unsolved challenges of clinical medicine. Despite refinements in suturing and stapling techniques for wound closure, physicians continue to struggle with the problem of leakage from internal wounds. For example, gastrointestinal wound closures have a leakage rate of 3-15%, resulting in dire clinical consequences (infection, hemorrhage, etc.) and a 2-3% mortality rate for patients undergoing gastrointestinal surgeries. There is great demand for tissue adhesives to augment or replace sutures and staples for internal wound repair.

Several types of tissue adhesives have been developed and commercialized, but these adhesives have generally not been successful due to biocompatibility or performance problems; for example, cyanoacrylate-based adhesives release toxic degradation products, and fibrin sealants exhibit poor adhesion and pose a risk of viral infection. We have developed the ActaMax™ sealant family of hydrogel tissue adhesives, formed by reacting an oxidized polysaccharide with a water-dispersable multi-arm polyether amine. Specifically, we have developed tissue adhesives composed of two components: dextran aldehyde and multi-arm PEG amine, which undergo a Schiff base reaction to form a crosslinked hydrogel (Figure 1). This two-component tissue adhesive system crosslinks in water, cures rapidly (<1 min) at room temperature, adheres to moist tissue, and degrades hydrolytically. The adhesive is non-cytotoxic to fibroblasts, and does not induce inflammatory activation of macrophages. The adhesive is also advantageous in that it does not pose a risk of viral contamination, and is based on polymers with a long history of clinical use as plasma expanders (dextran) and drug delivery agents (PEG).

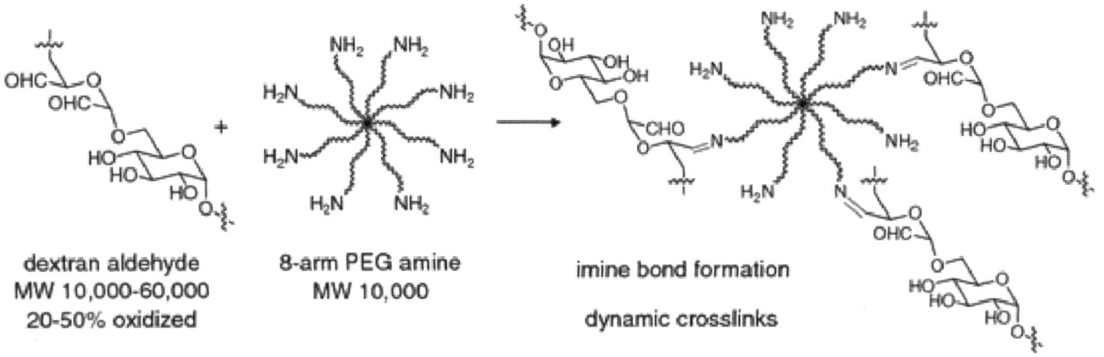

Figure 1. Foundation chemistry for polysaccharide-based tissue adhesives. The oxidized polysaccharide dextran aldehyde reacts with an 8-arm star PEG amine to form a crosslinked hydrogel network.

The foundation chemistry based on dextran aldehyde and PEG-amine allows fine-tuning of sealant properties to meet surgeon needs for specific clinical targets. ActaMax™ sealant demonstrates strong tissue adhesion (as strong as or better than that of all current commercial internal sealants in a burst pressure test). Additionally, ActaMax™ sealant demonstrates tissue adhesion that is substantially stronger than that of fibrin, the leading incumbent sealant. Our sealant is well-tolerated in short-term and long-term intestinal painting studies; the sealant remains on the target site with no injury to adjacent tissue. A multi-organ scouting study demonstrated that ActaMax™ tissue adhesive was efficacious in a variety of clinical procedures (Figure 2) including aortic graft, aortic puncture, aortic anastomosis, graft puncture, cardiac painting, cardiac puncture, coronary artery incision, intestinal anastomosis, liver lobectomy, splenectomy, lung lobectomy, and hernia patch attachment. ActaMax™ tissue sealant is a versatile bioadhesive platform that can be utilized in multiple surgical applications.

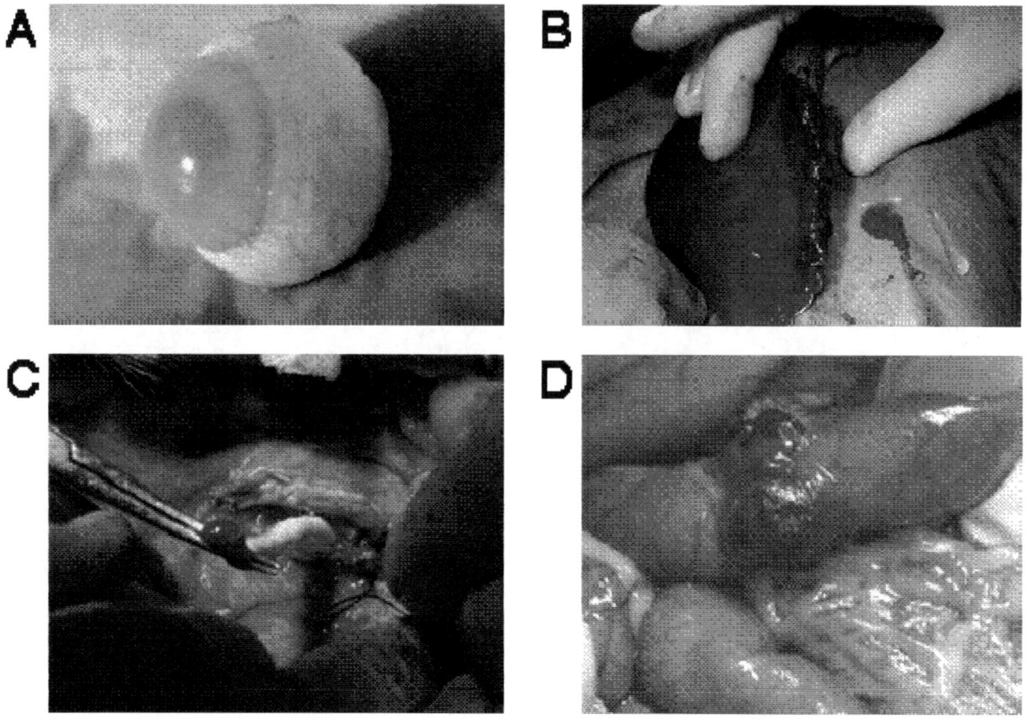

Figure 2. Use of ActaMax™ tissue adhesive for (A) cataract incision closure; (B) hemostasis following liver lobectomy; (C) vascular anastomosis closure; and (D) gastrointestinal anastomosis closure.

References:

S. K. Bhatia, S. D. Arthur, H. K. Chenault, G. K. Kodokian, "Interactions of Polysaccharide-Based Tissue Adhesives with Clinically Relevant Macrophage and Fibroblast Cell Lines," *Biotechnology Letters* (2007), 29:1645-1649.

S. K. Bhatia, S. D. Arthur, H. K. Chenault, G. D. Figuly, G. K. Kodokian, "Polysaccharide-Based Tissue Adhesives for Sealing Corneal Incisions," *Current Eye Research* (2007), 32:1045-1050.

207e Controlled Axon Regeneration In the Spinal Cord

Jason Coleman and Anthony Lowman. Chemical and Biological Engineering, Drexel University, 3141 Chestnut St., Philadelphia, PA 19104

Spinal cord injury (SCI) affects over 250,000 Americans a year and current clinical solutions focus primarily on improving quality of life. SCI results from a physical tear or compression of spinal tissue causing damage to axons, potentially resulting in various levels of paralysis. One of the most promising and clinically relevant strategies for axon reconnection is surgical implantation of a peripheral nerve graft (PNG) to bypass the injury sight. The main limitation towards the success of a PNG is the inability of the nerve to reach and reconnect with existing nerves. Growth factors, such as BDNF and NT-3 have been shown to help promote axon growth and regeneration post injury. By developing novel nanoparticulate delivery systems, a controlled release of these trophic factors can be created. Preliminary biodistribution studies have shown that nanoparticles stereotaxically injected into the spinal tissue remain at the sight of injection. Injections of nanoparticles loaded with growth factors can be used to create a concentration gradient by injecting at several locations around the injury site as well as the PNG. The release profiles of the growth factors from the particles are modified to allow for the creation of the gradient. The gradient will allow for the directional cues necessary for connection of the implanted PNG to injured axons. In order for reconnection to occur, the PNG must grow towards the existing nerves for several weeks. Thus, the release rates of growth factors from the particles must range from hours to weeks. Release rates from the particles are controlled by varying the type of polymers used, the fabrication process and the degree and method of crosslinking within the particles. Examples of polymers used for particle formation include hydrophobic polymers such as poly (lactic acid) (PLA) and poly (lactic-co-glycolic acid) (PLGA) and hydrophilic polymers such as poly (ethylene glycol dimethacrylate) (PEGDM) and poly (vinyl alcohol) (PVA). Several types of emulsion fabrication process are used and polymer crosslinking is performed by both physical and chemical means. Nanoparticles will be modified to maximize growth factor loading, control release, maintain growth factor bioactivity, minimize macrophage uptake, and ultimately, cause directional nerve growth and reconnection in vivo.

207f Towards the Rational Design of Biodegradable Controlled Release Vehicles

Sam N. Rothstein, William Federspiel, and Steven Little. Department of Chemical Engineering, University of Pittsburgh, 1237 Benedum Hall, Pittsburgh, PA 15261

Controlled release technology presents solutions to the problems of patient compliance and sub-optimal dosing that often plague modern pharmaceuticals. This technology is commonly manifested as biodegradable matrices that gradually release drug as the structure degrades and dissolves, obviating the need for removal. As therapeutics, these matrices, which form the basis of a $110 billion dollar industry, have simplified dosing schedules from daily to monthly or even biannual treatments. However, the design of controlled release therapeutics remains largely heuristic despite nearly 30 years of research, resulting in substantial development costs in the form of exploratory, in vitro release studies. At present, thousands of medications that could offer improved performance as controlled release formulations await development.

We have recently reported the first, broadly applicable, predictive model for biodegradable matrices composed of the most widely used polymers (e.g. polyesters and polyanhydrides). Unlike prior attempts to predict release, the published model uses readily measured design parameters to calculate the in vitro release profiles for a wide variety of agents, including small molecules, peptides and proteins without regression. As all these parameters can be controlled or set with standard microparticle fabrication methods (single/double emulsion processing, melt casting, ect.), it should be possible to produce formulations to match particular release profiles.

As one example, the model suggests that it is possible to precisely control the magnitude and timing of subsequent doses in a burst-lag-burst release profile (e.g. the duration of the lag phase). Such a profile has significant potential to replicate proven vaccine dosing schedules without repeat injections. To date, formulations where the timing of controlled release bursts match the desired vaccine dosing schedule have proven elusive. Using the published model, design parameters were computed for microspheres based on a common murine vaccination schedule. When fabricated to these specifications, release assays showed that dosing from these microspheres did indeed match the desired schedule within the bounds of experimental error.

As another example, theoretical predictions from this model also suggest that a sustained, constant rate of drug delivery may be obtained from bulk eroding polyester microspheres. This release behavior was previously thought to only be obtainable using surface eroding systems. Efforts are currently underway to experimentally confirm such precise specifications for microparticle formulations.

In conclusion, initial experimental evidence shows that our newly published model can serve as a powerful tool for the inexpensive and rapid design of select therapeutics. Using this model to guide matrix fabrication, it may be possible to both improve upon current formulations and also enable the design of new formulations to address clinical needs in the area of controlled release.

250a Effects of Multi-Valent Ionic Interactions on Polyelectrolyte Brushes

Robert Farina and Matthew Tirrell. Chemical Engineering, University of California Santa Barbara, Santa Barbara, CA 93101

The focus of this work is on the effects of multi-valent ions on the behavior of end-tethered polyelectrolyte brushes. The uses and applications of polyelectrolyte brushes are vast, appearing in both nature (e.g., surfaces of cartilage and mammalian lung interiors) and commercially (e.g., skin care products, shampoo, and surfaces of medical devices). This makes understanding the physical properties of these brushes important for many different applications. Using the Surface Forces Apparatus (SFA), we are able to study the forces of interaction between two polyelectrolyte brushes end-tethered to cylindrical surfaces. Depending upon the surrounding environment, which plays a critical role in brush behavior, the brushes we study can exhibit a wide range of physical properties.

In the presence of mono-valent counterions, polyelectrolyte brushes demonstrate purely repulsive forces when moved into contact. As the concentration of mono-valent ions in solution is decreased, electrostatic screening between the charged monomer segments of the polyelectrolyte chains is also decreased allowing the chains to extend further. These swollen chains at lower mono-valent salt concentrations also exhibit force versus distance profiles which reveal a softer brush. However, when multi-valent ions are added into solution, the force profiles portray a much more interesting story. The presence of multi-valent ions creates adhesive forces between the brushes after contact, which vary depending upon concentration. While the multi-valent counterions also display repulsive forces, it is the adhesion between brushes which allows for a much wider range of physical properties. The relevance of this work exists in the regularity at which multi-valent ions exist physiologically and in commercial applications.

250b Highly Conductive Ionic Liquid-Homopolymer Mixtures

Liang Gwee[1], David Salas-de la Cruz[2], Karen I. Winey[2], and Yossef A. Elabd[1]. (1) Chemical and Biological Engineering, Drexel University, 3141 Chestnut St., Philadelphia, PA 19104, (2) Materials Science and Engineering, University of Pennsylvania, 3231 Walnut Street, Philadelphia, PA 19104

Polymer electrolytes are of great interest due to their application in membrane-based separations, fuel cells, sensors, and actuators, where high ionic transport is desired for process efficacy. However, ionic conductivity is usually a strong function of water content. Therefore, dry or high temperature operation results in dehydration and consequently severe performance losses. This project aims to replace water as the medium for ionic transport with a class of organic salts commonly known as ionic liquids. In particular, room temperature ionic liquids are of interest due to their negligible vapor pressures, high degradation temperatures, large electrochemical windows, and high ionic conductivities. Successful implementation of ionic liquids as a substitute for water will allow for water-free operation of current polyelectrolyte membranes, expanding their range of operating conditions and utility.

In this study, a number of ionic liquid-homopolymer mixtures were investigated as a function of ionic liquid content. Several ionic liquids were explored, where the chemical compatibility with the polymer has a significant effect on the resulting ionic conductivity. For compatible (miscible) ionic liquid-homopolymer systems, the room temperature conductivity increased by seven orders of magnitude with increasing ionic liquid volume fraction. Also, the ionic liquid acts as a plasticizer, where there is a step change in the ionic conductivity at the ionic liquid content that corresponds with the glass transition temperature. For partially miscible and immiscible ionic liquid-homopolymer mixtures, the room temperature conductivity did not significantly change with increasing ionic liquid volume fraction. The phase behavior of these systems was confirmed with differential scanning calorimetry and scanning electron microscopy. Transport measurements will be coupled to static and in situ morphological studies. These results provide insights into the interactions between ionic liquid and polymers that will help with the design of new ionic liquid-polymer systems and new poly(ionic liquid)s to address key questions about ion transport.

250c Solute Induced Phase Transitions In Nafion

Jay B. Benziger[1], Paul W. Majsztrik[2], Christine Ranney[3], and Andrew B. Bocarsly[2]. (1) Department of Chemical Engineering, Princeton University, Princeton, NJ 08544, (2) Chemistry, Princeton University, Princeton, NJ 08544, (3) Chemical Engineering, Princeton University, Princeton, NJ 08544

The elastic modulus and creep of 1100 equivalent weight Nafion films were examined under environmental conditions of 20-90ºC and solute vapor activity of 0-1. At room temperature water and methanol plasticize Nafion reducing its elastic modulus and increasing its creep rate. However, at higher temperatures solvent absorption leads to a large reduction of the creep rate! The reversal of the solvent effect on the mechanical properties of Nafion between low and high temperature suggest that the solvents alter the microphase separation in Nafion. We propose that absorbed solvents alter the volume fraction of hydrophilic vs. hydrophobic phases in Nafion resulting in microphase evolution from dispersed, to cluster to cylindrical phases with increased solvent activity. The microphase evolution can dramatically alter the mechanical properties.

Web Page: http"//pemfc.princeton.edu

250d Polymers Incorporating Pi-Conjugated Small Molecules: Processing Characteristics, Impact on Electronic Properties, and Resistance to Ionizing Radiation

Robert J. Klein, John L. Schroeder, Shannon M. Cole, Michael E. Belcher, Phillip J. Cole, and Joseph L. Lenhart. Organic Materials, Sandia National Laboratories, P.O. Box 5800, MS-1411, Albuquerque, NM 87122

Polymers capable of specialized electronic properties, such as radiation resistance, anti-static coatings, enhanced breakdown strength, or tailored photo-electronic interactions, are needed for a new generation of organic devices. We have examined the method of impregnating polymer films by pi-conjugated small molecules from solution. This significantly modifies the electronic properties of the films, which we utilize for radiation tolerance from ionizing radiation. We have also thoroughly characterized the process using a physical model based on Fickian diffusion and solute partitioning, and determined a predictive model for the partitioning behavior using the activity coefficient and chi parameter.

Acknowledgement: Sandia is a multiprogram laboratory operated by Sandia Corporation, a Lockheed Martin Company, for the United States Department of Energy's National Nuclear Security Administration under contract DE-AC04-94AL85000.

Characterization of the Temperature Dependent Chemical and Mechanical Properties of a Diels-Alder Based Crosslinked Polymeric Material

Brian Adzima, Christopher J. Kloxin, Timothy F. Scott, and Christopher N. Bowman

A cross-linked polymeric material was formed by the Diels-Alder reaction between trisfuran and bismaleimide monomers. At elevated temperatures, chemical equilibrium shifts from the Diels-Alder reaction to the retro-Diels-Alder reaction, resulting in depolymerization. As a consequence, the material is both reversibly gelled and vitrified, existing as a liquid, polymer gel, or polymer glass, depending on the temperature and relative reaction progress. FTIR spectroscopy was used to characterize the conversion of furan and maleimide functionalities to the oxy-norbornene adduct. Equilibrium conversion of the furan and maleimide varied from 74 % at 85 degrees Celsius to 24% at 155 degrees Celsius, demonstrating significant reversion via the retro-Diels-Alder reaction. At lower temperatures vitrification and the subsequent mass transfer restrictions limit the extent of reaction rather than equilibrium. As a consequence the highest conversion realized was 87% at 45 degrees Celsius. The thermomechanical properties of the material were characterized by rheometry. Both the storage and loss moduli were found to decrease with increased temperature due to decreasing crosslink density, a result of the retro-Diels-Alder reaction. The gel point, as determined by the Chambon-Winters criterion, occurs at 92 degrees Celsius, corresponding to a 71% conversion of functional groups and consistent with the Flory-Stockmeyer equation for a stoichiometric mixture of A3 and B2 monomers.

Highly Selective Enzymatic Ring-Opening Polymerization: Syntheses and Characterizations of Thermoplastic Di-Block Co-Polyesters Containing Poly[(R)-3-Hydroxybutyrate] and Poly(ε-Caprolactone) Blocks

*Shiyao Dai and Zhi Li**

Department of Chemical & Biomolecular Engineering

National University of Singapore, 4 Engineering Drive 4, Singapore 117576

* Corresponding Author, E-mail: chelz@nus.edu.sg, Tel: +65-65168416, Fax: +65-67791936

ABSTRACT: Enzymatic modification of a microbial polyester was achieved by the ring-opening polymerization of ε-caprolactone (CL) with low-molecular weight telechelic hydroxylated poly[(*R*)-3-hydroxybutyrate] (PHB-diol) as initiator and Novozym 435 (immobilized *Candida antarctica* Lipase B) as catalyst in anhydrous 1,4-dioxane or toluene. The ring-opening polymerization was investigated at different conditions with two different types of PHB-diols: PHB-diol(P) containing a primary OH and a secondary OH end groups; and PHB-diol(M) consisting of 91% PHB-diol(P) and 9% PHB-diol containing two secondary OH end groups. The reactions were followed by GPC analyses of the resulting polymers at different time points, and the optimal conditions were established to be 70°C at a weight ratio of CL/enzyme/solvent of 8:1:24. The ring-opening polymerization of CL with PHB-diol(M) (M_n of 2380, NMR) at the molar ratio of 50:1 under the optimal conditions in 1,4-dioxane gave the corresponding poly[HB(56wt%)-*co*-CL(44wt%)] with M_n (NMR) of 3900 in 66% yield. Polymerization of CL and PHB-diol(P) (M_n of 2010, NMR) at the same condition in toluene gave the corresponding poly[HB(28wt%)-*co*-CL(72wt%)] with M_n (NMR) of 7100 in 86% yield. Both polymers were characterized by ^{1}H- and ^{13}C-NMR and IR analyses as *di*-block co-polyesters containing a PHB block with a secondary OH end group and a poly(ε-caprolactone) (PCL) block with a primary OH end group. NMR analyses and control experiments suggested no formation of random co-polymers and no change of the PHB block during the reaction. The enzymatic ring-opening polymerization was selectively initiated by the primary OH group of PHB-diol, whereas the secondary OH group remained as an end group in the final polymers. The thermal properties of the *di*-block poly(HB-*co*-CL)s were analyzed by DSC, with excellent T_g values for the elastomer domain: poly[HB(56wt%)-*co*-CL(44wt%)] with M_n (NMR) of 3900 demonstrated a T_g of -57°C, T_m of 145, 123, and 53 °C; and poly[HB(28wt%)-*co*-CL(72wt%)] with M_n (NMR) of 7100 gave a T_g of -60°C, T_m of 147 and 50 °C. Thus, the selective enzymatic ring-opening polymerization with PHB-diol as macro-initiator provides a new method for the preparation of PHB-based block copolymers as biomaterials with good thermoplastic properties and novel structures containing functional end groups.

Introduction

Poly[(*R*)-3-hydroxyalkanoates] (PHAs) are biodegradable and biocompatible polymers produced by a wide variety of microorganisms, with potential application in environmental, agricultural, marine, and biomedical fields.[1-6] Microbial poly[(*R*)-3-hydroxybutyrate] (PHB) is the most prominent polyester in the PHA family and can be easily produced in large quantity. PHB may be useful for drug delivery or tissue engineering, but its application as thermoplastic material is rather limited partially due to the high melting temperature (T_m) of 175-177°C and high glass transition temperature (T_g) of 4°C.[6] Many methods have been used to prepare PHB-based co-polymer to improve the thermoplastic properties. Bacterial syntheses by feeding with different substrates allowed for the preparation of random co-polyesters containing PHB and other PHA such as poly[(*R*)-3-hydroxyvalerate(HV)],[7] poly[(*R*)-3-hydroxypropionate],[8] poly[(*R*)-3-hydroxyhexanoate],[9] or poly[(*R*)-3-hydroxydecanoate].[10] Bacterial production of poly(HB-*co*-HV) containing 30% block units was also reported.[11] Chemical modification of PHB led to the preparation of block-co-polyesterurethane[12-15] and block-co-polyesters[16-17] containing the hard PHB blocks and other soft blocks such as poly(ε-caprolactone) (PCL),[12-13,17] poly[(*R*)-3-hydroxyoctanoate],[14,16] or poly(ethylene glycol)[15] blocks. A random co-polyester poly(HB-co-CL)[18] were also prepared by chemical modification of PHB. Some of these co-polymers demonstrated good thermoplastic properties. Nevertheless, microbial syntheses generally give random co-polyesters with relatively high production costs; and chemical modifications involve the use of toxic chemicals or catalysts.

We are interested in the enzymatic modification of microbial polyesters such as PHB to prepare block-co-polymers with good thermoplastic properties for biomedical application. Enzyme catalysis is highly chemo-, regio-, and stereo-selective, thus being a useful tool for the preparation of polymers with novel structures. Enzyme catalysis is *non*-toxic, which is very attractive for the preparation of polymeric biomaterials. Many enzymes have been used for the preparation of natural or unnatural materials,[19] and lipase-catalyzed ring-opening polymerizations (ROP) represent prominent examples.[20-22] Enzymes such as *Porcine pancreatic* lipase,[20,23-24] *Pseudomonas sp.* lipase,[25] and *Candida antarctica* lipase B (CALB)[26-30] have been successfully used for the ROP of lactones with alcohols,[20,23-25] diols,[29] and polyols[30] as initiators to prepare polyesters. Nevertheless, enzymatic ROP has not yet been applied for the modification of microbial polyesters.

Previously Suter,[12,13] we,[14,16] and others[15] successfully utilized low-molecular-weight telechelic hydroxylated PHB (PHB-diol)[12] as hard segment for chemical preparation of thermoplastic block-co-polymers. PHB-diol containing a primary and a secondary OH end group might also be a suitable initiator for enzymatic ROP. While both OH end groups were reacted in chemical polymerization,[12-16] the primary and the secondary OH end groups may show different reactivity in a lipase-catalyzed ROP. Thus, it could be possible that only the primary OH of the PHB-diol initiates the ROP giving a *di*-block copolymer with the unreacted secondary OH group as an end group. On the other hand, PCL is a biodegradable and biocompatible soft material with a T_m of 60°C and T_g of -60°C.[31] Although random co-polyester poly(HB-co-CL) prepared from PHB and PCL *via* enzyme-catalyzed transesterification[32-33] did not show improved elastic properties, incorporation of PCL block into PHB-based block co-polymers should significantly improve the elastic properties, which has been demonstrated in the chemically prepared block co-polyesterurethanes.[12-13] Recently, we have explored, for the first time, the lipase-catalyzed ROP of ε-caprolactone with PHB-diol to prepare block co-polyester poly(HB-co-CL). Here we report our results on this new and selective enzymatic ROP, the preparation of the corresponding block copolymers, the structural analysis of the novel *di*-block co-polyesters, and the characterization of physical properties of the polymers.

Experimental Section

Materials. Novozym 435 (immobilized *Candida antarctica* lipase B, 10000 PLU/g) was purchased from Novozymes. Telechelic hydroxylated poly-[(*R*)-3-hydroxybutrate], PHB-diol(M) (M_n of 3000, GPC, single peak), was a gift from Dr. P. Neuenschwander at ETH Zurich. Microbial PHB (>99%), dibutyltin dilaurate (95%), diglyme (99.5%), ε-caprolactone (99%), 1,4-dioxane (99.8%), and toluene (99.8%) were purchased from Aldrich. Ethylene glycol (99%), chloroform (GC, >99%), and *n*-hexane (HPLC, 99%) were obtained from Merck. Novozym 435 and PHB-diol were dried under vacuum at 50°C for 12 hours before use. ε-Caprolactone was freshly distilled over CaH_2 (83°C, 1.7 mmHg). 1,4-Dioxane and toluene were dried by refluxing over Na/benzophenone under argon.

Telechelic hydroxylated poly-[(*R*)-3-hydroxybutrate], PHB-diol(P): PHB-diol (P) was prepared by using the known procedure for the preparation of PHB-diol (M)[12] with large excess of ethylene glycol to PHB. Transesterfication of PHB (8.0 g) and ethylene glycol (40 mL) in the presence of dibutyltin dilaurate (118 mg) in diglyme (32 mL) was performed at 135 °C for 2 h. The mixture was then poured into cold water, and the precipitate was collected and washed by water (50 ml x 3) to remove the excess ethylene glycol. The crude product was dissolved in chloroform followed by precipitation at 4 °C by the addition of *n*-hexane. The product was separated by filtration and dried at 50°C under vacuum for 12 h to give 6.2 g of PHB-diol (P) in 76% yield. The molecular weight (M_n) was determined by GPC as 3700 g/mol with M_w/M_n of 1.38. The physical properties were determined by DSC with T_m of 149 and 134°C and T_g of -5.0°C. The chemical structure was identified by ^{1}H-NMR as a PHB-diol with a primary OH group on one end and a secondary OH group on the other end.

General procedure of enzymatic ring-opening polymerization of ε-caprolactone with PHB-diol: Novozym 435 (20-160 mg immobilized enzyme) and PHB-diol(M) (M_n of 3000, GPC; 88-212 mg) or PHB-diol(P) (M_n of 3700, GPC; 33-186 mg) were added to a dry schlenk tube containing a magnetic stirring bar and activated 4Å molecular sieves, and then dried at 50°C under vacuum for 12 h. Under argon atmosphere, the freshly distillated ε-caprolactone (400-600 mg) and freshly dried 1,4-dioxane or toluene (1.2 g-3.6 g) were added into the schlenk tube using a dry syringe. The mixtures were stirred under argon atmosphere at room temperature, 50°C, or 70°C and samples (50 μL) were taken at regular time intervals and analyzed by GPC. The reaction was stopped at 8-48 h, 10 mL chloroform was added, and the enzyme was removed by filtration. The solvent in the filtrate was removed under reduced pressure, and the product was dissolved in chloroform and then precipitated by adding *n*-hexane or methanol. The results are summarized in Table 1-2.

Preparation of poly[(*R*)-3-hydroxylbutyrate(56wt%)-*co*-ε-caprolactone(44wt%)]: According to the procedure described above, reaction of PHB-diol(M) (M_n of 3000, GPC; 212 mg) and ε-caprolactone (409 mg) with Novozym 435 (40 mg immobilized enzyme) as catalyst was carried out in 1,4-dioxane (1.6 g) at 70°C, and samples (50 μL) were taken at regular time intervals and analyzed by GPC. After 48 h polymerization, the reaction was terminated by the addition of 10 mL chloroform followed by the removal of Novozym 435 through filtration. 1,4-dioxane and chloroform were removed by evaporation under reduced pressure. The raw product was dissolved in 2 mL chloroform, treated with 18 mL methanol, and then precipitated at 4°C for 12 h. After removal of the solvent by filtration, the precipitates were dried at first by evaporation under vacuum and then in a vacuum oven at 50°C for 24 h. This gave 410 mg (66% yield) of the polymer. The molecular weight M_n was determined by GPC as 5400 (M_w/M_n of 1.63), and the structure was analyzed by NMR and IR as *di*-block co-polyester poly(HB-*co*-CL) with two different OH end groups. The ratio of PHB and PCL block was established as 56/44 (wt/wt) based on M_n (NMR) of PHB-diol(M) and the polymer. The physical properties were determined by DSC and WAXD with T_m of 145°C, 123°C, and 53°C, T_g of -57°C, and crystallinality of 19%.

Preparation of poly[(*R*)-3-hydroxylbutyrate(28wt%)-*co*-ε-caprolactone(72wt%)]: According to the procedure described above, reaction of PHB-diol(P) (M_n of 3700, 185 mg) and ε-caprolactone (632 mg) with Novozym 435 (76 mg immobilized enzyme) as catalyst was carried out in toluene (1.8 g) at 70°C, and samples (50 μL) were taken at regular time intervals and analyzed by GPC. After 16 h polymerization, the reaction was terminated by the addition of 10 mL chloroform followed by the removal of Novozym 435 through filtration. Toluene and chloroform were removed under reduced pressure with a rotary evaporator. The raw product was dissolved in 2 mL chloroform, treated with 18 mL methanol, and then precipitated at 4°C for 12 h. After removal of the solvent, the precipitates were dried at first by evaporation under vacuum and then in a vacuum oven at 50°C for 24 h. This gave 702 mg (86% yield) of the polymer. The molecular weight was determined by GPC as 7900 (M_w/M_n of 1.90), the structure was analyzed by NMR and IR as *di*-block co-polyester poly(HB-*co*-CL) with two different OH end groups. The ratio of PHB and PCL block was established as 28/72 (wt/wt) based on M_n (NMR) of PHB-diol(P) and the polymer. The physical properties were determined by DSC with T_m of 149°C and 54°C, and T_g of -61°C.

Gel permeation chromatography (GPC). Molecular weight analysis (M_n and polydispersity index M_w/M_n) was performed by using a Waters instrument, with Waters 510 pump, Waters 410 refractive index detector, and Waters HR4E, HR5E and HR6 columns placed in series. THF was used as the eluent at a flow rate of 1.0 mL/min at 30°C. Sample concentration was about 0.1% (w/v) and the injection volume was 100 μL. Polystyrene standards with molecular weights of 1310, 2970, 13900, 30200, 197000 and 696000 g/mol were used to generate a calibration curve.

Nuclear magnetic resonance (NMR). ^{1}H-NMR (500 MHz) and ^{13}C-NMR (125 MHz) spectra were recorded with a Bruker AMX500 NMR instrument in $DMSO_{d6}$ at 333K. Chemical shifts were referred to TMS at 0 ppm.

Differential scanning calorimetry (DSC). The thermal properties of polymers were measured on a Mettler Toledo DSC 822 system. Nitrogen was used as purge gas with a flow rate of 30 ml/min. Samples of 10 mg were prepared in aluminum foils, where the aluminum weights of the sample and reference were closely matched. The samples were heated from room temperature to 180 °C with a heating rate of 20 °C/min, cooled down to -100 °C with a cooling rate of -20 °C/min, and heated again from -100 °C to 180 °C at a heating rate of 20°C/min. T_m and T_g of the samples were obtained from the second heating curves.

Wide angle X-ray diffraction (WAXD). The crystallinity analysis was preformed with a SHIMADZU 6000 X-ray diffractometer with Cu Kα radiation at 40 kV and 30 mA in a 2θ range of 5-40 ° at scanning speed of 1.2 °/min.

Fourier transform infrared spectrophotometer (FTIR). IR spectra of the polymers were analyzed with a SHIMADZU FTIR-8400 system using potassium bromide (KBr) pressing.

Results and Discussion

Enzymatic ring-opening polymerization of ε-caprolactone with PHB-diol(M)

Reaction conditions. The enzymatic ring-opening polymerization of ε-caprolactone (CL) was initially investigated with PHB-diol (M) as an initiator (Scheme 1). PHB-diol(M) is a telechelic hydroxylated PHB with a M_n of 3000 g/mol (GPC) prepared by transesterification of PHB and ethylene glycol.[12] Novozym 435 [immobilized *Candida antarctica* lipase B (CALB)] was chosen as catalyst, as it is well known with high catalytic activity and good solvent resistance for the ROP of lactones.[21-22] The polymerization temperatures were examined from room temperature to 70°C, since Novozym 435 has the highest catalytic activity at around 70°C.[21-22]

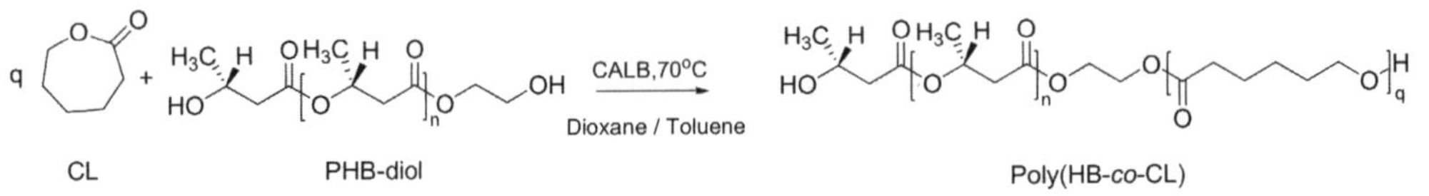

Scheme 1 Enzyme-catalyzed ring-opening polymerization of ε-caprolactone with PHB-diol

1,4-Dioxane and toluene were known solvents for enzymatic ROP,[21-22] and they have a boiling point of 103°C and 110°C, respectively, and good solubility for PHB-diol, thus being selected as solvents in our experiments. To avoid water-initiated ring-opening polymerization, solvent was dried before use and the reaction was carried out at anhydrous conditions under Argon atmosphere. In order to study the effect of different reaction conditions on the polymerizations, a set of experiments were designed with different weight ratios of CL/solvent, CL/enzyme, CL/PHB-diol and different temperature (Table 1). The reactions were performed in 1,4-dioxane and followed by taking samples at different time points to determine the molecular weight by GPC. As an example, the course of polymerization in experiment 11 in Table 1 was shown in Figure 1: the number average molecular weight (M_n) of the samples taken at 4 h, 8 h, 24 h and 48 h reached 3000, 4700, 4900, 6700, and 8400 g/mol, respectively. The molecular weight and yield of polymers under different reaction conditions are summarized in Table 1.

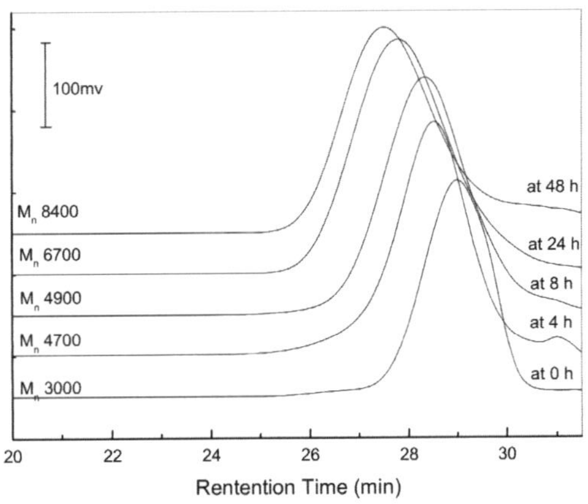

Figure 1. GPC chromatograms of block poly(HB-*co*-CL) formed at different time from experiment 11 in Table 1.

Table 1. Ring-opening polymerization of ε-caprolactone with PHB-diol(M) catalyzed by Novozym 435 in dioxane

No.	CL:**E** [a] wt:wt	CL:diol mol:mol	CL:**Sol** [b] wt:wt	Temp. °C	Time h	M_n (GPC)[c] g/mol	M_w/M_n	Yield %	CL Conv. %	Code
1	5:1	100:1	1:4	70	48	5500	1.65	63	53	
2	10:1	100:1	1:4	70	48	4500	1.56	57	47	
3	20:1	100:1	1:4	70	48	4000	1.48	51	40	
4	8:1	100:1	1:3	70	48	6700	1.68	63	54	
5	8:1	100:1	1:6	70	48	4300	1.53	64	55	
6	8:1	100:1	1:9	70	48	3400	1.38	64	55	
7	8:1	100:1	1:3	70	48	6700	1.79	64	53	A
8	8:1	100:1	1:3	50	48	4100	1.66	65	56	
9	8:1	100:1	1:3	RT	48	3600	1.55	69	59	
10	8:1	50:1	1:3	70	32	5400	1.63	66	49	B
11	8:1	75:1	1:3	70	48	8400	1.66	58	45	C

a: **E** for immobilized enzyme, Novozym 435. b: **Sol** for solvent, Dioxane c: Polystyrene was used as standards in GPC measurement

In experiment 1 to 3 (Table 1), three different ratios of CL/enzyme (E) were studied at 70°C with fixed ratios of CL/PHB-diol and CL/Solvent, and polymerization were shown in Figure 2(i). The highest molecular weight of the resulting product was obtained at CL/E in a weight ratio of 5:1. Within the range of 20-5:1, the molecular weight of the resulting polymers increased with catalyst amount. To reduce the amount of enzyme involved while achieving a high molecular weight of the polymer, the ratio of CL/E of 8/1 was used in the rest of experiments. In experiment 4-6 (Table 1), CL/solvent was examined in the weight ratio of 1:3, 1:6 and 1:9 at 70°C with fixed ratios of CL/PHB-diol and CL/E. As shown in Figure 2(ii), the use of less solvent resulted in higher molecular weight of the polymers, probably due to higher concentration of reacting species. On the other hand, the ratio of CL/solvent should not be too small, since sufficient amount of the solvent is required to dissolve PHB-diol. The effect of the reaction temperature on the polymerization was investigated in the experiment 7-9 (Table 1) at rt, 50°C and 70°C, respectively. Figure 2(iii) clearly demonstrated that higher temperature resulted in higher molecular weight of the polymers. These results are in close correlations to the reported best temperatures of 65-70 °C for ROP with Novozym 435. Finally the effect of the molar ratio of CL/PHB-diol was investigated in experiment 7, 10, and 11 (Table 1) at 70°C with fixed ratios of CL/E (8:1) and CL/Solvent (1:3). As shown in Figure 2(iv), the highest molecular weight of the resulting polymer was obtained with a molar ratio of CL/PHB-diol in 75:1. The use of more PHB-diol could create more opportunity for the opening of CL. However, this could be in competition with the chain growth by CL, resulting possible decrease of the molecular weight of final products. From these experiments, the best conditions examined so far for achieving highest molecular weight was 70°C, a weight ratio of CL/enzyme/solvent of 8:1:24, and a molar ratio of CL/PHB-diol of 75:1.

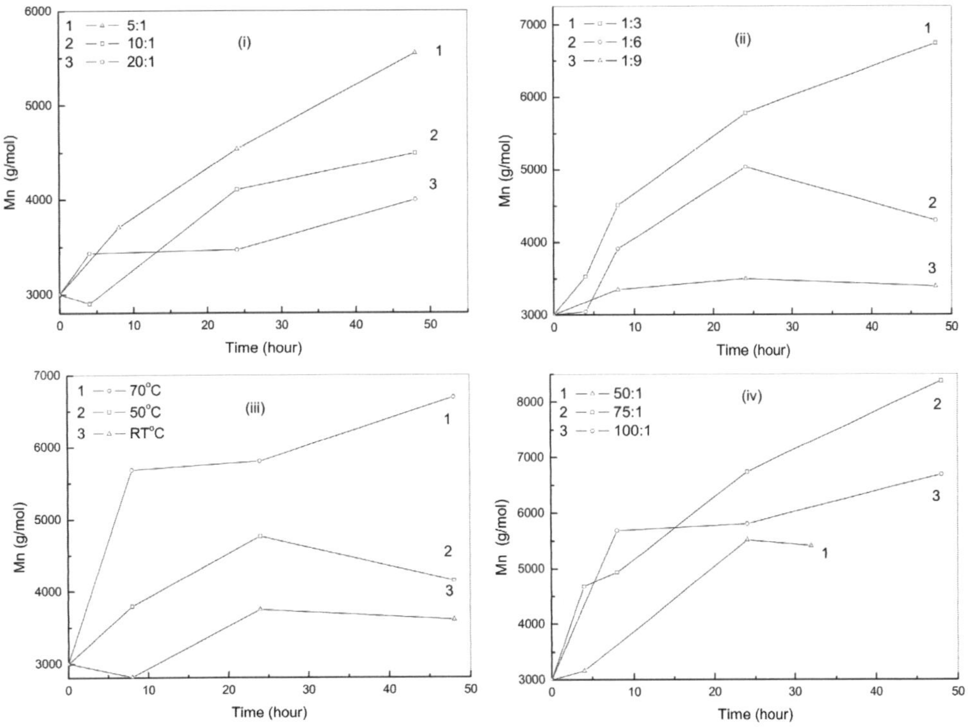

Figure 2. Enzyme-catalyzed ring-opening polymerization of ε-caprolactone with PHB-diol(M) under different reaction conditions: (i) different ratio of CL/Enzyme; (ii) different ratio of CL/Solvent; (iii) different reaction temperature; (iv) different ratio of CL/ PHB-diol(M).

Polymer isolation. The polymerization products from all experiments in Table 1 were isolated by adding chloroform to the reaction mixture, removing enzyme through filtration, evaporating 1,4-dioxane and chloroform at reduced pressure, and precipitating in chloroform/*n*-hexane or chloroform/methanol (1:9). Drying the precipitates under high vacuum at 50°C for 24 h gave the corresponding block co-polyesters with M_n (GPC) of 3600-8,400g/mol in 51-69% yields and 40-59% CL conversion, respectively.

Structure analysis. The chemical structure of PHB-diol(M) was determined by ^{1}H-NMR spectrum in Figure 3(i). Two different structures A and B were found in PHB-diol(M). All signals were assigned to the different protons in structure A and B according to the literatue.[12] Two different proton signals for primary OH and secondary OH were observed at 4.58 ppm (*c* proton) and 4.45ppm (*d* proton), respectively. The molar ratio of structures A and B was estimated as 91%:9% based on the intensities of *h* proton and *u* proton, and the number of repeating unit n was deduced as 23.6 based on the ratio of the intensities of *m* and *u* protons. The M_n of PHB-diol(M) can be thus established as 2380 g/mol which is comparable with the M_n of 3000 g/mol determined by GPC. In fact, the M_n obtained from NMR is more reliable, since the M_n determined from GPC was based on the use of polystyrene as standard.

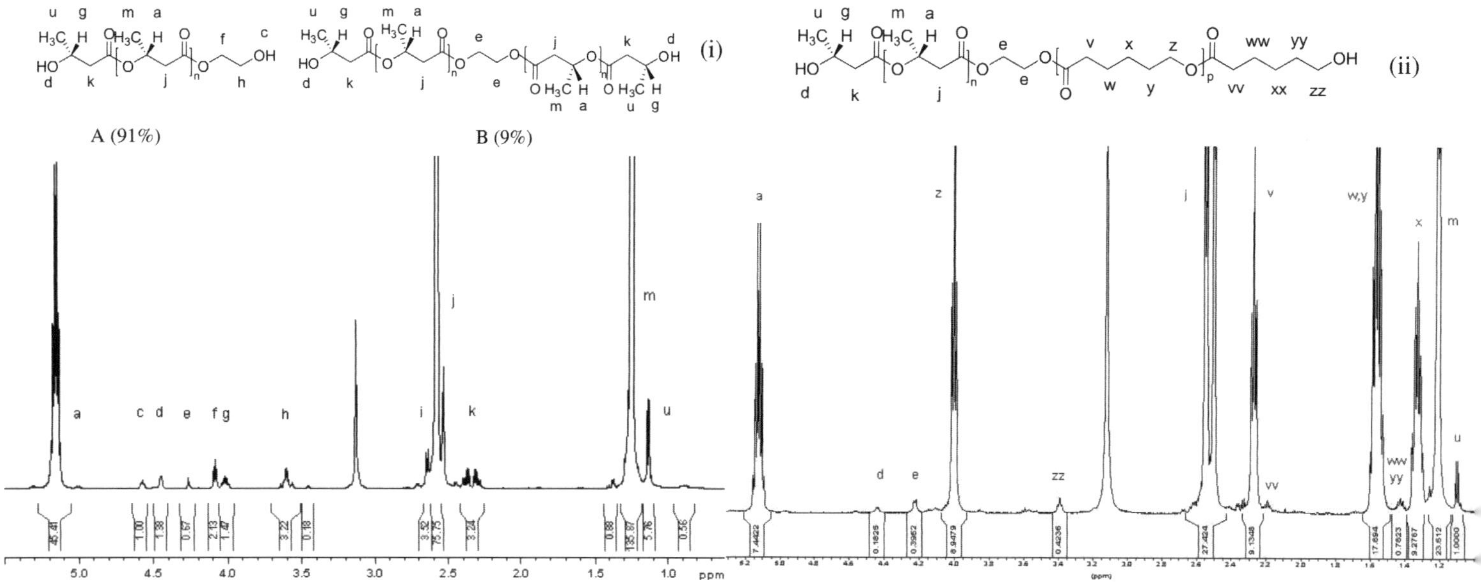

Figure 3. ^{1}H-NMR spectra in DMSO$_{d6}$ at 333K: (i) PHB-diol(M); (ii) Poly[HB(56wt%)-*co*-CL(44wt%)] (sample B in Table 1).

The chemical structures of the polymers were determined by NMR and IR spectra. The ^{1}H-NMR spectrum of sample B (Table 1) was shown in Figure 3(ii). The *c* proton (primary OH, δ = 4.58 ppm) of PHB-diol(M) disappeared and the *d* proton (secondary OH, δ = 4.45ppm) was clearly observed. This indicates that the primary OH end group of PHB-diol(M) was more reactive and hence totally reacted, while the secondary OH end group is less reactive thus remaining in the polymer. The signal of *h* proton (δ = 3.56ppm) and *f* proton (δ = 4.10ppm) disappeared, which suggested again that all primary OH groups of PHB-diol(M) was reacted. The polymerization was further evidenced by the increased intensity of *e* proton. While reaction with a primary OH end group of PHB-diol(M) did not change the ratio of *m*/u protons, polymerization with a secondary OH end group would transform a *u* proton into a *m* proton, thus increasing the ratio of *m*/*u* protons. From Figure 3 (ii), the ratio of *m*/*u* in the polymer was calculated as 23.5 which was the same as the *m*/*u* ratio for the starting material PHB-diol(M). Thus it is unlikely that the secondary OH end group was reacted in the ROP. As a result, the polymer is a *di*-block. The signals of PCL block were also assigned in Figure 3 (ii). With the ratio of *z*/*a* of 1.20:1, the number of repeating unit p in PCL block was calculated as 14.2. The molecular weight of poly(HB-*co*-CL) could thus be established as 3900 g/mol. This value is smaller than the M_n of 5400 g/mol determined by GPC. But, as it was mentioned above, the M_n obtained from NMR is more reliable.

The block copolymer structure was further evidenced by the ^{13}C-NMR spectrum in Figure 4. In the area of 160-180 ppm, only two signals at 168 and 172 ppm were observed and they were assigned to the carbonyl groups in PHB block and PCL block, respectively. From previous report[18] and our control experiment of lipase-catalyzed transesterification of PLC and PHB-diol, random polymer poly(HB-*co*-CL) contains at least two different types of carbonyl groups which absorbed at 169 and 171 ppm between the two signals at 168 and 172 ppm. In the ^{13}C-NMR spectrum of our polymer, no such signals can be detected. This indicated no random polymer formed during the polymerization.

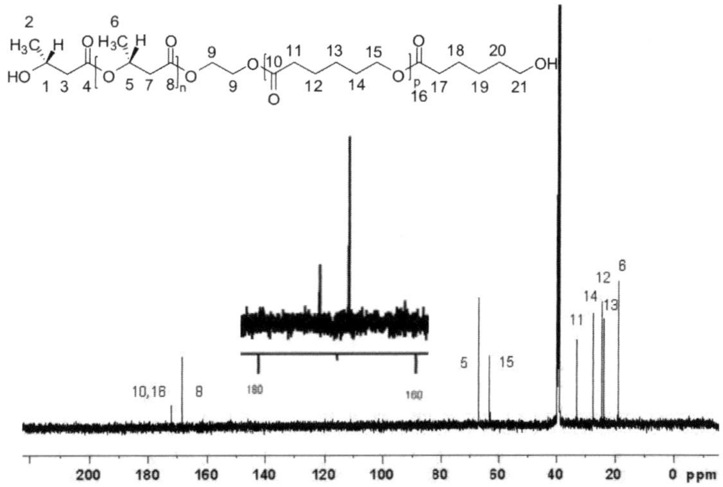

Figure 4. ^{13}C-NMR spectra of Poly[HB(56wt%)-*co*-CL(44wt%)] (sample B in Table1) in $DMSO_{d6}$ at 333K.

In the IR spectra in Figure 5(i) and (iii), the primary and the secondary terminal OH group of the starting material PHB-diol(M) showed two different absorptions at 3437 cm^{-1} and 3538 cm^{-1}. There were also two absorptions at 3438 cm^{-1} and 3533 cm^{-1} for the OH groups in the block co-polymer. This further confirmed that poly(HB-*co*-CL) is a *di*-block polymer containing a primary and a secondary OH group. In the case of A-B-A (PCL-PHB-PCL) *tri*-block structure, the OH end groups would be the same, thus giving only one absorption peak in the IR spectrum.

The weight percentage of PHB and PCL block in the co-polymer was calculated using the repeating unit *n* and *p* for PHB and PCL block, respectively, obtained from NMR analysis. In the case of the polymer prepared above, *n* is 23.5, thus the molecular weight for PHB block is 86 x (23.5+1) + 1 + 60 = 2168; p is 14.2, therefore, the molecular weight for PCL block is 114 x (14.2 +1) + 1 = 1734. Based on these results, it can be deduced that the *di*-block polymer contains 56wt% of PHB block and 44wt% of PCL block. This method was used to calculate the weight ratio of the two blocks for all *di*-block polymers in this study.

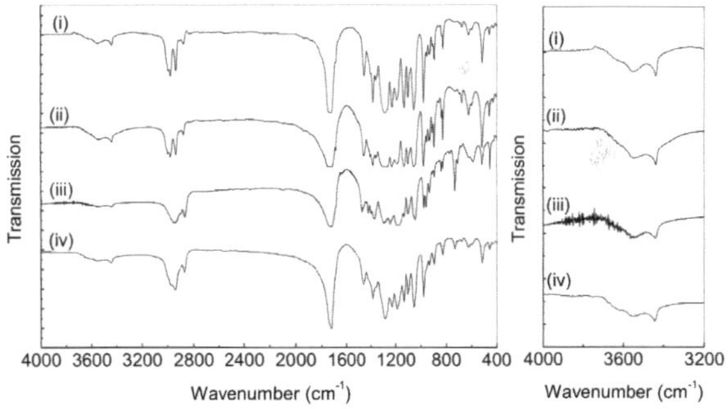

Figure 5. IR spectra of different polymers: (i) PHB-diol (M); (ii) PHB-diol (P); (iii) Poly[HB(56wt%)-*co*-CL(44wt%)] (sample B in Table 1); (iv) Poly[HB(28wt%)-*co*-CL(72wt%)] (sample D in Table 2).

Enzymatic ring-opening polymerization of ε-caprolactone with PHB-diol(P):

PHB-diol(P). For the easy investigation of the selectivities of the primary and secondary OH group of PHB-diol in the enzymatic ROP and for the easy preparation and characterization of the corresponding *di*-block co-polymer with simpler structure, PHB-diol that contains 100% structure A with a primary and a secondary OH end groups was designed as the initiator. This type of PHB-diols with M_n between 2000-4000 g/mol was prepared by transesterification of PHB and ethylene glycol according to the same procedure for the preparation of PHB-diol(M)[12], but with large excess of ethylene glycol to PHB. Their desired chemical structures were confirmed by ^{1}H-NMR analysis. As an example, the ^{1}H-NMR spectrum of PHB-diol(P) with M_n of 3700g/mol (GPC) was shown in Figure 6(i). The primary OH and secondary OH group absorbed at 4.58 ppm (*c* proton) and 4.45 ppm (*d* proton),[12] respectively, with nearly equal intensities. Similar to the case of PHB-diol(M), the molar ratio of structures A and B in PHB-diol(P) was estimated based on the intensities of *h* proton from the primary OH end and *u* proton from the secondary OH end as 100:0. The number of repeating unit n was calculated as 21.7 according to the ratio of *m*/*u* absorption intensity, which gave a M_n of 2010g/mol for PHB-diol(P). The primary and secondary OH groups of PHB-diol(P) showed two different absorptions at 3437 cm^{-1} and 3538 cm^{-1} in the IR spectrum in Figure 5(ii).

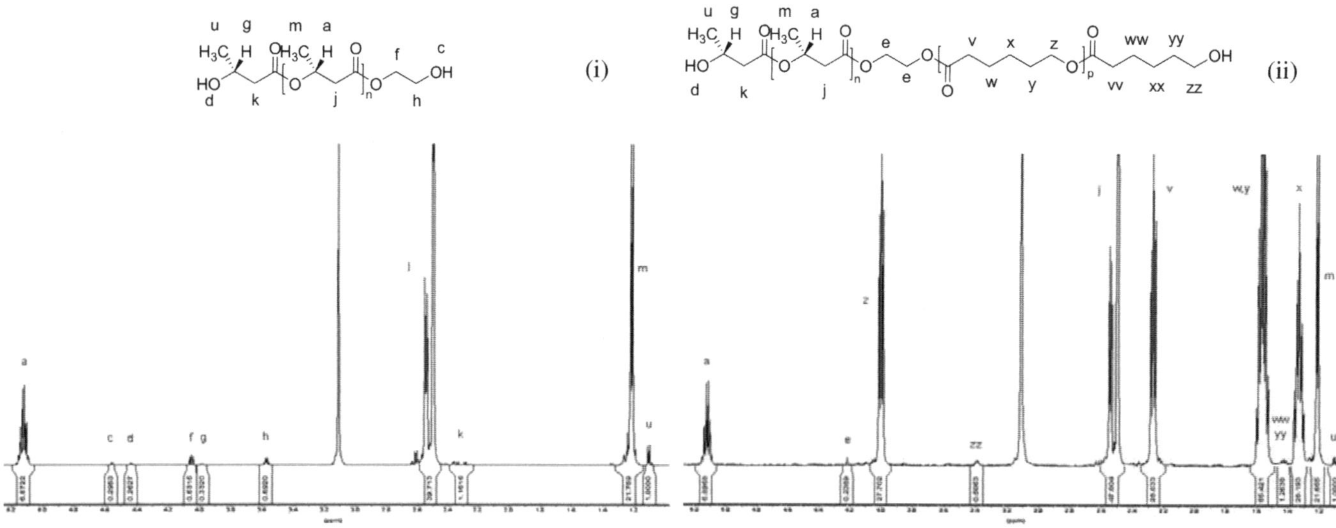

Figure 6. ^{1}H-NMR spectra in $DMSO_{d6}$ at 333K: (i) PHB-diol (P); (ii) Poly[HB(28wt%)-*co*-CL(72wt%)] (sample D in Table 2).

The physical properties of the new type of PHB-diols were characterized by DSC, and the T_m and T_g were found to be dependent on the M_n: PHB-diol (P) with M_n of 3700g/mol (GPC; M_w/M_n of 1.38) has T_m of 134 and 149°C and T_g of -5°C, while PHB-diol (P_8) with M_n of 2300g/mol (GPC; M_w/M_n of 1.17) has T_m of 115 and 129°C and T_g of -11.9°C. Both of them are good hard segment for block co-polymer syntheses. For the preparation of biomaterials with the possibility of sterilization, T_m of the polymers needs to be above 120°C. Therefore, PHB-diol (P) is an appropriate hard block for the preparation of such type of materials.

Ring-opening polymerization with PHB-diol(P). New batches of lipase-catalyzed ROP of CL with PHB-diol(P) (M_n of 3700, GPC) as the initiator were carried out in dioxane (experiments 12-15 in Table 2). The optimal conditions established for PHB-diol(M) such as 70 °C, CL:E in 8:1 wt ratio, and CL:Sol in 1:3 wt ratio were used for the ROP. Reaction for 16 h with a molar ratio of CL:PHB-diol(P) in 50:1 gave a block co-polymer with M_n (GPC) of 5500 g/mol. The molecular weight achieved here is similar to that obtained in the ROP with PHB-diol(M). Further increase of the molar ratio of CL:PHB-diol(P) to 100:1 did not increase the M_n of the final polymer. Changing the CL:E ratio from 8:1 to 4:1 did not result in big change of the polymer molecular weight, either. In experiment 16, ROP of CL and PHB-diol(P_8) with M_n(GPC) of 2300g/mol was carried out at similar conditions, affording a polymer with M_n (GPC) of 5500g/mol. To further increase the molecular weight of the polymer, toluene was examined as solvent (experiments 17-21). The ROP of CL with PHB-diol(P) were initially performed at the optimal conditions established for dioxane: reaction with CL and PHB-diol(P) in 100:1 and 50:1 molar ratio, respectively, gave the corresponding polymers with M_n of 8900 and 7900 g/mol, respectively. The reaction conditions were also varied in other experiments, which did not improve the molecular weight of the polymer. Nevertheless, the polymers prepared in toluene have higher M_n than those prepared in dioxane. This is possibly due to the different log P value of the solvents. Toluene with log P of 2.5 is possibly less toxic to the enzyme than 1,4-dioxane with log P of -0.42. All the polymers were isolated in 69%-90% yield with 64-88% CL conversion by using the same methods descript for the ROP with PHB-diol (M).

Table 2. Ring-opening polymerization of ε-caprolactone with PHB-diol(P) in dioxane or toluene

No.	Sol[a]	CL:**E**[b] wt:wt	PHB-diol	CL:PHB-diol mol:mol	CL:**Sol** wt:wt	Temp. °C	Time h	M_n (GPC) g/mol	M_w/M_n	Yield %	CL Conv. %	Diblock %	T_m °C	T_g °C	Cod
12	D[c]	8:1	P	100:1	1:3	70	16	5300	1.56	73	69		51.3 / 132.3 / 148.3	-57.5	
13	D	8:1	P	50:1	1:3	70	16	5500	1.55	79	73		52.0 / 135.7 / 150.0	-57.7	
14	D	4:1	P	100:1	1:3	70	12	5200	1.75	88	86		50.0 / 128.7 / 145.7	-60.0	
15	D	4:1	P	50:1	1:3	70	48	4200	1.57	69	64	100	48.3 / 123.3 / 141.0	-59.1	
16	D	4:1	P8[e]	75:1	1:3	70	48	5500	1.75	76	73	97	48.7 / 99.0 / 116.0	-55.4	
17	T[d]	8:1	P	100:1	1:3	70	16	8900	1.93	89	87		56.7 / 154.3	-57.3	
18	T	8:1	P	50:1	1:3	70	16	7900	1.90	86	82	100	50.0 / 147.0	-60.0	D
19	T	4:1	P	100:1	1:3	70	12	7900	1.86	90	88		50.0 / 153.3	-58.8	
20	T	16:3	P	100:1	1:4	70	8	7900	1.98	87	85		52.0 / 153.3	-60.3	
21	T	16:3	P	100:1	1:6	70	8	5600	1.95	79	76	98	54.0 / 149.0	-61.0	E

a: **Sol** for solvent, Dioxane. b: **E** for immobilized enzyme, Novozym 435. c: D for dioxane. d: T for toluene. e: P8 for PHB-diol (P8) with M_n (GPC) of 2300 g/mol.

Structure Analysis. Polymers from experiments 15, 16, 18, and 21 were analyzed by ^{1}H-NMR and they all contain *di*-block poly(HB-*co*-CL) in 97%-100%. The ^{1}H-NMR spectrum of sample D (experiment 18) from Table 2 was given in Figure 6(ii). The *di*-block structure was clearly evidenced by the absorption of u proton at the end of PHB block and the *zz*, *ww*, *yy* protons at the end of PCL block. The intensity ratio of *ww* and *yy* signals (4 protons,1.45 ppm) and *u* signal (3 protons, 1.10 ppm) was 1.26:1, which corresponds to a ratio of the two ends of the polymers about 0.97:1. New signal of *e* proton appeared at 4.22 ppm, indicating the formation of polymer. From the ^{1}H-NMR spectrum, the ratio of *m/u* signal intensities of the polymer was determined as 21.7 which is exactly the same as that for PHB-diol(P). As previously described, the formation of *tri*-block polymer would increase the ratio of *m/u*. Therefore, it can be concluded that there is no *tri*-block structures in the polymer sample D. The ratio of the absorption intensities of *z* proton in PCL block and *a* proton in the PHB block was determined as 4.02:1, thus the *p* in the

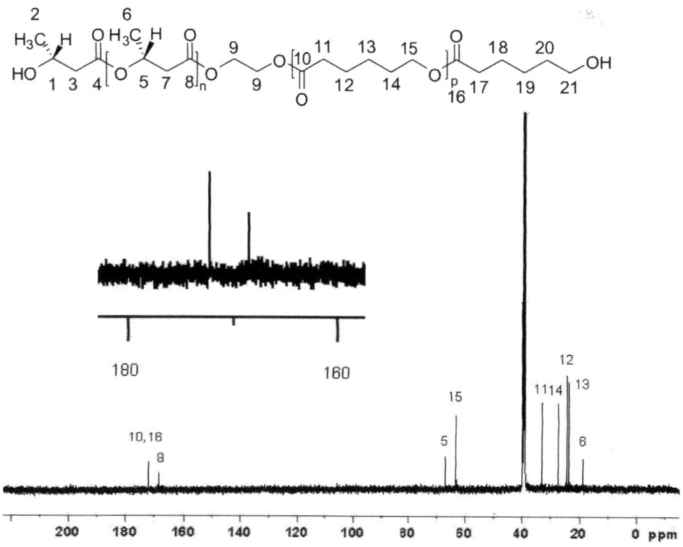

Figure 7. ^{13}C-NMR spectrum of poly[HB(28wt%)-*co*-CL(72wt%)] (sample D in Table 2) in $DMSO_{d6}$ at 333K.

PCL bock can be deduced as 43.6. This led to the establishment of the polymer M_n as 7100g/mol. Calculation based on the intensities of *m* and *x* protons gave a M_n of 6900g/mol for the polymer. These values are in good correlation to the M_n measured by GPC (7900g/mol).

The ^{13}C NMR spectrum of sample D in Table 2 was shown in Figure 7. In the area of 160-180 ppm, the two signals observed are the carbonyl groups of the PHB and PCL block. No other carbonyl signals were detected, suggesting no random copolymers formed during ROP. The *di*-block structure was further confirmed by two different absorption of the primary and the secondary end OH groups at 3438 cm^{-1} and 3533 cm^{-1} in the IR spectrum of sample D shown in Figure 5(iv). Based on NMR analysis, n = 21.7 and p = 43.6, thus the ratio of PHB and PCL block was established as 28/72 (wt/wt) for the polymer.

Mechanism of PHB-diol initiated ring-opening polymerization of ε-caprolactone.

Is the polymerization really initiated by the PHB-diol? Is it possible that PCL is first formed by water-initiated ring-opening polymerization and then it reacts with PHB-diol *via* transesterification to give the co-polymer? To answer these questions, control experiments were carried out in 1,4-dioxane with CL and novozym 435 at 70°C and at the same ratio of CL/enzyme/solvent as that used for the synthesis of sample B in Table 1. Reaction for 16 h without pre-drying gave PCL with M_n of 12000g/mol (GPC), while reaction under anhydrous conditions for 16 h afforded a PCL with M_n of 5000g/mol (GPC). From these experiments, the possible formation of PCL in our polymerization reactions can not be excluded. Based on the known mechanism, even only a very small amount of water could initiate the polymerization to give PCL. Further control reaction was to examine the possible transesterification between PHB-diol and PLC. Reaction of PHB-diol(M) (M_n of 3000 g/mol) and PCL (M_n of 5000 g/mol) in 1:1 molar ratio with Novozym 435 as catalyst was performed in dioxane under the anhydrous conditions. The M_n of the reaction mixture was 4400g/mol (GPC) at the beginning and dropped to 2200g/mol (GPC) at 48 h. The product was isolated by the same procedure used for ROP and analyzed by ^{1}H- and ^{13}C-NMR. The *m/u* signal ratio of the product in the ^{1}H-NMR spectrum was drop to 16.9 from 23.6 of PHB-diol(M), suggesting the degradation of PHB-diol. This must be caused by the Novozyme 435-catalyzed transesterification between PCL and PHB-diol, since no such degradation was observed in the same system without enzyme. In the

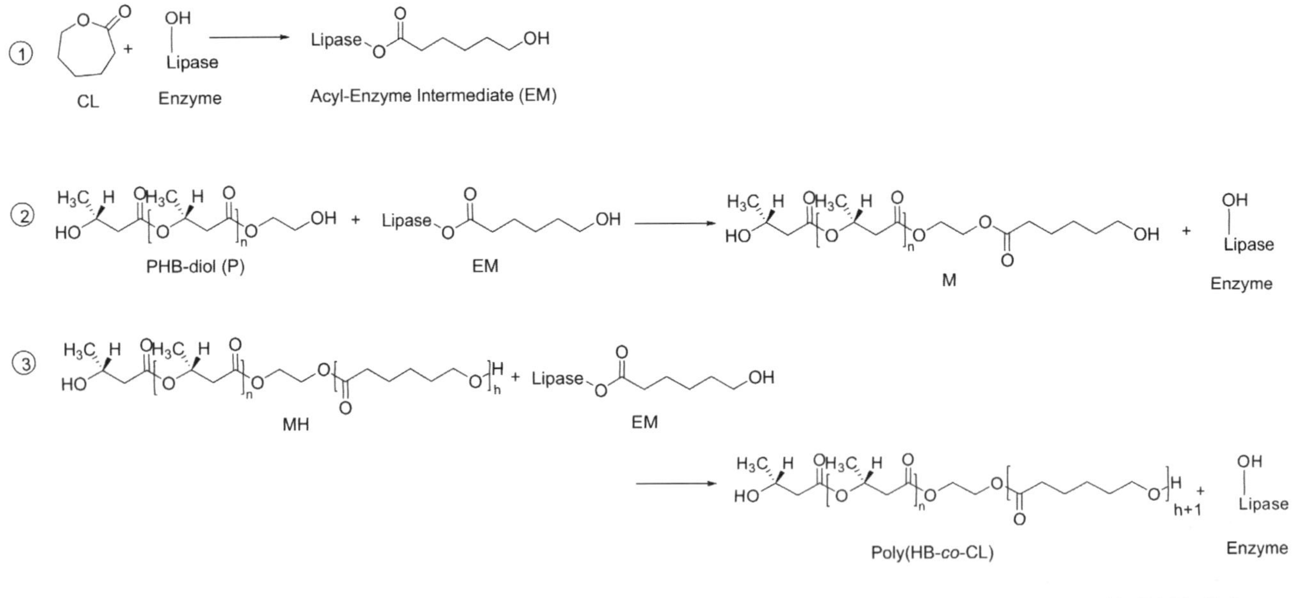

Scheme 2. Mechanism of enzyme-catalyzed ring-opening polymerization of ε-caprolactone with PHB-diol.

^{13}C-NMR, in addition to the signals at 168 and 172 ppm of the carbonyl group of the PHB and PCL block, two signals were observed at 169 and 171 ppm. These were the absorptions of the different type of carbonyl groups of random co-polymer. Thus, a mechanism of formation of PCL followed by transesterification with PHB-diol would generate co-polymers with decreased M_n, decreased *m*/*u* signal ratio in ^{1}H- NMR, and new signals in ^{13}C- NMR spectrum for random polymer units. Since all these phenomena were not observed in our polymer preparation, we can exclude such mechanism for our reaction.

The mechanism of enzyme-catalyzed ROP of CL was well known.[22] Accordingly, the steps of our ROP was proposed in Scheme 2. The first step is the opening of CL with lipase to give an acyl-enzyme intermediate (EM). Afterwards, the intermediate is reacted with an OH group of PHB-diol to form new molecule M and release the lipase. As the primary OH group is more reactive and has less stereo-hindrance than the secondary OH group, it is preferentially reacted with lipase to give compound M with a secondary OH end group from PHB block and a primary OH end group from CL. While the first two steps can be repeated, the compound MH could also react with EM to prolong the chain and release lipase. Also in this step, the primary OH group of CL/PCL in compound MH should be more reactive than the secondary OH group of PHB block, thus generating a *di*-block polymer with a secondary OH group at end of PHB block and a primary OH group at the end of PCL block. Although we can not exclude the possibility of initiating ROP of CL by the trace amount of water in the system such as in the enzyme hydration shell, the primary OH group of PHB-diol should be more reactive than water, thus effectively suppressing the water-initiated ROP. Even if PCL monomer and oligomer could be formed by water-initiated ROP, they could not react with the existing PHB-diol, its derivatives M and MH, and Poly(HB-co-CL) *via* enzymatic transesterification, since the enzyme active center is occupied through the formation of EM with CL in large excess. The only possibility remained for PCL monomer and oligomer is to react with EM for the elongation of the PCL chain, which is in competition with the elongation of Poly(HB-*co*-CL). In fact, only very small amount of PCL monomer and oligomer was observed in our experiments, and these side products were easily removed by precipitating Poly(HB-*co*-CL) in chloroform/*n*-hexane (1:9).

Physical properties of *di*-block poly[(*R*)-3-hydroxybutyrate-*co*-ε-caprolactone]s

The melting temperature (T_m) and glass transition temperature (T_g) of several poly(HB-*co*-CL)s, PCL, and PHB-diols were measured by DSC. The data were summarized in Table 2 and 3. As shown in Figure 8, T_m and T_g were determined from the second heating curve. T_m of PHB-diol(P) was found to be 134°C and 149°C, and T_m of PCL with an M_n(GPC) of 12000 g/mol was 57°C. Poly[HB(55wt%)-*co*-CL(45wt%)] (sample E) and poly[HB(28wt%)-*co*-CL(72wt%)] (sample D) showed T_m from both PHB and PCL blocks, with the values of 147-149 °C and 50-54°C. T_m of the PHB block is a broad peak in DSC, and in some case it was split into two peaks. T_g of PHB-diol(P) and the PCL was determined to be -5°C and -63°C, respectively. The polymers D and E showed a T_g of -60°C and

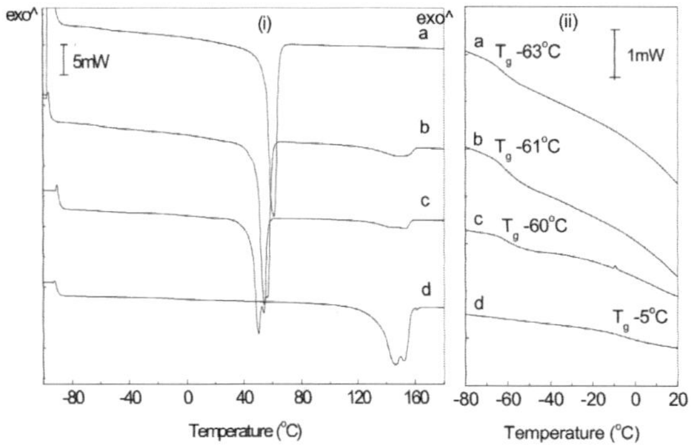

Figure 8. DSC spectra of (a) PCL with M_n 12000 (GPC), (b) Poly[HB(28wt%)-*co*-CL(72wt%)] (sample D in Table 2), (c) Poly[HB(55wt%)-*co*-CL(45wt%)] (sample E in Table 2), and (d) PHB-diol(P): (i) The second heating curve; (ii) The enlarged second heating curve.

-61°C, respectively, which was obviously the T_g from the PCL block. The T_g of PHB block was not detectable in all polymers prepared in this study. As shown in Table 3, all polymers containing 44-74wt% of PCL block have a T_g between -57°C and - 61°C for the elastomer domain. Thus, introduction of the PCL block significantly improved the thermal properties of block co-polymers for potential thermoplastic application.

Table 3. Physical properties of selected poly(HB-*co*-CL)s, PHB-diol and PCL

Polymer	Code	Initiator	M_n (GPC) g/mol	M_n (NMR) g/mol	HB:CL[a] wt%	T_m °C	T_g °C
PHB-diol(M)			3000	2380		137 / 147	-4
PHB-diol(P)			3700	2010		134 / 149	-5
Poly(HB-*co*-CL)	A	PHB-diol(M)	6700	6100	32:68	54 / 120 / 139	-57
Poly(HB-*co*-CL)	B	PHB-diol(M)	5400	3900	56:44	53 / 123 / 145	-57
Poly(HB-*co*-CL)	C	PHB-diol(M)	8400	7500	26:74	53 / 121 / 140	-56
Poly(HB-*co*-CL)	D	PHB-diol(P)	7900	7100	28:72	50 / 147	-60
Poly(HB-*co*-CL)	E	PHB-diol(P)	5600	3400	55:45	54 / 149	-61
PCL		water	12000			57	-63

a: The ratio was calculated based on NMR analysis.

In Figure 9, the DSC curves of different samples were compared, including PCL, PHB-diol(M), a mixture of PHB-diol(M) and PCL with weight ratio of 1:1, and poly[HB(56wt%)-*co*-CL(44wt%)] (sample B). While the 1:1 mixture demonstrated exactly the same T_m values as those of PHB-diol and PCL, the block co-poly(HB-*co*-CL) showed lower T_m than PHB-diol and PCL. This confirmed once again the formation of block-copolymer.

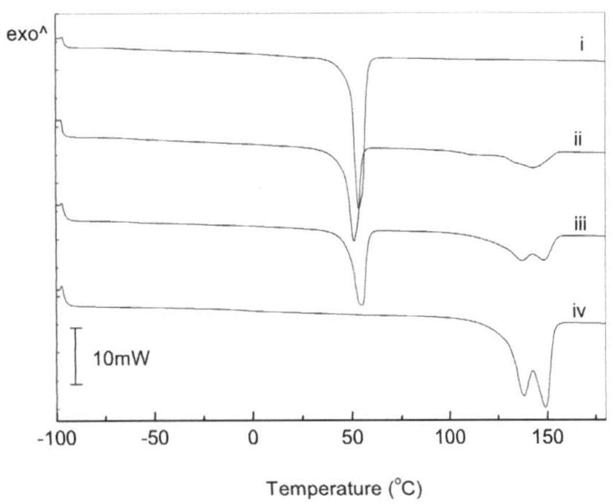

Figure 9. DSC spectra of different polymers: (i) PCL with M_n of 5000; (ii) Poly[HB(56wt%)-*co*-CL(44wt%)] (sample B in Table 1); (iii) Mixture of PHB-diol(M) and PCL with M_n of 5000 (GPC); (iv) PHB-diol(M).

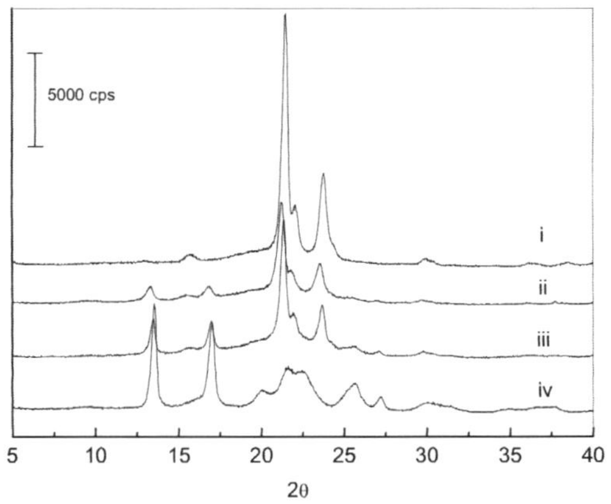

Figure 10. WAXD spectra of different polymers: (i) PCL with M_n of 12000 (GPC); (ii) Poly[HB(26%)-*co*-CL(74%)] (sample C in Table 1); (iii) Poly[HB(56wt%)-*co*-CL(44wt%)] (sample B in Table 1); (iv) PHB-diol(M).

The crystallinity of the polymers was investigated by the WAXD. As shown in Figure 10, poly[HB(56wt%)-*co*-CL(44wt%)] (sample B) and poly[HB(26wt%)-*co*-CL(74wt%)] (sample C) showed very similar patterns to that of PHB-diol and PCL, but with decreased intensity: two major peaks from PCL at 2θ of 21.4° and 2θ of 23.8° and two major peaks from PHB-diol at 2θ

of 13.6° and 2θ of 17.0° were founded in these copolymer. This indicated that the crystalline structures of the copolymers were similar to each of the two homopolymers. Thus, the polymer should show T_m from both PHB and PCL blocks. The crystallinities of poly(HB-*co*-CL)s, PHB-diol, PCL were estimated based on the ratios of crystalline peak area and amorphous peak area by using the affiliated software of SHIMADZU 6000 X-ray diffractometer. While PHB-diol and PCL showed crystallinities of 21% and 26% respectively, the two *di*-block poly(HB-*co*-CL)s sample B and C demonstrated a slightly decreased crystallinity of 19%.

Conclusion

Novel *di*-block co-polyesters containing PHB and PCL blocks were synthesized in high yield, for the first time, by enzymatic ring-opening polymerization of ε-caprolactone with PHB-diol. The structures of the *di*-block polymers with two different OH end groups were established by IR, ^{1}H-NMR and ^{13}C-NMR analyses. Poly(HB-*co*-CL)s with 44-74%(w/w) PCL demonstrated good thermal properties with T_g of about -60°C and T_m of 120-149°C and 50-54°C, being potentially useful thermoplastic biomaterials. Incorporation of PCL into the PHB-derived polyesters significantly improved the T_g of the materials.

Low molecular weight PHB-diol with a primary and a secondary OH end group was proven to be very useful for highly selective ring-opening polymerization, being the first example of using a telechelic macro-diol containing ester groups as an initiator. The primary OH end group of PHB-diol was found to initiate the ring-opening polymerization of ε-caprolactone, while the secondary OH end group was not reacted thus remaining as an end group of the final polymer. No enzymatic transesterification of PHB-diol and PCL happened and no random co-polymers formed during the polymerization. Optimal enzymatic polymerization conditions were established for the preparation of block co-polymers with different block ratio. This type of novel and selective ring-opening polymerization provides with new synthetic methods for preparing novel *di*-block co-polymers with functional end groups which could also be modified for other applications.

Acknowledgement

This work was support by National University of Singapore through research grants 279-000-194-112/133 and 279-000-225-112. We thank Dr. P. Neuenschwander at ETH Zurich for providing the sample of PHB-diol(M).

References and Notes

(1) Doi, Y. *Microbial polyester* VCH: New York, **1990**.

(2) Sudesh, K.; Abe, H.; Doi, Y. *Progress in Polymer Science*, **2000**, 25, 1503.

(3) Zinn, M.; Witholt, B.; Egli, T., *Advanced Drug Delivery Reviews*, **2001**, 53, 5.

(4) Sudesh, K.; Doi, Y.; in *Handbook of Biodegradable Polymers*, Rapra: Shawbury, 2005, Ch. 7, 219.

(5) Valappil, S. P.; Misra, S. K.; Boccaccini, A. R.; Roy, I. *Expert Review of Medical Devices*, **2006**, *3*, 853.

(6) Mobley, D. P. *Plastics from Microbes*, Hanser: Munich, **1994**

(7) Holmes, P. A.; Wright, L. F.; Colins, S. H. *Eur. Pat. Appl.* 005249 **1981**; *Eur. Pat. Appl.* 0069497 **1983**

(8) Shimamura, E.; Scandola, M.; Doi, Y. *Macromolecules* **1994**, *27*, 4429.

(9) Doi, Y.; Kitamura, S.; Abe, H. *Macromolecules* **1995**, *28*, 4822.

(10) Abe, H.; Doi, Y.; Fukushima, T.; Eya, H. *Int. J. Biol. Macromol.* **1994**, *16*, 115.

(11) Pederson, E. N.; McChalicher, C. W. J.; Srienc, F., *Biomacromolecules* **2006**, *7*, 1904.

(12) Hirt, T. D.; Neuenschwander, P.; Suter,U. W., *Macromol. Chem. Phys.* **1996**, *197*, 1609.

(13) Hirt, T. D.; Neuenschwander, P.; Suter, U. W.; *Macromol. Chem. Phys.* **1996**, *197*, 4253.

(14) Andrade, A. P.; Neuenschwander, P.; Hany, R.; Egli, T.; Witholt, B.; Li, Z. *Macromolecules* **2002**, *35*, 4946.

(15) Li, X.; Loh, X.J.; Wang, K.; He, C.B.; Li, J. *Biomacromolecues* **2005**, *6*, 2740.

(16) Andrade, A. P.; Witholt, B.; Chang, D. L.; Li, Z. *Macromolecules* **2003**, *36*, 9830.

(17) Reeve, M. S.; McCarthy, S. P.; Gross, R. A. *Macromolecues* **1993**, *26*, 888.

(18) Impallomeni, G.; Giuffrida, M.; Barbuzzi, T.; Musumarra, G.; Ballistreri, A. *Biomacromolecules* **2002**, *3*, 835.

(19) Santaniello, E.; Ferraboschi, P.; Grisenti, P.; Manzocchi, A. *Chem. Rev.* **1992**, *92*, 1071.

(20) Knani, D.; Gutman, A. L.; Kohn, D.H. *J. of Polym. Sci.:Part A: Polym. Chem.* **1993**, *31*, 1221.

(21) Gross, R. A.; Kumar, A.; Kalara, B. *Chem. Rev.* **2001**, *101*, 2097.

(22) Kobayashi, S.; Uyama, H.; Kimura, S. *Chem. Rev.* **2001**, *101*, 3793.

(23) MacDonald, R.T.; Pulapura, S. K.; Svirkin, Y.Y.; Gross, R.A. *Macromolecules* **1995**, *28*, 73.

(24) Henderson, L.A.; Svirkin, Y.Y.; Gross, R.A.; Kaplan, D.L.; Swift, G. *Macromolecules* **1996**, *29*, 7759.

(25) Dong, H.; W, H.D.; Cao, S.G.; Shen, J.C. *Biotechnology Letters* **1998**, *20*, 905.

(26) Cordova, A.; Iversen, T.; Hult, K. *Polymer* **1999**, *40*, 6709.

(27) Kumar, A.; Gross, R.A. *Biomacromolecules* **2000**, *1*, 133.

(28) Peeters, J.W.; Leeuwen,O.V.; Palmans, A.R.A.; Meijer,E.W. *Macromolecules* **2005**, *38*, 5587.

(29) Srivastava, R.K.; Albertsson, A.C.; *Macromolecules* **2006**, *39*, 46.

(30) Hans, Marc; Gasteier, Peter; Keul, Helmut; Moeller, Martin *Macromolecules* **2006**, *39*, 3184.

(31) Domb,A. J.; Kost, J.; Wiseman, D. M. *Handbook of biodegradable polymers,* Harwood: Amsterdam, **1997**.

(32) Kumar, A.; Gross, R.A.; *J. Am. Chem. Soc.* **2000**, *122*, 11767.

(33) Chen, C.; Fei, B.; Peng, S.; *J. Polym. Sci.:Part B: Polym. Phys.* **2002**, *40*, 1893.

250g Mechanical Properties of Semicrystalline Multiblock Copolymers

Manas R. Shah and Venkat Ganesan. Department of Chemical Engineering, The University of Texas at Austin, CPE 3.402, 1 University Station, Code C0400, Austin, TX 78712

The structure development in semicrystalline/rubbery multiblock copolymers involves complex interplay between two self-ordering mechanisms – microphase separation and crystallization of one of the components. Experiments have suggested connection between the mechanical properties and the molecular architecture of such linear copolymers, especially the role of bridging conformations in multiblock copolymers. We present a theoretical study to evaluate the bridging/looping fractions in these multiblock copolymers as a function of the morphology and the molecular architecture. We model the non-crystalline (rubbery) component as a flexible Gaussian chain and the crystalline component as a semiflexible chain with a temperature dependent rigidity and a favorable tendency to form parallel bonds. We calculate the domain spacing in lamellar phases of diblock copolymer where one flexible chain is attached to the semiflexible chain and compare the scaling exponents with existing scaling theories for semicrystalline diblock copolymers. Using self-consistent field theory, the bridging fractions of the various domains in these complex multiblock copolymers (triblock and pentablock) are evaluated as a function of the sequence of the chains and other parameters in the system namely, volume fraction of each domain and the rigidity of the semiflexible chains.

271a In Vivo Delivery of siRNA Targeting Tnf-α With Polyketal Nanoparticles

Sungmun Lee, Stephen Yang, Chen-Yu Kao, and Niren Murthy. Biomedical Engineering, Georgia Institute of Technology, 315 Ferst Drive, Atlanta, GA 30332

Small interfering RNA (siRNA) has great potential for the treatment of a variety of inflammatory diseases due to its ability to inhibit the expression of inflammatory genes. However, the clinical potential of siRNA has been limited by drug delivery problems. In this presentation, we present a new delivery vehicle for siRNA composed of polyketal nanoparticles. Polyketals are acid sensitive polymers which are designed to hydrolyze in the acidic environment of the phagosome and osmotically disrupt it. Polyketal-siRNA nanoparticles, targeting TNF-α, were fabricated by an oil in water emulsion method, generating nanoparticles that had average sizes of 800 nm - 900 nm. Cell culture experiments with macrophages demonstrated that polyketal nanoparticles dramatically improved the efficacy of TNF-α siRNA, in 10% serum. In vivo experiments with siRNA polyketal nanoparticles, targeting TNF-α, demonstrated that they were capable of reducing liver toxicity induced by LPS-mediated inflammation. Based on these results, we anticipate numerous applications of polyketal nanoparticles for the delivery of siRNA.

Combinatorial Design of Multiantibody-Targeted Immunolipopolyplex Nanoparticles of Oligonucleotides for Leukemia Therapy

Yuan Yuan[1,2], Bo Yu[1,2], Chaofang Yue[2], Chee Guan Koh[1,2], Andrew Morss[3], Raj.Muthusamy[4], John Byrd[4], Greg Lafyatis[3], Michael Paulaitis[2] and L. James Lee[1,2]*

The Ohio State University, 140 West 19th. Avenue, Columbus, OH 43210 (1) Nanoscale Science and Engineering Center for Affordable Nanoengineering of Polymeric Biomedical Devices, (2) Department of Chemical and Biomolecular Engineering, The Ohio State University,, (3) Department of Physics, (4) Comprehensive Cancer Center

Due to the sequence-specific hybridization and highly effective inhibiting translation of specific mRNA, oligonucleotides (ONs) such as antisense ODN and siRNA-based therapy has made considerable development and represented a new era of cancer therapy in the past several years. Unmodified ONs are vulnerable to exo- and endonucleases and are rapidly degraded *in vivo*. To address these drawbacks, polymeric and liposomal nanoparticles have been used as carriers to deliver ONs to down-regulate targeted mRNAs. Our group has successfully synthesized ON-containing immunolipopolyplex nanoparticles (ILPs) for the therapy of Chronic Lymphocytic leukemia (CLL), which is the most common adult leukemia in the world. To further improve the selectivity and specificity to target tumor cells and alleviate the side effects, multiantibody-conjugated ILPs were developed for B-CLL therapy through a combinatorial design. A microfluidics based antibody microarray with high throughput was used to study the combinatorial effects of antibodies for specific cell binding. Optical tweezer was applied to quantify the dynamic interactions and binding force of cell surface receptors and antibodies. The synergistic/additive cell internalization and bcl-2 down-regulation of ILPs targeted by selected antibody combination at different molar ratios were investigated by confocal microscopy, flow cytometry and Western blotting. Good agreement was observed between the microarray and ILPs results for both cell lines and patient cells.

271c Utilizing pH-Responsiveness and Mixed-Amine Ratio to Create An Enabling Technology for Effective Gene Delivery

Louisa R. Carr and Shaoyi Jiang. Chemical Engineering, University of Washington, Benson Hall, Box 351750, Seattle, WA 98195

A key limitation to the development of human gene therapy is the lack of safe, efficient, and controllable methods for gene delivery. Current research focuses on non-viral gene delivery agents, but these generally lack the required efficacy, cause toxicity due to both the materials used and nonspecific delivery, and are unstable in vivo when administered systemically. The engineering of specialized vehicles capable of overcoming various gene delivery barriers is critical to achieving successful gene transfection. This work describes efforts to create an integrated platform based on pH-responsive zwitterionic carboxybetaine methacrylate (CBMA) polymers. Zwitterionic materials are superlow-fouling in blood, biodegradable, readily tunable, and biomimetic (nontoxic). The key attribute of CBMA-ester polymers that makes them ideal for gene delivery lies in the fact that they are positively charged in their native form when they condense DNA, but they become zwitterionic CBMA in an acidic environment (as in the endosome). This zwitterionic state repulses the DNA to unpackage the nanoparticle with a remaining biomimetic CBMA coproduct. Here we synthesize ester-based CBMA polymers with mixtures of differently-substituted amines and report an optimal ratio that balances the advantages of secondary, tertiary and quaternary amines.

In the first part of this work, we verified the nonfouling properties and pH sensitivity of mixed-amine copolymers. Secondary amines are potentially proton sponging, tertiary amines are proton sponging and positively charged under slightly acidic conditions for DNA condensation, and quaternary amines are positively charged under all pH conditions to ensure DNA condensation. We measured the nonspecific protein adsorption, DNA binding and release, and the pH responsiveness of CBMA-ester polymer brushes containing mixtures of secondary/tertiary and tertiary/quaternary amine groups. In the second part of this work, the biophysical properties of CBMA-ester polymers containing different ratios of secondary/tertiary amines and tertiary/quaternary amines were studied, as were the corresponding nanoparticles. Specifically, we examined how changing the amine content of the polymers affects the polymer's ability to condense DNA into a nanoparticle, the resulting nanoparticle size and surface charge, the pH responsiveness of the bulk polymer and its corresponding nanoparticle, and the buffering ability of the nanoparticles. Finally, we have successfully transfected cells with the luciferase plasmid, and here we compared the transfection efficiencies of mixed-amine CBMA-ester polymers. We have correlated the transfection efficiencies with the biophysical characteristics and chemical properties of the polymers. Here we identify tertiary/secondary and tertiary/quaternary amine ratios that most effectively balance endosomal escape and nanoparticle packaging to induce COS-7 cells to express luciferase.

271d The Extracellular Matrix Environment Modulates Non-Viral Gene Transfer to Mouse Mesenchymal Stem Cells

Anandika Dhaliwal and Tatiana Segura. Chemical and Biomolecular Engineering, University of California-Los Angeles, 5531 Boelter Hall, 420 westwood boulevard, Los Angeles, CA 90095

The extracellular matrix environment modulates non-viral gene transfer to mouse mesenchymal stem cells

Genetically modified bone marrow derived mesenchymal stem cells (MSCs) have proven to be efficient cell carriers for local or systemic delivery of therapeutics for the treatment of diseases as well as for growth factors to augment tissue formation. Moreover, the effective transfection of MSCs inside tissue engineering constructs would transcend as an ideal approach to guide tissue formation in vitro and in vivo. However, the inability to efficiently transfect MSCs has limited these potential therapeutic applications. Most of the studies involving the enhancement of gene transfer have typically centered on improving the delivery vector systems, while the cellular microenvironment in which the cell resides and the cell itself are typically neglected. In the present study, the effect of structural extracellular matrix proteins, fibronectin (FN), laminin (LM), vitronectin (VT), collagen I (CL-I), and matrigel (MG), on non-viral gene transfer to cloned mouse MSCs (D1) is studied. In our gene transfer protocol cells are seeded on the immobilized structural proteins and DNA/PEI polyplexes are added post cell binding and spreading. Gene transfer was found to be significantly affected by the protein to which the cells were bound, with cells plated in MG having a statistically higher transgene expression than any other protein. In contrast, cells plated on collagen I resulted in statistically lower gene transfer efficiency than any other protein coating.

To determine the reason for MG enhancing and CL-I reducing the level of gene transfer we analyzed the morphology and proliferation for cells seeded on the ECM proteins. Although D1 cells were able to attach on all ECM proteins, cell spreading and the resultant morphology was different. For example cells plated on FN and LM resulted in well spread star-shaped cells with visually more actin stress fibers than cells seeded on CL-I, VT, and MG, while cells seeded on MG resulted in very elongated spindle cell morphology. Cells seeded on BSA coated wells (control) did not spread significantly. Surprisingly, although there were differences in cell morphology and spreading, cell proliferation was similar for all proteins except MG. Based on the above results it appears that gene transfer to D1 cells is affected by the cellular microenvironment were the cells reside, with ECM proteins that induce actin stress fibers and elongated spindle like cells resulting in more efficient gene transfer than ECMs that result in moderately spread cells. Proliferation rate was not able to explain the differences in gene transfer efficiency, with cells having the same proliferation rate resulting in drastically different gene transfer efficiency (e.g. FN vs CL-1) and cells with slower proliferation rate resulting in similar gene transfer rate as cells that had a faster proliferation rate (e.g. MG vs FN). We believe that the cellular microenvironment can be manipulated and engineered to achieve optimal non-viral gene transfer. The investigation of the effects of the cellular microenvironment on non-viral gene transfer will result into a comprehensive understanding of the limitations of gene transfer, which could further improve the current methods and vectors used for non-viral gene delivery.

271e Extracellular Matrix Protein Orientation by Adsorption on Self-Assembled Monolayers Controls Nonviral Gene Delivery

Beth A. Duensing, Biomedical Engineering Program, Biological Systems Engineering, University of Nebraska-Lincoln, 231 LW Chase Hall, Lincoln, NE 68583-0726 and Angela K. Pannier, Biological Systems Engineering, University of Nebraska-Lincoln, 231 LW Chase Hall, Lincoln, NE 68583-0726.

To date, most efforts to understand and improve the efficiency of nonviral gene delivery have focused on altering the physicochemical properties of delivery systems and developing new delivery strategies. The importance of the cellular microenvironment in achieving successful nonviral gene transfer has not been thoroughly examined, though proteins of the extracellular matrix (ECM) mediate cell adhesion and serve to regulate cellular behavior that may, in turn, dictate the cellular responsiveness to nonviral gene delivery. Self-assembled monolayers (SAMs) of alkanethiols on gold were used to control adsorption of ECM molecules and thus control the architecture of the extracellular environment presented to cells. SAMs of different surface chemistries were used to modulate the structure and activity of adsorbed ECM molecules to investigate how cells adhered to these ECM-immobilized SAMs respond to nonviral gene delivery. As an example ECM, fibronectin was immobilized to both hydrophobic and hydrophilic SAMs and mouse fibroblasts (NIH/3T3) were seeded onto the fibronectin- immobilized SAMs. DNA complexes (formed with plasmid encoding luciferase complexed with polyethylenimine or Lipofectamine 2000) were subsequently delivered as a bolus and transfection was assayed at 24 and 48 hours. Transfection on ECM-immobilized SAMs was measured using luciferase assays and compared to transfection in cells adhered to traditional polystyrene surfaces. Fibronectin immobilized on hydrophobic SAMs resulted in increased transfection levels relative to hydrophilic SAMs or control polystyrene surfaces. Furthermore, transfection was shown to increase as the concentration of fibronectin on the hydrophobic SAM was increased, indicating that hydrophobic interactions between the SAM and fibronectin resulted in a protein orientation that enhanced the ability of the cells to respond to gene delivery. A greater understanding of the ability of the extracellular environment to modulate gene delivery will enable the development of controlled and efficient nonviral delivery methods that incorporate appropriate ECM-cell interactions. In particular, controlling the architecture of the ECM could serve as an approach to enhance the ability of cells within a tissue engineering scaffold to internalize DNA, where gene delivery can present chemical factors to guide tissue formation.

271f In Vitro Gene Delivery with Non-Cytotoxic Cationic Oligopeptide Amphiphiles

Nikken Wiradharma, Drug Gene and Protein Delivery; Chemical and Biomolecular Engineering, Institute of Bioengineering and Nanotechnology; National University of Singapore, 31 Biopolis Way, #04-01, The Nanos, Singapore 138669, 21 Lower Kent Ridge Road, Singapore 119077, Singapore, Singapore, Yen Wah Tong, Department of Chemical and Biomolecular Engineering, National University of Singapore, 4 Engineering Drive 4, Singapore, 117576, Singapore, and Yi Yan Yang, Drugs, Proteins and Genes Delivery, Institute of Bioengineering and Nanotechnology, 31 Biopolis Way #04-01, Singapore, Singapore.

In this study, cationic and amphiphilic oligopeptides with a sequence A12H5K10 and A12H5K15 were designed and tested as a non-viral gene vector. The amphiphilic oligopeptides exhibit a critical micelle concentration at around 0.9 and 1.1 mg/mL, respectively, indicating their ability to self-assemble into micelle form in aqueous environment. The formation of the micellar nanoparticles were also confirmed by scanning electron microscopy. Both peptides were able to condense DNA at small N/P ratio, with a size of around 300-400nm, and zeta potentials ranging from +7-18 mV. We also hypothesized that the pre-formation of micelle before complexation increases cationic surface density, hence providing stronger DNA binding capacity of the vector. As such, control study with a peptide without hydrophobic block, H5K10, was also performed to evaluate this hypothesis. In addition, DNAse degradation of DNA-bound peptide complex also provided strong evidence on this hypothesis. To evaluate the efficacy of the cationic peptide amphiphiles, cytotoxicity test and luciferase transfection test were performed for all the three peptides against HepG2, HEK293, and 4T1 cell lines. All these studies were performed in comparison with PEI as the positive controls. The results showed that our designed oligopeptides were able to induce in vitro gene expression at about one order of magnitude less than PEI, if not comparable, with significant reduction in cellular cytotoxicity.

271g Cationic Core/Shell Nanoparticles Self-Assembled from Cholesterol-Conjugated Oligopeptides as An Efficient Gene Delivery Vector

Fanny Tandiono, Xin Dong Guo, Chuan Guan Tan, Ding Yue Khor, Majad Khan, and Yi Yan Yang. Drugs, Proteins and Genes Delivery, Institute of Bioengineering and Nanotechnology, 31 Biopolis Way #04-01, Singapore, Singapore

To avoid adverse immunogenic response invoked by viral vectors, increasing non-viral alternatives including peptide-based carriers have been proposed. In this paper, cationic and amphiphilic cholesterol conjugated oligopeptides containing a histidine block (5 or 10 residues) and an arginine block (10 residues), i.e. Chol-H5R10 and Chol-H10R10 were designed and synthesized. They were self-assembled into cationic core/shell nanoparticles in aqueous solution at concentrations of 17.8 and 28.2 mg/L respectively. These cationic nanoparticles condensed DNA efficiently. After being complexed with DNA, the particles had small average size of about 200 nm and high zeta potential of about 60 mV at higher N/P ratios (molar ratio of histidine and arginine content in peptide to phosphorus content in DNA). The nanoparticles induced much higher luciferase expression level in both human embryonic kidney HEK293 and human liver carcinoma HepG2 cell lines than the peptides without cholesterol because the formation of nanoparticles increased the local density of cationic charge, leading to more efficient DNA binding and higher zeta potential of DNA complexes. The luciferase expression level mediated by Chol-H10R10 nanoparticles was almost 10 times higher than that induced by Chol-H5R10 nanoparticles probably because an increased length of histidine led to higher capacity of endosomal escape and stronger DNA binding ability. The luciferase expression level induced by Chol-H10R10 nanoparticles was comparable to that mediated by PEI at the optimal N/P ratio (i.e. 10), but the cytotoxicity of nanoparticles was lower. In conclusion, Chol-H10R10 nanoparticles may be a promising gene delivery carrier.

271h Covalent Tethering of Plasmid DNA for Substrate-Mediated Gene Delivery

Kory M. Blocker[1], Kristi L. Kiick[2], and Millicent O. Sullivan[1]. (1) Chemical Engineering, University of Delaware, 150 Academy Street, Newark, DE 19716, (2) Materials Science and Engineering, University of Delaware, 201 DuPont Hall, Newark, DE 19716

Gene delivery enables mammalian cell manipulation necessary for disease treatment, tissue development, and comprehension of biochemical functions and cellular response.[1] Currently, non-viral gene delivery is accomplished by bolus delivery of cationic polymer-complexed DNA, which is limited by mass transport and deactivation processes. For tissue engineering applications, incorporation of DNA onto biomaterial scaffolds may avoid these limitations and increase cell transfection by maintaining a high concentration of DNA in the cellular microenvironment.[2] Surface immobilization of DNA also allows spatial targeting of cells.[2,1] Presently, DNA immobilization to a surface is dependent upon molecular interactions between DNA packaging materials and the surface, and requires careful design to support cellular uptake of the DNA.[3] As an alternative, covalent binding of the DNA to a substrate via a labile peptide sequence would allow surface immobilization and greater control over cellular transfection. Our design focuses on chemical functionalization of plasmid DNA to form the vector-surface covalent bond. This is accomplished using a peptide nucleic acid (PNA) and peptide-based surface tethering system. PNA is a DNA analog that sequence-specifically hybridizes with DNA to form a highly stable conjugate but does not interfere with the DNA transcriptional activity. DNA-PNA-peptide (DPP) conjugates may be directly linked to a variety of biomaterial surfaces. The use of coupling peptides that include cell adhesive and matrix metalloproteinase-1 (MMP-1) degradable sequences[4] enables release and uptake of the complexes in a cell-responsive manner.

Formation of the DPP conjugate was validated using agarose gel electrophoresis and peptide estimation assays. After formation, the conjugates were attached to self-assembled monolayer-functionalized gold model surfaces.[5] Atomic force microscopy was used to validate attachment and to observe the structures of the attached conjugates. The MMP-1 sensitivity of the coupling peptide has been studied with reverse phase HPLC. We are currently exploring the kinetics of conjugate release upon exposure to enzyme. After further validation of the tethering system via surface analysis techniques as well as cellular transfection studies, this chemistry will be translated from 2D model scaffolds to 3D tissue engineering scaffolds, and will be employed to promote the correct cellular responses necessary for tissue engineering.[6]

1) Gersbach, C.A. et al. Biomaterials 2007; 28; 5121-5127.

2) Segura, T. et al. Journal of Controlled Release 2003; 93; 69-84.

3) Jang, J. et al. Journal of Biomedical Materials Research Part A 2005; 77A; 50-58.

4) Lutolf, M.P. et al. Advanced Materials 2003; 15; 888-892.

5) Pannier, A.K. et al. Acta Biomaterialia 2005; 1; 511-522.

6) Segura, T. et al. Biomaterials 2005; 26; 1575-1584.

272a Atomic Layer Deposition Surface Modified Porous Polymer for Tissue Engineering Applications

Xinhua Liang[1], Aaron D. Lynn[1], David M. King[2], Stephanie J. Bryant[1], and Alan W. Weimer[3]. (1) Chemical and Biological Engineering, University of Colorado, 1111 Engineering Drive, Campus Box 424, Boulder, CO 80309-0424, (2) Department of Chemical Engineering, University of Colorado, 1111 Engineering Dr., Campus Box 424, Boulder, CO 80309, (3) Department of Chemical and Biological Engineering, University of Colorado, 1111 Engineering Drive, Boulder, CO 80309-0424

The overall research goal is to develop novel porous polymer/ceramic composite materials for bone tissue engineering applications. Towards this goal, the main objective of this study is to demonstrate the feasibility of using atomic layer deposition (ALD) to coat the surface of porous polymers for bone scaffolds. As a proof of concept, highly porous poly(styrene-divinylbenzene) (PS-DVB) particles were employed and were successfully coated with alumina and titania films at the atomic level in a fluidized bed reactor using a low-temperature scalable particle ALD process. Alumina and titania films were deposited by the alternating reactions of trimethylaluminum (TMA) and H2O at 33 °C, and titanium tetrachloride (TiCl4) and H2O2 (50 wt.% in H2O) at 100 °C, respectively. Analytical characterization revealed that conformal alumina and titania films were grown on internal and external polymer particle surfaces. ALD provides a controllable method for reinforcing porous polymeric structures while modifying surface properties. The ALD layers can improve the bioactivity and cell adhesion of the polymer substrate. The bioactive behavior of the composites was assessed by the formation of bone-like apatite on the material surfaces. The improved cell adhesion was demonstrated by protein adsorption of the composites from fetal bovine serum (FBS) as well as NIH/3T3 cell attachment and spreading on particle surfaces.

272b Antibacterial Silver-Containing Nanocomposites for Bone-Defect Repair

Oliver D. Schneider[1], Stefan Loher[1], Robert N. Grass[1], Tobias J. Brunner[1], and Wendelin J. Stark[2]. (1) Department of Chemistry and Applied Biosciences, ETH Zurich, Wolfgang-Pauli-Strasse 10, Zurich, 8093, Switzerland, (2) Chemistry and Applied Biosciences, ETH Zurich, Wolfgang-Pauli-Str. 10, ETH Hönggerberg, Zurich, 8093, Switzerland

In reconstructive surgery, bioresorbable implant materials are in great demand for the repair of bone defects. Amorphous calcium phosphate nanoparticles (ATCP) [1] with biodegradable poly(lactide-co-glycolide) (PLGA) proved to be a nanocomposite suitable for this purpose. However, bone reconstruction often suffers from high bacteria concentration in surrounding tissues. To overcome this problem, 1-2 nm Ag containing scaffolds inhibiting antibacterial properties were introduced.

In vitro bioactivity tests of these wool-like electrospuncomposite PLGA/ATCP (80 : 20) containing 0.5 wt% silver showed rapid hydroxyapatite deposition on the nanocomposite within 2 days. Antibacterial tests using *E. coli* demonstrated a strongly prolonged antibacterial effect of the scaffolds containing finely dispersed silver on TCP if compared to current clinically used methods based on soaking the scaffolds with a tetracycline solution prior to implantation. These results indicate the potential for a broader use of silver in flexible implant materials and suggest applications in non-load bearing bone defects of lesions with high exposure to microorganisms, for example in dental surgery.

Figure: Scanning electron microscopy images after incubation in suspension of *E. coli* at 37 °C under shaking conditions: (a) PLGA/TCP 80 : 20 and (b) silver doped nanocomposite after 8 h immersion both showed small holes on the fibres; (c) silver containing nanocomposite after 160 h immersion (*E. coli* containing simulated body fluid changed almost every 12 h) showing the deposition of a hydroxyapatite layer.

References

[1] S. Loher, W.J. Stark, M. Maciejewski et al., Chem Mater 2005;17(1):36-42.

[2] O.D. Schneider, S. Loher, T.J. Brunner, L. Uebersax, M. Simonet, R.N. Grass, H.P. Merkle, W.J. Stark, J. Biomed Mater Res B, 2007, 84B(2), 350-62.

[3] O.D. Schneider, S. Loher, T.J. Brunner, P. Schmidlin, W.J. Stark, J. Mater. Chem., (2008)

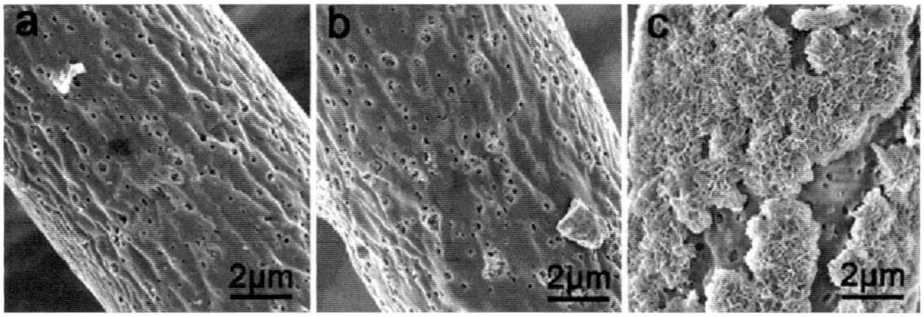

272c In Situ Synthesis of Hydroxyapatite in Block Copolymer Hydrogels

David M. Griffin and Surita R. Bhatia. Department of Chemical Engineering, University of Massachusetts, 159 Goessmann Laboratory, 686 North Pleasant St., Amherst, MA 01003

Nanocomposites of hydroxyapatite in polymeric matrices have a number of important applications in bone and cartilage tissue engineering. Most studies to date have focused on systems where the polymeric matrix is a homopolymer gel or block copolymer solution. Use of a block copolymer hydrogel may offer additional control over crystallite size and morphology. Here we report on our efforts to synthesize hydroxyapatite in situ in block copolymer gels. Inorganic ceramic nanocomposites were formed by diffusion of calcium ions into a phosphate-containing Pluronic® F127 hydrogel. Initial pH of the gel prior to nucleation strongly influenced the final crystal structure and morphology. Nucleation at near physiological pH preferentially produced a highly crystalline calcium phosphate phase on the millimeter scale whereas mineral growth at alkaline conditions was found to produce hydroxyapatite crystals in the micrometer size range. Calcium phosphate mineral phases were determined by powder x-ray diffraction (XRD) and substantiated through energy dispersive x-ray spectroscopy (EDS). Morphology of the composites was examined using scanning electron microscopy (SEM). Rheological studies on these systems show that higher elastic moduli are obtained with composites prepared using the in situ synthesis technique as compared to conventionally prepared polymer-hydroxyapatite composites. The biomimetic nature of our investigation suggests that composites formed by this in situ technique my have significant biomaterial and drug delivery applications.

FIBER REINFORCED HYDROGELS AS A SYNTHETIC MENISCUS REPLACEMENT

Julianne L. Holloway, Giuseppe R. Palmese, Anthony M. Lowman
Chemical and Biological Engineering, Drexel University, Philadelphia, PA

The meniscus is a semi-lunar fibrocartilage disc present in the knee that is important primarily for weight bearing and shock absorption.[1-3] Injuries in the meniscus occur frequently and over time lead to degeneration of the knee joint.[4] Most injuries occur in the inner two-thirds of the meniscus, which is avascular and generally not surgically repairable. Initially, the meniscus was regarded as an unnecessary appendage and was completely removed at the first sign of injury. In the past few decades, however, total meniscectomy has been shown to lead to severe joint degeneration. Partial meniscectomy is currently the most common treatment for a meniscal injury; however, degradation of the joint still occurs. Allograft transplantation has been used to replace injured menisci and is capable of alleviating pain and swelling; however, the long-term results of this procedure are still unknown. Availability, donor matching, and the possibility of transferring diseases limit allograft transplantation in its use as well.[1,3]

Current research is attempting to regenerate the meniscus using tissue-engineering concepts and to evaluate the use of autografts and allografts as a replacement meniscus. Tissue-engineered scaffolds show promise; although, poor initial mechanical properties limit the use of these scaffolds as complete meniscal replacements. Autografts and allografts are primarily ill suited to replace the meniscus due to size mismatches between the donor tissue and the original tissue. Limited research has been performed on the development of a synthetic meniscus. The selection of the biomaterials to use for the implant has shown to be a significant issue.[1]

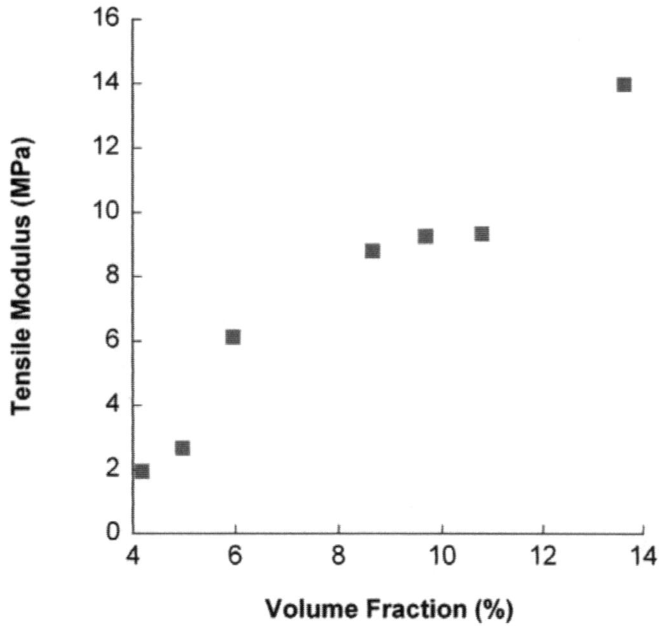

Figure 1 Tensile modulus of polypropylene reinforced hydrogels as a function of fiber volume fraction

Fiber reinforced hydrogels have been selected as a possible biomaterial capable of withstanding the forces present in the knee during loading. Several different fiber types including polypropylene and ultra high molecular weight polyethylene (UHMWPE) have been chosen for reinforcement. Preliminary work synthesizing and characterizing these composites has shown significant increases in tensile modulus by incorporation of polypropylene fibers within the hydrogel matrix. Figure 1 shows a tensile modulus of approximately 14 MPa at a polypropylene fiber volume fraction of 14% compared to the hydrogel tensile modulus of 0.3 MPa.

Environmental scanning electron microscopy (ESEM) was used to look at the interface between the fiber and hydrogel. As shown in Figure 2, circular voids are present around the polypropylene fibers. These voids are believed to be a result of poor adhesion between the fiber and hydrogel matrix.

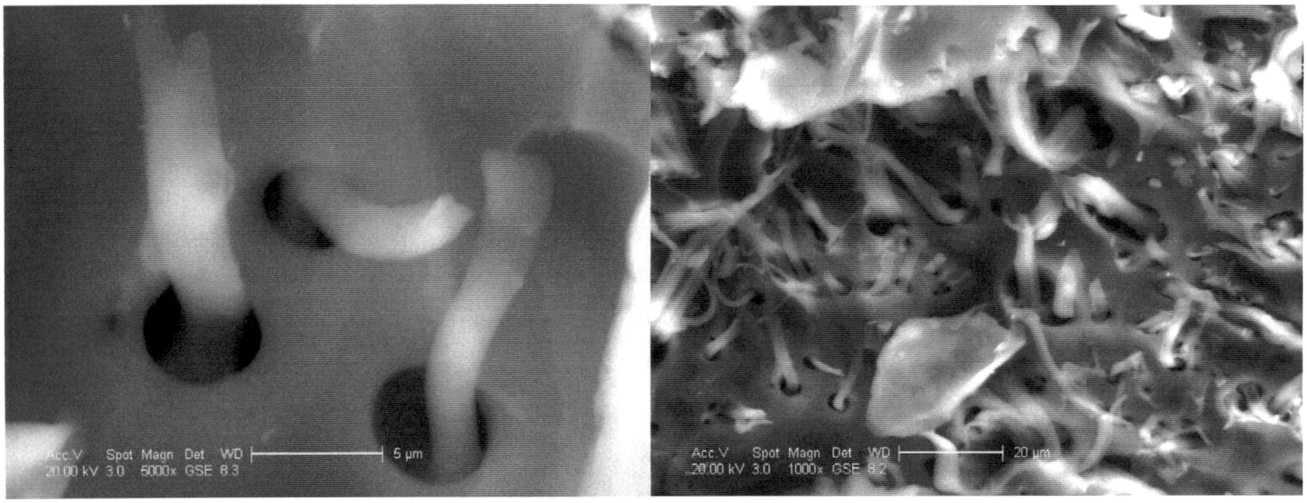

Figure 2 ESEM images of polypropylene reinforced hydrogel composites

Additional research was performed on composites synthesized using UHMWPE fibers; however, all of the samples tested in tension experienced interlaminar failure in between 5% and 15% strain. The tensile modulus before failure was approximately 20.8 MPa at a fiber volume fraction of 10%.

Polypropylene reinforced hydrogels successfully satisfy the loading conditions of the meniscus in areas that experience minimal loading, such as at the surface and radially. UHMWPE reinforced hydrogels show promise for the areas of the meniscus that experience increased loads, such as circumferentially. However, continued research needs to examine the impact of fiber orientation and interfacial strength in order to minimize the occurrence of failure.

1. Buma, P.; Ramrattan, N. N.; van Tienen, T. G.; Veth, R. P. H., Tissue engineering of the meniscus. *Biomaterials* **2004,** 25, (9), 1523-1532.
2. Rodeo, S. A.; Kawamura, S., Form and Function of the Meniscus. *Orthopaedic Basic Science* **2007**, 175-189.
3. Rath, E.; Richmond, J. C., The menisci: basic science and advances in treatment. *British Journal of Sports Medicine* **2000,** 34, (4), 252-257.
4. Rodkey, W., Basic biology and response to injury. *Instr Course Lect* **2004,** 49, 189-93.

272e Mechanical Performance and Microstructure of Biomimetic PEG-Agarose Interpenetrating Networks as Determined by Dynamic Mechanical Analysis and AFM

Joseph Lomakin, Chemical and Petroleum Engineering, Kansas University, The University of Kansas, 4132 Learned Hall 1530 W. 15th Street, Lawrence, KS 66045, Michael Detamore, Dept. of Chemical & Petroleum Engineering, University of Kansas, 1530 W. 15th St., 4132 Learned Hall, Lawrence, KS 66045, and Stevin H. Gehrke, Departments of Chemical & Petroleum Engineering and Pharmaceutical Chemistry, University of Kansas, 1530 W 15th Street, Lawrence, KS 66045-7609.

The combination of rigid macromolecules embedded in a ductile matrix is common to a wide variety of biological structural composites such as the mammalian extracellular matrix, insect cuticle and crustacean exoskeleton. Novel synthetic materials that mimic this biological organization have recently been shown to possess exceptional mechanical properties with Young's modulus and toughness values orders of magnitude greater than that of individual composite components.

In keeping with this biological design principle, an interpenetrating network (IPN) of PEG-diacrylate in agarose is synthesized and investigated as a novel composite material mimetic of the insect cuticle. The cuticle is largely composed of chitin fibers interpenetrated with a protein matrix. In this synthetic construct, agarose is selected to mimic the role of chitin and the PEG-diacrylate to mimic the role of the cuticular proteins. These materials were also selected because of their utility in a variety of biomedical applications. The interpenetrating networks are formed by photopolymerization of thermally gelled agarose gels soaked in aqueous solution of PEG-diacrylate (700 Da). The resulting IPN hydrogels as well as single networks of crosslinked PEG-diacrylate and agarose are investigated with respect to their compressive mechanical performance via transient and dynamic mechanical analysis. The toughness of initial formulations of agarose-PEG-DA gels has been determined to be 100 times greater than that of agarose and 5 times greater than that of a crosslinked PEG-diacrylate network. The biological potential of the synthesized networks is further established by demonstrating effective chondrocyte encapsulation with high long term cell viability within the novel biomaterial.

The mechanical properties obtained in macroscopic studies depend on the nanoscale material structure and mechanical properties as probed by submerged AFM. AFM is used to determine pore sizes and microscale organization of agarose and PEG-DA polymers. The heterogeneity of compression modulus on the micron scale is assessed, as this is considered an important parameter underlying the exceptional mechanical properties of interpenetrating networks.

Optimization of the nanofibrous structure of non-woven mats of electrospun biodegradable nanocomposites using Response Surface Methodology

A. Tsimpliaraki[1,2], S. Svinterikos[1], S. I. Marras[2], I. Zuburtikudis[2*] and C. Panayiotou[1]

[1]*Department of Chemical Engineering, Aristotle University of Thessaloniki, Thessaloniki, Greece*

[2]*Department of Industrial Design Engineering, TEI of Western Macedonia, Kozani, Greece*

Abstract

Electrospun mats of polymer-clay nanocomposites were prepared in order to study the influence of material and process parameter settings on their morphology. The polymer solution concentration, the flow rate of the injected solution and the organically-modified-clay content of the polymer matrix were the factors chosen to be investigated according to a design of experiments (DoE) within the context of response surface methodology (RSM). The developed quadratic models and the individual and coupling effect of the three factors examined are given. The results suggest that the dominant parameter affecting mats' morphology is polymer solution concentration and that a broader range in the factor settings, especially for concentration, should be used in a subsequent optimization.

Introduction

Presently, nanofiber manufacturing is one of the key advancements in nanotechnology. For this reason, electrospinning has been widely used as an alternative technique in fabrication of mats consisting of fiber diameters ranging from few nanometers to few micrometers. This is a markedly simple and cost effective process which operates on the principle that the solution is extracted under a high electric field. The final fibrous structure can be tailored by altering the material and process parameters involved, such as the concentration of the polymer solution, the applied voltage, the flow rate, the diameter and angle of the spinneret, the spinning distance, the solution's viscosity and conductivity, just to mention a few [1]. Furthermore, the combination of their nanometric dimensionality, high surface area, porosity, flexibility and mechanical integrity makes the electrospun fibers suitable for several value-added applications, such as tissue engineering, wound dressing, clothing protection, nanoscale and biological adsorption, filter and membrane technology.

During the last decade various polymers, most of which dissolved in solvents, but also some in melt form, have been successfully electrospun into ultra fine fibers [2]. However, there are few studies regarding the electrospinning of polymer/layered silicate

* Corresponding author:

Tel.: +30-24610-40161; fax: +30-24610-39682, e-mail address: izub@teikoz.gr (I. Zuburtikudis) nanocomposites [3-9] and particularly of biodegradable-polymer based nanocomposites, in spite of the fact that they are environmentally friendly with potential applications both as filters and as reinforcing components in other composite systems [2].

In the present work, a novel polymer – clay nanocomposite system is studied. A biodegradable aliphatic polyester was used as a polymer matrix, where organically modified montmorillonite is incorporated. Membranes comprised of various structures of the nanocomposite material were subsequently prepared via electrospinning according to a design of experiments (DoE). An investigation was conducted to identify and quantify the important electrospinning material and processing parameters that affect the morphology of the collected mats and to recognize the experimental settings domain useful for a subsequent optimization.

Experimental

Materials and methods

The biodegradable aliphatic polyester PBSA by the commercial name 'Bionolle 3001' was supplied by Showa Highpolymer Co., Ltd (Tokyo, Japan). Bionolle 3001 is a copolyester of succinic acid (S), adipic acid (A) and 1,4 butanediol (B) with a composition ratio 40/10/50, respectively. The nanofiller used, was organically modified montmorillonite (Cloisite 25A) and was purchased from Southern Clay Products (Texas, USA), while dichloromethane (CH_2Cl_2) was obtained from Sigma-Aldrich.

Initially, the nanocomposite materials were prepared in bulk as films according to the same solution casting routine followed in our previous studies [10,11]. The films obtained were dried in vacuum at 40 ^oC for 24 h. The organoclay content of the produced films was ranging from 0 to 9 wt%. The electrospinning set up used for nanofibres manufacturing was homemade. Dichloromethane was used as a solvent for the preparation of the solutions from the nanocomposite stock materials previously prepared.

The membrane structures were explored by Scanning Electron Microscopy (SEM) using a JEOL microscope, (mod. JSM-840A) and processed by the appropriate software. The generation of the D-optimal design and the subsequent regression analysis was attained with the help of MODDE™ 8.0 software.

Experimental design and responses

The DoE used included 25 experimental runs, four of which were pure replicates (the experimental space investigated is shown in Fig.1). Three variables were chosen as controllable factors; flow rate (qualitative factor), solution concentration and clay loading in polymer nanocomposite (quantitative factors).

Three responses were defined in order to achieve a more reliable representation of the experimental outcomes. A full picture of each sample's morphology is pursued by the ratio of the average bead and fiber diameters, D_{bead}/ D_{fibre}, and the number surface density of beads, N_{bead}, as well as the average fiber diameter, D_{fibre}.

Results and discussion

Morphology

SEM photographs of the electrospun mats resulted in five different morphologies depicted at Fig.1: Fibers and a small amount of spindle-like beads (Exp.No4), fibers with a large amount of spindle-like beads (Exp.No16), pearls-on-string morphology (Exp.No25), only beads (Exp.No9) and, finally, there were two experiments (Exp.No13&18) for which the solution was cast from the spinneret on the collector without previous solvent evaporation. This was observed at low solution concentrations and clay loadings and at very high flow rates. Therefore, it was impossible to take a measurement for any of the three responses at these experimental runs. That reduced the 25 runs to 23.

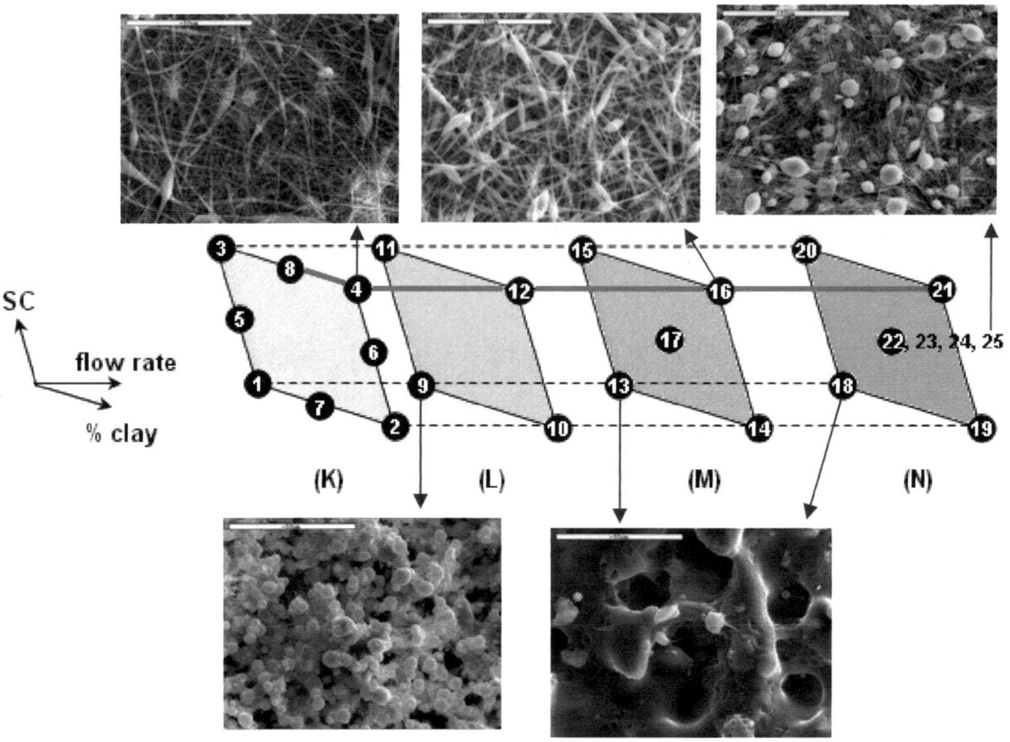

Fig. 1. The five indicative morphologies observed overall by SEM at various combinations of factor settings; scale bar at 50 μm. The numbers correspond to the 25 experiments conducted.

In general, SEM observations indicate that the higher the polymer solution concentration and clay loading, the more uniform, bead free and fibrous the structure tends to be.

Regression analysis

Multiple linear regression analysis (MLR) was carried out to evaluate the contribution of the three factors to the morphology of the nanocomposite collected mats and to recognize the proper experimental settings space for optimizing their fibrous structure. It was found that the three measured responses ($y_1 \equiv D_{fibre}$, $y_2 \equiv N_{bead}$, $y_3 \equiv D_{bead}/D_{fibre}$) were not normally distributed and were skewed with a tail to the right. They were, therefore, transformed appropriately so that MLR could be conducted. The MLR analysis resulted in three quadratic model equations, one for each of the three responses.

Fig. 2 shows some results from the statistical analysis which refer to the goodness of fit (R^2) and of prediction (Q^2) of the quadratic equations for the transformed responses and to their validity and reproducibility, after the model has been refined for the insignificant terms present due to the 4-level qualitative factor of flow rate.

The Q^2 value of the refined model for Y_3 is small (<0.5) and the difference between R^2 and Q^2 is well far from the region 0.2 - 0.3. This suggest that the refined model for Y_3 cannot be trusted [12], although its model validity value (=0.5973) and its reproducibility (=0.8721) could be accepted. Thus, only the Y_1 and Y_2 model equations are taken into account in what follows.

(a)

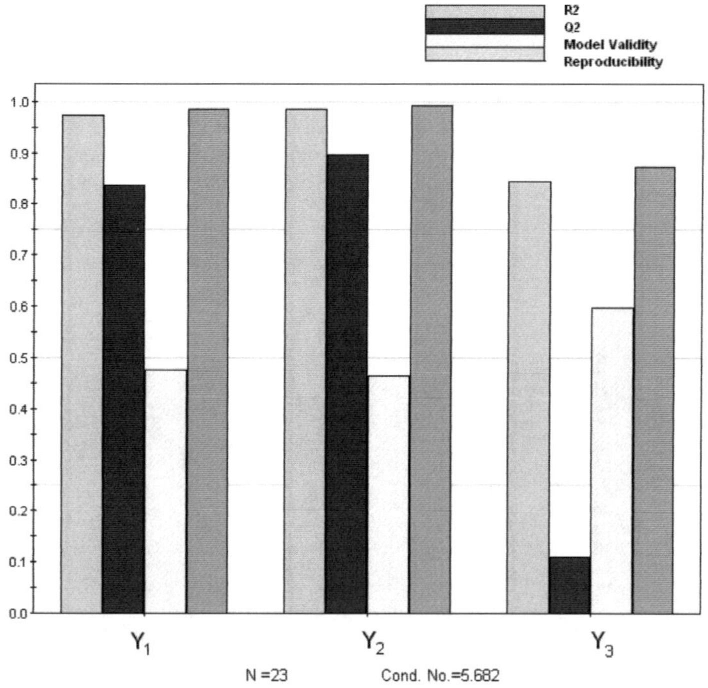

(b)

Response	R^2	Q^2	Model Validity	Reproducibility
Y_1	0.9730	0.8361	0.4762	0.9846
Y_2	0.9845	0.8970	0.4639	0.9915
Y_3	0.8443	0.1111	0.5973	0.8721

Fig. 2. Model check for each of the three transformed responses based on results from regression analysis after model refinement: (a) graphical representation, (b) summarized on a table.

Based on these results, the important factors in the electrospinning of the polymer nanocomposite can be recognized and optimal conditions can be obtained for predetermined fiber diameters and morphologies of the electrospun mats.

Important factors

In order to find the extent of the impact of the three factors to these responses, the plots of coefficients were obtained where the importance of the inputs (coefficients) can be ranked and presented in a bar chart as shown in Fig. 3. Since Y_1 and Y_2 have different ranges, the coefficients are normalized by dividing them with the standard deviation of their respective response. In absolute values, the higher the bar of a term, the more important the term is.

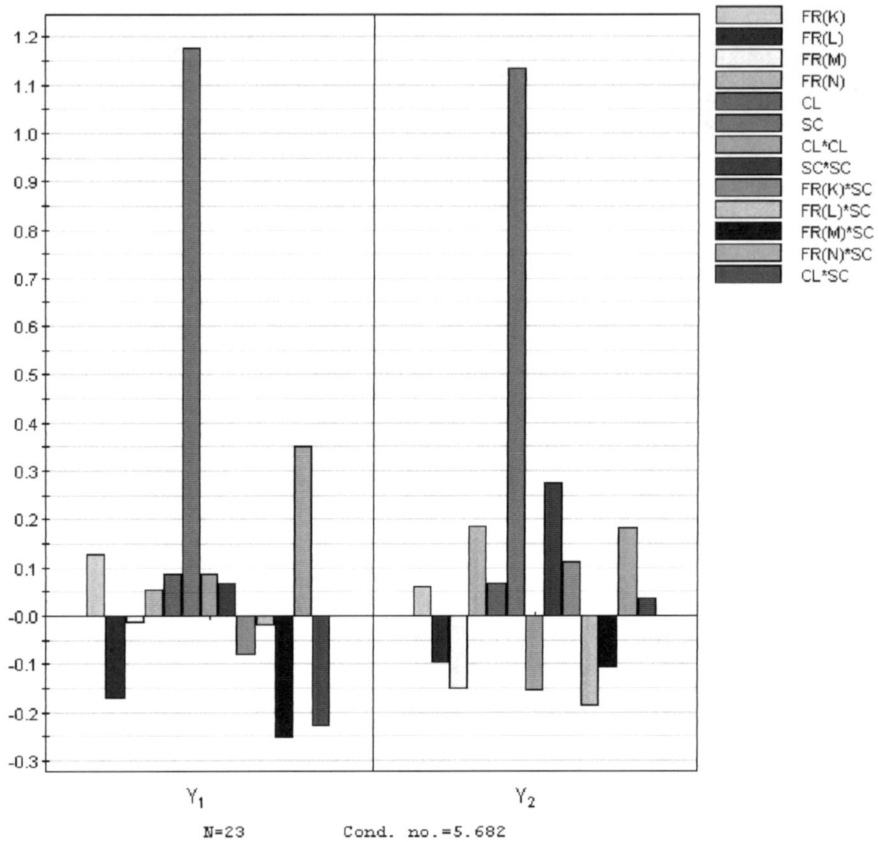

Fig. 3. The normalized coefficients plots for responses Y_1 and Y_2.

The most striking feature of the obtained plots (Fig.3) is that for both responses Y_1(=transformed D_{fiber}) and Y_2(=transformed N_{bead}), polymer solution concentration (SC) is the eminent parameter. This great impact of SC on the responses overshadows the effect of the other terms and possibly if SC values were chosen within a different and less wide range, then other trends might be revealed.

Response surface results

The contour plots for the two non-transformed responses D_{fiber} and N_{bead} were calculated with the aid of the corresponding model equations and were found similar for all four discrete levels of FR. The similarity of results for the four values of flow rate, which correspond to a tenfold increase in the flow rate, do suggest that this factor (FR) does not play any role on the fiber diameters and on the number surface density of beads. At Fig.4 the response surfaces for two flow rates are given.

For FR at the K-level (=0.5 ml/h), fibers at the 150 nm-scale of diameter are formed, when SC is low. Furthermore, when SC is low and below 7.5% w/v, the bead number density is high. Therefore, a mat is produced with fibers in the 150 nm-scale and with plenty of beads. On the other hand, the clay content in nanocomposite has only a minor effect on the diameter of the produced fibers (D_{fiber}), since the contours are almost parallel to each other and almost parallel to the horizontal axis, which is the clay content axis (the plot consists of almost horizontal strips of different colour that represents a different fiber diameter value). The contour plot for the beads number density (N_{bead}) is similar. It also consists of almost horizontal different-colour strips and therefore, the clay content has a minor effect on the number of the produced beads, as well. In general, all contour plots conjointly show that fiber diameters and number of beads per μm^2 are affected by the polymer solution concentration in opposite ways: increase of SC results in increase in D_{fiber} and decrease in N_{bead}. If fibers at the 500 nm-scale are desired, the SC has to be high, over 10% w/v and even higher. At these values of SC, a very small number of beads will be formed in accordance with the experimental findings.

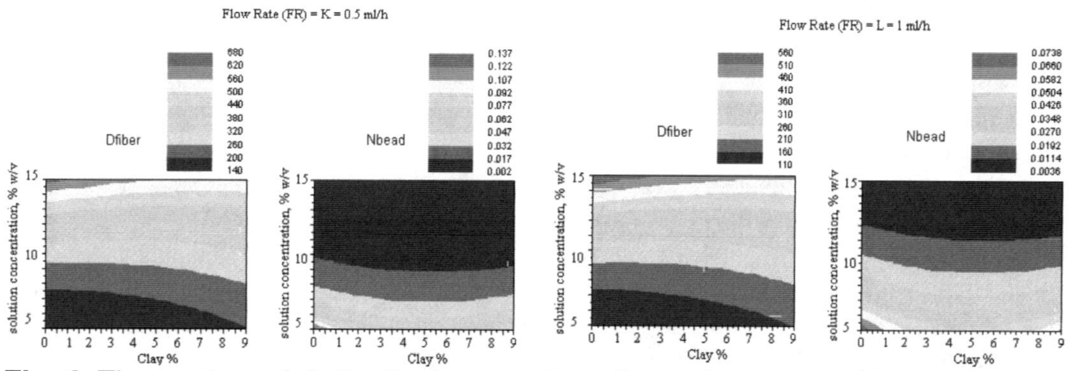

Fig. 4. The contour plots for the two non-transformed responses D_{fiber} and N_{bead} as a function of SC and CL at the 4 levels of FR.

For FR at the L-level (=1.0 ml/h), the same picture evolves. In this case, however, the window of very low bead number density is narrower than that for FR at the K-level. This only suggests that an even higher than 12% w/v solution concentration is necessary, if we are to produce fibers with just a few beads and diameters in the 500 nm-scale.

All the above response surface results are in good agreement with the experimental findings and show the way for further experimentation. The polymer solution concentration ought to be set in a more extended region; higher concentrations than the ones used here should be employed.

Conclusions

A statistically sound quadratic model was built for the fiber diameter and the bead number density of the electrospun non-woven mats of a biodegradable-polymer nanocomposite following the DoE methodology and using a 25-run D-optimal design. The conclusions of the present study can be stated as it follows:

1. The model showed and the experimental findings ensured that the most crucial factor for the morphology of the electrospun mats is polymer solution concentration.
2. The clay content in the nanocomposite has a minor effect on the fiber diameters and the bead number density, when compared to the effect of the solution concentration at the processing conditions chosen, while the flow rate does not play any role, despite its wide value range.
3. Fiber diameters and number of beads per μm^2 are affected by the polymer solution concentration in opposite ways.
4. Further experiments with a broader range in the experimental settings (particularly in solution concentration) should be designed incorporating even more factors such as the applied voltage and/or the room relative humidity in order to improve the process.
5. Optimization of the mat's morphology can be achieved by putting specific goals regarding the desired mat morphology and by choosing the experimental settings domain with the help of the results presented here.

References

1. Berkland C, Pack DW, and Kim K. Biomaterials 2004;25(25):5649-5658.
2. Huang ZM, Zhang YZ, Kotaki M, and Ramakrishna S. Composites Science and Technology 2003;63(15):2223-2253.
3. Ji Y, Li BQ, Ge SR, Sokolov JC, and Rafailovich MH. Langmuir 2006;22(3):1321-1328.
4. Fong H, Liu WD, Wang CS, and Vaia RA. Polymer 2002;43(3):775-780.
5. Ristolainen N, Heikkila P, Harlin A, and Seppala J. Macromolecular Materials and Engineering 2006;291(2):114-122.
6. Daga VK, Helgeson ME, and Wagner NJ. Journal of Polymer Science Part B-Polymer Physics 2006;44(11):1608-1617.
7. Wang A, Hsieh AJ, and Rutledge GC. Polymer 2005;46(10):3407-3418.
8. Lee YH, Lee JH, An IG, Kim C, Lee DS, Lee YK, and Nam JD. Biomaterials 2005;26(16):3165-3172.
9. Marras SI, Kladi KP, Tsivintzelis L, Zuburtikudis I, and Panayiotou C. Acta Biomaterialia 2008;4(3):756-765.
10. Marras SI, Tsimpliaraki A, Zuburtikudis I, and Panayiotou C. Journal of Physics: Conference Series 2007; 61, 1366-1370.
11. Marras SI, Tsimpliaraki A, Zuburtikudis I, and Panayiotou C. Polymer Engineering and Science, in press.
12. Eriksson L, Johansson E, Kettaneh-Wold N, Wikstrom C, and Wold S. Design of Experiments – Principles and Applications, ISBN 91-973730-0-1, 2000.

276a Microfluidic Hydrogel Cultures for Analysis of Oxygen-Dependent Angiogenic Signaling by Tumor Cells

Nak Won Choi[1], Daniel J. Brooks[2], Kang-Yeol Park[3], Lawrence J. Bonassar[2], Claudia Fischbach-Teschl[2], and Abraham D. Stroock[1]. (1) School of Chemical and Biomolecular Engineering, Cornell University, Ithaca, NY 14853, (2) Department of Biomedical Engineering, Cornell University, Ithaca, NY 14853, (3) Samsung Cheil Industries Incorporation, Uiwang-si, South Korea

Tumors consist of spatially and functionally distinct niches which may result from spatiotemporal variation in oxygen concentration (pO_2) within growing tumors. Cells located in the periphery of tumors are well nourished by diffusion of oxygen and nutrients from adjacent blood vessels (up to ~ 200 μm), whereas cells located in the center of the tumors are exposed to varying levels of hypoxia. We have recently shown that up-regulation of certain angiogenic factors (e.g. vascular endothelial growth factor) is likely due to low pO_2 in central hypoxic niches, whereas other angiogenic factors (e.g., interleukin-8) are only partially controlled by this aspect. However, further investigation with spatiotemporal control of pO_2 in tissues is required to more explicitly clarify the role of pO_2 in angiogenic factor secretion. In this study, we have designed and explored microfluidic tumor models to recreate and analyze distinct tumor niches by varying pO_2. Microchannels were embedded within alginate-based 3-D tumor models to allow for spatiotemporal control of pO_2 and subsequent analysis of angiogenic factor secretion. Our strategy includes: 1) mathematical modeling and experimental approaches to establish design requirements for microfluidic tumors, 2) a lithographic technique to build functional microfluidic structures within alginate hydrogels seeded with OSCC-3 (oral squamous carcinoma cells), 3) confirmation of our ability to control pO_2 with oxygen-sensing microbeads, and 4) characterization of angiogenic factor secretion as a function of spatiotemporal variations in pO_2.

276b 'marker of Self' Proteins on Synthetics - Exploiting the Body's Own System of Distinguishing Foreign from Self

Richard Tsai[1], Pia Rodriguez[1], and **Dennis E. Discher**[2]. (1) University of Pennsylvania, Room 129 Towne Building, 220 South 33rd Street, Philadelphia, PA 19104-6393, (2) Chemical and Biomolecular Engineering, University of Pennsylvania, Room 129 Towne Building, 220 South 33rd Street, Philadelphia, PA 19104-6393

Any particle injected or surface implanted in us or any other mammal must contend with Macrophages that have – for eons – swept up invading bacteria, yeast, and other pathogens. At the same time, Macrophages leave our own 'Self' cells alone. The Foreign vs Self difference certainly does NOT reside in steric repulsion by the glycocalyx, which some have argued is well-mimicked by the modern synthetic polymer PEG. We have sought to address how Macrophages specifically recognize Self and to define its physicochemical limits, and we have focused on a cell-surface protein found on all of our own cells. We demonstrate a two-step procedure for recognizing intruders that helps avoid misdirected attacks. In the first step, which is well known, Macrophages adhere and begin engulfing objects studded with antibodies or plasma complement proteins that bind interlopers and will also bind to the body's own cells. However, before a macrophage engulfs its target, it also gropes and checks for the second form of identification on all self cells, the protein CD47. The ~100 aa extracellular domain of CD47 is sufficient to induce a macrophage to disengage a cell from the same species or even a synthetic particle decorated with this domain. We will detail the divergence in Foreign vs Self adhesive signaling mechanisms, the dependence on protein densities, and the length scale dependence for this fundamental facet of immunocompatability.

276c Microengineered Co-Culture Substrates: Investigating Cell-Cell Interactions Using Selectively Degradable Hydrogels

Andreia S. Ribeiro[1], Filipa Maia[2], Ozlem Yasar[3], Binil Starly[3], and Jennie B. Leach[1]. (1) Chemical and Biochemical Engineering, University of Maryland Baltimore County, 1000 Hilltop Circle, Baltimore, MD 21250, (2) Departamento de Engenharia Química, Faculdade de Engenharia da Universidade do Porto, Porto, 4000, Portugal, (3) School of Industrial Engineering, University of Oklahoma, Norman, OK 73019

Interactions between neighboring cells regulate cell behavior as well as tissue function. Tools to investigate heterologous cell-cell interactions in vitro are therefore critically important for the development of new diagnostic and drug screening tools. Layered manufacturing based technologies have enabled the development of novel tools to study cell-cell and cell-material interaction for applications in tissue engineering. Among these techniques, microfabrication and stereolithography provide powerful means to pattern cells, but typically rely on immobilized adhesive domains that permit stronger affinity of one cell type for a specific biochemical group relative to a second cell type. While successful, these techniques generally cannot be applied to co-cultures of similar cell types (e.g., two types of neurons) that do not demonstrate selective adhesion to biochemical cues. In this work, we aim to demonstrate a new co-culture patterning approach based on the selective degradation of patterned gels thus allowing the culture of any two cell types for investigating cell-cell interactions in vitro. Using maskless projection micro-stereolithography (μSL), we developed defined multi-component micropatterned architectures of poly (ethylene glycol) dimethacrylate (PEG-DM) and carboxymethylcellulose-aminoethyl methacrylate (CMC-MA) on glass substrates. Since both polymer surfaces are resistant to cell and protein adsorption, the first cell type can only adhere to the free glass spaces. The degradation of the CMC-MA patterns by the enzyme cellulase provides new adhesive regions on the substrate where a second cell type can adhere. This technique allows control of cell-cell contact and spacing between two different cell types at the microscale and will be used as a new tool to investigate heterologous cell-cell interactions in co-culture systems.

276d Surface Functionalization for Selective Cell Attachment

Anka N. Veleva[1], Daniel Heath[2], Cam Patterson[3], and Stuart L. Cooper[2]. (1) Department of Chemical and Biomolecular Engineering, North Carolina State University, 911 Partners Way, Box 7905, Raleigh, NC 27695, (2) Department of Chemical and Biomolecular Engineering, The Ohio State University, Columbus, OH 43210, (3) Department of Medicine, The University of North Carolina at Chapel Hill, 8200 Medical Biomolecular Research Bldg., 103 Mason Farm Road, Chapel Hill, NC 27599

A common approach to direct cell and tissue responses at synthetic surfaces is to modify the material to mimic the extracellular matrix. Typically, short recognition sequences presenting cell binding motifs found in components of the extracellular matrix (e.g., the RGD or the YIGSR peptides) are immobilized on the material to promote cell adhesion via ligand-receptor interactions. Although surface functionalization remains one of the most promising strategies, the clinical utility of this approach is limited by the need to develop materials that are optimized to promote adhesion of a specific cell type. Towards this end, we designed and implemented a high-throughput protocol for screening a phage display peptide library to identify and select novel peptide ligands that bind with high affinity and specificity to human blood outgrowth endothelial cells (HBOEC). Here we demonstrate the use of the phage display-selected-HBOEC-specific peptides as a tool to direct and modulate endothelial cell (EC) behavior with a focus on designing functional biomaterials intended for use in cardiovascular applications. First, we ensured that our peptide ligands did not interfere with EC function as tested by proliferation, migration, tube formation, and response to vascular endothelial growth factor. Second, peptides that supported EC function were incorporated into methacrylic terpolymers via chain transfer free radical polymerization. The HBOEC-specific peptide, TPSLEQRTVYAK, when covalently coupled to a terpolymer matrix, retained binding affinity towards HBOEC in a serum-free medium. Under the same binding conditions, the attachment of human umbilical vein endothelial cells (HUVEC) was limited, thus establishing HBOEC specificity. The findings from this work could facilitate the development of autologous cell therapies with which to treat cardiovascular disease.

276e Polymeric Biomaterial Tuned to the Adhesion of Human Blood Outgrowth Endothelial Cells

Daniel Heath, Chemical and Biomolecular Engineering, Ohio State University, 140 W. 19th Ave, Koffolt Labs, Rm 125A, Columbus, OH 43210, Anka N. Veleva, Department of Chemical and Biomolecular Engineering, North Carolina State University, 911 Partners Way, Box 7905, Raleigh, NC 27695, John J. Lannutti, Department of Material Science and Engineering, Ohio State University, Columbus, OH 43210, Cam Patterson, Department of Medicine, The University of North Carolina at Chapel Hill, 8200 Medical Biomolecular Research Bldg., 103 Mason Farm Road, Chapel Hill, NC 27599, and Stuart L. Cooper, Department of Chemical and Biomolecular Engineering, The Ohio State University, 140 West 19th Avenue, Columbus, OH 43210.

Thrombus undermines the function of many blood contacting biomedical devices. We hypothesize that a surface which specifically adheres human blood outgrowth endothelial cells (HBOECs) will scavenge these adult stem cells from circulation and lead to the development of a confluent, functioning, and blood compatible endothelial cell layer. A biostable methacrylic terpolymer synthesized from hexyl methacrylate (HMA), methyl methacrylate (MMA), and methacrylic acid (MAA) was used in this research [1]. The mechanical properties of the material were modulated by controlling the molar ratio of HMA and MMA in the polymer while MAA provided acid functionality for post-synthesis derivitizations [2,3]. The biological properties of the base polymer have been enhanced through the introduction of topography [4], peptide ligands [2,3,5], and non-fouling character. Electrospinning generated fibrous terpolymer constructs with random or aligned fiber orientations [4]. Peptide ligands which specifically bind HBOECs were discovered through phage display screening and have been covalently incorporated into the biomaterial [5,6]. Resistance to protein adsorption and unwanted cell adhesion will be achieved through copolymerization of methacrylate monomers with non-fouling pendant groups: poly(ethylene oxide) (PEO), phosphorylcholine (PC), and quaternary ammonium/ sulfonic acid (QS).

Electrospinning was used to generate fibrous terpolymer constructs with both random and aligned fiber orientations with the electrospinning enhancing the cytocompatibility of the material as observed through increased cell adhesion [4]. The HBOEC-specific ligands have been incorporated into the polymer through novel chain transfer chemistry [5]. The peptide ligand TPSLEQRTVYAK retained its biofunctionality after covalent attachment to the polymer and was illustrated to increase the adhesion of HBOECs on films in media not enriched with serum [6]. Currently we are creating polymers with higher ligand densities to maximize the cellular response. Polymers containing PEO pendant groups have been synthesized as well as the QS-methacrylate monomer. Results on non-fouling polymers will be presented.

In conclusion, a polymer system with tunable mechanical properties has been synthesized. This material has been electrospun into random and aligned fiber constructs and biofunctionalized with cell-specific peptide ligands. The topography introduced through the electrospinning process and the covalently incorporated ligands have each increased the cytocompatibility of the material towards HBOECs. Currently we are maximizing HBOEC specificity through increasing the density of ligands in the material and by incorporating non-fouling character to inhibit the adsorption of unwanted proteins and the adhesion of undesired cell types. These results support the use of this polymer system as a cardiovascular biomaterial.

References

[1] Veleva AN, et al, JBMR. 2005; 74A: 117 – 123.

[2] Fussell GW, et al, Biomaterials. 2004; 25: 2971 – 2978.

[3] Fussell GW, et al, JBMR. 2004; 70A: 265 – 273.

[4] Veleva AN, et al, JBMR. Submitted.

[5] Veleva AN, et al, Biotechnology and Bioengineering. 2007; 98.1: 306 – 312.

[6] Veleva AN, et al, Biomaterials. Submitted.

276f Effect of Biomimetic Substrate Microtopography on Intestinal Epithelial Cell Behavior

Lin Wang[1], Shashi K. Murthy[1], Gilda A. Barabino[2], and Rebecca L. Carrier[1]. (1) Chemical Engineering Department, Northeastern University, 360 Huntington Ave., Boston, MA 02115, (2) Biomedical Engineering Department, Georgia Institute of Technology, Atlanta, GA 30332

The inner surface of the small intestine contains a monolayer of epithelial cells resting on a basement membrane, a specialized form of the extracellular matrix (ECM)), which supports and facilitates the growth and differentiation of epithelial cells. The basement membrane has a convoluted topography consisting of finger-like projections (villi) with deep invaginations (crypts) between the villi. The migration, proliferation, differentiation, and function of small intestinal epithelial cells vary with position relative to these structures. Thus, crypt-villus micro-architecture may induce changes in cell phenotype. Understanding the role of micron-scale topographic factors will provide a foundation for intestinal tissue engineering research using biomimetic scaffolds. In present study, crypt and villus structures were simplified as a micro-well array and a micro-pillar array, respectively. Both poly-dimethylsiloxane (PDMS) and type I collagen cell culture substrates with designed topography were fabricated using soft lithography technique. Caco-2 cells were cultivated on both substrates. Cell morphology, viability, proliferation, and differentiation were evaluated.

It was found that the microtopography affects Caco-2 morphology and migration. Cells cultured in PDMS substrates with 50 um and 100 um diameter wells had much more distinct protrusions than the cells cultured on flat substrates, and generally extended their protrusions towards vertical sidewalls. Cells tended to avoid attachment to the sharp bottom corner of well structures, often bridging this corner in attaching both to the bottom and side walls of the micro-wells. It was also observed that on the collagen substrates, cells tended to elongate along the top boundaries of wells. Substrate topography also influenced cell differentiation; the expression of cell differentiation markers (i.e. alkaline phosphatase, aminopeptidase N, and pipetidyl peptidase IV) varied with substrate topography. Especially in the case of alkaline phosphatase, cells grown on collagen substrates with 100 um-well array had the highest alkaline phosphatase activity, comparing with flat, 50 um-well, and 500 um-well collagen substrates.

276g Effect of Structural Variations In Porous Scaffolds on Cell Growth

Pooja Iyer and Sundararajan V. Madihally. Chemical Engineering, Oklahoma State University, 423 Engineering North, Oklahoma State University, Stillwater, OK 74078-5021

The demand for organs or tissue transplants is more than the supply. This has created the need to find alternate structures that can mimic the properties of the body tissues. These alternate structures which are artificially engineered should be porous, degradable, bindable to the target organ and conducive to cell growth in human body. The first three factors have been widely studied and reported in literature. Cell growth in both chitosan and chitosan-gelatin tissues need to be studied and compared in greater detail. The emphasis needs to be on reproducing the growth of these tissues in human body and recreating conditions that are similar to the human body. This includes the use of serum free media to grow the cells.

In our experiments, chitosan and chitosan-gelatin scaffolds, with 2D and 3D architectures were constructed to determine the growth of human fibroblasts on these porous structures, which were cultured using serum free media. These structures were first quantified for mechanical properties and results showed a broad range of elastic modulus (2 Pa to 2,000,000 P). Hence they were utilized to assess the effect of material properties in the absence and presence of cell-binding domains. Human foreskin fibroblasts seeded on these structures in serum free medium were analyzed using MTT assay. Further, the cytoskeletal structure of the tissue was analyzed using actin staining. Collagen assay was also performed to investigate the viability of these tissues. The analysis was done on the 1st day, 4th day and 7th day after cell seeding.

In preliminary experiments, cytoskeletal organization of human fibroblasts on chitosan structures was found to be minimal. Chitosan gelatin scaffolds demonstrated better organization of the cells on them as indicated by pictures taken after actin staining. Experiments were also carried out on chitosan hydrogels to investigate if the change in structure could aid in better cell growth than in the 3D porous chitosan structure. The results indicate little effect on cytoskeletal structure confirming the fact that a blend of chitosan gelatin is needed to aid in cell growth when conditions similar to those found in the human body is encountered.

310a Compositional Tuning of Bimetallic Nanoparticles for Low Temperature Carbon Nanotube Growth

Wei-Hung Chiang and R. Mohan Sankaran. Chemical Engineering, Case Western Reserve University, 10900 Euclid Avenue, Cleveland, OH 44106-7217

The unique physical properties of carbon nanotubes (CNTs) have attracted a great deal of interest for potential applications. Despite the availability of a wide range of synthesis techniques, the structure of CNTs has not been controlled sufficiently during growth. In chemical vapor deposition, transition metals are used to catalyze nucleation and growth of CNTs. However, the role of the catalyst in CNT growth remains unknown. Here, we present a systematic investigation of the catalytic properties of Fe and Ni nanoparticles towards CNT growth. In addition to pure monometallic catalysts, our unique synthetic approach allows the preparation of bimetallic nanoparticles which exhibit lower activation energies and growth temperatures. Nanometer-sized metal nanoparticles are produced in a continuous-flow, gas-phase process based on an atmospheric-pressure microplasma. Microplasmas are miniaturized versions of direct-current glow discharges formed between a hollow cathode with a microcavity (d~180 microns) and a counter electrode. Vapor-phase precursors such as nickelocene or ferrocene are introduced into the microplasma and decomposed non-thermally by electron impact to nucleate particles homogenously in the gas phase. Particle growth is limited to the small reactor volume (less than 1 nL) created by the microplasma geometry. As a result, ultrasmall nanoparticles (1-3 nm mean diameter) with narrow size distributions and tunable chemical composition are synthesized. To grow CNTs, a mixture of acetylene (C_2H_2) and hydrogen (H_2) gases are added to the particle flow and heated in a tube furnace at fixed temperatures between 300 and 650°C. As-grown nanoparticles and nanotubes are continuously monitored in real time using a differential mobility analyzer. Kinetic studies have been performed to determine activation energies of 119 and 73 kJ/mol for monometallic Fe and Ni nanoparticles, respectively. In comparison, NixFe1-x bimetallic nanoparticles are observed to have dramatically different activation energies as low as 28 kJ/mol. Additionally, the nanoparticle alloys are found to catalyze CNT growth at 300°C, significantly lower than temperatures normally required for monometallic catalysts. These results suggest that the bimetallic catalyst particles are characterized by synergistic effects during the nucleation and growth of CNTs. A possible mechanism for CNT growth based on the diffusion of carbon on the catalyst particles will be presented. Microcharacterization of the catalyst particles and CNTs by high-resolution TEM, X-ray diffraction, and micro Raman spectroscopy will also be discussed.

310b Design of a High Throughput Microwave Plasma Reactor for Bulk Production of Metal Oxide Nanowires

Jeong H. Kim[1], Vivekanand Kumar[2], and Mahendra K. Sunkara[2]. (1) Chemical Engineering, University of Louisville, 106 Ernst Hall, Department of Chemical Engineering,, University of Louisville, Louisville, KY 40292, (2) Department of Chemical Engineering, University of Louisville, 106 Ernst Hall, Department of Chemical Engineering,, University of Louisville, Louisville, KY 40292

A novel microwave plasma reactor which can produce bulk quantity of metal oxide nanowires and nanoparticles has been successfully developed. A custom made wave applicator which produces high density of plasma jet discharge in a large volume has been designed. Low cost, very fast, large scale production of different metal oxide nanowires can be accomplished using this reactor.

310c Modeling Plasma-Surface Interactions and Their Role In Inducing Structural Transitions In Materials

Dimitrios Maroudas, Department of Chemical Engineering, University of Massachusetts, 686 N Pleasant Street, Amherst, MA 01003

Plasma-surface interactions play a crucial role in determining the structure and properties of thin films and nanostructures grown by plasma deposition processes. Such thin films and nanostructures of electronic materials are used widely in numerous technological applications. For example, thin films of hydrogenated amorphous silicon (a-Si:H) and nanocrystalline silicon grown by plasma-enhanced chemical vapor deposition from silane-containing discharges are important materials in the fabrication of optoelectronic and photovoltaic devices. Film properties, such as crystallinity and surface roughness, are determined by the surface reactivities and mobilities of reactive species from the plasma that impinge on the growth surfaces; these properties are key factors in determining device performance.

In this presentation, we discuss a hierarchical modeling approach for the fundamental understanding of plasma-surface interactions and the first-principles-based simulation of plasma deposition and other plasma treatment processes. The approach combines first-principles density functional theory calculations of surface reaction and diffusion phenomena that occur during deposition or post-deposition plasma treatment, molecular-dynamics simulations of radical-surface interactions for radicals originating from the plasma, and kinetic Monte Carlo simulations of the long-time dynamics that govern film growth during plasma deposition processes. Representative results are presented for a-Si:H growth under conditions that make the silyl radical the dominant deposition precursor. The results are used to discuss issues of surface reactivity during deposition, film smoothening mechanisms, and film surface composition as a function of temperature.

We place special emphasis on understanding structural transitions induced upon processing of plasma deposited thin films and nanostructures with dihydrogen plasmas. Specifically, we discuss the mechanism of hydrogen-induced crystallization of a-Si:H films when they are exposed to hydrogen atom fluxes, which triggers an amorphous-to-nanocrystalline phase transformation. Furthermore, we present recent results of interactions of multi-walled carbon nanotubes (MWCNTs) with H atoms from a dihydrogen plasma and demonstrate a H-induced MWCNT-to-diamond transition that leads to formation of diamond nanocrystals even at room temperature.

3-Dimensional Modeling and Simulation of Surface Roughening under Plasma Etching

Wei Guo, **Hiroyo Kawai** and **Herbert H. Sawin**
Department of Chemical Engineering
Massachusetts Institute of Technology, Cambridge, MA 02139, USA

Abstract -Mixing-layer kinetics model was built to describe plasma-surface interaction and incorporated into a 3-Dimensional (3-D) Monte Carlo feature profile simulator. Etching yields and surface composition were modeled in 3-D simulator and compared to experimental data and kinetics modeling results. Surface roughening on poly-Si, SiO_2 were simulated in the 3-D simulator as a function of etching chemistry, ion incidence angle and amount of etching time. Experimental data and 3-D simulation results were in good quantitative agreement in terms of etching yield, roughness level and surface morphology, suggesting the incorporated mixing-layer kinetics model is able to accurately account for the chemistry taking place on various substrates and plasma chemistries.

Introduction

Plasma etching has broad applications on manufacturing processes [1] and much effort has been devoted to the near-surface chemical and physical modification in order to understand the plasma-surface interactions. Various kinetics models have been proposed to elucidate the reaction mechanism during plasma etching and yet few of them are able to deal with the complicated chemistry such as oxide etching in fluorocarbon chemistry. Meanwhile, the development of feature scale simulator is important to reduce the time and cost of process development and optimization. In this paper a generic kinetics model was built and incorporated into a 3-Dimensional feature scale simulator. Various feature evolution was simulated and compared to experimental observations.

Translating Mixed-Layer Kinetics Model

A number of assumptions were made in order to develop the translating mixed-layer kinetics model. First, there is a layer on top of the substrate formed by ion bombardment, in which all the atoms are well mixed and randomly bonded to each other. This mixing layer can move up or down depending upon deposition or etching is dominant, as shown in Fig. 1.

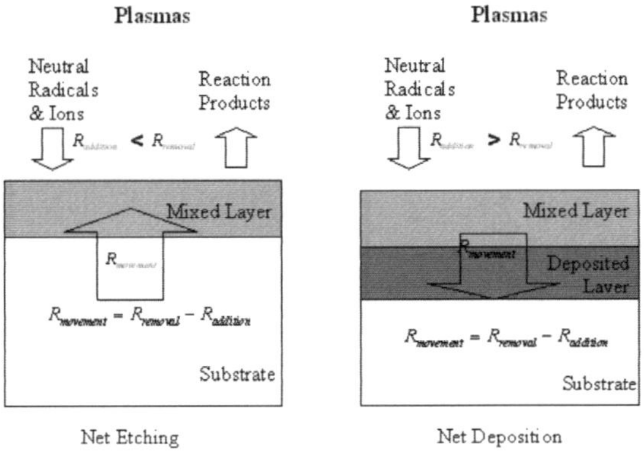

Fig. 1. Schematic of the translating mixed layer. It moves up or down during deposition or etching. Mass is conserved in the layer and substrate acts as the source or drain of the materials.

Second, concentrations of the surface chemical complexes are calculated based upon this random mixing assumption and the surface composition. Nearest bonding probability is defined as the probability for any two species sitting in neighbor. The formula is shown as follows:

$$J_{i\_j} = \frac{b_i x_i \times b_j x_j}{\left(1+\delta_{i,j}\right)\sum_{k=1}^{N} b_k x_k} \quad (1)$$

in which J_{i_j} is the nearest bonding probability, $\delta_{i,j}$ is the Kronecker delta-function, x_i is the composition of the ith species in the mixed layer as normalized by the total number of atoms in the mixed layer, b_i is the maximum number of bonding neighbors for the ith species. With the nearest bonding probability, the concentrations of any chemical complexes can be calculated and the reaction rates expressed as functions of those concentrations.

A subset of reactions were chosen to represent the overall reaction mechanism based upon both the experimental evidence of the primary products, the independence that the experimental data can be fitted, as well as the completeness to fully reflect the removal mechanism. All the important reactions in plasma are considered, including ion incorporation, neutral adsorption, physical sputtering, ion-induced etching, dangling bond creation and densification. Based on these removal and addition reaction rates, time differential equations of surface compositions are expressed as follows:

$$\frac{dx_{Si}}{dt} = Si_{Addition} - Si_{Removal}$$
$$\frac{dx_{Cl}}{dt} = Cl_{Addition} - Cl_{Removal} \qquad (2)$$
$$\frac{dx_{V}}{dt} = V_{Addition} - V_{Removal}$$

where x_{Si} is the surface composition of silicon, $Si_{Addition}$ and $Si_{Removal}$ are the summation of all the silicon addition and removal reaction rates (similar for chlorine and vacancy). Here the rate coefficients will be determined by fitting the experimental data. Unlike the other kinetics models which obtain steady-state solutions by setting the derivatives to zero, we numerically integrated those differential equations to get the physically reasonable solutions. This was accomplished by using JACOBIAN®, which is a robust program at solving differential equations accurately in mathematical and physical sense. We also fitted the rate coefficients by matching up the simulation results with the experimental data.

Incorporation of kinetics model into 3-dimensional Monte Carlo simulator

When the kinetics model is incorporated into the 3-D profile simulator, the concept of the mixing layer is approximated in the cellular domain as shown in Fig. 2: when an ion strikes a cell, first it is evaluated whether to incorporate or scatter according to the scattering probability as a function of ion species, incident angle and energy using Monte Carlo methods. If the ion is to be incorporated, the struck cell and its neighboring cells are mixed and compositions are averaged. Then all the ion-induced reactions take place with the reaction rates calculated using the averaged compositions and products removed from the struck cell.

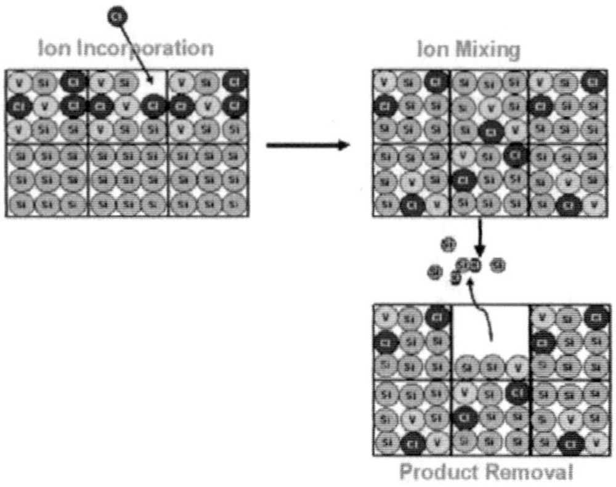

Fig. 2. Schematic of translating the kinetics into the 3-D profile simulator. When an ion strikes the surface, ion bombardment triggers ion mixing locally and the

composition is averaged among all cells surrounding the struck cell. The reaction kinetics and product removal are dependent upon this averaged composition.

In order to simulate the feature at off-normal angle, the angular dependence and curvature dependence effect was included in this paper. According to sputtering theory, the etch rates depend on ion angle of incidence. The angular dependence varies greatly according to process conditions such as gas composition and substrate materials, which can be characterized as either physical sputtering or ion-enhanced etching, as shown in Fig 3. Curvature dependence was proposed by Bradley and Harper [2] to explain the dependence of the surface curvature on the local sputtering yield. They proposed the more energy is deposited onto a surface with a positive curvature than a surface with a negative curvature and that results in a high sputtering yield at the bottom of valleys than at the top of the hills on a given surface, as shown in Fig. 4.

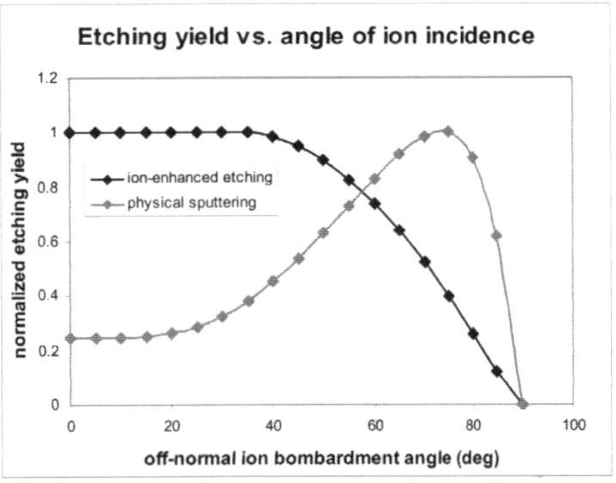

Fig. 3. Angular dependences of ion-enhanced etching and physical sputtering on etching yield. The ion-enhanced etching (Cl_2 plasma) yield remains constant up to a certain off-normal angle and decreases monotonically at high off-normal incidence, whereas the physical sputtering yield increases with off-normal angles of incidence and peaks at around 60° to 75°.

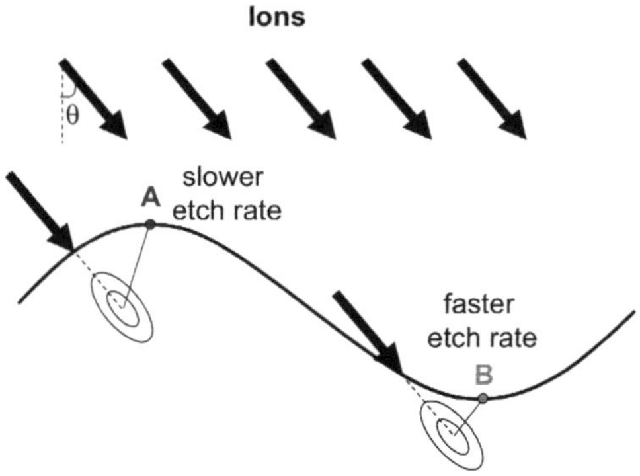

Fig. 4. B-H model of curvature-dependent etching. When the ions bombard the surface at off-normal incidence, the amount of energy deposited at B is larger than at A because the distance from the center of energy distribution contour to the surface is smaller for point B.

Results and Discussion

Results will be discussed in two parts. First, etching yields of poly-Si in Cl/Ar^+ and Cl_2 plasmas was modeled using the kinetics model and the 3-D MC profile simulator with the kinetics incorporated. The modeled etching yields and compositions will be compared to the experimental ones to demonstrate the kinetics model is able to account for surface process quantitatively. Second, Si sputtering in Ar^+ was simulated at different ion incidence angle to explore the transition of surface morphology. The effect of etching time on surface morphology was captured and compared to experimental observations.

A. Etching yields of Si in Cl/Ar^+

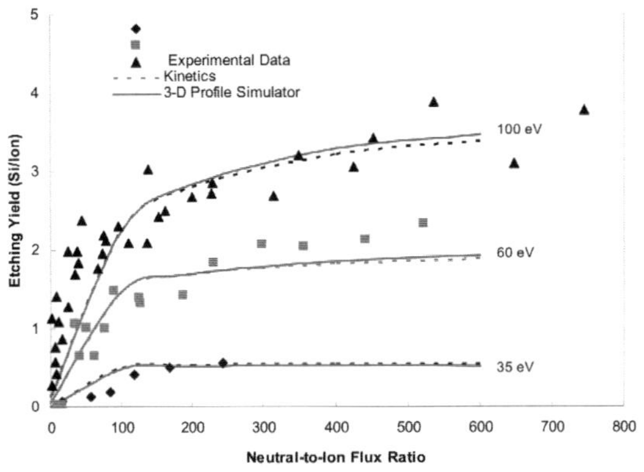

(a) Etch yield vs. neutral-to-ion flux ratio at E=35 eV, 60 eV, 100 eV.

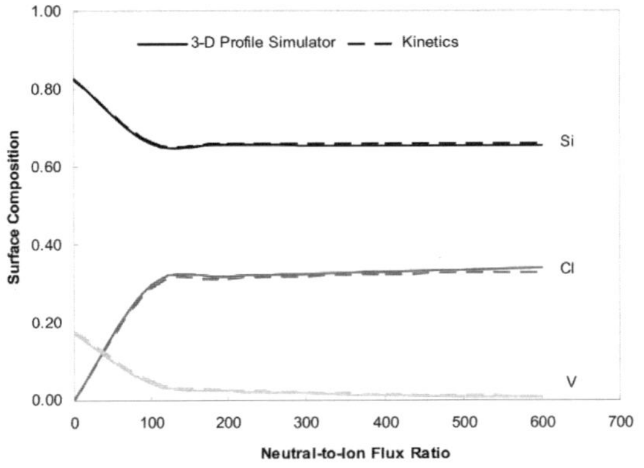

(b) Surface composition vs. neutral-to-ion flux ratio at E=60eV.

Fig. 4. Poly-Si etching in in Cl/Ar^+ and comparison of experiments (dots), translating-layer kinetics modeling (dash lines) and 3-D MC profile simulation (solid lines).

Fig. 4 (a) compares the etching yields from the kinetics modeling, 3-D MC profile simulation and the experimental data for Si etching in Cl/Ar^+ discharge at different conditions. Kinetic results closely match the experimental data, suggesting the lumped reaction set as well as the fitted parameters are able to carry on the chemistry. The 3-D MC profile simulator results fall right on top of the kinetic simulations as well as the experimental data, validating the accuracy of the translation from kinetics modeling to 3-D feature scale simulation. Fig. 4 (b) shows the steady-state surface coverage. Excellent agreement between the kinetics and the 3-D profile simulating suggests the kinetics has been translated completely.

B. Etching yields of Si in Cl_2

Si etching in Cl_2 plasmas was predicted using parameters determined from beam scattering experiments. Experiments conducted by Vitale *et al* [3] showed that the chlorine gas plasmas was composed of 10% Cl and 90% Cl_2 radicals and the ion flux was composed of 30% Cl^+ and 70% Cl_2^+ ions. The measured etching yield was found to be between silicon etching yield in Cl/Cl^+ and Cl_2/Cl_2^+ cases, as shown in Fig. 5(a).Identical parameters were used to those fitted for silicon etching in Cl/Cl^+ beams and Cl_2/Cl_2^+ beams. As shown in Fig. 5(a), the model predicted the dependence of etching yield on square root of ion energy, which closely matches the experimental data, indicating the model is able to predict conditions that were not fitted.

Surface compositions of silicon, chlorine and vacancy as a function of ion bombardment energy are shown in Fig. 5(b). The vacancy concentration increases and the chlorine concentration decreases at higher ion bombardment energy, consistent with experimental observations.[4] Comparisons of surface composition also indicate that the silicon/chlorine concentration in Cl_2 plasmas are in between that of silicon etching in Cl/Cl^+ and Cl_2/Cl_2^+ beams.

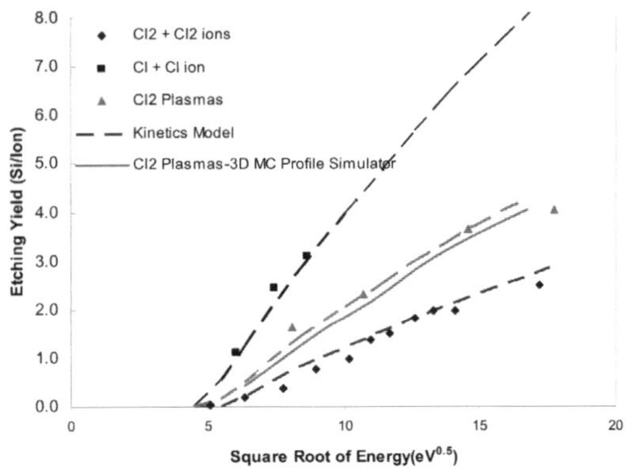

(a) Etching yield vs. square root of energy.

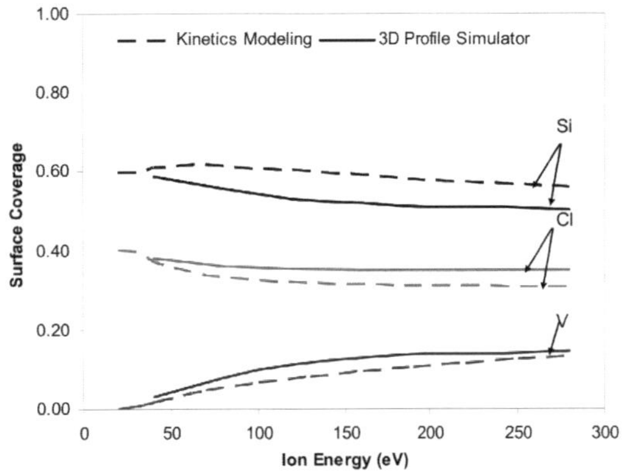

(b) Surface composition vs. ion energy

Fig. 5. Poly-Si etching in Cl discharge including silicon etching in Cl/Cl^+ beams (diamond), in Cl_2/Cl_2^+ beams(square) and in Cl_2 plasmas (triangle) and comparison of experiment (dots), the mixing-layer kinetics modeling (dashed lines) and 3-D MC profile simulation (solid lines). Neutral-to-ion flux ratios are all 500.

C. Profile simulation of sputtering of Si by Ar^+ bombardment

Physical sputtering of Si by Ar^+ was studied using the profile simulator. The sputtering yields were assumed to follow those determined from the angular and local curvature dependences and the energy of the Ar^+ ions was kept constant at 250 eV. The surface was bombarded at different ion incidence angles to examine its effects on the surface morphology. Fig. 6 shows the contour plot of the surface morphology of Si after sputtering at different off-normal angles. The surface was etched to a depth of 400 nm for an area of 250 x 250 nm^2. The vertical scale is ±35 nm. At 45° off-normal incidence, wave-like patterns develop transverse to the ion beam direction. The average wavelength of these wave-like patterns after etching to a depth of 400 nm is about 70 nm, with amplitude of about 10 nm. At 75° off-normal incidence, the structures extend in the direction parallel to the ion beam. Experimental data were shown in Fig. 6. (c) and (d). The transverse striation that appear on Si at 60° off-normal angle shown in (c) is consistent with the simulation results shown in (a). The parallel striation developing on Si at 75° off-normal angle is also consistent with the simulation result shown in (b). That suggests this simulator is able to capture the physical sputtering process at various off-normal ion bombardment with the angular and curvature dependence included.

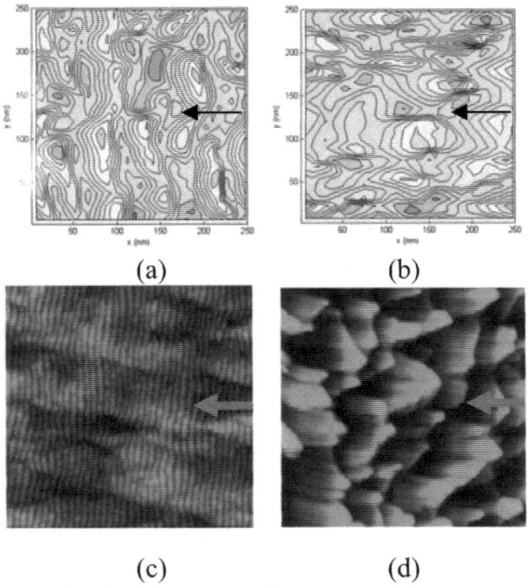

Fig. 6. Comparison of experimental and simulation results of single crystal silicon after sputtering in Ar plasma. a) AFM image of single-crystal Si after sputtering at 60° off-normal angle for 400 nm, b) AFM image of single-crystal Si after sputtering at 75° for 200 nm. Domain is 1μm by 1 μm for both

AFM images. c) simulation result for single-crystal Si after sputtering at 45° for 400 nm, d) simulation result for single-crystal Si after sputtering at 75° for 400 nm. Domain is 250 nm by 250 nm.

D. Transient surface morphology with the amount of etching

Single-crystal Si sputtering in Ar^+ was simulated as a function of amount of etching. Fig. 7(a), (b) and (c) shows the ripple growth on Si after removing 100 nm, 200 nm and 300 nm materials at 75°. It can be seen that mild ripples transverse to the ion beam appear first and develop into the grooves parallel to ion direction with etching. It is consistent with the experimental observation for single-crystal Si in Ar^+ sputtering after removing similar thickness, as shown in (d), (e) and (f).

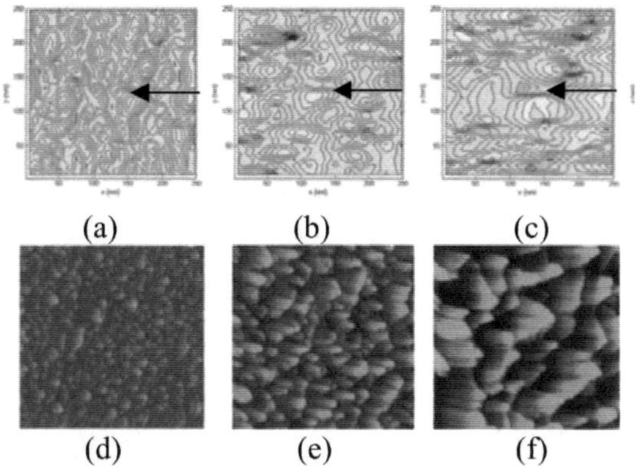

Fig. 7. Comparison of experimental and simulation results of single-crystal Si sputtering in Ar plasma at 75 ° off-normal angle for different amount of time. (a)-(c) Simulation of Si surface after removing (a) 100 nm b) 200 nm c) 300 nm. (d)-(f) Experimental AFM images of Si surface after removing (d) 40 nm (e) 80 nm (f) 200 nm.

E. SiO_2 etching in C_4F_8 plasma

The reaction set and rate coefficients determined from the kinetics model for SiO_2 etching in C_4F_8 plasma was incorporated into the 3-D simulator in a similar fashion described previously. The surface roughening was simulated and compared to the experimental observation as shown in Fig. 8. At a neutral-to-ion flux ratio of 5, the reactive species is mainly ion thus the etching mechanism is close to sputtering. Therefore, the simulated surface morphology is transverse to ion bombardment angle when 80 nm is etched. Experimentally, the surface also appears to the transverse to ion beam, as shown in Fig. 8 (b). 250 nm x 250 nm area from the top-left corner of the AFM image was selected in Fig. 8 (b) and re-plotted in a contour fashion in order to compare to the simulation in (a). It can be seen that the amplitude and frequency of the feature in experiment is close to that simulated in (a), suggesting the simulator is able to predict SiO_2 in C_4F_8 plasma at different off-normal angles. It is believed as etching persists the pattern is at an intermediate state between transverse to parallel striation.

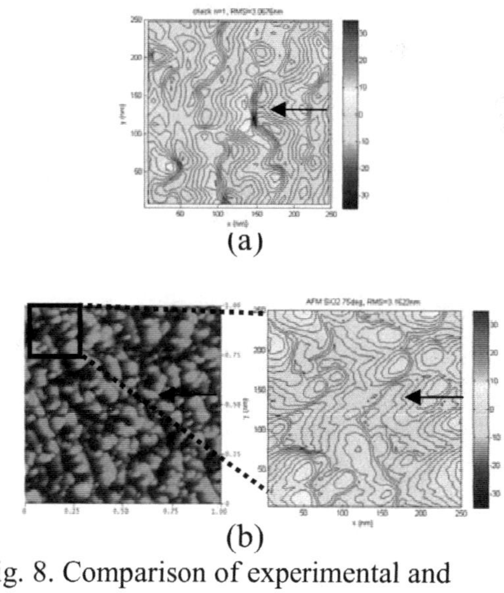

Fig. 8. Comparison of experimental and simulation for oxide etching in C_4F_8 plasma. (a) Simulation result of oxide etching in C_4F_8 plasma. Neutral/Ion flux ratio=5, Ion energy =350 eV. Domain is 250 nm by 250 nm. RMS =3.1 nm after 80 nm etched. (b)

On the left, experimental AFM images for oxide etching in C_4F_8 plasma. Neutral/Ion flux ratio=5, Ion energy =350 eV. Domain is 1μm by 1 μm. On the right, the contour plot of the upper-left corner of the AFM image. Domain is 250 nm by 250 nm. RMS =3.2 nm after 82 nm etched.

Conclusions

In this paper a mixing-layer kinetics model was built to describe plasma-surface interaction and incorporated into a 3-Dimensional (3-D) Monte Carlo feature profile simulator. Surface roughening on poly-Si, SiO_2 were simulated in the 3-D simulator as a function of etching chemistry, ion incidence angle and amount of etching time. Experimental data and 3-D simulation results were in good quantitative agreement in terms of etching yield, roughness level and surface morphology, suggesting the incorporated mixing-layer kinetics model is able to accurately account for the chemistry taking place on various substrates and plasma chemistries. The transition from transverse to parallel striation at different ion angles were captured with the profile simulator by combining the curvature-dependent sputtering with surface diffusion suggested by B-H model, through which the impinging ions deliver more energy to the surface in depressions relative to elevations. It was demonstrated experimentally and in modeling that the ripple formation is sensitive to the amount of etching: transverse striation on single-crystal Si at 60° ion angle gave way to parallel striation as etching persisted.

Acknowledgement

Work was supported by Semiconductor Research Corporation (SRC) and Numerica Technology for Jacobian program.

References

[1] J. W. Coburn and H. F. Winters, *Annual Review of Materials Science* **13**, 91-116 (1983).

[2] R. M. Bradley, J. M. E. Harper, *J. Vac. Sci. Tech. A*, 6, 2390-2395 (1988)

[3] S. A. Vitale, H. Chae, and H. H. Sawin, *J. Vac. Sci. Tech.A* **19**, 2197-2206 (2001).

[4] J. P. Chang, A. P. Mahorowala, and H. H. Sawin, *in* "International workshop on basic aspects of nonequilibrium plasmas interacting with surfaces, Vol. 16, pp. 217-224. *AVS*, 1998

310e Simulation of Profile Evolution In Shallow Trench Formation by Plasma Etching

John Hoang, Chemical and Biomolecular Engineering Department, UCLA, Los Angeles, CA 90095 and Jane P. Chang, Department of Chemical and Biomolecular Engineering, University of California at Los Angeles, Los Angeles, CA 90095.

In this work, a Monte Carlo based feature scale model was developed to accurately portray the profile evolution during shallow trench isolation etch (STIE) in chlorine based plasmas. A novel surface representation eliminates the artificial surface flux fluctuations due to the highly sloped sidewall features under simulation and the discrete cell nature of the simulation domain. It also enables a precise calculation of the surface normal, which dictates the trajectory of the reflected reactive species that control the profile evolution. The number of particles simulated is estimated from the depth and width of the etched profiles determined by scanning electron microscopy (SEM), with the assumption that the etch processes occur at high neutral-to-ion flux ratios. Through a set of carefully planned design of experiments (DOE) in which the effects of plasma density and plasma chemistry were assessed, the model was shown to accurately predict key features of STIE profiles, including microtrenching, mask faceting, and sidewall tapering, as a result of changing neutral-to-ion ratio, the mean ion energy, ion energy and/or angle distribution function.

A two-dimensional numerical fluid model was developed to investigate the dual-coil and dual-feed reactor design on the radial profiles of plasma species, namely etch products and positive ions. The dual-coil parameter was determined to be effective in tailoring the radial ion flux profile at pressures higher than 20 mT, while the dual-feed parameter was shown to alter the etch product transport in the convection-dominant flow regime. Coupling of the reactor scale model to the feature scale model allowed investigation of subtle yet important changes in the etched feature profile from the center to the edge of the wafer. This hybrid model suggests that the radial decrease in the etch depth from wafer center to edge, seen from a set of DOE, is caused by an inherent net neutral-to-ion ratio decrease. In addition, the increase in the silicon sidewall angle from wafer center to edge can be qualitatively explained by a decrease in the concentration of the etch products. To study the local variations at the die/meso scale, the simulation domain is expanded to study the effects of etch product distributions at the die level.

310f Molecular Dynamics Simulations of Plasma-Surface Interactions: Nanoscale Feature Etching on a Silicon Substrate

Joseph J. Végh and David B. Graves. Department of Chemical Engineering, University of California, Berkeley, 201 Gilman Hall, Berkeley, CA 94720

Classical molecular dynamics (MD) simulations have been employed to model the fabrication of very small (< 5 nm) features in silicon. As scale-down continues in the semiconductor and other thin-film device industries, fundamental understanding of etch mechanisms at these smaller scales is becoming increasingly important for process design and control. In our MD approach, confined beams of ions and radicals have been used to approximate a perfect masking layer, using either cylindrical (hole) or rectangular (trench) geometries with a nominal confinement size of ~2 nm. Bombarding species (chosen from typical plasma processing gases) include Ar^+ and CF_x^+ ions, as well as F and CF_x radicals. The nano-feature evolution is tracked over several thousand impacts, allowing direct examination of the etch process as a function of ion/radical fluence. By varying the ratios of the incoming species, various physical and chemical contributions can be examined and their corresponding roles in the etch process and sidewall formation can be studied. For example, we show that ion bombardment alone is sufficient to create a dense, mixed damaged/passivation layer along the sidewalls of the feature being etched that is ~1 nm thick. However, for some bombarding chemistries, radical deposition along the feature also plays an important role in determining the final sidewall composition and subsequent etch characteristics. We elucidate which mechanisms are most critical under the various conditions studied. Additional simulations have been carried out to examine the effects of an explicit masking layer (i.e. amorphous carbon deposited on top of the silicon) on both the final geometry of the feature and to overall feature size limitations. For example, ions impinging on the masking layer's sidewalls can lead to sputtering of the masking material into the feature, where it mixes into the substrate layer during subsequent bombardment, changing the effective etch rate. Further, the mask can collapse at the very small length scales examined, limiting the transport of radicals and ions to the substrate material. Mask erosion under certain chemistries can lead to loss of critical dimension control or mask impurity incorporation into the substrate. Future developments to capture more realistic phenomena (such as surface charging effects, longer-than-MD-timescale diffusion effects, etc.) are also discussed, including the parallelization of the code to allow for atomistic simulation of larger scale features.

310g Stochastic Differential Charging and Its Effects on Charging Damage and Feature Profile Evolution during Plasma Processing

Eunsu Paek, Dept. of Chem. Eng., The University of Texas at Austin, 1 University Station C0400, Austin, TX 78712 and Gyeong S. Hwang, Chemical Engineering, Dept. of Chem. Eng.,The University of Texas at Austin, 1 University Station Stop C0400, Austin, TX 78712.

Differential surface charging is often a serious drawback in application of plasma techniques to defining high aspect ratio structures, necessary for the manufacturing of modern microelectronic devices. It is well established that charges can accumulate on the exposed insulating surfaces of patterned structures during plasma exposure, due to the directionality differences between impinging ions and electrons. This, in turn, gives rise to electric fields which can alter the trajectory, flux, and kinetic energy of incident ions, often resulting in undesirable side effects in the plasma-assisted processes. Earlier theoretical studies have focused on describing the mean behavior of surface charge densities and potential distributions on patterned dielectric surfaces at the micronscale or larger. However, as device feature sizes shrink into the nanometer scale regime, the influence of an individual charge transferred to the surface will be larger, leading to an increase in the variability of potentials within the charging area. This leads to the question of whether a true steady-state-like behavior will be reached for high aspect ratio dielectric structures with small absolute dimension or will large oscillations in potential lead to essentially stochastic behavior. In this presentation, we will show our simulation results of plasma bombardment of high aspect ratio dielectric structures, which demonstrate stochastic charging behavior arising as absolute dimension decreases from 500 nm to 50 nm. We will also discuss how the stochastic changing influences charging damage and feature profile evolution during plasma etching and ashing.

310h Coupling Gas Phase and Surface Reaction Kinetics In C_4F_8 and SF_6 Plasmas Used for Si and SiO_2 Etching

George Kokkoris[1], **Evangelos Gogolides**[1], Andy Goodyear[2], and Mike Cooke[2]. (1) Institute of Microelectronics, NCSR Demokritos, Terma (End) of Patriarhou Gregoriou St., Aghia Paraskevi, 15310, Greece, (2) Plasma Technology, Oxford Instruments, North End, Yatton, Bristol, United Kingdom

C_4F_8 plasma has been widely used for dielectric etching in microelectronics fabrication.[1] It is also met in the area of micro-electro-mechanical systems (MEMS) fabrication, where in combination with SF_6 plasma is used for deep Si etching during the Bosch[2] process. In particular, SF_6 plasma is used for the etching of Si and C_4F_8 plasma is used to deposit a fluorocarbon (fc) film which protects the sidewalls of the etched Si structure and sustains the anisotropy of the etching process. C_4F_8 has been also used for plasma enhanced chemical vapor deposition of fc films,[3] which are exploited either as low-k films or in bio-fluidic applications.[4]

Several models (0D, 1D, and 2D) for C_4F_8 plasmas[5-8] have been reported, while there is a lack of models for SF_6 plasmas[9] in low pressure conditions. None of the models has focused on the interaction of the gas phase with the reactor surfaces. The importance of the interactions with the reactor walls increases as the constraints for manufacturing become stricter; the interaction with the walls can affect the reproducibility of the process.[10]

In this work, a 0D or global type[11] model for C_4F_8 and SF_6 plasmas is combined with a surface reaction model. It consists of a) mass balances for all species (neutral or charged) except electrons, b) charge neutrality equation, and c) energy balance for the electrons. The reaction set in the gas phase comprises several types of reactions such as electron impact reactions (e.g. ionization, dissociation, attachment), ion – ion recombination, and neutral – neutral recombination; it is assembled based on literature sources for cross sections (electron impact reactions) and rate coefficients. The surface model describes the interaction of the species produced in the gas phase with the reactor walls (e.g. sticking, recombination, deposition); it is phenomenological, it is assembled so as to take into account pertinent experimental measurements, and its formulation is the surface site balances on the reactor surfaces. The rate coefficients of the surface model are fitted to the available experimental measurements.

The combined model, not only takes into account the effect of surface reactions in the densities of species in the gas phase, but allows the calculation of derived outputs which extend the potential experimental measurements for the validation of global models. In particular, it allows the calculation of the pressure rise after the ignition of the discharge (compared to that before) which links to the degree of dissociation of the parent gas. The combined model also allows the calculation of the effective sticking coefficients of the species which signify the net consumption of the species on the reactor surfaces and are the values measured in the experiments. Finally, it allows the calculation of the deposition rate and the ratio of F/C of the fc film deposited on the reactor surfaces (C_4F_8 plasma).

The results of the combined model compare well with experimental measurements concerning pressure rise as well as densities of F atoms, CF_2 and CF radicals, and ion flux in an inductively couple plasma reactor. Summarizing the results for C_4F_8 plasma, the pressure rise and the densities of CF and CF_2 radicals are compared with experimental measurements in Figures 1a and 1b respectively. In addition, the parent gas (C_4F_8) is vastly dissociated, CF_4 dominates after 1000 W, CF_3 radical is predicted to be produced on the reactor walls, and the fc film deposited on the reactor walls is mainly due to deposition of light fc radicals; Concerning the results for SF_6 plasma, the loading phenomenon[12] observed during Si etching by SF_6 plasma is predicted.

The combined model for C_4F_8 and SF_6 plasmas is also coupled with a previously developed simulation framework[13,14] for feature scale etching, which links the output of a global model to the shape of the etched feature (e.g. trench or hole), in order to investigate the effect or the operational parameters of the reactor on the shape of the features during Si and SiO_2 etching.

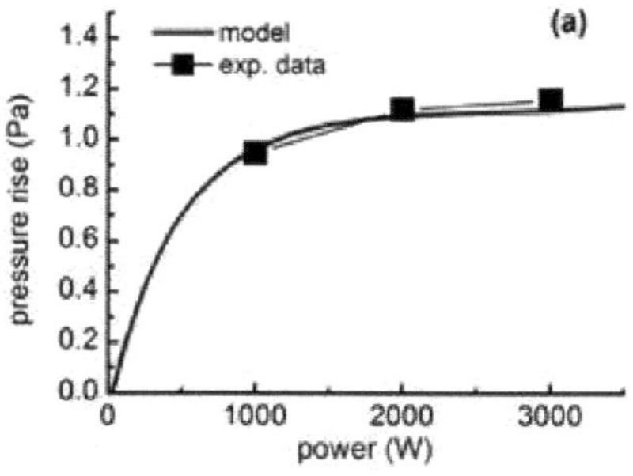

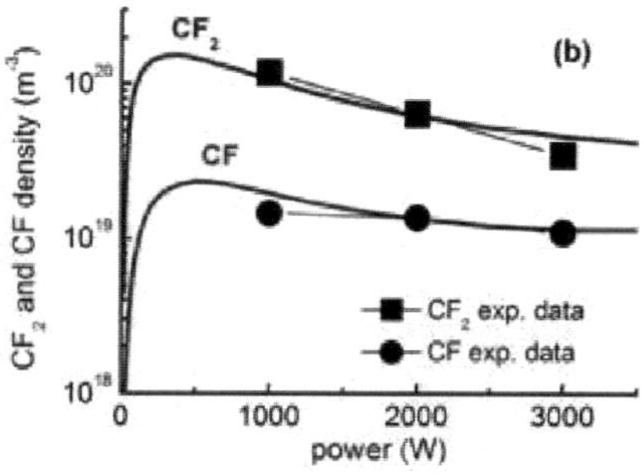

Figure 1. Comparison of model results with experimental measurements for C_4F_8 plasma. Pressure before plasma ignition is 1.29 Pa and gas temperature is 300 K. a) Pressure rise. b) Densities of CF_2 and CF radicals.

References

1. H. H. Doh, J. H. Kim, S. H. Lee, and K. W. Whang, J. Vac. Sci. Technol. A **14**, 2827 2834 (1996).
2. F. Laermer and A. Schilp, German Patent DE 4241045 (25 May 1994).
3. K. Endo, K. Shinoda, and T. Tatsumi, J. Appl. Phys. **86**, 2739-2745 (1999).
4. P. Bayiati, A. Tserepi, P. S. Petrou, K. Misiakos, S. E. Kakabakos, E. Gogolides, and C. Cardinaud, Microelectron. Eng. **84**, 1677-1680 (2007).
5. S. Rauf and P. L. G. Ventzek, J. Vac. Sci. Technol. A **20**, 14 (2002).
6. A. V. Vasenkov, X. Li, G. S. Oehrlein, and M. J. Kushner, J. Vac. Sci. Technol. A **22**, 511-530 (2004).
7. D. Bose, S. Rauf, D. B. Hash, T. R. Govindan, and M. Meyyappan, J. Vac. Sci. Technol. A **22**, 2290-2298 (2004).
8. G. I. Font, W. L. Morgan, and G. Mennenga, J. Appl. Phys. **91**, 3530-3538 (2002).
9. C. Riccardi, R. Barni, F. De Colle, and M. Fontanesi, IEEE Trans. Plasma Sci. **28**, 278-287 (2000).
10. G. Cunge, B. Pelissier, O. Joubert, R. Ramos, and C. Maurice, Plasma Sources Sci. Technol. **14**, 599-609 (2005).
11. C. Lee, D. B. Graves, M. A. Lieberman, and D. W. Hess, J. Electrochem. Soc. **141**, 1546-1555 (1994).
12. R. A. Gottscho, C. W. Jurgensen, and D. J. Vitkavage, J. Vac. Sci. Technol. B **10**, 2133-2147 (1992).
13. G. Kokkoris, A. Tserepi, A. G. Boudouvis, and E. Gogolides, J. Vac. Sci. Technol. A **22**, 1896-1902 (2004).
14. G. Kokkoris, A. G. Boudouvis, and E. Gogolides, J. Vac. Sci. Technol. A **24**, 2008 (2006).

318a Yield Stress of Ethylene-Methacrylic Acid Copolymers and Ionomers

Robert C. Scogna and Richard A. Register. Chemical Engineering, Princeton University, Princeton, NJ 08544-5263

Introduction

Ethylene-methacrylic acid (E/MAA) copolymers were first developed over four decades ago. It was quickly realized that the MAA groups could be partially or fully neutralized with a metal cation – such as sodium, magnesium, zinc, etc. - to form a class of extraordinarily tough semicrystalline "ionomers" [1]. The improved mechanical properties of these ionomers has led to their widespread use as packaging films, scratch resistant coatings and as an interlayer in shatter-resistant glass windows. Despite decades of commercial utilization, the structure-processing-property relationships for these materials are only now being explored [2-4]. The aim of this investigation is to reveal the microstructural origins of the yield stress (σ_y) for E/MAA copolymers and ionomers.

Experimental

E/MAA copolymers and ionomers were provided by DuPont Packaging and Industrial Polymers. The copolymer base resins contained 6, 11.5, 15 and 19 wt% MAA (2.0, 4.1, 5.5 and 7.1 mol % MAA, respectively). The degree of neutralization varies between 0 and 83%, or 0 to 2.6 wt% Na. As-received pellets were melt-pressed into 0.2-0.5 mm thick sheets at 140-160 °C in a PHI hot press and quenched to room temperature. The sheets were then stored in a desiccator at room temperature for three weeks. ASTM D1708 dogbones were stamped from these sheets for tensile testing. DMTA samples measuring approximately 5 x 22 mm^2 were cut from the same sheets to preserve the thermal history. Uniaxial tensile stress-strain curves were obtained with an Instron Model 5865, equipped with an Environmental Chamber 3111 retrofitted to control the testing temperature to within 0.3 °C with a cycle time of 1 second. Samples were tested at temperatures between 0 and 35 °C. Crosshead speeds ranged from 8.5×10^{-3} to 8.5 mm/s, producing initial strain rates of 3.8×10^{-4} to 3.8×10^{-1} s^{-1}. Subambient temperatures were achieved by introducing a small quantity of dry ice into the chamber. DMTA measurements were performed at 1 Hz on a TA Instruments RSA 3, using the film fixture. Data were collected every 5 °C for an effective heating rate of about 10 °C/min.

Results and Discussion

Tensile tests of E/MAA copolymers and ionomers were performed over a range of temperatures and strain rates. The yield stress was taken as the stress exerted on the sample at a strain that corresponds to the intersection of the Young's modulus line and a line fit to the pseudo-linear region immediately following the yield point. This construction allows for the determination of σ_y even when a local maximum in stress – corresponding to the initiation of necking – is not observed.

As expected, σ_y increases as the temperature is lowered and as the strain rate is increased. For the unneutralized E/MAA copolymers, two regimes of behavior are observed. At low strain rates, σ_y is a very weak function of strain rate and is well-described by the polyethylene crystal plasticity expression of Shadrake and Guiu [5,6],

$$\sigma_c = \frac{K(T)B}{\pi r_0} \exp\left(\frac{-2\pi \Delta G^*}{K(T)B^2 l_c} - 1 \right) \quad (1)$$

where K(T) is the shear modulus of the slip plane, B is the magnitude of the Burgers vector, r_0 is the dislocation core radius, l_c is the crystal thickness and ΔG^* is the strain rate-dependent energy barrier for the nucleation of a [001] screw dislocation from one of the lateral faces of a polyethylene crystal. At higher strain rates, however, σ_y starts to increase very quickly with strain rate due to incomplete relaxation of the amorphous phase. DMTA tests reveal that, for the E/MAA copolymers, the incomplete relaxation corresponds to the glass transition temperature, T_g. The tensile stress contribution, σ_{re}, from a single incomplete relaxation process is given by the Ree-Eyring model [7],

$$\frac{\sigma_{re}}{T} = \frac{R}{v} \sinh^{-1}\left(\frac{\dot{\varepsilon}}{\dot{\varepsilon}_0} \exp\left[\frac{\Delta H}{RT} \right] \right) \quad (2)$$

where R is the gas constant, v is the activation volume, $\dot{\varepsilon}$ is the applied strain rate, $\dot{\varepsilon}_0$ is a constant pre-exponential factor and DH is the activation energy. Thus, when σ_{re}/T is plotted against $\ln(\dot{\varepsilon})$, this regime appears as a straight line. Overall, we can describe the temperature and strain rate dependence of the E/MAA copolymer σ_y by the simple superposition of the two contributions so that,

$$\frac{\sigma_y}{T} = \frac{\sigma_c}{T} + \frac{\sigma_{re}}{T} \quad . (3)$$

Eq. (3) also works well to describe the σ_y of E/MAA ionomers. However, over the range of conditions studied, the amorphous phase was not completely relaxed when the degree of neutralization was greater than 0.4 wt% Na. At this critical degree of neutralization, the stiff "regions of restricted mobility" [8] surrounding each ionic aggregate are large (or numerous) enough to impinge upon each other and nearby polyethylene crystals. The percolation of the stiff material through the amorphous phase is responsible for the elevated yield stress. Additional neutralization above the critical degree results in minimal gains in the measured yield stress; there is little advantage in neutralizing to stoichiometric equivalence.

References

[1] R.W. Rees, D.J. Vaughn, Polym. Prepr. Am. Chem. Soc., Div. Polym. Chem., 6, 296 (1965).

[2] K. Wakabayashi, R.A. Register, Polymer, 46, 8838 (2005).

[3] K. Wakabayashi, R.A. Register, Macromolecules, 39, 1079 (2006).

[4] R.C. Scogna, R.A. Register, Polymer, 49, 992 (2008).

[5] L.G. Shadrake, F. Guiu, Philos. Mag., 34, 565 (1976).

[6] L.G. Shadrake, F. Guiu, Philos. Mag. A., 39, 785 (1979).

[7] T. Ree, H. Eyring, J. Appl. Phys., 26, 793 (1955).

[8] A. Eisenberg, B. Hird, R.B. Moore, Macromolecules, 23, 4098 (1990).

318b Nanofilled and Nanoporous Epoxies: Processing Routes, Size-Scale Transitions, and Mechanical Properties

Robert J. Klein, John L. Schroeder, and Joseph L. Lenhart. Organic Materials, Sandia National Laboratories, P.O. Box 5800, MS-1411, Albuquerque, NM 87122

There are many routes to the implantation of nano-scaled fillers or pores in polymers, but the baseline impact of the neat material properties on processing and the resulting mechanical properties are poorly understood. Two systems are examined here: high-modulus fillers in epoxies, and porous epoxies templated via phase-separation of a removable small molecule. In the filled systems, we utilize alumina and nickel-fillers, with and without surface treatments, to examine both the rheological impact of high volume fractions of filler as the size scale decreases, and the mechanical properties in the cured system, focusing on fracture mechanisms and toughness. In the porous systems, proper selection of the small molecule chemistry and volume fraction leads to spherical pores in the cured epoxy on a size scale ranging from microns to nanometers.

Acknowledgement: Sandia is a multiprogram laboratory operated by Sandia Corporation, a Lockheed Martin Company, for the United States Department of Energy's National Nuclear Security Administration under contract DE-AC04-94AL85000.

318c Dynamics of Polymer Nanocomposites Under Stress

Robert A. Riggleman, Gregory N. Toepperwein, and Juan J. de Pablo. Department of Chemical and Biological Engineering, University of Wisconsin, Madison, 1415 Engineering Drive, Madison, WI 53706

Several recent simulation and experimental studies have explored the effects of deformation on the dynamics of glassy materials, and both have demonstrated that the segmental dynamics in glassy polymers can be enhanced by up to two orders of magnitude. However, the origins of the enhanced dynamics remain largely unknown. Furthermore, it is not known how the observed changes are affected by aging processes. In this study, we have performed extensive molecular dynamics simulations of a pure polymer glass and a polymer nanocomposite glass under stress in the nonlinear creep regime. We measure the dynamics using the bond autocorrelation function and show that the effective relaxation times are reduced during the deformation by up to two orders of magnitude. The dynamics are most enhanced during periods of high strain rate, and we also find the largest changes in the features of the potential energy landscape during periods of fast dynamics and high strain rate. Additionally, we find that the relaxation times in the nanocomposite have approximately the same dependence on strain rate as the pure polymer at a constant temperature despite being farther into the glassy state. The results of our simulations are compared to those of fluorescence recovery after photobleaching. The agreement between theory and experiment is highly satisfactory, giving credence to the conclusions extracted from our molecular simulations.

MULTI-LENGTH SCALE STRUCTURE OF SEGMENTED PEG-BASED IONOMERS

Wenqin Wang
University of Pennsylvania, Philadelphia, PA, USA

Shichen Dou, Gregory J. Tudryn, Ralph H. Colby,
Pennsylvania State University, University Park, PA, USA

Karen I. Winey
University of Pennsylvania, Philadelphia, PA, USA

Introduction

Polymeric solid electrolytes have been subject of intensive research for potential applications in electrochemical devices, especially lithium-ion batteries. The most widely studied system is based on poly(ethylene glycol) (PEG), which can effectively solvate a variety of alkali and alkaline cations.[1] Although mixtures of polymers with salts show reasonable conductivity at room temperature,[2] they suffer from the undesirable electrode polarization due to anion aggregation at the cathode surface.[3] Single-ion conductors with anions fixed to polymer backbone are able to resolve this problem and achieve a cation transference number of 1.[1] Unfortunately, the single-ion conductors studied to date show far lower conductivity than the bi-ion conductor. Understanding the structure and cation transport mechanism in single-ion conductors is critical for designing future cation transport materials with high conductivity.

In this study, the structure and ionic association behavior of PEG-based ionomers are investigated as a function of PEG segment length, cation type, and temperature. A series of sulfonated polyester ionomers with well-defined PEG spacer lengths are synthesized by melt polycondensation. Multiple-angle X-ray scattering data reveal the morphology of the ionomers at different length scales. The effects of temperature on the nanoscale structures of these ionomers are investigated by in situ X-ray scattering over a wide temperature range.

Experimental

PEG-based ionomers were prepared by a two-step catalyzed melt transesterification of PEG oligomer diol with a mixture of Dimethyl 5-sulfoisophthalate sodium salt and dimethyl isophthalate to give random ionomers of varying ion content. The molecular weight of PEG spacer was 400, 600, 900, or 3000, and the cation was exchanged from sodium to cesium by dialysis. The concentrated ionomer solutions were freeze-dried after diafiltration and then vacuum-dried at 120 ^{o}C to constant mass. The ionomers are denoted as PEGx-yM, where x is the molecular weight of PEG oligomer, y is the percent of sulfonated phthalates over total amount of phthalates in ionomers, and M represents the type of cation (Na or Cs). PEG600-

50% Na designates the ionomers with PEG spacer length of 600 g/mol and 50% sulfonated phthalates. Note that y is not the neutralization level of the acid and theoretically all the acids in different ionomers are neutralized with metal cations.

The multi-angle X-ray scattering system (MAXS) used Cu X-ray from a Nonius FR 591 rotating-anode generator operated at 40 kV and 85 mA. The bright, highly collimated beam was obtained via Osmic Max-Flux optics and pin collimation in an integral vacuum system. The scattering data were collected using a Bruker Hi-Star multiwire detector with a sample to detector distance of 11, 54 and 150 cm. Temperature scanning was performed from room temperature to 150 °C with a step size of 25 °C and a heating rate of 10 °C/min. The ionomers were then cooled back down to room temperature at cooling rate of 10°C/min. The samples were held at each temperature for 10 min before starting the X-ray data collection. The 2-D data reduction and analysis were performed using Datasqueeze software.[4]

Results and Discussion

Figure 1 shows the room temperature X-ray scattering results for five fully sulfonated PEG-based ionomers over a wide range of scattering angle ($q = 0.1$ to 18 nm^{-1}). The morphological features observed in the X-ray scattering data of these ionomers are summarized in Table 1. The ionomers are amorphous when the PEG spacer length is short (400 or 600 g/mol). As the PEG segment length increases to 900 or 3000 g/mol, the polymer chain starts to crystallize. The crystallinity of fully sulfonated polyester ionomers based on PEG length of 900 g/mol is highly dependent on the thermal history, while the fully sulfonated polyester ionomer based on PEG with M = 3000 g/mol is always semi-crystalline. The crystalline reflection peaks at 13.6 nm^{-1}, 16.5 nm^{-1} and 18.5 nm^{-1} correspond well to the main crystalline reflections of typical PEO crystals having a helical chain conformation and a monoclinic lattice.[5] Because the lattice parameter remains unchanged in the PEG-based ionomers, we conclude that the ions are excluded into the amorphous phase. When the counterion is sodium, there is no "ionomer peak" in the expected angular range at room temperature. In contrast, when the cation is exchanged from sodium to a higher atomic number species such as cesium, a broad peak appears at ~7 nm^{-1}. The origin of this peak is not well-understood. The angular position of the peak is much higher than the typical ionomer peaks observed in various ionomer systems (1-5 nm^{-1}) and it is also much broader than typical ionomer peaks. Small-angle X-ray scattering shows the usual ionomer upturn at low q (roughly with intensity scaling as q^{-2}) for all of the ionomers. The semi-crystalline ionomers (with PEG $M \geq 900$) also show three small angle peaks superimposed on this power law. The peak positions relative to the first peak position are 1:2:3, which might indicate a layered structure. Typical semicrystalline polymers usually show only one crystalline lamellae-lamellae scattering peak due to a lack of ordering and the limited contrast between the amorphous and crystalline domains. In these semicrystalline PEG-based ionomers, the existence of multiple scattering peaks is probably the result of higher contrast between the crystalline domains from which the cations are excluded and the amorphous domains containing the cations. For example, although PEG900 – 100% Cs has less crystallinity shown in the wide-angle regime, the small-angle scattering peaks are sharper than PEG3000 – 100% Na, because the higher atomic number cation Cs^+ enhances the amorphous/crystalline contrast relative to the lower atomic number cation Na^+.

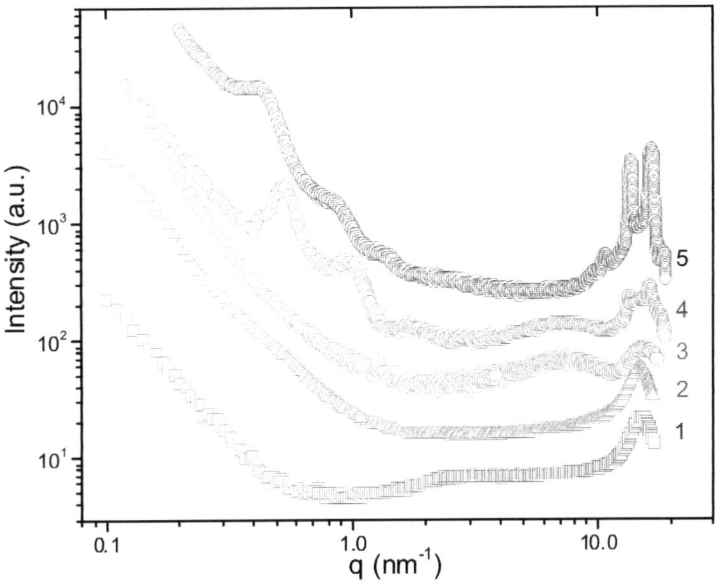

Figure 1. X-ray scattering intensity vs. scattering vector q for fully sulfonated PEG-based ionomers neutralized by monovalent cations. Log-log scale is used. Sample numbers correspond to materials listed in Table 1.

Table 1. Morphological features observed in X-ray scattering of PEG-based ionomers.

Sample	**Materials**	**$q<2\ nm^{-1}$**	**$2<q<10\ nm^{-1}$**	**$q>10\ nm^{-1}$**
1	PEG400 – 100% Na	Upturn	No Peak	Amorphous
2	PEG600 – 100% Na	Upturn	No Peak	Amorphous
3	PEG600 – 100% Cs	Upturn	Broad Peak at 7 nm^{-1}	Amorphous
4	PEG900 – 100% Cs	Upturn & three scattering peaks	Broad Peak at 7 nm^{-1}	Semicrystalline
5	PEG3000 – 100% Na	Upturn & three scattering peaks	No Peak	Semicrystalline

To further understand the morphology in our ionomers systems, *in situ* X-ray scattering was performed over a wide temperature range to investigate the effects of temperature on the crystallinity and ionic association behavior. High-temperature X-ray scattering on the sodium salt of the fully sulfonated reveals a new peak in the angular range of 2-3 nm^{-1}, reminiscent of conventional ionomers. Figure 2 shows a representative variable-temperature scattering data for PEG600 -100% Na. For the fully sulfonated ionomers neutralized by cesium cations, the broad peak at 7 nm^{-1} gradually diminishes as the temperature increases, and a new peak also appears at $q \sim 2\text{-}3\ nm^{-1}$. The intensity of the new peak decreases as the sulfonation level decreases and the peak appears broader and asymmetric due to the overlap with the upturn.

There is no new peak appears at high temperature for materials with 0% sulfonation, which is expected. As the temperature increases, the amorphous halo of PEG-based ionomers shifts to the lower-angular position in all ionomers, which is caused by the thermal expansion of the polymer. The morphology of ionomers is reversible when the system is cooled down, and the scattering profile at room temperature after cooling is identical to that of room temperature before heating (Figure 2). The appearance of "ionomer peak" at high temperature is attributed to the formation of new ionic aggregates. A mechanism is proposed to explain the ionic association behavior as a function of temperature. These results will provide new insights into the development of advanced single-ion conductors for lithium-ion batteries.

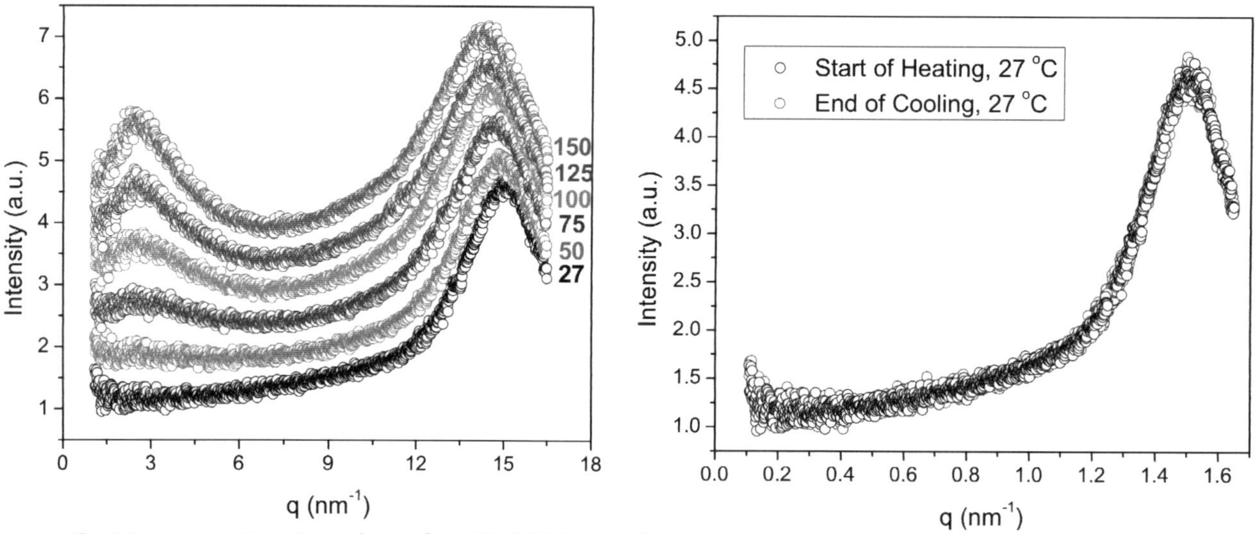

Figure 2. X-ray scattering data for PEG600 – 100% Na as a function of temperature: 27, 50, 75, 100, 125, 150 °C (left). Room temperature X-ray scattering data for PEG600 -100% Na before and after heating (right).

References

1. Xu, Kang (2004), "Nonaqueous Liquid Electrolytes for Lithium-Based Rechargeable Batteries," *Chem. Rev.*, 104, 4303.
2. Zhang, Shihai; Runt, James (2004), "Segmental Dynamics and Ionic Conduction in Poly(vinyl methyl ether)-Lithium Perchlorate Complexes," *J. Phys. Chem. B*, 108, 6295.
3. Dou, Shichen; Zhang, Shihai; Klein Robert. J.; Runt, James; Colby, Ralph H. (2006), "Synthesis and Characterization of Poly(Ethylene Glycol)-Based Single-Ion Conductors" *Chem. Mater.* 18, 4288-4295.
4. Heiney, Paul A. (2005), "Datasqueeze: A Software Tool for Powder and Small-Angle X-Ray Diffraction Analysis" *Comm Powder Diffr Newsletter*, 32, 9-11.
5. Zhu, Lei; Cheng, Stephen Z. D. *et al.* (2000), "Crystallinization Temperature-Dependent Crystal Orientations within Nanoscale Confined Lamellae of a Self-Assembled Crystalline-Amorphous Diblock Copolymer", *J. Am. Chem. Soc.*, 122, 5957-5967.

318e Effect of Polymer Mobility on Conductivity of Single-Ion Conductors

Kokonad Sinha and Janna K. Maranas. Chemical Engineering, The Pennsylvania State University, 115 Fenske Laboratory, University Park, PA 16802

Scientists are turning to the use of polymers as substitutes for liquid electrolytes in lithium ion batteries, because of their mechanical flexibility and non-toxic properties. Physical mixtures of lithium salt and poly(ethylene oxide) (PEO + $LiClO_4$) are commonly chosen because they have potential for high ionic conductivities. However, high mobility of ions in these mixtures results in electrode polarization, which affects battery performance. To isolate the effect of the cation and to reduce the obstacle of concentration polarization, the anion is chemically incorporated into the backbone of the polymer, thereby rendering it immobile. These single-ion conductors are called ionomers.

It is a widely accepted theory that in polymer electrolytes, ion conduction is aided by the segmental mobility of the polymer. The cation electrostatically coordinates with 5-7 ether oxygen atoms and these cations "hop" from one polymer segment to another by making and breaking temporary "cross-links". This study determines how the mobility of the ionomer affects the ion conductivity and how the ion content affects segmental mobility in return.

Neutron scattering experiments have been conducted on ionomers with varying degrees of ionization. We observed that there is an optimum ion content that has the maximum ionomer mobility and the highest conductivity. This can be explained by the fact that when ion content increases, there are more ions present to contribute to conductivity, but electrostatic coordination of cations with ether oxygen atoms reduces the polymer mobility.

Neutron scattering experiments further show that polymers with ions have a second process, which is absent in polymers with no ion concentration at all. This second process is slower that than the segmental motion of the polymer and it constitutes of making and breaking of electrostatic cross-links. This feature becomes more prominent with increasing ion content.

On comparison of the timescales for segmental mobility and the second process we noted that for PEO +$LiClO_4$ mixtures, this second process is significant in improving conductivity. However, for ionomers, this process does not play a major role in the mechanism of conduction. This comparison in terms of mobility as a function of ion concentration gives a better insight for understanding the fundamental mechanism of ion conduction in ionomers. This will enable us to choose an ion loading that gives optimum performance.

We measured ionomer mobility using the High-Flux Backscattering Spectrometer [HFBS] and the Disk Chopper Time-of-Flight Spectrometer [DCS] at the NIST Center for Neutron Research in Gaithersburg, Maryland. DCS primarily detects segmental motion of the polymer, which lies in a timescale between 0.1 ps and 40 ps. HFBS detects motion on a timescale between 240 ps and 2 ns, and this primarily constitutes of the second process. Mobility is measured for ionomers with from 0%, 5%, 10%, 25%, 50% and 100% ionization, at temperatures of 298K, 325K and 350K.

318f Aggregation In Dilute Solutions of High Molar Mass Poly(ethylene) Oxide and Its Effect on Polymer Turbulent Drag Reduction

Abhishek M. Shetty and Michael J. Solomon. Chemical Engineering, University of Michigan, 2300 Hayward Street, 3074 H.H Dow Bldg, Ann Arbor, MI 48109

Poly(ethylene oxide) (PEO) is used in wastewater treatment, oil recovery, drilling fluid stabilization, as an additive in pharmaceutical formulations and in drag reduction. Aqueous solutions of PEO are in a state of molecular aggregation, which is hypothesized to be the source of their anomalous rheology and high drag reduction capability. Validation of models for prediction of the viscoelastic behavior of PEO solutions is hindered by the lack of appropriate experimental data. In this work, we apply methods of dynamic light scattering (DLS) and fluid mechanics to quantitatively establish the role of aggregation in the turbulent drag reduction of high molar mass poly(ethylene oxide) solutions. By means of DLS, we show that the aggregation state of dilute solutions of high molar mass PEO WSR-301 (Mw ~ 4 x 10^6 g/mole) can be manipulated by addition of the chaotropic salt guanidine sulfate (GuS) or the divalent salt magnesium sulfate (MgSO4). In aqueous solution, we find $\Gamma \sim q^{2.8 \pm 0.1}$, where Γ is the relaxation rate and q is the scattering vector. This scaling is consistent with internal motions of a large coil or aggregate. The above results provide molecular level evidence that support earlier rheological inferences (Tirtaatmadja et al., Physics of Fluids, 18, 043101, 2006) that the structure of WSR-301 at ~ 10-100 ppm concentrations is not consistent with the picture of an isolated polymer coil in a dilute solution. Addition of salt progressively decreases the scaling to $\Gamma \sim q^{2.0 \pm 0.1}$(at 0.5M) consistent with center of mass diffusion of isolated coils. We furthermore find that manipulating the aggregation state of PEO in this way shifts the critical condition for onset of turbulent drag reduction at dilute concentrations by a factor of 2.5. This critical condition is inversely proportional to the viscoelastic relaxation time of the polymer solution. The implication is that the aggregation state and the turbulent drag reduction behavior of PEO are strongly correlated. This correlation definitively confirms prior speculation (Dunlop et al., Nature, 249, 1974) that the high molar mass PEO commonly used in literature studies of turbulent drag reduction is in a state of aggregation.

322a a Study of Diffusion and Release in Porous Alumina

Bradley R. Gordon[1], Charles E. Luckett[2], Douglas S. English[2], Sheryl H. Ehrman[1], and Daniel D. Lim[2]. (1) Department of Chemical and Biomolecular Engineering, University of Maryland, College Park, MD 20742, (2) Department of Chemistry and Biochemistry, University of Maryland, College Park, MD 20742

A model is presented for the diffusion and release of Doxorubicin (Dox) and Sulforhodamine 101 (SR101) in porous alumina. The diffusion of Dox and SR101 were studied using fluorescence recovery after photobleaching (FRAP) at pH's 5.0 and 7.4 over the temperature range 273K to 298K. Data from these experiments were fitted to first-order kinetics to predict the effects of pH and temperature on the adsorption of Dox and SR101. It was observed that the diffusion of Dox is dependent upon the pH whereas the release is not. It was further found that the diffusion of SR101 is independent of pH and that the release is temperature dependent. These observations suggest that the pH induced charge interactions between the diffusing species and the alumina surface play a significant role in the uptake and release from porous alumina.

Infusion of Fluid into Powder Beds

David B. Todd[1], **Costas G. Gogos**[1], **Ming-Wan Young**[1], **Linjie Zhu**[1], **Bainian Qian**[2] and **Chen Wan**[3], (1) Polymer Processing Institute, NJIT, GITC Bldg 3901, Newark, NJ 07102, (2) Rohm and Haas Electronic Materials, 451 Bellevue Road, Newark, DE 19713, (3) Sabic Innovative Plastic, 9226 DuPont Road, PO Box 68, Newark, DE 26181

Abstract

When it is desirable that a product contain the maximum volume percentage of selected solids within a fluid matrix, which in turn gels or solidifies, it may be desirable to consider an infusion process. An infusion process permits achievement of a higher solids loading than can be obtained utilizing a pourable premixed slurry. However, infusing a viscous fluid into a bed of fine powder is a daunting task. The wide size distributions and non-uniform shapes frequently encountered with many powders make the classical equations for fluid flow through packed beds difficult or impossible to use. An alternate approach is proposed whereby the mechanism is characterized as flow through parallel capillary channels, rather than just as flow over the particle surfaces. A simple test system can be used to define the effective capillary size, from which the pressure and time required to infuse a fluid into powder mixtures can be predicted as a function of the viscosity of the fluid and the geometry of the vessel into which it is being pumped. The predictability is particularly important when the fluid is a polymerizable monomer with only a narrow operating window before it sets up, such as some potting compounds and energetic applications. Examples are given with various fluid/powder systems.

Introduction

The current art of manufacturing energetic products, such as propellants and explosives, usually involves mixing of the ingredients for long periods to achieve a uniform distribution of the solid ingredients in a binder matrix that is still pourable but eventually slowly polymerizes or gels into a solid mass. The slurry paste becomes very viscous because of the generally high loading of solids. The increase in viscosity caused by the solids loading frequently sets the maximum solids content that can be processed. The high viscosity of the mixture requires substantial torque to blend. The power to mix combined with long mixing times result in inefficient mixing and high power consumption, and can lead to potentially undesirable and dangerous heat buildup.

For example, mixing for the solid propellants such as used in the space shuttle program typically contain 12.6 % by volume aluminum powder as fuel (10 to 20 microns), 63.5 % ammonium perchlorate as oxidant, generally bimodal size distribution (about 20 microns and about 300 microns), and 23.9 % polymerizable mixture of prepolymer and plasticizer. Mixing the above formulation may be done in large planetary change-can mixers.

The recent ACE™ wet manufacturing process described by Gogos et al [1] involves similar energetic products for the opposite end of the size spectra, namely long narrow metal tubes known as burster tubes. The ACE process achieves higher solids loading by replacing the slurry-mix-and-pour step by preconditioning the solids and dry filling the burster tube with the solid mixture, followed by infusing and curing of the binder fluid. The work described herein concerns the time and pressure required to achieve complete infusion before gelling occurs.

Pressure Drop through Packed Beds

Flow of fluids through packed beds has been well investigated by the Ergun [2] equation in the creeping flow regime.

$$\frac{\Delta p}{L} = \frac{p_1}{p}\left[150\frac{(1-\varepsilon)^2}{\varepsilon^3}\frac{\mu U_1}{d_{sv}} + 1.75\frac{1-\varepsilon}{\varepsilon^3}\frac{\rho_{g1}U_1^2}{d_{sv}}\right] \approx 150\frac{(1-\varepsilon)^2}{\varepsilon^3}\frac{\mu U_1}{d_{sv}^2} \quad (1)$$

Where, μ, viscosity of the fluid, E, fraction void, Δp, pressure drop across bed, d_{sv}, surface volume of the packing particles, U, superficial velocity, ρ, fluid density, L, bed depth.

Since the superficial flow velocity $U_1 = \varepsilon\frac{dL}{dt}$, thus infusion time t can be rearranged as:

$$t = \left\{75\frac{(1-\varepsilon)^2}{\varepsilon^3}\frac{\mu}{\Delta p \times d_{sv}^2}\right\}L^2 = kL^2 \quad (2)$$

Appling the Ergun equation to predict the time *vs.* infusion length for packed beds of widely differing particle sizes and shapes is difficult. Figure 1, 2 and 3 show size distributions representive of two of the solids used in these tests.

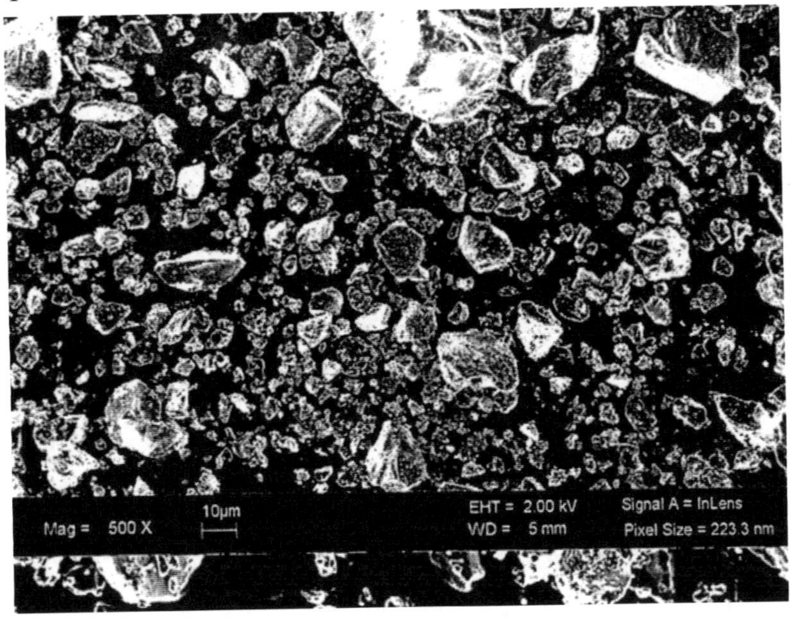

Figure 1 Photomicrograph of Dechlorane Plus 515 particles

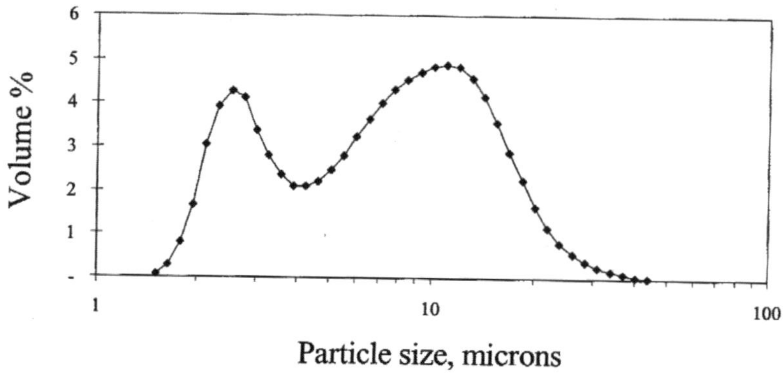

Figure 2 Particle size distribution of Dechlorane Powders

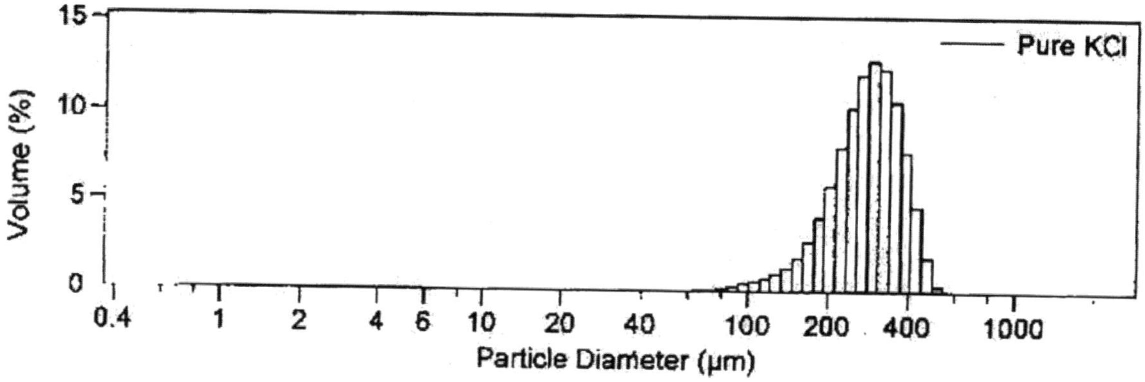

Figure 3 Particle size distribution of KCl crystals

Instead of trying to describe the flow pattern based on the sizes of the particles around which the binder fluid has to flow, we decided to simplify the model for any packed bed of powder by characterizing the flow as many parallel passages with an equivalent capillary diameter, Φ, which can be determined from the same Δp *vs.* L data. For a Newtonian fluid, the flow in a capillary channel follows the formula:

$$\frac{\Phi \Delta p}{4L} = \left(\frac{8V}{\Phi}\right)\mu \quad (3)$$

in which Φ is the effective capillary diameter, and V is the velocity of the fluid

$$V = \frac{dL}{dt} \quad (4)$$

where t is the time. Equation (3) would be further converted into:

$$dt = \frac{32\mu L dL}{\Phi^2 \Delta p} \quad (5)$$

Integrating Equation (5) along the filled length L results in:

$$\Delta t = \frac{16\mu L^2}{\Phi^2 \Delta p} \quad (6)$$

Rearranging Equation (6), the effective capillary diameter Φ can be calculated.

$$\Phi = \left[\frac{16\mu L^2}{(\Delta t)(\Delta p)}\right]^{0.5} \quad (7)$$

Testing Procedure and Results

A test device was assembled (Figure 4) which could hold a metal burster tube or its polycarbonate equivalent. Tube dimensions are listed in Table 1, and test conditions in Table 2. With packed powder tubes, plots of time *vs.* the square of infusion-filled length in the plastic tubes are shown in Figures 5 to 11. The data indicate that the filling time is 3 to 10 times longer than that predicted by the Ergun equation. Table 1 also indicates the hypothetical capillary diameter Φ calculated from Equation (7). Figures 12 and 13 illustrate the time and length observations for both 100 and 300 microns glass beads and for both sizes of the plastic burster tube.

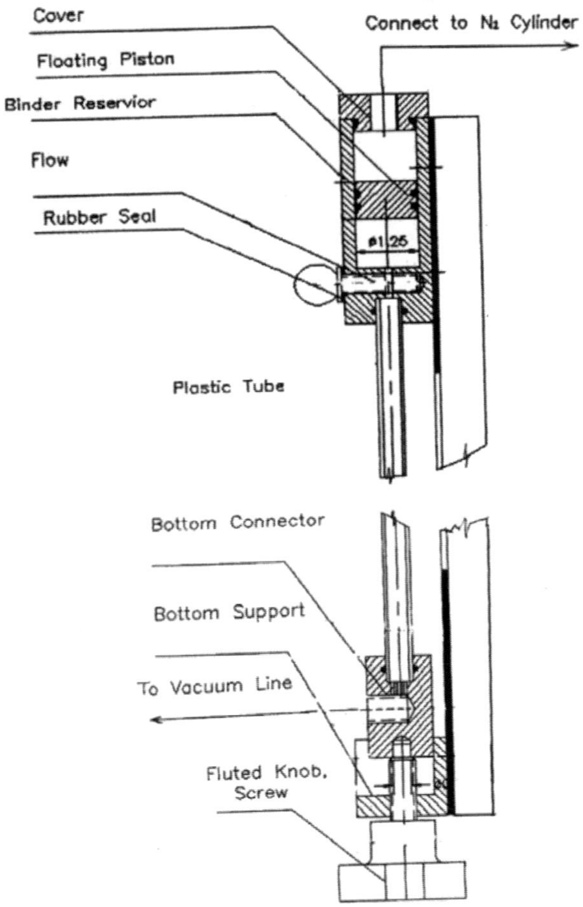

Figure 4 Infusion test device

Table 1 Burster Tube Dimensions

Designation	1 D (mm)	L/D
120-Aluminum	11.11	31.1
155-Steel	15.49	33.2
Plastic-Small	9.52	36.3
Plastic-Large	15.88	32.4

Table 2

Test	Tube Diam (mm)	Solids	%	Shape	Fluid	Viscosity (mPa·s)	Pressure (kPa)	Fraction Void	d_{sr} (mm)	Φ (mm)	Figure
1	9.5	KCl	100	Crystal	SC-B	62	80	0.37	280	30	5
2	9.5	KCl	100	Crystal	SC-B	62	827	0.37	280	29	6
3	9.5	KCl Dechlorane	80 20	Cube Irreg	MMA	0.55	827	0.27	280 23	2.5	7
4	9.5	KCl Dechlorane	80 20	Cube Irreg	SC-A+B	47	1184	0.32	280 23	4.5	8
5	9.5	KCl	100	Cube	SC-A+B	47	1255	0.36	220	49	9
6	9.5	KCl	100	Cube	SC-A+B	47	1360	0.37	220	55	10
7	15.9	KCl	100	Cube	SC-A+B	47	1017	0.37	220	55	11
8	9.5	Glass	100	Beads	Lube	35	680	0.36	100	15	12
9	9.5	Glass	100	Beads	Lube	35	680	0.36	300	52	12
10	15.9	Glass	100	Beads	Lube	35	1296	0.36	100	16	13
11	15.9	Glass	100	Beads	Lube	35	1316	0.36	300	53	13

*Notes: Smooth-cast 327 consists of two components, A and B to form polyurethane. Lube is a non-reactive lubricant.

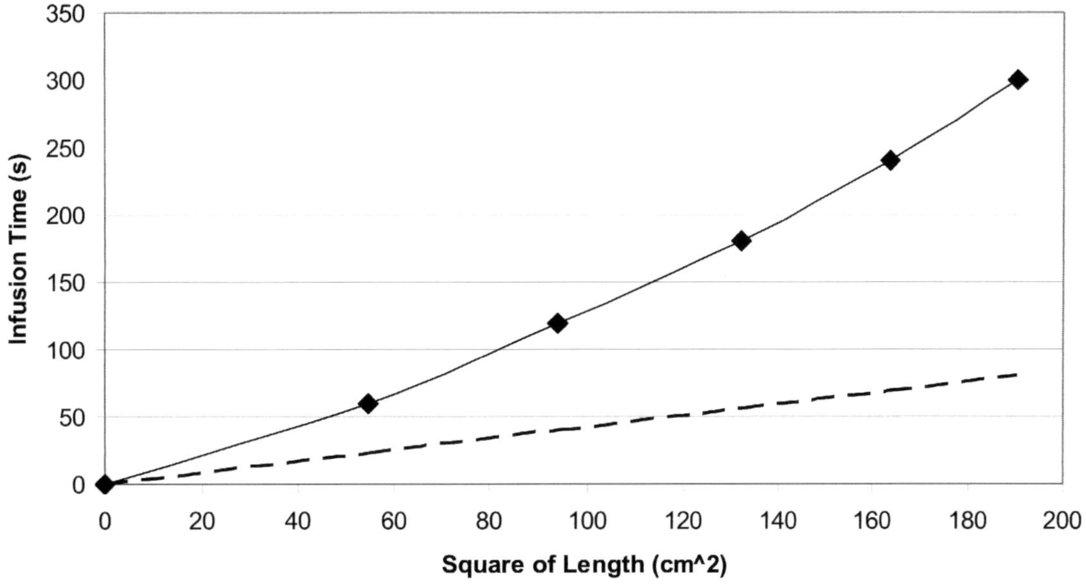

Figure 5 Infusion length *vs.* time for Test 1 with KCl crystals and Smooth-cast B fluid at 80 kPa Δp (-■- experimental, --- predicted by the Ergun equation)

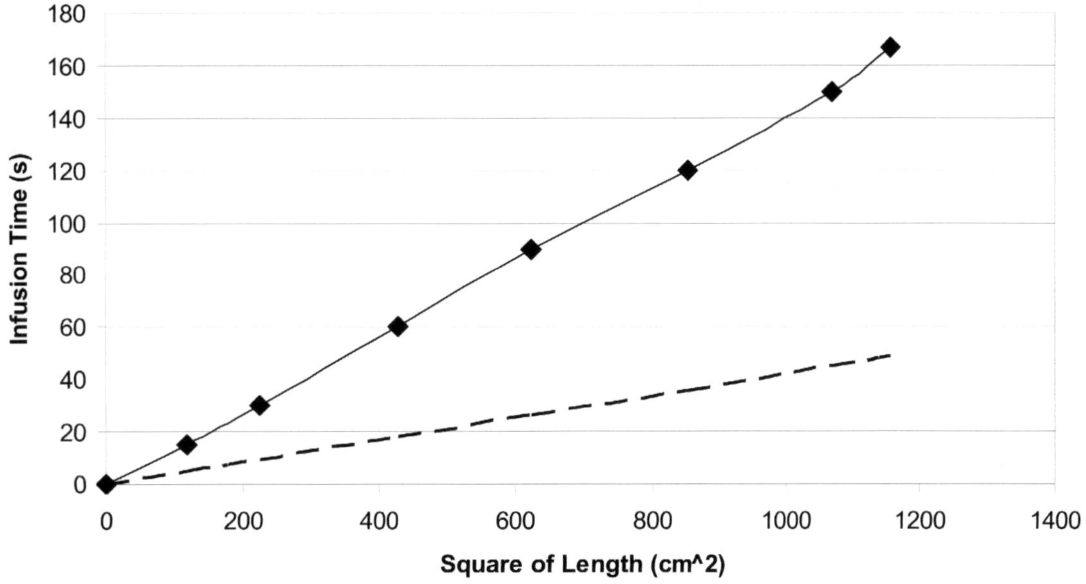

Figure 6 Infusion length *vs.* time for Test 2 with Smooth-cast B fluid at 827 kPa Δp (-■- experimental, --- predicted by the Ergun equation)

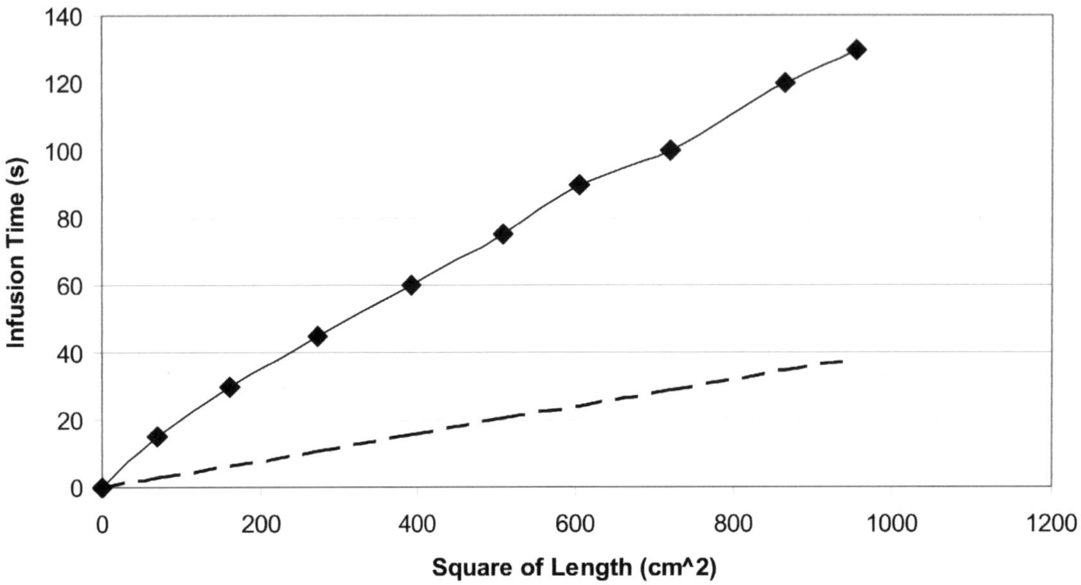

Figure 7 Infusion length *vs*. time for Test 3 for 80 % KCl and 20 % dechlorane with methyl methacrylate fluid at 827 kPa Δp (-■- experimental, --- predicted by the Ergun equation)

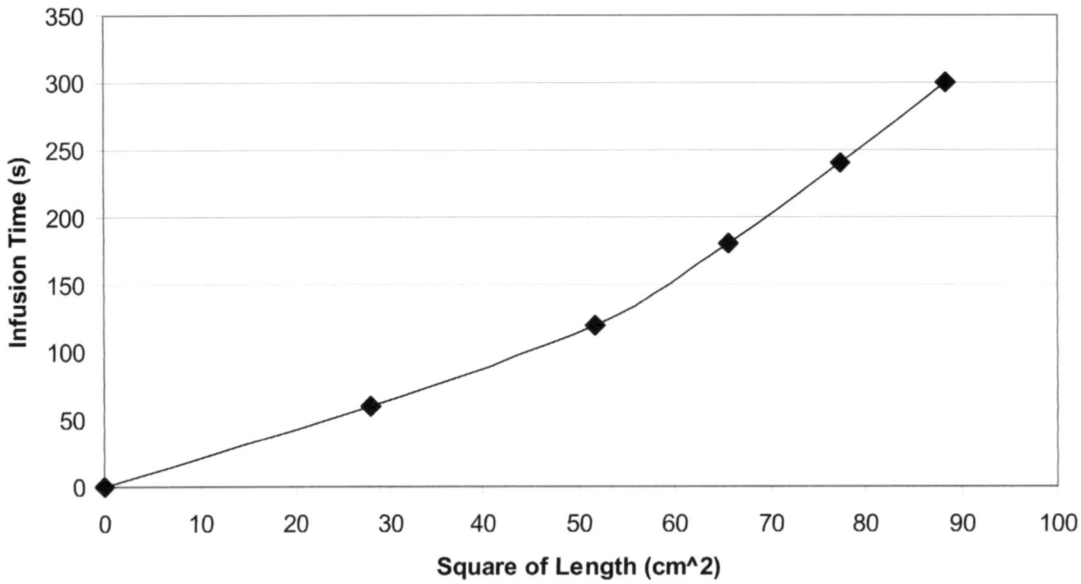

Figure 8 Infusion length *vs*. time for Test 4 for 80 % KCl, 20 % dechlorane with reacting Smooth-cast 327 A and B components

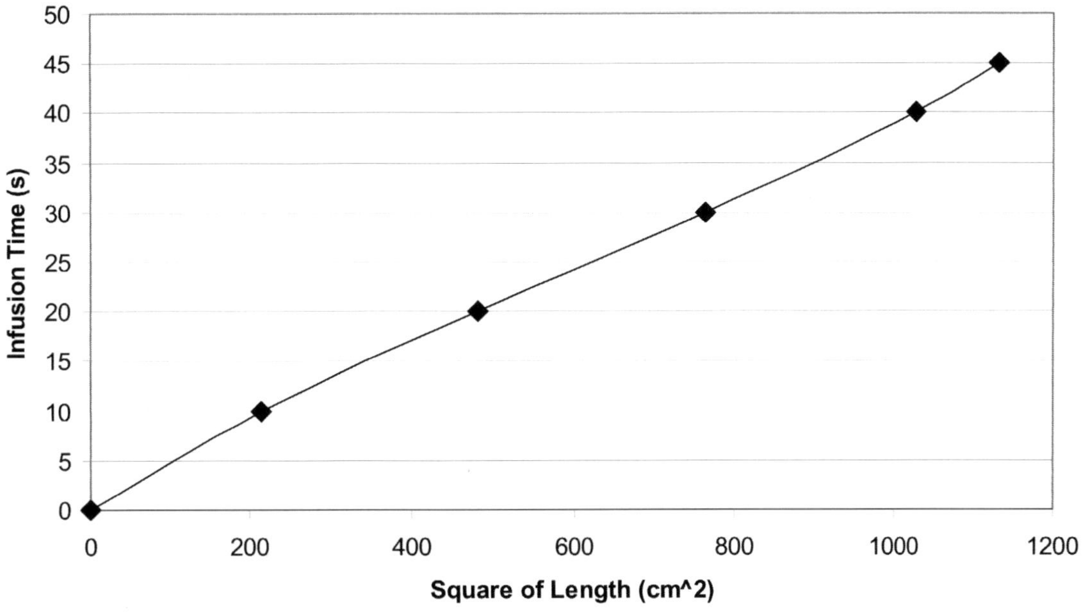

Figure 9 Infusion length *vs*. time for Test 5 with KCl crystals and Smooth-cast 327 binders

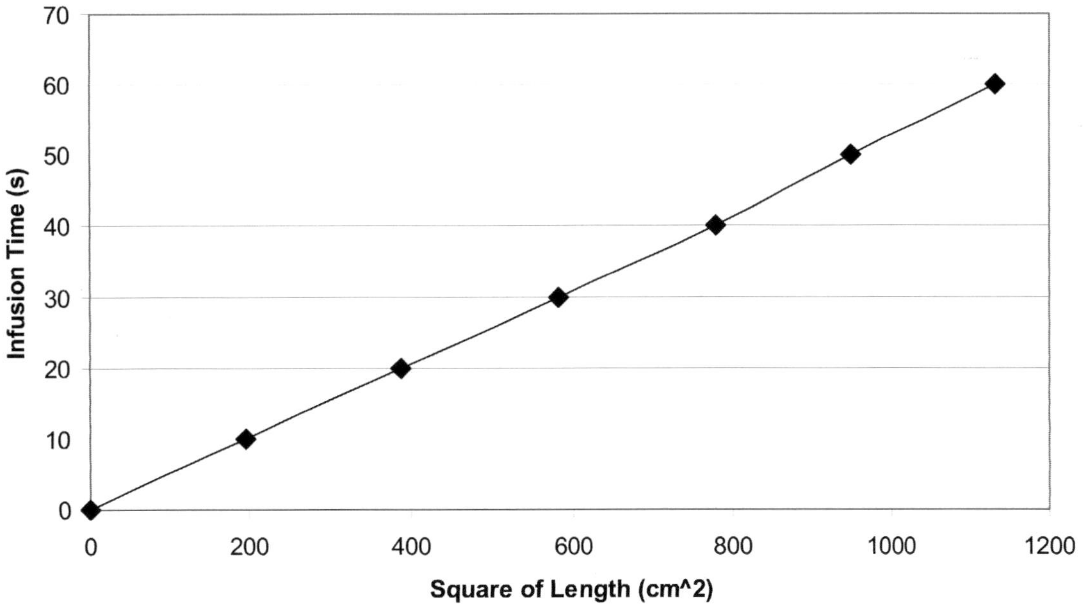

Figure 10 Infusion length *vs*. time for Test 6 for Smooth-cast 327 in 120 mm burster tube

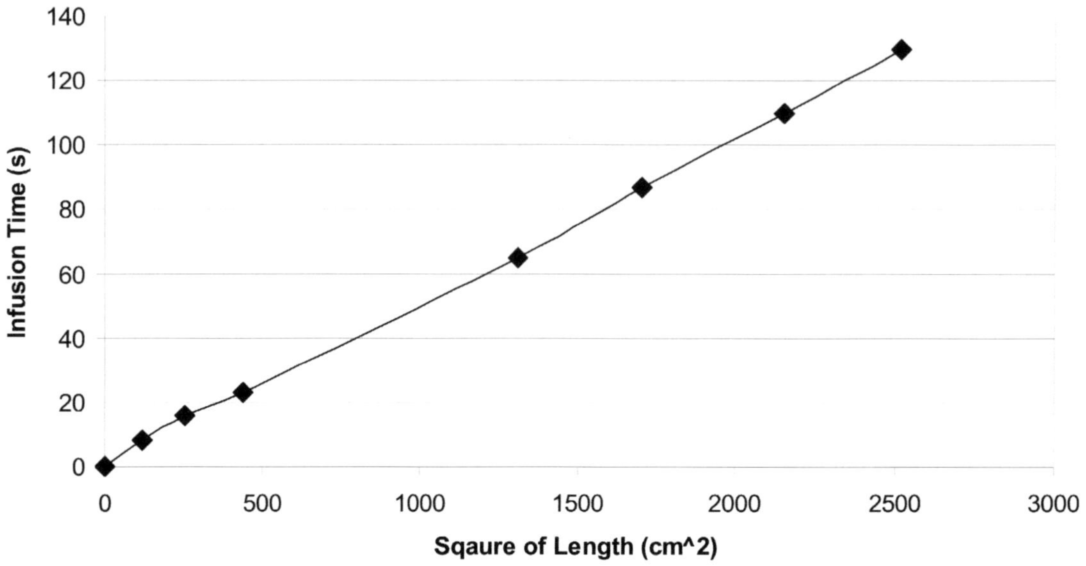

Figure 11 Infusion length *vs.* time for Test 7 for Smooth-cast 327 in 155 mm burster tube

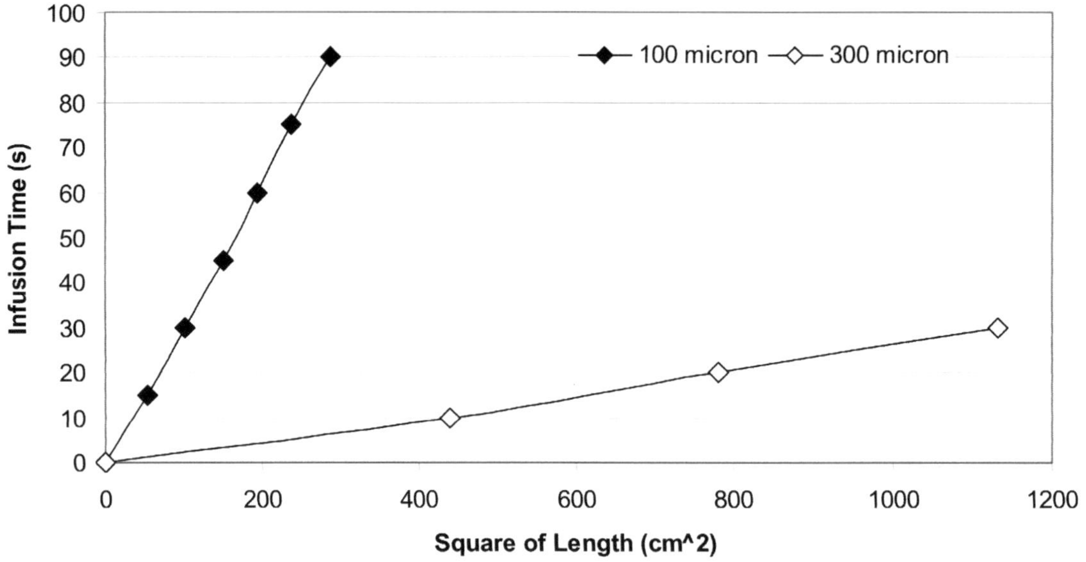

Figure 12 Infusion length *vs.* time for Test 8 with lubricant flow in 120 mm burster tube filled with glass beads

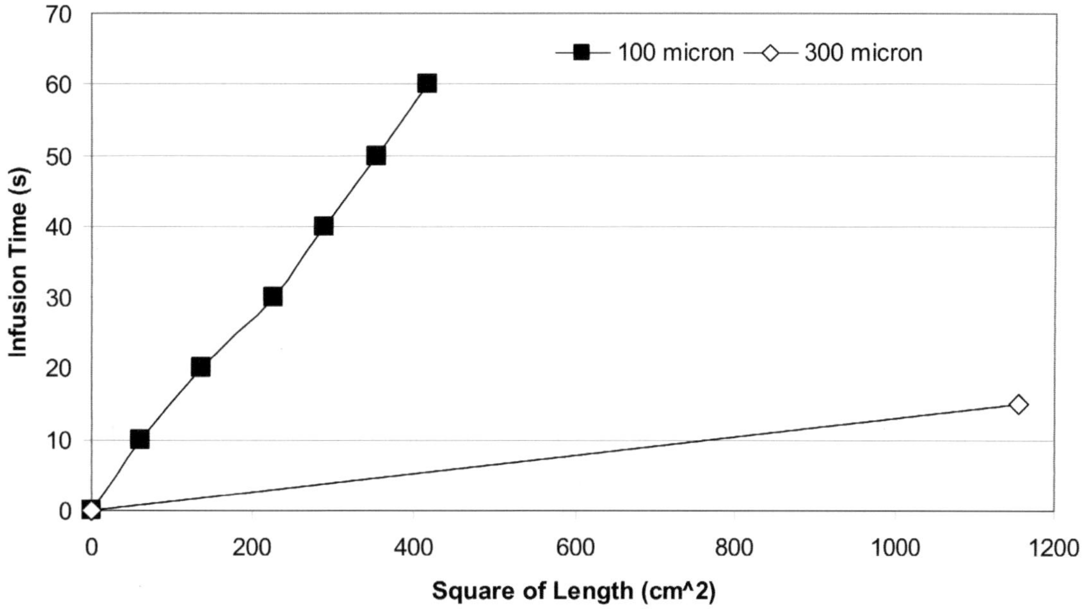

Figure 13 Infusion length *vs.* time for Test 10 lubricant flow through a bed of 100 micron glass beads in a 15.9 mm plastic tube

Conclusion

A simple model based on a simple test device and procedure has been developed to determine pressure (Δp) - time (t) - penetration depth (L) for the infusion of a fluid into mixed beds of powders. From the observed Δp-t-ΔL data, a hypothetical equivalent capillary Φ is calculated:

$$\Phi = \left[\frac{16\mu L^2}{(\Delta t)(\Delta p)}\right]^{0.5}$$

Testing can be done at room temperature with a known viscosity liquid, and extrapolated for equivalent packed beds of mixed powders for other viscosities and processing conditions.

The hypothetical capillary diameter Φ is about $0.15d$ to $0.18d$ for a bed of uniform diameter (d) spheres. For beds of non uniform powders, Φ is sensitive to sizes and shapes of the particulates, and may range from about $0.1d_{sv}$ to $0.3d_{sv}$, depending on how the average diameter is defined.

Acknowledgement

This work was partially supported by ARDEC at the Picatinny Arsenal

References

1. Costas G. Gogos, Ming-Wan Young, Linjie Zhu, Chen Wan, Raj Dave, David Todd "A Novel Net Shape Manufacturing Process for Concentrated Polymer-Particulate Composite Products", SPE ANTEC, Milwaukee, May 6 [2008]

2. Ergun, S., Chem Eng Prog, 48, 89 [1952]

322c Mechanically Robust Nanoparticle Stabilized Transparent Liquid Marbles

Prasad Bhosale, Center for Energy Systems Research, Tennessee Technological University, Cookeville, TN 38505, Mahesh V. Panchagnula, Dept of Mechanical Engineering, Tennessee Technological University, Cookeville, TN 38505, and Holly A. Stretz, Department of Chemical Engineering, Tennessee Technological University, Cookeville, TN 38501.

Liquid marbles are non-wetting liquid drops coated with hydrophobic powder. The presence of the particles on the free surface causes the effective liquid-solid interfacial tension to decrease substantially resulting in near non-wetting conditions. A common necessity of micro-fluidic devices is to be able to easily and controllably manipulate minute quantities of liquids and to be able to cause chemical reactions between such quantities of liquids. Liquid marbles have been shown to possess the potential to deliver upon these promises. However it is known that the ability to use liquid marbles in microfluidic devices is currently limited by their mechanical robustness. Here we present, remarkably robust transparent liquid marbles of glycerol and water under both compressive and tensile loading conditions formed using commercially available nanoparticulate fumed silica treated with hydrophobic agent, Hexamethyldisalazane (nHFS) and micron size polytetrafluoroethylene powder (µPTFE). ESEM and mass transfer data are used to investigate the cause of this robustness. Liquid marble evaporation data indicates that the nanoparticulate fumed silica marbles are more robust due to increased particle-particle interaction. During the evaporation process convective flows within the evaporating drop causes hydrophobic particles to self-organize and form a robust 2-dimensional array on the liquid-vapor interface. We also report translucent liquid marbles if not entirely transparent formed using nHFS particles. Transparent liquid marbles allow the contents of the liquid marbles to be probed optically during transport.

322e Microfluidic Assembled Lipopolyplex for Antisense Oligonucleotide Delivery

Chee Guan Koh, Department of Chemical and Biomolecular Engineering, The Ohio State University, 140 West 19th Avenue, Columbus, OH 43210, Xulang Zhang, NSF Center for Affordable Nanoengineering of Polymeric Biomedical Devices, The Ohio State University, 140 West 19th Avenue, Columbus, OH 43210, and L. James Lee, Chemical and Biomolecular Engineering, The Ohio State University, 125 Koffolt Labs, 140 W 19th Ave, Columbus, OH 43210.

For therapeutics applications, small and uniformly sized nanoparticles, polymer-based (polyplex) and lipid-based (lipopolyplex), are important for successful and efficient in vitro and in vivo gene/drug delivery. Nanoparticles, e.g. liposomes, have been studied extensively as nanocarriers because of their biocompatibility and ability to incorporate drug molecules. Conventional bulk mixing (BM) method of producing liposomes requires mixing of multiple reagents and postprocessing steps such as sonication or membrane extrusion to produce small and uniformly sized nanoparticles. Therefore, in this study, microfluidic devices and novel micromixers were developed for assembling lipopolyplex nanoparticles. The size, surface charge, morphology, and gene/drug encapsulation of the nanoparticles were characterized using dynamic light scattering, zeta potential measurement, atomic force microscopy, cryo-TEM, and gel electrophoresis. Nanoparticles encapsulating antisense oligonucleotides synthesized were evaluated for downregulation of Bcl-2, an antiapoptotic protein commonly overexpressed in many cancer cells.

TiO_2 Hollow Nanofibers Templated by Electrospun Polyethylene Oxide (PEO) Aqueous Solutions

Shinsuke Nagamine, Yoshitaka Tanaka, Masahiro Ohshima*
Department of Chemical Engineering, Kyoto University, Katsura Campus, Nishikyo-ku, Kyoto 615-8510, Japan
* Tel: +81-753832686, Fax: +81-753832646, E-mail: nagamine@cheme.kyoto-u.ac.jp

Abstract

A new and simple method of fabricating TiO_2 hollow nanofibers which employs electrospinning was developed. An aqueous solution of polyethylene oxide (PEO) was electrospun and thread-like liquid jets were introduced directly into a titanium tetraisopropoxide (TTIP) / hexane solution. This induced the formation of a core – sheath structure due to the rapid hydrolysis and condensation of TTIP at the interface between the PEO solution and TTIP solution. As a result, mat-like samples were obtained, which was composed of nanofibers with a diameter of several hundreds nanometers and a length of several tens to hundreds micrometers. Hollow spherical micron particles with a diameter of several micrometers and thick hollow fibers could also be synthesized by varying the molecular weight and concentration of PEO solution. It was suggested that the product morphology was governed by two parameters relating to properties of PEO solution, Berry number $C[\eta]$ and κ / γ, representing the degree of overlap of polymer chains and the strength of electric charge repulsion against the surface tension, respectively.

Introduction

Hollow microparticles and microfibers of inorganic materials have a wide range of applications, e.g. microcapsules for controlled release, catalysts, sensors, and electric devices. The most effective strategy for preparing the hollow structure is to utilize the precipitation of solids at the interfaces, such as spray drying (liquid droplet – gas interface), the emulsion method (water – oil interface stabilized by surfactant), and the solid templating method (solid particle – liquid interface). We have previously developed another simple synthetic method for creating TiO_2 hollow microparticles, what we call "spray hydrolysis". In this method, water is atomized and introduced into an organic solution containing titanium tetraisopropoxide (TTIP) as the precursor of TiO_2 [1, 2]. The rapid hydrolysis and condensation of TTIP at the interface of the water droplets – organic phase leads to the formation of a TiO_2 shell and keep the droplets from coalescing. The merits of this method are as follows: (1) Spherical hollow microparticles can be produced quickly at ambient temperature and pressure; (2) The synthesis process is free from contamination and there is no complicated post-treatment to remove additives such as surfactants.

Our present purpose is to produce hollow fibers based on the concept of spray hydrolysis. As a simple synthetic method of micro and nanofibers, electrospinning has attracted considerable attention and is well studied [3]. Hollow fibers of metal oxides [4, 5] were prepared using co-axial electrospinning, where two immiscible liquids were simultaneously electrospun from the co-axial double-tube spinneret. Another synthetic method for creating hollow fibers which also employs electrospinning is to use preformed polymer fibers as templates [6]. This method consists of three steps: (1) the formation of template polymer fibers by electrospinning, (2) the coating of the fibers with the target solid, and (3) the removal of the template fibers by the thermal treatment.

In this study, a new and simple method of fabricating TiO_2 hollow nanofibers is proposed. This method consists of electrospinning a viscous aqueous solution of polyethylene oxide (PEO) and the direct introduction of elongated jets into the organic solution of TTIP. The core-sheath structure is formed via the hydrolysis of TTIP at the interface between the aqueous PEO solution and the TTIP solution. This method may appear similar to the above-mentioned method which uses the electrospun fibers as templates. The difference is that our method enables the formation of the thread-like templates and the TiO_2 core-sheath structure in a single process. The product morphology would reflect that of electrospun jet, which strongly depends on the properties of PEO solution, such as viscosity, surface tension, and electric conductivity. Therefore the effects of those properties determined by molecular weight and concentration on the product morphology were investigated.

Experimental Section

PEO (M_W = 500,000, 2,000,000, and 4,000,000), TTIP and hexane were purchased from Wako Chemical, Japan and were used as supplied. Aqueous solutions of PEO with different concentrations (C = 0.1 – 10 wt%) were prepared. Viscosities of PEO solutions were measured using capillary viscometer. The measurements of surface tensions were performed by Wilhelmy plate method using Kyowa CBVP-A3. Electric conductivities were measured using conductance meter (HORIBA ES-51). The measurements of these properties were conducted at 298 K.

The electrospinning of these PEO solutions was conducted using a commercial electrospinning device (Kato-Tech NEU). The schematic of experimental setup is illustrated in Fig. 1. A PEO solution was loaded in a downward plastic syringe with a steel spinneret (inner diameter: 0.8 mm). A TTIP solution in hexane (1.0 mol L^{-1}) in a petri dish was located on the ground electrode. The distance from the tip of spinneret to the top of the TTIP solution is fixed at 11.4 cm. The PEO solution was then fed at the rate of 3.8 cm^3 h^{-1} and a positive voltage of 10 kV was applied to the spinneret. The jetted PEO

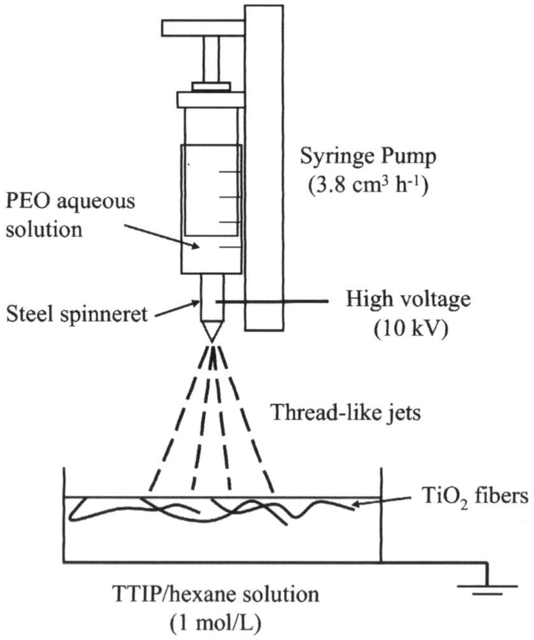

Fig. 1. Schematic illustration of sample preparation process.

solution was introduced into the TTIP solution, resulting in the immediate formation of a mat-like structure that floated on the surface of the TTIP solution. The products were recovered by filtration, washed with hexane, and dried at 363 K for several hours. A part of sample was calcined at 773 K to remove PEO completely and to convert it to crystalline TiO_2. The products were observed by a field emission scanning electron microscope (FE-SEM, JEOL JSM-6340FS).

Results and Discussion

Fig. 2a shows a photograph of a sample prepared with a PEO having M_W = 500,000 and C = 5 wt%. The SEM observation reveals that the mat-like sample is mainly composed of fibers, although micrometer-sized spheres are also present as by-products (Fig. 2b). The diameters of fibers are on a submicron scale and the lengths are greater than tens of micrometers. The outer morphologies of the fibers are a little larger when compared to those of PEO nanofibers prepared by the conventional electrospinning method. However, the inner structure significantly differs from that of PEO fiber, that is, the fibers possess a hollow core-sheath structure as shown in the magnified SEM image of fractured fibers in Fig. 2c. This result suggests that the TiO_2 sheaths were formed by the hydrolysis of TTIP at the interface between the elongated PEO aqueous solution and the organic TTIP solution. An interesting feature is that the hollow structure is formed without the removal of PEO. This is probably because the template was composed mostly of water and had a low concentration of PEO. After drying, the volume of precipitated PEO was too small to fill the space inside the sheath of TiO_2, which led to the creation of hollow structure. Fig. 2d presents the SEM image of calcined sample. The hollow fibrous structure is maintained even after calcination.

The products with other morphologies were also obtained by varying M_W and C. Fig. 3a shows SEM image of sample prepared with M_W = 500,000 and C = 0.5 wt%. The main products are the spherical particles with a diameter of 10 to 20 micrometers and smaller distorted

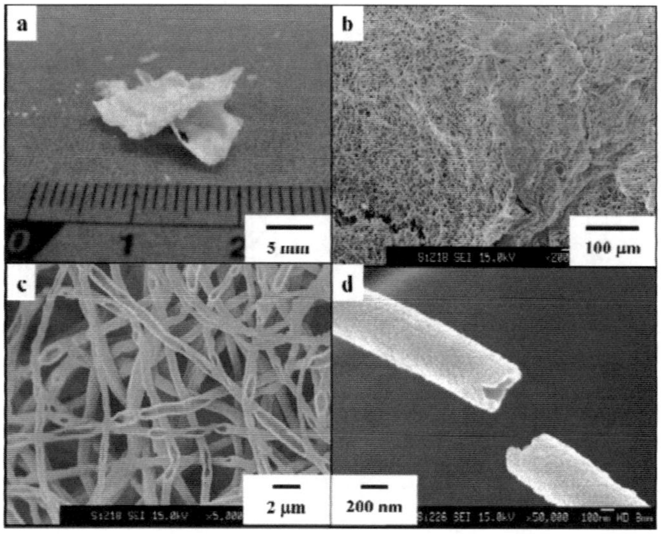

Fig. 2. (a) a photograph of a mat-like morphology and (b, c) SEM images of hollow nanofibers prepared with M_W = 500,000 and C = 5 wt% at different magnifications. (d) a SEM image of hollow nanofibers calcined at 773 K.

Fig. 3. SEM images of products prepared with (a, b) M_W = 500,000 and C = 0.5 wt%, (c, d) M_W = 2,000,000 and C = 5 wt% at different magnifications.

particles. The magnified SEM image reveals the formation of hollow core-shell structure in the larger particles (Fig. 3b). In this case, the projected PEO solution could not take the thread-like shape due to low viscosity, and the generated spherical droplets served as templates to yield hollow particles. Hollow fibers with a diameter of several micrometers were also obtained with M_W = 2,000,000 and C = 5 wt% (Fig. 3c, d). During the electrospinning process of this PEO solution, it was observed that a single thread-like jet with a visible thickness traveled continuously from the tip of spinneret to the TTIP solution. Probably the strong entanglement of PEO chains inhibited the disruption of the jet of the PEO solution, and led to the formation of thick hollow fibers. The dependence of the product morphology on M_W and C is summarized in Fig. 4. As a general tendency, the low M_W and low C lead to the formation of spherical particle. As C increases, the product morphology turns to hollow fiber. Hollow nanofiber could be obtained at M_W = 500,000, while thick fibers were formed at the higher M_W.

The morphology of electrospun jet is determined by the properties of spinning solution. Intuitively, when the surface tension is low and the charge density, relating to the electric conductivity, is high, the charged droplet tends to deform to increase surface area. If the viscosity is sufficiently high, the elongation occurs, while at the low viscosity, the droplet is disrupted to smaller droplets. The results of surface tension measurements are shown in Fig. 5. The surface tension γ is almost independent of C and M_W in the tested range. As shown in Fig. 6, the electric conductivity κ continues to increase with C for M_W = 500,000, while it stops increasing at 0.5 wt% for M_W = 2,000,000 and 4,000,000. Here κ / γ is regarded as a rough index representing the electrostatic repulsion against the surface energy.

Fig. 7 shows the viscosity η greatly increasing with C and M_W. The viscosity reflects the configuration of polymer chains. The spinnability of PEO solution is considered to be determined by the degree of overlap of polymer chains. It is known that the concentration C^* at

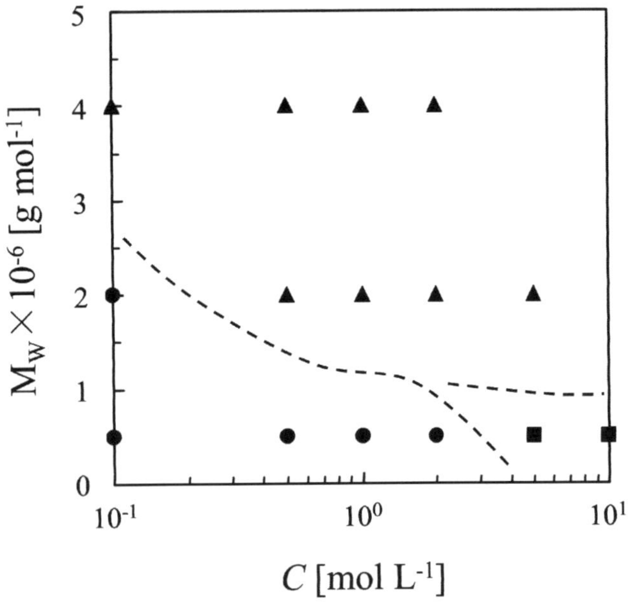

Fig. 4. The dependence of product morphology on molecular weight M_W and concentration C of PEO. (●) hollow particle, (▲) thick hollow fiber, and (■) hollow nanofiber.

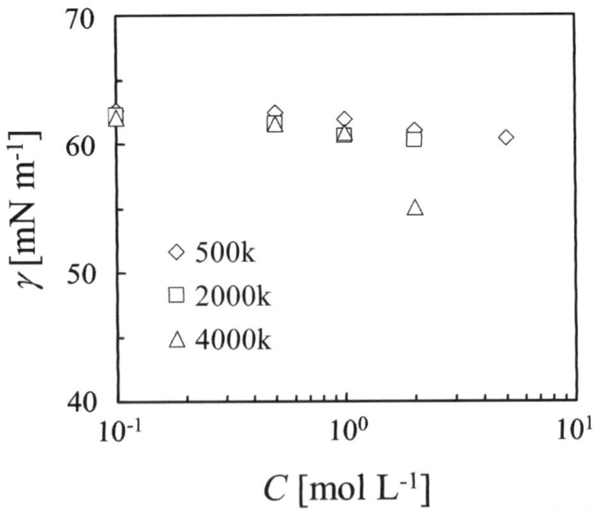

Fig. 5. Plot of surface tensions γ of PEO solutions with C for different M_W.

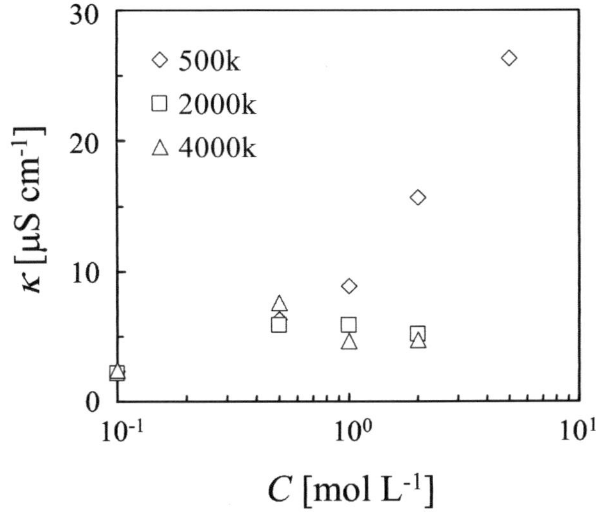

Fig. 6. Plot of electric conductivities κ of PEO solutions with C for different M_W.

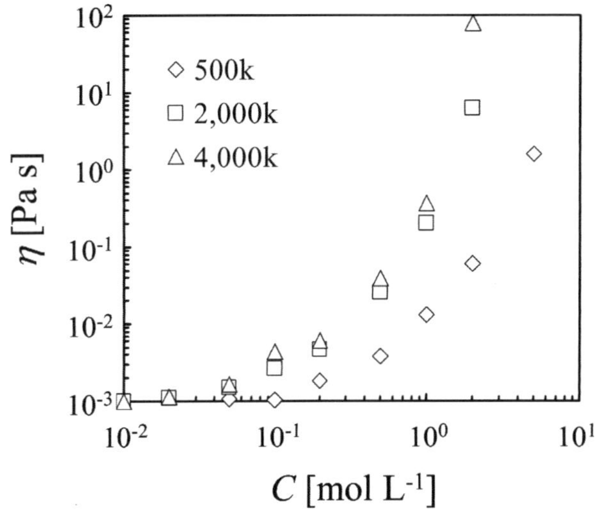

Fig. 7. Plot of viscosities η of PEO solutions with C for different M_W.

which overlap starts is associated with the intrinsic viscosity $[\eta]$ by the following equation [7].

$$C^* \sim [\eta] \quad (1)$$

$[\eta]$ can be calculated by applying the Huggins equation to the dependence of η on C.

$$\eta / \eta_S = 1 + [\eta]C + k([\eta]C)^2 + \cdots \quad (2)$$

Here Berry number *Be* defined by the following equation is introduced as an index of the degree of overlap of polymer chains.

$$Be = [\eta]C \quad (3)$$

Fig. 8 shows the dependence of product morphology on *Be* and κ / γ. This map could be successfully partitioned into three zones associated to the morphologies. When *Be* was large, overlap of polymer chains induced the formation of fiber. When κ / γ was large, the electric charge repulsion surpassed the surface tension. As a result, the disruption of droplet led to the production of thin nanofibers or spherical particles.

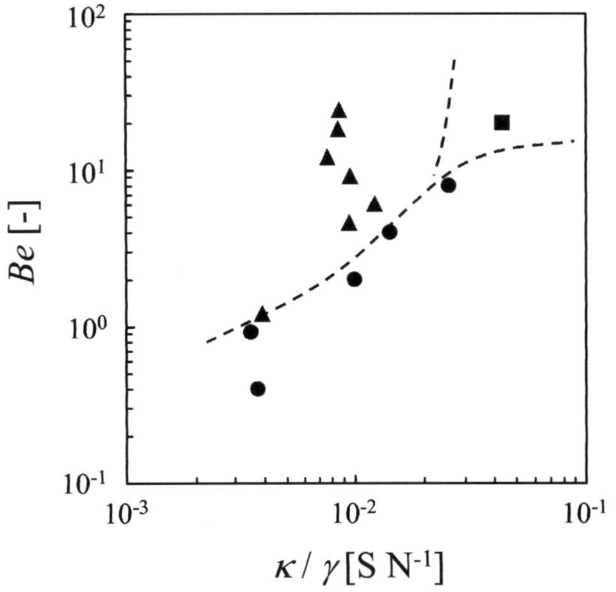

Fig. 8. The dependence of product morphology on Berry number $Be = [\eta]C$ and κ / γ. (●) hollow particle, (▲) thick hollow fiber, and (■) hollow nanofiber.

Conclusion

A novel and simple method to synthesize hollow TiO_2 nanofibers was developed, which based on electrospinning and an interfacial sol-gel reaction. The introduction of the electrospun PEO aqueous solution into the TTIP / hexane solution resulted in the formation of TiO_2 hollow nanofibers. Hollow microparticles and thick microfibers were also obtained by varying the molecular weight and the concentration of PEO. The dependence of product morphology on the properties of PEO solution could be summarized in a diagram using *Be* and κ / γ as indexes. Spherical particles were produced with small *Be*. As *Be* increased, the products tended to be fibers due to the overlap of polymer chains. When κ / γ was large, strong electric charge repulsion caused the disruption of droplets and therefore thin nanofibers were produced.

References

1. S. Nagamine, A. Sugioka, and Y. Konishi, *Mater. Lett.*, 2007, **61**, 444.
2. S. Nagamine, A. Sugioka, H. Iwamoto, and Y. Konishi, *Powder. Technol.*, 2008, **186**, 168.
3. D. Li, J. T. McCann, Y. Xia, and M. Marquez, *J. Am. Ceram. Soc.*, 2006, **89**, 1861.
4. I. G. Loscertales, A. Barrero, M. Márquez, R. Spretz, R. Velarde, and G. Larsen, *J. Am. Chem. Soc.*, 2004, **126**, 5376.
5. D. Li and Y. Xia, *Nano Lett.*, 2004, **4**, 933.
6. M. Bognitzki, H. Hou, M. Ishaque, T. Frese, M. Hellwig, C. Schwarte, A. Schaper, J. H. Wendorff, and A. Greiner, *Adv. Mater.*, 2000, **12**, 637.
7. P. Gupta, C. Elkins, T. E. Long, G. L. Wilkes, *Polymer*, 2005, **46**, 4799.

322g Reconfigurable Microfluidics with Metallic Containers

Hongke Ye, Jung-Rae Park, and **David H. Gracias**. Department of Chemical and Biomolecular Engineering, Johns Hopkins University, 221 Maryland Hall, 3400 N. Charles Street, Baltimore, MD 21218

We describe a microfluidic scheme based on remotely controlled self-assembled containers that allows spatio-temporal control over nanoliter scale chemical reactions. We discuss finite element simulations of the inductive coupling of radio frequency radiation to the containers; this coupling enables remotely triggered release of chemical reactants. We demonstrate on-demand chemical release from stationary and mobile containers patterned with different porosities. We also explore reactions between chemicals released from two containers that form liquid products and deposit solid precipitates. We argue that these remotely controlled metallic containers provide an attractive platform for carrying out reconfigurable microfluidics "without channels".

331a Cell Detachment from Porous Poly(L-Lactic Acid) Scaffolds Cultured Under Flow Perfusion for Bone Tissue Engineering

Samuel VanGordon[1], Roman S. Voronov[2], Dimitrios V. Papavassiliou[3], and Vassilios I. Sikavitsas[3]. (1) Bioengineering, University of Oklahoma, 100 E. Boyd St., SEC T-335, Norman, OK 73019, (2) School of Chemical, Biological, and Materials Engineering, University of Oklahoma, 100 E. Boyd, SEC Room 311, Norman, OK 73019, (3) Chemical, Biological, and Materials Engineering, University of Oklahoma, Sarkeys Energy Center, 100 East Boyd Street, Norman, OK 73019

Cell adherence is crucial to long term culturing of cells seeded on porous scaffolds using flow perfusion. Development of tissues from osteoblasts cultured in flow perfusion systems is important for the growth of tissue engineered bone grafts. Mechanotransduction of preosteoblastic cells from fluid shear forces has been found to promote cell differentiation and the production and calcification of extracellular matrix (ECM). High fluid shear also causes the disengagement of cells from scaffold surfaces immediately subsequent to seeding. These same high fluid shear forces also instigate detachment of cells from developed ECM that is secreted from cells at later culturing points. Measuring cell detachment numbers and creating detachment profiles for porous 3-dimensional (3D) cellular constructs will shed light on how cells cultured under flow perfusion can have their disengagement from scaffolds reduced while exposing them to ample stimulatory shear forces.

Characterization of different fluid shear rates on the detachment of preosteoblastic mesenchymal stem cells (MSC) seeded and cultured on porous 3D poly(L-lactic acid)(PLLA) scaffolds using a perfusion bioreactor is preformed in this study. Using oscillatory flow perfusion to homogeneously distribute cells, MSC's were seeded on scaffolds. These scaffolds were then cultured under unidirectional flow of 0.15mL/min for time periods up to 12 days using osteoblastic media. After each time period, cells were detached from scaffolds using increasing fluid flows. Serial collection of detached cells was used to generate a detachment profile. Critical flow rates were established for each culture time period analogous to a different cell microenvironment where significant cell disengagement occurred.

PLLA scaffolds made by the solvent casting with substrate leaching technique have random internal architectures. These random architectures create complex shear force environments that cannot be accurately characterized by simple bulk mathematical calculations. To characterize the internal shear environment, mathematical modeling using Lattice Boltzmann method was employed. Mathematical modeling was also coupled with the detachment of cells from flat nonporous PLLA surfaces under equivalent shear forces to those exerted in the 3D microenvironments of the scaffolds under flow perfusion.

331b Comparing the Attachment and Growth of Bone Cells on Chitosan Bound by Two Silane Molecules to Titanium for Use In Joint Replacements

Holly J. Martin, Dave C. Swalm School of Chemical Engineering, Mississippi State University, 330 Swalm - President's Circle, Mississippi State, MS 39762, Kirk H. Schulz, Vice President for Research and Economic Development, Mississippi State University, 617 Allen Hall, Mississippi State, MS 39762, and Joel D. Bumgardner, Biomedical Engineering Department, Herff College of Engineering, University of Memphis, Memphis, TN 38152.

Metals that are used for implantation, such as titanium, are chosen because of advantageous mechanical properties, such as strength, weight, and durability. These metals also possess the ability to be passivated, meaning the implant is unreactive with the surrounding physiological fluids. Unfortunately, osseointegration, or the attachment and growth of bone cells, does not occur on metal implants, preventing the integration of the implant into the bone. By attaching bioactive materials, such as calcium phosphate/hydroxyapatite [1,2] or biological molecules [3-5], osseointegration can be achieved [6].

One implant coating that has been shown to be bioactive is chitosan, a de-acetylated form of chitin [7]. Found in the exoskeletons of shellfish and insects, chitin is the second most abundant form of polymerized carbon in nature, behind only cellulose [7,8]. Chitosan is being investigated as an implant coating because it is non-toxic and the by-products produced are considered a normal part of metabolism [8,9]. In addition to the non-toxic nature, chitosan is cationic, or positively charged, meaning it will attract negatively charged proteins and cells [10]. This attraction has been shown to encourage the attachment and growth of bone cells [10]. Along with the attraction, chitosan is useful as an implant coating because the biopolymer encourages proper bone formation [11]. The cells responsible for bone production retain the desired cell shape, which influences cell-specific functions [11].

At Mississippi State University, methods to improve the bond between chitosan and titanium are being studied. A two step process to deposit chitosan consisted of two titanium surface treatments, passivation and piranha, followed by the deposition of triethoxsilylbutyraldehyde (TESBA) [12]. A three step process to deposit chitosan also consisted of the two titanium surface treatments followed by the deposition of aminopropyltriethoxysilane (APTES) and gluteraldehyde [13]. Four treatment combinations were produced, by varying the titanium surface treatment and the silane deposited. The titanium surface was documented following each reaction step using X-Ray Photoelectron Spectroscopy, which showed that more silane was deposited on the piranha treated surface as compared to the passivated titanium surface [12,13]. The chitosan deposited using one of the four treatment combinations were also documented using XPS, which showed no differences in the films [14]. Other bulk properties, including hardness and elastic modulus, were unchanged from films deposited on glass slides [14]. An increase in the adhesion strength of the chitosan coating to the titanium surface was seen as compared to other methods of deposition, although no differences were seen between the four treatment combinations [7,14].

The individual steps of the reaction series have been documented using XPS, while the effects of the reaction series on the mechanical properties have been documented using nano-indentation and tensile testing [12-14]. Some research on the effects of the silane molecule on the biological properties of chitosan has been performed. Using APTES to bond either chitosan [7] or other biological molecules [4,5], no decrease in the ability of the bone cells to attach to the coating was seen [4,5,7]. However, these studies were conducted using the 5% water – 95% ethanol mixture as a carrier for APTES [4,5,7]. By changing the solvent from a 5% water - 95% ethanol mixture to toluene, the adhesion strength between the chitosan coating and the titanium surface was increased at least 10 fold [14]. However, the effects of changing the solvent to toluene on the biological properties of chitosan have not been determined. Since toluene is considered mutagenic, there is a chance that any trapped solvent could negatively affect the attachment and growth of the bone cells. The effects of changing the silane deposited from APTES to TESBA have also not been studied. The research presented will cover the effects of bonding chitosan to the titanium surface using two silanes and toluene as the solvent.

[1] Y. Yang, C.M. Agrawal, K.H. Kim, H. Martin, K. Schulz, J.D. Bumgardner, J.L. Ong. *Journal of Oral Implantology*, 29, 6, 270-277, 2003.

[2] R.J. Friedman, T.W. Bauer, K. Garg, M. Jaing, Y.H. An, R.A. Draughn. *Journal of Applied Biomaterials*. 6, 231-235, 1995.

[3] A. Nanci, J.D. Wuest, L. Peru, P. Brunet, V. Sharma, S. Zalzal, M.D. McKee. *Journal of Biomedical Materials Research*, 40, 324-335, 1998.

[4] D. A. Puleo. *Journal of Biomedical Materials Research*, 37, 222-228, 1997.

[5] D. A. Puleo. *Journal of Biomedical Materials Research*, 29, 951-957, 1995.

[6] B.D. Ratner and A. S. Hoffman. In: *Biomaterials Science: An Introduction to Materials in Medicine*. California: Academic Press, Inc., 1996, Foreword, 1-8.

[7] J.D. Bumgardner, R. Wiser, P.D. Gerard, P. Bergin, B. Chestnutt, M. Marini, V. Ramsey, S.H. Elder, J.A. Gilbert. *Journal of Biomaterials Science: Polymer Edition*, 14, 5, 423-438, 2003.

[8] G. Haipeng, Z. Yinghui, L. Jianchun, G. Yandao, Z. Nanming, Z. Xiufang. *Journal of Biomedical Materials Research*, 52, 285-295, 2000.

[9] M. Prasitsilp, R. Jenwithisuk, K. Kongsuwan, N. Damrongchai, P. Watts. *Journal of Materials Science: Materials in Medicine,* 11, 773-778, 2000.

[10] C. Jarry, C. Chaput, A. Chenite, M.A. Renaud, M. Buschmann, J.C. Leroux. *Journal of Biomedical Materials Research (Applied Biomaterials)*, 58, 127-135, 2001.

[11] A. Lahiji, A. Sohrabi, D. S. Hungerford, C. G. Frondoza. *Journal of Biomedical Materials Research*, 51, 586-595, 2000.

[12] H.J. Martin, K.H. Schulz, J.D. Bumgardner, K.B. Walters. *Applied Surface Science*, 254, 4599-4605, 2008.

[13] H.J. Martin, K.H. Schulz, J.D. Bumgardner, K.B. Walters. *Langmuir*, 23, 6645-6651, 2007.

[14] H.J. Martin, K.H. Schulz, J.D. Bumgardner, J.A Schneider. *Thin Solid Films*, 2008, In Press.

331c Macroporous Cyclic Acetal Hydrogels for Orbital Floor Regeneration

Martha W. Betz[1], John F. Caccamese[2], Domenick P. Coletti[2], John J. Sauk[3], and John P. Fisher[1]. (1) Fischell Department of Bioengineering, University of Maryland, College Park, MD 20742, (2) Oral Maxillo Surgery, University of Maryland Dental School, Baltimore, MD 21201, (3) Diag Sci & Pathology, University of Louisville Dental School, Louisville, KY 40202

This work investigates a cyclic acetal hydrogel construct for orbital floor regeneration. Orbital floor injuries are a devastating form of craniofacial trauma. Current clinical treatments, including implantation of plastics or metals, are often inadequate due to loss of function as well as poor aesthetics. These concerns have led us to investigate tissue engineering approaches for the treatment of orbital bone defects. Cyclic acetal biomaterials may be preferred as they hydrolytically degrade to form diol and carbonyl primary degradation products, which should not affect the local acidity of the implant or phenotypic function of a delivered cell population. Specifically, EH-PEG hydrogels composed of the cyclic acetal monomer 5-ethyl-5-(hydroxymethyl)-β,β-dimethyl-1,3-dioxane-2-ethanol diacrylate (EHD) and hydrophilic poly(ethylene glycol) diacrylate (PEGDA) were investigated. Previous work has demonstrated that EH-PEG hydrogels support long term viability of encapsulated bone marrow stem cells (BMSCs), and can deliver bone morphogenetic protein-2 (BMP-2) to an orbital floor defect model to support new bone growth.[1,2]

Macroporous EH-PEG biomaterials have been created previously with non-water soluble polymers using techniques including porogen-leaching. Macroporosity within hydrogels facilitates both molecular diffusion and cell migration, and thus should promote osteogenic signaling among cells within EH-PEG hydrogels. This study investigates the influence of fibronectin, an extracellular matrix protein, on osteogenic cell signaling and differentiation in porous EH-PEG hydrogels. Human mesenchymal stem cells (hMSCs) were loaded in porous EH-PEG hydrogels with fibronectin concentrations of 0, 10, and 100 μg/mL gel. At days 1, 4, 8, and 12 total RNA was isolated and reverse transcribed. Quantitative real-time PCR was completed on BMP-2 and BMP-2 receptors to analyze osteogenic signal expression, as well as alkaline phosphatase and osteocalcin to analyze differentiation; corresponding protein expression was analyzed by enzyme based assays as well as immunohistochemistry. Results show increased cell attachment and spreading with fibronectin concentration throughout the study. Nevertheless, the inclusion of macropores within a hydrogel does impact the strength of the biomaterial. As the purpose of the orbital floor is to maintain the orbital contents, approximately 43 grams[3], our scaffold must provide the necessary mechanical support. Therefore, studies were completed to assess the material strength of porous EH-PEG hydrogels. To improve the strength of the construct, a thin layer of crosslinked EHD was bound in the center of the construct creating a three-layer scaffold. Compression testing was performed in addition to three-point bending, which emulates physiological orbital stresses, using an Instron 5565 mechanical tester. Results demonstrate improved strength with the addition of the EHD layer. Overall, this work develops both the biomolecular and mechanical properties of a novel EH-PEG hydrogel for orbital bone regeneration.

References:

1. MW Betz, PC Modi, JF Caccamese, DP Coletti, JJ Sauk, and JP Fisher. Cyclic Acetal Hydrogel System for Bone Marrow Stromal Cell Encapsulation and Osteodifferentiation. Journal of Biomedical Materials Research, Part A. In Press.

2. MW Betz, JF Caccamese, DP Coletti, JJ Sauk, and JP Fisher. Orbital Floor Regeneration Using Cyclic Acetal Hydrogels. Journal of Biomedical Materials Research, Part A. In Press

3. Haug, RH, Nuveen E, Bredbenner T. An evaluation of the support provided by common internal orbital reconstruction materials. Journal of Oral and Maxillofacial Surgery 1999;57(5):564-70

331d The Effects of Material Properties of Hydrogels for Cartilage Tissue Engineering

Kara L. Spiller and Anthony Lowman. Chemical and Biological Engineering, Drexel University, 3141 Chestnut Street, Philadelphia, PA 19104

Currently no clinical successful technique exists to halt the degeneration of articular cartilage due to a cartilage defect. Despite extensive research into the use of hydrogels as scaffolds for cartilage repair, controversy exists regarding optimal material properties. Poly (vinyl alcohol) hydrogels have been modified to improve cellular adhesion by the physical or covalent addition of extracellular matrix proteins like fibronectin. The porosity of these hydrogels has also been controlled up to 65%, with pore sizes ranging from 50 to 100 um, and the effects of porosity on the migration of bone marrow stromal cells (BMSCs) throughout the hydrogels are being investigated. In order to examine the effects of mechanical properties on the response of BMSCs, it was important to ensure that the initial distribution of cells throughout the hydrogels was homogenous. Therefore, hydrogels based on poly(n-isopropylacrylamide) have been developed, which are thermally responsive in that they are liquid below body temperature and form stiff gels above body temperature. Cells can be incorporated into the hydrogels by stirring at room temperature and subsequently raising the temperature of this cell-hydrogel mixture, so that cells are physically entrapped and uniformly distributed throughout the matrix. The mechanical properties of this hydrogel system can be varied by the addition of methacrylic acid. The elastic modulus of the gels has been controlled from 100 to 500 kPa, depending on the relative amounts of monomers. The effects of mechanical properties on the behavior of BMSCs, in terms of gene expression and deposition of extracellular matrix proteins, is currently being investigated in both static cell culture and under dynamic loading conditions.

In the US alone, roughly 1.4 million patients undergo operations requiring arterial prostheses annually. Tissue engineering represents a potential means to construct functional small diameter vascular grafts in situations where autologous tissue is unavailable and conventional synthetic materials fail. While initial results with many of the tissue engineered vascular grafts (TEVGs) constructed to date are very encouraging, the potential for thrombosis, hyperplasia, and mechanical failure has limited their success. The relatively poor mechanical properties of TEVGs are believed to result primarily from insufficient extracellular matrix (ECM) deposition by and the inappropriate phenotype of associated smooth muscle cells (SMC).

In the present study, we characterize the effects of novel inorganic-organic hydrogel scaffolds generated by the photo-cure of hydrophobic star polydimethylsiloxane ($PDMS_{star}$) (inorganic polymer) and hydrophilic linear poly(ethylene oxide) (PEO) on SM progenitor cell differentiation and ECM production. Initial studies have shown these hybrid scaffolds to be biocompatible, non-thrombogenic, and highly elastic. In addition, they display unique microstructural, biochemical, and biomechanical properties that can be precisely tuned over a broad range, enabling systematic exploration of the effects of scaffold material properties on cell behavior.

Twelve hydrogel $PDMS_{star}$-PEO hydrogel formulations were prepared by varying the MW of the $PDMS_{star}$ arms and the weight ratio of $PDMS_{star}$ to PEO, resulting in scaffolds covering a broad range of initial microstructures, hydrophobicities, and mechanical properties. These alterations in materials properties had a significant impact on cellular differentiation and ECM production. Specifically, progenitor cells demonstrated increasing levels of differentiation with increased scaffold hydrophilicility and decreasing scaffold modulus. Similarly, the levels of collagen and elastin production varied significantly with hydrogel $PDMS_{star}$ concentration. From these results, we have identified a material property range which appears to evoke both high ECM deposition and SMC differentiation toward TEVG optimization.

331f Biodurable Reticulated Elastomeric Matrix as Scaffolds for Tissue Engineering

Craig Friedman, Yong Song, Arindam Datta, **Lawrence Lavelle**, and Rujul Majmundar. Biomerix Corporation, 41 West 57th Street, New York, NY 10019

Introduction: A suitable matrix for tissue engineering (TE) scaffold is a biocompatible material that allows for attachment, growth and proliferation of cells to regenerate new tissues for repair of chronic wounds [1,2]. Two areas that could benefit from using TE scaffolds with appropriate functionalities are the reconstruction of chronic rotator cuff (RC) tears that are associated with high rates of failure and the repair of various hernias in the abdominal walls with reduced complications [3,4]. The primary goal of both repairs is adequate initial strength of fixation to allow appropriate rehabilitation while the reparative process and eventual repair, reattachment and integration of the damaged or weakened tissue is completed.

In case of chronic RC tears, sometimes the defect persists as the tear cannot be completely repaired to its bony insertion and the common mode of failure of the suture tendon construct, which is normally repaired by sutures and suture anchors, can be attributed to poor biology and biomechanics of the compromised musculotendinous unit. It is believed that augmentation of such a defect with appropriate scaffold material will allow tissue ingrowth and reorganization leading to better integration of tendon and mineralized tissue and possibly reduced re-tear rates. In case of hernia, complaints such as seromas, deceased abdominal mobility and adhesion are known to occur even when the current meshes provide adequate mechanical support for resisting intra-abdominal pressures. It is believed that flexible scaffolds that will allow for higher tissue ingrowth and improved biointegration will be able to alleviate the situation[5].

The objective of the study is to investigate whether the use of an appropriate biocompatible TE scaffold in the augmentation of a chronic RC tendon or the repair of an abdominal hernia will result in tissue ingrowth, exhibit favorable histology and increase biomechanical strength of the healed tissue.

Materials and Methods: The scaffold used in these repairs comprises a matrix that is a crosslinked polycarbonate polyurethane-urea. The matrix is biodurable and elastomeric with a reticulated morphology consisting of an open celled, porous structure with an interconnected network of cells and pores. The network of cells forms a 3-dimensional spatial structure or void phase, which is interconnected via the open pores that provide connectivity between the cells. It has a large void content around 95%, which together with its reticulated morphology permits a high degree of cellular proliferation and tissue ingrowth throughout the porous scaffold. The matrix is biostable resisting degradation in hydrolytic, enzymatic and oxidative environments. 1-D fibers or 2-D meshes can be used to reinforce the matrix to enhance its mechanical performance, if necessary, when used as a surgical mesh for reconstructive surgery.

In the first study, chronic defects were created by detaching the infraspinatus tendons from the humerus in twenty-four skeletally mature female sheep weighing 60 to 100 Kg. The tendons were reattached to the humerus using a suture/anchoring technique mimicking a traditional RC repair. In 12 sheep, 2mm thickness patches of the matrix (containing a grid of PET sutures) were augmented on top of the repaired site by the use of sutures and anchors. In the remaining 12 sheep, reattachment was performed with no augmentation. All animals were sacrificed after 12 weeks and analyzed for histology and biomechanical testing.

In the second study, host tissue response was studied by placing patches (measuring 1 cm x 1 cm in size) of the matrix to repair defects (hernias) in the abdominal wall of rats. The defect left part of the abdominal fascia and the peritoneum intact, removing all layers of the internal and external abdominal oblique muscles. The test article was then implanted to replace the excised muscle and was surrounded by native muscle tissue, subcutaneous tissue, and fascia. Each implant group consisted of 4 rats and which were sacrificed at various intervals up to 16 weeks.

Results and Discussion: The properties of the matrix are as follows: density is 0.056gm/cc; average cell sizes are 300 to 400 μ m; tensile strength is 0.35 MPa; elongation to break > 175 %; tensile modulus is 0.20 Mpa; compression strength @ 50% strain is 0.01 MPa; Darcy permeability is greater than 500 (compared to under 10 for unreticulated porous foam of same density); and the material recovers to 90% of the original height in less than 80 seconds after being compressed at 50% strain for 2 hours.

For the sheep study, the ultimate force and ultimate stress for the augmented shoulders were 1328±427 N and 4.2±2.3 MPa compared to 762±474 N and 2.1±1.4 MPa for the shoulders without scaffold augmentation, demonstrating that both ultimate force and stress were significantly greater for the shoulders repaired with the use of a TE scaffold patch. Histological evaluation (H&E stained) showed that the patch supported tissue ingrowth (Fig. 1) along with tendon healing and repair. The bone tendon interface showed expected normal cellular healing at the repair site and the matrix was incorporated into the infraspinatus tendon by multiple fibrous interdigitation and into the bone through a mature fibrocartilaginous transition zone.

In the rat abdominal study, minimal inflammatory response with increasingly organized connective tissue was observed over time from histological evaluation (Fig. 2). Multinucleate cells were observed within the graft material as time progressed, a response that is typical for a slowly resorbed or nonresorbable implanted material. There was high degree of tissue infiltration into the patch without necrosis of the surrounding tissue. Additionally, there was minimal scar tissue formation and adhesion within the abdominal space. The healing response within the matrix and the surrounding tissue was superior to the typical response for other soft tissue meshes

Conclusions: Biodurable reticulated elastomeric matrix as a TE scaffold shows favorable tissue ingrowth with no adverse effects. This matrix as a TE scaffold supports collagenous and fibrovascular tissue ingrowth throughout the voids in the scaffold, enabling the healing and repair of damaged tissue. These scaffolds have been shown to increase the mechanical reinforcement of a chronic RC repair and solicit favorable healing of defects in the abdominal wall. This healing response is characterized by the interdigitation of organized dense collagenous tissue and bone for RC repair and characterized by organized connective tissue for abdominal wall defect repair.

References:

1.Liu, X and Ma, P.X, Annals Biomed. Eng, 32 (3), 477, 2004.

2.Peter, S.J., Miller, M.J., Yasako, A.W., Mikos, A.G., J. Biomed. Mater. Res., 43, 422 , 1998

3.Gerber C., Fuchs B. and Hodler J., J Bone Joint Surg, 82,505, 2000.

4. Harryman, Mack, Wang, J Bone Joint Surg,73, 1991.

5. Klosterhalfen, B., Junge, K. and Klinge, U., Exprt Rev. Med. Devices 2 (1), 2005.

Figures:

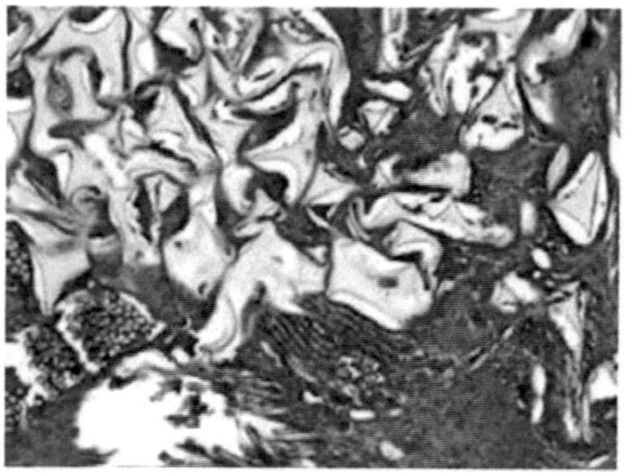

Figure 1

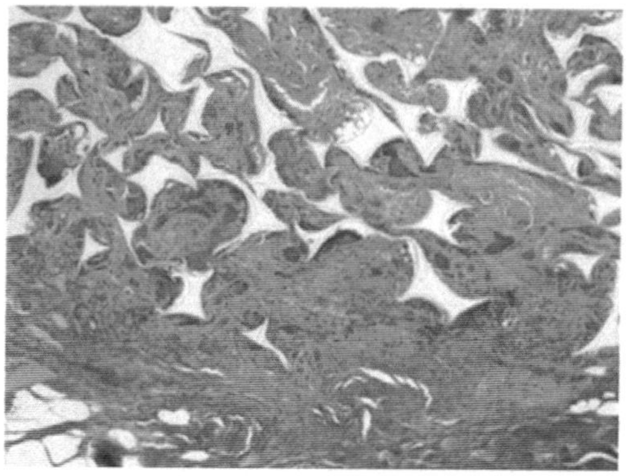

Figure 2

Scaffolds Covalently Immobilized with VEGF and Angiopoietin-1 to Promote Angiogenesis In Engineered Cardiac Tissues

Loraine L.Y. Chiu[1] *and Milica Radisic*[1,2,3]

[1] *Department of Chemical Engineering and Applied Chemistry, University of Toronto, Toronto, Ontario, Canada*
[2] *Institute of Biomaterials and Biomedical Engineering, University of Toronto, Toronto, Ontario, Canada*
[3] *Heart & Stroke/Richard Lewar Centre of Excellence, University of Toronto, Toronto, Ontario, Canada*

Introduction:

Cardiovascular diseases have been the main cause of death in Canada for the past 25 years, and accounted for more than 30% of deaths in 2004 [1]. Myocardium infarction causes irreversible damage to the heart, and the heart becomes incapable of regeneration due to the non-proliferative nature of terminally differentiated adult cardiomyocytes. As a result, there is motivation to engineer cardiac tissue patches *in vitro* by culturing cardiac cells in a scaffold matrix, after which these patches can be grafted into the diseased heart to regenerate the injured myocardium.

One of the recent challenges in cardiac tissue engineering is that functional blood supply is needed for constructs larger than 100-200 microns in thickness in order to provide sufficient oxygen and nutrients. Oxygen transport only occurs up to a distance of 100μm under normal medium diffusion [2]. Cell density and viability is compromised under limited oxygen supply [3]. One solution is to design a physiologically interactive replacement consisting functional blood vessels for the injured vascular tissue. By inducing vascularization within the engineered tissue *in vitro*, limited transport capacity of oxygen and nutrients into the tissue can be overcome, thus improving its survival both *in vitro* and *in vivo* [4].

Neovascularization requires coordination of multiple endothelial growth factors, receptors and intracellular signalling pathways. Amongst these angiogenic growth factors, vascular endothelial growth factor (VEGF) and angiopoietin-1 (Ang1) have been shown to be the most important regulators of blood vessel formation. Past studies have shown that VEGF promotes the formation of new capillary vessels, while Ang1 induces the maturation and stabilization of new vessel networks, suggesting a complementary relationship between these two angiogenic factors [5].

In this study, we immobilized VEGF and Ang1 in collagen scaffolds to protect growth factors against cellular inactivation and digestion, and to allow highly sustained and localized activity. We hypothesize that VEGF and Ang1, when covalently immobilized in collagen scaffolds, work together to first form new vessels, and then to stabilize these vessels.

Materials and Methods:

A metal borer was used to cut circular collagen scaffolds with diameter of 7mm and thickness of 2mm from a sheet of porous collagen sponge (Ultrafoam collagen sponge, Davol, 1050050). Both mouse recombinant VEGF-165 (Cell Sciences, CRV014B) and human recombinant Ang1 (R&D Systems, 923-AN-025) were immobilized into the scaffolds using EDC chemistry (**Figure 1A**). EDC chemistry was performed by immersing the collagen scaffolds into 150μL sterile filtered solution of EDC (N-(3-dimenthylaminopropyl)-N'-ethylcarbodiimide hydrochloride, Sigma, E7750) and Sulfo-NHS (N-hydroxysulfosuccinimide, Pierce Chemicals, 24510) in a 96-well plate (EDC/sulfo-NHS concentrations of 24mg/60mg per 1mL 1M PBS). The activation of the collagen scaffolds was allowed to proceed for 20 minutes at room temperature. The scaffolds were then removed from the solution, and immersed into 100μL solution of a) 1μg/mL VEGF (VEGF+PBS), b) 1μg/mL Ang1 (Ang1+PBS), c) 0.5μg/mL VEGF and 0.5μg/mL Ang1 (1/2VEGF+1/2Ang1), or d) 1μg/mL VEGF and 1μg/mL Ang1 (VEGF+Ang1) in PBS. The reaction was allowed to proceed for 1 hour at room temperature. For control samples (PBS control), collagen sponges were immersed in PBS at room temperature. Afterwards, the scaffolds were immersed in fresh PBS for 8 times to wash away the uncrosslinked VEGF and/or Ang1, EDC and sulfo-NHS. After washing in PBS, the scaffolds were incubated in culture medium for 30 minutes and dried on autoclaved Kim-wipes prior to cell seeding.

For *in vitro* studies, the collagen scaffolds were transferred to a clean 24-well plate after scaffold preparation. 50000 H5V endothelial cells were seeded onto the scaffolds in 10μL culture medium. The scaffolds were incubated for 40 minutes at 37°C for cells to attach. After incubation, 1mL fresh culture medium was added to each well. The samples were cultured for 3 or 7 days with a 100% change and collection of culture medium on Day 3, Day 5 and Day 7 (**Figure 1B**). Collected culture medium was stored at -20°C for further lactate and glucose assays. The samples were analyzed for final cell number by XTT assay, lactate production rate by lactate assay, glucose consumption rate by glucose assay, and cell density at various heights of the scaffold by cryosectioning and DAPI staining. The effective VEGF and Ang1 concentrations were quantified by ELISA.

For preliminary *in vivo* studies, the collagen scaffolds immobilized with solutions of a) 0μg/mL VEGF (control), b) 0.5μg/mL VEGF (low VEGF), and 2μg/mL VEGF (high VEGF) in PBS were surgically inserted into the right ventricular outflow tract (RVOT) of adult rat hearts to repair transmural defects. At Week 1 and Week 4, the biomaterials were analyzed for patch thickness, cellular density and angiogenesis (i.e. vessel density and vessel diameter).

A

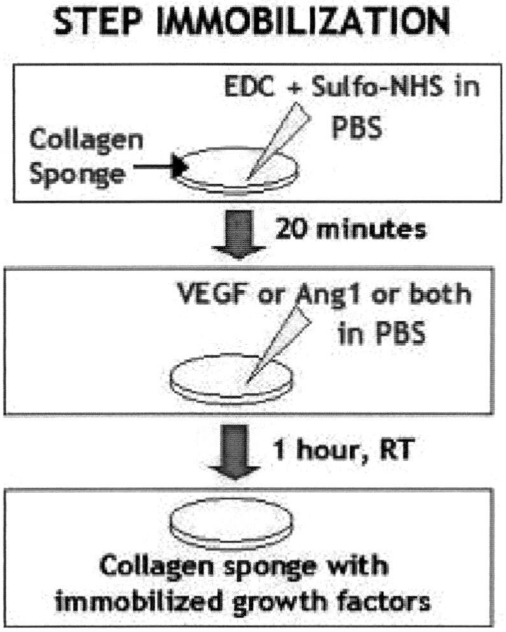

B

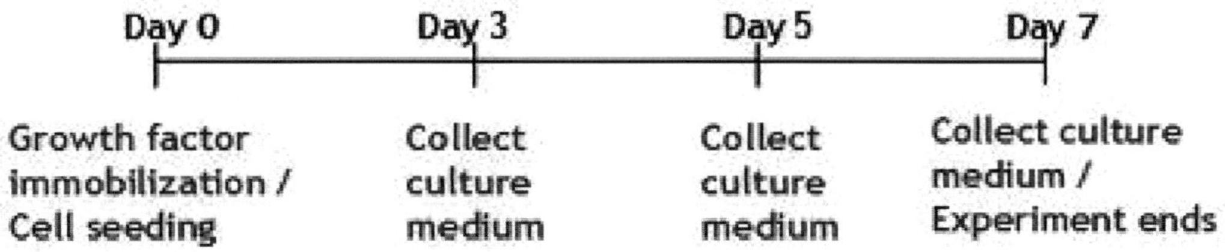

Figure 1: Experimental details. (A) Experimental procedure for step immobilization of VEGF and Ang1 on collagen scaffolds. **(B)** Experimental timeline for *in vitro* cultivation of H5V endothelial cells on collagen scaffolds.

Results and Discussion:

The amount of VEGF immobilized on the scaffolds, as determined by VEGF ELISA, was found to be 1.13±0.44ng for 1/2VEGF+1/2Ang1 group and 2.33±0.36ng for VEGF and VEGF+Ang1 groups. The amount of VEGF immobilized on the scaffolds was 1.46±0.11ng for low VEGF group and 11.37±0.20ng for high VEGF group. The amount of Ang1 immobilized on the scaffolds, as determined by Ang1 ELISA, was found to be 0.30±0.04ng for 1/2VEGF+1/2Ang1 group and 0.59±0.10ng for Ang1 and VEGF+Ang1 groups.

High endothelial cell density within the scaffold is the first step in achieving vascularization in a tissue construct. *In vitro* results (**Figure 2A**) showed that collagen scaffolds with immobilized VEGF (VEGF+PBS, 1/2VEGF+1/2Ang1 and VEGF+Ang1 groups), had increased final cell number (1.6-fold, 1.8-fold and 1.7-fold respectively) compared to PBS control at Day 3 (one-way ANOVA, P = 0.0213). This is consistent with the cell proliferative effect of VEGF. Although Ang1+PBS group had a 1.4-fold higher cell number than PBS control, it did not show statistically significant difference in final cell number compared to PBS control. This is due to the fact that Ang1 is responsible for endothelial cell survival rather than cell proliferation.

Scaffolds were also cryosectioned and stained with DAPI to count number of cell nuclei at various heights (**Figure 2B, 2C**). At Day 3, groups with immobilized VEGF and/or Ang1 all showed higher cell density (**Figure 2C**) than PBS control (two-way ANOVA, $P < 0.0001$). Interestingly, at 1400μm from the top of the scaffold, only VEGF+PBS and VEGF+Ang1 groups (and not Ang1+PBS and 1/2VEGF+1/2Ang1 groups) showed significantly higher cell density than PBS control (**Figure 2C**), thus indicating a dependency of cell infiltration on immobilized VEGF concentration.

Lactate production rate (**Figure 2D**) and glucose consumption rate (**Figure 2E**) both increased from Day 3 to Day 7 for all experimental groups *in vitro*. More importantly, VEGF+PBS, Ang1+PBS, 1/2VEGF+1/2Ang1 and VEGF+Ang1 groups all showed higher lactate production rates (**Figure 2D**) compared to PBS control at Day 7 ($P < 0.0001$ two-way ANOVA; $P < 0.05$, $P < 0.05$, $P < 0.001$ and $P < 0.01$ respectively Bonferroni post-tests). Ang1+PBS and 1/2VEGF+1/2Ang1 groups also had higher lactate production rates compared to PBS control at Day 5 (Bonferroni post-tests, $P < 0.05$ for Ang1+PBS, $P < 0.01$ for 1/2VEGF+1/2Ang1). VEGF+PBS, Ang1+PBS, 1/2VEGF+1/2Ang1 and VEGF+Ang1 groups all showed higher glucose consumption rates (**Figure 2E**) compared to PBS control at Day 7 (2.8-fold, 1.4-fold, 2.7-fold and 2.5-fold respectively). Specifically, VEGF+PBS and VEGF+Ang1 groups showed significantly higher glucose consumption rate than PBS control at Day 7 (two-way ANOVA Bonferroni post-tests, $P < 0.05$).

For preliminary *in vivo* studies (data not shown), sponges were immobilized with low dose VEGF (500ng/mL VEGF) and high dose VEGF (2μg/mL VEGF), and used in the adult rat RVOT. It was found that the patch thickness was significantly increased with high dose VEGF compared to both PBS control and low dose VEGF at Week 1. The patch thickness was also significantly increased with low dose VEGF and high dose VEGF compared to PBS control at Week 4. The increase in patch thickness was not due to increased cellular density, since there was no statistically significant difference in cell density amongst groups at both Week 1 and Week 4. Rather, trichrome stained images and corresponding image analysis showed that blood vessel density was increased in high dose VEGF patches compared to PBS control at Week 1 and compared to both PBS control and low dose VEGF patches at Week 4.

The modification of cardiac patches with immobilized VEGF seems to increase blood vessel density in the biomaterial over time, thus increasing the thickness of the implanted VEGF patch. This is consistent with the angiogenic effect of VEGF. However, the vessel diameters were not significantly different between high VEGF and control sponges, suggesting a need for an additional growth factor such as Ang1 for vessel stabilization. Future work includes an *in vivo* study on the effect of co-immobilized VEGF and Ang1.

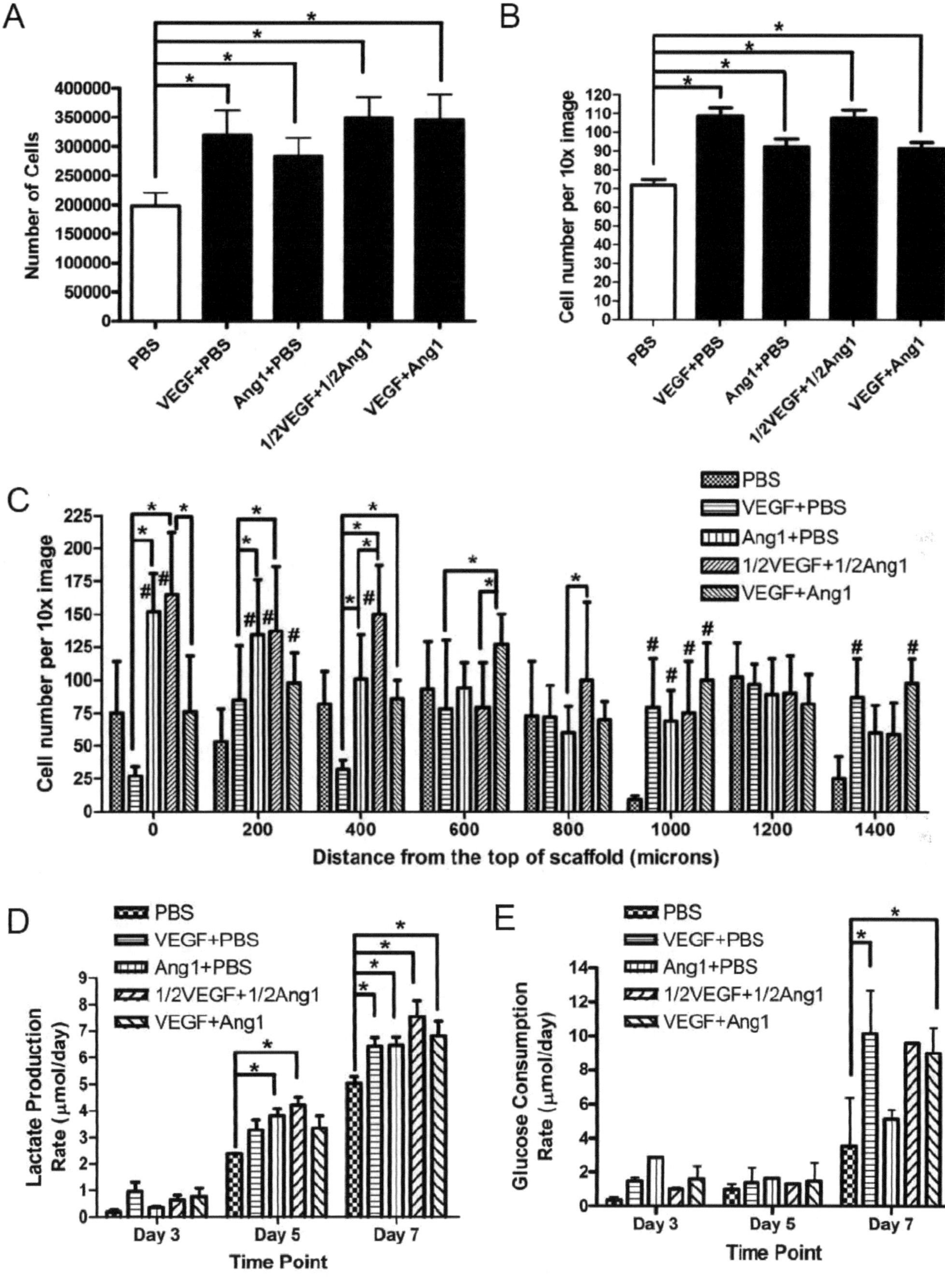
A
Number of Cells
400000
350000
300000
250000
200000
150000
100000
50000
0
PBS
VEGF+PBS
Ang1+PBS
1/2VEGF+1/2Ang1
VEGF+Ang1
B
Cell number per 10x image
120
110
100
90
80
70
60
50
40
30
20
10
0
PBS
VEGF+PBS
Ang1+PBS
1/2VEGF+1/2Ang1
VEGF+Ang1
C
PBS
VEGF+PBS
Ang1+PBS
1/2VEGF+1/2Ang1
VEGF+Ang1
Cell number per 10x image
225
200
175
150
125
100
75
50
25
0
0
200
400
600
800
1000
1200
1400
Distance from the top of scaffold (microns)
D
PBS
VEGF+PBS
Ang1+PBS
1/2VEGF+1/2Ang1
VEGF+Ang1
Lactate Production Rate (μmol/day)
9
8
7
6
5
4
3
2
1
0
Day 3
Day 5
Day 7
Time Point
E
PBS
VEGF+PBS
Ang1+PBS
1/2VEGF+1/2Ang1
VEGF+Ang1
Glucose Consumption Rate (μmol/day)
14
12
10
8
6
4
2
0
Day 3
Day 5
Day 7
Time Point

Figure 2: Comparison of individually immobilized and co-immobilized growth factors *in vitro*. (A) XTT assay indicating final cell numbers in collagen scaffolds at Day 3. **(B)** Averaged cell density at Day 3. **(C)** DAPI staining indicating cell density at various heights. **(D)** Lactate production rate. **(E)** Glucose consumption rate. * denotes statistically significant difference, # denotes statistically significant difference compared to PBS control ($P < 0.05$; one way ANOVA with post-hoc Tukey test for cell number and averaged cell density; two-way ANOVA with Bonferroni post tests for cell density, lactate production rate, and glucose consumption rates)

Conclusions:

The covalent immobilization of combined growth factors VEGF and Ang1 in collagen sponges leads to endothelial cell proliferation and survival, and increased cell metabolism *in vitro*. More importantly, angiogenic effect was evident when collagen scaffolds with immobilized VEGF were used in preliminary *in vivo* studies to repair the right ventricular outflow tract of adult rat hearts. Further studies will be performed to investigate the effect of combined immobilized growth factors on angiogenesis *in vivo*.

Acknowledgments:

This study was supported by NSERC Canada Graduate Scholarship (to L.L.Y.C.) The authors thank Dr. Ren-Ke Li and Yasuo Miyagi for performing the surgical component of the preliminary *in vivo* studies.

References:

1. Statistics Canada. Mortality, Summary List of Causes (2004), 2007.
2. Morritt AN, Bortolotto SK, Dilley RJ, Han X, Kompa AR, McCombe D, Wright CE, Itescu S, Angus JA, Morrison WA, "Cardiac tissue engineering in an in vivo vascularized chamber", Circulation 115:353-360, 2007.
3. Radisic M, Malda J, Epping E, Geng W, Langer R, Vunjak-Novakovic G, "Oxygen gradients correlate with cell density and cell viability in engineered cardiac tissue", Biotechnology and Bioengineering 93:332-343, 2006.
4. Shen YH, Shoichet MS, Radisic M, "Vascular endothelial growth factor immobilized in collagen scaffold promotes penetration and proliferation of endothelial cells", Acta Biomaterialia 4:477-489, 2008.
5. Kim SH, Kiick KL, "Heparin-mimetic sulfated peptides with modulated affinities for heparin-binding peptides and growth factors", Peptides 28:2125-2136, 2007.

331h Making of a New Liver Using Decellularized Livers as Scaffolds

Basak E. Uygun[1], Alejandro Soto-Gutierrez[1], Carley Shulman[1], Korkut Uygun[2], François Berthiaume[3], and Martin Yarmush[4]. (1) Surgery, Center for Engineering in Medicine, Massachusetts General Hospital, Harvard Medical School, 51 Blossom Street, Boston, MA 02114, (2) Center for Engineering in Medicine, Harvard Med School, 51 Blossom Street, Boston, MA 02114, (3) Center for Engineering in Medicine, Massachusetts General Hospital, Harvard Medical School, 51 Blossom Street, Boston, MA 02114, (4) Research, Center for Engineering in Medicine, 51 Blossom Street, Boston, MA 02114

Orthotopic liver transplantation is the only clinically proven effective treatment for end-stage liver disease and there are currently ~100,000 patients on the organ transplant waiting list in the US, a number that far exceeds the supply of available organs, and that continues to grow ~5% each year. One of the most promising strategies that is being explored as means for addressing this critical shortage is constriction of bioartificial tissue and organs (which aims to manufacture tissue and organ analogues in vitro). In this study, we aimed to reform liver tissue using perfusion decellularized rat livers as scaffolds. Rat livers were harvested from cadaveric donors and decellularized by perfusion with 1% sodium dodecyl sulfate. The removal of the cellular component from the liver was confirmed by hematoxylin and eosin staining of histology sections of the decellularized specimen. The decellularized livers were recellularized with adult primary hepatocytes (100 million cells, 90% viability) isolated by perfusion incubation over 60 minutes and the recellularized livers were cultured in organ perfusion system for 24 hours. At the end of the seeding period, it was found that 80 million cells were retained in the organ and cell viability was approximately 70%. Viability of the cells during perfusion culture was monitored by measuring lactate dehydrogenase activity in the perfusion medium. Hepatic metabolic function was measured by analysis for urea and albumin secretion in the medium. At the end of the 24 hr perfusion culture, histology analysis of the recellularized liver samples revealed that cells were distributed uniformly around the openings which used to constitute liver vasculature. Hepatocytes were found to be positive for glycogen after periodic acid-Schiff staining. There were very minimal apoptotic cells as confirmed by TUNEL staining. The results showed that hepatocytes successfully grafted in the decellularized livers using organ perfusion system. Future work will include demonstration of long-term function of hepatocytes after repopulating the decellularized liver.

332a Smart Packaging: Localized Drug Delivery to Treat Ovarian Cancer

Eva Christabel Williams, Ryan Toomey, and Norma Alcantar. Dept. of Chemical and Biomedical Engineering, University of South Florida, 4202 E. Fowler Ave - ENB 118, Tampa, FL 33620

Localized drug delivery is emerging as an effective technique due to its ability to administer therapeutic concentrations and controlled release of drugs to cancer sites in the body. It also prevents the contact of harsh chemotherapy drugs to healthy regions in the body that otherwise would become exposed to current treatments. In this project, we have developed a model drug delivery system consisting of non-ionic surfactant vesicles (niosomes) packaged within a biodegradable, temperature and pH sensitive hydrogel network. Our first step has been to characterize the behavior of the noisomes when exposed to environments that mimic body fluids. Characteristics such as diffusion and mass transfer of 5,6-Carboxyfluorescein, which is a dye with similar physical properties as therapeutic drugs for cancer, is used to determine release rate from the niosomes. Depending on the conditions to which individual niosomes are exposed, the release rate can be controlled to last from 24 hours to more than 144 hours. The hydrogel provides another layer of control and a stable environment for the niosomes. For this we used a cross linked hydrogel (chitosan) network into which the niosomes were embedded. The niosome-hydrogel system, which is a liquid at room temperature, starts gelling once inside the body. In- vitro release rate studies proved that drugs continued to be released appreciably even after 3 months of the experiment. One of the systems that we are targeting with this study is intraperitonial cavities after ovarian cancer is discovered and removed to increase the life span of patients. Single administration rates have been tested and compared to local delivery via an external catheter. We found that our system lasted longer than catheter administration leading to less frequent administration and also resulted in reduced toxicity. Our results will help in the development of a low cost and improved method for drug delivery with application to intracavitary ovarian cancer treatment and other cancer types.

332b Anti-HER2/neu Liposomes for Triggered Content Release In Ovarian and Breast Cancer Cells

Shrirang Karve, Ali Alaouie, and Stavroula Sofou. Othmer-Jacobs Department of Chemical and Biological Engineering, Polytechnic University, 6 Metrotech center, Brooklyn, NY 11201

Solid tumors still account for the majority of cancer mortality. For such cases of advanced disease, nanometer-sized long-circulating drug delivery carriers could provide enhanced accumulation and retention of therapeutics at the tumor sites relative to normal organs, and, following internalization by cancer cells, they could increase drug bioavailability by triggered release of encapsulated contents.

In this work, we describe PEGylated anti-HER2/neu liposomes composed of rigid bilayer membranes containing only lamellar forming lipids that are triggered to form phase-separated domains in a pH-dependent manner resulting in fast and extensive content release. These liposomes were evaluated in serum supplemented media (60% v/v serum). Anti-HER2/neu liposomes with encapsulated fluorescent compounds were also evaluated for binding, internalization and content release in HER2/neu overexpressing ovarian cancer cells, and breast cancer cells, and were compared to compositions similar to FDA approved formulations. Our studies demonstrate the potential of the proposed liposomes as delivery carriers to increase the drug bioavailability in metastatic tumors.

332c An In Vivo Examination of the Vesosome, a Novel Multi-Compartment Drug Carrier

Benjamin J. Wong[1], Shelley Esakoff, and Joseph A. Zasadzinski[2]. (1) Department of Chemical Engineering, University of California, Santa Barbara, Santa Barbara, CA 93106, (2) Chemical Engineering, University of California, Santa Barbara, Santa Barbara, CA 93106

An optimal drug delivery vehicle should circulate in the body long enough to reach the site of illness or disease and also localize itself at the desired site to consequently deliver its contents at a rate appropriate for maximum therapeutic benefit. It should also possess a large drug loading capacity and retain its contents over the course of treatment. While liposomal systems have experienced success with extending circulation, content retention and controlled release remain problematic. The vesosome – a large lipid bilayer enclosing many smaller liposomes – is the most suitable candidate for addressing these issues. The external lipid bilayer offers a second barrier of protection for interior components and also serves as the anchor for active targeting components. Furthermore, internal compartmentalization permits customization of separate environments for multiple therapeutics and release triggers, highlighting the vesosome's potential as a single site, single dose, multiple component drug treatment.

The processes of vesosome formation and functionalization are based on the phase behavior exhibited by homogeneous lipid mixtures. Although slight changes in lipid composition, such as the addition of PEG-lipids or fluorescently-labeled lipids, may have a large impact on the structures that form, vesosome formation can still be achieved using heterogeneous lipid mixtures, a success vital to moving the vesosome toward in vivo studies. A variety of microscopy techniques, including freeze-fracture and cryo transmission electron microscopy as well as fluorescence and confocal microscopy, are used to characterize these functionalized drug carriers.

To assess the viability of the vesosome as a drug carrier, the *in vivo* lifetimes and biodistribution of various formulations were examined in mice. Our work examines how these properties are affected by lipid composition and the addition of other functional components, including ones for controlled release and active targeting.

332d PR_B Peptide-Labeled Liposomes for Targeting Prostate Cancer Cells

Döne Demirgöz, Ashish Garg, and **Efrosini Kokkoli**. Department of Chemical Engineering and Materials Science, University of Minnesota, 421 Washington Ave. SE, Minneapolis, MN 55455

Integrin alpha(5)beta(1) is expressed on several types of cancer cells, including prostate cancer, and plays an important role in tumor growth and metastasis. The ability to target the integrin alpha(5)beta(1) using an appropriate drug delivery nanovector can significantly help in inhibiting tumor growth, reducing tumor metastasis, and decreasing deleterious side effects associated with different cancer therapies. The goal of this study is to design peptide-functionalized stealth liposomes (liposomes covered with polyethylene glycol (PEG)) that will specifically target the integrin alpha(5)beta(1) and effectively release their encapsulated load intracellularly. The targeting moiety is the PR_b peptide, that has been designed in our lab, and has been shown to specifically recognize and bind to alpha(5)beta(1) expressing cells. The peptide-functionalized nanoparticles have been characterized with NMR, TEM, and DLS, as well as in vitro studies with prostate cancer cells. In this study we have demonstrated that by optimizing the amount of PEG and PR_b on the liposomal interface we can engineer nanovectors that bind to human prostate cancer cells in a specific manner and are internalized through alpha(5)beta(1)-mediated endocytosis. RGD-targeted stealth liposomes bind to prostate cancer cells and internalize, but are less specific and have much lesser efficiency than PR_b targeted stealth liposomes. We have further shown that the PR_b liposomes upon internalization are capable of releasing their therapeutic payload intracellularly. The proposed targeted delivery system has a great potential to deliver therapeutic agents directly to cancer cells, in an efficient and specific manner.

332e Release of Model Macromolecules from Self-Assembling Peptide Hydrogels

Monica C. Branco[1], Norman J. Wagner[1], Darrin J. Pochan[2], and Joel P. Schneider[3]. (1) Department of Chemical Engineering, University of Delaware, 100 Academy St., Newark, DE 19716, (2) Department of Materials Science and Engineering, University of Delaware, 100 Academy St., Newark, DE 19716, (3) Department of Chemistry and Biochemistry, University of Delaware, 100 Academy St., Newark, DE 19716

Advances in biotechnology techniques have led to the rapid development of small protein and antibody therapeutics. However, several limitations remain in the preparation and delivery of these drugs due to the susceptibility of proteins to degrade during storage and upon administration. To address this problem, hydrogels have been used as delivery devices for these protein drugs. We have designed a class of self-assembling peptides that undergo triggered hydrogelation in response to physiological pH and salt conditions (pH 7.4, 150 mM NaCl). These peptides adopt a random coil conformation in aqueous pH 7.4 solutions and are freely soluble. However, when a physiological relevant concentration of NaCl is added, the peptides fold into a β-hairpin, and subsequently, self-assemble to form a rigid hydrogel stabilized by non-covalent cross-links. For these peptides, it is possible to control the folding and assembly kinetics to form hydrogels with different rigidities. These changes affect the porous morphology within the hydrogel system, and subsequently influence the rate of macromolecular diffusion within the peptide fibrillar network. Another unique characteristic of these hydrogels is that under applied shear, the hydrogel will shear-thin into a low-viscosity gel; however, the gel quickly resets and recovers its initial mechanical rigidity after the applied shear is removed. This property allows hydrogels encapsulating therapeutics to be administered via syringe to target sites for delivery. This study focuses on determining the mass transport properties of model probes from self-assembled MAX8 hydrogels before and after delivery through a syringe. Probes of varying molecular weights and pIs were chosen to assess the effect of fibrillar charge density and hydrogel mesh size on their release properties from these self-assembled networks. Probes are added into solution during folding and self-assembly, yielding gels with macromolecules directly and homogeneously encapsulated into the network.

332f Targetted Delivery of Doxorubicin into Human Liver Cancer Cells Encapsulated with D-Galactose-Peptide Amphiphile

Yen Wah Tong, Department of Chemical and Biomolecular Engineering, National University of Singapore, 4 Engineering Drive 4, Singapore, 117576, Singapore, Nikken Wiradharma, Drug Gene and Protein Delivery; Chemical and Biomolecular Engineering, Institute of Bioengineering and Nanotechnology; National University of Singapore, 31 Biopolis Way, #04-01, The Nanos, Singapore 138669, 21 Lower Kent Ridge Road, Singapore 119077, Singapore, Singapore, Shao Qiong Liu, Drug, Gene and Protein Delivery, Institute of Bioengineering and Nanotechnology, 31 Biopolis Way, #04-00, Singapore, 130889, Singapore, and Yi Yan Yang, Drugs, Proteins and Genes Delivery, Institute of Bioengineering and Nanotechnology, 31 Biopolis Way #04-01, Singapore, Singapore.

Liver cancer cells have been reported to have an over-expression of asialoglycoprotein receptor, which were able to recognize D-galactose residue. In this study, we have designed and synthesize an amphiphilic peptide-galactose conjugate to self-assemble into core-shell micelle nanoparticle with a small hydrodynamic radius of around 200nm. The micelle was used to encapsulate anticancer drug, doxorubicin, as a model drug. Encapsulation efficiency, loading level, and in vitro release profile of the peptide-galactose conjugate was studied as a novel and non-cytotoxic carrier of drugs to target cells with over-expression of asialoglycoprotein receptors. To evaluate the targeting capability of this carrier, cytotoxicity study of doxorubicin-loaded into D-galactose-peptide conjugate was performed against human liver carcinoma, HepG2 cell lines, under competitive binding with free D-Galactose present in the culture media. As a control study, we had also provided similar experiments against human breast cancer, PC3 cell lines, which does not have an over-expression of glycoprotein receptors.

Implantable hydrogel beads entrapping PLGA-paclitaxel microspheres: Exploring the effects of near-zero-order drug release for intracranial chemotherapy

Abstract ID: 123910

Sudhir H. Ranganath, Alvin Yang, Ying Ying Chan, Jinghan Huang, William B. Krantz, Chi-Hwa Wang

Department of Chemical and Biomolecular Engineering, National University of Singapore, 4 Engineering Drive 4, Singapore 117576

INTRODUCTION

Local intracranial chemotherapy using drug-loaded microspheres has shown considerable potential in many pre-clinical studies [1,2]. However, its translation into clinical trials has been hindered by two major disadvantages: implantability and excessively high initial drug release. Intra-tumoral injection of microspheres (suspended in saline) reduces the efficiency of implantation due to the expelling of microspheres from the target site because of high interstitial pressure in the tumor. Also, owing to the concentration of drug near the surface, the microspheres exhibit a high initial drug release that can cause neurotoxicity. In an effort to overcome these limitations, we have developed alginate hydrogel beads that entrap PLGA-paclitaxel microspheres.

Thus in this study, the main objectives were to design, develop and evaluate a novel composite implant that couples the sustained release properties of PLGA-paclitaxel microspheres with the intracranial implantability advantages of alginate matrix for post-surgical chemotherapy to treat malignant glioma. PLGA-paclitaxel microspheres were fabricated using Electrohydrodynamic Atomization (EHDA) since this process was reported to be very efficient in controlling the microsphere size, morphology and monodispersity (3). An electrospray technique in dripping mode was utilized to entrap the microspheres in alginate hydrogel beads owing to its ease of operation and reported monodispersity (4). Furthermore, the effect of various parameters such as gelation time, gelation bath concentration and microsphere loading on in vitro drug release was studied followed by cell culture studies (evaluation of cytotoxicity against C6 glioma cells). Studies of drug release sustainability of the hydrogel beads, apoptosis against C6 glioma cell in vitro, and in vivo anti tumor efficacy in a subcutaneous tumor model are ongoing. These studies will be followed by in vivo intracranial studies such as drug bio-distribution and treatment efficacy (tumor volume and survival analysis) and are warranted.

EXPERIMENTAL METHODS

Fabrication of PLGA-Paclitaxel Microspheres by EHDA Process

Fabrication of PLGA-paclitaxel microspheres was performed using EHDA as reported earlier (3).

Fabrication of Alginate Beads Entrapping PLGA-Paclitaxel Microspheres by Electrospray

The electrospray process as depicted in Fig. 1 was employed to entrap PLGA-paclitaxel microspheres in alginate beads. The setup constituted a high voltage generator, syringe pump with syringe and a metal tip (tapered conical section made of copper sheet) and a grounded stage. The metal tip was connected to the syringe and used instead of a needle to avoid clogging by the microspheres. A petri dish filled with $CaCl_2$ solution was placed on the grounded stage. PLGA-paclitaxel microspheres were suspended in deionized water (with 0.1% (w/v) of Tween 80) to form a uniform dispersion. Sodium alginate was added to the dispersion at 1% (w/v) and then completely dissolved to result in a viscous suspension. The suspension was then pumped at 1 ml/min using the syringe pump with the metal tip connected to a high voltage generator at 6 kV. Due to the high voltage at the metal tip, the suspension broke into monodisperse droplets and was dripped directly into $CaCl_2$ solution. The alginate in the droplets instantaneously underwent gelation in $CaCl_2$ solution resulting in calcium crosslinked hydrogel beads entrapping microspheres in the matrix. In addition, hydrogel beads were also fabricated by manually dripping without the use of the electrospray and were used to compare with the beads fabricated by the electrospray.

The following preparative variables were evaluated in an effort to understand their effect on drug release profiles: Gelation time (1, 5 and 15 min), $CaCl_2$ concentration (0.5, 1, and 2% w/v) and microsphere loading (50, 80, and 90% w/w). Following gelation, the beads were washed with deionized water (to remove non-crosslinked Ca^{2+} ions) and freeze-dried for at least 7 days to remove water and residual DCM. The beads were sterilized by exposing to UV light for 12 hours before using for cell culture and animal studies. Table 1 shows the different formulations fabricated and evaluated.

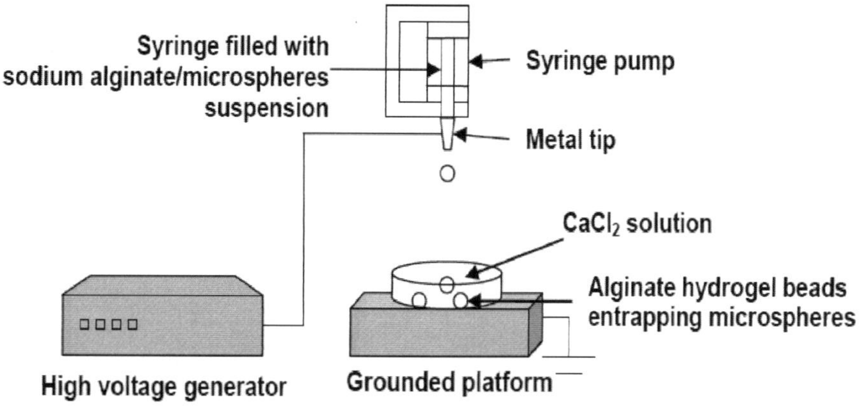

Figure 1: Schematic representation of microsphere entrapment in hydrogel beads using electrospray.

Table 1. Preparative Variables for Hydrogel Bead Fabrication and Respective Formulations

Preparative Variables	Formulations								
	H50	M50	L50	H80	M80	L80	H90	M90	L90
Microsphere Loading (% w/v)	50	50	50	80	80	80	90	90	90
Gelation Time (min)	15	5	1	15	5	1	15	5	1
$CaCl_2$ Concentration (% w/v)	2.0	1.0	0.5	2.0	1.0	0.5	2.0	1.0	0.5

RESULTS AND DISCUSSION

Physicochemical Characterization of Hydrogel Beads

Morphology, Size Distribution, Uniformity of Microsphere Entrapment and Loading Efficiency

The beads were prepared using two types of dripping methods, manual dripping (MD) and electrospray dripping (ED), after which they were characterized and compared. Manual dripping yielded polydisperse non-spherical beads with a non-uniform distribution of the microspheres within them (data not shown). The ED method yielded highly monodisperse spherical beads with a uniform microsphere distribution within them as revealed by the SEM images in Figure 2. Also, from Table 2 we can observe that the hydrogel beads possess acceptable monodispersity. However, the ED method yielded hydrogel beads with better uniformity in microsphere distribution compared to the MD method; hence, for further studies beads fabricated using the ED method were used.

Table 2. Physicochemical Characterization of Hydrogel Beads Fabricated by Electrospray

Formulations	Average Bead Diameter (mm)	Loading Efficiency (%)
50% microsphere loaded beads	1.61 ± 0.041	144.2 ± 11.07
80% microsphere loaded beads	1.65 ± 0.057	73.5 ± 11.02
90% microsphere loaded beads	1.68 ± 0.046	52.1 ± 3.77

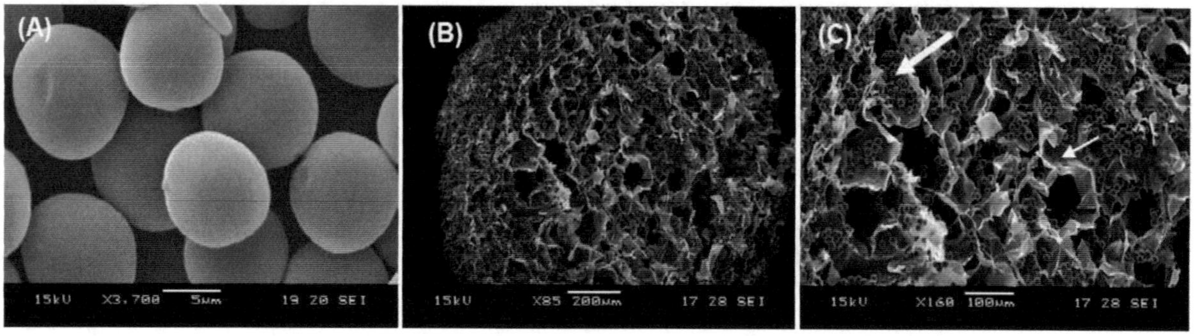

Figure 2: SEM images of hydrogel beads entrapping PLGA-paclitaxel microspheres. (A) PLGA-paclitaxel microspheres; (B) & (C) Alginate hydrogel beads entrapping PLGA-paclitaxel microspheres. Big arrow shows microspheres, while small arrow shows alginate matrix.

Thermal Analysis of Hydrogel Beads

Differential scanning calorimetery (DSC) was performed in order characterize the solid-state properties of paclitaxel encapsulated in the alginate gel beads by thermal analysis on the melting of crystalline regions and phase transitions. Polymer-drug crystallinity of PLGA microspheres in alginate is a key concern related to rate-controlled delivery and drug delivery efficiency. It has been established that molecular dispersions give better release profiles as compared with particulate dispersions of the same drug [5]. The characteristic DSC thermographs were obtained for the individual components of the bead. The thermograms for pure calcium alginate, PLGA 50:50 and pure crystalline anhydrous paclitaxel were analyzed (data not shown) and then compared with the thermograms of the hydrogel beads.

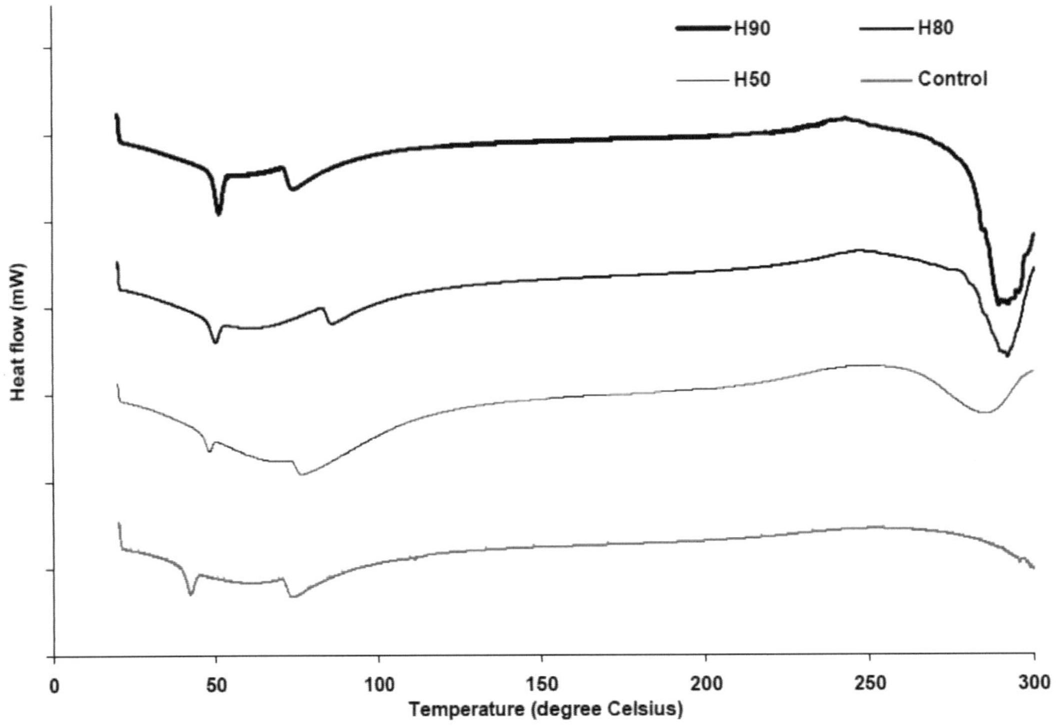

Figure 3: Thermograms of H90, H80, H50 and control beads (no paclitaxel).

According to the DSC thermograms in Fig. 3 the endotherms for the pure and anhydrous paclitaxel were missing in all the formulations. This indicated that the drug was in an amorphous phase in the polymer matrix after the EHDA and electrospray dripping processes. Decreasing the proportion of microspheres to alginate resulted in thermograms that resembled those of the pure calcium alginate. Furthermore, an increase in glass transition temperature was noted as increasing proportions of drug-loaded microspheres were used in each formulation. As the percentage of microsphere loading increased, the increase in T_g may be explained by the additional endothermic overlap, representing the energy required to overcome the microstructure of the polymer which had developed during the EHDA process [6]. These results suggested that paclitaxel was in an amorphous state that was ideal for drug diffusion.

In vitro Characterization of Hydrogel Beads

Hydrogel Bead Disintegration Study

The disintegration time of hydrogel beads containing the entrapped microspheres is affected by gelation time, calcium chloride concentration in the gelling bath and microsphere loading per bead. Experiments were performed by varying the three parameters to evaluate the disintegration time.

When subjected to monovalent ions in the buffer solution, the beads were disintegrated by the substitution of mainly sodium ions with crosslinked calcium ions. Longer exposure of beads to the counter ion solution (Fig. 4A) and exposure of beads to a more concentrated counter ion solution (Fig. 4B) increased the extent of cross-linking resulting in stronger beads. This increased the disintegration time under the simulated conditions. However, in Fig. 4C, as the microsphere loading increased, both the effects of calcium concentration and gelation time on disintegration time diminished. Reduced slopes were observed as the microsphere loading increased from 50 to 90%; this was due to the decreasing amount of alginate content and lower extent of calcium crosslinking. These data suggested that by controlling the three parameters, disintegration time of the beads can be controlled.

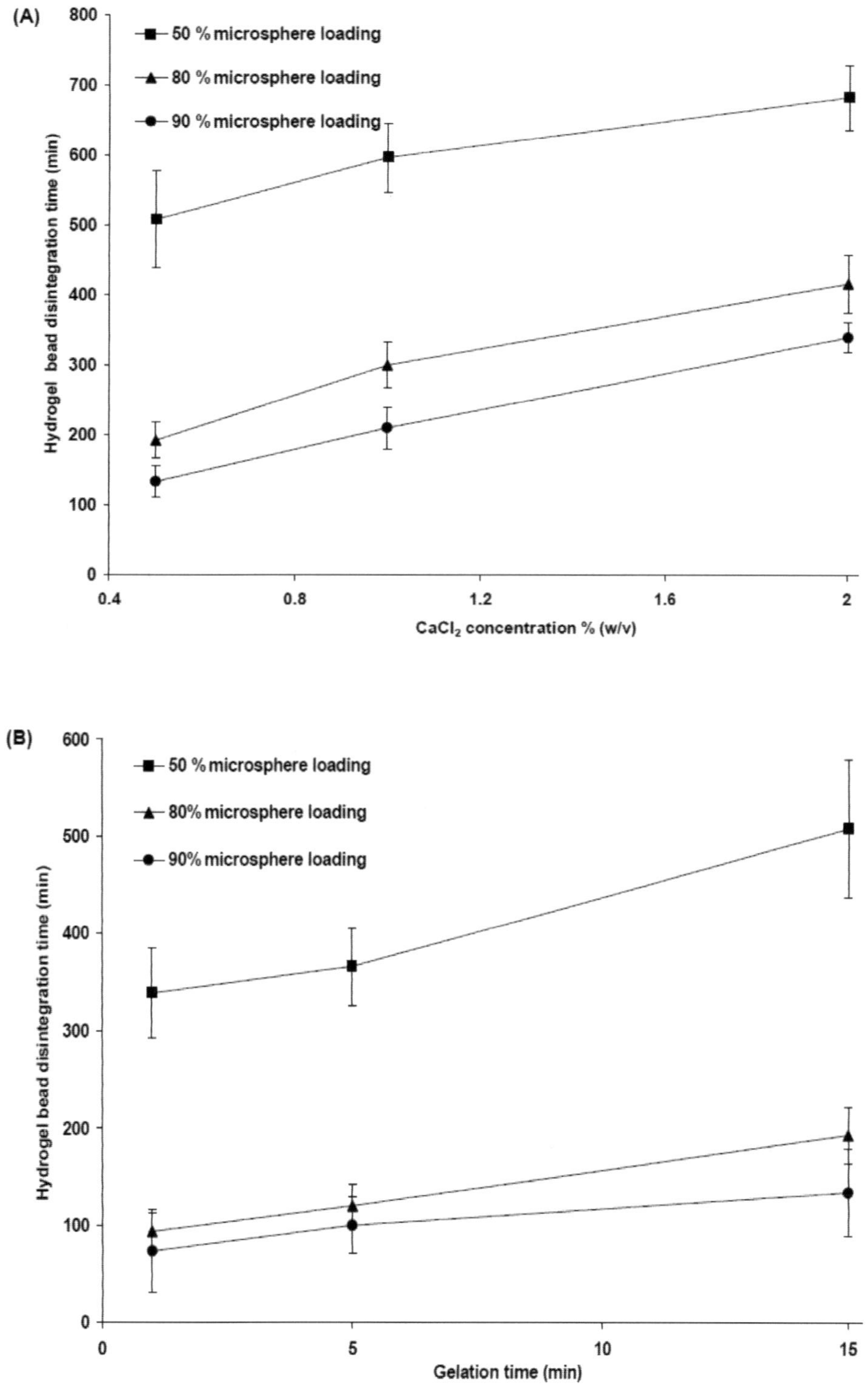

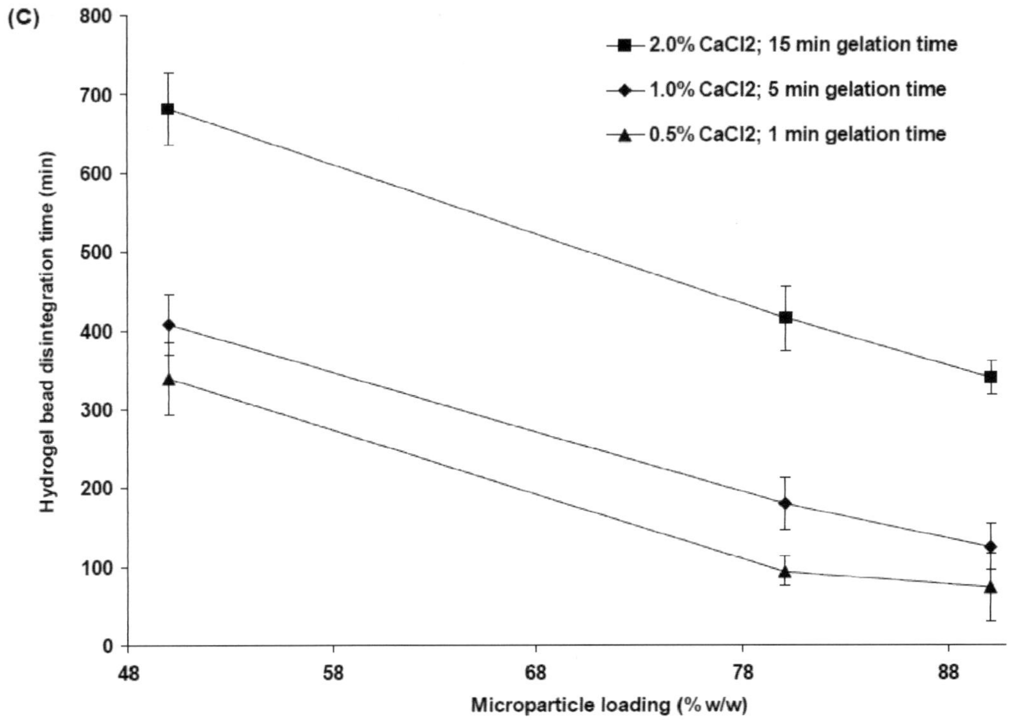

Figure 4: (A) Effect of gelation bath concentration on bead disintegration time (at 15 minute gelation time); (B) Effect of gelation time on bead disintegration time (at 2% w/v gelation bath concentration); (C) Effect of microsphere loading on bead disintegration time.

In vitro Paclitaxel Release from Hydrogel Beads

The nine formulations stated in Table 1 were used in this study to evaluate paclitaxel release under simulated physiological conditions in vitro. Figure 5 depicts the release profiles of the formulations.

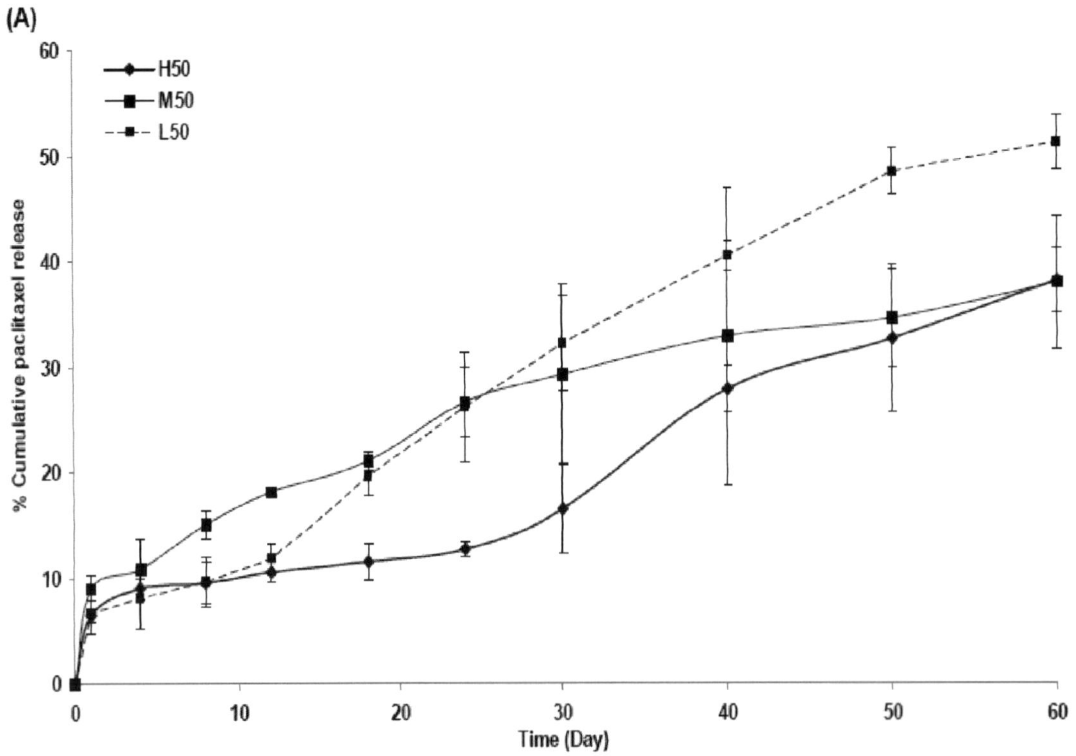

(B)

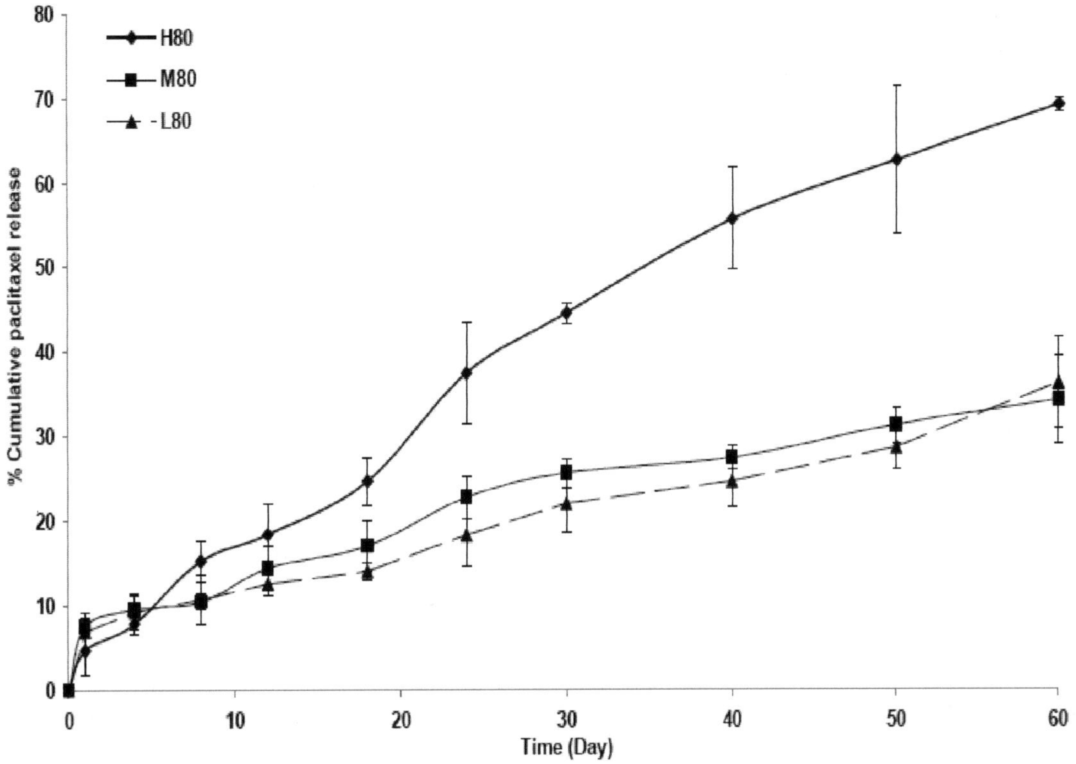

(C)

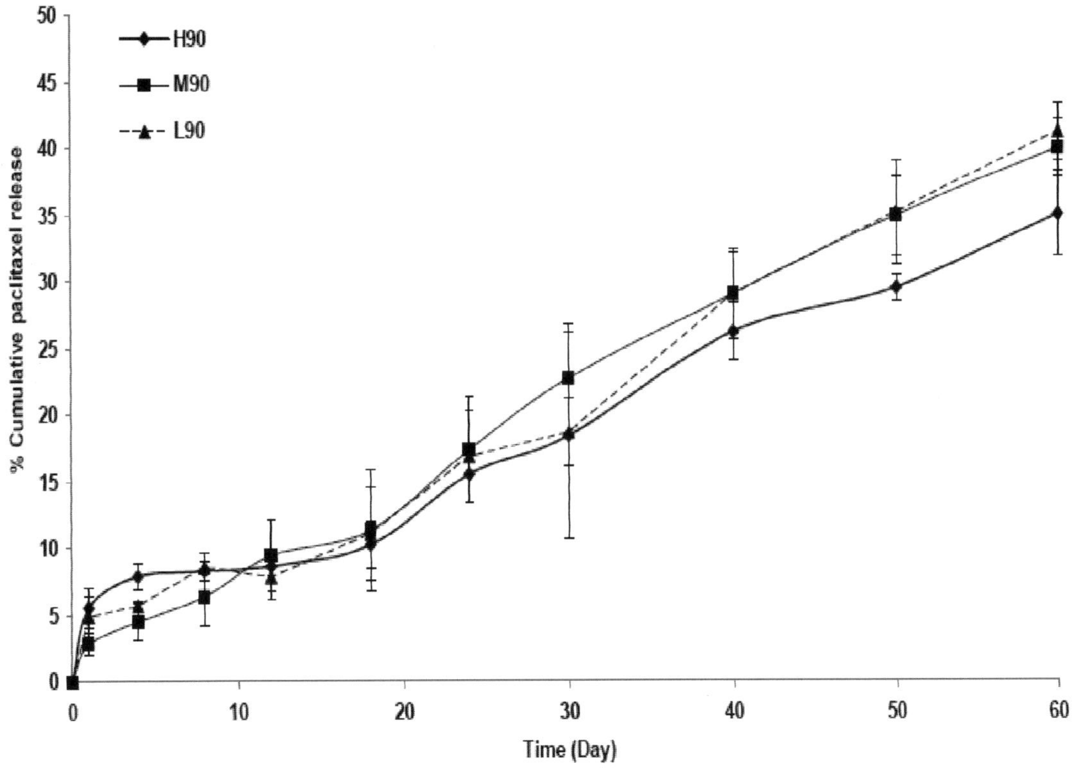

Figure 5: In vitro paclitaxel release from (A) 50 % microsphere loaded beads; (B) 80 % microsphere loaded beads; (C) 90 % microsphere loaded beads under specified conditions in Table 1.

From Figure 5 it can be noted that as the microsphere loading increases, a decrease in the drug initial burst results. The reason could be that at lower microsphere loading the microspheres were less compactly distributed and had good contact with the release buffer. Whereas in case of the higher microsphere loaded beads, due to densely distributed microspheres the initial burst was minimized. The later part of the release profile for all the formulations was near zero order. Although 50% microsphere loaded beads exhibited the highest initial burst, the release rate later was relatively slow. This could be due to the barrier of non-disintegrated alginate matrix through which the drug has to diffuse out. Whereas the 90% microsphere loaded beads exhibited a slow release rate probably because the higher compaction resulted in low penetration of buffer into the bead. Of the three formulations, the 80% microsphere loaded beads exhibited the highest release rate (especially H80) with near zero-order kinetics; this may be because this formulation has the optimum loading (optimum microsphere compaction and degree of crosslinking) that results in low initial burst and higher release rate with zero-order kinetics. Hence, the H80 and M80 formulations were screened for the subsequent in vitro cell culture and in vivo animal studies. Also, H80 and M80 having different release rates would be useful to neurosurgeons for applying them for patient-specific glioma chemotherapy.

In vitro Cytotoxicity Against C6 Glioma Cells

The H80 and M80 beads were evaluated against C6 glioma cells in vitro for their cytotoxic potency using IC_{50} values. IC_{50} refers to the concentration of the drug required to inhibit 50% of cell growth in vitro against the control sample. The C6 glioma cells were treated with control (no drug), H80 and M80 beads at different paclitaxel concentrations for three days. Figure 6 shows the decrease in cell viability after three days of incubation with varying paclitaxel concentrations. The M80 has an IC_{50} value of 16 µg/ml while the H80 has 37 µg/ml. This difference could be explained by the in vitro release profiles. After 3 days, the M80 exhibits a higher release (around double) of drug compared to the H80 and thus is more potent. However the IC_{50} values of both M80 and H80 correspond to that of Taxol®, which is reported to be about 30 µg/ml for a three-day incubation period [3]. In fact, the M80 outperformed Taxol® while the H80 had a similar IC_{50} value.

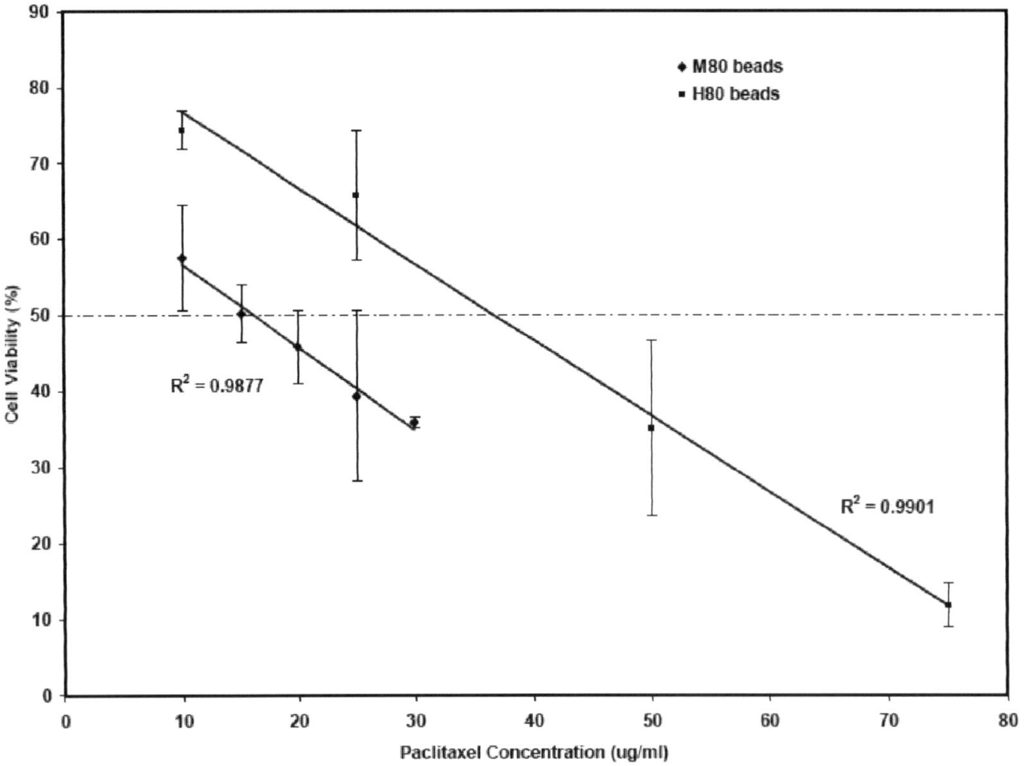

Figure 6: IC_{50} values for H80 and M80 beads used to treat C6 glioma cells in vitro for 72 hours.

These values indicate the cytotoxic potency of the H80 and M80 beads against C6 glioma cells in vitro. Studies of the drug release sustainability of the hydrogel beads, apoptosis against C6 glioma cell in vitro, and in vivo anti tumor efficacy in a subcutaneous tumor model are ongoing. These studies will be followed by in vivo intracranial studies such as drug bio-distribution and treatment efficacy (tumor volume and survival analysis) and are warranted.

CONCLUSIONS

The entrapment of microspheres was done using two types of dripping methods, manual dripping (MD) and electrospray dripping (ED). Manual dripping yielded polydisperse non-spherical beads with a non-uniform distribution of the microspheres within them. In the ED method highly monodisperse spherical beads with a uniform microsphere distribution within them were obtained. Hence, for further studies, beads fabricated using the ED method were used. The hydrogel bead disintegration study revealed that with higher gelling bath concentration and gelation time the beads disintegrate slowly, whereas with an increase in microsphere loading the disintegration time decreased. *In vitro* release of paclitaxel from nine formulations revealed that 80% microsphere loaded beads provided the optimum release profile with low initial burst with a higher release rate and near zero-order kinetics. The study of the IC_{50} value for the H80 and M80 in comparison with Taxol® against C6 glioma cells showed that in vitro, the beads were more potent. In vivo intracranial studies focusing on paclitaxel bio-distribution in the brain, survivability analysis and tumor volume through bio-imaging are ongoing and will give more insight into the performance of the hydrogel beads.

REFERENCES

[1] Naraharisetti PK, Ong BYS, Xie JW, Lee TKY, Wang CH, Sahinidis NV. In vivo performance of implantable biodegradable preparations delivering paclitaxel and etanidazole for the treatment of glioma. Biomaterials 2007;28(5):886-894.

[2] Li KW, Dang W, Tyler BM, Troiano G, Tihan T, Brem H, et al. Polilactofate microspheres for paclitaxel delivery to central nervous system malignancies. Clin Cancer Res 2003;19:3441-3447.

[3] Xie J, Marijnissen JCM, and Wang CH. Microparticles developed by electrohydrodynamic atomization for the local delivery of anticancer drug to treat C6 glioma in vitro. Biomaterials. 27:3321-3332 (2006).

[4] Xie J, and Wang CH. Electrospray in the dripping mode for cell microencapsulation. J Colloid Interf Sci. 312:247-255 (2007).

[5] Freiberg S, Zhu XX. Polymer microspheres for controlled drug release. Intl. J. Pharm., 282:1-18, 2004.

[6] Dubernet C. Thermoanalysis of microspheres. Thermochimica Acta, 248: 259-269, 1995.

370a An Analysis of the Deposition Mechanisms Involved during Self-Limiting Growth of Metal Oxides by Pulsed PECVD

Michael T. Seman[1], David N. Richards[2], and **Colin A. Wolden**[2]. (1) CMD Research LLC, 400 Corporate Circle, Unit B, Golden, CO 80401, (2) Chemical Engineering, Colorado School of Mines, 1500 Illinois, Golden, CO 80401

Self-limiting deposition (~ Å/pulse) of several metal oxides (Al2O3, TiO2, ZnO, TiO2) has been achieved by pulsed plasma-enhanced chemical vapor deposition (PECVD). In this process the metal precursor and O2 are delivered continuously to a PECVD reactor while the rf power is pulsed at low frequency (~ 1 Hz). The net deposition rate of pulsed PECVD can exceed that of continuous wave operation, and the quantity of impurities was dramatically attenuated. The growth mechanism of alumina synthesis from trimethyl aluminum (TMA) was investigated by comparing the results from pulsed PECVD with those of plasma-enhanced atomic layer deposition (PE-ALD). For both processes the rate/cycle saturated with ~200 L of TMA exposure. At 165 °C a rate of 1.37 Å/cycle was obtained using PE-ALD. For pulsed PECVD the rate scaled linearly with the TMA partial pressure, and its extrapolation was in good agreement with PE-ALD. The results suggest that deposition in pulsed PECVD involves an ALD component which is supplemented by PECVD growth, and that the contribution of the latter may be tuned using the TMA partial pressure. Experiments using patterned wafers supported this hypothesis. Conformal coatings were observed within 10:1 aspect ratio trenches using pulsed PECVD, however the deposition rate on the surface of these substrates was greater than within the trench. The ratio between the two corresponds well to the ratio of rates obtained from pulsed PECVD and PE-ALD on planar substrates. With cycle times < 1 s, net rates > 30 nm/min were obtained by pulsed PECVD while retaining high quality and digital control.

370b Kinetic Monte Carlo Simulations of Surface Growth during Plasma Deposition of Silicon Thin Films

Sumeet C. Pandey[1], Tejinder Singh[1], and Dimitrios Maroudas[2]. (1) Department of Chemical Engineering, University of Massachusetts - Amherst, 213 Goessmann Laboratory, 686 North Pleasant Street, Amherst, MA 01003-3110, (2) Department of Chemical Engineering, University of Massachusetts, 686 N Pleasant Street, Amherst, MA 01003

Plasma-assisted deposition of hydrogenated amorphous silicon (a-Si:H) thin films from silane-containing discharges is used extensively in large-area electronics, optoelectronics, and photovoltaics for fabrication of solar cells, thin-film transistors for flat panel displays, and detectors for medical imaging. Important properties of these films include their hydrogen content, surface composition, roughness, and crystallinity, which are governed by the interactions between the film surface and chemically reactive species arriving at the surface from the plasma. Prediction of surface and film properties, as determined by such plasma-surface interactions, requires properly developed coarse-grained dynamical simulations capable of capturing effectively surface length scales and deposition time scales. Toward this end, in this presentation, we report results of surface growth during Si plasma deposition according to kinetic Monte Carlo (KMC) simulations based on a first-principles database for radical-surface and adsorbed radical-radical interactions as identified from molecular-dynamics (MD) simulations of a-Si:H film growth.

We present KMC simulation results under growth conditions that render the SiH_3 radical the dominant deposition precursor. The transition probabilities for the various kinetic events accounted for in the KMC simulations are based on first-principles density functional theory (DFT) calculations on the H-terminated Si(001)-(2x1) surface. The relevant surface transport and reaction processes have been identified in the MD simulations and include SiH_3 diffusion, SiH_3 chemisorption and insertion into Si-Si bonds, surface H abstraction reactions, surface hydride dissociation reactions, as well as SiH_4 and Si_2H_6 desorption into the gas phase. In our DFT analysis, the H-terminated Si(001)-(2x1) surface is considered as a representative model of local chemical environment and atomic coordination on the growing film surfaces. Our DFT calculations are conducted within the generalized gradient approximation (GGA) and employ plane-wave basis sets, ultra-soft pseudopotentials, slab supercells, and the nudged elastic band method for determining optimal surface reaction/diffusion pathways and the corresponding activation energy barriers.

The predicted surface hydride, SiH_x (x = 1,2,3), composition and overall hydrogen content are in very good agreement with experimental *in situ* composition measurements using attenuated total reflection Fourier transformed infrared (ATR-FTIR) spectroscopy on a-Si:H films deposited under similar growth conditions. The temperature dependence of the surface hydride composition and the trends that have been captured, over a range of deposition temperatures, are fully consistent with reported experimental observations. Si trihydrides are predicted to be the dominant surface hydride species at low temperatures ($T \leq 373$ K), while dihydrides become dominant at 500 K and monohydrides become the most abundant surface species at higher temperatures ($T \geq 640$ K). The surface hydrogen concentration is predicted to decrease drastically over the temperature range from 350 to 500 K and to remain almost constant over the range from 500 to 640 K. The surface dangling bond coverage is found to be practically independent of temperature and quite low, on the order of $\sim 10^{-2}$.

370c Investigation of the Growth Mechanism during Plasma-Assisted Deposition of a-C:H

Bhavin N. Jariwala[1], Cristian V. Ciobanu[2], and Sumit Agarwal[1]. (1) Department of Chemical Engineering, Colorado School of Mines, 351C Alderson Hall, 1500 Illinois Street, Golden, CO 80401, (2) Division of Engineering, Colorado School of Mines, 1610 Illinois St., Golden, CO 80401

We have employed classical molecular-dynamics (MD) simulations, based on the modified extended Brenner potential, to study the structure and properties of hydrogenated amorphous carbon (a-C:H) thin films. The structure and properties of these films are defined by the sp^2-to-sp^3 hybridization ratio and the H content. In particular, interaction of H generated in the plasma results in local and overall transformations due to reactions such as, insertion into C-C bonds, surface H abstraction, bulk diffusion, and hydrogen-induced etching. In this presentation, we will discuss the mechanism and energetics of these reactions and show how they affect the film structure.

To study the radical-surface interactions, we have developed a procedure for creating realistic a-C:H thin films starting with diamond(001) slabs. Due to the density difference between a-C:H (1.7-2.2 g/cm^3) and diamond (3.52 g/cm^3) a certain number of atoms were randomly removed from the diamond(001) slab. The resulting structure was amorphized at 1000 K, hydrogenated, relaxed and then thermalized at 700K to create the a-C:H films. We have also formulated a scheme to characterize the sp^2-to-sp^3 hybridization ratio of these films based on the coordination number and the dangling bonds for each carbon atom, and the values obtained agree well with the range of values reported in the literature for experimentally deposited a-C:H films. Furthermore, the sp^2-to-sp^3 hybridization ratio and the H content in these films can be varied by tuning their initial density. H-induced reactions were observed by impinging H atoms on the surface of these realistic a-C:H films at 700 K. The detailed energetics of the various reaction pathways along with the structure and energies of the corresponding equilibrium and transition-state configurations will be presented.

370d Remote Atmospheric Pressure Plasma Activation of Polymers

Eleazar Gonzalez II[1], Michael Barankin[1], Peter C. Guschl[2], and Robert F. Hicks[1]. (1) Chemical & Biomolecular Engineering, University of California, Los Angeles, Box 951592, Los Angeles, CA 90095-1592, (2) Surfx Technologies, Culver City, CA 90232

The surfaces of polyethylene terephthalate (PET), polyethylene naphthalate (PEN), and polyphenylene sulfide (PPS) were activated remotely using an atmospheric pressure oxygen and helium plasma. The afterglow of the plasma, containing O atoms, and to a lesser extent metastable oxygen (1Δg O2) and ozone, came into contact with the polymer surface for periods of time ranging from 0 to 400 milliseconds. Changes in the surface energy and chemical composition of the surfaces were determined by contact angle measurements and x-ray photoelectron spectroscopy (XPS). The decrease in water contact angle followed an exponential decay indicative of a Langmuir adsorption process. For both PET and PEN, x-ray photoelectron spectroscopy revealed an 18 to 29% decrease in the area of the C 1s peak located at 285 eV, which is attributable to the aromatic carbons. The C 1s peak area corresponding to the C=O increased by 11.4 and 23.6% for PET and PEN, respectively, while the C 1s peak area resulting from the C-O bond increased by approximately 5% for both polymers. These results demonstrate that the O atom radicals generated in the remote plasma oxidize the aromatic rings in the polymer chain. The reaction is believed to occur by H abstraction from the polymer surface by the O radical, followed by O2 addition to the resulting aromatic radical. The peroxy species then decomposes to from ketones and alcohols. The surface of PPS was also oxidized, despite not containing any ethylene carbons as found in PET and PEN. It is believed that the O radicals attack the aromatic rings and abstract a hydrogen atom, which has not been seen before in literature. The results of this study will be presented at this meeting.

370e Thin-Film Deposition on Nanoparticles and Nanowires In Low-Pressure Plasma

Anaram Shahravan and Themis Matsoukas. Chemical Engineering, Penn State University, Department of Chemical Engineering, Fenske Lab, Room 210, University Park, PA 16802

Film deposition on particles is of great interest for several potential applications, ranging from catalysts, to nanocomposites, to sensors. Currently, this is mainly accomplished in wet environments using the tools of colloid science. Plasma deposition, although a well-established technology for microelectronics, has not been successfully adapted to particles. This work presents a new process that extends the usage of plasma deposition to finely dispersed materials such as micron- and sub-micron particles, and nanowires. We accomplish this by taking advantage of electrostatic effects that allow us to suspend particles in the plasma for extended periods of time. This trapping is now understood to involve an interplay of forces whose main component is electrostatic interaction and gravity. While the collective behavior of these so-called “dusty” plasmas have generated theoretical interest, we view these systems as an analogue to a "fluidized" bed with the ability to electrostatically confine micro and nano-particles, much smaller than the size that can be fluidized by conventional methods based on fluid drag.

In this talk we present results of film deposition on silica particles and on nanowires of various sizes in a plasma reactor in parallel-plate configuration driven by RF generator. Particles are introduced into the reactor via hollow stem of the top electrode and they become trapped in the sheath, a few mm above the bottom electrode. Vertical confinement is accomplished by force balance between the weight of the particles, and electrostatic repulsion from the electrode. Horizontal confinement is achieved by introducing asymmetry between the top and bottom electrodes. Various organic gases and vapors are introduced in the plasma and supply the coating material. Under the influence of the plasma, these organic molecules fragment and produce amorphous hydrogenated carbon (a-C:H) which is deposited on the surface of the suspended particles. Monodisperse silica particles of various sizes are studied for their ability to remain confined and to receive uniform deposition. We find that the thickness of the films is a linear function of time and is of the order of a few nanometers per minute. Analysis of the distribution of the deposited films further suggests that the sheath region where trapping occurs is an inhomogeneous environment that exhibits a wide, quasi exponential distribution of deposition rates. While the majority of particles appear to be trapped in regions of low deposition, substantially higher deposition rates are also observed. We discuss the implications of these findings on the feasibility of converting this semi-batch coating process into one that is continuous.

370f Multiferroic BiFeO3 Thin Films Deposited by MOCVD Method

Manish Singh, Dept of Chemical Engineering, University of Illinois at Chicago, 810 S Clinton St, Chicago, IL 60607, **Yi Yang**, Chemical Engineering, University of Illinois at Chicago, 810 S Clinton St, Chicago, IL 60607, and Christos G. Takoudis, Chemical Engineering and BioEngineering, University of Illinois at Chicago, 810 S Clinton St, Chicago, IL 60644.

BiFeO3 (BFO) is a novel multiferroic material which can couple both ferroelectric and antiferromagnetic effects in the same phase. This coupling provides an extra degree of freedom in design of devices and novel applications never thought to be possible before. Potential device applications include high speed low power electrically controlled magnetic memory elements, electrically tunable microwave devices and highly sensitive magneto sensors.

Most of the known single phase bulk magnetoelectric materials do not exhibit strong magnetoelectric coupling, have Néel or Curie temperatures far below room temperature, and are often difficult to grow in thin film form. Among the several known single-phase magnetoelectrics, BiFeO3 is known to be the only material that exhibits ferroelectromagnetism at room temperature. Recently, large magnetoelectric effects were reported in thin epitaxy-strained BiFeO3 films. This strong and robust ferroelectric and magnetic ordering in BiFeO3 provides an excellent opportunity for design of multifunctional oxides using it as a host.

Obtaining pure BiFeO3 with desirable properties is a challenge and several deposition approaches including pulsed laser deposition, molecular beam epitaxy, sputtering and solution based techniques have been used. Metalorganic chemical vapor deposition (MOCVD) can deposit conformal thin films over large areas with the right stoichiometry, high yield and high throughput at low cost making it ideal for fabrication of thin film-based devices. Strangely, compared to other deposition techniques, there have been extremely few reports on the use of MOCVD for multiferroic films. The choice of precursors is critical and herein we report MOCVD of crystalline BiFeO3 films on platinized silicon substrates using n-butylferrocene, triphenylbismuth and oxygen. Thermogravimetric analysis shows the suitability of these two precursors for depositing BiFeO3. Surface morphology was studied using scanning electron microscope. Composition analysis using x-ray photoelectron spectroscopy revealed that the films were stoichiometric BiFeO3 and did not have any carbon or organic impurities. X-ray diffraction studies indicated that BiFeO3 films were phase pure. Electrostatic force microscopy indicated that the film had polarizable domains that showed no deterioration in polarization over time long after electric poling.

370g Interesting Characteristics of a Vapor-Deposited Silica Thin Film

A. Anderson and W. Robert Ashurst. Chemical Engineering, Auburn University, 212 Ross Hall, Auburn, AL 36849

The deposition of silica layers from the vapor phase hydrolysis of silicon tetrachloride will be discussed. These silica layers, which support the formation of organic monolayers formed from standard organosilane chemistries, can be deposited onto a wide variety of surfaces, including silicon, cellulose, zinc selenide, polystyrene, and poly (methyl methacrylate), offering the potential to drastically expand the types of materials that silane surface modification chemistry can be applied to. This talk will present an overview of our studies on this silica layer, including morphological and spectral characterization, dehydration and rehydration properties, interactions with water, and interactions with bacterial cells.

379a Tunable Polymer Networks Based on Specific Hydrogen Bonding Motifs

Kamlesh P. Nair, School of Chemistry and Biochemistry, Georgia Institute of Technology, Atlanta, GA 30332, Marcus Weck, Department of Chemistry, New York University, New York, NY 10003, and **Victor Breedveld**, Chemical and Biomolecular Engineering, Georgia Institute of Technology, 311 Ferst Drive, Atlanta, GA 30332-0100.

We have investigated the rheology of polymer networks in which the interpolymer crosslinks are based on controlled hydrogen bonding between polymer side groups and small-molecule crosslinking agents. Hydrogen bonds are noncovalent interactions, like metal coordination, Coulombic and dipole-dipole interactions, and therefore reversible under the appropriate environmental conditions. Hydrogen bonds are readily broken at elevated temperatures and can thus be used to form reversible, thermosensitive polymer networks.

Another unique property of hydrogen bonds as intermolecular interactions is that multiple bonds are often combined in specific molecular recognition motifs. Thymine and cyanuric acid are both well-known examples of compounds that form strong, multiple hydrogen bonds with receptor molecules that possess complementary acceptor-donor motifs. Thymine, one of the bases in DNA, forms 2-point hydrogen bonds with adenine in DNA, but can also form 3-point hydrogen bonds with diaminotriazines. Cyanuric acid forms 3-point hydrogen bonds with melamine or diaminotriazine, and even stronger 6-point hydrogen bonds with Hamilton wedge receptors.

For this study, a variety of co- and terpolymers was synthesized via ring-opening metathesis polymerization (ROMP) of three norbornene-based monomers with different functional side chains: an inert alkane (C8), cyanuric acid, and thymine. The alkane-based spacer monomers were used to dilute the functional groups and enhance solubility of the polymers in 1-chloronaphthalene. Selective ditopic crosslinking molecules were also synthesized: one with two Hamilton wedge motifs and another with two diaminotriazine groups. The objective was to create a tunable system in which the strength of the network can be manipulated by selective crosslinking different polymer side chains. As controls, we also used monotopic Hamilton wedge and diaminotriazine, which should only be able to bind to one side chain at a time and therefore not act as crosslinkers.

Shear and oscillatory rheology will be shown for solutions of spacer/cyanuric acid and spacer/thymine copolymers in the absence and presence of the small molecule hydrogen-bonding receptors. It was found that, depending on the combination of polymer and crosslinking agent a wide range of rheological properties could be obtained, from highly viscous fluid to strongly elastic gel. The variability of rheological properties reflects significant variations in the underlying molecular structure, which are currently not fully understood. The tunability of polymer networks based hydrogen bonding will be highlighted by presenting the rheology of solutions of spacer/thymine/cyanuric acid terpolymers in the presence of all cross-linking agents in varying combinations and concentrations. We were able to manipulate the rheology of the terpolymer solution by taking advantage of the competitive binding of the cyanuric acid and thymine residues with the various cross-linking agents.

379b Structure and Rheology of Fibrin Networks as Probed with Small Angle Scattering

Danilo C. Pozzo and Kathleen Weighandt. Chemical Engineering, University of Washington, 357 Benson Hall, Box 351750, Seattle, WA 351750

Small angle scattering experiments with neutrons (SANS, USANS) and light (SALS) are used to probe the structure of fibrin networks (blood clots) over a broad range of length scales (1-20,000 nm). SANS probes the rich internal structure of the fibers (protein arrangement, fiber porosity) while USANS and SALS provide valuable information on the structure of the bifurcating fibrin network (e.g. correlation length). In our experiments, we are also able to probe the mechanical properties (rheology) of the clots while simultaneously obtaining structural information through scattering. This allows us to correlate the structure of these important materials to their unique elastic and strain-hardening properties. Furthermore, we also discuss changes occurring in fibrin networks that are formed under flow. It is found the polymerization of the network in a shearing environment results in significant alterations of its structure and rheology.

379c Characterization of the Network Structure of Molecularly Imprinted Polymers by Determination of Kinetic Chain Length Distribution

Vishal D. Salian and Mark E. Byrne. Biomimetic & Biohybrid Materials, Biomedical Devices, and Drug Delivery Laboratories, Department of Chemical Engineering, Auburn University, Auburn, AL 36849-5127

Living/controlled polymerization has been reported to enhance the binding characteristics of imprinted polymer networks, which has been hypothesized to be due to improved structural homogeneity of the binding sites and a global energy minimum of the spatial arrangement of polymer chains. Macromolecular memory within imprinted networks are due to two synergistic effects: (i) shape specific cavities that match the template molecule, which provide stabilization of the chemistry in a crosslinked matrix, and (ii) chemical groups oriented to form multiple non-covalent complexation points with the template. Increased potential for the growing polymer network and template binding complexes to reach a global energy minimum can lead to further memorization of the chain conformation and enhance binding parameters in both highly crosslinked and weakly crosslinked polymers. This work demonstrates the effect of template molecules that non-covalently interact with functional monomers on the distribution of the resulting linear polymer chains in both conventional free-radical polymerization and living/controlled polymerization. Kinetic chain length distributions were studied in both conventional and living/controlled free-radical polymerization reactions by varying functional monomer/template and iniferter/conventional photoinitiator ratios. Results highlight as the strength of the monomer template interaction and the monomer/template ratio increase, there is a more pronounced effect on the distribution of polymer chains. Since crosslinked polymers are insoluble and cannot be analyzed by size exclusion chromatography, it is difficult to apply most commonly used techniques. It is possible, however, to characterize the degradation products of these networks. Therefore, imprinted poly (methacrylic acid-co-ethylene glycol dimethacrylate) (poly (MAA-co-EGDMA)) networks were synthesized at various crosslinking densities and porcine liver esterase (PLE) was used to selectively catalyze the hydrolysis of the EGDMA ester linkages to form smaller bi-acid polymer molecules. These molecules were then characterized by HPLC to yield structural and kinetic chain length distribution of the polymers. In addition, binding studies performed on the imprinted polymer networks yielded the imprinting efficiency of the polymer. The results confirm that improved structural homogeneity of the binding sites and a global energy minimum of the spatial arrangement of polymer chains lead to enhanced binding parameters.

Web Page: www.eng.auburn.edu/~byrneme/

Structural characterization of protein-imprinted gels using lattice Monte Carlo simulation

Liora Levi and Simcha Srebnik*

Department of Chemical Engineering, Technion – Israel Institute of Technology, Haifa, Israel 32000.

Key words: molecular imprinting: proteins; simulation; Monte Carlo; polymer gel

Abstract

Molecular imprinting finds relevance and importance in a range of applications including the development of new chemical sensors, separation technologies, and selective binding particularly for catalytic and bioreactors. While molecular imprinting technology has found success when imprinting small molecules, progress in molecular imprinting of biomacromolecules, such as polypeptides and proteins, is hindered, in part due to their large size and complex functionality. Unlike with small molecules, imprinting of biomacromolecules results in poor performance of the imprinted gel, characterized by significant nonspecific binding of molecules of similar shape and size to the template. Our research in this field aims to characterize the structure of protein-imprinted gels in an effort clarify the effects that lead to poor protein imprinting.

The determination and study of the three dimensional molecular structure of such supramolecular assemblies is a challenging problem. To gain a general understanding of the process, we focus on simplified coarse-grained lattice models that are known to capture some generic and universal properties of polymer gels and biopolymers. The protein and imprinting matrix are modeled using coarse-grained lattice techniques, where gelation is simulated using a modified kinetic gelation algorithm. Structural and functional properties are probed using protein diffusion within the gel. Different gelation conditions are considered and analyzed for imprinting performance. The effects of protein size, hydrophobicity, charge, and flexibility on gel structure are studied.

In this work, we focus on a bulk imprinting method using radical polymerization of hydrogels, in an effort to reveal the problematic points in the polymerization process, which ultimately result in low imprinting efficiency. We relate to the work of Ou et al. [1] and Kimhi et al. [2], who studied molecular imprinting of small globular proteins (lysosyme and cytochrome c) in polyacrylamide gels with electrostatic functional groups. We will present the method and results of the structural, energetic and functional properties of these imprinted gels and compare them with experimental observations.

* Corresponding author: simchas@technion.ac.il. Current address: Fischell Department of Bioengineering, University of Maryland, College Park, MD 20742

Molecular imprinting is a process in which organic or inorganic polymeric materials that are able to recognize specific substrates are fabricated [3]. These molecularly imprinted polymers (MIPs) are formed by cross-linking of functional monomers in the presence of a template molecule. When the template is removed, it leaves cavities that are complementary to the target molecule in size, shape, and functional group orientation. These cavities will preferentially rebind in specific locations to the template molecule and not to similar molecules on account of the matching location of the functional groups [3-5].

Molecular imprinting of bio-macromolecules and proteins in particular, is a challenging area of research that is currently receiving a great deal of attention [6, 7]. As a well-established technology in the field of synthetic molecular recognition, it offers a generic, robust and cost-effective alternative to existing biological recognition techniques such as monoclonal antibodies [8]. Selective protein binding is very useful in isolation, extraction, biosensors and other issues applied in laboratory practice [8], procedures that are almost entirely dependent on antibodies, which are extremely expensive [6, 8]. Protein imprinted polymers (PIPs) can be used as the solid phase for chromatographic analysis and purification of proteins, and can also provide researchers with a low-cost easily obtainable method for studying the fundamental interactions occurring during biological recognition processes [9]. The potential application of such materials could also extend beyond the laboratory by serving as a tool to control biological systems. For example, artificial receptors, which are capable of recognizing proteins in aqueous media, would allow for creating materials and devices capable of mimicking natural processes and could potentially be used in medical diagnostic applications and clinical analysis [9].

While imprinting of small molecules has advanced significantly over the years and a number of companies now sell tailor-made imprints, efforts to generate imprints of protein targets have been far less successful [6]. The reason for this can be attributed to several factors related to the properties of the protein templates. A key contribution to poor specificity is inherent to the non-covalent approach (the most widely used method for MIP fabrication due to its simplicity [3, 10]), where heterogeneity of the binding sites cannot be avoided mainly as a result of the relative instability of template-monomer complex and the limited choice of polymerization conditions that will favor monomer-template assembly, even for small imprint molecules [4, 5, 10, 11]. Furthermore, while the majority of molecular imprinting relies on hydrogen and ionic bonding for recognition and therefore takes place in apolar, organic solvents that maximize electrostatic interactions [7, 8], proteins obligate the use of water as the reaction solvent, which further diminishes the power of molecular recognition. Moreover, water limits the choice of monomers, since many monomer types commonly used in MIP preparations are insoluble in water [8]. The large number of potential recognition sites on the protein also favor nonspecific binding. An additional drawback of the large size of the protein is poor mass transport, and permanent entrapment within traditionally highly cross-linked MIP gels. Reducing the degree of cross-linking results in highly porous and thus less selective gels, in addition to poor mechanical properties [7, 8] .

The following experimental works exemplify the unimpressive protein selectivity of PIPs achieved so far in comparison with MIPs. Ou et al. [1] created acrylate based Lysosyme-imprinted polymer by free radical bulk polymerization using charged functional monomers, with imprinting factor (binding of protein to imprinted polymer /

binding of protein to non-imprinted polymer (NIP)) of 1.83-3.38. However, rebinding was not strictly template-specific as the polymer also bound albumin. Furthermore, more than 25% of the original template remained in the polymer, a situation that impairs selectivity as it reduces diffusion efficiency, diminishes the number of free recognition sites within the polymer, and may lead to protein aggregation during rebinding. Similar results were obtained by Kimhi and Bianco-Peled [2], in their study of PIPs using cytochrome c and lysozyme on lysozyme-imprinted polymer. Although absorption in the imprinted gel showed preference towards the template protein, it is far less significant compared to MIPs. Takeuchi et al. [12] created an array of imprints, which together were used to determine unique binding profiles for a set of analyte proteins. Though every imprint bound its own template protein with the highest affinity, the selectivity was modest (at most 2.5-fold). By using acidic or basic proteins and monomers it was anticipated to find differences in the strength of interactions accordingly. However, no clear trend was observed. In comparison, Yavuz et al. [13] performed non-covalent imprinting of cholesterol in polymeric particles using bulk polymerization and showed that adsorption capacity of cholesterol by MIP particles was 8 fold higher than by the control NIP particles, while Wang et al. [14] prepared a cholic acid-imprinted polymer exhibiting imprinting factor of 7.9.

In view of the difficulties associated with the production of efficiently selective PIP [6, 8], the need to plan new imprinting strategies for proteins arises. For this purpose, one must first reveal the problematic points in the polymerization process of current methods that result in low imprinting efficiency. For example, does the protein-imprinted, fairly loose, gel retain the form of the imprinted protein pore following protein extraction and if not, does it have a "memory" of the template protein such that it reconstructs the pore upon template absorbance? How does the number of recognition sites in the imprinted cavity affect the PIP specificity, and does higher number of linked sites necessarily result in higher recognition and better selectivity? Moreover, Could imprinting particular sites on the protein surface promote or inhibit specific recognition? What is the effect of protein flexibility on PIP recognition and selectivity? Does the lack of molecular rigidity play an important part in lowering the efficiency of the imprinted polymer? What is the effect of protein size and shape on imprinting efficiency? How does charge density and charge distribution on the protein shell affect recognition? How do gel properties such as chain flexibility, cross-linker concentration, volume fraction, the fraction of functional monomers etc., affect recognition and PIP selectivity?

The most common polymer system for PIP production is free radical initiated vinyl polymerization, which is easy to prepare using a broad range of commercially available monomers and cross-linkers [3, 10]. The kinetic gelation model (KGM) describes the chemistry of irreversible polymer gelation and relates to the rules of initiation, propagation, termination and radical trapping [15-17]. KGM models generally describe the formation of a rigid network without relating to the flexibility of the resulting structure. Verdier et al. [18, 19] present a MC method for studying polymer chain dynamics and equilibrium configurations. In their model the polymer is represented by a self-avoiding walk on a simple cubic lattice. Trial repositioning of chain particles is performed by several distinct types of motion (end rotation, crankshaft, kink jump, etc.), while obliging a fixed bond length.

We use a three-dimensional lattice MC simulation to model the complexation between a template protein and charged functional monomers. This stage is followed by KGM simulating radical reaction mechanism [20-22] of gelation in presence of the template protein. The occupied lattice sites in the model represent protein residues and polymer particles. The monomers are bifunctional and either neutral, negatively or positively charged, and the cross-linkers are all tetrafunctional and neutral. In the current model, the protein is modeled as a rigid cube composed of a fixed number of lattice sites, each representing one amino acid unit. Charged protein residues are randomly distributed on the protein shell according to charge density parameter. The electrostatic functionality for charged monomers and charged protein residues is taken to be unity. Reversible electrostatic bonds are allowed to form between two oppositely charged functional monomers, or between a functional monomer and a protein residue with opposite charges. Both covalent and electrostatic links can only form between two nearest lattice neighbors. No statistical condition limits the formation of covalent or electrostatic links; however, the covalent bonding is of first priority. Hence, a particle or a cluster will first implement all covalent bonding opportunities and only then will have the option to link electrostatically. However, a pair of covalently linked monomers cannot link electrostatically as well.

Initially, particles (monomers, cross-linkers and protein) are randomly distributed amongst lattice sites achieving a chosen volume fraction, ϕ. Each simulation is divided into two main stages: The first is a prepolymerization stage, during which energetic equilibrium is achieved via particle movement and electrostatic interactions. The energy of the system is given by, $E = b\varepsilon$, where ε is the interaction energy of a single electrostatic bond, and b is the number of electrostatic bonds in the system. It was our aim to first isolate the effect of protein-monomer complexation, which is determined by electrostatic attraction and is presumed to dictate the functionality of the imprinted pore. For completeness, electrostatic repulsion between similarly charged elements should also be included. However, since the system reaches equilibrium before polymerization this addition is not expected to affect the number of recognition sites but should diminish the number of unfavorable interactions within the pore. The second simulation stage begins with fast initiation, as radical "signatures" are randomly distributed amongst polymer particles [23], and continues with propagation, as covalent bonding occurs. The process ends either when all polymer particles are connected to the main cluster (gelation is completed) or when the number of clusters does not change during a sufficient number of iterations, which means that all radicals are either terminated or trapped. Since this simulated process is fairly rapid, we can be certain that the last situation assures the ending of polymerization.

During propagation, each radical molecule randomly links to one of its nearest neighbors, thus starting the formation of a chain of covalently linked particles. Only a radical molecule can covalently bond to another molecule, and by doing so transfers the radical signature to it. Once no options of further linking are left, cluster displacement moves take place. At the end of each movement, another series of covalent bonding is performed, as well as an update of electrostatic bonding around the moved cluster. This order of events is significant only computational and does not affect the final gel structure.

In each attempted move, one particle or cluster is randomly selected as a candidate for displacement. Clusters move as unities similar to single particles and currently there are no independent movements of their inner chains. The probability P that a cluster will attempt to move depends on its size such that $P = 1/M^{\gamma}$, where M is the number of particles composing the cluster and γ is a parameter that correlates the degree to which particle size affects its diffusion [24-29]. Once a particle has been selected, the direction of the movement is randomly chosen out of the six nearest neighbor lattice directions and an overlap check is performed. If no overlap is found, the displacement takes place with probability P' that depends on energetic calculations as follows. The energy difference between the current and proposed conformation is calculated according to $\Delta U = \varepsilon\,(b_2 - b_1)$, where ε is the interaction energy of a single electrostatic bond, b_1 and b_2 are the number of electrostatic bonds in the current and proposed configurations, respectively. Moves are accepted according to the Metropolis criterion [30].

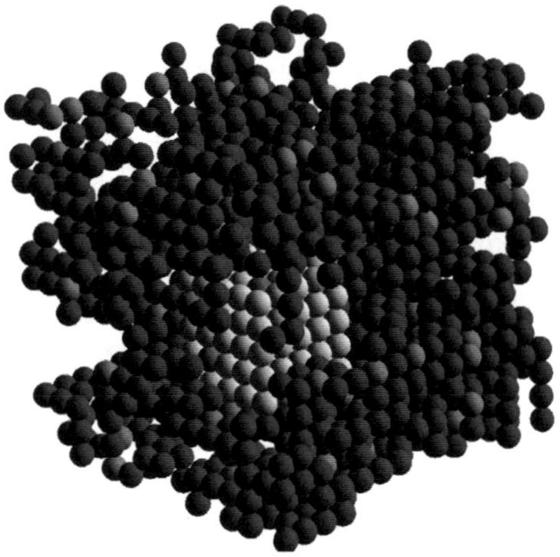

Figure 1. Illustration of protein imprinted polymer. Different colors indicate different particle types: monomers (blue), crosslinkers (pink), protein residues (beige), and protein residues that are linked to functional monomers (red). Simulation parameters correspond to $\phi = 0.4$, cross-linkers percent is 7.6%, protein charge density is 0.6, fraction of charged monomers is 0.6.

In order to study PIP selectivity it is necessary to simulate the processes of protein extraction and re-adsorption, which require simulations of protein diffusion into and out of a swollen gel, and is the subject of our continuing research. In this work, we focus on properties of the dry gel, which facilitates the investigation of the imprinted pore structure and functionality, and provides an indication on the optimal potential of protein recognition. A sample configuration of the protein-imprinted gel is illustrated in Figure 1, for a cubic lattice containing 15^3 sites. Lattice size L=15 was found to be sufficiently large so that lattice size effects are negligible [31].

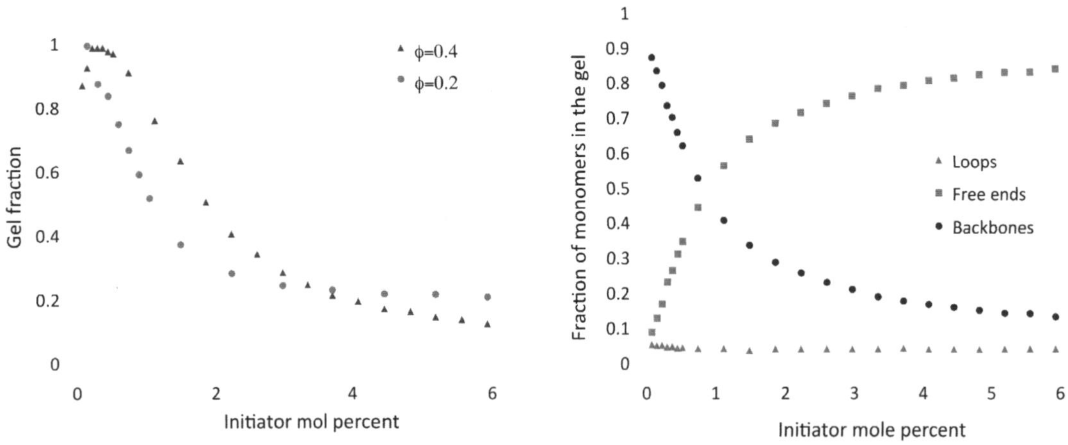

Figure 2. Effect of initiator concentration on gelation at $\phi = 0.2$ and 0.4 (left) and gel structure for $\phi = 0.4$ (right).

It is shown in Figure 2 that the initiator molar ratio influences gelation by changing the final gel fraction (GF) and gel composition; GF represents the fraction of particles (monomers and crosslinkers) that belong to the main cluster. For each particle volume fraction, ϕ, studied an optimal range of initiator concentration in which the GF value is maximal (Figure 2a) and the resulting gel is mostly composed of backbone chains (Figure 2b). It is striking that the initiator molar ratios used in experimental works of protein imprinting via acrylate-based bulk radical polymerization [1, 2] correspond to the optimal range found in the current results.

The effect of polymer volume fraction and cross-linker molar ratio on the resulting gel structure, shown in Figure 3, exemplify a similar trend with both parameters such that when increasing ϕ or cross-linker ratio the number of chains increases and chain length decreases. Finally, we note that while at the dilute protein concentrations studied, our results indicate that gelation in the presence of the protein does not affect the final gel structure, the presence of the protein creates cavities in the dry gel that are of the order of protein size in dense gels (high ϕ), that are unlikely to form otherwise.

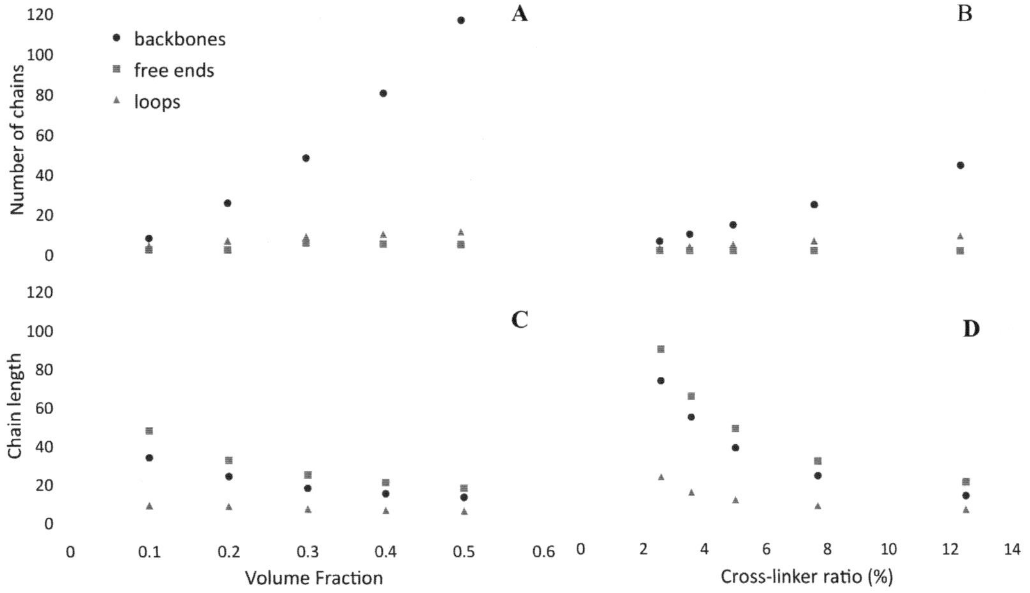

Figure 3. Effect of volume fraction and cross-linker molar ratio on gel structure. Cross-linker ratio in A and C is 1/13; ϕ in B and D is 0.2.

Figure 4 illustrates the effect of monomer volume fraction on pore size and the number of recognition sites present on pore walls (normalized by pore size) and on the average fraction of linked protein residues during diffusion of the protein within the cavity. Both these factors are linearly correlated with the ϕ, while the fraction of average links out of all potential recognition sites increases rapidly with ϕ and reaches a plateau at $\phi > 0.5$ when the cavity approaches protein size. The large values of pore size are due to percolation of pores in the system that occurs in low particle densities [32, 33]. Ou et al. showed in their experimental work that increasing ϕ leads to increase in the imprinting efficiency. This trend was explained by elevated stability of the protein-monomers complex at high polymer concentrations [1], and is directly related to the enveloping effect of the imprinted pore at higher values of ϕ.

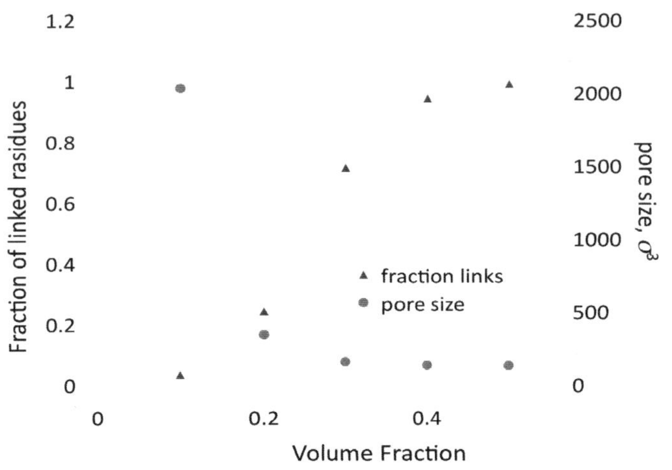

Figure 4. Effect of ϕ on pore size (circles) and linking characteristics of the template protein to the imprinted pore (triangles).

Along with the linked fraction of charged residues, we also calculated the average fraction of protein surface that interacts with the gel. **Figure 5** compares these values for an imprinted pore with that of a randomly formed pore within a nonimprinted polymer. It can be clearly seen that the imprinted pores recognize the protein via more directions and by larger number of recognition sites than the non-imprinted ones. In another simulated experiment that gives an indication on PIP selectivity, the template protein is removed from the gel following polymerization and replaced by other proteins of the same size but which differ from the template protein either by charge distribution alone or by charge density as well.

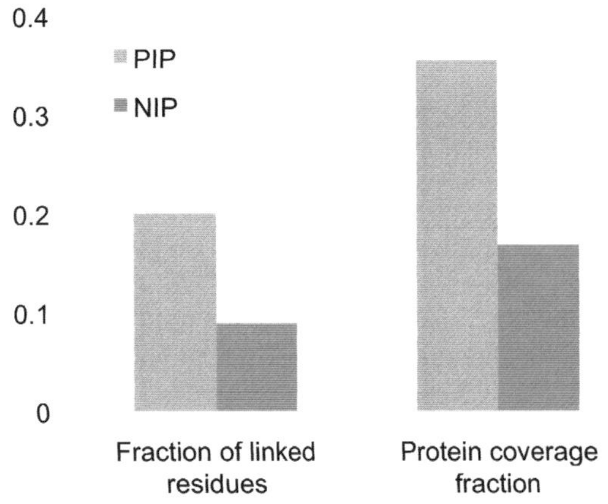

Figure 5. Effect of imprinting on the linked fraction of charged protein residues and on the fraction of protein coverage by the imprinted pore walls ($\phi = 0.4$).

Figure 6 shows a comparison of the fraction of linked residues of template and non-template proteins with various charge densities in such diffusion experiments. As expected, we find that when considering an identical charge density on the protein shell, the template protein has a higher number of linked recognition sites than a non-template protein, suggesting that charge distribution on the protein shell affects the imprinting process. However, at varying charge densities it seems that the main influencing factor on the fraction of linked residues is the charge density on the protein shell. Irrespective of the charge density on the template protein, a highly charged protein will have more recognition sites on the imprinted pore walls than a poorly charged one, even if the latter is the template protein itself. This result corroborates the suggestion that multiplicity of recognition sites may impair selectivity [8] and provides an explanation for the differences in PIP recognition of lysosyme and cytochrome c, as was indicated by experiments [2].

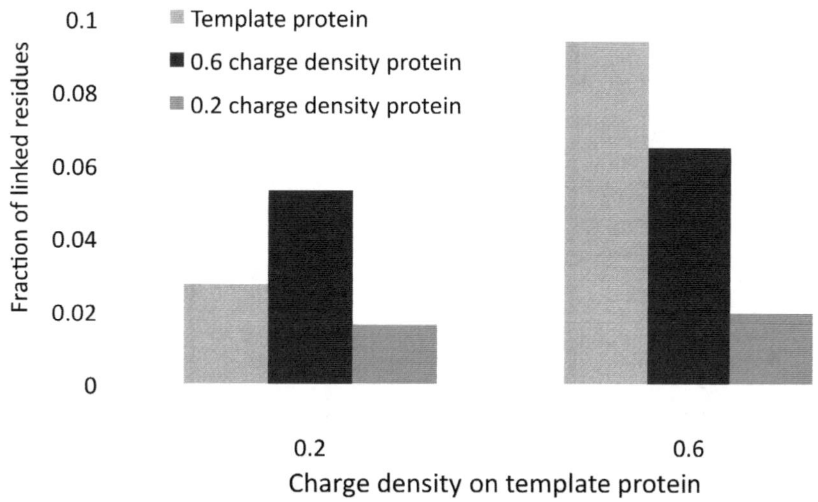

Figure 6. Linking characteristics of template and non-template proteins with different charge densities ($\phi = 0.4$).

Our preliminary results on the structure and functionality of the imprinted pore appear to provide an indication on the potential recognition properties of the PIP, even in its dry (non-swollen) state. In view of experimental data, it can be postulated that at least some of the trends observed by our presented pore structure analysis will be corroborated by the results of more rigorous diffusion simulations. Future stages of the proposed research will include simulations of swelling, protein extraction and absorption (via diffusion within the gel), and polymer and protein flexibility. Such simulations will indicate whether it is possible to improve PIPs performance and selectivity and will shed some light on the prominent issues on which one should focus in order to enable these improvements for practical applications.

Acknowledgements

We thank the Russell Berrie Nanotechnology Institute and Israel Science Foundation for partial support.

References:

1. Ou, S.H., et al., Analytica Chimica Acta, 2004. **504**(1): p. 163-166.
2. Kimhi, O. and H. Bianco-Peled, Langmuir, 2007. **23**(11): p. 6329-6335.
3. Alexander, C., et al., Journal of Molecular Recognition, 2006. **19**(2): p. 106-180.
4. Maier, N.M. and W. Lindner, Analytical and Bioanalytical Chemistry, 2007. **389**(2): p. 377-397.
5. Holthoff, E.L. and F.V. Bright, Analytica Chimica Acta, 2007. **594**(2): p. 147-161.
6. Hansen, D.E., Biomaterials, 2007. **28**(29): p. 4178-4191.
7. Bossi, A., et al., Biosensors & Bioelectronics, 2007. **22**(6): p. 1131-1137.
8. Turner, N.W., et al., Biotechnology Progress, 2006. **22**(6): p. 1474-1489.
9. Janiak, D.S. and P. Kofinas, Analytical and Bioanalytical Chemistry, 2007. **389**(2): p. 399-404.
10. Zhang, H.Q., L. Ye, and K. Mosbach, Journal of Molecular Recognition, 2006. **19**(4): p. 248-259.
11. Jiang, X.M., et al., Analytical and Bioanalytical Chemistry, 2007. **389**(2): p. 355-368.
12. Takeuchi, T., D. Goto, and H. Shinmori, Analyst, 2007. **132**(2): p. 101-103.
13. Yavuz, H., et al., International Journal of Biological Macromolecules, 2007. **41**(1): p. 8-15.
14. Wang, Y.J., et al., Polymer, 2007. **48**: p. 5565-5571.
15. Henthorn, D.B. and N.A. Peppas, Industrial & Engineering Chemistry Research, 2007. **46**(19): p. 6084-6091.
16. Wen, M., L.E. Scriven, and A.V. McCormick, Macromolecules, 2003. **36**(11): p. 4140-4150.
17. Wen, M., L.E. Scriven, and A.V. McCormick, Macromolecules, 2003. **36**(11): p. 4151-4159.
18. Verdier, P.H. and W.H. Stockmayer, Journal of Chemical Physics, 1962. **36**: p. 227.
19. Verdier, P.H., Journal of Chemical Physics, 1966. **45**(6): p. 2122-&.
20. Herrmann, H.J., D.P. Landau, and D. Stauffer, Physical Review Letters, 1982. **49**(6): p. 412-415.
21. Bansil, R., H.J. Herrmann, and D. Stauffer, Macromolecules, 1984. **17**(5): p. 998-1004.
22. Ghiass, M., A.D. Rey, and B. Dabir, Polymer, 2002. **43**(3): p. 989-995.
23. Manneville, P. and L. de Seze, in *Numerical methods in the study of critical phenomena*, I. Della Dora, J. Demangeot, and B. Lacolle, Editors. 1981, Springer-Verlag: Berlin. p. 116-124.
24. Gimel, J.C., T. Nicolai, and D. Durand, Physical Review E, 2002. **66**(6): p. 64105.
25. Meakin, P. and J.M. Deutch, Journal of Chemical Physics, 1984. **80**(5): p. 2115-2122.
26. Meakin, P., Physical Review Letters, 1983. **51**(13): p. 1119-1122.
27. Meakin, P., Journal of Colloid and Interface Science, 1984. **102**(2): p. 491-504.
28. Meakin, P., T. Vicsek, and F. Family, Physical Review B, 1985. **31**(1): p. 564-569.
29. Kolb, M., R. Botet, and R. Jullien, Physical Review Letters, 1983. **51**(13): p. 1123-1126.
30. Metropolis, N., et al., Journal of Chemical Physics, 1953. **21**(6): p. 1087-1092.
31. Levi, L. and S. Srebnik, Macromolecular Symposia, submitted.
32. Jan, N., Physica A, 1999. **266**: p. 72.
33. Jan, N., D. Stauffer, and A. Aharony, Journal of Statistical Physics, 1998. **92**: p. 325.

379e Influence of Network Structural Modifications on the Dynamic Relaxation Characteristics of Crosslinked Poly(ethylene oxide) Copolymer Membranes

Douglass S. Kalika[1], Jeffrey J. Richards[1], Victor A. Kusuma[2], and Benny D. Freeman[2]. (1) Chemical and Materials Engineering, University of Kentucky, 177 Anderson Hall, Lexington, KY 40506-0046, (2) Chemical Engineering, The University of Texas at Austin, 10100 Burnet Rd., Bldg 133, Center for Energy and Environmental Resources, Austin, TX 78758

The structure and dynamic relaxation characteristics of a series of crosslinked poly(ethylene oxide) copolymer networks based on the UV photopolymerization of bisphenol A ethoxylate diacrylate [BPA-EDA] crosslinker have been investigated. Rubbery amorphous networks with high ethylene oxide content have recently been identified as promising membrane materials for the selective removal of carbon dioxide from light gas mixtures. By strategic control of crosslink density and the introduction of selected pendants along the network backbone it is possible to modify both the solubility and free volume characteristics of the membranes and thereby tailor them for specific gas separations.

Polymer networks were synthesized by UV photopolymerization of commercially-available BPA-EDA crosslinker. Crosslink density within the networks was controlled in two ways: (i) variation of crosslinker molecular weight, with the diacrylate crosslinkers encompassing 4, 8 and 30 total ethylene oxide segments, respectively; and (ii) copolymerization of selected monofunctional ethylene oxide acrylates into the network backbone. Copolymerization results in the insertion of fixed-length pendants into the network structure and a corresponding reduction in crosslink density. Variation in the size and end group associated with these pendants provides additional variables by which to tune the chemical composition, architecture and free volume of the resulting membrane networks.

In this study, dynamic mechanical analysis and broadband dielectric spectroscopy have been used to investigate the sub-glass and glass-rubber viscoelastic relaxation characteristics of the networks in order to elucidate the underlying motional origins of each relaxation and their correlation with the structural and morphological features of the materials. These complementary techniques provide detailed information with respect to relaxation intensity and breadth, and the influence of crosslink constraint on the local and segmental motions inherent to the materials. Based on these studies, a series of design rules have been established to guide the formulation of the networks for optimum gas transport performance. The gas transport characteristics of the BPA-EDA networks are compared to other families of crosslinked PEO membranes with similar properties, and the overall performance of these materials are interpreted within the context of the selectivity-permeability "upper bound" concept.

379f Mechanical Properties of the Poly (HEMA-co-NVP) Hydrogel: Molecular Dynamics Simulation Approach

Seung Soon Jang[1], Seung Geol Lee[2], Giuseppe Brunello[1], David Bucknall[3], and Hannah Lee[3]. (1) School of Materials Science and Engineering, Georgia Institute of Technology, 771 Ferst Drive, Love Building, Atlanta, GA 30332-0245, (2) School of Polymer, Textile and Fiber Engineering, Georgia Institute of Technology, 801 Ferst Drive, MRDC1, Atlanta, GA 30332-0295, (3) Polymer, Textile, and Fiber Engineering, Georgia Institute of Technology, 801 Ferst Drive, MRDC1, Atlanta, GA 30332-0295

Poly (hydroxyethyl methacrylate-co-N-vinyl-2-pyrrolidone), poly(HEMA-co-NVP) is a well-balanced biomaterial that has high hydrophilicity and less toxicity on the support of proper mechanical strength. In this study, we investigated the mechanical and transport properties of poly(HEMA-co-NVP) using fully-atomistic molecular dynamics simulation. After equilibrating the poly(HEMA-co-NVP) network structure in the presence or absence of water, we deformed the systems by applying the uniaxial compression mode up to 80 % in our 2-ns MD simulation. In comparison to the reference system consisting of only linear chains in the absence of crosslinking point, the crosslinked poly(HEMA-co-NVP) network shows higher level of stress as a function of strain. We also investigated the effect of monomeric sequence consisting of HEMA and NVP on the mechanical property of the hydrogel by comparing the random sequence and blocky sequence. Interestingly, the system with the random sequence showed higher stress than that with the blocky sequence. We will discuss the structural aspects affecting the mechanical properties.

379g Fast Dynamics of Semiflexible Chain Networks of Self-Assembled Peptides

Monica C. Branco[1], Florian Nettesheim, Joel P. Schneider[2], and Norman J. Wagner[1]. (1) Department of Chemical Engineering, University of Delaware, 100 Academy St., Newark, DE 19716, (2) Department of Chemistry and Biochemistry, University of Delaware, Lammont Dupont Lab, Newark, DE 19716

The cytoskeleton within cells is comprised of several biopolymers, which include actin, microtubules, and intermediate filaments. Rheological and spectroscopy techniques show that these polymer chains are semiflexible and form networks with unique viscoelastic properties, such as high storage moduli at low monomer concentrations and significant strain hardening at low strain. Recently, synthetic polymers have been developed with similar semiflexible characteristics. One example is a synthetic β-hairpin peptide analogue, MAX1, which undergoes triggered self assembly at the nanoscale to form a physically crosslinked network of fibrils with a defined cross-section. These peptides adopt a random coil conformation in aqueous solutions and are freely soluble. However, when subjected to a stimulus, such as physiological pH and ionic strength, the peptides fold into a β-hairpin, and subsequently, self-assemble to form a structurally rigid hydrogel stabilized by non-covalent cross-links. MAX8 is an alternate peptide sequence derived from MAX1, in which faster folding and self assembly kinetics are observed at the same buffer conditions, resulting in more rigid gels at the same concentration of peptide. Here, we present the first quantitative neutron spin echo (NSE) measurements of self-assembling peptide hydrogel networks to study their dynamics on the nanometer and nanosecond length and time scales. These measurements demonstrate that these self-assembled peptide fibrils can be described by the theory of semiflexible chains on these length and time scales. Alteration of peptide sequence does not affect the nanoscale dynamics, but does significantly affect the macroscopic rheology. We conjecture that the difference arises from the rate of assembly and the degree of branching, rather than a change in the fibril structure or persistence length.

396a Viscoelastic Modeling of Porous Matrices Used In Tissue Engineering

Rahul D. Mirani and Sundararajan V. Madihally. School of Chemical Engineering, Oklahoma State University, 423 Engineering North, Stillwater, OK 74078

Many naturally obtained matrices such as small intestinal sub mucosa and synthetic matrices are utilized in tissue regeneration to support the in growth of cells to be colonized. Recent advances have shown the elastic property of the porous structure play a significant role in cell colonization. However, mechanical properties of the porous structures are not well understood, particularly the viscoelastic properties. In this study we explore the viscoelastic properties of both the natural matrix SIS and synthetic matrices made of chitosan/gelatin supported by 50:50 PLGA layer. First, the uniaxial tensile tests done on the composite in hydrated medium at 37°C at a controlled cross head speed of 10 mm per minute and a fixed load of 100 N show that it can be elongated up to nearly 400 % of its initial length, nearly 8 times more than SIS.

In order to understand the behavior of the composite material under relaxation, stress relaxation tests are performed. The specimen is subjected to a predetermined amount of increasing ramp strain inputs, consisting of a loading part wherein the strain is applied at the rate of 3.125 % per second for 16 s and a relaxation part where the material is allowed to relax for 100 seconds. Strain rates ranging from 0.9375 % to 3.125 % for two different relaxation times of 60 s and 100 s were used for the test. Similar experiments are performed on the SIS and the output stress relaxation behavior was compared. A lower range of the strain rate had to be used for the SIS because it fails at the higher ranges. It was also observed that the SIS has a higher maximum stress value which shows that it accumulates more stress than the composite. Also for a similar time range the composite relaxed to a much greater extent than the SIS and accumulated less stress. Thus the composite was able to relax more naturally as compared to the SIS with less deformation in the long run. To quantify the relaxation behavior of the matrices, the Quasi Linear Viscoelastic (QLV) model is used and the data obtained from the model is compared to the experimentally obtained stress values for both the loading as well as the relaxation part of the stress relaxation curve. The QLV model is modified to incorporate the fact that the loading was done in a finite amount of time and not instantaneously as assumed by Fung YC. The data obtained from feeding the model equation in MAT LAB matched the experimentally obtained data with an average deviation of 3%. Thus the QLV model is successful in quantifying the nonlinear behavior of the composite.

Enhancing the Biomechanical Properties of Chitosan Scaffolds for Tissue-Engineered Heart Valves

Mohammad Zaki Albanna[1], **Therese H. Bou-Akl**[2], **Henry L. Walters III**[2], and **Howard W. T. Matthew**[1, 3]

(1) Department of Biomedical Engineering, Wayne State University, 5050 Anthony Wayne Drive, Detroit, MI 48202,

(2) Children's Hospital of Michigan, 3901 Beaubien Blvd., Detroit, MI 48201.

(3) Department of Chemical Engineering and Materials Science, Wayne State University, 5050 Anthony Wayne Drive, Detroit, MI 48202

Introduction:

Previous tissue engineering approaches to create heart valves constructs have been limited by the structural immaturity and mechanical properties. Tissue engineered heart valves constructs have been successfully implanted at the pulmonary position in lambs where the construct is subjected to smaller loads compared to the aortic position. However, the constructs still lack the mechanical integrity needed to withstand systemic pressures. Therefore, several studies aim to improve the biomechanical properties and tissue formation of the engineered heart valve via mechanical conditioning of the cell-seeded scaffold in a bioreactor system. Chitosan is a biocompatible, biodegradable linear polymer, and due to its particular properties, it is currently being used in a number of tissue engineering applications. It offers a wide variety of mechanical properties depending on the molecular weight (MW), degree of deacetylation (DD), and molecular architecture [1]. Our research seeks to enhance the biomechanical properties of chitosan-based heart valve tissue engineered scaffolds. In this study, our approach involves embedding various weight percentages of high molecular weight (HMW) chitosan fibers in a porous heart valve scaffold made out of medium molecular weight (MMW) chitosan. Cell attachment and viability is also investigated throughout this study.

Materials and Methods:

Chitosan fibers were made out of 1.5 wt% HMW chitosan dissolved in 0.2 M acetic acid by extruding the solution through a 26-gauge (i.e. 0.45 mm diameter) teflon catheter directly into 10 vol% ammonia solution. The gelled chitosan fibers were then air-dried at room temperature.

Test scaffolds were made out of 2 wt% MMW chitosan dissolved in 0.2 M acetic acid. Chitosan fibers (2-10 mm length, 40 µm diameter) were blended into the solution at four different weight percentages (0, 20, 30, and 40 wt %). Scaffolds were generated by freezing the fiber-containing solutions in 60 mm Petri dishes at -70°C, followed by lyophilization for 24 hours, and rehydration in 1% ammonium hydroxide an then washed with Phosphate Buffer Saline (PBS). Pore sizes ranged from 180 to 240 µm. Each scaffold was cut into 2x5x20 mm rectangular samples. Mechanical properties of the samples were evaluated by monotonic tensile testing. All samples were tested in a fully hydrated condition. The maximum load, ultimate tensile strength (UTS), maximum strain, and modulus of elasticity were evaluated.

To evaluate the biocompatibility of the fiber material, HMW chitosan films were cast with a different thickness in culture wells. The desired volume (i.e. 50 µl, 150 µl, or 250 µl) of 1.5 wt% HMW chitosan was distributed homogenously in 35 mm culture wells and left to dry overnight. After drying, the films were neutralized by 0.2 M sodium hydroxide, and then washed with PBS. The films were sterilized by immersion in 70% ethanol overnight and then washed with sterile PBS. The films were then seeded with valvular interstitial cell (VIC's) at a density of 8000 cell/cm^2. The cells were cultured in M199 medium that containing Fetal Bovine Serum (FBS) and Gentamycin. Cell attachment, viability and growth was then followed by phase contrast microscopy and Calcein AM fluorescence at regular intervals.

Results and Discussion:

Tensile testing results showed that addition of 40 wt% chitosan fibers (2-3 mm length) improved scaffold tensile strength approximately 3 fold from 45±18 KPa to 139.5±40 KPa (Figure. 1), and the Modulus of Elasticity from 0.07±0.021 MPa to 0.42±0.07 MPa. Statistical analysis (student t-tests-paired two samples for mean) confirmed the significance of these improvements ($p<0.004$). The values for the ultimate tensile strength and the modulus of elasticity were substantially lower than those reported in the literature for the aortic valve. Stradins and colleagues [2] reported, based on their experimental results, that the tensile strength of aortic valve cusps ranges from 0.32±0.04 MPa to 1.74±0.29 MPa, while the modulus of elasticity ranges from 1.98±0.15 MPa to 15.3±3.8 MPa respectively.

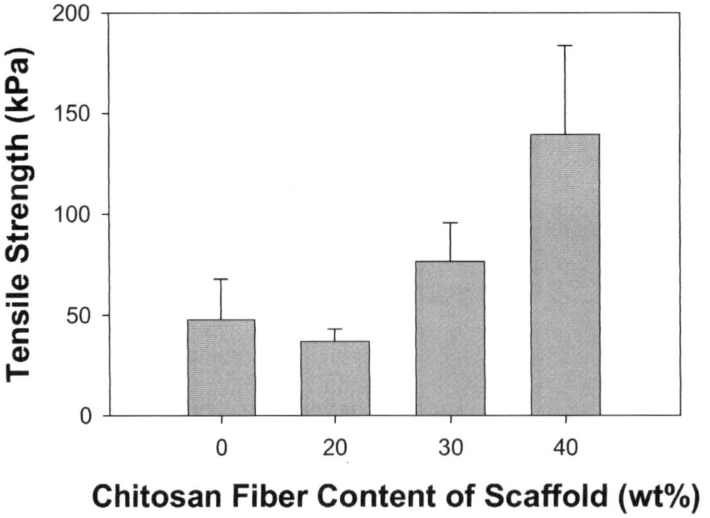

Figure. 1: Effect of chitosan fiber content on scaffold strength.

Figure 2 shows Calcein AM fluorescence microscopic images after 3 days of VIC's cultured on 3 chitosan cast films with increasing thickness. The thicker films show greater cellular attachment than the thinner ones, and in each condition, most of the cell attachment occurred at the periphery of the wells. Attachment of cells at the center of the well was either minimal or absent. Since the cast films are known to be thicker near the container walls (a result of the fluid surface tension phenomenon), this observation suggests that the difference in cell adhesion may reflect a film thickness-induced change in film microstructure of chitosan membranes between the center and periphery of the well. This thickness effect makes the chitosan more amorphous towards the periphery. The reduced crystallinity may alter either the amount or conformation of adhesion proteins adsorbed onto the film surface, thus enhancing the cellular attachment.

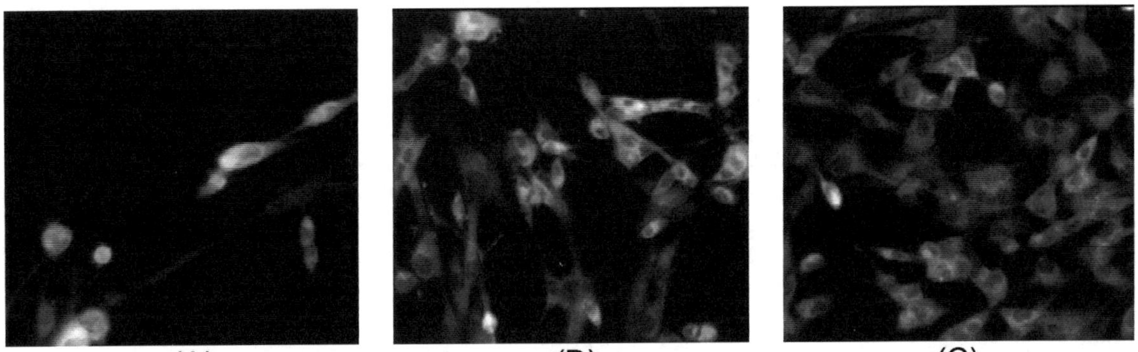

(A) (B) (C)

Figure. 2: Phase contrast microscopic images at 4X magnification of Chitosan membranes with different thicknesses depending on the added volume as follow: (A): 50µl, (B): 150 µl, and (C) 250 µl

These findings of biomechanical properties and the cell studies show that fiber reinforcement of porous chitosan scaffolds can significantly improve baseline scaffold properties.

Acknowledgements

This research was partially funded by the Perry Foundation through a grant to the Cardiovascular Surgery Department of Children's Hospital of Michigan.

References

[1] Madihally SV, Matthew HW, Porous chitosan scaffolds for tissue engineering. Biomaterials, 20(12), 1133-42, June 1999

[2] Stradins et al. Comparison of biomechanical and structural properties between human aortic and pulmonary valve. European Journal of Cardio-thoracic Surgery 26(2004)634-639.

396c Plasma Enhanced Chemical Vapor Deposited Poly (2-hydroxyethyl methacrylate) for Fabricating a Degradable, Biocompatible Intestinal Tissue Culture Substrate

Courtney A. Pfluger, Chemical Engineering, Northeastern University, 342 Snell Engineering Center, Boston, MA 02115, Rebecca L. Carrier, Department of Chemical Engineering, Northeastern University, 342 Snell Engineering Center, 360 Huntington Avenue, Boston, MA 02115, and Daniel D. Burkey, Northeastern University, 342 Snell Engineering Center, 360 Huntington Avenue, Boston, MA 02115.

It has been shown that, in the study of oral drug transport, epithelial cells cultured on a chemically biomimetic substrate more accurately reflect in vivo cellular behavior [1,2]. It is expected that biomimetic nano- and micro-topology will also influence intestinal epithelial cell phenotype, and specifically the expression of proteins important to drug uptake and metabolism, because these proteins are expressed differentially along the villus-crypt axis of the intestinal basement membrane in vivo [3]. To test this hypothesis, we are fabricating biomimetic molds of sections of the small intestine and examining the attachment, migration and protein expression of intestinal epithelial cells on the biomimetic scaffold.

The topology of the intestinal basement membrane was replicated using a plasma CVD technique to create a film with micron scale thickness and with micro- and nano-meter scale features on a section of pig intestine to be ultimately used as a cell culture scaffold. The biodegradable polymer deposited was poly(2-hydroxyethyl methacrylate) (pHEMA). pHEMA's porosity can be controlled by adjusting the degree of cross-linking, making it ideal for drug transport studies and as a material for tissue scaffolds [4]. FT-IR analysis confirmed the existence of key functional groups in poly(2-hydroxyethyl methacrylate) after plasma deposition. Carbonyl peak intensity confirmed film cross-linking at varying degrees. Cross-linking density affected the degree of swelling and extent of degradation of the film over time.

Analysis of cell attachment, growth, and metabolic activity demonstrated substrate biocompatibility for intestinal epithelial cell culture. Comparisons of cell attachment on CVD cross-linked pHEMA verses other culture substrates such as collagen, PDMS (alone and coated with fibronectin), and glass were performed. Cells adhered to high cross-linked pHEMA just as well as collagen, polystyrene and PDMS coated with fibronectin and better than to glass and PDMS alone. This confirms that cross-linked pHEMA can be used as a scaffold with the same cell adhesion properties as polystyrene, collagen, and PDMS. CVD pHEMA offers the advantage over these materials of forming conformal films with biomimetic topography in a dry, one step process.

In summary, we have demonstrated a plasma CVD process for fabricating a biocompatible pHEMA polymer film suitable for cell culture and demonstrated that we can alter its properties via variable cross-linking. We have shown that cross-linked pHEMA can be deposited onto the intestinal basement surface and retain micro- and nano- scale topological features. The next step will be to use the cross-linked pHEMA conformal, biomimetic scaffold for cell culture and protein expression experiments, and compare resulting cultures to those on flat cell culture substrates.

1. P. Artursson, Critical Reviews in Therapeutic Drug Carrier Systems 1991, 8(4), 305-330.

2. B. H. Stewart, O. H. Chan, N. Jezyk, D. Fleisher, Advanced Drug Delivery Reviews 1997, 23(1 3), 27-45.

3. N. Cigna, C. Nicoleeti, A. Durand, J. C. Chaix, T. Giardina, J. Perrier, Cell Biology International 2007, 31, 966-973.

4. H. G. Pryce Lewis, K.K.S. Lau, Y. Mao, K.K. Gleason, Annual Technical Conference Proceedings - Society of Vacuum Coaters 2005, 48th, 90-94.

396d Designing 3D Photopolymer Gels to Regulate Biomechanical Cues

Garret Nicodemus, Idalis Villanueva, and **Stephanie J. Bryant**. Chemical and Biological Engineering, University of Colorado, Campus Box 424, Boulder, CO 80309

Synthetic hydrogels fabricated from photopolymerization are attractive for tissue engineering due to the ability to finely tune their macroscopic properties, incorporate biological functionalities, and encapsulate cells. In designing hydrogels for tissue engineering applications where mechanical forces are prevalent, the gel structure and its chemistry will play a key role in translating the applied mechanical forces into biochemical cues. Therefore, the goal of the present study was to exploit the attractive features of synthetic hydrogels to elucidate the role of gel structure and chemistry in regulating biomechanical cues for cartilage tissue engineering.

Photopolymerized poly(ethylene glycol) (PEG) hydrogels were employed as a base system to study cell response to loading in the absence of cell-material interactions. To study mechanoreceptors, matrix analogs (e.g., RGD, a common cell adhesion ligand) were systematically incorporated into the PEG gels. Chondrocytes were encapsulated in PEG and PEG+RGD hydrogels and subjected to dynamic compressive strains. Cell response was measured by anabolic and catabolic gene expression as well as matrix production.

In the absence of loading, anabolic gene expression by chondrocytes encapsulated in PEG gels was stimulated (e.g., aggrecan expression increased 6-fold) while catabolic gene expression was significantly inhibited (e.g., matrix metalloproteinase-3 decreased by 2 orders of magnitude) by day 10 of culture. When PEG constructs were subjected to dynamic loading (0.3 Hz, 1 hr on/1hr off for 12 hrs, 12 hrs off), both anabolic and catabolic gene expression were enhanced (e.g., aggrecan and MMP-3 increased by 2.2 and 7-fold, respectively, by day 7 of culture) suggesting that loading may induce remodeling of the newly deposited tissue. The addition of RGD into PEG gels up to 0.4 mM RGD did not affect chondrocyte phenotype as measured by the ratio of collagen II/I in the absence of loading. The addition of 0.4 mM RGD led to a significant reduction in proteoglycan synthesis compared to PEG gels in the absence of loading. However, when PEG+RGD gels were subjected to dynamic loading (0.3 Hz, continuous for 2 days), proteoglycan synthesis was stimulated 2-fold with 0.4 mM RGD compared to unloaded PEG+RGD gels suggesting that RGD is acting as a mechanoreceptor to stimulate matrix production.

Together our results demonstrate that loading and the incorporation of matrix analogs have a profound effect on chondrocyte response. We are currently exploring a range of loading regimes and other matrix analogs (e.g., glycosaminoglycans) toward elucidating biomechanical cues that stimulate tissue growth.

396e Multifunctional Tissue Engineering Scaffolds Via Organic Sol-Gel Chemistry

Hugh Lippincott[1], Poonam Borgaonkar[2], Sachin Sharma[2], Ming Chen[3], Sankha Bhowmick[3], and **Daniel F. Schmidt**[2]. (1) Biomedical Engineering & Biotechnology, University of Massachusetts Lowell, 1 University Avenue, Lowell, MA 01854, (2) Plastics Engineering, University of Massachusetts Lowell, 1 University Avenue, Lowell, MA 01854, (3) Mechanical Engineering / Biomedical Engineering & Biotechnology Program, University of Massachusetts Dartmouth, 285 Old Westport Road, Dartmouth, MA 02747

The need for multifunctional tissue engineering scaffolds with well-defined structure-composition-properties relations cannot be underestimated. Current methodologies for producing scaffolds generally involve attempts to produce pore structures appropriate for the applications in mind in a limited range of compositions. While effective, it is difficult to prepare materials with arbitrary chemistries or independently vary composition and structure via most of these techniques, making the establishment of structure-composition-properties relations difficult.

We present a new approach that attempts to address some of these issues through the application of organic sol-gel chemistry. Here, materials are prepared via polymer network formation in solution, with polymerization-induced phase-separation giving rise to fully open, interconnected pore networks. Both polyester and polyurethane chemistries are explored, with materials analyzed via mercury intrusion porosimetry, scanning electron microscopy, dynamic mechanical analysis and cell culture studies with NIH 3T3 mouse fibroblasts.

First, we briefly review our past work to show that porous polymer networks with a range of compositions may be produced via this technique, pore size and porosity may be varied independent of composition, and cytocompatibility may be achieved with an appropriate combination of structure and composition [1]. Second, we demonstrate that the cytocompatible pore structures of natural porous bodies may be replicated in our sol-gel systems via a modified replamineform process [2]. Finally, we show that useful shape-memory characteristics and therapeutic agents may be incorporated into these scaffolds via appropriate compositional choices. In sum, we believe that this approach may be promising as a simple and versatile means of producing multifunctional tissue engineering scaffolds and better understanding their performance.

1. 'A Flexible Approach to the Preparation of Polymer Scaffolds for Tissue Engineering', P. Borgaonkar, S. Sharma, M. Chen, S. Bhowmick, D. F. Schmidt, Macromolecular Biosciences, 7, 2, 2007, 201-207.

2. 'Carbonate Minerals as Precursors of New Ceramic, Metal, and Polymer Materials for Biomedical Applications', J. N. Weber, E. W. White, Minerals Science and Engineering, 5, 2, 1973, 151-165.

A Fiber-Reinforced, Large Vessel, Chitosan Scaffold for Pediatric Applications

Irina S. Robu[1], Henry L. Walters III[2], Howard W.T. Matthew[1]
[1]Department of Chemical Engineering & Materials Science, Wayne State University, Detroit, MI
[2]Department of Cardiovascular Surgery, Children's Hospital of Michigan, Detroit, MI

Introduction

Congenital heart and large vessel defects are a significant cause of morbidity and mortality in pediatric patients. Current vessel replacement options lack the capacity to grow and remodel in response to patient needs. This limitation is a major problem for pediatric patients who may require multiple surgical procedures before reaching adulthood. In this study, we explored the prospect of a tissue engineered large vessel graft using polysaccharide biomaterials. Our scaffold design was based on the use of chitosan as the main structural biopolymer, with cell adhesion ligands provided by blending Type I collagen with the chitosan. In addition, we explored the effects of covalently grafting various glycosaminoglycans to the scaffold surfaces to promote binding and presentation of growth factors and extracellular matrix proteins. The aim of the present study is to evaluate a tissue engineered blood vessel in which the extracellular matrix production, collagen and cellular growth can be enhanced and produced similar to natural artery.

Materials and Methods

Scaffold Fabrication

Porous tubular scaffolds with an inner diameter of 12 mm, an outer diameter of 14 mm, and a length of 7 cm were produced by freeze drying a solution of 1.5 wt% chitosan and 0.15 wt% type I collagen in 0.2 M acetic acid within an annular mold containing preformed chitosan fibers. After freeze-drying, scaffolds were rehydrated through an ethanol series and then finally equilibrated in PBS. To improve the cell proliferation and attachment, the scaffolds were then mounted in a perfusion system and surface-derivatized by perfusing carbodiimide-activated heparin through the porous structure. The final heparin immobilization level was 0.2 mg heparin per mg of chitosan. Scaffolds were then washed with PBS and sterilized by immersion in 80 % ethanol for 24 hours. They were then washed with PBS and equilibrated overnight with serum-supplemented culture medium.

Cell Seeding and Scaffold Bioreactor Culture

Porcine aortic smooth muscle cells (SMC) were suspended in 20 ml of culture medium and 30 million cells were seeded into each scaffold by sealing on end of the scaffold and perfusing the cell suspension into the lumen and out through the porous wall. The cell suspension was recirculated for 30 min at a flow rate of 50 ml/min. This resulted in the wall-entrapment of most of the cells. The scaffolds were then placed in 10 cm culture plates and incubated in culture medium for 4 days. The culture medium was MCDB 131 supplemented with amphotericin B, FGF-2, EGF, insulin, 10% FBS and penicillin-streptomycin. During perfusion culture, the medium was additionally supplemented with 0.3 mM ascorbic acid 2-phosphate and 0.1 mM retinoic acid. The seeded scaffolds were then transferred to a perfusion bioreactor system where pulsatile flow through the construct lumen was initiated at a flow rate of 50 ml/min for 4 days. The flow rate was subsequently increased to 100 ml/min and the culture medium supplementation with 0.1 mM retinoic acid and 0.3 mM ascorbic acid was initiated and maintained for 12 days. Samples were excised from the perfused vessel scaffold at intervals and used to evaluate cell viability and distribution.

Scaffold Analyses

Cell viability was evaluated in scaffold samples using using Calcein red-orange fluorescence.
For scanning electron microscopy (SEM), the scaffold samples were fixed in 2.5 % glutaraldehyde in PBS, dehydrated through an ethanol series, and then flash frozen in liquid nitrogen and lyophilized. After sputter coating with gold samples were examined under an SEM. For histology, scaffolds were

fixed in 10 % buffered formalin, dehydrated and embedded in paraffin. Sections 8-10 μm in thickness were cut and stained with hematoxylin and eosin.

Results and Discussion

Scanning electron microscopy (Figure 1) revealed that the luminal surfaces of heparin-chitosan-collagen after 16 days of growth were covered with smooth muscle cells that were oriented in the direction of the flow.

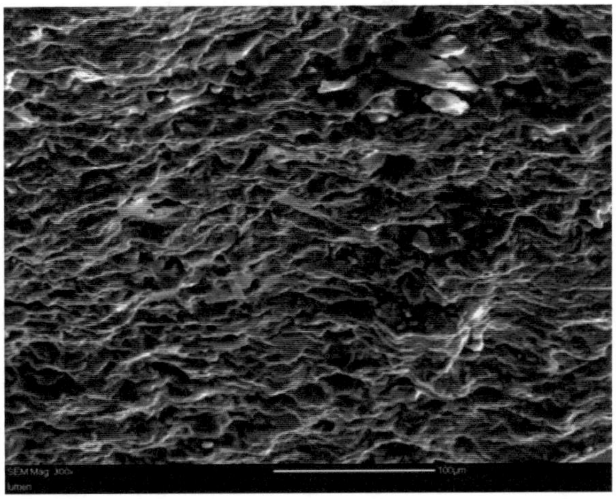

Figure 1. SEM of lumenal surface of heparin-chitosan-collagen scaffolds seeded with porcine aortic smooth muscle cells.

Histology (Figure 2) of the heparin-chitosan-collagen scaffold show clusters of cells present within the scaffold, with the highest cell densities occurring near the scaffold periphery. However, coverage was incomplete and a substantial increase in the amount of cells is needed to fill the scaffold void space.

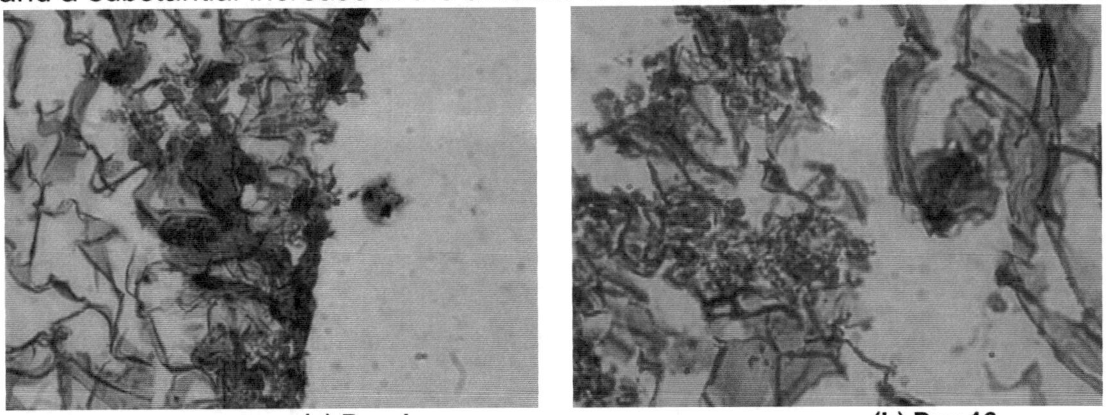

(a) Day 4 **(b) Day 16**

Figure 2. H & E stained sections of the scaffold on Day 4 of the culture showing SMC near the scaffold periphery

Calcein red-orange fluorescence showed spreading and allignment of SMC. In particular, the cells appeared to align in the direction of flow (Figure 3). shows SMC forming clusters within the scaffold and their viability increased. Figure 3c and 3d shows the cells lining up with the flow and forming arranged clusters. The viability examination of the cells showed that heparin-chitosan-collagen is a viable scaffolding material for engineering large diameter vascular grafts.

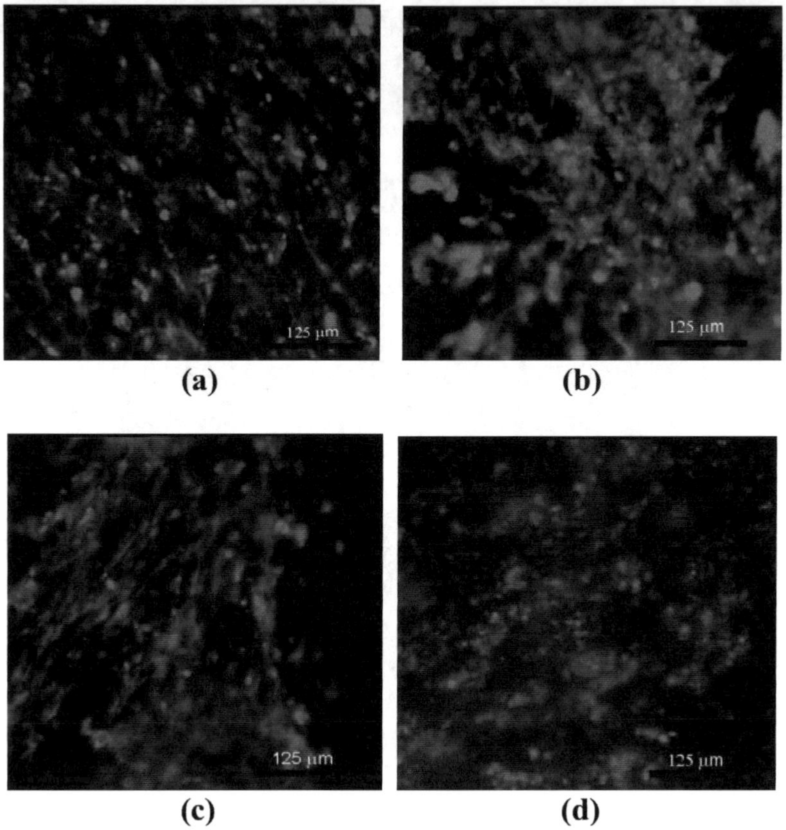

Figure 3. Calcein red-orange fluorescence indicating cell viability scaffold samples taken from the perfusion system on (a) day 2, (b) day 4, (c) day 13 and (d) day 16.

Conclusions

Scanning electron microscopy showed the presence of SMC on the luminal surface, histology examination showed the radial distribution of cells and their penetration within the heparin-chitosan-collagen graft. The graphs were examined regularly for cell viability, and showed high cell viability and aligment of SMC cells in the direction of the flow. We believe that the perfusion seeding produced better cell distribution within the scaffold. However, higher seeding densities are needed to accelerate tissue formation and extracellular matrix deposition.

Acknowledgements

This research was partially funded by the Perry Foundation through a grant to the Cardiovascular Surgery Department of Children's Hospital of Michigan.

396g Accelular Dermis Promotes Neovascularization and Epidermal Regeneration: Implications for Wound Healing

Liana M. Lugo, Surgery, University at Buffalo, State University of New York, Buffalo, NY 14203 and **Stelios T. Andreadis**, Chemical and Biological Engineering, SUNY-Buffalo, 908 Furnas Hall, Buffalo, NY 14260-4200.

Despite significant advances in management of severe wounds such as burns and chronic ulcers, autologous split-thickness skin grafts are still the gold standard of care. The main problems with this approach include pain and discomfort associated with harvest of autologous tissue, limited availability of donor sites and the need for multiple surgeries. Although tissue engineering has great potential to provide alternative approaches for tissue regeneration several problems have hampered progress in translating technological advances to clinical reality. Specifically, engineering of skin substitutes requires long culture times and delayed vascularization after implantation reduces graft survival. To address these issues we developed a novel two-prong strategy for tissue regeneration in vivo: (i) vascularization of acellular dermal scaffolds by infiltration of angiogenic factors; (ii) generation of stratified epidermis by direct delivery of epidermal keratinocytes in vivo. Using athymic mouse as a model system we found that gene delivery from acellular human dermis enhanced the density and diameter of infiltrating blood vessels. Surprisingly, cell delivery resulted in fully stratified epidermis with normal spatiotemporal expression of differentiation markers including keratin-10, involucrin and fillagrin. This strategy promoted tissue regeneration in vivo with no need for engineering skin tissue in vitro and therefore, it may be useful for treatment of major wounds when skin donor sites are scarce and rapid wound coverage is required.

396h Characterizing the Effect of 3D Porous Structure on Flow Properties In Tissue Engineering Scaffolds

Benjamin J. Lawrence and Sundararajan Madihally. School of Chemical Engineering, Oklahoma State University, 423 Engineering North, Stillwater, OK 74078

Biomaterials for tissue engineering are commonly constructed from degradable 3D porous matrices which provide support and guidance for cellular colonization and ingrowth. In a reactor in order to regenerate large sections of tissue required for bladder, skin, or blood vessel regeneration the cell seeded construct must be matured in a bioreactor. The flow distribution within the bioreactor affects cellular colonization and extracellular matrix (ECM) production by controlling the distribution of nutrients as well as the shear stress profiles within the reactor. Changes to the porous structure alter both the pressure required for perfusion through the material and the shear forces present within the structure. Therefore, this study focused how changing the features of the material's 3D porous architecture affected both the pressure drop across the structure and the shear stresses within the structure. The flow distribution, pressure drop, and shear stresses within a thin rectangular reactor (2.5 cm × 8 cm × 0.2cm) were evaluated using computational simulations and bioreactor experiments. The simulations were performed using commercially available software, CFX 11 (ANSYS Inc, Canonsburg, PA.) and/or Comsol 3.3 Multiphysics (COMSOL, Inc., Burlington, MA). In order to account for both viscous and drag forces, flow through the porous structure was modeled using the Brinkman Equation, which characterizes the porous structure using pore size and the number of pores per unit area. The effect of matrix construction was modeled by changing both pore size and pores per area, the number of pores per area increases as the pore size shrinks, and the effect of cellular ingrowth and ECM production was modeled by keeping the pores per area constant while shrinking the pore size. A variety of values were tested ranging from 200 μm pores, 15 pores/mm^2 to 10 μm pores, 1500 pores/mm^2. The values centered on the experimentally measured value of 85 μm pores, 120 pores/mm^2 (determined from chitosan scaffolds freeze dried at -80°C). Both computational and experimental results showed expected flow profiles with increased pressure drop and shear stress as pore size decreased. This modeling strategy allows better understanding of the effect of 3D porous architecture on bioreactor properties, and shows promise in enhancing the design of tissue engineering materials.

420a Design and Assembly of Anisotropic Particles: The Shapes of Things to Come (With a Little Help from Computer Simulation)

Sharon C. Glotzer, Chemical Engineering and Materials Science and Engineering, University of Michigan, 2300 Hayward Street, Ann Arbor, MI 48109

Materials comprised of "designer" nanoscale building blocks organized into structures with novel functionality represent the next frontier for materials research and development. Breakthroughs in synthesis and fabrication techniques over the past few years have produced an exquisite library of building blocks of nearly arbitrary shape and composition. Still, control of the assembly of these building blocks into target structures is required for these new "atoms" and "molecules" to form the foundation for next-generation materials in applications ranging from energy to medicine. Theory, modeling and simulation are critical tools in this quest. Needed for progress are useful schemes for classifying anisotropy and its impact on assembly; insight into synthesis and fabrication methods for producing designed particles; efficient computational methods for predicting assemblies of building blocks with arbitrary anisotropy; and new methods for characterizing the degree and type of order in assembled phases of anisometric shapes. In this talk* **, we present examples of computer simulation studies of self-assembly of a rich array of structures from "patchy" and "tetnered" nanoparticles and colloids that demonstrate how various anisotropy attributes may be exploited and combined for self-assembly. We describe a new computational algorithm that rapidly approximates ground state structures of strongly-interacting particles with arbitrary anisotropy, and which shows how small changes in particle patchiness and shape can produce large changes in macrostructure. Finally, we introduce new order parameters schemes for characterizing the local translational and orientational order of assembled phases formed from complex shapes.

* In collaboration with A.H.Akbari, E.R. Chan, T. Chen, P. Ghorai, M.A. Horsch, C.R. Iacovella, E. Jankowski, A.S. Keys, T.D. Nguyen, C. Phillips, A. Santos, T. Sheh, C. Singh, S. Teich-McGoldrick, X. Zhang, Z.L. Zhang, and N.A. Kotov, M.J. Solomon, and F. Stellacci.

**Funded by DoE, NSF, DoEd, AFOSR, NASA, and the McDonnell Foundation

420b Computational Nanotribology - Understanding Lubrication at the Nanoscale Using Molecular Simulation

Peter T. Cummings, Yongsheng Leng, and Hugh Docherty. Department of Chemical Engineering, Vanderbilt University, Nashville, TN 37235-1604

To be added

420c Plasma-Surface Interactions and the Control of Nanostructure

David B. Graves, UC Berkeley, Department of Chemical Engineering, Berkeley, CA 94720

Low temperature, nonequilibrium ionized gas plasmas are powerful tools for creating nanostructures on surfaces, especially over large areas. However, plasma-surface interactions are notoriously complex with many species interacting synergistically over disparate length and time scales. In this talk, I will review some of the opportunities and challenges in creating and controlling surface nanostructures with plasmas, with a focus on selected topics including nano-scale features in plasma etch and the challenges of using organic polymer masks for etching.

420d Synthetic Scaffolds for Tissue Engineering

Antonios G. Mikos, Departments of Bioengineering and Chemical and Biomolecular Engineering, Rice University, 6100 Main Street, MS-142, Keck Hall, Room 116, Houston, TX 77005-1892

Recent developments in biomaterials technology have enabled the production of biodegradable, cytocompatible tissue engineering scaffolds that can be tailored with appropriate mechanical and biological signals to restore damaged tissue in specific sites in the body. For instance, our laboratory has developed a class of fumarate-based polymers for the controlled delivery of cells and/or bioactive agents to improve tissue repair. We have applied this technology to fabricate biodegradable synthetic scaffolds from a hydrophilic fumarate-based polymer, oligo(poly(ethylene glycol) fumarate), and load bearing scaffolds from a hydrophobic fumarate-based polymer, poly(propylene fumarate). These synthetic scaffolds have been widely investigated as injectable carriers for marrow derived stem cells and/or controlled bioactive factor delivery to promote regeneration of orthopedic tissues. Additionally, composites of fumarate-based polymers and nanomaterials have been explored to provide a means of increasing the mechanical integrity of synthetic tissue engineering scaffolds. This talk will present examples of fumarate-based materials from our laboratory to illustrate significant recent advances in synthetic scaffold technology for tissue engineering.

420e The Road to Low-Cost and High-Efficiency Solar Cells Via Self-Assembled Nanomaterials

Hugh W. Hillhouse, School of Chemical Engineering and Energy Center at Discovery Park, Purdue University, 480 Stadium Mall Drive, West Lafayette, IN 47907

Developing economic and green methods to supply our future energy needs is perhaps the grand challenge of our time. Due to its abundant and distributed supply, solar energy may play a key role in this revolution. However, limitations in cost and efficiency have hindered solar photovoltaic energy conversion from supplying a large fraction of our energy. The presentation will focus on our progress towards solving the key challenges to decrease the cost and increase the efficiency of photovoltaic energy conversion by developing new nanomaterials and devices. In particular, I will discuss our recent developments of a new low-cost route to solar cells based on nanocrystals and a new method for forming nanowire solar cells designed to break the Shockley-Queisser limit (33%, the upper limit of energy conversion for a conventional single junction solar cell). For the latter, the formation of monodisperse diameter low-bandgap semiconductor quantum wires is a key challenge. However, as will be shown in the presentation, this can be overcome by new methods of nanofabrication based on self-assembly.

427a Dispersion of Semiconductor Nanoparticles In a Polymer Matrix:a Fluorescence Energy Transfer Study

Shyam V. Vaidya[1], M. Lane Gilchrist[1], Charles Maldarelli[2], and **Alexander Couzis**[1]. (1) Chemical Engineering, City College and the Graduate Center of the City University of New York, Steinman Hall, 140th St @ Convent Ave, New York, NY 10031, (2) Levich Institute and Chemical Engineering, The City College of New York, The School of Engineering, T-1M15Steinman hall, 140th street and Convent avenue, New York, NY 10031

The focus of this presentation is the development of optically barcoded polymer beads for use in high-throughput, multiplexed screening applications such as protein microarrays or flow cytometry. Luminescent semiconductor nanoparticles (or quantum dots (QDs)) with different emission wavelengths (colors), and incorporated in different compositions in polystyrene (PS) beads are used to define an optical barcode. The incorporation is undertaken by copolymerizing the PS beads with hydrophobically capped, core-shell, CdSe/ZnS QDs, using a spraying suspension polymerization procedure. Confocal laser scanning microscopy (CLSM) and transmission electron microscopy (TEM) images of the beads indicate that the QDs are segregated into inclusions distributed throughout the bead. The segregation of the QDs inside the polymer beads is due to enthalpy and entropy-driven rejection of the QDs from polymerizing loci as the polymerization proceeds. CLSM and fluorimeter measurements of the emission spectra of PS beads embedded with multicolored QDs in varying concentrations are reported which verify that distinguishable optical ratio-metric barcodes, derived from the spectral scans of the barcoded beads, can be obtained by this technique.

Comparison of the emission profiles of the barcoded beads with that of the same QDs dispersed in styrene indicates Electronic Luminescence Energy Transfer from the lower wavelength QDs to the higher wavelength QDs, providing evidence that the QDs are situated within nanometers of each other in the inclusions. The energy transfer limits our ability to obtain significant number of ratio-metric barcodes for use in multiplexing applications. Therefore, we obtained a reduction in the energy transfer by separation of the segregating QDs from one another during the polymerization process. This separation was achieved by highly cross-linking the PS beads using divinylbenzene (DVB) as one of the monomer units. A diffused luminescence from the CLSM luminescence images of the beads containing DVB confirms the non-aggregation of the QDs in presence of DVB during the polymerization. TEM images of the bead interiors also indicate that the QDs are uniformly distributed in the polymer matrix. CLSM emission profiles of these barcoded beads in comparison with that of the beads without DVB show a significant reduction in the energy transfer.

427b Characterization of Biocompatible Polymer Composites

James J. Lee and Gerold A. Willing. Chemical Engineering, University of Louisville, Eastern Parkway, Louisville, KY 40292

Several polymer composites for actuating devices that are responsive to an electric current and also is biocompatible were studied. These composites may be functionally used as actuators in biomedical devices or robotics. Hydrogels, electroactive polymers (EAPs), and gold nanoparticles were utilized in the development of several biocompatible polymer composites. The hydrophilicity, conductivity, response to electric conduction, strain, thermal stability, and degree of cross-linking of several hydrogel and EAP monomers were characterized. The hydrophilic nature of each homopolymer was characterized to determine the how well the material may function as a part of the biocompatible polymer composite. The mechanical response of each homopolymer was characterized by applying an electrical current. Conduction across the EAP and hydrogel polymer may be enhanced, directed, and controlled by doping it with gold nanoparticles. The size, concentration, and distribution of the gold nanoparticles in the EAP and hydrogel monomers were studied. Several combinations of hydrogels and EAPs with the gold nanoparticles were then characterized to develop several electrically actuating biocompatible polymer composites.

Understanding and characterizing the homopolymer and copolymers with and without the gold nanoparticles provided a preliminary step in engineering a system to design and develop an EAP-hydrogel copolymer composite. The characteristics will be used to later develop mathematical relationships between the water solubility and the conductivity of the monomers, EAP-hydrogel copolymers, and EAP-hydrogel copolymer composites. The mathematical models will be used to design a structured, scientific approach to develop an EAP-hydrogel copolymer composite with the desired biocompatible, mechanical, and electrical properties. The potential impact of this polymer composite's characterization and mathematical model may aid in development of various biocompatible EAP-hydrogels for multipurpose uses in artificial muscles, biomimetic robotic machines, sensors and actuators from the macro to the nano scale, and smart materials that can change according to environmental conditions.

427c Hierarchical Polymer-Based Nanocomposites for Electro-Magnetic Interference (EMI) Shielding

J. J. Huang, Plastics Engineering and NSF Nanoscale Science and Engineering Center for High-rate Nanomanufacturing, University of Massachusetts, One University Avenue, Lowell, MA 01854

EMI, also called electromagnetic pollution, has become an increasingly important problem due to the increasing use of electronic, computing and telecommunication equipment. Many efforts have been made in reducing and/or minimizing the interference by employing polymeric materials (which are made conductive) in place of conductive metallic housing, because of the advantages of polymers, such as lighter weight, lower cost, and design freedom. By employing conductive nanofillers, we recently fabricated polymer-based nanocomposites for EMI shielding applications. The formed nanocomposites show superior shielding property coupled with improved mechanical properties. The enhanced shielding property is attributed to significant surface areas of these nanofillers as shown by electron microscopy.

427d Carbon Fiber Based Multifunctional Energy Storage Structural Composites

Tony Pereira[1], **Zhanhu Guo**[2], Simon Nieh[3], and H. Thomas Hahn[1]. (1) Mechanical and Aerospace Engineering, University of California Los Angeles, 420 Westwood Plaza,, Eng IV, 48-121, Los Angeles, CA 90095, (2) Chemical Engineering Department, Lamar University, Beaumont, TX 77710, (3) Front Edge Technology, Inc., 13455 Brooks Drive, Baldwin Park, CA 91706

Developing power composite structures capable of energy harvesting and storage in addition to load bearing requires the integration of materials dissimilar physical property [1-3]. Recent progress is reported towards the fabrication of new multifunctional composites with the ability to store electrical energy in addition to single unifunctional load bearing.

In this presentation, the process to embed all-solid-state thin-film lithium energy cells into a carbon fiber reinforced plastic (CFRP) is reported. The performance of the integrated power composites is investigated. The mechanical properties of the multifunctional composite (modulus and strength) under quasi-static uniaxial loading conditions did not significantly deviate from those of the standard fully cured pure CFRP. The embedded energy cells in the multifunctional structure charged and discharged at baseline levels with force loadings of up to about 50% of the CFRP tensile strength.

References

[1] T. Pereira, R. Scaffaro, S. Nieh, J. Arias, Z. Guo and H. T. Hahn, Journal of Micromechanics and Microengineering, 2006, 16, 2714-2721.

[2] T. Pereira, R. Scaffaro, Z. Guo, S. Nieh, J. Arias and H. T. Hahn, Advanced Engineering Materials, 2008, 10(4) 393-399.

[3] T. Pereira, Z. Guo, S. Nieh, J. Arias and H. T. Hahn, Composites Science and Technology, 2008, 68, 1935-1941.

Web Page: www.seas.ucla.edu/~pereira/

Development of Remendable Polymer Networks with Thermally Reversible Bonds

Amy M. Peterson & Giuseppe R. Palmese

Materials that can recover properties following mechanical failure are desirable, as they offer increased durability and safety. Mendable materials are particularly desirable for load-bearing applications in which material failure is costly and dangerous.

Inspiration for remendable materials comes from nature and there is a great interest in developing biomimetic processes by a number of functionalities can be applied to synthetic materials. Biomimicry in remendable materials has led to composites with vascular networks that "bleed" resin when a crack forms.[1]

Two approaches for healing polymer networks have captured much attention. In one method, polymer networks are made to self-heal by adding microcapsules filled with uncured resin [2-8]. Upon fracture the microcapsules rupture releasing the resin, which hardens to heal the crack. The other method relies on the inherent reversibility of bonds designed into a polymer network [9-26]. The reversible nature of these linkages allows for network remodeling at the damaged site. We have investigated a composite approach to self-healing that combines many advantages of healing via encapsulation and healing via reversible bonds. This methodology allows for localized healing by including a secondary phase that induces healing through thermoreversible bonding while maintaining the mechanical and physical properties of the base thermoset.

Epoxy-amine thermosets are high modulus materials that are used in composites for their mechanical strength. We report on the development of two healing systems for epoxy-amine thermosets based on the thermoreversible Diels-Alder reaction of furan and maleimide. In one, crack healing of a traditional epoxy-amine thermoset is induced by thermally reversible crosslinking of a secondary phase. In the other, furan functionalization of an epoxy-amine thermoset allows for in situ crack healing of this thermoset with a bismaleimide solution. Both phenomena occur at room temperature and minimal pressure and significant load recovery is possible multiple times in a given location.

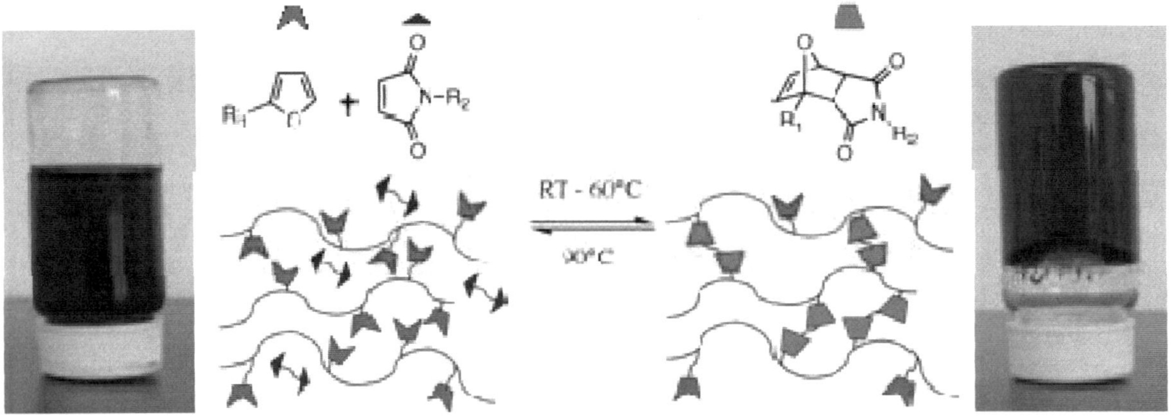

Figure 1. Thermoreversible crosslinking gel: crosslinking chemistry (top center), a schematic of reversible network (bottom center), and photographs showing liquid (left) and gel (right).

In this work the thermoreversible Diels-Alder reaction of furan and maleimide was selected to impart remendability. Specifically, the crosslinks of the gel are based on the Diels-Alder reaction between maleimide groups on a bismaleimide and pendant furans on a linear polymer. A schematic representation of this network is shown in Figure 1 with maleimide moieties as triangles, furan moieties as notched trapezoids and Diels-Alder adducts as trapezoids. Also shown in Figure 1 are the Diels-Alder reaction and pictures of vials containing the liquid system at 90°C and the gel at room temperature (RT). This form of the Diels-Alder reaction is reversible, forming a ring structure at room temperature and reforming the respective diene and dienophile between 60 and 90°C.

Load recovery in the furan-functionalized network is postulated to be the result of both physical and chemical bonding across the crack surface. Physical bonding is caused by solvent-mediated swelling and subsequent interlocking of crack surfaces, while chemical bonding results from the Diels-Alder reaction of furan and maleimide.

References

1. Toohey, K.; Sottos, N. R.; Lewis, J.; Moore, J.; White, S. R., Self-healing materials with microvascular networks. *Nat Mater* **2007,** 6, (8), 581-585.
2. White, S. R.; Sottos, N. R.; Geubelle, P. H.; Moore, J. S.; Kessler, M.; Sriram, S. R.; Brown, E. N.; Viswanathan, S., Autonomic healing of polymer composites. *Nature* **2001,** 409, (6822), 794-797.
3. Kessler, M.; Sottos, N. R.; White, S. R., Self-healing structural composite materials. *Composites Part A: Applied Science and Manufacturing* **2003,** 34, (8), 743-753.
4. Liu, X.; Sheng, X.; Lee, J. K.; Kessler, M. R., Synthesis and characterization of melamine-urea-formaldehyde microcapsules containing ENB-based self-healing agents. *International Conference on Smart Materials and Nanotechnology in Engineering. Edited by Du* **2007,** 6423, 112.
5. Brown, E. N.; White, S. R.; Sottos, N. R., Retardation and repair of fatigue cracks in a microcapsule toughened epoxy composite—Part II: In situ self-healing. *Composites Science and Technology* **2005,** 65, (15-16), 2474-2480.
6. Li, H.; Wang, R.; He, X.; Liu, W.; Hao, H., Preparation and characterization of self-healing poly (urea-formaldehyde) microcapsules. *International Conference on Smart Materials and Nanotechnology in Engineering. Edited by Du* **2007,** 6423, 98.
7. Yin, T.; Zhou, L.; Zhi Rong, M.; Qiu Zhang, M., Self-healing woven glass fabric/epoxy composites with the healant consisting of micro-encapsulated epoxy and latent curing agent. *Smart Mater. Struct.* **2008,** 17, 5019.
8. Blaiszik, B.; Sottos, N. R.; White, S. R., Nanocapsules for self-healing materials. *Composites Science and Technology* **2008,** 68, (3-4), 978-986.
9. Craven, J. M., Cross-linked Thermally Reversible Polymers Produced from Condensation Polymers with Pendant Furan Groups Cross-linked with Maleimides. *Patent* **1969,** 3,435,003, 5.
10. Kennedy, J. P.; Castner, K. F., Thermally reversible polymer systems by cyclopentadienylation. II. The synthesis of cyclopentadiene-containing polymers. *Journal of Polymer Science: Polymer Chemistry Edition* **1979,** 17, (7), 2055-2070.
11. Salamone, J. C.; Chung, Y.; Clough, S. B.; Watterson, A. C., Thermally reversible, covalently crosslinked polyphosphazenes. *Journal of Polymer Science Part A: Polymer Chemistry* **1988,** 26, (11), 2923-2939.
12. Chujo, Y.; Sada, K.; Saegusa, T., Reversible gelation of polyoxazoline by means of Diels-Alder reaction. *Macromolecules* **1990,** 23, (10), 2636-2641.
13. Canary, S. A.; Stevens, M., Thermally reversible crosslinking of polystyrene via the furan-maleimide Diels-Alder reaction. *J. Polym. Sci. A Polym. Chem.* **1992,** 30, (8), 1755-1760.
14. Sijbesma, R., Reversible Polymers Formed from Self-Complementary Monomers Using Quadruple Hydrogen Bonding. *Science* **1997,** 278, (5343), 1601-1604.
15. Nijenhuis, K., Thermoreversible Networks. *Advances in Polymer Science* **1997,** 130, 51-57.
16. Fall, R.; Ward, T. Puncture Reversal of Ethylene Ionomers - Mechanistic Studies. 2001.

17. Gheneim, R.; Perez-Berumen, C.; Gandini, A., Diels-Alder Reactions with Novel Polymeric Dienes and Dienophiles: Synthesis of Reversibly Cross-Linked Elastomers. *Macromolecules* **2002,** 35, (19), 7246-7253.
18. Liu, Y.; Hsieh, C.; Chen, Y., Thermally reversible cross-linked polyamides and thermo-responsive gels by means of Diels–Alder reaction. *Polymer* **2006,** 47, (8), 2581-2586.
19. Kavitha, A. A.; Singha, N. K., A tailor-made polymethacrylate bearing a reactive diene in reversible diels-alder reaction. *J. Polym. Sci. A Polym. Chem.* **2007,** 45, (19), 4441-4449.
20. Liu, Y.; Chen, Y., Thermally Reversible Cross-Linked Polyamides with High Toughness and Self-Repairing Ability from Maleimide- and Furan-Functionalized Aromatic Polyamides. *Macromolecular Chemistry and Physics* **2007,** 208, (2), 224-232.
21. Cordier, P.; Tournilhac, F.; Soulié-Ziakovic, C.; Leibler, L., Self-healing and thermoreversible rubber from supramolecular assembly. *Nature* **2008,** 451, 977.
22. Bergman, S.; Wudl, F., Mendable polymers. *J. Mater. Chem.* **2008,** 18, (1), 41.
23. Varley, R.; van der Zwaag, S., Ionomers as Self Healing Polymers. In *Self Healing Materials*, 2007; pp 95-114.
24. Kalista, S. J.; Ward, T. C., Thermal characteristics of the self-healing response in poly(ethylene-co-methacrylic acid) copolymers. *Journal of The Royal Society Interface* **2007,** 4, (13), 405-411.
25. Murphy, E. B.; Bolanos, E.; Schaffner-Hamann, C.; Wudl, F.; Nutt, S. R.; Auad, M. L., Synthesis and Characterization of a Single-Component Thermally Remendable Polymer Network: Staudinger and Stille Revisited. *Macromolecules* **2008,** 41, (9), 3169-3174.
26. Chen, X.; Dam, M. A.; Ono, K.; Mal, A.; Shen, H.; Nutt, S. R.; Sheran, K.; Wudl, F., A Thermally Re-mendable Cross-Linked Polymeric Material. *Science* **2002,** 295, (5560), 1698-1702.

AUTHOR INDEX

AUTHOR INDEX

AUTHOR INDEX

AUTHOR INDEX

AUTHOR INDEX

AUTHOR INDEX

AUTHOR INDEX

AUTHOR INDEX

AUTHOR INDEX

AUTHOR INDEX

AUTHOR INDEX

AUTHOR INDEX

AUTHOR INDEX

AUTHOR INDEX

AUTHOR INDEX

AUTHOR INDEX

AUTHOR INDEX

AUTHOR INDEX

Materials Engineering and Sciences Division

Presentations at the 2008 AIChE Annual Meeting

Philadelphia, Pennsylvania
16 - 21 November 2008

Volume 2 of 2

ISBN: 978-1-61567-223-3

Printed from e-media with permission by:

Curran Associates, Inc.
57 Morehouse Lane
Red Hook, NY 12571

Some format issues inherent in the e-media version may also appear in this print version.

Printed by Curran Associates, Inc. (2009)

AIChE
3 Park Avenue
New York, NY 10016-5991

Phone: (203) 702-7660
Fax: (203) 775-5177

www.aiche.org

Additional copies of this publication are available from:

Curran Associates, Inc.
57 Morehouse Lane
Red Hook, NY 12571 USA
Phone: 845-758-0400
Fax: 845-758-2634
Email: curran@proceedings.com
Web: www.proceedings.com

TABLE OF CONTENTS

VOLUME 1

VOLUME 2

436a Draw Ratio Enhancement in Non-Isothermal Melt Spinning

Balram Suman and Satish Kumar. Chemical Engineering and Materials Science, University of Minnesota, 151 Amundson Hall, 421 Washington Ave. SE, Minneapolis, MN 55455

Non-isothermal melt spinning of materials having a step-like viscosity variation with temperature is studied in this work. A set of nonlinear equations is used to describe the fiber behavior and to obtain the draw ratio, the square of the ratio of the fiber diameter at the entrance to that at the exit of the fiber-spinning device. The fluid-flow equation is based on a slender-jet approximation, and external heating and cooling have been accounted for with a one-dimensional model in order to obtain the fiber temperature and viscosity along the fiber length. The model is similar to that used by Wylie et al. [J. Fluid Mech. 570, 1 (2007)], but accounts for inertia, shear stress at the fiber surface, surface tension, gravity, cooling, and larger heating rates. Steady-state analysis reveals that the draw ratio increases with an increase in the pulling force, passes through a maximum, and then starts increasing again, resulting in three possible pulling forces for the same draw ratio. However, linear stability analysis reveals that depending on the strength of heating and/or cooling, at most two of the steady states are stable. The stability analysis also predicts complicated oscillatory and non-oscillatory dynamical behavior as the pulling force varies. Nonlinear simulations reveal that an unstable system always tends to limit-cycle behavior. Systems predicted stable by the linear stability analysis are also stable for large-amplitude perturbations. External heating is found to dramatically enhance the draw ratio of the melt-spinning process. The addition of a cooling section suppresses the draw ratio, but this can be compensated for with a higher heating strength.

436b The Mechanics of Stable Electrospinning Polymer Jets and Consequences for Measurement of Polymer Extensional Viscosity

Matthew H. Helgeson[1], Kristie N. Grammatikos[2], Norman J. Wagner[3], and Joseph Deitzel[1]. (1) University of Delaware, Newark, DE 19711, (2) Chemical Engineering, University of Delaware, 150 Academy Street, Newark, DE 19711, (3) Department of Chemical Engineering, University of Delaware, 150 Academy St., Newark, DE 19716

Electrospinning has gained significant attention as a method to produce polymer and composite nanofibers for a variety of emerging applications. However, realization of such applications will require simplified models for the design and control of electrospinning processes that explicitly account for the extensional rheology of electrospinning fluids. Here, we develop a simplified model based on slender body electrohydrodynamic (EHD) theory of stable jets that includes extensional strain hardening. The EHD model is analyzed during the startup of uniaxial extension, yielding predictions for the kinematics of electrospun jets solely in terms of operating conditions and independently measurable fluid properties. Experimental observations using high speed videography and velocimetry of polyethylene oxide-water jets validate scaling predictions of the model, enabling the reduction of an entire jet profile to several key parameters for comparison across different materials and operating conditions. Furthermore, the measurements allow for determination of the apparent extensional viscosity of the spinning fluid at extension rates experienced during electrospinning, which far exceed the measurable rates of conventional extensional rheometers. Finally, the theory enables correlation of electrospun fiber diameter by dimensional analysis of the EHD model that enables prediction of the electrospun fiber morphology solely from operating conditions and polymer spinning solution properties.

436c Nozzle-Less Fiber Nanospinning – a Rapid and Efficient Method for Fabrication of Micron Diameter Fibers

Stoyan K. Smoukov and Orlin Velev. Department of Chemical and Biomolecular Engineering, North Carolina State University, 911 Partners Way, EB1, Raleigh, NC 27695

Synthetic microfibers for non-woven textiles are predominantly made by methods that involve extrusion through a nozzle to form the fiber. A technology for wet-spinning of microfibers would greatly expand the range of microfiber feedstocks, and possible microfiber products. We will report a new nanospinning method where fibers are formed by shear extension and anti-solvent based precipitation in bulk solution. Instead of using nozzles, the method relies on combined capillary instabilities and shear extension to form the fibers. This is desirable for the incorporation of various particlulate additives in composite fibers, which is problematic for traditional wet-spinning due to clogging of the nozzles. Compared to electrospinning, the process is straightforward to scale up. It is applicable to a wide range of polymers: we demonstrate fibers from recycled, renewable and biodegradable polymers. The fibers of diameter ~1-2 microns could find application in fine filtration, medical wound dressings, better barrier materials, and tissue scaffolds.

436d Structure and Rheology In a Model Non-Polar Clay Nanocomposite

Saswati Pujari[1], Wesley R. Burghardt[1], Marie-Claude Heuzey[2], Christophe Mobuchon[2], and Pierre J. Carreau[2]. (1) Department of Chemical & Biological Engineering, Northwestern University, Evanston, IL 60208, (2) Chemical Engineering Department, École Polytechnique de Montréal, CREPEC, Montreal, QC H3C3A7, Canada

The rheology of polymer-clay nanocomposite materials is strongly affected by flow-induced changes in fluid structure. For the purposes of elucidating rheological characteristics particularly associated with the dispersed clay, recent research has demonstrated the virtues of studying idealized model materials employing polymer matrices that are essentially Newtonian in the absence of clay. In this work, we follow up on recent studies of Mobuchon and coworkers [Rheol Acta 46:1405 (2007)] on an organically modified clay (Cloisite® 15A, Southern Clay Products) dispersed in a Newtonian polybutene matrix, studying a similar sample using in situ small-angle x-ray scattering to study flow-induced orientation of the clay particles. In this work we have studied a sample with 2 wt% loading of clay. This sample was studied in an annular cone and plate x-ray capable shear cell that allows interrogation of the degree and direction of particle orientation in the flow-velocity gradient plane. Particle orientation was observed to progressively increase with increasing shear rate. Upon cessation of shear flow, orientation was found to partially relax, to an extent which depended on the previous shear rate: higher residual particle orientation was found after cessation from higher shear rates. These trends appear to be closely connected with evolution of the sample's viscoelasticity observed in mechanical rheology following low cessation. Orientation and rheology are also compared for other transient shear flow protocols, such as step-change in shear rate and shear flow reversal.

436e Rheological Characterization of Supercritical CO2 Processed Polystyrene-Clay Nanocomposites

Robert Bellair[1], Mihai Manitiu[2], Steven E. Horsch[2], Esin Gulari[3], and Rangaramanujam M. Kannan[2]. (1) Department of Chemical Engineering and Materials Science, Wayne State University, 5050 Anthony Wayne Dr., Suite 1100, Deroit, MI 48202, (2) Chemical Engineering and Materials Science, Wayne State University, 5050 Anthony Wayne Dr., Suite 1100, Deroit, MI 48202, (3) Clemson University, Clemson, SC

Polymer-clay nanocomposite research is largely focused on improving material properties by maximizing polymer-clay interactions and increasing particle surface area through delamination of clay platelets. A novel method utilizing supercritical CO_2 ($scCO_2$) has recently been shown to be effective in separating clay platelets with or without polymer present, and can be used to disperse them in a polymer matrix.[1] By investigating the rheological changes that occur upon $scCO_2$ processing, we can gain a better understanding of the physical process of delamination. High molecular weight polystyrene and montmorillonite with varied organic modifiers were processed with a co-solvent under multiple processing conditions. Use of a co-solvent is a convenient way to reduce processing temperatures, allowing us to avoid thermal degradation of clay modifier, and enabling direct comparison of our nanocomposite properties with solution blended analogs.

Results indicate that $scCO_2$ processing produces a dramatic rheological improvement at low frequencies, over both the neat polymer and solution blended benchmark samples, even at 2 wt% nano-clay. As much as a 3 order of magnitude improvement in storage modulus is observed despite a lack of change in the x-ray diffraction peak of the clays, suggesting that a bi-modal distribution of dispersed platelets and small tactoids exist in the $scCO_2$ processed materials. The supercritical fluid process appears to produce dispersion that is also dependent on the chemical species on the nano-clay organic modifier. Finally, the rate at which the system is depressurized as well as the number of times the clay is processed in CO_2 have a strong effect on rheological enhancement with better results produced by faster depressurization rates and a larger number of CO_2 depressurizations. These results are analyzed in the scope of interpreting the clay structural changes that occur due to the catastrophic depressurization of CO_2 in our processing method.

[1] Horsch, S.; Serhatkulu, G.; Gulari, E.; Kannan, R. M. *Polymer* **2006**, 47, (21), 7485-7496.

Web Page: www.kannangroup.com

436f Interfacial Adhesion and Strain Hardening in Polymer Nanocomposites

Krishnamurthy Jayaraman, Chemical Engineering & Materials Science, Michigan State University, 2527 Engineering Building, East Lansing, MI 48824-1226 and Tanmay J. Pathak, Chemical Eng. & Materials Science, Michigan State University, 2527 Engineering Building, East Lansing, MI 48824-1226.

The effect of interfacial adhesion on strain hardening behavior in extensional flow has been investigated for intercalated polymer-clay nanocomposites with a small loading of nanofiller. The nanocomposites were prepared with a functionalized polypropylene (linear chains) as the matrix and organoclay at 5wt% loading with and without silane treatment. The strain hardening behavior of the nanocomposite melts was evaluated in uniaxial extensional flow at 10°C and 15°C above the melting point of 145°C and at four different strain rates with the Extensional Viscosity Fixture mounted on a TA-ARES instrument. Strain hardening of the melt was observed in both nanocomposites but to a lesser extent in the case of silane treated organoclay. The strain hardening behavior of the nanocomposite melts was described with the molecular stress function model of Wagner et al. (2001, 2006) and the maximum value of tube diameter contraction, f^2_{max} fitting the extensional viscosity data was found for each case. Since the composites were intercalated, the interfacial adhesion between polymer matrix and nanoclay could be evaluated experimentally with the help of stress strain curves from tensile tests at 5°C and 10°C below the melting point – cf. Piggott.

Wagner, M.H., Rubio, P. and Bastian, H., J. Rheol. 45(6), 1387 (2001)

Piggott, M.R. Ch 8 in Load Bearing Fibre Composites, Pergamon Press 1980

437a Investigation of NMR Signal Loss during Hydrolytic Polycondensation of Organoalkoxysilanes

Jyothirmai Ambati and Stephen E. Rankin. Chemical and Materials Engineering Department, University of Kentucky, 177 Anderson Hall, Lexington, KY 40506

Kinetic studies of the polymerization of bis(triethoxysilyl)ethane in ethanol/acidic water solutions suffers from NMR signal loss for certain molar ratios. Since the solutions remain macroscopically clear, we speculate that loss of signal could be due to microphase separation or drastic changes in NMR parameters of the polymerizing species during NMR characterization. For instance, while particles smaller than about 50 nm would be expected to undergo rotational diffusion on a time scale that would allow motional averaging for static NMR [C. Mayer, Progr. Nuc. Magn. Reson. Spec. 2002, 40, 307], the spin-lattice relaxation time (T_1) of nuclei on the interior of a rigid particle can be large enough that saturation leads to loss of signal. However, we have evidence that solid particle formation is not responsible for the loss of signal that we observe; for a strongly acidic solution of bis(triethoxysilyl) ethane, ethanol and water, we observed that NMR signal is regained after certain time. This phenomenon was previously detected by Rankin et al. [J Phys. Chem. A, 1999, 103, 4233] during ^{29}Si NMR kinetic study of trimethylethoxysilane but not studied in detail. Here, we investigate polymerizing organoalkoxysilane solutions exhibiting NMR signal loss starting with solubility studies of trimethylethoxysilane and hexamethyldisiloxane in ethanol/acidic water mixtures with the goal of quantifying the relationship between the kinetics of hydrolytic condensation and the loss of signal of hydrolyzed and condensed species. The observation of microphase separation in these systems has important implications for zeolite, Stoeber particle, and mesoporous (e.g. MCM-41) materials nucleation.

437b Kinetic Modeling of Single-Site Olefin Polymerization with Multi-Response Data: Even Models with Many Parameters Cannot Fit An Elephant

Krista A. Novstrup[1], Grigori A. Medvedev[1], Nicholas E. Travia[2], Corneliu Stanciu[2], Jeffery M. Switzer[1], Thomas A. Manz[1], W. Nicholas Delgass[1], Mahdi M. Abu-Omar[2], and James M. Caruthers[1]. (1) School of Chemical Engineering, Purdue University, Forney Hall of Chemical Engineering, 480 Stadium Mall Drive, West Lafayette, IN 47907-2100, (2) Department of Chemistry, Purdue University, 560 Oval Drive, West Lafayette, IN 47907-2084

Single-site catalysts are revolutionizing polyolefin production because they enable more precise control of the polymer's molecular architecture, which in turn controls the physical properties of the polymer. The ability to design single-site catalysts requires fundamental understanding of the polymerization, including detailed knowledge of the governing kinetic mechanisms and values for all of the rate constants. This task is difficult because for each of the individual steps in the polymerization (i.e. activation, initiation, propagation, and chain transfer) there are a number of mechanistic possibilities, resulting in potentially thousands of different polymerization models. Hence, it would seem that any number of proposed kinetic models would be capable of describing the polymerization, because of the large number of kinetic parameters associated with each potential model. On the contrary, we have found that for a rich set of multi-response data (i.e. evolution of monomer concentration, molecular weight distribution, different types of terminal groups via NMR, etc) it is actually very difficult to fit all the data with a mechanistically based kinetic model – one cannot fit the proverbial elephant no matter how many parameters if the proposed mechanism, and hence associated kinetic equations, is in error. Thus, quantitative kinetic modeling provides an essential tool for determining detailed mechanisms.

The power of quantitative kinetic modeling in combination with rich multi-response experimental data will be demonstrated here for polymerization of 1-hexene with the [rac-$(C_2H_4(1\text{-Ind})_2)$ZrMe][MeB$(C_6F_5)_3$] single-site catalyst system. Experimental batch polymerization data include (i) monomer concentration versus time profiles via ^{1}H-NMR, (ii) time evolution of the molecular weight distribution, (iii) terminal vinylene and vinyledene concentrations via ^{1}H-NMR, (iv) structure of the polymer attached to catalysts obtained by quenching with deutero-methanol with subsequent ^{2}H-NMR analysis and (v) electronic structure of the metal-counterion complex from UV-VIS. This set of rich multi-response data, in combination with quantitative kinetic analysis, indicates that the mechanisms proposed in the literature for this catalyst are inadequate to describe all the observations. A new mechanism with a dormant site that can reactivate is required to describe the data. This communication will demonstrate how this approach can be used to discriminate between possible mechanisms, where the researcher can be confident in a kinetic model once it is able to fit all the multi-response data.

437c Kinetics of Carbon Accelerated Radical Polymerization (CARP)—the Role of Diffusional Limitations

Ruohua Xiong, Fouad Teymour, and Hamid Arastoopour. Department of Chemical and Biological Engineering, Illinois Institute of Technology, Chicago, IL 60616

The catalytic effect of carbon on free radical polymerization, mediated by benzoyl peroxide initiation, has been previously reported by our group [1,2]. These studies have concluded that the carbon nanoparticles are involved in a mechanism that catalyzes the dissociation of the initiator on their surface. Consequently, the polymerization proceeds at temperature much below those normally used for the monomers considered. For methyl methacrylate, near complete conversion is achieved in less than three hours at 50 °C, and polymerization can be even carried out at refrigerator temperatures overnight. The current study aims at a systematic quantification of the accelerated kinetics, with the broader objective of elucidating the catalytic reaction mechanism. Kinetic data from free radical polymerization of different vinyl monomer such as glycidyl methacrylate (GMA), methyl methacrylate (MMA), 2-hydroxyethyl methacrylate (HEMA) at different reaction parameters are compared and discussed. The effect of carbon content is studied and elucidated, as an initial increase of catalytic activity is observed at low carbon percentages followed by a decrease after a maximum is reached; this results from the competition between acceleration of dissociation and inhibition of polymerization, which becomes appreciable at higher levels. In all these systems, the formation of nanocomposites with superior dispersion of the carbon nanoparticles in the polymer matrix is observed; results on the characterization of these composites will be presented elsewhere.

It was interesting to note that the catalytic activity of the carbon particles is retained when these are present in a rubber formulation. Pulverized recycled rubber tire microparticles were tested and shown to catalyze the polymerization. As a matter of fact, the initial discovery of this catalytic effect occurred during the preparation of composites based on the pulverized rubber particles. Value-added materials can be obtained by synthesizing interpenetrating polymer networks (IPNs) based on pulverized rubber particles. Screening of the various additives in the rubber formulation for their effect on polymerization clearly showed that only carbon black is responsible for the catalysis. It was, however, interesting to note that the magnitude of the catalytic effect (normalized to the proper carbon content) is higher in the presence of rubber microparticles than when carbon nanoparticles are used. After excluding any possibility of another chemical catalytic activity effect, it was concluded that the near-perfect dispersion of the carbon black nanoparticles in the rubber formulation matrix creates a compartmentalization effect that interferes with the termination of radicals attached to one particle with those on other particles, thus severely impacting the rate of termination and consequently accelerating the net rate of polymerization. This of course is similar to the Tromsdorff effect observed in many radical polymerization systems as the viscosity of the medium increases with conversion. In experiments aimed at confirming this hypothesis, we conducted the polymerization in the presence of the carbon nanoparticles and varying amounts of poly(methyl methacrylate) added to the formulation. Results confirm the synergetic coupling between the reactive catalytic effect and the diffusional limitation effect. Quantification of the kinetics is reported for the proposed polymerization systems.

References:

1. Fouad Teymour, Sung Hun Kim, Amit Sachdeva, CARP: Carbon Accelerated Radical Polymerization —Free Radical Polymerization On Carbon Nanoparticles. Polymer Reaction Engineering VI, Engineering Conferences International, May 21-26, 2006, Halifax, Nova Scotia, Canada.

2. Ruohua Xiong, Fouad Teymour, and Hamid Arastoopour, Carbon Black Acts as Catalyst in Making Amphiphilic IPNs Based on Pulverized Rubber Particles, AIChE 2006 Annual Meeting, San Francisco, CA.

437d Well Defined Nanoparticles from Controlled Inverse Miniemulsion Polymerization

Genggeng Qi[1], Christopher W Jones[1], and F. Joseph Schork[2]. (1) Chemical & Biomolecular Engineering, Georgia Institute of Technology, 311 Ferst Drive, Atlanta, GA 30332-0100, (2) Department of Chemical & Biomolecular Engineering, University of Maryland, 2113D Chemical Engineering Bldg 090, College Park, MD 20742

Well defined hydrophilic or amphiphilic polymer latexes and polymer composites in submicron size have important applications in various fields such as drug delivery, industrial catalysis, and coatings. However, these well defined polymer latexes and composites are usually difficult to produce in large scale due to the issue of mixing and heat transfer. To overcome the above problems, RAFT inverse miniemulsion polymerization is applied here as a unique approach to synthesize these nanosized hydrophilic polymer latexes and composites. Moreover, the polymers derived from RAFT inverse miniemulsion polymerizations maintain their "livingness" and can be easily functionalized. The kinetic behavior of inverse RAFT miniemulsion polymerization is studied and effects of reaction parameters on inverse miniemulsion polymerizations are investigated. Responsive nanoparticles are synthesized and the applications of RAFT inverse miniemulsion polymerization are also discussed.

437e Mathematical Modeling of a Dispersive Suspension Polymerization

Carla Luciani[1], Yunju Jung[1], Joong Jin Han[2], and **Kyu Yong Choi**[2]. (1) Chemical and Biomolecular Engineering, University of Maryland, Bldg 090, College Park, MD 20742, (2) Department of Chemical and Biomolecular Engineering, University of Maryland, College Park, College Park, MD 20742

Porous particles exhibit a variety of applications. Recently, a novel synthetic method for producing porous polymer particles has been developed. It is based on the phase separation process that takes place inside monomer droplets along a suspension polymerization. In this paper, a mathematical model that describes a dispersive suspension polymerization of methyl methacrylate is presented, and the evolutions of conversion, phase volumes, and polymer molecular structure are predicted. The phase separation point and the distribution of species between phases are calculated through the Flory-Huggins theory. In each phase, the kinetic mechanism involves thermal and chemical initiation, propagation, chain transfer to the monomer, and termination reactions. Gel effect and volume contraction are also considered by the model. The global results are used as the input of another model for estimating the morphology of the polymer particles. This last model assumes that an intra-particle instantaneous local distribution of species is the main responsible for the morphology development. Due to this local distribution, a fraction of the polymer chains is produced homogeneously, yielding a dense structure; while the other fraction precipitates yielding a less dense structure. The proposed model requires to solve global and local mass balances, chemical potential balances, fluid-dynamic correlations, and population balances. It is adjusted and validated with experimental results.

437f Kinetic Study of the Factors Affecting the Compositional Gradient along Copolymer Chains

Lin Wang and Linda J. Broadbelt. Department of Chemical and Biological Engineering, Northwestern University, 2145 Sheridan Rd. Tech E136, Evanston, IL 60208-3120

A gradient copolymer has a gradient in the repeat units comprising the backbone of the polymer arranged from predominantly monomer A to predominantly monomer B along the copolymer chain. Because of the gradual change of the composition along the copolymer chains, gradient copolymers exhibit distinct physical properties compared to those of random or block copolymers of the same composition. Furthermore, different sequence distributions along gradient copolymer chains may lead to markedly different physical properties, suggesting that gradient polymers can be tailored for specific applications by tuning the sequence distribution. Thus, it would be valuable to be able to predict the specific sequence distribution of gradient copolymers given different monomer pairs and synthesis conditions.

To address this challenge, we developed kinetic Monte Carlo (KMC) models, which track molecules instead of concentration, in order to track the explicit sequence distribution for each copolymer chain. In addition, KMC models permit the explicit molecular weight distribution (MWD) and chemical composition distribution (CCD) of gradient copolymers to be tracked, which is not possible with moment-based continuum models. We developed these KMC models in the context of the nitroxide-mediated controlled radical polymerization (NM-CRP) of two different monomer systems: styrene/4-acetoxystyrene and methyl methacrylate/t-butyl methacrylate. The simulated MWD and CCD were compared to experimental data from matrix-assisted laser desorption/ionization-time of flight mass spectroscopy (MALDI-TOF-MS). The effects of different synthesis conditions on the MWD, CCD and formation of the compositional gradient along the copolymer chains were studied. Finally, experimental methods for obtaining information about the sequence distribution to compare to the detailed KMC output were explored.

437g Reaction Analysis of "Living/controlled" Polymerization Techniques Used to Enhance Binding Characteristics of Highly Crosslinked Imprinted Polymer Networks

Asa D. Vaughan and Mark E. Byrne. Biomimetic & Biohybrid Materials, Biomedical Devices, and Drug Delivery Laboratories, Department of Chemical Engineering, Auburn University, Auburn, AL 36849-5127

Imprinted polymer networks with intrinsic binding properties are materials that have a particular "trained" affinity for a specific target molecule. The "trained" macromolecular memory within these polymer networks are due to two synergistic effects: (i) shape specific cavities that match the template molecule, which provide stabilization of the chemistry in a crosslinked matrix, and (ii) chemical groups oriented to form multiple non-covalent complexation points with the template. Kinetics analysis of the polymerization reaction show increases in the chemical controlled polymerization mechanism with the use of "living" polymerization techniques. Binding parameter analysis in conjunction with kinetic parameter analysis indicates that the formation of the binding site takes place during the chemical controlled propagation mechanism of the polymerization reaction. Extending the chemical controlled propagation mechanism via the reversible termination reaction allows for fewer frustrations within the network. Binding parameter research of ethyladenine-9-acetate (EA9A) imprinted poly(methacrylic acid-co-ethylene glycol dimethacrylate) (poly(MAA-co-EGDMA)) networks synthesized via "living" polymerization techniques show increases in template loading capacities and binding affinities compared to standard UV-initiated free radical polymerization. Results from binding studies show a 63% in template loading capacity in EA9A imprinted poly(MAA-co-EGDMA) networks with an increase in fractional double bond conversion, and a 85% increase in template binding affinity for EA9A imprinted poly(MAA-co-EGDMA) networks with equivalent double bond conversions. These results highlight reaction methodologies to tailor imprinted networks and modify the binding characteristics. This will enhance the application potential of "artificial" recognition materials for use in sensors, diagnostic devices, chromatography, and drug delivery carriers.

447a Simulation Studies on the Tensile Properties of Polymer Networks with Heterogeneous Microstructure

Bernardo M. Aguilera-Mercado, Claude Cohen, and Fernando A. Escobedo. School of Chemical and Biomolecular Engineering, Cornell University, Ithaca, NY 14853-5201

Coarse-grained molecular models of polymer networks were used to study how energetic, kinetic, and entropic effects can be harnessed to manipulate the network architecture and the tensile properties. We investigated first the impact of chain length polydispersity and its kinetic implications on the mechanical properties of bimodal end-linked networks, which are formed by endlinking two sets of telechelic linear polymer chains of different molar mass that are chemically identical. Molecular simulations of the end-linking reaction and network deformation have been used to elucidate, in particular, the origin of the enhancement of the mechanical properties (like toughness) seen experimentally in end-linked bimodal polymer elastomers. The impact of the system composition (mol% of short chains) on the network microstructure was monitored by obtaining detailed network topological characterizations that include the identification of network imperfections (i.e., pendant chains and loops), elastic chains, and short-chain clusters. The responses upon deformation of these topological chain types and their contribution to the network elasticity were also quantified through their segment orientation, end-to-end distances and gyration radii. Our results suggest that, at low concentrations of short chains (relatively to their pervaded volume), the kinetics of the end-linking reaction leads to the formation of a significant amount of network structural inhomogeneities (i.e., short-chain clusters) and defects (mostly short-chain loops), which considerably decrease the toughness and modulus of the material. Our results further identify the conditions at which optimal toughness is expected. Uni-axial stretch experiments, and network swelling of bimodal Poly(dimethylsiloxane) (PDMS) networks compared well with the simulation results.

Additionally, the tensile properties of idealized networks with highly regular topologies (corresponding to bi-continuous phases) were also simulated. These networks, formed by regularly connected chains with few or no entanglements, are expected to have tunable elastic modulus and toughness—depending upon network topology, chain stiffness, and molecular weight—and high ultimate strain and stress values due to their structure regularity. These regular networks exhibit a particular step-wise elastic response and multiple ordering transitions (as seen in some extremely tough natural materials such as the organic adhesive inside the abalone shells, muscle protein "titin", and spider silk). In addition, possible ways of obtaining these tri-dimensional regularly connected networks—inspired in recent experimental works on the fabrication of bi-dimensional regular networks—are being explored through simulations.

447b Flexoelectric Networks from Bent-Core Nematic Liquid Crystal Polymers

Rafael Verduzco[1], Martin Chambers[2], Phillip F. Britt[3], Antal Jákli[4], Samuel Sprunt[2], and James T. Gleeson[2]. (1) Center for Nanophase Materials Science, Oak Ridge National Laboratory, 1 Bethel Valley Road, PO Box 2008, MS 6494, Oak Ridge, TN 37831, (2) Department of Physics and Liquid Crystal Institute, Kent State University, Kent, OH 44242, (3) Center for Nanophase Materials Sciences and Chemical Science Division, Oak Ridge National Laboratory, Building 4500S, S-144, MS-6129, P.O. Box 2008, Oak Ridge, TN 37831-6197, (4) Liquid Crystal Institute, Kent State University, Kent, OH 44242

The flexoelectric effect is an electro-mechanical phenomenon that arises in nematic liquid crystals (LCs) whereby a macroscopic polarization develops in response to deformations of the liquid crystal alignment [1]. This effect has potential technological application in devices such as strain gauges, micropower generators, and nano-actuators that utilize an electro-mechanical response to operate. Recently, Harden and coworkers discovered that nematic bent core LCs exhibit a flexoelectric coefficient more than three orders of magnitude larger than in previously studied calamitic nematic LCs, paving the way for electro-mechanical devices that utilize the flexoelectric effect [2]. In order to develop practically viable flexoelectric materials, it is necessary to incorporate this type of nematic LC between flexible substrates or in an easily processible polymer matrix. Towards that goal, we synthesized reactive bent core LCs that can be used to prepare bent core polymers, elastomers, and gels. While previous studies have focused on reactive bent-core mesogens and bent-core LC networks that display smectic phases [3], we focus on nematic bent-core polymer networks. We prepared two different types of reactive bent core LCs with either an alkene or acrylate functionality. Monofunctional bent-core LCs with a reactive alkene group can be used to make aligned nematic elastomers using the well-known Finkelmann method [4], and these elastomers can either be studied in the bulk or swollen with bent core LC to investigate their flexoelectic behavior. Following a method demonstrated for calamitic LCs, monofunctional bent-core monomers with an acrylate functionality can also be dissolved in a nematic solvent and polymerized in the presence of a crosslinker to make bent-core LC networks and gels. Furthermore, monofunctional acrylate bent core LCs can be polymerized to make well-defined side-group liquid crystalline polymers. We present the synthesis and characterization of reactive, nematic bent-core LC mesogens.

[1] Meyer, R. B. Phys. Rev. Lett. 22, 918 (1969).

[2] Harden, J, Mbanga, J., Éber, N., Fodor-Csorba, K., Sprunt, S., Gleeson, J. T., and Jákli, A. Phys. Rev. Lett. 97, 157802 (2006).

[3] (a) Keum, C. D., Kanazawa, A. Ikeda, T. Adv Mater., 13, 321 (2001). (b) Sentman, A. C., Gin, D. L. Angew. Chem. Int. Ed.42, 1815 (2003). (c) Barberá, J. Gimeno, N., Monreal, L. Piñol, R. Blanca Ros, M., Serrano, J. L. J. Am. Chem. Soc., 126, 7190 (2004).

[4] (a) Küpfer J., Finkelmann H. Makromol. Chem., Rapid Commun. 12, 717 (1991). b) Küpfer, J., Finkelmann, H. Macromol. Chem. Phys., 195, 1353 (1994).

[5] (a) Urayama K., Arai Y. O., Takigawa, T. Macromolecules 38, 3469 (2005) (b) Urayama K., Okuno Y., (b) Kawamura T., Kohijiya S. Macromolecules 35, 4567 (2002).

447c Stress-Strain Behavior of Smectic Main-Chain Elastomers

Ronald C. Hedden, Daniel M. Lentz, and Harshad P. Patil. Materials Science & Engineering, Penn State University, University Park, PA 16802

Liquid crystalline elastomers (LCE) are rubber-like polymer networks that usually contain rigid, rod-like chemical units (mesogens). Because of spontaneous orientational or positional ordering of the mesogens, LCE exhibit remarkable deviations from ordinary rubber elasticity, leading to fascinating physical phenomena such as spontaneous shape changes triggered by electric fields or heating. As their exceptional dynamic mechanical response distinguishes LCE from all other elastomers, recent efforts have focused on understanding and controlling the molecular factors governing their static and dynamic mechanical response at both low and high strains.

Smectic LCE are those that form a layered mesophase having inter-layer spacings of about 30 Å. Crosslinking a smectic polymer in the absence of an aligning field produces a polydomain elastomer containing numerous randomly oriented microdomains. Under uniaxial tension, polydomain smectic elastomers are well known to undergo a transition to a globally oriented monodomain state. The mechanism for this transition has been debated, as domain rotations and/or transient disordering of domains have each been suggested to occur in response to applied strain. New studies of smectic elastomers having mesogens embedded in the polymer backbone (main-chain elastomers) suggest a three-stage deformation process. At low strains, amorphous (non-smectic) material deforms readily, with minimal change to the size or orientation of smectic domains. At intermediate strains, disordering of smectic microdomains via unfolding of hairpin structures is the dominant mechanism for elongation. At very high strains, elastic chains approach the finite extensibility limit, and layer buckling becomes a significant mechanism for elongation. Because of crude chain-folding in the polydomain state, the stress-strain behavior of smectic elastomers bears similarities to the cold drawing of semicrystalline polymers. A yield stress is found in plots of the nominal stress vs. strain, the value of which increases with increasing domain size. The observed yield stress is shown to result from a "necking" instability, which we have reported in a smectic LCE for the first time.

447d Monte Carlo Simulation of Liquid Crystalline Elastomers

Brian T. Gettelfinger[1], Raj Shekar[2], and Juan J. de Pablo[2]. (1) Department of Chemical and Biological Engineering, University of Wisconsin-Madison, Madison, WI 53706-1691, (2) Chemical and Biological Engineering, University of Wisconsin-Madison, Madison, WI 53706-1691

Liquid crystalline elastomers exhibit a variety of intriguing properties. In these materials, the optical and elastic properties of liquid crystals are combined with polymer chains to create a novel phase of matter. When polymerized under strain, liquid crystalline elastomers exhibit interesting anisotropic mechanical properties that can be used in a variety of applications, including sensors. Theoretical and computational studies of liquid crystalline elastomers have been limited. In our work, we have developed a model for liquid crystals, elastomers, and liquid crystalline elastomers that is capable of describing the structure and properties of these materials over a wide range of length scales. The model relies on a self-consistent field formulation of the system. The model is solved by resorting to a Monte Carlo approach that facilitates description of interfaces and nanoscale inclusions. The validity of the model is illustrated by describing the behavior of bulk liquid crystals and polymeric molecules at liquid crystal interfaces. Results are then presented for liquid crystalline elastomers in the bulk.

447e Cavitation Rheology and Fracture Behavior of Polyacrylamide Hydrogels

Santanu Kundu, Department of Polymer Science and Engineering, University of Massachusetts-Amherst, 120 Governors Drive, Amherst, MA 01003 and Alfred J. Crosby, Polymer Science and Engineering Department, University of Massachusetts-Amherst, 120 Governors Drive, Amherst, MA 01003.

Cavitation rheology is a new characterization technique for the measurement of mechanical properties on small length scales, e.g. 10 -100 µm, at any arbitrary location within a soft material. The technique involves growing a cavity at the tip of a syringe needle and monitoring the pressure of the cavity at the onset of instability. This critical pressure is directly related to the local modulus of the material. We used this technique to characterize the network mechanics of polyacrylamide hydrogel materials, a common material used in many biological applications. Mechanical properties were characterized for the gels with different dimensions and different moduli, which were obtained by varying initial monomer to water ratio. With the increase of monomer concentration, a transition from cavitation to fracture was observed and the results indicate that the critical pressure for fracture initiation also scales with the elastic modulus of a gel. These results provide a quantitative foundation for the extension of this technique to in vivo characterization of biological tissues.

447f Well-Defined Polymeric Structures Based on Novel Degradable Cross-Linkers

Efrosyni Themistou, Department of Chemical and Biological Engineering, University at Buffalo - The State University of New York (SUNY), 303 Furnas Hall, Buffalo, NY 14260-4200 and Costas S. Patrickios, Department of Chemistry, University of Cyprus, P. O. Box 20537, Nicosia, 1678, Cyprus.

Non-linear, branched polymers have attracted significant attention due to their unique structure and properties. Of particular interest to our group are polymer networks (soft materials – gels) whose branching points are cleavable under appropriate chemical or physical stimuli. The properties of polymer networks containing degradable crosslinks undergo dramatic changes upon degradation and branch removal, thus rendering such polymers ideal in biomedical and electronics applications (as hydrolyzable biomaterials for controlled drug delivery and tissue engineering, and thermolyzable polymers for removal and re-attachment of components, respectively). The investigation of these changes for chemically different branching points and under various degradation conditions constitutes a field of active research.

Our work is the first to address the synthesis via a "controlled" polymerization method, group transfer polymerization (GTP), characterization and degradation of polymer networks and star polymers based on cross-linkers degradable under different conditions. The novelty of our work lies on: (1) the use of newly synthesized novel degradable cross-linkers, bearing different labile groups; and (2) the control of the architecture, molecular weight and polydispersity of the synthesized polymers – neat cross-linker networks, randomly cross-linked networks, and cross-linked star polymer model networks and their constituent star polymer nanoobjects – and their degradation products.

The successful synthesis and degradation (hydrolysis or thermolysis) of such well-defined polymer networks was established with a variety of experimental techniques including Gel Permeation Chromatography (GPC), Nuclear Magnetic Resonance (NMR) spectroscopy, Dynamic and Static Light Scattering (DLS and SLS), and Thermogravimetric Analysis (TGA). Structure-property relationships were established.

447g The Impact of Sol Molecular Weight on the Mechanical and Adhesive Properties of Polymer Gels

Randy A. Mrozek[1], Phillip J. Cole[2], and Joseph L. Lenhart[1]. (1) Sandia National Laboratories, PO Box 5800 MS 1411, Albuquerque, NM 87185, (2) NNSA Satellite Programs, Sandia National Laboratories, P.O. Box 5800, MS-1411, Albuquerque, NM 87122

The addition of non-reactive soluble material (sol) to a polymer network can have a dramatic impact on the resulting physical properties. The degree of property change is dependent on the sol quality and loading. We are interested in determining the impact of the sol size on the mechanical and adhesive properties of the gel. To investigate the impact of the sol, a model silicone system with the ability to control the crosslink density was used. The incorporated sols have a molecular weight range from ~1,000 to ~ 400,000 g/mol. The initial results from a model network with a low crosslink density indicate a distinct change when the molecular weight of the sol is below and above the molecular weight of entanglement. The scaling factors for sol sizes below the molecular weight of entanglement approach the theoretical scaling factor of 2.3 determined for networks formed in the presence of a theta solvent. Sols with molecular weight higher then the molecular weight of entanglement exhibit a decreased scaling factor, indicating a reduced influence on the storage modulus, and a more pronounced frequency dependence. The decreased scaling factor is attributed to the higher molecular weight sols ability to entangle with the polymer network and contribute to the modulus. The frequency dependence is likely the result of the longer reptation times of the higher molecular weight sols resulting in a higher effective storage modulus at high frequencies. The study is currently being expanded to network gels that have crosslink densities at and below the molecular weight of entanglement. We anticipate that the results of this study will aid in the design of gel materials to determine the optimum sol loading and molecular weight at a given crosslink density to provide tailored adhesive and mechanical performance.

462a Fine-Tuning Notch Signaling to Promote Angiogenesis

Lan Cao, Praveen Arany, Yuan-Shuo Wang, and David J. Mooney. School of Engineering and Applied Sciences, Harvard University, Cambridge, MA 02138

Promoting angiogenesis is both a critical aspect for tissue regeneration and a potential effective therapy for cardiovascular diseases. Inspired by the recent findings that systemic introduction of Notch inhibitors reduced blood flow to tumor tissues by forming excessive yet dysfunctional vasculature, we proposed that precisely controlled partial and local Notch inhibition might enhance regional neovascularization, by altering the responsiveness of local endothelial cells and/or progenitors to angiogenic stimuli. We tested our hypothesis with in vitro assays with human endothelial cells, as well as in a rodent hindlimb ischemia model. In vitro, an optimal level of inhibition of Notch signaling by gamma secretase inhibitor (GSI) significantly increased the proliferation, migration and sprout formation of human endothelial cells (derived from diabetic and non-diabetic subjects) in response to VEGF stimulation. In vivo, delivery of an appropriate combination of GSI and VEGF from an injectable alginate hydrogel system to ischemic hindlimbs led to a greater recovery of blood flow than VEGF or GSI alone; perfusion levels reached 80% of the normal level by week 4 with combined GSI and VEGF delivery. Strikingly, direct intramuscular or intraperitoneal injection of GSI did not result in the same level of improvement, suggesting that the extended presence of GSI (gel delivery) is important for its activity. Potential side effects of GSI delivery were also investigated by examining the differentiation of crypt cells in the small intestine. The optimal dose of GSI delivered from alginate hydrogels did not show any adverse effects, in contrast to systemic introduction of GSI. Altogether, these results suggested a new approach to promote angiogenesis by fine-tuning Notch signaling, and may provide new options to treat patients with diseases, such as diabetes, that can diminish angiogenic responsiveness.

462b Induction of Angiogenesis In Tissue Engineered Scaffolds for Bone Repair: A Combined Gene Therapy-Cell Transplantation Approach

Ehsan Jabbarzadeh, Chemical Engineering, California Institute of Technology, 1200 E. California Blvd., Pasadena, CA 91125 and Cato T. Laurencin, Department of Orthopaedic Surgery, Chemical Engineering, and Biomedical Engineering, University of Virginia, Charlottesville, VA 22904.

One of the fundamental principles underlying tissue engineering approaches is that newly formed tissue must maintain sufficient vascularization to support its growth. Efforts to induce vascular growth into tissue engineered scaffolds have recently been dedicated to developing novel strategies to deliver specific biological factors that direct the recruitment of endothelial cell (EC) progenitors, and their differentiation. The challenge, however, lies in orchestration of the cells, appropriate biological factors and optimal factor doses. This study reports a novel approach as a step forward to resolve this dilemma by combining an ex-vivo gene transfer strategy and EC transplantation. The utility of this approach was evaluated using three dimensional (3-D) poly(lactide-co-glycolide) (PLAGA) sintered microsphere scaffolds for bone tissue engineering applications. Our goal was achieved by isolation and transfection of adipose derived stromal cells (ADSCs) with adenovirus encoding the cDNA of vascular endothelial growth factor (VEGF). We demonstrated that the combination of VEGF releasing ADSCs and ECs results in marked vascular growth within PLAGA scaffolds. We, thereby, delineate the potential of ADSCs to promote vascular growth into biomaterials.

462c Assessment of Biotin-Streptavidin Surface Stability In the Presence of Various Proteases for Controlled Release of DNA Polyplexes

Talar Tokatlian and Tatiana Segura. Chemical and Biomolecular Engineering, University of California, Los Angeles, 420 Westwood Blvd., 5531 Boelter Hall, Los Angeles, CA 90095

Non-viral gene delivery is an ideal approach to guide tissue regeneration through the delivery of genes that encode for bioactive signals. However, gene delivery efficiency and controlled release of DNA from tissue engineering scaffolds have proven to be major limitations in their widespread application. One possible solution to address both of these issues is to immobilize and release DNA near the proximity of the targeted cells. Immobilization of polyplexes to cell binding surfaces has been shown to enhance non-viral gene transfer, and through the modulation of the tether degradation rate, one could control the polyplex release rate resulting in controlled transgene expression. Our group has proposed that through the immobilization of DNA/PEI complexes to biomaterials via protease sensitive tethers, release can be targeted to cells that express these proteases. In our proposed approach the cationic polymer used to complex DNA is modified with protease sensitive peptide tethers. These tethers also contain a functional group (e.g. biotin) that can be used to immobilize the polyplexes to a cell-binding surface that has an affinity to the functional group (e.g. streptavidin). Biotin-streptavidin systems have been used widely for immobilization and purification of proteins and have been previously used for the immobilization of biotinylated polyplexes. However, to date there have been no studies conducted regarding the stability of streptavidin surfaces in the presence of various buffers and proteases. We conducted a thorough study to analyze the robustness of streptavidin surfaces and in preventing release of fluorescently labeled biotinylated nanoparticles in the presence of phosphate buffered saline (PBS), conditioned cell media, matrix metalloproteinase 2 (MMP-2), collagenase I and trypsin. Particles were immobilized to a streptavidin surface and incubated at 37°C for up to 128 hours with the solution of interest. Surface fluorescence was quantified using a Typhoon scanner and particle release was directly correlated to biotin-streptavidin stability. After an initial burst release of loosely bound particles over the first 20 hours, the surface proved to be stable in PBS, mesenchymal stem cell (MSC) conditioned media, 50ng/ml MMP-2, and 1.25mg/ml collagenase I, averaging only 10%, 8%, 8%, and 15% particle release from the surface over the following 100 hours, respectively. Only trypsin proved to be detrimental to the surface stability with a sharp burst release followed by steady continued release. Overall, incubation in trypsin resulted in 78% of the particles being released from the surface over 120 hours. However, surfaces incubated in trypsin for 48 hours post 120 hours of incubation in PBS showed only 19% release indicating that prolonged incubation in PBS may have aided in further stabilizing the biotin-streptavidin bond. To ensure that the biotin-streptavidin bond was also stable in the presence of cells, multiple cell types (HeLa, MSC, M202, and PAE) were incubated on top of the bound particles and particle internalization was qualitatively assessed for up to 5 days using a fluorescence microscope. Extensive particle internalization would indicate the surface is not stable for cell culture. No extensive particle internalization was observed for any cell type even when the cells were found to express high levels of proteases enzymes. The immobilization of DNA/PEI polyplexes via a biotinylated peptide tether to a protease resistant streptavidin surface would ensure that release kinetics would be dominated by tether sensitivity to the proteases and not surface instability. We have shown here that the biotin-streptavidin surface is an ideal 2D system to study DNA/PEI polyplex release and cell internalization kinetics.

462d Extracellular Matrix Composition Influence on Gene Delivery In Hydrogels

Jackie A. Shepard and Lonnie D. Shea. Chemical and Biological Engineering, Northwestern University, 2145 Sheridan Road, Evanston, IL 60208

Hydrogels are a central component of many approaches to engineer functional tissue replacements. These hydrogels exhibit mechanics similar to native tissue, and cell adhesion and matrix degradation can be readily manipulated. More recently, gene delivery from hydrogels has been investigated as a means to modulate the local microenvironment. We are investigating the design of hydrogels for gene delivery, and this report addresses the role of the adhesive environment on transfection. Previous reports from our laboratory have indicated that fibronectin significantly enhanced gene transfer in the traditional 2D cell culture. To regulate cell adhesion in a hydrogel, we employ a 4-arm PEG backbone, with bifunctional degradable crosslinker, and monofunctional ECM-based adhesion sites. Fibrin hydrogels were employed as a reference hydrogel, as it is a natural material that supports cell adhesion. PEG hydrogels were formed with RGD peptides, fibronectin, or the fibronectin fragment (Fn 7-10). The obtainable ranges of mesh sizes calculated from swelling experiments were between 14 to 27 nm, which would be expected to retain DNA complexes within the network until sufficient degradation allowed for release. Cell migration is supported within the hydrogels, indicating the ability to adhere to the matrix and degrade the hydrogel. Release profiles using radiolabeled DNA indicated that 20% of the encapsulated DNA complexes were released by burst effect, followed by a sustained release for 20 days. Confocal microscopy was used to visual complexes within the hydrogel, and indicated complexes were stable without aggregation. Three configurations of cell and DNA encapsulation were investigated to simulate the multiple uses of hydrogels. The first two approaches involve encapsulating cells within the hydrogel to simulate the use of the hydrogel for *in vitro* tissue models or a cell transplantation vehicle. These two approaches involve a mixing order of i) cells, DNA, and precursor solution or ii) DNA, precursor solution, and cells, followed by gelation. Finally, DNA and precursor solution were mixed, gelled and cells seeded on top, which simulates cell infiltration into the hydrogel. The first two approaches produced substantially higher cell transgene expression relative to the final approach. ECM has widely been employed to regulate cell function, and these studies indicate that the ECM identity and density can also regulate transgene expression. Our results demonstrate the ability to achieve transfection of cells *in vitro* while supporting cell proliferation and infiltration, which may be useful for a range of applications.

462e Hydrogel Composite Materials for Long-Term Neurotrophin Delivery In Neural Prostheses

Ning Han, Chemical and Biomolecular Engineering, Ohio State University, 125 Koffolt Laboratories, 140 West 19th Ave., Columbus, OH 43210, Jed Johnson, Material Science and Engineering, Ohio State University, Columbus, OH 43210, John J. Lannutti, Department of Material Science and Engineering, Ohio State University, Columbus, OH 43210, and Jessica O. Winter, Chemical and Biomolecular Engineering/Biomedical Engineering, Ohio State Unversity, 140 W 19th Ave, Columbus, OH 43210.

Long-term implanted electrode arrays have a significant importance in neural prosthetics for recording the electrical signals from nearby neurons and generating electrical signals to stimulate nearby tissue. Therefore, the density of neurons and their proximity to the electrode sites play an important role in the electrode performance. In the past decades, biodegradable and biocompatible polymers have been used as electrode coatings to minimize the acute and chronic inflammatory response and to preserve neurons near the recording sites. Since poly(ethylene glycol) (PEG) is nontoxic, non-immunogenic, and inert to most biological molecules, such as proteins, block copolymers of PEG have been studied in our group to form hydrogels as the scaffolds that can support or stimulate neuron growth. The results of our previous work demonstrated that the hydrogels could be used as a short-term (1~3 weeks) controlled delivery system for neurotrophins (the growth factors which can stimulate the neurite extensions) in vitro. To further reduce the acute immune response (usually lasts for 6~8 weeks) caused by the insertion of electrodes and to enhance integration of neurons with electronic circuitry, we have constructed composite drug delivery systems comprised of polymeric hydrogels (i.e., poly(ethylene glycol)-poly (lactic acid), (PEGPLA) or poly(ethylene glycol)-polycaprolactone, (PEGPCL)), and other vehicles, such as biodegradable poly(lactic co-glycol acid) (PLGA) microspheres or polycaprolactone (PCL)-based electrospun nanofibrous scaffolds, to provide neurotrophins at a predetermined rate for 2~3 months in vivo. The stable and sustained release of neurotrophins can remarkably enhance the attraction, attachment and restoration of neurons around the chronically implanted electrodes.

462f Release of Insulin Like Growth Factor-1/Green Fluorescent Protein Plasmids from PEG Coated Porous Eh Networks

Erin E. Falco, Department of Chemical and Biomolecular Engineering, University of Maryland, College Park, MD 20742, J. Scott Roth, Department of General Surgery, University of Maryland Medical School, Baltimore, MD 21201, and John P. Fisher, Fischell Department of Bioengineering, University of Maryland, 3238 Jeong H. Kim Building, College Park, MD 20742-2111.

Incisional hernias are a common clinical problem occurring in up to 10% of all patients undergoing abdominal procedures. Primary closure, synthetic biomaterials, as well as xenografts and allografts have been used to treat incisional hernia defects. Despite these approaches, the incidence of hernia recurrence ranges from 24% to 54%. To address this high recurrence rate, we propose an incisional hernia treatment that utilizes a functional biomaterial developed for skeletal muscle regeneration. To this end, the use of a porous cyclic acetal biomaterial (EH network), based on 5-ethyl-5-(hydroxymethyl)-β,β-dimethyl-1,3-dioxane-2-ethanol diacrylate, coated with poly(ethylene glycol) (PEG) to deliver non-viral lipoplexes encoding for insulin like growth factor 1 (IGF-1) and green fluorescent protein (GFP) in vitro was investigated. Porous EH networks were fabricated using a leachable porogen strategy. Briefly, the EHD monomer was combined with 7wt% benzoyl peroxide (BP) in acetone, 70wt% NaCl and the accelerant N,N-dimethyl-p-toluidine (DMT). The slurry was cast in a glass mold to form porous EH sheets. PEG was then reacted onto one side of the EH network using the same radical initiator solution. To characterize these scaffolds, rectangles with 14x3 mm dimensions were made. These rectangles were then placed in 7.4 pH PBS for 1, 4, 8, 14, 21, 28, 56, 84 days. At each time point the scaffolds were weighed and tested in tension to determine the mass lost and yield strength. Since results indicate that these scaffold have clinically relevant degradation times and mechanical strength, additional scaffolds were fabricated, washed and sterilized under UV light overnight. To demonstrate their utility as a plasmid carrier, IGF-1/GFP plasmids were constructed from the IGF-1 gene which was isolated from human skeletal myoblasts. Plasmids were complexed with Lipofectin (Invitrogen) in a ratio of 1:7. Networks were loaded with three different complex concentrations and the complexes were released into growth media. Released plasmids were analyzed for both the plasmid DNA and lipid concentrations using picogreen and an ammonium ferrothiocyanate solution respectively. Bioactivity of the released plasmid was evaluated through the tranfection of human skeletal myoblasts. Results showed that the networks were able to sustain the release of the lipoplexes over a 48 hours time period with the complexes remaining bioactive.

An integrated biodegradable, immunoprotective, dual nanoporous capsule for cell-based therapies

Xulang Zhang,[a] Hongyan He,[a, b] Chi Yen,[a, b] Winston Ho,[b] and L. James Lee[a, b]

[a] NSF Nanoscale Science and Engineering Center for Affordable Nanoengineering of Polymeric Biomedical Devices, The Ohio State University, Columbus, Ohio 43210

[b] Department of Chemical and Biomolecular Engineering, The Ohio State University, Columbus, OH, USA

1. Introduction

Cell-based therapeutic strategy has been around for two decades. Primary cells or cell line with specific secretion can be used as "seed" cells for therapeutic function, such as pancreatic B cell secreting insulin for diabetes therapy [1]. Immunoprotection of transplanted cells and tissue by size-based semipermeable membranes is an attractive approach for cell-based therapy to treat patients with such diseases as Alzheimer's, diabetes mellitus, hepatic disease, etc. [2-5]. Such cell-based devices are thought to hold great promise in applications requiring site-specific and sustainable drug delivery of cell-synthesized molecules. To design the immunoprotection devices, the semipermeable membrane plays a critical role. Typically, its pore size must be in a few nanometers to allow diffusion of small nutrient molecules but to prevent the passage of large immune molecules to achieve high immunoisolation. Diffusion chamber is a device used in such applications. In medical and pharmaceutical fields, the drug-loaded chambers are capable of controlling the release rate and keeping the concentration within the therapeutic windows for a long time [6]. Micromachined silicon membranes containing arrays of channel-like nanopores and alumina nanopore membranes produced by anodization have been used in biocapsules for the immunoisolation of transplanted pancreatic islet cells to treat diabetes [7-9]. The precisely controlled pore size of ~10 nm effectively prevents cellular and humoral immune species from penetrating through the membranes. These pores, on the other hand, allow sufficient passage of oxygen and nutrients to maintain cell viability and permit the reasonable release of insulin. However, if implanted, these devices must be surgically removed after use. Inexpensive polymers such as poly (ε-caprolactone) (PCL), on the other hand, can be fully biodegradable. A variety of low-cost techniques are available for the fabrication of polymeric micro/nanostructures, including nanopore membranes [10].

In this work, an immunoprotective and biodegradable delivery system combining two nanoporous capsules was developed for releasing cell-secreted product (Figure 1(a)). Figure 1(b) shows the schematic of microcapsules for cell encapsulation. Microencapsulation is a technology that encloses chemicals, drugs, enzymes, and cells within a membrane. A microcapsule is a hollow sphere with a diameter that can be controlled in the range of 200-1500 μm and a biocompatible semipermeable membrane with a surface pore size of 5-200 nm. These pores allow the bidirectional diffusion of nutrients, oxygen, secreted therapeutic products, and waste, but can prevent the penetration of high molecular weight substances from the microcapsule. Due to the 3-D microenvironment in microcapsules, transplanted cells, can grow well and provide therapeutic effect to treat a wide range of diseases from endocrine disorders and central nervous system diseases [11, 12]. Figure 1(c) describes the fabrication protocol for the large nanoporous capsule including the formation of a chamber, design of the release gate, and device assembly. As a cell-based release system, the chamber and the membrane need to provide sufficient space for good cell viability and proper nanopores for nutrient and oxygen diffusion and immunoprotection. The outer surface of the chamber membrane also needs to be biocompatible to prevent fibrosis and other inflammatory responses for in vivo applications. In this study, we investigate the cell growth profile and the secreted product release in the integrated device. For comparison, the same measurements were also carried out in individual capsules.

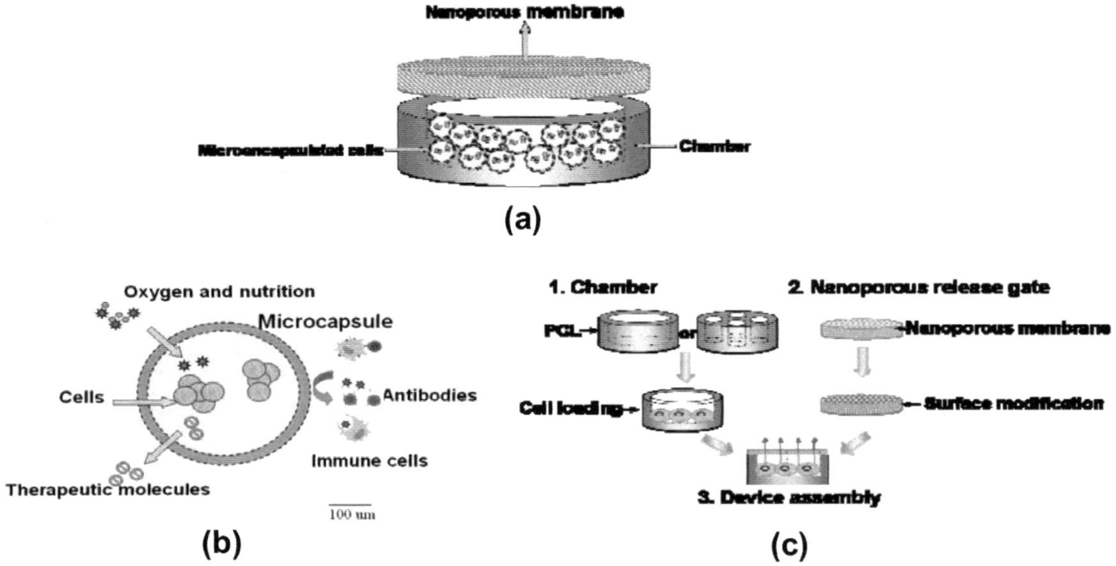

Figure 1. Schematic diagram of biodegradable nanoporous membrane systems. (a) integrated device; (b) microcapsule; (c) PCL large capsule with nanoporous gate.

2. Materials and Methods

2.1 Chemicals and reagents

Alginate sodium (low viscosity), poly-L-lysine (MW15,000~30,000), PCL (MW 70,000), lysozyme (MW 14,700), poly(ethylene glycol) diacrylate (PEGDA), and monothioglycerol were purchased from Aldrich Chemicals (Milwaukee, WI). 1,4-

dioxane was obtained from Mallinckrodt Chemicals (Philipsburg, NJ). 2-methoxyethanol (ACS reagent, ≧93%) was purchased from Sigma-Aldrich (St. Louis, MO). Citrate sodium was provided by Fluka Biochemika. EnzChek® Lysozyme assay kit, Phosphate Buffered Saline (PBS).

2.2 Device design and fabrication

Figure 1(c) describes the fabrication protocol for the PCL nanoporous capsule. This miniature device consists of two parts: a nanoporous membrane gate and a chamber for loading mES cells or microcapsules. PCL nanoporous membranes were prepared via the combination of thermally and nonsolvent induced phase separations [13]. PCL was first dissolved and stirred well in a diluent to form a 20 wt. % solution at 50°C for 2 hr. The diluent was a mixture of dioxane as solvent and 2-methoxyethanol as nonsolvent with a ratio of 15 to 65 by weight. After complete dissolution, the polymer solution was cast over a Teflon plate and immersed in a coagulation bath, i.e. water, at 5°C. After 6 hr, the resulting membrane was peeled from the Teflon plate and dried in a vacuum oven for 24 hr at room temperature. To prevent the fibrosis formation, hydrophilic PEG molecules were grafted on the PCL surface by using plasma technique [14].

An array of wells, each 4.5 mm in diameter and 1.8 mm in height, was first patterned on an aluminum plate using CNC machining. Then, a mixture of poly (dimethyl siloxane) (PDMS) resin and its curing agent (SYLGARD 184, Dow Corning) was poured onto the patterned mold. After 2 hr of curing at 70°C, the PDMS mold with an inverse pattern (i.e. pillars) was peeled off for hot-embossing of the PCL chamber. The PCL powders were first added into another aluminum mold patterned with slightly larger wells.

The mold was placed on a hotplate and set at 140°C for melting PCL. The PDMS mold was then aligned with the aluminum mold and the PDMS pillars were pushed into the molten PCL in the wells. After 5 min, the setup was cooled down to below 40°C and the embossed chambers were taken out from the mold. After the PCL chambers were grafted with PEG molecules using the same procedure, the inside chamber was coated with 1.0 wt. % gelatin solution for mES cell culture.

Figure 1(b) shows the schematic of microcapsules. AP microcapsules were prepared as described previously with modification [15]. Briefly, mES cells (2×10^6 cells/mL) were suspended in 2 % w/v sterilized sodium alginate and extruded through a 27-gage needle into 100 mM $CaCl_2$ using an electrostatic droplet generator (NISCO, Sweden) to form calcium alginate gel beads. The voltage used was 10 KV and the distance between the needle tip and the solution level was 1.5 cm. The gel beads were incubated with 0.05 % w/v poly-L-lysine to form alginate-poly-L-lysine membrane around the surface. The membrane-enclosed gel beads were further suspended in 55 mM sodium citrate to liquefy the alginate gel core. Resulted AP microcapsules were 300-400 μm in diameter. Microcapsules with mES cells were cultured in a 24-well plate at 37°C in 5% CO_2 in the mES maintenance medium. The integrated device was formed by loading microcapsules containing cell in the PCL capsule before bonding with nanoporous membrane (shown in Figure 1(a)).

2.3 *In vitro release performance*

To investigate the release performance through nanoporous membranes, a hydrophilic protein, lysozyme, was selected as a model drug and lysozyme diffusion through PCL membranes was conducted in a Costar Transwell Test Setup. A modified PCL membrane was cut into a disc shape of 1.4 cm in diameter and bonded to the bottom of the Transwell insert by the bioadhesive to separate the two chambers. After the adhesive was completely solidified, 0.3 mL of a 1.0 mg/mL lysozyme solution was pippeted into the donor chamber and allowed to diffuse into the receptor chamber containing 3.7 mL PBS (Phosphate Buffered Saline) solution with 0.1% sodium azide. These diffusion experiments were subjected to constant shaking at 100 rpm. At various time intervals, the donor chamber was placed in a new receptor chamber containing 3.7 mL fresh buffer. Samples containing diffused lysozyme were analyzed using EnzChek[@] Lysozyme assay kit. To investigate lysozyme diffusion through microcapsule surface, empty microcapsules were placed in 1.0 mg/mL lysozyme solution. The remaining lysozyme concentration in the solution was measured at different time intervals.

2.4 Scanning Electron Microscopy (SEM) for membrane morphology

To visually examine the surface morphology of nanoporous PCL membranes, a Hitachi Model S-4300 SEM was used to analyze the porous structure. The air-dried samples were loaded on the surface of an aluminum SEM specimen holder and sputter coated with gold for 40 s (Pelco Model 3 Sputter Coater) before observation. A working distance about 8-10 mm, an accelerating voltage of 10 KV, and a chamber pressure of

10^{-8} Torr were found to be suitable for obtaining high-resolution images. The magnification used in this study varied from 2000× to 120,000× depending on the pore structure.

3. Results and Discussion

3.1 SEM results of PCL membrane

Figure 2 shows the SEM images of the PCL membranes with or without surface treatment. PCL membrane has an asymmetric structure consisting of a thick porous support layer (bottom) and a thin and dense layer (top). The thick support layer contributes to the mechanical stability of the membranes. During the casting step, the solvent molecules located at the bottom of the cast film had difficulty for evaporation and

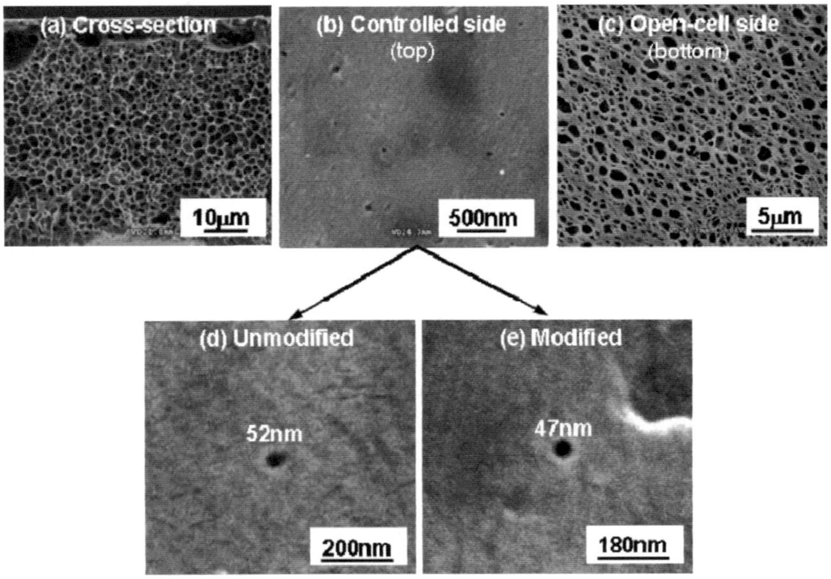

Figure 2. SEM images of PCL membrane formed with 20 wt% in mixture of dioxane and 2-Methoxyethanol with a ratio of 15 to 65 by weight (evaporation time: 60 sec, temperature of water bath: 5 °C): (a) cross-section, (b) controlled side from top view, (c) open-cell side from top view, (d) controlled side of unmodified PCL membrane using higher magnification, and (e) controlled side after PEG grafting.

could easily diffuse into the water bath for liquid-liquid diffusion, leading to micro-sized open cells (shown in Figure 2(c)). On the other hand, the thin and dense layer has nano-sized pores and is used for rate controlling in membrane separation processes and controlled drug delivery. On the dense side image in Figure 2(b), the average pore size is around 50nm before PEG grafting (shown in Figure 2 (d)). After PEG molecules grafted on the PCL surface using plasma technique, the membrane showed a slightly lower pore size. This can be clearly seen from SEM image in Figure 2(e). For microcapsule alginate and poly-L-lysine membrane, the cut-off molecular weight is around 72,000 Da (data not shown).

3.2 In vitro diffusion through nanoporous membranes

Lysozyme was selected as a model drug to demonstrate the releasing profile with molecular weight of 14,700 Da. Figures 3 compared the release pattern in the two types of capsules. For microcapsules, lysozyme released very fast in the first 3-4 hr because the surface area to volume ratio is very large and the diffusion distance is very short implying that microcapsules were not suitable for controllable and sustained release of drug. On the other hand, the lysozyme diffusion through the PCL membrane increased very slowly.

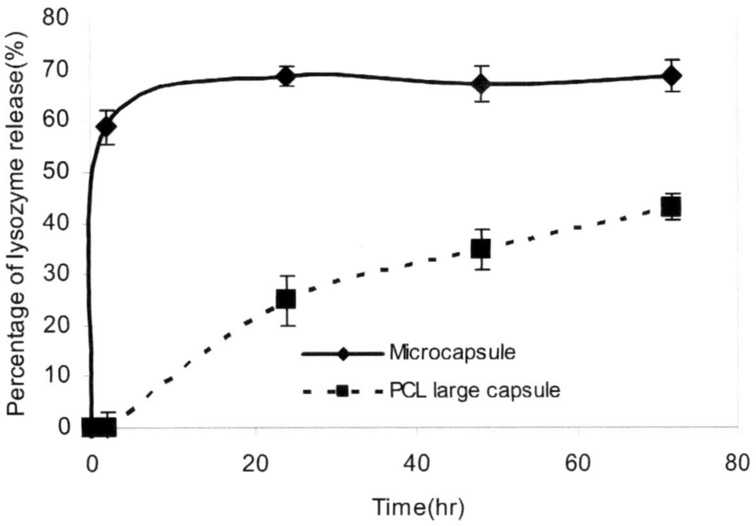

Figure 3. Lysozyme diffusion through the nanoporous membrane in microcapsule and PCL large capsule systems.

4. Summary

In summary, a new biodegradable cell-based device has been developed based on the integration of two techniques: microcapsules and surface coated PCL capsule. Microcapsules provide a 3-D microenvironment for spatial cell growth with good viability and proliferation. Coating biocompatible and hydrophilic PEG gelatin on the PCL surface could mediate the inflammatory response, minimize the fibrosis formation, and maintain the release performance. Our ongoing work is to test AP and PCL nanoporous membrane immunoprotection and controlled release functions for therapeutic cell secreting products. Hopefully, the dual nanoporous construct provides a unique way to allow superior cell growth, immuoprotection, fibrosis prevention and controllable release of secreted product in a biodegradable device.

Acknowledgements

This work is funded by the National Science Foundation sponsored Nanoscale Science and Engineering Center for Affordable Nanoengineering of Polymeric Biomedical Devices (NSEC-CANPBD) at The Ohio State University.

References:

[1] Kofman AV, Kharazova AD, Pinaev GP. Gene- and cell-based therapy of diabetes: between bench and bedside. Tsitologiia 2002; 44:1157-1177.

[2] Aebischer P, Winn SR, Galletti PM. Transplantation of neural tissue in polymer capsules. Brain Res 1988; 448:364–368.

[3] Colton CK. Implantable biohybrid artificial organs. Cell Transplant 1995;4:415–436.

[4] Desai TA, Hansford DJ, Ferrari M. Micromachined interfaces: new approaches in cell immunoisolation and biomolecular separation. Biomol Eng 2000; 17: 23-36.

[5] Robert PL, William LC, Willem M. Microcapsules and composite microreactors for immunoisolation of cells, US Patent No: 402209, 1995.

[6] LaVan DA, McGuire T, Langer R. Small-scale systems for *in vivo* drug delivery. Nat. Biotech. 2003; 21:1184-1191.

[7] Desai TA, Chu WH, Rasi G. Microfabricated biocapsules provide short-term immunoisolation of insulinoma xenografts. Biomed microdevices 1999;1:131-138.

[8] Walczak RJ, Boiarski A, West T, Shapiro J, Sharma S, Ferrari M. Long-term biocompatibility of NanoGATE drug delivery implant. NanoBiotech 2005;1:35-42.

[9] LaFlamme KE, Mor G, Gong D, Tempa T, Fusaro VA, Grimes CA, et al. Nanoporous alumina capsules for cellular macroencapsulation: transport and biocompatibility. Diabetes Technol Ther 2005;7: 684-694.

[10] Lin WJ, Lu CH. Characterization and permeation of microporous poly(ε-caprolactone) films. J Mem Sci 2002; 198:109-118.

[11] Orive G, Hernandez RM, Gascon AR, Igartua M, Pedraz JL. Survival of different cell lines in alginate-agarose microcapsules. Eur J Pharm Sci 2003;18:23-30.

[12] Jeon Y, Kwak K, Kim S, Kim Y, Lim J, Baek W. Intrathecal implants of microencapsulated xenogenic chromaffin cells provide a long-term source of analgesic substances. Transplant Proc 2006;38: 3061-3065.

[13] Yen C., He H., Lee L.J., Ho W. Synthesis and characterization of polycaprolactone nanoporous membranes for liquid-phase controlled release. J Mem Sci. (submitted).

[14] Carlisle ES, Mariappan MR, Nelson KD, Thomes BE, Timmons RB, Constantinescu A, et al. Enhancing hepatocyte adhesion by pulsed plasma deposition and polyethylene glycol coupling. Tissue Eng 2000; 6: 45-52.

[15] Ma X, Vacek I, Sun A. Generation of alginate-poly-Llysine-alginate (APA) biomicrocapsules: the relationship between the membrane strength and the reaction conditions. Artif Cells Blood Subtit Immobil Biotechnol 1994; 22: 43-69.

463a Direct Relationship Between Enhanced Gene and Matrix Protein Expression by Osteoblasts Exposed to Bioactive Glass Ions

Venu G. Varanasi[1], Nicole Shabnam Barkhordar[1], Eduardo Saiz[2], Peter Loomer[1], Bernadette Ancheta[1], Nakako Uritani[1], Antoni Tomsia[2], Sally J. Marshall[1], and Grayson W. Marshall[1]. (1) University of California, San Francisco, 707 Parnassus Avenue MS 0758, San Francisco, CA 94143, (2) Lawrence Berkeley National Laboratory, 1 Cyclotron Road MS 62-203, Berkeley, CA 94720

Introduction and Specific Aim¶ Currently, titanium (Ti) implants are used successfully for tooth replacements in mandibular or maxillary bone (86% and 76% success rate, respectively [1]) and for bone replacement in the cranium (nearly 93% success rate [2]). However, faster bone integration and improvements in implant success rate and longevity are still needed. A new family of bioactive glasses (50-59 wt. % SiO_2) may enhance the osteointegration potential of Ti [3-15]. The bioactive glasses acts as a functionally graded interface that remains well adherent to titanium [15] while also forming a hydroxyapatite surface layer and releasing calcium and silicon ions to osteoblasts for direct bone bonding [14, 16].¶ Interestingly, it has been discovered that these ions released to osteoblasts may play an active role in osteoblast function. Foppiano and colleagues [17] and Hench and colleagues [18-27] found these ions enhance osteoblast gene expression, yet, no significant effort on downstream matrix protein production has been established. **Thus, this study tested the hypothesis that bioactive glass ions enhance osteogenesis and that a direct connection exists between enhanced gene expression and enhanced matrix protein synthesis.** The key osteogenic markers to be investigated here are collagen type 1 alpha 1 (Col1α1) and alpha 2 (Col1α2), core binding factor a (Cbfa1/Runx2), alkaline phosphatase (ALP), and osteocalcin.¶ **Materials and Methods**¶ Melt-derived experimental bioactive glass (6P53-b) and commercial Bioglass (45S5) were used in this study by soaking glass specimens in cell culture medium to make glass conditioned medium (Table 1). The 6P53 b and 45S5 glass conditioned mediaum were measured for their ion concentrations using inductively coupled plasma mass spectrometry (ICP MS) prior to addition to mouse osteoblast progenitors (MC3T3 E1.4, at confluence in multi-well plates). The cells were induced to differentiate by adding ascorbic acid (50 μ g mL^{-1}) to GCM and control media. Cells were lysed for both mRNA and protein for sample analysis. Gene expression (Col1α1, Col1α2, Runx2, osteocalcin) was measured using quantitative reverse transcription-polymer chain reaction (qRT PCR) and amplification curve results were analyzed using a sigmoidal curve fitting method [28]. Protein expression (ALP, osteocalcin) was measured using enzyme-linked immunosorbent assay (ELISA). Statistical comparisons ($p < 0.05$ for statistical significance) were made between GCM- and control-treated cells using two-way ANOVA (independent variables: time, treatment; dependent variables: gene, protein expression) and Tukey's for individual group comparisons. **Results** It was observed that increased silicon and calcium concentrations were contained in 45S5 and 6P53-b GCM as compared to control media (α-MEM) (Table 2). GCM treatment enhanced osteocalcin expression (Figure 1). The expression of osteocalcin peaked (Figure 1A) for cells treated with ascorbic acid and 6P53 b GCM (40x control after 10 days) and 45S5 GCM (70x control after 6 days). These significantly increased osteocalcin protein levels correlated with the increased osteocalcin mRNA levels within 7 days of treatment. While MC3T3-E1.4 cells differentiated, osteocalcin gene expression reached a maximum after 5 days in 6P53 b GCM (14x control) and after 3 days in 45S5 GCM (19x control) (Figure 1B). Other markers measured included: Runx2 levels were approximately 2x control after 7 days exposure to GCM, which confirmed the results of Foppiano and colleagues [17]. Alkaline phosphatase activity was also 2x control for both GCM treated cells after 10 days of exposure. Finally, Col1α1 expression was enhanced after 1 day of exposure (45S5 GCM: 3x control; 6P53 b GCM: 4x control) and throughout the course of differentiation (day 5, Col1α2, 45S5 GCM: 3.15x control; 6P53 b GCM: 2.35x control).¶ **Conclusions**¶ This study tested the hypothesis that bioactive glass ions enhance osteogenesis and that there is a direct connection between enhanced gene expression and protein matrix production. The increased gene expression was seen for collagen type 1, Runx2, and osteocalcin, which fits into the known sequence of events of osteoblast differentiation in that collagen type 1 expression is followed by Runx2 expression and these two markers are necessary for downstream expression of osteocalcin and alkaline phosphatase [29]. These results are significant because to this point, no connection has been made to the enhanced gene expression by osteoblasts exposed to bioactive glass ions and their enhanced downstream matrix protein expression. Future research will explore the underlying extracellular and intracellular mechanisms that cause enhanced differentiation and impact mineralized-tissue formation.¶ **References**¶ [1] R. Adell, U. Lekholm, B. Rockler and P. I. Branemark, Int J Oral Maxillof 10 (6), 387-416 (1981).¶ [2] B. A. Miles, D. P. Sinn and G. G. Gion, J Craniofac Surg 17 (5), 889-897 (2006).¶ [3] E. Saiz, M. Goldman, J. M. Gomez-Vega, A. P. Tomsia, G. W. Marshall and S. J. Marshall, Biomaterials 23) 3749–3756 (2002).¶ [4] A. Pazo, E. Saiz and A. P. Tomsia, Acta Mater 46 (7), 2551-2558 (1998).¶ [5] T. Oku, K. Suganuma, L. R. Wallenberg, A. P. Tomsia, J. M. Gomez-Vega and E. Saiz, J Mater Sci-Mater M 12 (5), 413-417 (2001).¶ [6] S. Lopez-Esteban, E. Saiz, S. Fujino, T. Oku, K. Suganuma and A. P. Tomsia, J Eur Ceram Soc 23 (15), 2921-2930 (2003).¶ [7] J. M. Gomez-Vega, E. Saiz, A. P. Tomsia, T. Oku, K. Suganuma, G. W. Marshall and S. J. Marshall, Adv Mater 12 (12), 894-898 (2000).¶ [8] J. M. Gomez-Vega, E. Saiz, A. P. Tomsia, G. W. Marshall and S. J. Marshall, Biomaterials 21 (2), 105-111 (2000).¶ [9] J. M. Gomez-Vega, E. Saiz and A. P. Tomsia, Journal of Biomedical Materials Research 46 (4), 549-559 (1999).¶ [10] J. M. Gomez-Vega, A. Hozumi, E. Saiz, A. P. Tomsia, H. Sugimura and O. Takai, Journal of Biomedical Materials Research 56 (3), 382-389 (2001).¶ [11] S. Fujino, H. Tokunaga, E. Saiz and A. P. Tomsia, Mater Trans 45 (4), 1147-1151 (2004).¶ [12] S. Fujino, K. Morinaga, E. Saiz and A. P. Tomsia, Glass Sci Technol 75, 221-226 (2002).¶ [13] S. Foppiano, S. J. Marshall, E. Saiz, A. P. Tomsia and G. W. Marshall, Acta Biomaterialia 2:133-142 (2006).¶ [14] S. Foppiano, S. J. Marshall, G. W. Marshall, E. Saiz and A. P. Tomsia, J Biomed Mater Res A 71A (2), 242-249 (2004).¶ [15] D. R. Bloyer, J. M. Gomez-Vega, E. Saiz, J. M. McNaney, R. M. Cannon and A. P. Tomsia, Acta Mater 47 (15-16), 4221-4224 (1999).¶ [16] V. G. Varanasi, T. Vallortigara, P. M. Loomer, E. Saiz, A. P. Tomsia, S. J. Marshall and G. W. Marshall, presented at the Materials Research Society Symposium Proceedings, San Francisco, CA, 2006 (unpublished).¶ [17] S. Foppiano, S. J. Marshall, G. W. Marshall, E. Saiz and A. P. Tomsia, Acta Biomaterialia 3 (5), 765-771 (2007).¶ [18] I. D. Xynos, M. V. J. Hukkanen, L. L. Hench and J. M. Polak, J Pathol 187, 38a-38a (1999).¶ [19] I. D. Xynos, M. V. J. Hukkanen, J. J. Batten, L. D. Buttery, L. L. Hench and J. M. Polak, Calcified Tissue Int 67 (4), 321-329 (2000).¶ [20] I. D. Xynos, A. J. Edgar, M. Ramachandran, L. D. K. Buttery, L. L. Hench and J. M. Polak, J Pathol 193, 31a-31a (2001).¶ [21] I. D. Xynos, A. J. Edgar, L. D. K. Buttery, L. L. Hench and J. M. Polak, Journal of Biomedical Materials Research 55 (2), 151-157 (2001).¶ [22] I. D. Xynos, A. J. Edgar, L. D. K. Buttery, L. L. Hench and J. M. Polak, Biochem Bioph Res Co 276 (2), 461-465 (2000).¶ [23] J. R. Jones, P. Sepulveda and L. L. Hench, Journal of Biomedical Materials Research 58 (6), 720-726 (2001).¶ [24] J. R. Jones and L. L. Hench, Curr Opin Solid St M 7 (4-5), 301-307 (2003).¶ [25] L. L. Hench and J. M. Polak, Science 295 (5557), 1014 (2002).¶ [26] R. C. Bielby, R. S. Pryce, L. L. Hench and J. M. Polak, Tissue Eng 11 (3-4), 479-488 (2005).¶ [27] R. C. Bielby, I. S. Christodoulou, R. S. Pryce, W. J. P. Radford, L. L. Hench and J. M. Polak, Tissue Eng 10 (7-8), 1018-1026 (2004).¶ [28] H. Qiu, K. Durand, H. Rabinovitch-Chable, M. Rigaud, V. Gazaille, P. Clavere and F. G. Sturtz, Biotechniques 42 (3), 355-362 (2007).¶ [29] G. Z. Xiao, R. Gopalakrishnan, D. Jiang, E. Reith, M. D. Benson and R. T. Franceschi, J Bone Miner Res 17 (1), 101-110 (2002).¶

Table 1. Composition (wt. %) of Bioglass (45S5) and experimental bioactive glass (6P53-b)

	SiO_2	Na_2O	K_2O	MgO	CaO	P_2O_5
45S5 (Mo-Sci)	45.0	24.5			24.5	6.0
6P53-b (LBL)	52.7	10.3	2.8	10.2	18.0	6.0

Table 2. Ion Extract Concentration (mg L^{-1})

	Si^{4+}	Na^{+}	K^{+}	Ca^{2+}	Mg^{2+}	PO_4^{3-}
α-MEM	-	4300	325.0	49.0	37.6	28.8
45S5+ α-MEM	47.9 ± 10.4	4168 ± 404	299.8 ± 26.6	69.8 ± 14.0	27.8 ± 2.5	23.6 ± 4.7
6P53b+ α-MEM	33.4 ± 3.8	4458 ± 186	298.3 ± 34.9	57.1 ± 2.8	32.7 ± 5.4	32.6 ± 9.3

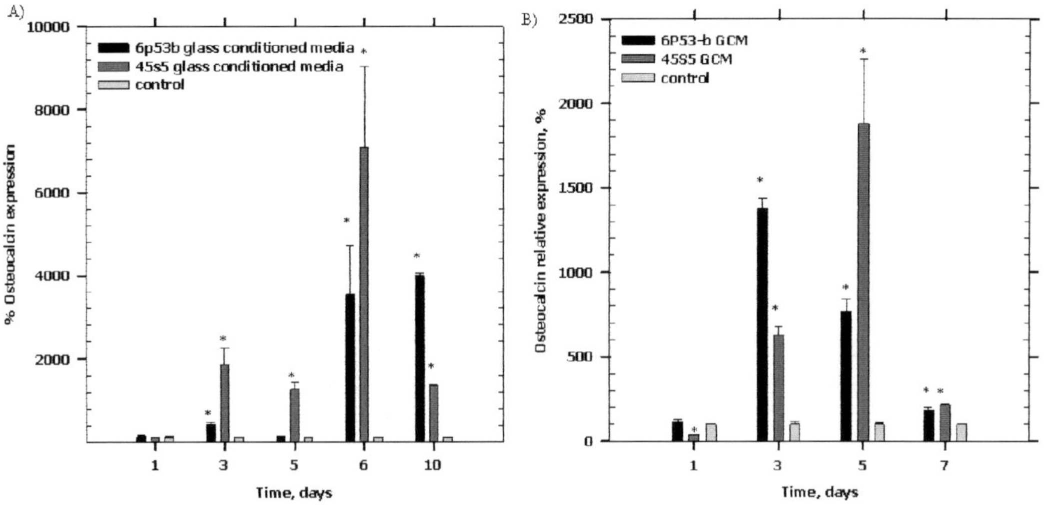

Figure 1. Osteocalcin expression after exposure to GCM and control treatment. A) Osteocalcin protein expression over 10 days of differentiation. B) Osteocalcin mRNA expression over 7 days of differentiation. (ANOVA, $p < 0.05$, * denotes statistical significance).

463b The Design of Wettability of Antifouling Materials on the Basis of the Molecular Properties of Marine Fouling Bacteria

Bong-Jae Park, Travis Haines, and **Nehal I. Abu-Lail**. School of Chemical Engineering and Bioengineering, Washington State University, 118 Dana Hall, P.O. Box 642710, Pullman, WA 99164

Our ability to design marine antifouling materials is severely limited by the lack of molecular-scale studies on how heterogeneities of the surface properties of micro-fouling bacteria affect their design. Motivated by the lack of such studies, we aimed at optimizing the wettability of antifouling materials necessary to repel fouling bacteria from marine submerged surfaces. Our approach was based on using atomic force microscopy (AFM) to quantify the nanoscale adhesion forces between a Gram-negative fouling bacterium (Acidovorax sp.) and surfaces that vary in their wettability in 0.5 M NaCl to represent the ionic strength of marine water. To vary the wettability, we decorated gold-coated AFM cantilevers with self assembled monolayers (SAMs) of ù-substituted alkanethiolates that ranged from hydrophilic (OH-SAMs) to hydrophobic (CH_3-SAMs). Our results clearly showed that the nanoscale adhesion forces of Acidovorax sp. to hydrophilic OH-terminated SAMs (0.491 ± 0.245 nN, n = 709) were significantly lower than adhesion forces measured between the same bacterium and CH_3-terminated SAMs (2.289 ± 1.352 nN, n = 771) (Mann-Whitney Rank Sum Test, P<0.001). Our preliminary results suggest that antifouling materials should be designed using hydrophilic materials to repel marine bacteria. Future work will quantify adhesion forces to a range of surface wettabilities created using mixed SAMs.

463c Fibrin-Infiltrated Small Intestine Submucosa as a Scaffold for Tissue Engineered Vessels Using Hair-Follicle Derived Smooth Muscle Progenitor Cells

Hao-Fan Peng[1], Jin Yu Liu[1], Juhee Han[1], Daniel D Swartz[2], and Stelios T. Andreadis[1]. (1) Chemical and Biological Engineering, SUNY at Buffalo, 916 Furnas Hall, buffalo, NY 14260, Buffalo, NY 14260, (2) Buffalo Children's Hospital, SUNY at Buffalo, 916 Furnas Hall, buffalo, NY 14260, Buffalo, NY 14260

Our laboratory recently demonstrated that fibrin-based tissue engineered vessels (TEVs) can be prepared from bone marrow derived smooth muscle and endothelial cells. These cells demonstrated high proliferation potential and contractile function as well as great ability for matrix remodeling after implantation as interpositional grafts into the jugular veins of an ovine animal model. More recently we obtained contractile smooth muscle progenitor cells from ovine hair follicles (HF-SMPC), thereby identifying another source of stem cells for vascular tissue engineering. In this paper we aimed at improving the strength of TEVs while retaining some of the functional advantages of fibrin. To this end, we constructed a vascular matrix composite comprised of small intestinal submucosa (SIS) infiltrated with fibrin. Scanning electron microscopy showed that adhesion and spreading of HF-SMPCs were enhanced on the surface of fibrin-infiltrated compared to control SIS. In addition, fibrin promoted migration of cells into the matrix and increased proliferation significantly after 2 weeks in culture. At that time, fibrin-SIS was significantly compacted and cells expressed and secreted collagen and elastin. Notably, despite the increased strength of SIS, HF-SMPC were able constrict the matrix in response to vasoactive agonists. Taken together our data showed that fibrin-SIS composite enhanced cellular function and improved the mechanical properties of TEVs, suggesting that this composite of natural biomaterials may be used for arterial implantation of HF-SMPC based TEVs.

463d Cell-Level Selectivity from Engineered Placement of Binding Groups

Maria M. Santore[1], Jun Zhang[1], Surachate Kalasin[2], Vincent Rotello[3], and Sudhanshu Srivastava[3]. (1) University of Massachusetts Amherst, Conte Polymer Laboratory, 120 Governor s Drive, Amherst, MA 01003, (2) Department of Physics, University of Massachusetts, Amherst, MA 01003, (3) Chemistry, University of Massachusetts, Goessmann Lab. 686 North Pleasant St., Amherst, MA 01002

A common strategy to promote cell adhesion to biomaterials employs peptide sequences which impart molecular level specificity to cell binding. As many cells present similar surface molecules, however, molecular-level specificity does not always translate to cell-level specificity. The current contribution demonstrates the power of control over the placement of binding groups on a surface to achieve selectivity. Biomaterials are being developed to adhere /kill bacteria while maintaining compatibility with mammalian cells. A key principle being exploited is a cell-level adhesion threshold which develops when surface repulsions are added to molecular-scale attractions between the surface and targets. This talk demonstrates how this threshold arises from the fine tuning of interfacial forces. Different thresholds for different cells allows materials to be designed to sharply discriminate targets.

463e Endogenous Growth Factor Gene Expression Profile of Bone Marrow Stromal Cells on 3D Porous Poly(Propylene Fumarate)/Hydroxyapatite Nanocomposites

Kyobum Kim, Department of Chemical and Biomolecular Engineeringt, University of Maryland, 2113 Chemical and Nuclear Eng Bldg, College Park, MD 20742, Minal Patel, Fishcell Department of Bioengineering, University of Maryland, 3740 Jeong.H.Kim Engineering Bldg, College Park, MD 20742, David Dean, Department of Neurological Surgery, Case Western Reserve University, 10900 Euclid Avenue, Cleveland, OH 44106, Antonios Mikos, Department of Bioengineering, Rice University, 6100 Main Street, MS-142, Houston, TX 77005, and John P. Fisher, Fischell Department of Bioengineering, University of Maryland, 3238 Jeong H. Kim Building, College Park, MD 20742-2111.

Hydroxyapatite (HA) nanoparticles have been investigated for improving the surface properties of tissue engineered scaffolds due to their structural similarity to organic substances and inorganic minerals in bone. Photocrosslinked poly(propylene fumarate) (PPF) is a promising bone-tissue engineering scaffold due to its biodegradability and mechanical strength. Incorporation of HA with PPF would be a promising strategy to create an osteoconductive nanocomposite material and to enhance cellular responses, including endogenous osteogenic signal expression. We hypothesize that the intracellular signaling mechanisms among a transplanted cell population are largely influenced by the surrounding environment. To examine this relationship, we have investigated the effect of HA nanoparticle concentration in PPF on the osteogenic signal expression among rat bone marrow stromal cells (BMSCs) cultured on a 3D porous PPF/HA scaffold. Our specific hypothesis is that an increase in HA nanoparticle concentration would promote osteogenic signaling by augmenting the surface roughness and hydrophilicity, and therefore enhancing BMSC attachment and function. BMSCs were isolated from rats then cultured for 2 wks on PPF/HA scaffold (0, 10, 20, and 30 wt% HA in PPF, 6 mm in diameter and 1.5 mm in thickness). Cell attachment on the scaffold surfaces was demonstrated by scanning electron microscopy. Osteodifferentiation was tracked by alkaline phosphates and osteocalcin mRNA expression by quantitative reverse transcriptase - polymerase chain reaction (qRT-PCR). Endogenous osteogenic signals of interest include bone morphogenetic protein-2, transforming growth factor-β1, platelet-derived growth factor-A, and fibroblast growth factor-2. This work demonstrates a relationship between tissue engineered construct properties with HA nanoparticles and osteogenic signal expression as well as the possibility of regulating the signal expression level by biomaterial modulation.

463f Effect of Glass-Forming Matrices on Phospholipid Bilayers during Biopreservation. A ^{31}P NMR Line Shape Simulation Study

Pragati Jain, Sabyasachi Sen, and Subhash Risbud. Chemical Engineering and Materials Science, University of California, Davis, One Shields Ave, Davis, CA 95616

It is well known that certain sugars have the ability to preserve and stabilize membranes, proteins and other living cells in the dry state for extended periods of time. Certain crustaceans, plant seeds and other organisms can even survive dehydration for several years, and resume metabolism upon rehydration. Such organisms accumulate large amounts of disaccharides in their cells, which gives them the capability of undergoing anhydrobiosis, the ability to survive complete dehydration. Other sugars such as lactose, in addition to sucrose and trehalose, are used commercially for freeze-drying biomaterials and products used in the pharmaceutical industry. There are contrasting views on the mechanism that enables disaccharides to stabilize and preserve living membranes. The two principal theoretical models present in the literature to explain this phenomenon are the water replacement model and the vitrification model. Structural and dynamical studies of the atomic-scale interaction of these sugars with the lipid head groups are required to understand the role of these vitreous sugars in the stabilization of biomembranes. ^{31}P nuclear magnetic resonance (NMR) spectroscopy is especially useful in this regard as it can give direct quantitative information about the orientation and dynamics of the lipid head groups. In a previous study we have reported temperature dependent ^{31}P NMR spectra of pure, freeze-dried dipalmitoylphosphatidylcholine (DPPC) lipid membrane and of DPPC in a wide range of glass-forming matrices. Three glass-formers were selected for this study: trehalose is a disaccharide with $T_g = 115°C$, glucose is a monosaccharide with $T_g = 30°C$ and hydroxyethyl starch (HES) is a carbohydrate polymer with $T_g = 110°C$. Changes in ^{31}P NMR line shapes as a function of temperature were qualitatively related to the effects of various glass-forming matrices on the mobility of lipid phosphate head groups. However, quantitative studies to determine the effects of glass-forming additives on the orientational and dynamical properties of lipid head groups had not been attempted. Here we report the results of quantitative simulations of temperature dependent ^{31}P NMR line shapes of these systems acquired at temperatures ranging between 25°C and 80°C. The orientational and dynamical parameters of the DPPC lipid head groups as obtained from these simulations are shown to provide key information about the structural and phase change behavior of the lipid membrane and to lead to a better understanding of the role of sugars in biostabilization. It may be noted here that the selection of these three glass-formers was driven by the significant differences between trehalose and glucose in terms of their T_g and by the contrasting behavior of HES with respect to the sugars in terms of the absence of hydrogen bonding with DPPC in the latter. Such a selection ensures that both the water replacement model and the vitrification model can be adequately tested. Simulations of these line shapes with a model of rotational diffusion of the PO_4 headgroups yield dynamical and orientational information regarding the mobility of these head groups as a function of temperature in the presence of two different sugars and one starch. The gel-to-liquid crystalline phase transition of DPPC is characterized by a sudden onset of rapid uniaxial rotational diffusion of the PO_4 headgroups. The temperature of this onset of rapid rotational motion shifts to higher values in the presence of glass-forming sugars such as trehalose and glucose. The order of the shift in the transition temperature is glucose > trehalose > HES ≈ pure DPPC while the order of Tg for the glass formers is: trehalose ≈ HES > glucose. Moreover, only trehalose and glucose are found to be able to preserve the initial orientation of the diffusion axes of the lipid headgroups during phase transformation while DPPC/HES behaves in a similar way as pure DPPC. The dynamical and orientational rigidity of the lipid headgroups imparted by the trehalose and glucose matrices are consistent with the presence of hydrogen bonding between the lipid and the sugars and lend support to the water replacement hypothesis of biopreservation.

463g Intracellular Uptake and Morphological Changes of Thermoresponsive and Degradable Nanogels

Gauri P. Misra[1], Mofya Diallo[1], and Tao L. Lowe[2]. (1) Surgery, Penn State University College of Medicine, 500 University Drive, Hershey, PA 17033, (2) Departments of Surgery, Bioengineering, and Materials Science and Engineering, Pennsylvania State University, 500 University Drive, Hershey, PA 17033

There is a growing interest in polymeric nanogels for drug delivery and diagnostic applications because of unable chemical and physical properties [1]. Our group has been working on a novel polymeric nanogel system having both thermoresponsive and degradable properties [2]. The nanogels were synthesized by UV photopolymerization at 45 °C using N-isopropyl acrylamide (NIPAAm) and a cross-linkable macromer, 2-hydroxyethylmethacrylate-lactate-dextran of varying mass ratios. The size and morphology, and hydrodynamic radii of the nanogels were studied by atomic force microscopy and dynamic light scattering, respectively. Time-dependent interaction of these nanogels with rat retinal precursor (R28) and human glioblastoma (U87) cells were investigated using atomic force microscopy, laser confocal microscopy, and flow cytometry techniques. The MTT cell viability data indicated non-toxic nature of the nanogels at 100 µg/mL concentration. The nanogels with more hydrophilic macromer component showed higher cell viability compared with the nanogels with a high NIPAAm content. Both confocal microscopy and flow cytometry studies confirmed uptake of the nanogels by the R28 and U87 cells. The uptake of the one week degraded nanogels decreased in the case of R28 cells while the opposite occurred in the case of U87 cells. The AFM images revealed morphological changes in the nanogels in culture media after one week. The nanogels in R28 culture medium exhibited a strong association between the nanogels with time leading to an increase in size while the U87 medium did not show such behavior. These studies suggest that cell type and morphological changes of nanogels are important factors affecting cellular interaction of nanogels.

1. N.A. Peppas, Intelligent therapeutics: biomimetic systems and nanotechnology in drug delivery, Advanced Drug Delivery Reviews 56, 1529 (2004).

2. Xiao Huang, Gauri Misra, Amit Vaish, Bryan Sutermaster, and Tao Lu Lowe, Novel Nanogels with Both Thermoresponsive and Hydrolytically degradable Properties, Macromolecules, 2008 (in press).

481a Graphene Stabilized Copper Nanoparticles as An Air-Stable Substitute for Silver and Gold In Low-Cost Ink-Jet Printable Electronics

Norman A. Luechinger[1], Evagelos K. Athanassiou[2], and Wendelin J. Stark[2]. (1) Chemistry and Applied Biosciences, Swiss Federal Institute of Technology (ETH Zurich), Wolfgang-Pauli-Str. 10, ETH Hönggerberg, Zurich, 8093, Switzerland, (2) Chemistry and Applied Biosciences, ETH Zurich, Wolfgang-Pauli-Str. 10, ETH Hönggerberg, Zurich, 8093, Switzerland

Metal nanocolloids have received tremendous attention for a wide range of applications. Metal oxidation, however, has traditionally restricted the range of useful nano-metals to silver, platinum, palladium and gold. Technical low cost applications such as ink jet printing of flexible conductors and radio frequency identification (RFID) tags, however, require manufacturing costs of typically less than one cent per tag. The most evident candidate to substitute silver or gold would be copper, but nano-copper rapidly oxidizes under ambient conditions. We demonstrate how deposition of graphene bi- or trilayers on copper can be realized on a technical scale and enables full protection of the copper metal core up to 200 °C under humid air. The protected copper can then be used in domestic ink jet printers for preparation of conductive patterns under ambient conditions. References [1] N. A. Luechinger, N. Booth, G. Heness, S. Bandyopadhyay, R. N. Grass, W. J. Stark, Advanced Materials (2008), in press. [2] N. A. Luechinger, S. Loher, E. K. Athanassiou, R. N. Grass, W. J. Stark, Langmuir, 23, 3473-7 (2007). [3] N. A. Luechinger, E. K. Athanassiou, W. J. Stark, in review (2008) Figure: Flame spray synthesis (a) as a low-cost process for the production of metallic copper nanoparticles (b). Transmission electron micrograph (c) and corresponding model sketch (d) of an air-stable individual copper nanoparticle with a graphene coating. Photographs of ink-jet printed conductive copper patterns showing operating red LED's (e, f).

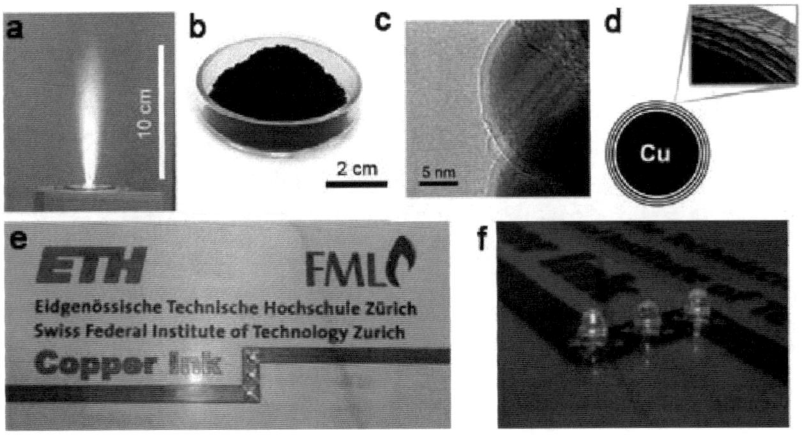

481b Spectroscopical Characterization of Thin Film Deposition Processes: Protective and Conversion Coatings

Jorge E. Gatica[1], Andrew R. Snell[1], Szabolcs Sofalvi[1], and Wilfredo Morales Sr.[2]. (1) Chemical and Biomedical Engineering, Cleveland State University, 455 Stilwell Hall, 2121 Euclid Avenue, Cleveland, OH 44115-2214, (2) National Aeronautics and Space Administration, Glenn Research Center, Cleveland, OH 44135-3191

With the raise of environmental awareness and the renewed importance of environmentally friendly processes, surface pre-treatment processes based on chromates have been targeted by the United States Environmental Protection Agency. Indeed, this process has been subject to regulations under the Clean Water Act as well as other environmental initiates, and there is today a marked movement to phase the process out in the near future. Therefore, there is a clean need for new advances in coating technology that could provide practical options for replacing present industrial practices. Depending on the final application, such coatings might be required to be resistant to corrosion and/or act as chemically resistant coatings. Multi-oxide conversion coating (nanometer scale) films were successfully grown onto aluminum alloys from precursors based on aryl phosphate esters. These films were primarily characterized by X-ray Photoelectron Spectroscopy (XPS) and Fourier Transform Infrared Spectroscopy (FTIR) techniques. Thermal analysis methods were used to formulate a model describing the chemical vapor deposition (CVD) process leading to conversion coatings on solid substrates. The results presented illustrate the potential of this technology to replace chromate-based processes. Results showing the effectiveness of these films as a protective barrier for corrosive environments and as enhancement of adhesion properties of metallic substrates are presented.

481c Capping Ligand Effect on CdSe Quantum Dot Langmuir-Blodgett Monolayer Stability

Chander Radhakrishnan[1], **Michael K. F. Lo**[1], Miguel Garcia-Garibay[2], and Harold G. Monbouquette[1]. (1) Chemical and Biomolecular Engineering, University of California, Los Angeles, 420 Westwood Plaza, 5531 Boelter Hall, Los Angeles, CA 90095-1592, (2) Department of Chemistry and Biochemistry, UCLA, 607 Charles E. Young Drive East, Los Angeles, CA 90095-1569

The monolayer coating of amphiphilic compounds onto surfaces is commonly achieved via the Langmuir-Blodgett (LB) technique. Such a technique exploits the hydrophobic and hydrophilic properties of amphiphiles to self-orient into a monolayer at the air-water interface with the hydrophilic head groups associated with water and the hydrophobic tails directed into the air. While LB monolayers of fatty acids or phospholipids have been well established, little is known about LB films of quantum dots (Qdots). This study successfully demonstrated that stable cadmium selenide (CdSe) Qdot Langmuir monolayers were formed at the air-water interface. The effect of Qdot capping ligands [tri-n-octylphosphine oxide (TOPO), dodecanethiol (DDT), and dithiocarbamates (DTC)] on the stability of the Langmuir monolayer was investigated in detail, and a theoretical model was used to describe Qdot desorption. Pressure-controlled experiments indicated that TOPO-capped Qdots formed the most stable Langmuir monolayers. The effect of deposition pressure on the transfer of DDT-capped CdSe Qdots onto a smooth silicon (100) substrate was correlated with the observed density of Qdots via atomic force microscopy (AFM). Finally, DDT-capped Qdots (3.4 nm) were LB deposited on an azide-terminated, self-assembled monolayer substrate, which was subsequently illuminated. Partial photocatalytic reduction of azides to amines by deposited CdSe Qdots was confirmed by XPS. Controlled deposition of Qdots followed by photocatalytic reduction of azide to amine under light irradiation can potentially be used to synthesize chemical patterns on surfaces with a tight pitch.

481d Enhanced Field-Effect Mobility for Solution-Processable Organic Thin Film Transistors by Surface Modification with Organosilanes

Lin Jiang[1], Jie Zhang[2], Daniel R. Gamota[2], and Christos G. Takoudis[3]. (1) Chemical Engineering, University of Illinois at Chicago, 810 S Clinton St, Chicago, IL 60607, (2) Motorola, 1301 E. Algonquin Rd, Schaumburg, IL 60196, (3) Chemical Engineering and BioEngineering, University of Illinois at Chicago, 851 S. Morgan St. (M/C 063), Room 218 Science and Engineering Offices (SEO), Chicago, IL 60607

In Organic Thin Film Transistors (OTFTs), the performance is profoundly affected by organic semiconductor crystal formation or organic structure ordering. As it is known, the ability of conjugated molecules to transport charge due to the π-orbital overlap of neighboring molecules provides their semiconducting and conducting properties. The self-assembling or ordering of these molecules enhances this π-orbital overlap and is the key to improvements in carrier mobility. Since the semiconductor materials are deposited on dielectric layer, the surface chemical and mechanical properties of dielectric materials do affect the alignment and the crystal formation of semiconductor. In this work, we present methods about dielectric surface modification to obtain preferred functional group and promote semiconductor ordering on dielectric surface.

OTFTs were made on heavily doped silicon wafer and the thermally grown SiO_2 was modified by organosilanes which have different functional groups, such as –SH, $-C_6H_5$. Those functional groups help the assembling of semiconductor molecular on dielectric surface. A field effect carrier mobility of 0.01 $cm^2/V \cdot s$ was obtained for OTFTs with solution processable bis(triisopropylsilylethynyl) (TIPS) pentacene. X-ray photoelectron spectroscopy was used to characterize the chemical composition of modified dielectric surface. More ordered semiconductor crystal structure was shown by atomic force microscopy after surface treatment.

Nanocomposite Optical Filters Under Extreme Strain

Thad Druffel, Omar Buazza, Matt Lattis, Scott Farmer*

Optical Dynamics, 10106 Bluegrass Parkway, Louisville, KY, USA 40299-2202

Abstract

Nanocomposites are created by doping host polymers with nanoparticles that typically have higher or lower refractive indices. The ability to tailor the mechanical and optical performance of these composites has led to their increased use in transparent materials. Nanocomposites maintain the elastic properties of the binding polymers and exhibit infinite refractive index tunability between the limits of the system. These unique properties provide distinct benefits for multilayer thin-film optical filters. Because the nanoparticles are dispersed in a fluid or bound in a polymer matrix in use, toxicity risks that may be associated with raw particles are reduced. Using a stable dispersion of titanium dioxide nanoparticles and a UV-curable monomer, we were able to design and produce several quarter-wave filters that were capable of strains of up to 20 percent.

Introduction

Nanocomposites using organic polymers and inorganic nanoparticles will play an important role in applications requiring high visible transparency and mechanical flexibility. A unique class of materials is created when the elastic properties of polymers are combined with the inherent hardness of crystalline nanoparticles. Transparent polymers are generally more flexible than glass materials and exhibit a moderate range of optical, electrical, and mechanical properties. The incorporation of inorganic nanoparticles in the polymer system significantly increases the physical parameter range beyond that of the host polymer and allows material properties to be engineered for specific applications. Of particular interest to our group are thin-film nanocomposites on polymer substrates in which strain domains are well matched.[1,2]

True nanoparticles are significantly smaller than visible light wavelengths and, therefore, generally do not cause significant optical scattering when incorporated in a polymer. Careful nanoparticle selection can influence refractive index, electrical conductivity, UV absorption, magnetism, and a host of other properties in the finished nanocomposite. The elastic properties of the binding polymers can be maintained while improving abrasion resistance through the inclusion of high-hardness nanoparticles. This provides opportunities to improve transparent materials that require clarity and abrasion resistance.

Thin-film coatings of less than 100 nm to tens of microns in thickness are important in many polymer substrate applications, but the conditions under which these systems are used must often be limited to maintain film and substrate integrity. Since mainstream coatings are often ceramics applied using vacuum deposition, they normally have strain domains that are significantly different from the polymer substrate. Under large strains these coatings tend to crack and induce objectionable optical scattering that is quantifiable as haze. These cracks allow other types of damage to easily propagate from them as well. The processing temperatures used to apply these coatings also leave behind large intrinsic stresses that further limit their utility due to significant changes in the ultimate strength of the coated substrate.

Here we demonstrate a thin-film optical filter composed of a multilayer nanocomposite undergoing strains in excess of 20 percent. These filters were designed to have a peak reflectance at a chosen wavelength in the visible region. As strain is applied to the system the peak wavelength changes, but the intensity of the reflectance remains constant. This confirms that the films are not failing at large strains and demonstrates a performance unmatched by traditional vacuum deposition techniques.

Background

Our group has been investigating systems that use nanoparticles in UV-curable monomers. For our typical applications, the nanoparticle stabilization methods must be compatible with solvents and monomers that work well in a spin coating process.

Our nanocomposite thin films consist of metal-oxide nanoparticles and a UV-cured acrylate polymer that acts as a binder. The nanoparticles are used to both engineer the refractive indices of the individual layers and to improve the mechanical properties of the film. The nanoparticles are initially suspended in a solvent, along with an acrylate monomer and a photoinitiator. To ensure transparency and avoid excessive light scattering, the primary particle size must be much less than 100 nm and the suspension must be properly stabilized to avoid particle agglomeration.

The refractive index of a layer is controlled by adjusting the volume ratio of nanoparticles and monomer. For a layer using high-index particles in a lower index monomer, the refractive index limits occur at the index of the pure monomer (the index minimum in this case) and at the index achieved with approximately 60 volume percent nanoparticle loading (the index maximum in this case). This corresponds well with the theoretical close packing of spheres, and we have noted that the refractive index and modulus of the film reach their maxima at this loading.

Quarter-Wave Stacks

High-reflectance filters are often created with a stack of alternating high and low refractive index layers, with each layer having an optical thickness of ¼ wave.

The reflectance of an uncoated planar surface from a wave perpendicular to it is

$$R = \left(\frac{n_0 - n_{sub}}{n_0 + n_{sub}} \right)^2 \tag{1}$$

where

R = reflectance
n_o = index of refraction of air (1.00)
n_{sub} = index of refraction of the substrate

If PMMA is the substrate (n=1.5 at 550 nm), then the amount of incident light reflected by the uncoated surface is 4%. In this case the reflectance is purely a function of the refractive index difference between the substrate and air (it should be noted that the refractive index of the substrate varies with wavelength and, therefore, the reflectance does as well).

A coating of multiple thin films of different refractive indices on the substrate can be used to interfere with the reflected waves. For a film having ¼ wave optical thickness the reflectance can be computed as

$$R = \left(\frac{(n_0 - Y)}{(n_0 + Y)} \right)^2 \tag{2}$$

In this case the refractive index of the substrate has been replaced by the admittance of the surface (Y) and is a ratio of the total tangential magnetic and electric fields as described by Macleod.[3] The admittance for a system of i alternating high and low-index films is

$$Y = \left(\frac{n_{high}^{(i+1)}}{n_{sub} n_{low}^{(i-1)}} \right) \tag{3}$$

The first three equations can be used to model the magnitude of the reflectance for a multilayer filter. Continuing with the case of a dual refractive index system, the width of the filter's notch is a function of the ratio of the refractive indices of the layers

$$\Delta g = \frac{2}{\pi} \sin^{-1} \left(\frac{n_{high} - n_{low}}{n_{high} + n_{low}} \right) \lambda \tag{4}$$

where

Δg is the half-width of the notch

From this equation it can be seen that as the ratio of refractive indices increases, the width of the filter's notch will do the same.

These relationships allow filters to be designed and their simplified response at a specified design wavelength (λ) to be modeled. Each layer's optical thickness (d) is made equal to $\lambda/4$, with optical thickness defined as the product of the physical thickness and the refractive index of the material (refractive index indicating how much the speed of light is reduced in the material compared to the speed of light in a vacuum). In cases where layer thickness is not equal to a quarter wave, computation of the reflectance is more laborious and generally requires the use of a computer.

Stress in Thin Films

The deformation of a system of films is controlled by the stiffness of each of the individual layers. The stiffness of each layer is the product of the modulus of elasticity and its cross-sectional area.

$$K = EA \tag{5}$$

where
K = stiffness
E = modulus of elasticity
A = cross sectional area (thickness x width)

Quarter-wave optical films have thicknesses on the order of 100 nm. To create a viable optical element, these films are generally laid down on a substrate with a thickness of at least a millimeter. This represents a thickness difference of three to five orders of magnitude. While the modulus of elasticity of ceramic films is generally one to two orders of magnitude larger than that of a plastic substrate, the stiffness of the film will still be several orders of magnitude smaller than the substrate because the films are so thin. Additionally, ceramic films are typically limited to strain domains of approximately 1 percent, while many polymers can withstand very large strains. This means the films will not significantly contribute to the overall stiffness of the article and they will be quickly strained beyond failure. Externally applied loads will ultimately be borne almost entirely by the polymer substrate.

What this means in real-world applications is that plastic substrates coated with thin-film filters using vacuum deposited ceramic materials are typically unable to resist large strains induced by thermal and mechanical forces. As a result, the brittle ceramic films will either buckle under compressive loads or crack under tensile loads. These brittle failures on the surface will cause excessive light scattering and will reduce visible transparency. Therefore, engineering applications for thin-film dielectric coatings on polymers tend to be limited. A nanocomposite film, on the other hand, has a strain domain that is closely matched to the polymer; therefore, it can undergo large induced strains without failure.

Experimental

The individual layers in the stack were spin coated onto a substrate using a machine by Optical Dynamics. This well-understood technique controls layer thickness by balancing the centrifugal forces of a developing thin film with the viscous forces that increase as evaporation takes place. The repeatability of this method is extremely high as long as the coating environment is controlled such that the evaporation rate stays constant. This method can also be extended to coat surfaces with roughness on the order of several microns.

After the solvent evaporates, a 50-150 nm film of a UV-curable monomer and nanoparticles remains. In this case the monomer is a trimethylolpropane triacrylate (TMPTA). The film is then cured using a pulsed xenon UV lamp, leaving a polymer nanoparticle composite. Subsequent layers are then added on top of the previous layer to build the filter.

The physical thickness of the layers was determined by applying and curing the individual films on a hard substrate (glass). Part of the film was scratched from the surface of the glass and the step height was determined using a profilometer (model number XP2 by Ambios Corporation). The accuracy of the method was confirmed by measuring thickness in several locations and on multiple layers.

The optical response of the final article was measured using a contact probe spectrophotometer model F20 by Filmetrics. For a given quarter-wave stack design of alternating index layers the refractive index of the layers is computed using equations (2) and (3) and the physical thickness (as measured by profilometry). The software package TFCalc, by Software Spectra, Inc., was used to confirm the results of the equations. A simple method of determining the refractive index of the films is to build a multilayer stack of high and low refractive index materials and model their response using a computer (once the physical thicknesses have been established).

The titanium dioxide dispersion was synthesized in our laboratory using a hydrothermal process. The titania nanoparticles have a mean diameter of approximately 20 nm and are functionalized to improve their stability in an alcohol. The functional groups on the particles also aid in their adhesion to the polymer matrix, which prevents cracking of the film under large strains.

Demonstration of multilayer film flexibility was accomplished by casting the film onto a highly flexible, thin substrate and then applying strains to the article. The first method was to simply bend the substrate/film. In order to apply larger strains we clamped the article in an apparatus that allows pressurized water to be introduced into a cavity formed by the thin, flexible substrate and the apparatus. Since the apparatus is much more substantial, the pressurized water inflates the substrate/film. This method introduces an isotropic strain in the plane of the film.

Results

In the case of the high-index layer, the introduction of the inorganic particles creates a non-uniform refractive index across the visible spectrum, complicating analysis of the film. To verify the refractive index of the nanocomposites, we first built a 9-layer reflective filter composed of alternating high and low refractive index films. Because the finished filter provides significant physical thickness and a large, highly responsive peak at its fundamental wavelength, the optical properties of the high-index layer can be accurately determined using thin-film modeling software to match the measured response of the filter.

In this study we used an anatase form of titanium dioxide nanoparticles to engineer the refractive index of the high-index layer. The refractive index of cured TMPTA (the binding polymer of the layer) is approximately 1.48, and the anatase titanium dioxide is approximately 2.2. With 60 volume percent particle loading, the theoretical refractive index is 1.92; however, surface modification of the nanoparticles decreases their refractive index.

The low index utilized silicon dioxide and TMPTA, which have similar refractive indices and very flat refractive index distributions. This makes the finished film relatively easy to characterize. By depositing nine successive layers of this film alone we determined that the thickness of the individual layers is 91.6 nm and the refractive index is 1.475.

The thickness of the 9-layer reflective stack was measured to be 825 nm, which yields a thickness of 92.4 nm for the high-index layer (according to the measurements above). This system was modeled in TFCalc using a flat dispersion of 1.48 for the refractive index of the low and a mixture of 1.48 and anatase TiO_2 for the high. The best fit to the observed reflectance spectrum yields a refractive index for the high of 1.885 (figure 1). As a check, we also ran a sensitivity analysis in which we allowed the thicknesses of the coatings to vary by +/- 5 nm. The results show that the optical response of the filter fits well using these parameters.

This system was then tested for flexibility by simply bending the substrate/filter at a very short radius. This is shown in figure 2. It can be observed that the optical clarity of the system has not been impacted by the induced strains. No visible damage was noted to the article. In fact, this system was also subjected to very large thermal loads in which the article was moved between a boiling water bath and an ice water bath with no visible signs of coating degradation.

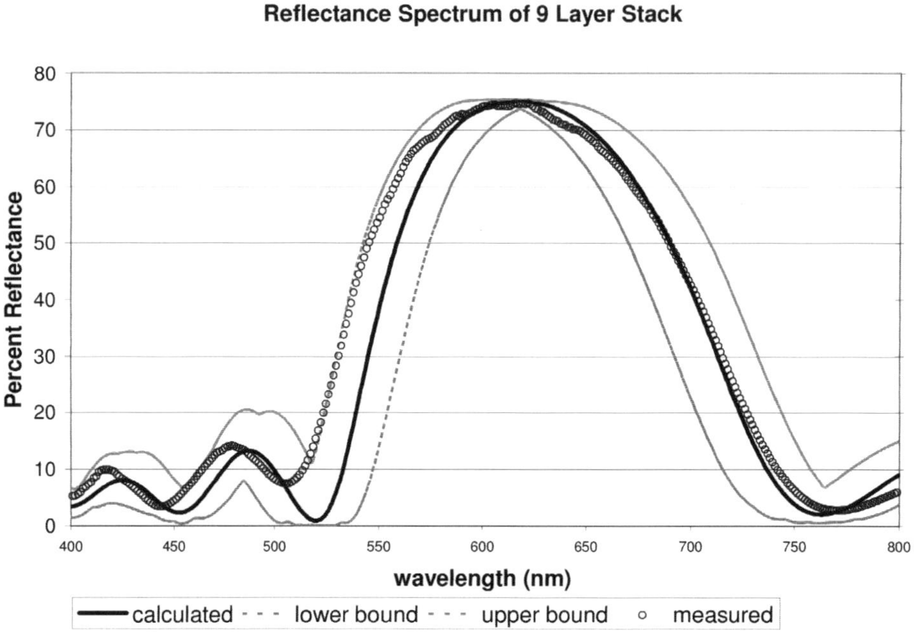

Figure 1: Measured reflectance spectrum of a 9-layer alternating high (TiO2 nanocomposite) and low (SiO2 nanocomposite) index stack compared to a modeled stack bounded by +/- 5nm thickness variations.

Figure 2: A 9-layer nanocomposite thin-film coating on a polymer substrate demonstrating high flexibility and transparency.

To further study these nanocomposite films, we built an apparatus that would allow for the application of an in-plane stress. By applying hydrostatic pressure to one side of the substrate, we cause it to inflate and produce a tensile strain. The coating/substrate was strained up to 20 percent and the reflectance spectrum was measured. As expected, the system did not fail while undergoing these strains, as shown by the response of the reflectance (figure 3). In fact, we observed no significant change to the height of the reflectance response, indicating that the refractive indices of the layers were maintained. This information could be used to determine the Poisson's ratio of the thin films.

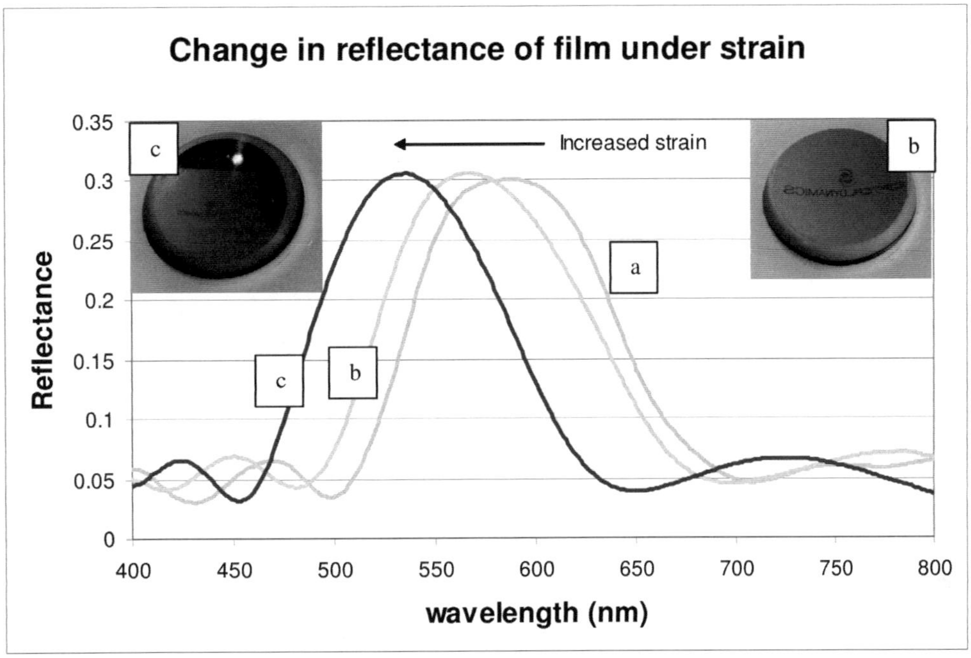

Figure 3: Demonstration of nanocomposite flexibility by stretching a thin film. Pressure was applied on one side of the article inducing a uniaxial strain in the plane of the films. The reflectance response of the films at 0, 10%, and 20% strain are shown graphically. The unstrained film has a peak response at approximately 590 nm and is shown by curve (a). At a 10% strain (b) the spectrum is shifted to about 560 nm and the reflected color is green. At a 20% strain (c) the spectrum's peak is shifted to about 530 nm and the reflected color of the mirror is blue. In each case the peak reflectance of the mirror stays constant at 30%.

Conclusion

Thin-film coatings of less than 100 nm to tens of microns in thickness are important in many polymer substrate applications, but the conditions under which these systems are used must often be limited to maintain film and substrate integrity. Since mainstream coatings are often ceramics applied using vacuum deposition, they normally have strain domains that are significantly different from the polymer substrate. Under large strains these coatings tend to crack and induce objectionable optical scattering that is quantifiable as haze. These cracks allow other types of damage to easily propagate from them as well. The processing temperatures used to apply these coatings also leave behind large intrinsic stresses that further limit their utility due to significant changes in the ultimate strength of the coated substrate.

Here we demonstrate a thin-film optical filter composed of a multilayer nanocomposite undergoing strains in excess of 20 percent. These filters were designed to have a peak reflectance at a chosen wavelength in the visible region. As strain is applied to the system the peak wavelength changes, but the intensity of the reflectance remains constant. This confirms that the films are not failing at large strains and demonstrates a performance unmatched by traditional vacuum deposition techniques.

References

[1] Krogman, K., T. Druffel, and M. Sunkara, Anti-reflective optical coatings incorporating nanoparticles. Nanotechnology, 2005. 16: pp. S338-S343.

[2] Druffel, T., Mandzy, N, Sunkara, M. and Grulke, E., Polymer nanocomposite thin film mirror for the infrared region. Small, 2008, 4, pp 459-461

[3] Macleod, H.A., Thin Film Optical Filters. 2001, London: Institute of Physics. 641.

481f Hollow-Fiber-Based Adsorbent System for CO2 Capture from Flue Gas

Ryan P. Lively[1], Ron R Chance[1], William J Koros[1], Harry W. Deckman[2], and B. T. Kelley[3]. (1) Chemical and Biomolecular Engineering, Georgia Tech, 778 Atlantic Dr NW, Atlanta, GA 30332, (2) ExxonMobil Research and Engineering, 1545 Route 22 East, Annandale, NJ 08801, (3) ExxonMobil Upstream Research Company, Houston, TX 77098

We describe a rapid temperature swing adsorption (RTSA) system based on hollow polymer fibers impregnated with CO2 sorbents and designed for the capture of CO2 from the flue gas streams of coal-fired power generation facilities. This post-combustion capture system relies on a two-bed configuration, each bed being composed of hollow fibers with sorbent particles dispersed in the porous polymer matrix. The system takes advantage of the hollow fiber morphology by passing cooling water through the bores during sorption to maximize sorption capacities, and steam through the bores during desorption to release CO2 to the collection system. The thin walled hollow fibers offer the advantage of rapid heat and mass transport , with less than 1 second response times expected for both under anticipated operating conditions. To ensure that the heating and cooling agent only thermally interacts with the fiber walls, a dense lumen layer is required on the interior of the fiber wall and must provide high resistance to molecular transport between the fiber sheath and the core. To probe the latter, nitrogen gas permeation experiments have been conducted. Incorporation of a fiber lumen layer composed of poly-vinylidene chloride yields a substantial reduction in N2 transport through the fiber wall, from about 8 x 10-8 mol/m2-s-Pa uncoated to about 6 x 10-13 mol/m2-s-Pa coated. Further reductions are anticipated with optimization of the composition and structure of the lumen layer. Rapid cycling is also critical for the technical and economic viability of the system. Gravimetric sorption experiments on the fiber sorbents yield a CO2 sorption halftime of about 7 seconds at a CO2 partial pressure of 7 psia. Subsequent sorption measurements and modeling show that much more rapid equilibration should occur in actual practice with the thermally moderated fibers having cooling water flowing in the bore. To explore the economic viability of this system and provide research guidance, an economic analysis of the system is being conducted, including capital expenditures, operating costs, and compression to a sequestration-ready stream. The flue gas stream is assumed to be received at ~1 atm and 40°C, and the additional 0.35 atm bed pressure drop is assumed to be supplied by additional draft fans. The cycle time is taken to be 25 seconds, which is much longer than the observed halftime for sorption, as well as the expected timescale for heat transfer in the fiber. Greater than 95% of the effluent CO2 is captured. The CO2-rich capture stream is pressurized to 2300 psia for underground sequestration. Though the details of the economic analysis are being continually refined, it is clear that the system has the potential for a carbon cost on CO2-avoided basis that is significantly less than systems based on liquid sorbents. This potential is enabled in part by the low pressure drop across the bed and rapid cycles, both of which are made possible by the thin fiber sorbents. Our current estimate for the parasitic load on a 400MW power station is about 120MW, with about half being consumed for CO2 compression. This hollow-fiber-based adsorbent system has many potential advantages over conventional and competing technologies, including low cost, no waste products, and applicability to retrofit scenarios. The system should also be applicable to many pre-combustion CO2 capture scenarios.

481g Repeated In Situ Crystallization: A Facile Route to Oriented Films of Metal-Organic Framework

Masaru Kubo, Watcharop Chaikittisilp, and Tatsuya Okubo. Department of Chemical System Engineering, The University of Tokyo, 7-3-1 Hongo, Bunkyo-ku, Tokyo 113-8656, Japan

As a new class of porous materials, metal-organic frameworks (MOFs), constructed from metal ions and organic ligands, have attracted considerable attention. The first step to apply unique features of MOFs such as diverse structures and flexible frameworks is to prepare MOFs as a highly uniform film. [Cu2 (pzdc)2(pyz)]n (CPL-1; pzdc = pyrazine-2,3-dicarboxylate; pyz = pyrazine) possessing one-dimensional channels (4 Å× 6 Å) [1] is considered here because of the unique characteristics of CPL-1. Very recently, thin films of several MOFs were prepared on surface-modified substrates by in situ crystallization and layer-by-layer assembly; however, CPL-1 films were not obtained by such methods. Thus, we have developed a new method "repeated in situ crystallization", combing the in situ crystallization and the secondary growth, to prepare continuous CPL-1 films. [2] In this concept, the crystals formed on the substrates act as seeds for the further growth. The procedure is as follow; the substrates were floated for 2h on the CPL-1 synthesis solution freshly prepared at RT, then the substrates were thoroughly rinsed with distilled water and dried. The procedure (float-wash-dry) was repeated for five cycles. The formation of CPL-1 films was confirmed by X-ray diffraction (XRD) and field emission scanning electron microscopy (FE-SEM). The XRD patterns mainly show the peaks attributed to (0k0) in all of the samples (one to five cycles), and the (0k0) peaks become more intense after repeating the procedure, clearly representing preferential b orientation. FE-SEM images show that the number of crystals increased after repeated growth. After the fifth cycle, no bare surfaces of the substrate were observed. In addition, the influence of the substrates was remarkable at the early stages of the formation of CPL-1 film, however, no difference of the substrates was observed after the fifth cycle. These results suggest that repeated in situ crystallization could increase the density of nuclei on the surface regardless of the chemical or physical natures of the substrates. Compared to previous procedure for preparing MOF films, the present procedure is simple and the obtained CPL-1 films are continuous and uniform. It is expected that the repeated in situ crystallization would be suitable for other MOFs.

Reference

1. S. Kitagawa et al., Angew. Chem. Int. Ed., 2004, 43, 2334.

2. M. Kubo et al., Chem. Mater., 2008, 20, 2887-2889.

481h Response Studies of Microcapsules to pH, Ionic Strength and Osmotic Pressure

Shyam Kadali, Department of Chemical and Biomolecular Engineering, Rice University, MS 362, 6100 Main Street, Houston, TX 77005, Hitesh G. Bagaria, Dept of Chemical and Biomolecular Engineering, Rice University, 6100 Main Street, Houston, TX 77005, and M. S. Wong, Departments of Biological & Chemical Engineering and Chemistry, Rice University, 6100 Main Street, Houston, TX 77005.

Abstract

Microcapsules find wide-ranging applications in pharmaceutical, food, and consumer industry by serving as miniature containers to store, deliver, and release substances. Nanoparticle-assembled capsules (NACs) are one such model structure, which we can prepare through nanoparticle (NP) assembly on polymer-salt aggregates [1-5]. In this polymer-salt aggregate templating route, a solution of linear cationic polymer like poly (allylamine hydrochloride) is combined with a multivalent anionic salt solution, to form a suspension of salt-bridged polymer aggregates. An aqueous suspension of NPs, such as silica NPs, is then added to form polymer-filled capsules, water-filled capsules or polymer/inorganic solid spheres, depending on the synthesis conditions. This electrostatically-driven route is attractive for scale-up of microcapsule synthesis because the formation reaction occurs in water at mild pH values and at room temperature.

NACs containing macromolecules such as enzymes and drugs are difficult to release unless one takes definite measures to unlock or break open the capsule. This measure can be achieved by tuning the responsive parameters such as pH, ionic strength and osmotic pressure. The objective of study is to gain an understanding of the behavior of NACs in response to pH, ionic strength and osmotic pressure. As an illustration, capsules can be made either polymer filled or water filled by changing pH between 3 and 10. On the other hand, outside this pH range NACs can be dissembled. Similar controlled studies were performed by changing the ionic strength with NaCl and changing the osmotic pressure [6] with poly (styrenesulfonate sodium salt) and poly(ethylene glycol). Understanding the sensitivity of NACs towards the various conditions would help in better understanding and perhaps tailoring their properties for drug delivery and controlled release.

Keywords: Microcapsule, hollow, nanoparticle, NACs

References:

[1] R. K. Rana et al., Adv. Mater. 17, 1145-1150 (2005)

[2] V. S. Murthy et al., J. Phys. Chem. B., 110 (51), 25619 -25627 (2006)

[3] J. Yu et al., Chem. Commun. 10, 1097-1099 (2006)

[4] J. Yu et al., Chem. Mater. 19(6), 1277 – 1284(2007)

[5] S. B. Kadali et al., Topic. Catal, accepted (2008)

[6] C. Gao, et. al., Eur. Phys. J. E., V5, 21(2001)

499a Protein Adsorption on Biodegradable Polyanhydride Microspheres

Brenda R. Carrillo-Conde[1], Alicia Garza[2], James Anderegg[3], and Balaji Narasimhan[1]. (1) Chemical and Biological Engineering Department, Iowa State University, 2129 Sweeney Hall, Ames, IA 50011, (2) Chemical Engineering Department, ITESM Campus Monterrey, Monterrey, 64849, Mexico, (3) Ames Laboratory, Iowa State University, 325 Spedding, Ames, IA 50011

In the past few decades, many efforts have focused on the study of biodegradable polymer microparticles (μPs) as effective drug carries with the purpose of developing protein delivery vehicles and single dose vaccines. In this context, polyanhydrides have shown to provide the right environment for protein stability and release. These surface-erodible chemistries have been shown to maintain the primary, secondary and tertiary structures of different proteins (i.e., ovalbumin, lysozyme, and tetanus toxoid), as well as their biological activity. In addition, in vitro release studies have showed that it is possible to control the release kinetics of proteins, ranging from days to months, by choosing the appropriate polyanhydride chemistry.

The biodistribution and biocompatibility of these carriers, and therefore their effectiveness as vaccine adjuvants, can be influenced by the adsorption of serum proteins in vivo; once μPs are administrated intravenously, they will first interact with plasma proteins. It has been suggested that adsorption of plasma proteins on the surface of microspheres can affect the release kinetics profiles provided by the microspheres. Particle surface properties (i.e.. hydrophobicity) play an important role in the amount of protein adsorbed. For this study, polymers based on sebacic acid (SA), 1,6-bis(p-carboxyphenoxy) hexane (CPH), and 1,8-bis(p-carboxyphenoxy)-3,6-dioxaoctane (CPTEG) anhydrides were synthesized. Cryogenic atomization method was used to fabricate microspheres based on CPH:SA and CPTEG:CPH that were characterized by SEM. Three model proteins from bovine serum plasma were chosen to study in vitro protein adsorption: bovine serum albumin (BSA), which is the most abundant protein in plasma; Immunoglobulin G (IgG), which is known to be an opsonin; and Fibrinogen (Fib), which in spite of its opsonization properties is considered a "sticky" protein. Interaction between these proteins and μPs was studied by X-Ray Photoelectron Spectroscopy (XPS) and SDS-PAGE. In addition, confocal microscopy was used to observe the distribution of adsorbed BSA-FITC and IgG-Cy5 on the surface of polyanhydride μPs. Differences in the amount of protein adsorbed were detected ass a function of: (i) copolymer composition, and (ii) the specific protein studied. A direct correlation was observed between polymer hydrophobicity and protein adsorbed. On the other hand, higher quantities of Fib and IgG were absorbed on the μPs. In vitro release studies were performed with ovalbumin (Ova)-encapsulated microspheres that were then incubated with fibrinogen; the results obtained from SDS-PAGE showed a reduction in the amount of Ova released from the microspheres when Fib is adsorbed on the surface.

It is important to considerer that the adsorption of plasma proteins can affect the in vivo behavior of these polymeric devices in different ways. For example, it has been suggested that the presence of opsonins like immunoglobulin G and fibrinogen can accelerate the uptake of these microparticles by phagocytes; while the presence of serum albumins can maintain the circulation of these carriers for longer periods of time after i.v. administration. Therefore, determination and knowledge of protein adsorption patterns may be useful to understand and optimize the in vivo performance of these drug delivery devices.

499b Injectable, In Situ Hardening Macromers for Bone Tissue Engineering

James D. Kretlow[1], Michael C. Hacker[1], Brandy B. Ma[1], Leda Klouda[1], and Antonios G. Mikos[2]. (1) Department of Bioengineering, Rice University, 6100 Main Street, MS-142, Houston, TX 77005, (2) Departments of Bioengineering and Chemical and Biomolecular Engineering, Rice University, 6100 Main Street, MS-142, Keck Hall, Room 116, Houston, TX 77005-1892

Injectable biomaterials are desirable in tissue engineering for many reasons, including the ability to fill defects of complex shapes through minimally invasive delivery methods and to deliver cells and growth factors during injection. Many injectable biomaterials are hydrogels, a class of crosslinked hydrophilic polymers characterized by a high water content, making them ideal carriers for encapsulated cells and growth factors. In bone tissue engineering applications the need for injectable or moldable materials is great as the surgical protocols prior to placement of an implant often call for extensive debridement, making scaffold prefabrication difficult or impossible. Hydrogels are often not ideal as they lack the mechanical strength necessary to fill and stabilize a bony defect.

In order to address this weakness, we have synthesized novel biodegradable macromers containing hydrophobic moieties as well as moieties promoting ionic crosslinking and subsequent hardening in situ. These materials were synthesized via radical polymerization from pentaerythritol diacrylate monostearate, a hydrophobic building block to mediate disperse interaction and mechanical stability, vinyl phosphonic acid, a calcium-binding moiety, and other acrylic monomers including N-isopropylacrylamide. The chemical structure of the macromers was verified using NMR spectroscopy and thermogravimetric analysis. Rheology was used to determine gelation properties and mechanical characteristics of the macromers in regular culture media as well as in the presence of high Ca++ solution. Dynamic light scattering was also used to investigate the interaction of these macromers with free calcium ions. Cytocompatibility of the macromers was also investigated.

Initial results demonstrate that we have synthesized macromers containing the radically polymerized monomers in ratios close to those of the monomer feeds. The materials are liquid at room temperature but undergo a rapid sol-gel transition at around 37ºC, making them a suitable candidate as injectable biomaterials. Rheological studies further indicate that in the presence of free calcium ions, this rapid temperature-dependent phase transition occurs at a lower temperature and that the resultant gel has a higher storage modulus than gels formed in the absence of free calcium ions. Direct contact cytocompatibility assays demonstrate that these materials have good cytocompatibility when compared to positive controls (tissue culture polystyrene).

In summary, we have synthesized novel, cytocompatible macromers that undergo a rapid sol-gel transition at temperatures close to body temperature and also interact with free calcium ions to further stabilize the gel. These materials warrant further investigation as matrices for cell delivery and bone tissue engineering applications.

Extensive Dark Cure from Controlled Polymerization Based on a Method Using Visible-Light Activated Initiator System

Dongkwan Kim, and Jeffrey W. Stansbury

University of Colorado-Denver, School of Dental Medicine, Department of Craniofacial Biology, Anschutz Medical Campus, PO Box 6511/Mail Stop 8310, Aurora, Colorado 80045

ABSTRACT

Free-radical photopolymerization has been widely used for polymeric biomaterials because of rapid reaction, excellent mechanical properties and the versatility available with a broad array of monomers. In contrast, cationic polymerization shows greater sensitivity to trace impurities and involves a much more restrictive selection of monomers; however, cationic polymerization allows significant dark curing because of long active center lifetimes. The dark curing process is one of the distinct differences between cationic and free-radical photopolymerization. In this contribution, we report the development of a photo-initiator system that allows extensive dark curing from visible light-initiated controlled radical polymerization. The photo-initiator is a radical-based, three-component initiator system that includes a light-absorbing photosensitizer (PS), an electron donor and an electron acceptor, which is usually an onium salt. In this system, 5,10,15,20-tetraphenyl-21H,23H-porphyrin zinc (Zn-tpp) is used as a PS, diazabicyclooctane (DABCO) is used as an electron donor and diphenyliodonium hexafluoroarsenate (DPI) is used as an electron acceptor. The three-component initiator system, Zn-tpp/DABCO/DPI produced a limiting conversion of 93.3 % conversion in 2-hydroxyethyl methacrylate (HEMA) upon full exposure time over 90 min. Real time conversion was monitored using near-infrared spectroscopy while photopolymerizations were conducted with a 100 W halogen lamp at an irradiance of 0.15 mW/cm^2.

Remarkably, three-component initiator system did not exhibit rapid termination of HEMA polymerization when the irradiation source was turned off. For example, only 30, 60 or 300 sec partial illumination coincided with conversion of 1.7 %, 4.3 % and 30.7 % respectively, but the final conversions reached 81.5 %, 84.7 % and 93.3 % respectively, at 90 min. These conversion kinetic results illustrate a unique controlled radical dark polymerization where the radical active centers are not terminated even when the light source is extinguished. Among many applications, this approach has practical significance for polymeric biomaterials including dental composite materials and bone cements where depth of cure and shadow cure are important considerations.

Introduction and Motivations

Free radical photopolymerization serves as the industry standard due its excellent efficiency across broad classes of vinyl-based monomers. However, the free radical process produces short propagating radical lifetimes due to termination reactions which means continuous initiation is required to maintain active polymerization. The result is a rapid cessation of polymerization when the irradiation source is extinguished or more simply, very limited dark cure potential. As such, complete polymerization requires either repeated incremental exposures or continuous irradiation throughout the course of polymerization. In contrast, cationic photopolymerization allows significant dark curing because of long active center lifetimes; however, compared with radical polymerizations, there are few monomers compatible with the cationic curing process[1]. The dark curing behavior is one of the distinct differences between cationic and free-radical photopolymerization. The prospect of removing this dark cure limitation from radical-based polymerizations with conventional monomers is highly significant since effective dark curing could be useful by allowing reduced processing times and lower initiator concentrations as well as achieving photocuring into shadow regions, enhanced depths of cure and photocuring of pigmented or highly filled systems using radical polymerizations.

Recently, we have identified visible-light activated three-component initiator systems that provide dramatic dark cure behavior with conventional monomers. The initiator system includes a light-absorbing photosensitizer, an electron donor and an electron acceptor[2-11]. Although a number of three-component initiator systems have been reported, no studies have attempted to provide extensive dark polymerization kinetic results using of three-component initiator systems. In this contribution we report a series of kinetic studies of radical dark polymerizations[12,13] of three-component initiator systems. We performed dark polymerization experiments as a function of iodonium salt counter anion size. For this research 5,10,15,20-tetraphenyl-21H,23H-porphyrin zinc (ZnTPP) dye was chosen as a photosensitizer (PS) due to the thermodynamic and kinetic considerations.

Experimental Methods

Materials.

The organic dye 5,10,15,20-tetraphenyl-21H,23H-porphyrin zinc (ZnTPP) was purchased from Aldrich Chemical Company and were used as received. The chemical structures of Zn-ttp and CQ are shown in Figure 1. The monomer 2-hydroxyethyl methacrylate (HEMA) was obtained from Aldrich and used as model monomer systems. 1,4-diazabicyclo[2.2.2]octane (DABCO) was purchased from Aldrich and were used as electron donors (DH). Four different iodonium salts were used as a function of counter anion size (e.g. Cl^-, PF_6^- AsF_6^- long chain structure with SbF_6^-) were purchased from Aldrich. All chemicals were used without further purification.

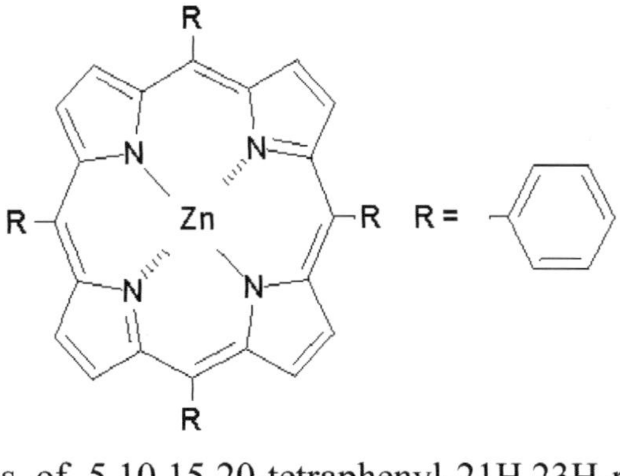

Figure 1. Chemical structures of 5,10,15,20-tetraphenyl-21H,23H-porphine zinc (Zn-tpp).

Methods

Real-time FT-near-infrared (NIR) spectroscopy.

The HEMA polymerization conversion profile was monitored *in situ* by FT-near-infrared (NIR) spectroscopy (Nicolet Nexus 670, Nicolet Instrument Corp., Madison, WI) equipped with an extended KBr beam-splitter and an MCT/A detector. To initiate photopolymerizations, a 385 - 800 nm light source (Oriel Model 77501 Fiber Optics Source, Stratford, CT) was used. The visible light source contains an adjustable iris, a manual shutter and a 100 W quartz halogen lamp outfitted with 3.2 mm diameter and 0.9 m length silica fiber optic cable. The output from the light source was passed through IR blocking filter to remove IR light (wavelengths greater than 800 nm). The filtered light irradiance reaching the sample was ~ 0.15 mW/cm^2, as measured by a calibrated diode array spectrometer. Samples with a variety of photoinitiator compositions were prepared in a rectangular mold made by glass slides at a thickness of 0.5 mm. To judiciously characterize kinetics, very mild reaction conditions were selected; all photopolymerizations were conducted at room temperature with ~ 0.15 mW/cm^2 incident light intensity. Samples with a series of photoinitiator component were place in a rectangular mold made by glass slides at a thickness of 0.5 mm. For all samples, [Zntpp] = 2.32 × 10^{-3} M [DABCO] = 0.25 M, [DPI] = 0.02 M in HEMA. All photopolymerizations were conducted at room temperature. The absorbance peak area method was used to calculate conversion. The =CH_2 first overtone absorption band at ~ 6167 cm^{-1} in the NIR region was used to characterize the concentration of the HEMA double bond. For each initiator system, at least three independent measurements were made, and the average values are reported.

Results and Discussion

Effect of selection of PS on the radical dark polymerization kinetics.

Extensive dark curing from a visible-light initiated controlled radical polymerization has been demonstrated where the radical active centers are not terminated even when the light source is extinguished. For example, when 5,10,15,20-tetraphenyl-21H,23H-porphyrin zinc (Zn-tpp) as a photosensitizer (PS), diazabicyclooctane (DABCO) as an electron donor and diphenyliodonium hexafluoroarsenate (DPI) as an electron acceptor are used as three-component initiator system, dramatic dark cure with a conventional methacrylate monomer was observed. Remarkably, the three-component initiator system did not exhibit rapid termination of 2-hydroxyethyl methacrylate (HEMA) polymerization when the irradiation source was turned off. Specifically, 30, 60 or 300 seconds partial illumination coincided with conversion values of 1.7 %, 4.3 % and 30.7 % respectively, at the time the curing light was extinguished; however, polymerization continued very efficiently in the dark and the final conversion values attained were 81.5 %, 84.7 % and 93.3 % respectively, at 90 min. The Zn-tpp/DABCO/DPI produced a limiting conversion of 93.3 % conversion in 2-hydroxyethyl methacrylate (HEMA) upon full exposure time over 90 minutes as shown in Figure 2. The same monomer polymerized with conventional photoinitiators provides very limited (< 5 %) additional dark cure conversion once the light exposure is ended due to efficient biradical termination mechanisms that rapidly consume the active radical centers.

We also report preliminary selection criteria of three-component initiator system and monomers for efficient controlled radical dark polymerizations. For example, significant improvement in conversion under the dark polymerization condition was obtained with photo-oxidizable PS, electron donors with high radical cation persistence, electoron acceptors with larger counter anions and in hydrogen abstractable monomers. Notably, the initiator systems including photo-reducible PS such as camphorquinone (CQ) and methylene blue (MB) as the PS did not produce any controlled radical dark polymerization behavior, which means that these initiator systems showed typical rapid cessation of polymerization under dark polymerization conditions as illustrated in Figure 3.

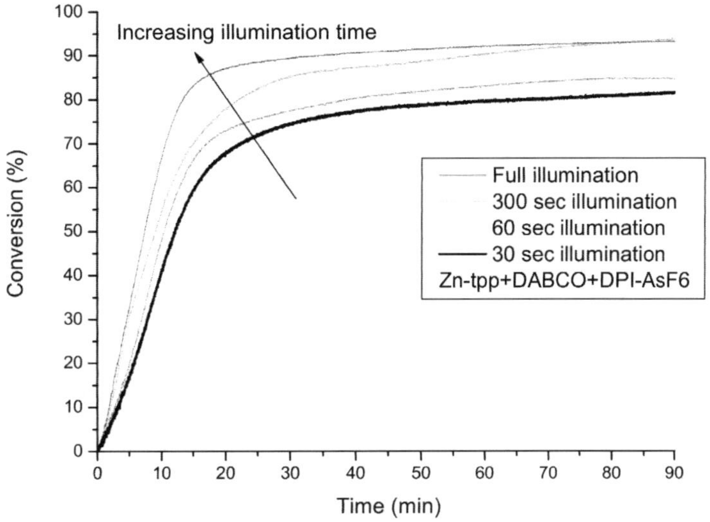

Figure 2. Conversions of HEMA polymerizations as a function of illumination time with Zntpp/DABCO/DPI-AsF_6, as measured by NIR at room temperature with an incident light intensity of 0.15 mW/cm^2. For all samples, [Zn-tpp] = 2.32×10^{-3} M, [DABCO] = 0.25 M and [DPI-AsF_6] = 0.02 M, in neat HEMA.

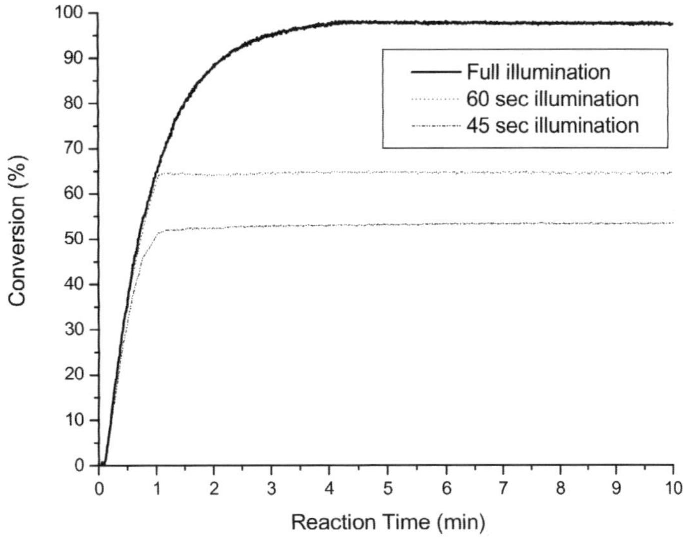

Figure 3. Conversions of HEA polymerizations as a function of illumination time with CQ/DABCO/DPI-AsF_6, as measured by NIR at room temperature with an incident light intensity of 440 mW/cm^2. For all samples, [CQ] = 0.5 M, [DABCO] = 0.25 M and [DPI-AsF_6] = 0.02 M, in neat HEA.

Effects of counter anion size and electron density of iodonium salts on the radical dark polymerization kinetics

The conversion data illustrated that size and electron density of counter aniom of iodonium salt plays an important role in the photopolymerization reaction under full illumination as shown in Figure 4. This conversion data showed both initiation and propagation step are dependent on both size and electron density of counter aniom of iodonium salt. For these reasons, ZnTPP/DABCO/DPI-AsF_6^- initiator system shows highest conversion (~ 90 % conversion at 20 min). As a result, as counter aniom size decrease, the time to reach final conversion increase and final conversion decreases. This data also showed electron density of counter anion influenced the conversions.

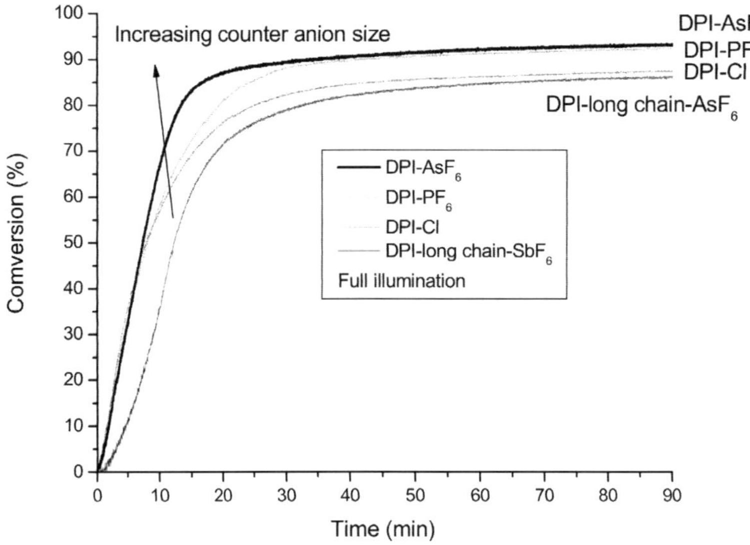

Figure 4. Conversions of HEMA polymerizations with Zn-tpp/DABCO/DPI-Xn, as measured by NIR at room temperature with an incident light intensity of 0.15 mW/cm^2. For all samples, [Zn-tpp] = 2.32×10^{-3} M, [DABCO] = 0.25 M and [DPI-Xn] = 0.02 M, in neat HEMA.

Under the 30 sec illumination and dark polymerization conditions, size and electron density of counter aniom of iodonium salt plays the same role in the both initiation and propagation step of dark polymerization reaction as shown in Figure 5. ZnTPP/DABCO/DPI-AsF_6^- initiator system shows the most dramatic conversion efficiency under the 30 sec illumination. The initiator system produced ~ 75 % conversion at 30 min under dark polymerizations with 30 sec illumination. However, as the counter anion size was decreased, final conversion obviously decreased. In addition, electron density of counter anion also influenced the conversions. From the data in Figure 4 and Figure 5, we observed that ZnTPP/DBCO/DPI-AsF_6^- initiator system is the most efficient initiator system under both full illumination and dark polymerizations with 30 sec illumination. Figure 5 illustrates the effect of size and electron density of counter aniom of iodonium salt on the conversion profile. Unlike cationic polymerizations, typical radical polymerizations do not exhibit long-lived life time of radical active centers. As a result dark polymerizations generally are not on the radical polymerizations. Therefore, these kinetic data are significantly important for radical polymerizations.

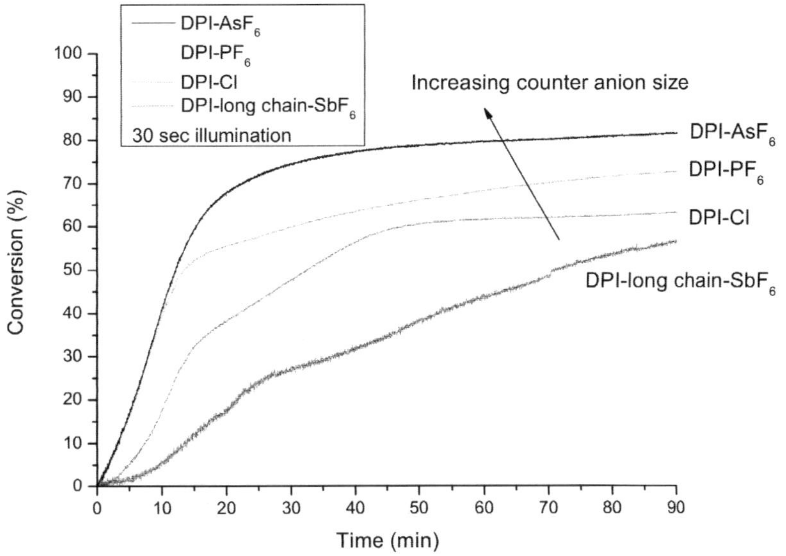

Figure 5. Conversions of HEMA polymerizations under 30 sec illumination time with Zntpp/DABCO/DPI-Xn, as measured by NIR at room temperature with an incident light intensity of 0.15 mW/cm^2. For all samples, [Zn-tpp] = 2.32×10^{-3} M, [DABCO] = 0.25 M and [DPI-Xn] = 0.02 M, in neat HEMA.

Conclusions

The photo-initiator is a radical-based, three-component initiator system that includes a light-absorbing photosensitizer (PS), an electron donor and an electron acceptor, which is usually an onium salt. In this system, 5,10,15,20-tetraphenyl-21H,23H-porphyrin zinc (Zn-tpp) is used as a PS, diazabicyclooctane (DABCO) is used as an electron donor and diphenyliodonium hexafluoroarsenate (DPI) is used as an electron acceptor. The three-component initiator system, Zn-tpp/DABCO/DPI produced a limiting conversion of 93.3 % conversion in 2-hydroxyethyl methacrylate (HEMA) upon full exposure time over 90 min. Real time conversion was monitored using near-infrared spectroscopy while photopolymerizations were conducted with a 100 W halogen lamp at an irradiance of 0.15 mW/cm^2. Remarkably, the three-component initiator system did not exhibit rapid termination of HEMA polymerization when the irradiation source was turned off. For example, only 30, 60 or 300 sec partial illumination coincided with conversion of 1.7 %, 4.3 % and 30.7 % respectively, but the final conversions reached 81.5 %, 84.7 % and 93.3 % respectively, at 90 min. These conversion kinetic results illustrate a unique controlled radical dark polymerization where the radical active centers are not terminated even when the light source is extinguished.

References

1. Oxman, J.D.; Jacobs, D.W.; Trom, M.C.; Sipani, V.; Ficek, B.; Scranton, A.B., *J. Polym. Sci. Part A: Polym. Chem.,* 2005, 43, 1747-1756.
2. Kim, D.; Scranton, A.B., *J. Polym. Sci. Part A: Polym. Chem.,* 2004, 42, 5863-5871.
3. Padon, K.S.; Scranton, A.B.; Pandalai, S. G., Ed.; *Transworld Research Network*: Trivandrum, India, 1999, Vol. 3, 369-385.
4. Padon, K. S.; Scranton, A. B., *J. Polym. Sci. Part A : Polym. Chem.,* 2000, 38, 3336-3346.
5. Padon, K. S.; Scranton, A. B., *J. Polym. Sci. Part A : Polym. Chem.,* 2001, 39, 715-723.
6. Oxman, J.D.; Ubel, F.A.; Larson, E.G., US Patent, 1989, 4,828,583.
7. Palazzotto, M.C.; Ubel, F.A.; Oxman, J.D.; Ali, Z.A.; US Patent, 2000, 6,017,660.
8. Oxman, J.D.; Jacobs, D.W., US Patent, 1999, 5,998,495.
9. Oxman, J.D.; Jacobs, D.W., US Patent, 2000, 6,025,406.
10. Erddalane, A.; Fouassier, J. P.; Morlet-Savary, F.; Takimoto, Y., *J. Polymer Sci., A: Polym. Chem.,* 1996, 34, 633-642.
11. Fouassier, J. P.; Wu. S. K., *J. Appl. Polm. Sci.,* 1992, 44, 1779-1786.
12. Lalevee, J.;.El-Roz, M.; Allonas, X.; Foussier, J. P. *Journal of Polymer Science Part A-Polymer Chemistry* 2007, 45: 2436-2442.
13. Kilambi, H.; Reddy, S. K.; Schneidewind, L.; Stansbury, J. W.; Bowman, C. N.. *Polymer* 2007, 48: 2014-2021.

499d Oligo(trolox) – Synthesis and Characterization of An Antioxidant Polymer Additive for the Suppression of Biomaterial Oxidative Stress

Paritosh Wattamwar, Chemical and Materials Engineering, University of Kentucky, 177 F. Paul Anderson Tower, Lexington, KY 40506 and Thomas Dziubla, Department of Chemical and Materials Engineering, University of Kentucky, Lexington, KY 40506.

Biodegradable polymers and copolymers like poly-lactic acid, poly-lactic-co-glycolic acid are considered to be biocompatible since their biodegradation results in residues that are already present in the body. However, recent studies have found out that local accumulation of both water-soluble and water-insoluble biodegradation products of such polymers trigger an inflammatory response which results in oxidative stress (Jiang, Su et al. 2007). We had proposed an idea of synthesizing biodegradable poly/oligomers having native antioxidant activity, biodegradation of which will result in release of active antioxidants, thereby attenuating the injury caused by oxidative stress and local inflammation.

In this work we polymerized trolox, a synthetic antioxidant, to form poly/oligo(trolox). Trolox is a water-soluble analogue of Vitamin E which scavenges free radicals and provides protection from mild forms of oxidative stress (Chow, Lynch et al. 1994). Polymer/oligomer was synthesized using Steglich Esterification (Wiener and Gilon 1986) and characterized using Fourier Transformed Infrared Spectroscopy (FTIR) and Gel Permeation Chromatography (GPC) was used to determine molecular weight. Effect of reaction conditions (e.g., reactant ratios and temperature) on extent of reaction, molecular weight of oligomer and antioxidant activity was studied. Polymer blending studies with standard degradable biomaterials (e.g., Polylactide and Poly(lactide-co-glyocolide)) was also performed.

References:

Chow, H. S., J. J. Lynch, 3rd, et al. (1994). "Trolox attenuates cortical neuronal injury induced by iron, ultraviolet light, glucose deprivation, or AMPA." Brain Res 639(1): 102-8.

Jiang, W. W., S. H. Su, et al. (2007). "Phagocyte responses to degradable polymers." J Biomed Mater Res A 82(2): 492-7.

Wiener, H. and C. Gilon (1986). "An Improved Method for the Catalytic Preparation of Tert-Butyl Esters of Carboxylic and Fatty-Acids." Journal of Molecular Catalysis 37(1): 45-52.

499e Hydrophobically Modified Alginate (HMA) for Drug Delivery and Tissue Engineering

Soumitra Choudhary and Surita R. Bhatia. Chemical Engineering, University of Massachusetts, Amherst, 686 North Pleasant Street, Amherst, MA 01003-9303

Alginate modified with hydrophobic groups was synthesized by an amide coupling reaction with alkyl amine (C8) at low pH. At or above critical concentration, HMA forms a physical gel in aqueous media. Unreacted guluronic units of alginate can be further crosslinked with divalent cations, such as Ca2+. Uniform gels were obtained by slow release of calcium ions from calcium-ethylene diamine tetra acetic acid complex with the addition of D-glucono-δ-lactone. Interplay of the two different gelation mechanisms, hydrophobic association and chemical crosslinking, enables us to tune the mechanical properties of the system. Solubility of lipophilic drugs can be improved comparatively due to preferential uptake of the drugs by micelles formed by hydrophobic moiety. Better release rates can be expected due to stronger crosslinked alginate units surrounding the hydrophobic rich domains. Small angle scattering analysis will be presented along with rheological data to determine the structure-property relationship of the resultant gels. These experiments will be compared with semi-interpenetrating polymer network of alginate and hydrophobically modified ethyl hydroxyl ethyl cellulose for possible biomedical applications

499f Long-Term Resistance of Novel Zwitterionic Surfaces to Biofilm Formation

Gang Cheng[1], Guozhu Li[1], and Shaoyi Jiang[2]. (1) Department of Chemical Engineering, University of Washington, Box 351750, Seattle, WA 98105, (2) Chemical Engineering, University of Washington, Benson Hall, Box 351750, Seattle, WA 98195

Microbial adhesion onto surfaces and subsequent formation of biofilm are critical issues for many biomedical and engineering applications. One of the most effective methods to prevent the formation of biofilm is to avoid or reduce the initial adhesion of bacteria to a surface. In order to suppress the attachment of microorganisms onto the surfaces and delay biofilm formation, the most commonly used method is to graft surfaces with hydrophilic polymers, such as poly(ethylene glycol) (PEG) derivatives and more recently zwitterionic polymers. In our previous work, we demonstrated that poly(sulfurbetaine methacrylate) p(SBMA) is highly resistant to nonspecific protein adsorption, and can dramatically reduce bacterial adhesion and biofilm formation, and a more recent study show that the two novel zwitterionic surfaces poly(carboxybetaine methacrylate) (pCBMA) and poly(carboxybetaine acrylic amide) (pCBAA) can resist nonspecific protein adsorption from 100% plasma and serum. The outstanding performance of zwitterionic polymers and PEG derivatives to resist protein adsorption and bacterial attachment is attributed to their intrinsically strong hydration via electrostatic interactions for pSBMA or hydrogen bonding interactions for PEG derivatives. Our hypothesis is that the water structures in the hydration may influence the long-term non-fouling performance of these hydrophilic polymers.

There are three objectives of this study. The first one is to understand the correlation between water structures in the hydration layer of PEG and zwitterionic polymers and the capacity of these polymers to resist biofilm formation. The second objective is to understand the effect of film thickness and surface density on the performance of zwitterionic surfaces to resist biofilm formation. The third objective is to identify the key factors which lead to the eventual fail of the non-fouling surfaces in the long-term biofilm formation study. In this work, pCBAA, pCBMA and poly(oligo(ethylene glycol) methyl ether methacrylate) (pOEGMA) were grafted from glass surfaces via atom transfer radical polymerization (ATRP) to achieve the highest efficiency to resist biofilm formation on top of these surfaces. The long-term accumulation (5 days) of Gram-negative Pseudomonas aeruginosa on these surfaces was studied using a laminar flow chamber, and the biofilm formation on these surfaces were observed and recorded in situ using a Nikon epifluorescent microscope at different time points. The result showed that pCBAA and pCBMA can efficiently delay the biofilm formation up to five days comparing to untreated glass. The effect of film thickness and density on the performance of zwitterionic surfaces to resist biofilm formation was studied by using the polymer brushes with different precisely controlled thickness. The stability of the polymer brushes in the complex medium is one of the key factors affecting the performance of polymer surface. After 5 days, biofilm was removed and the surfaces were analyzed by XPS. The effect of the stability of ester or amide bond on the performance of non-fouling polymer brushes is obtained by comparing pCBMA with pCBAA, and the contribution of functional group stability to non-fouling performance is obtained by comparing pOEGMA and pCBMA.

500a Transient Shear and Extensional Rheology and Nanostructure of Polymer Nanocomposites

Christopher Kagarise[1], Monon Mahboob[2], Koki Miyazono[1], Stephen E. Bechtel[2], and Kurt W. Koelling[1]. (1) Chemical and Biomolecular Engineering, The Ohio State University, 140 W 19th Ave, Columbus, OH 43210, (2) Mechanical Engineering, The Ohio State University, 140 W 19th Ave, Columbus, OH 43210

There has been much interest in carbon nanofiber (CNF) based polymer nanocomposites in recent years because of the potential to create materials with enhanced mechanical, thermal, and electrical properties while using much lower particle loadings than microscale fillers such as carbon black, carbon fibers, and glass fibers. The mechanical, thermal, and electrical performance properties of the nanocomposite are greatly dependent upon the nanoparticle orientation developed during processing and the dispersion of the particles within the polymer matrix. Thus it is desirable to be able to model and predict the evolution of particle orientation during processing. Accurate models provide the predictive capability necessary to design manufacturing processes that optimize the desired performance properties.

We have studied the rheological behavior of melt phase polystyrene/carbon nanofiber composites under transient shear and extensional flows. A constitutive model has been developed that predicts the rheology and nanostructure development with time during these two different types of flow. The model includes parameters that account for properties of the composite such as the rheology of the polymer matrix, nanoparticle shape, nanoparticle concentration, particle-particle interactions and polymer-particle interactions. To improve the model's predictions we have developed a method to experimentally measure the 3-dimensional structure of the nanofibers. Our nanostructure measurement technique allows us for the first time to be able to validate 3-D model predictions through comparison with 3-D experimental measurements.

500b Dynamics of Individual Chains In Linear Polyethylene Liquids Under Shear

Jun Mo Kim, David Keffer, Bamin Khomami, and **Brian Edwards**. Chemical and Biomolecular Engineering, The University of Tennessee, 419 Dougherty Hall, Knoxville, TN 37996-2200

Many previous studies investigated dynamics of polymer chains under shear. In this work, we examined dynamics of individual polymer chains in linear polyethylene liquids under shear using nonequilibrium molecular dynamics (NEMD) simulation and Brownian dynamics (BD) simulation. As expected, the probability distribution function of the end-to-end vector's magnitude follows Gaussian behavior at low shear rate. As the shear rate increases to intermediate values, individual polymer chains continue to be stretched on average, but also begin to rotate with the local fluid kinematics. Therefore, the end-to-end vector distribution begins to exhibit non-Gaussian behavior. Indeed, we observed a bimodal probability distribution function at intermediate and high values of the shear rate. The first peak of the distribution as low magnitudes of the end-to-end vector is associated with the rotational motion of the chains, and the peak at higher values of the end-to-end vector corresponds to the degree of chain elongation.

We also calculated time auto- and cross-correlation functions of each component of end-to-end vector with respect to itself and the other components, and extracted multiple time scales of the chain dynamics, associated with various physical mechanisms including the rotational motion of the chains. These time scales vary strongly with shear rate.

We compared NEMD simulation results with BD simulation results mapping atomistic configurations to the bead-rod model. Each mapped atomistic chain was also classified 'Stretched, Dumbbell, Half Dumbbell, Kink, Fold, Coil' according to its configuration and compared with BD results. Furthermore, we calculated the time that each chain spends in the four quadrants of its coordinate system, and found that an individual chain spends more time with positive orientation with respect to the direction of flow as the shear rate increases. For example, the chains spent 66% of the time at a positive orientation at the highest shear rate examined, while only 55% of the time at the lowest value of shear rate examined.

Web Page: clausius.engr.utk.edu/

500c High Resolution Shear Profile Measurements In Entangled Polymers

Keesha A. Hayes[1], Lynden A Archer[1], Mark R Buckley[2], and Itai Cohen[2]. (1) Chemical & Biomolecular Engineering, Cornell University, 120 Olin Hall, Cornell University, Ithaca, NY 14853, (2) Physics, Cornell University, 120 Olin Hall, Cornell University, Ithaca, NY 14853

We use confocal microscopy and particle image velocimetry to directly visualize polymer flow in a custom-built, planar-Couette rheometer. Polybutadiene (melt M_w = 200K, 788K) solutions spanning a range of entanglement densities ($8 \leq N/N_e \leq 56$) are seeded with 250-300 nm particle tracers and sheared in narrow gaps ~ 35 μ m. Compared to Couette and cone & plate geometries, our small aspect ratio (a_s < 1/143) parallel plate shear cell is a more rigorous setting in which to explore banding in entangled polymer flows. As the imposed shear rate increases, we find that not only does the velocity at the boundaries violate the no-slip condition, but that there are increasingly large differences between the imposed and measured rates. Despite these observations, importantly, the measured velocity profiles are generally linear (see Figure), even for shear rates well within the non-Newtonian shear regime. This finding disagrees with recent reports that shear banding is a characteristic flow response of entangled polymers, and instead points to interfacial slip as an important source of strain loss. The measured shear rates and shear stresses are used to characterize slip. We find two slip regimes; the transition to the highly non-linear second regime occurs at shear stresses comparable to the elastic modulus of the entangled polymer network. Our results are consistent both with theories for polymer slip and with published data obtained via other techniques. However, we also find that at very high dimensionless shear rates, a non-linearity in the velocity profile emerges near the moving plate. Motivated in part by this last result, we complement the confocal rheology with large-gap velocimetry using micron-sized tracers to investigate the limits and/ or range of applicability of the banding and interfacial slip phenomena. FIGURE: Velocity profiles for PBD200K, ϕ=0.4 solution; y/H=0 and y/H=1 are the top & bottom plates respectively

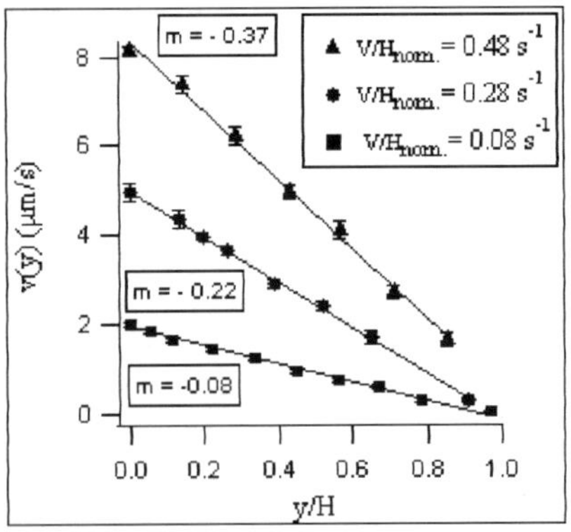

500d Computational Linear Rheology of Metallocene-Catalyzed High Density Polyethylene Predicted by Two Alternative Hierarchical Models

Xue Chen and Ronald G. Larson. Chemical Engineering, University of Michigan, 2300 Hayward St.3074 H.H. Dow Building, Ann Arbor, MI 48105

We compare the predictions of a "hierarchical model" for commercial long-chain-branched polymers, developed by Park and Larson with another publicly available model that incorporates similar physical mechanisms developed by Das et al. (2005). Both models include reptation, primitive path fluctuations of chain ends, and constraint release by "constraint release Rouse motion" or "dynamic dilution." The model of Das et al. also accounts for branches on branches, i.e., hyper-branching. These two models are applied to single-site metallocene catalyzed high density polyethylene (mHDPE) to predict the effect of long-chain branching on the linear viscoelastic properties. We generate the distributions of molecular weight and of long chain branching for mHDPE using a Monte Carlo simulation of Costeux et al. (2002), where the simulation parameters are obtained from the average molecular weight and the degree of LCB. Besides the algorithm differences described by Das et al. (2005), we compare the parameters needed in the two models. The prediction results show the differences between these two models in accuracy and efficiency.

500e Unraveling Entanglements and Polymer Rheology

Richard P. Wool, Department of Chemical Engineering and Center for Composite Materials, University of Delaware, 150 Academy St, Newark, DE 19716

The percolation model of entanglements (Wool 1993) makes unique predictions regarding the dynamics of polymer chains in the terminal relaxation zone of reptating linear polymer melts of molecular weight M. We show that the percolation entanglement model and the packing model are related through the dependence of the characteristic ratio C„V on the monomer molecular weight Mj as C„V ~ Mj1/2. When percolation occurs during relaxation of entangled chains in the terminal zone, we observe some unusual results. These include: (a) for homopolymers of molecular weight M >> Mc, reptating chains appear to be non-reptating as their ends and centers relax at the same rate during percolation. (b) During stress relaxation, the random coil dimensions Rg(//) and Rg(*f*Î) are predictably not fully relaxed when the stress and birefringence relax to zero. (c) The matrix molecular weight P effects on relaxation time *f*ä of the probe chain M >> Mc are as follows: Rouse-like dynamics is observed for reptating chains with *f*ä ~ PM2 when M>>P and (d) the relaxation time *f*ä ~ PoM3 when P>>M, in accord with reptation. These unusual results predicted by entanglement percolation are supported by significant experimental data from selectively deuterated polystyrene chains HDH, DHD and DPS. Entanglement Percolation complements Constraint release and dominates the Chain End Fluctuation mechanisms in the understanding of the dynamics of polymers in the melt and concentrated solutions.

500f Integrating Encapsulated Extensional Rheology and Step-Strain into the Methodology of the Multi-Mode Pom-Pom Model

Christopher D. McGrady, Christopher W. Seay, and Donald G. Baird. Chemical Engineering, Virginia Tech, 154 Randolph Hall, Blacksburg, VA 24061

Analysis of using encapsulated extensional rheology and step-strain to fit the non-linear parameters of the multimode pom-pom model is examined for a series of HDPE sparsely branched resins. The information available on these materials offers an excellent opportunity to compare our predictions to and aid in the possible determination of a more unique set of parameters. A novel encapsulation technique designed to overcome ductile failure in extensional flow shows promise for gathering data at higher strains, thereby improving model parameter fits. Differences in the behavior of the relaxation modulus at short times following step strain is seen as a function of LCB content.

503a Effect of Functionalization and Processing Parameters on SWNT Polypropylene Nanocomposites

Vinod K. Radhakrishnan and Virginia A. Davis. Department of Chemical Engineering, Auburn University, 212 Ross Hall, Auburn, AL 36849

The outstanding mechanical and thermal properties of single-walled carbon nanotubes (SWNTs) make them attractive reinforcing agents in fabricating the next generation composite. However, strong van der Waals attraction between the tubes results in the formation of aggregates leading to poor performance. Knowledge of the effect of fabrication parameters on the nanotubes dispersion and aggregate size is critical in manufacturing high performance nanocomposites. Our previous studies compared various pre-processing methods and showed that pre-processing along with extrusion conditions have a significant effect on the properties of melt extruded nanocomposites. In this study we focus solely on the effect of pre-processing parameters on the properties of the final nanocomposite.

We report the influence of fabrication parameters on the properties of 0.5 % vol. SWNT and functionalized SWNT polypropylene (MFI = 12) nanocomposites produced by mixing polypropylene in a mildly sonicated dispersion of SWNTs in chloroform followed by melt extrusion. Statistically designed experiments were employed to analyze the effect of polypropylene pellet size, sonication time and functional groups on the rheological and thermal properties of the nanocomposite. The level of SWNT dispersion was estimated by microscopy and Raman spectroscopy while TGA and DSC were used to observe variations in thermal properties. Rheological properties were investigated under controlled strain within the linear viscoelastic region of the nanocomposite. The results of the investigation show that initial dispersion of SWNTs in polypropylene has a large effect on the properties of the final nanocomposite.

503b Excellent Dispersion and Improved Properties of Polymer/multiwall Carbon Nanotube Nanocomposites Made Via a Two-Step Process of Solid-State Shear Pulverization Followed by Melt Mixing

Jun'ichi Masuda, Department of Chemical and Biological Engineering, Northwestern University, 2145 Sheridan Rd, TECH E136, Evanston, IL 60208 and John M. Torkelson, Department of Chemical and Biological Engineering and of Materials Science and Engineering, Northwestern University, 2145 Sheridan Rd, TECH E136, Evanston, IL 60208.

Since the discovery of carbon nanotubes (CNTs) more than a decade ago, polymer/CNT nanocomposites (NCs) have been extensively studied, because low levels of well-dispersed CNTs can yield NCs with enhanced mechanical properties, electrical conductivity, thermal conductivity, and flame retardancy. However, the structural nature of CNTs, which are commonly strongly aggregated due to their high specific surface area and heavy entanglements (the latter especially so with commercially available CNTs), has made it very difficult to employ conventional melt processing to achieve excellent dispersion of CNTs in a polymer matrix. Other processes that have been employed in selected polymer systems to achieve CNT dispersion include solvent blending and in-situ polymerization usually with the aid of surface functionalization and/or sonication, but these processes are limited for use in only a subset of polymer systems and are not commercially attractive routes for production of polymer/CNT NCs.

Here, we incorporate originally heavily entangled, multiwall CNTs of different dimensions (diameter and/or length) into polypropylene (PP) by a continuous process called solid-state shear pulverization (SSSP). With SSSP, deformation energy can be highly stored in the system and released by creating a new surface, which allows for effective pulverization of the system into fine powder. The dispersion of CNTs and NC properties are studied by scanning electron microscopy (SEM) as a function of the energy consumed during pulverization. After SSSP processing, SEM reveals an unusual structure in which loose CNT agglomerates are interpenetrated. We have found that these interpenetrated agglomerates can be further dispersed by additional melt mixing. A NC with 1 wt% MWCNT (30-50 nm outer diameter and aspect ratio of ~125) and fabricated by the two-step process of SSSP followed by melt mixing is shown via SEM to have CNTs that are almost fully debundled into individual tubes even under relatively low energy input during SSSP. This well-dispersed NC exhibits a greater than 50% increase in Young's modulus relative to neat PP. This enhancement in Young's modulus agrees well with predictions from the Halpin-Tsai model for a perfectly dispersed 99/1 wt% PP/MWCNT NC with a MWCNT aspect ratio of 125. Additionally, relative to neat PP, the NC exhibits a PP crystallization half-time that is reduced by more than 75 % and a thermal degradation temperature that is elevated by more than 25 °C.

In contrast, when we use the same conditions for our two-step SSSP plus melt mixing process on PP and 1 wt% of significantly thinner (< 8 nm outer diameter) and originally much more heavily entangled MWCNTs, we find that our hybrid product contains MWCNT agglomerates with diameters that are less than 500 nm. This nano-agglomerated 99/1 wt% PP/MWCNT NC exhibits properties that are slightly better than those of our other NC. However, because the aspect ratio of the thinner MWCNTs is greater than that of the thicker MWCNTs, the observed Young's modulus in the agglomerated NC is well below the value predicted from the Halpin-Tsai model. These results demonstrate that the ability to disperse CNTs in polymers is a function of their dimensions and the degree of CNT entanglements and that our novel, two-step process method must be accordingly tailored to the energy required to disperse each MWCNT system.

Studies are underway involving other CNT systems, optimization of our two-step SSSP plus melt mixing process approach, assessment of any damage of the PP chains and MWCNTs during our two-step process, and characterization of the electrical conductivity and rheological properties of our resultant NCs.

In-situ Densification of Combustion Synthesized Nickel Aluminide-Carbon Nanotube Composites from Nanoreactants

Lori J Groven and Jan A Puszynski
Department of Chemical and Biological Engineering
South Dakota School of Mines and Technology
501 East Saint Joseph, Rapid City, SD 57701

Abstract

In-situ densification and combustion synthesis was investigated as a unique processing technique for the formation of carbon nanotube reinforced nickel aluminides. Nano-sized reactants were utilized to form NiAl-CNT composites with densities up to 96% of the theoretical and nanosized grains were observed. For composites synthesized with 1-wt% single-walled carbon nanotubes a 30 % improvement in Vickers microhardness was observed over the monolithic material. The same improvement was not observed for multi-walled carbon nanotubes. It is demonstrated that with increased carbon nanotube loading the reaction is critically impacted leading to pre-melting and hence low product density, and a reduction in Vickers microhardness.

Introduction

Carbon nanotube (CNT) reinforced composites have been the focus of much research during the past few years. They are considered the ultimate reinforcement material with a Young's modulus of 1.2 TPa [1] for single-walled carbon nanotubes (SWCNTs) and 950 GPa [2] for multi-walled carbon nanotubes (MWCNTs). While the majority of research in this area has been focused on polymers; ceramics, metals and alloys have also been investigated. For ceramic, metal and alloys the processing methodologies have been those primarily used in traditional powder technology [3-5]. The most significant issue with these processing methods is a long term exposure to high temperatures which results in partial damage of the CNTs and also increases the vulnerability of the CNTs to interfacial reactions with the matrix material. Therefore, in order to retain the desired reinforcement properties the exposure to high temperatures must be minimized [6].

In-situ densification and combustion synthesis offers a route for the formation of dense composites reinforced with CNTs that may be completed within seconds compared to hours in conventional processing, therefore minimizing the long exposure time to high temperatures. This processing technique utilizes the exothermic heat of reaction to spontaneously transform reactants, when ignited, to desired products. With this synthesis technique a variety of reactant powder sizes can be utilized to yield products with nano to micron sized grains. While very high temperatures (>3000K) can be reaching during the combustion reaction, those temperatures are maintained for only a fraction of second [7]. By utilizing this method it is possible to synthesize a variety of ceramic and intermetallic materials that are reinforced with CNTs.

Experimental

The objective of this experimental study was to extend our previous research work [8] on combustion synthesis of CNT-reinforced materials by further investigating the high temperature reaction of Ni and Al with nano-sized reactants and varying additions of CNTs. For this work the mixing solvent, dispersant, and methodology are varied and we address what effect varying CNT addition has on the product morphology, the phase composition, and the Vickers microhardness of the composites.

The following raw materials were utilized: SWCNTs (Helix Materials, d=1.4 nm), MWCNTs (Cheap Tubes, d=8-15 nm), Ni powder (Argonide, d=100nm), and Al powder (Novacentrix, d=80nm). The stoichiometric reaction between nickel and aluminum was investigated. In the case of nano-sized aluminum there is a considerable aluminum oxide layer present (23% for 80 nm) and must be accounted for. The resulting material is a nickel mono-aluminide alumina composite.

Prior to varying the addition of CNTs into the reactive mixture several different solvents, dispersants, and mixing techniques were investigated. The following solvents were used: phenyl ethyl alcohol, ethanol, and isopropyl alcohol. Two different dispersants were tested: sodium dioctyl sulfosuccinate and 1C Coulter. Sonication with and without additional mechanical mixing was used. Mechanical mixing was achieved using a SPEX 8000 shaker with 3 mm stainless steel grinding media. The results and most effective methodology are discussed below.

Simultaneous combustion synthesis and densification experiments of these composites were conducted in a uniaxial press at a preheating temperature of 520 ºC and a densification pressure of 200 MPa as shown in Fig 1a.

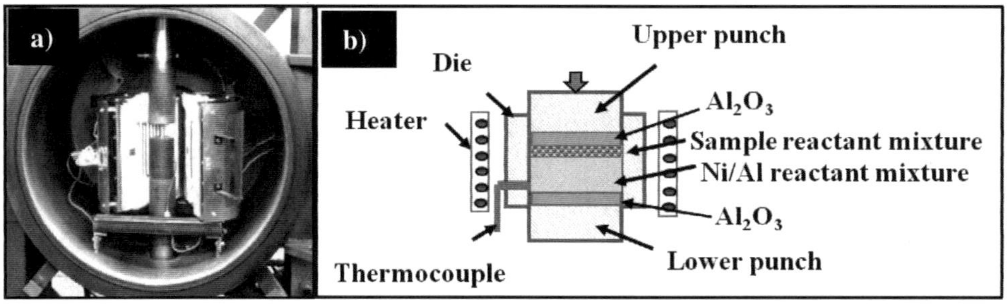

Fig. 1. *a) Experimental equipment and, b) die setup for the in situ combustion synthesis and densification of carbon nanotube loaded reactant mixtures to produce nanocomposites.*

For these experiments a 0.5" inconel graphite foil lined die and vacuum atmosphere were utilized. In these experiments a small amount (approx. 2 g) of reactant mixture containing CNTs was used. Due to a high level of heat losses, an additional chemical energy source had to be added in order to simulate the combustion process on a larger scale. This was accomplished by placing the reactant mixture under study above 10 grams of micron Ni/Al reactant mixture and placing 1 g Al_2O_3 on top and bottom as a non-conducting medium as shown in Fig 1b.

Vickers microhardness was measured with a Buehler Micromet 4 Hardness Tester using a 500 g load, with a 10 s dwell time. Samples were mounted and polished with SiC paper from 240-1200 grit, then followed with 1 µm and 0.5 µm Al_2O_3 suspensions prior to microhardness measurement. Density measurements were made using Archimedes principle.

Results

Mixing Methodology

In this work several different solvents, dispersants, and mixing methodologies were evaluated. These experiments were conducted for 1-wt % SWCNT addition. The effectiveness was based on the resulting Vickers microhardness as shown in Table 1. Initial synthesis for this system was conducted using phenyl ethyl alcohol (PEA) as the mixing solvent [8], as indicated by the low microhardness value this method led to low density samples and insufficient dispersion of the CNTs. The use of SDS as pre-dispersant also yielded poor results due to interaction of the dispersant with the Al nanopowder during the combustion process.

Table 1. *Vickers microhardness (GPa) values for composites synthesized with 1-wt% SWCNT addition.*

MIXING METHOD	SOLVENT	Ethanol	PEA	IPA
	DISPERSANT			
Sonication	none	2.92±0.51	2.74±0.40	
Sonication	SDS	2.34±0.27		
Sonication + 30min SPEX shaker w/media	none			3.48±0.17
Sonication + 30min SPEX shaker w/media	1C Coulter			4.62±0.13

It was found that the most effective mixing technique utilizes pre-dispersion of the SWCNTs with 1C Coulter as the dispersant, followed by mechanical mixing via the SPEX shaker. The dispersant is removed with vacuum treatment at 130ºC. All subsequent experiments were conducted using 1C Coulter as the dispersant with 20 minutes of sonication followed by 30 minutes of SPEX shaker mixing.

Product Morphology and Phase Composition

SEM was used to compare the product morphologies for composites synthesized with 0, 1, 2.5, and 5-wt% addition of SWCNTs. Samples were fractured and the interior morphology imaged. As shown in Fig 2 when 1-wt% to 5 wt% SWCNT addition is compared a significant difference is observed. For the 1-wt% material nearly full density (96 % TMD) was achieved and there are regions of melt as well as nano-sized grains observed.

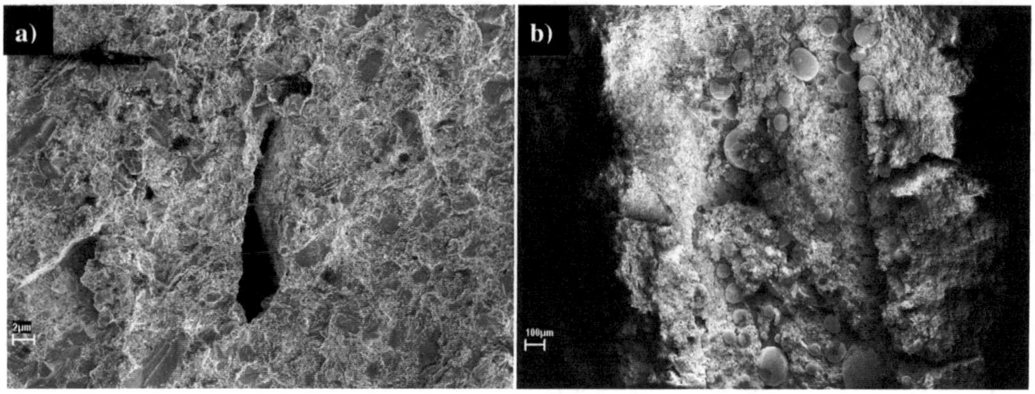

Fig.2. *Fractured surfaces of in-situ densification and combustion synthesized nickel aluminide composites; a)1-wt% and b)5-wt% SWCNT addition.*

The SWCNT addition can clearly be seen as the CNTs clearly bridge any residual porosity within the sample. For both the 2.5 and the 5-wt % SWCNT additions the product is very porous due to the pre-mature formation of product as indicated by the spherical particles. EDS of these spheres confirm that they are not an individual reactant melted but rather the desired product NiAl.

XRD patterns of the composites revealed that for all compositions that full conversion to the desired product, NiAl, is achieved as shown in Fig 3. In addition, no carbide peaks are detected.

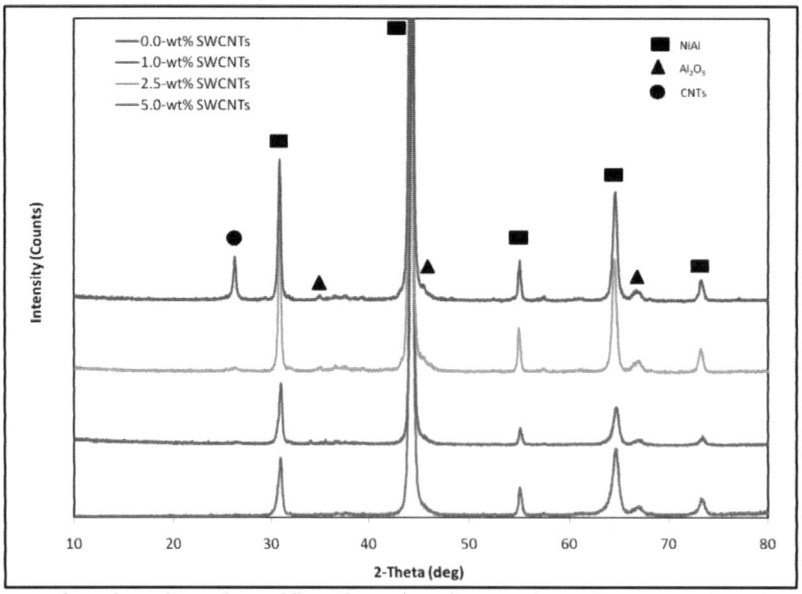

Fig 3. *XRD patterns for in-situ densification and combustion synthesized nickel aluminide composites.*

Microhardness

Vickers microhardness was measured for 0, 1, 2.5, and 5-wt% SWCNT additions and for 1-wt% MWCNT addition. Due to the pre-mature formation of product within higher CNT addition the resulting products exhibited low densities and therefore low microhardness values with increased CNT addition. As shown in Fig. 4 there is a 30 % improvement in the Vickers microhardness for composites synthesized with nano-sized reactants and 1-wt% SWCNT addition over that of monolithic.

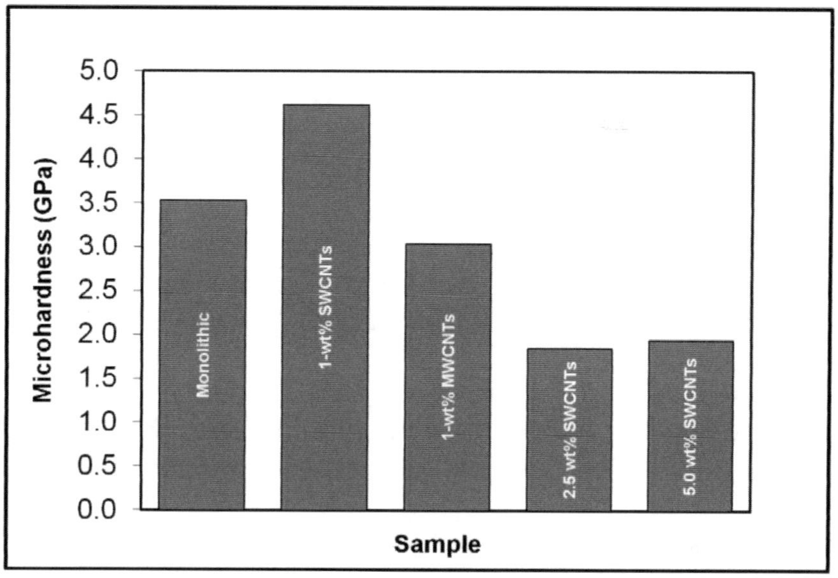

Fig. 4. *Vickers microhardness for in-situ densification and combustion synthesized nickel aluminide composites with varying CNT additions.*

This was not the case for composites synthesized with 1-wt% MWCNTs. For 1-wt% MWCNT addition there was a reduction rather than an improvement in the microhardness. This is due to the low product density achieved in this system. Representative product densities are listed in Table 2 for the different composites synthesized with this technique.

Table 2. *Representative product densities for CNT-reinforced NiAl composites.*

Sample	**0 wt% CNTs**	**1 wt% SWCNT**	**1 wt% MWCNT**	**2.5 wt% SWCNT**	**5 wt% SWCNT**
Density, g/cc	4.74	5.30	4.84	4.39	3.47

Summary

In summary this work has clearly demonstrated that in-situ densification and combustion synthesis coupled with CNTs offers a processing route to CNT-reinforced composites. The role of appropriate mixing methodologies in achieving composites with improved properties has been shown. In addition, the critical effect increased CNT addition has on the reactive processing of these types of materials was discovered and is presently being further examined from the perspective of increased thermal conductivity of the reactive mixtures.

Acknowledgment

The authors are thankful for the support of this work by the National Science Foundation with a National Science Foundation Graduate Research Fellowship.

References

[1] J.P. Salvetat, G.A. Briggs, J.M. Bonard, R.R. Bacsa, A.J. Kulik, T. Stöckli, et al., "Elastic and shear moduli of single-walled carbon nanotube ropes," *Phys Rev Lett*, vol. 82(5), 1999, pp. 944-7.

[2] F. Lupo, R. Kamalakaran, C. Scheu, N. Grobert, and M. Rühle, "Microstructural investigations on zirconium oxide–carbon nanotube composites synthesized by hydrothermal crystallization," *Carbon*, vol. 42, 2004, pp.1995–99.

[3] A. Peigney, Ch. Laurent, E. Flahaut, and A. Rousset, "Carbon nanotubes in novel ceramic matrix composites". *Ceram Int*, vol 26, 2000, pp. 677-83.

[4] T. Kuzumaki, O. Ujiie, H. Ichinose, and K. Ito, "Mechanical characteristics and preparation of carbon nanotube fiber-reinforced Ti composite," *Adv Eng Mater*, vol. 2(7) 2000, pp. 416–8.

[5] Tiny Tubes Boost for Metal Matrix Composites. *Met Powd Rep*, 2004, pp.40-3.

[6] K. Metenier, S. Bonnamy, F. Beguin, C. Journet, P. Bernier, et al., "Coalescence of single-walled carbon nanotubes and formation of multi-walled carbon nanotubes under high pressure treatments," Carbon, vol. 40, 2002, pp. 1765-73.

[7] A. Varma, A. S. Rogachev, A. S. Mukasyn, and S. Hwang, "Combustion Synthesis of Advanced Materials: Principles and Applications", Adv. in Chem. Eng., vol. 24, pp. 79-225, 1998.

[8] L.J. Groven and J.A. Puszynski, "Effect of Carbon Nanotube Addition on SHS Materials", International Journal of Self-Propagating High Temperature Synthesis, Vol. 16, No. 4, pp. 189–198, 2007.

503d Using External Fields to Control the Location of Nanoparticles In Block Copolymers: Experiments and Simulations

Vibha Kalra[1], Jinwoo Lee[2], Sergio Mendez[1], Fernando Escobedo[1], and Yong L. Joo[1]. (1) School of Chemical and Biomolecular Engineering, Cornell University, 120 Olin Hall, Ithaca, NY 14853, (2) Department of Chemical Engineering and School of Environmental Science & Engineering, Pohang University of Science and Engineering, xxx xxx xxx, Pohang, xxx, South Korea

Nanocomposite nanofiber mats (100-400 nm diameters) with internal self assembly used to tailor nanoparticle location have been fabricated. The fibers are obtained using coaxial electrospinning with symmetric poly(styrene-*b*-isoprene)/magnetite NP as the core and silica (synthesized using sol gel technique) as the shell. Silica shell provides thermal stability to the fibers and helps to anneal the materials at temperatures higher than polymer glass transition to obtain equilibrium, confined self assembly without destroying the fiber morphology. The magnetite NPs (4 nm) used in this work, are surface coated with oleic acid to provide slight selectivity towards isoprene domain while preserving their magnetic moment. All nanofiber samples are characterized using scanning electron microscopy (SEM), small angle x-ray scattering (SAXS) and transmission electron microscopy (TEM). A systematic study on effect of NP:polymer volume ratio on material self assembly has been conducted. One of our key findings is that strong deformation (~10000 s^{-1}) and fast solvent evaporation (200 nl/s) during electrospinning enables selective, uniform dispersion of magnetite NPs in the isoprene domain for as high as 20 wt% NPs with respect to isoprene in spite of strong magnetic inter-particle interactions. Magnetic properties are measured using superconducting quantum interference device (SQUID) magnetometer and all nanocomposite fiber samples exhibit superparamagnetic behavior.

To further understand the effect of flow conditions on nanoparticle location in block copolymers, we have performed coarse grained molecular dynamics simulations using dissipative particle dynamics (DPD) thermostat. We modeled and simulated the behavior of symmetric diblock copolymer/ nanoparticle systems under simple shear flow. We considered three systems such that; a) NPs have selective interactions with one block, b) NPs have equal interactions with both blocks and c) NP-NP interactions are dominant. We find that shear has a significant effect on prevention of NP aggregation in cases where NP-NP attraction is dominant and this effect of shear is a strong function of both nanoparticle size and polymer chain length. These results qualitatively match with our experimental results on PS-*b*-PI/magnetite NP nanofibers.

Nanocomposite materials are gaining popularity because of their unique properties. High interfacial area and the interactions at these interfaces between the different constituents, result in properties that are not present in the individual bulk components. Mullite is one of such nanocomposite material and finds wide applications because of its high creep resistance and refractory properties. It is a naturally occurring aluminum silicate and is often produced synthetically because of its rarity. One common method of producing mullite is to mix aluminum oxide and silicon oxide particles in liquid solvents and then firing the mixture at temperatures as low as 1250 °C. However, removing the suspending liquid can be expensive in terms of both time and resources. Therefore, the objective of this work was to use dry mixing methods to improve the quality of mixing of alumina and silica for the purpose of reducing the temperature for mullite formation and to increase % conversion of a mixture to mullite.

Various environmentally friendly dry mixing methods have been used to prepare nanoparticle mixtures of alumina (d_p = 13 nm) and silica (d_p = 16 nm) at the stoichiometric ratio of 3:2 for the purpose of forming mullite at sintering temperatures above 1250 °C. One such method is Magnetically Assisted Impact Mixing, which uses sub-millimeter magnets propelled by oscillating magnetic field to mix the nanopowders. The quality of mixing, which was determined by Energy Dispersive X-Ray Spectroscopy (EDS) was correlated to the amount of mullite formed as measured by X-ray powder diffraction (XRD) and differential scanning calorimetry (DSC).

Preliminary results correlating the quality of mixing to the amount of mullite formed after firing at 1550°C are shown below in Table 1. Mullite content was determined by quantitative XRD analysis using CaF_2 as an internal standard. The mullite content was correlated to the intensity of segregation (IOS), a mixing quality index proposed by Danckwerts [1], in order to demonstrate the effect of mixing on % of mullite formation. IOS ranges from 0 to 1, where 0 implies perfect homogeneity and 1 implies complete heterogeneity and was determined from EDS analysis. Lower IOS implies more uniform mixing and hence increases % conversion to mullite.

Table 1: Comparison of mixing quality, characterized by the intensity of segregation to the amount mullite formed at 1550 °C.

Sample	IOS	Mullite Content (%)
1	0.0017	~ 100
2	0.030	87
3	0.450	55

[1] Danckwerts, PV. The Definition and Measurement of some characteristics of Mixtures. Applied Science Research. 1952: 3;A, 279-296.

503f Fabrication, Characterization and Microwave Properties of Polyurethane Nanocomposites Reinforced with Magnetic and Ferroelectric Nanoparticles

Zhanhu Guo[1], Sang-Eui Lee[2], Amar B. Karki[3], David P. Young[3], and H. Thomas Hahn[2]. (1) Chemical Engineering Department, Lamar University, Beaumont, TX 77710, (2) Mechanical and Aerospace Engineering Department, University of California Los Angeles, Los Angeles, CA 90095, (3) Department of Physics and Astronomy, Louisiana State University, Baton Rouge, LA 70803

Polymer nanocomposites have been reported for applications in coating as UV shielding agent,1,2 electromagnetic interface (EMI) shielding,3,4 and microwave absorption industries5,6 due to their easy processability, low-cost manufacturing, and good adhesion to substrate. For microwave absorption, the fillers could be chosen from either magnetic or dielectric materials.7 Most of the polymer matrices are insensitive to microwave and only serve as binders between the nanoparticles. The hosting polymer matrix facilitates local stress transfer from the polymer matrix to the tougher nanoparticles and provides good adhesion between the composite and the substrate. The active component of absorption is the magnetic or dielectric fillers, which absorbs the microwave due to the interactive loss process of dielectric or magnetic dipoles of the fillers.8 Thus, composites with higher filler loading have great potential to increase the absorption ability. However, the subsequent manufacturing puts a limitation in the availability of high-quality nanocomposites with high filler loadings. High-loading can cause heavy-cracks or even bubbles, which will definitely decrease the mechanical properties.9-11 Toward achieving nanocomposites with high-loading and good mechanical performance, coupling agents have been widely utilized to improve the particle dispersion and to enhance the interaction between the nanoparticles and the polymer matrix.2,12

Recently, we have introduced a facile surface-initiated-polymerization (SIP) method to fabricate polyurethane nanocomposites reinforced with iron oxide nanoparticle13 or silicon carbide nanoparticles.14 Both show improved mechanical properties at high particle loading and the nanoparticle materials have an effect on the mechanical behaviors of the polymer. The superparamagnetic iron oxide nanoparticles were observed to behave as ferrimagnetic due to the weaker dipolar interparticle interaction induced by the nonmagnetic polymer spacer. Ferroelectric ceramic barium titanate possesses a high dielectric constant with a great potential for energy storage supercapacitor15 and dielectric microwave absorber applications.16 The addition of two different kinds of nanoparticles could enhance the overall performance of the polymer nanocomposites by introducing the individual constitute nanoparticles with different properties.

In this presentation, surface-initiated-polymerization (SIP) method has been used to fabricate polyurethane nanocomposites reinforced with two different nanoparticles, i.e. magnetic Fe_2O_3 nanoparticles and ferroelectric barium titanate with high dielectric constant. The thermal stability of the nanocomposite was qualitatively characterized by TGA and the chemical interaction between nanoparticles and polymer was characterized by DTG. Magnetic properties of the nanocomposites with different -Fe2O3 particle loadings and two different nanopartilces were investigated in a 9-Tesla Physical Properties Measurement System (PPMS) by Quantum Design . The effect of high dielectric constant ceramic barium titanate nanoparticles and ferromagnetic Fe2O3 nanoparticles on the microwave properties (permittivity and permeability) of the polyurethane nanocomposites was investigated according to volume fraction and frequency of 10 MHz to 1 GHz. The effective complex permittivity and permeability of the polymer nanocomposites were measured using the transmission/ reflection (TR) method based on Nicolson-Ross algorithm,17-19 which is calculated from the measured scattering parameters (S parameters). The measurement was carried out with an Agilent E4991A impedance analyzer, combined with an Agilent 16453A dielectric material test fixture and an Agilent 16454A magnetic material test fixture, respectively, under a sweep of 10 MHz – 1 GHz. Specimens of permittivity measurement are cylinders with a diameter of 17.5 mm, and those for permeability are hollow cylinders with an inner diameter of 10 mm and an outer diameter of 17.5 mm. The thickness of all the samples ranges from 0.5 mm to 2.0 mm. The predicted microwave properties from Bruggeman's equation were consistent with measured data, except the real permittivity of Fe2O3/BaTiO3/PU. The volume average method (VAM) usually used for fiber-reinforced composites with reinforcements in thickness directions was applied in our nanocomposite system. The predicted real permittivity by VAM was found in a better agreement with the measured data than that predicted by Bruggeman's equation. This indicates that VAM can be used to predict microwave properties of nanocomposites with multiphases.

References:

1 D. Sun, N. Miyatake, and H.-J. Sue, Nanotechnology 18, 215606 (2007).

2 Z. Guo, S. Wei, B. Shedd, R. Scaffaro, T. Pereira, and H. T. Hahn, Journal of Materials Chemistry 17, 806-813 (2007).

3 C.-S. Zhang, Q.-Q. Ni, S.-Y. Fu, and K. Kurashiki, Composites Science and Technology 67, 2973-2980 (2007).

4 J. Gass, P. Poddar, J. Almand, S. Srinath, and H. Srikanth, Advanced Functional Materials 16, 71-75 (2006).

5 V. B. Bregar, IEEE Transactions on Magnetics 40, 1679-1684 (2004).

6 Z. Guo, S. Park, H. Thomas Hahn, Suying Wei, M. Moldovan, A. B. Karki, and D. P. Young, Journal of Applied Physics 101, 09M511 (2007).

7 X. F. Zhang, X. L. Dong, H. Huang, Y. Y. Liu, W. N. Wang, X. G. Zhu, B. Lv, J. P. Lei, and C. G. Lee, Applied Physics Letters 89, 053115 (2006).

8 S. M. Abbas, A. K. Dixit, R. Chatterjee, and T. C. Goel, Materials Science and Engineering: B 123,

167-171 (2005).

9 X. Zhang and L. C. Simon, Macromolecular Materials and Engineering 290, 573-583 (2005).

10 Z. Guo, K. Lei, Y. Li, H. W. Ng, and H. T. Hahn, Composites Science and Technology 68, 1513-1520 (2008).

11 R. M. Rodgers, H. Mahfuz, V. K. Rangari, N. Chisholm, and S. Jeelani, Macromolecular Materials and Engineering 290, 423-429 (2005).

12 Z. Guo, T. Pereira, O. Choi, Y. Wang, and H. T. Hahn, Journal of Materials Chemistry 16, 2800-2808 (2006).

13 Z. Guo, S. Park, S. Wei, T. Pereira, M. Moldovan, A. B. Karki, D. P. Young, and H. T. Hahn, Nanotechnology 18, 335704 (2007).

14 Z. Guo, T. Y. Kim, K. Lei, T. Pereira, J. G. Sugar, and H. T. Hahn, Composites Science and Technology, 164-170 (2008).

15 I. Burn and D. M. Smyth, Journalf of Materials Science 7, 339-343 (1972).

16 M. P. McNeal, S.-J. Jang, and R. E. Newnham, Journal of Applied Physics 83, 3288-3297 (1998).

17 A. M. Nicolson and G. F. Ross, IEEE Transaction Instrument Measurement IM-19, 377-382 (1970).

18 A. M. Nicolson, IEEE Transactions on Instrum. Meas. 17, 395-402 (1968).

19 C. Brosseau and P. Talbot, IEEE Transactions on Dielectrics and Electrical Insulation 11, 819-832 (2004).

Web Page: www.seas.ucla.edu/~zhanhu

524a Prediction of Thermodynamic Compatibility Between Self Associating Poly(Ethylene Oxide)-*B*-Poly(ε-Caprolactone) and Two Water Insoluble Drugs Using *In Silico* Methods

Sarthak K. Patel, Chemical & Materials Engineering, University of Alberta, 536 CME Building, Edmonton, AB T6G 2G6, Canada, Afsaneh Lavasanifar, Faculty of Pharmacy and Pharmaceutical Sciences, University of Alberta, 4119 Dent/Pharm Centre, Edmonton, AB, Canada, and Phillip Choi, Chemical&Materials Engineering, University of Alberta, 536 CME Building, Edmonton, AB T6G 2G6, Canada.

Recently, drug formulations formed from polymer excipients has attracted considerable attention as a novel nanoscopic vehicle for the encapsulation and the controlled delivery of water insoluble drugs. Polymer micelles used for such an application are formed by self association of amphiphilic block copolymers in an aqueous environment. The compatibility between the drug and the hydrophobic block of a block copolymer determines the encapsulation efficiency of the micelles, release profile and even the stability of the delivery system. Currently, most of the drug formulations proceed by "trial and error" method with no distinct method to predict the right combination of block copolymers and drugs to give all the desired functional properties. This is simply because such drug delivery systems involve complex intermolecular interactions and geometric fitting of molecules of different shapes. So, in the context of block copolymer design process, quantification and prediction of the interactions between potential block copolymers and the target drug are of great importance. Computer simulations of copolymer/drug mixtures, to predict these interactions between them, can provide an alluring alternative to the cumbersome and expensive "trial and error" type of formulation studies.

In the present work, we report the use of molecular dynamics (MD) simulation to predict the solubility of two water insoluble drugs, i.e., fenofibrate and nimodipine, in a series of self associating PEO-b-PCL block copolymers with combinations of blocks with different molecular weights. The solubility predictions based on the MD results were then compared with those obtained from solubility experiments and those obtained by the commonly used group contribution method (GCM). The results showed Flory-Huggins interaction parameters computed by the MD simulations to be consistent with the solubility data of the drugs/PEO-b-PCL systems while those calculated by the GCM deviate significantly from the experimental findings. We have also accounted for the possibility of drug solubilization in the PEO shell and assessed the effect of both hydrophilic (PEO) and hydrophobic (PCL) drug solubility in PEO-b-PCL micelles. Our results confirm that not only the interaction energy potential but also the local molecular arrangement plays a vital role in the accurate predictions of solubility parameters and Flory-Huggins interaction parameters.

524b New Shapes for Drug Delivery: Persistent Circulation of Filomicelles Opens the Dosage Window for Tumor Shrinkage

Dennis E. Discher, Takamasa Harada, and Shenshen Cai. Chemical and Biomolecular Engineering, University of Pennsylvania, Room 129 Towne Building, 220 South 33rd Street, Philadelphia, PA 19104-6393

Shape effects of drug delivery vehicles are largely unexplored, especially in vivo, but flexible cylinders have recently been shown to circulate in mice and rats longer than spheres of the same composition. Degradable worm-like filomicelles are shown here to have significant advantages for drug delivery in vivo. Not only is drug loading of the hydrophobic drug paclitaxel (tax) substantially increased with a cylindrical geometry compared to sonication-generated spherical micelles of the same copolymer, but intravenous injection of tax-loaded filomicelles nearly doubles the maximum tolerated dose in mice. In tumor-bearing mice, the higher dose of tax produced greater and more sustained tumor shrinkage. Apoptosis with filomicelles was also more pronounced in the tumor than in non-tumor organs, and drug biodistribution confirmed the enhanced tumor-selectivity of delivery. Flexible worm-like filomicelles thus increase both dosage and tumor-selective effects in vivo relative to spherical micelles and thus appear promising as anticancer drug delivery systems.

Stimuli-responsive hydrogels are attractive as components of microfluidic device because the swelling response of hydrogel systems is typically diffusion limited, and thus hydrogels at small scales are more efficient. In addition, hydrogels are easy to fabricate in microfluidic device and eliminate the need of integrated electronics, controls, and power source that are required for active components. Here, we demonstrate the concept of using external stimuli to actuate temperature-responsive hydrogel valves. Externally actuated valves are attractive for applications like pulsatile drug release from implanted device, where the dose can be varied easily by variation of external stimuli.

Previously, we have developed magnetic nanocomposites of temperature responsive poly (N-isopropylacrylamide) hydrogels by addition of superparamagnetic Fe_3O_4 particles. The composites respond to alternating magnetic field (AMF) and have been demonstrated for matrix type of externally actuated pulsatile drug release systems. The rate of hydrogel collapse and recovery are key parameters in design of hydrogel as a valve. In this study, the systems were characterized for remote heating on application of AMF and AMF triggered collapse. The effect of AMF was characterized for different particle loadings and different nanocomposite film thickness. Heat and mass transfer analysis was conducted to correlate the film thickness with kinetics of AMF triggered heating, collapse, and subsequent recovery. The gels with higher thickness collapsed faster because heat transfer dominates the collapse. On the contrary, gels with higher thickness recovered slower because mass transfer dominates the recovery process. A simple model was proposed that explains and predicts the collapse and recovery process with hydrogel thickness.

In this study, NIPAAm based hydrogel composites were incorporated as valves in a ceramic microdevice. Flow of room temperature water resulted into hydrogel swelling and the valve closed. The valve was demonstrated for ON-OFF flow control with variation in temperature using a hot plate. The characterization of time needed for valve to open/close was also conducted with a pressure sensor. The goal of this study was also to look at remote actuation capabilities of this valve. Application of AMF to the device selectively heated the valve and resultant temperatures were imaged by an IR camera. As the temperature of matrix increased above the lower critical solution temperature (LCST), the hydrogel collapsed, and opened the valve. This process showed very good reproducibility for multiple ON-OFF cycles. Hence, hydrogel nanocomposites were demonstrated as remote controlled valves.

524d "Living/controlled" Molecular Imprinting Polymerization Techniques to Enhance Drug Delivery Networks

Asa D. Vaughan and Mark E. Byrne. Biomimetic & Biohybrid Materials, Biomedical Devices, and Drug Delivery Laboratories, Department of Chemical Engineering, Auburn University, Auburn, AL 36849-5127

Enhanced drug loading and extended release in hydrogels can be achieved by molecular imprinting techniques, which involve the formation of a pre-polymerization complex between the template therapeutic and functional monomers by non-covalent chemistry. In this work, thin polymer films have been synthesized via templated mediated polymerization resulting in macromolecular recognition or "trained" recognition for diclofenac sodium, a non-steroidal anti-inflammatory drug, within the resulting network. Creating molecular memory within polymer chains can significantly delay release and is a novel, generalizable way to further control and extend release and increase loading in addition to current techniques. Also, "living/controlled" polymerizations techniques have been used in the creation of these polymer gels and significantly enhanced template loading and delayed transport. Diclofenac sodium imprinted poly(diethylaminoethyl methacrylate-co-2-hydroxyethyl-methacrylate-co-polyethylene glycol200dimethacrlyate) (poly(DEAEM-co-HEMA-co-PEG200DMA)gels synthesized via "living/ controlled" methods show a two-fold increase in the duration of release compared to diclofenac sodium imprinted gels synthesized via standard free-radical polymerization techniques and a four-fold increase in the duration of release compared to the non-imprinted control gel. Also, diclofenac sodium loading capacities increased two-fold from (1.87 ± 0.30 to 3.54 ± 0.25) x 10-2 mmole/g while retaining similar template binding affinities when comparing conventional free-radical polymerization to living polymerization. Structural analysis of the imprinted gels also demonstrated that "living/controlled" polymerization techniques decreased the average mesh size from 30.3 ± 1.7 to 19.7 ± 2.1 A. This directly correlated with the enhanced loading and delayed release of the imprinted structure and is the first time living polymerization strategies have been shown to lead to more effective binding cavities within a flexible hydrogel network. The release profiles from imprinted gels demonstrated less Fickian behavior and moved closer toward a zero-order release profile with both "living/controlled" and standard free-radical imprinting polymerization techniques. Utilization of "living" polymerization techniques within the synthesis of loosely crosslinked polymers thin films augment the tailorability of the release profiles and enhance the resulting loading capacity which significantly enhances the potential of hydrogel drug delivery carriers. Characterization analysis of the network structure of the hydrogel carrier in terms of molecular weight between crosslinking points, mesh size, and diffusion studies provides an aid to optimizing the design and begins to answer fundamental questions on the nature of the recognition and extended delivery in imprinted hydrogels on the chain level.

524e Polyketals: A New Acid-Sensitive Biomaterial for Drug Delivery

Sungmun Lee, Stephen Yang, Michael Heffernan, and Niren Murthy. Biomedical Engineering, Georgia Institute of Technology, 315 Ferst Drive, Atlanta, GA 30332

There is currently great interest in developing acid-sensitive materials for drug delivery because of their ability to hydrolyze in the acidic environment of phagosomes and release drugs into the cytoplasm. In this presentation, we introduce a new acid-sensitive polymer for drug delivery, poly(cyclohexane-1,4-diyl acetone dimethylene ketal) (PCADK), which is ideal for treating inflammatory diseases because of its biocompatible degradation products. PCADK was synthesized using the acetal exchange reaction between 1,4-cyclohexanedimethanol and 2,2-dimethoxypropane. PCADK hydrolyzes into innocuous compounds, 1,4-cyclohexanedimethanol and acetone, which have FDA approval for human use. The hydrolysis half-life of PCADK is 24.1 days at pH 4.5 and over 4 years at pH 7.4, which is suitable for targeting chronic diseases. The therapeutic enzyme superoxide dismutase (SOD), which scavenges reactive oxygen species, was encapsulated into PCADK-based microparticles using a double emulsion procedure. Cell culture experiments demonstrated that PCADK-based microparticles significantly improved the ability of SOD to scavenge reactive oxygen species produced by macrophages. We anticipate the use of PCADK to treat a wide variety of chronic inflammatory diseases such as lung fibrosis, arthritis, and congestive heart failure.

Local intracranial drug delivery using biodegradable PLGA-paclitaxel micro/nano-fiber implants to treat malignant brain tumors

Abstract ID: 123898

Sudhir H. Ranganath, Chi-Hwa Wang

Department of Chemical and Biomolecular Engineering,
National University of Singapore, 4 Engineering Drive 4, Singapore 117576

INTRODUCTION

Malignant brain tumors (gliomas) characterized by aggressive proliferation, inexorable local recurrence and invasive nature are among the most recalcitrant cancers to treat in the human body. This necessitates the urgent need for the development of implantable, biocompatible and biodegradable delivery devices carrying cytotoxic and radiosensitizing drugs for post-surgical synergetic chemo-radiotherapy against malignant gliomas. Most of the polymeric implants delivering paclitaxel (a cytotoxic and radiosensitizing drug) developed are either micro/nano-particles, compressed micro/nano-particle discs or wafers. However, due to very high interstitial pressure in the brain/tumor, intratumoral injection of drug loaded microparticles will always run the risk of being expelled out of the target site thus negating the advantages of local delivery [1]. The other disadvantage of using microparticles is the high initial burst due to the presence of the drug on the surface which might lead to undesired neurotoxicity. On the other hand, drug loaded compressed discs and wafers have low surface area to volume ratio available for polymer degradation and drug diffusion and hence would result in low drug release rates and undesired secondary burst [2].

In an effort to improve local therapy against glioma in terms of avoiding high initial burst and providing drug release sustainability and implantability, we have developed novel poly (D,L-lactide-co-glycolide) (PLGA) micro/nano-fiber implants bearing paclitaxel by Electrohydrodynamic Atomization process (EHDA). To demonstrate proof of concept, PLGA-paclitaxel fiber discs and sheets were fabricated and characterized. The performance of the formulations in sustaining drug release and cytotoxicity were also evaluated *in vitro* through apoptosis study on C6 glioma cells. In addition, the formulations were tested against subcutaneous C6 glioma tumors in BALB/c nude mice. These studies allowed a comparison between fiber discs and sheets in terms of implantability, compactness and drug release profiles which showed that the discs were better suited. Subsequently, to better understand the contributions of the co-polymer and the fiber diameter towards the drug release, six fiber disc dosage forms with varying co-polymer and fiber diameter were fabricated and characterized in terms of in vitro release profiles.. Also, the best dosage forms for further in vivo studies were selected which exhibited small initial burst, relatively zero-order release rate and sustainability over extended periods.

EXPERIMENTAL METHODS

PLGA was dissolved in a mixture of dichloromethane (DCM) and dimethylformamide (DMF) at 30% (w/v) concentration. 9.1% (w/w) paclitaxel was then dissolved in the solution and the drug-polymer solution was pumped at a predetermined rate using a syringe pump, forming a bead of solution at the tip of syringe. A high voltage difference (12-20 kV) was applied between the nozzle (27 G needle with diameter of 0.34 mm), a negative potential and a grounded collection target. As the jet broke up into fibers from the Taylor Cone, DCM was evaporated giving rise to relatively dry fibers which were subsequently spun on the aluminum foil wrapped rotating shaft until multilayered fiber mat was obtained.

Paclitaxel-PLGA fiber discs were obtained by punching the fiber mats into 3 x 1 mm discs. The mats were cut into 5 x 5 mm sheets to obtain paclitaxel-PLGA fiber sheets. PLGA 85:15 was used to fabricate microfiber disc (MFD) and sheet (MFS), whereas PLGA 50:50 was used for submicrofiber disc (SFD) and sheet (SFS). The dosage forms were freeze dried for approximately one week to remove residual DCM. The discs were sterilized using Co-60 gamma irradiation at a dose of 15 kGy before using for in vitro and in vivo studies. To compare the effect of co-polymer and fiber diameter on drug release profile, PLGA 85:15 co-polymer was used to fabricate microfiber (E1), sub-microfiber (E2) and nano-fiber (E3) discs with fiber diameters of about 3.1 µm, 930 nm and 100 nm respectively. Similarly, PLGA 50:50 was used to fabricate microfiber (F1), sub-microfiber (F2) and nano-fiber (F3) with the same diameters as above.

RESULTS AND DISCUSSION

In vitro Characterization

Fiber Diameter, Distribution, Morphology and Paclitaxel Encapsulation Efficiency

In the proof of concept study, the reason for choosing two different PLGA co-polymers was that PLGA 50:50 having higher hydrophilic glycolic acid (G) content is known to degrade at a faster rate than PLGA 85:15 owing to faster hydrolysis. This difference in bulk degradation rates was utilized to develop faster and slower releasing formulations. Additionally, by varying the fiber diameter for the two formulations the difference in surface area to volume ratio was availed to aid faster or slower release. SEM images in Fig.1 suggest that electrospun fibers were formed with good monodispersity in fiber diameters which is confirmed in Table 1.

Table 1. Electrospinning parameters and characterization of Paclitaxel-PLGA microfibers

Samples	DCM/DMF Ratio % (v/v)	Flow rate (ml/h)	Voltage difference between nozzle & ground (kV)	Distance from needle tip to shaft (cm)	Fiber mean diameter	Paclitaxel encapsulation efficiency (%)	Paclitaxel loading (%) w/w
PLGA 85:15 Microfiber (MF)	80/20	6.00	12 -16	15	3.5 ± 0.32 μm	98 ± 4.9	8.92 ± 0.49
PLGA 50:50 Submicrofiber (SF)	70/30	0.30	16 - 18	15	930 ± 35 nm	94 ± 0.57	8.60 ± 0.06

Fig. 1 also confirmed that the surface of the fibers were smooth with high encapsulation efficiency of paclitaxel. Such a non-woven fiber was hypothesised to provide higher surface area to volume ratio for drug release when compared to conventional drug delivery discs/wafers made of either compressed drug-polymer mixture or compressed drug-polymer microparticles.

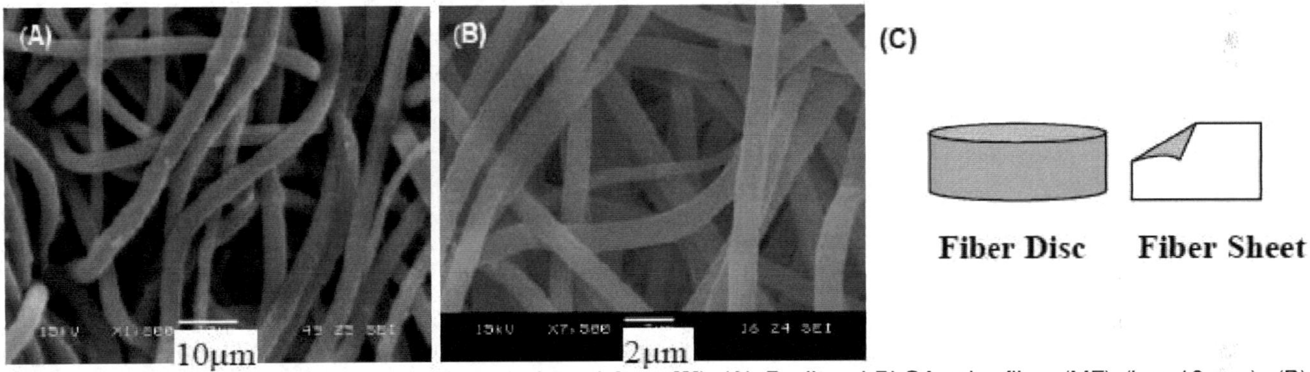

Figure 1: SEM images of electrospun fibers (*adopted from [3]).* (A) Paclitaxel-PLGA microfiber (MF) (bar 10 μm); (B) Paclitaxel-PLGA submicrofiber (SF) (bar 2 μm). (C) Depiction of fiber disc and fiber sheet.

Later, from the two formulations (MF and SF), four dosage forms (MFD, SFD, MFS and SFS) were prepared as described earlier. A comparative study was performed on the polymer degradation, in vitro drug release, apoptotic activity and in vivo subcutaneous tumor inhibition from discs and sheets.

In vitro Release of Paclitaxel from Fiber Discs/Sheets

This study was aimed at quantifying the % cumulative release of paclitaxel from the four dosage forms under simulated physiological conditions over an extended period of time. The *in vitro* release profiles shown in Fig.2 suggest that paclitaxel release was sustained for more than 80 days for all the dosage forms. While the fiber discs showed nearly zero order release kinetics with slight initial burst (6% for MFD and SFD after 9 days), the fiber sheets (MFS and SFS) exhibited a relatively high initial burst (16% for SFS and 10% for MFS after 9 days) followed by a zero-order release. For MFD, after day 7 the release rate was almost linear and after day 80, 26% of total paclitaxel was released. For SFD, till day 31 a slow release rate was observed followed by a slight increase in release rate up to day 80. After day 80, around 35% of total paclitaxel was released from SFD. SFD and MFD also exhibit two slightly different release rates. We can notice that for SFD, the release profile from day 0 till day 35 is strikingly similar to the release profile from day 30 to day 70. Whereas for MFD, after day 64 a slight burst is seen which was similar to its initial burst.

The results from the release profile confirm the fact that SF has a higher release rate than MF for the most obvious reason of faster bulk degradation rate due to higher glycolic acid content in PLGA 50:50 compared to PLGA 85:15. This correlates well with the GPC results (not shown). MFD and SFD in our study demonstrate low initial burst probably due to the disc geometry and multilayered structure of fibers, where as

MFS and SFS due to less compact structure exhibit a relatively higher initial burst. Furthermore, SFD and MFD were made by multilayered fiber sheets, thus not all the sheets were equally exposed to release buffer. This multilayered structure made them exhibit two slightly different release rates. The reason for this was as the old layer degraded and exposed the new layer; a slight burst was seen followed by a constant release which went on till most of the polymer had degraded. Even though the discs exhibited two different release rates, they were not much different in magnitude in comparison to the biphasic release seen in the fiber sheets.

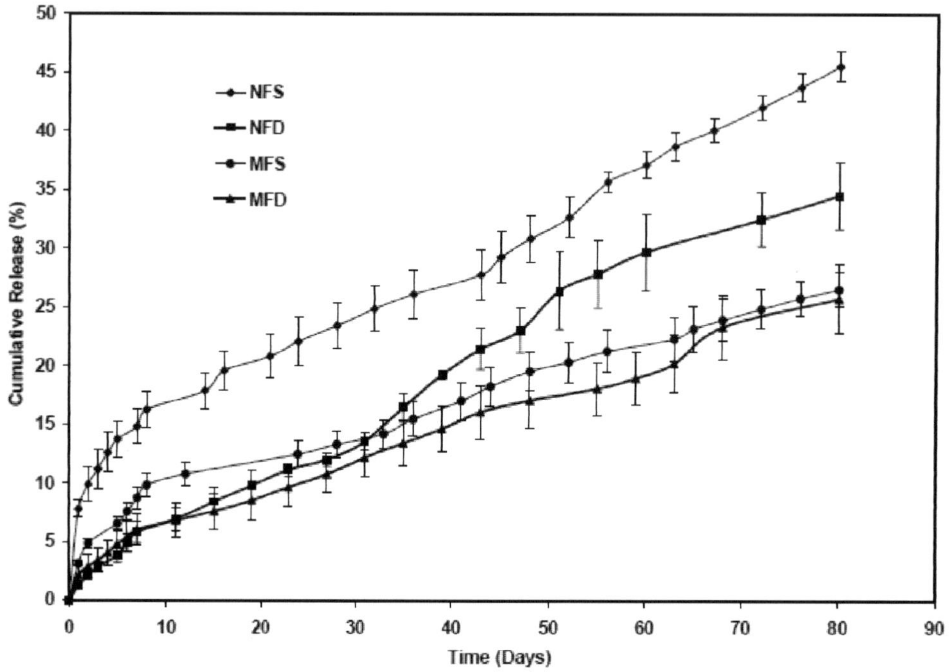

Figure 2: In vitro paclitaxel release from different dosage forms (a*dopted from [3]).* (◆) SFS; (■) SFD; (●) MFS; (▲) MFD. Each data point represents the average of triplicate samples and error bars represent standard deviation.

Other paclitaxel loaded polymeric discs/wafers like PLGA spray dried microparticle discs [2], MPEG 2000-DSPE liquid crystalline cubic phases [4], polilactofate microsphere discs [5] exhibit significantly lower release rates probably because of lower surface area to volume ratio available for polymer degradation and drug diffusion. On the other hand, other dosage forms like PLGA micro/nano-fiber films [6], PLGA microparticles by EHDA and spray drying [1], polilactofate microspheres [5], PCPP-SA microsphere discs [7] and docetaxel loaded PCPP-SA microsphere discs [8] exhibit very high initial burst in comparison to fiber discs and sheets. The large burst in some cases was due to fast degrading polymer like PCPP-SA [8] or due to surface bound drug release [1, 5]. The large burst has been reported to cause sporadic toxicity among experimental animals [7]. In comparison, the fiber discs and sheets have the advantages of low initial burst, near constant drug release rate and longer drug release sustainability which might be crucial for treating recurrent glioma. This provides the flexibility in selecting dosage forms with different drug release rates according to patient specific glioma chemotherapy (based on residual tumor post-surgery).

In vitro Cellular Apoptosis

This study was designed to demonstrate the long term paclitaxel release sustainability at a cellular level. Paclitaxel induced apoptosis was quantified for C6 glioma cells treated with MFD, SFD, MFS and SFS and compared with blank and Taxol® controls for sustained apoptotic activity. Fig.3 shows quantitative results of extent of apoptosis (caspase-3 activity) after days 4, 8 and 12. Fig.3A and 3B show low level of apoptosis in C6 glioma cells treated with blank discs/sheets, where as large apoptotic activity was observed on day 4 for Taxol® groups. For the fiber treated groups, apoptotic activity slightly higher than that of blank discs and considerably lower than that of Taxol® controls was observed. Once Taxol® was washed off and fresh media was replaced on day 3, the cells started recovering which is reflected in increased cell density (data not shown) and decreased apoptotic activity after days 8 and 12 as seen in Fig.3A and 3B. In contrast, for the fiber treated groups the cell density continued to decrease (data not shown) and apoptotic activity continued to increase after days 8 and 12. Also, SFS treated group showed the highest apoptotic activity after 12 days, whereas MFD showed the least among the fiber treated groups.

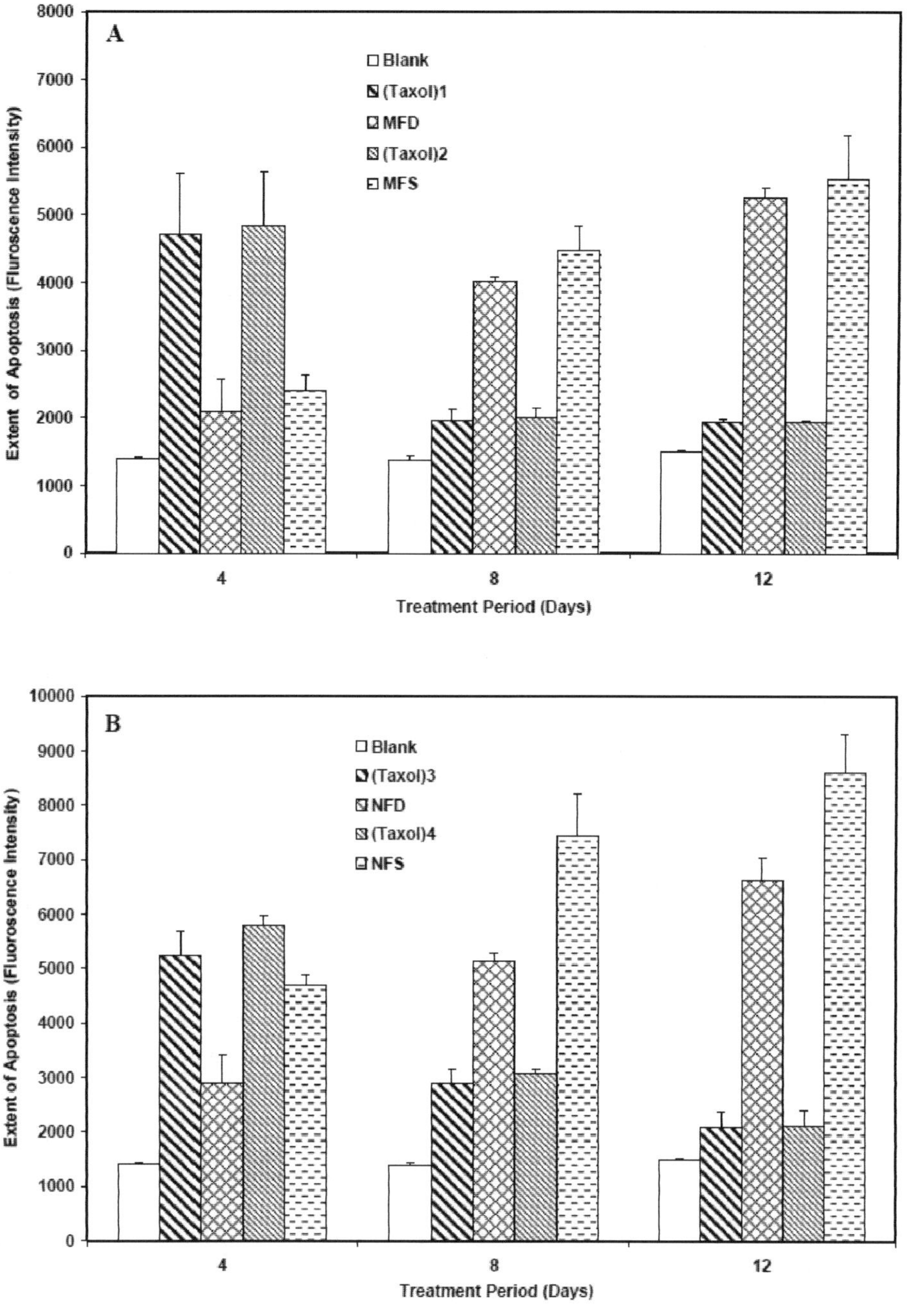

Figure 3: Extent of apoptosis in C6 glioma cells in vitro treated with different groups after 4, 8 and 12 days (a*dopted from [3])*. (A) Blank, (Taxol)$_1$, MFD, (Taxol)$_2$, MFS; (B) Blank, (Taxol)$_3$, SFD, (Taxol)$_4$, SFS. (Taxol)$_1$, (Taxol)$_2$, (Taxol)$_3$, (Taxol)$_4$ represent equivalent amounts of commercial Taxol® corresponding to paclitaxel release from MFD, SFD, MFS and SFS after 12 days (from *in vitro* release profiles). Each data point represents the average of three samples and error bars represent standard deviation.

Cellular recovery observed in Taxol® treated groups was indicative of the limitations of systemic drug administration because it only provided short and acute exposure due to low terminal half life of paclitaxel. But for the fiber treated groups, due to sustained release of paclitaxel at therapeutic concentrations, we could observe continued decrease in cell number even after 12 days (data not shown). Sustained increase in apoptotic activity for the fiber treated groups was indicative of continued paclitaxel induced apoptosis. Furthermore probably due to higher release rate from submicrofiber formulations, higher apoptotic activity was observed compared to microfiber treated group. This result also confirms GPC and *in vitro* release results.

In vivo Characterization

In vivo Subcutaneous Tumor Inhibition

This study was aimed at evaluating the efficacy of MFD, SFD, MFS and SFS in comparison with placebo fiber discs and commercial Taxol® in inhibiting subcutaneous C6 glioma tumor in BALB/c nude mice after surgical implantation. It was observed from Fig.4 that for animals with placebo discs the tumor grew unabatedly (more than 2000 mm^3) till day 24 after which the animals had to be sacrificed owing to animal ethics consideration and IACUC regulations on maximum allowable tumor size (~ 1.7 cm in diameter). Tumor volume inhibition (TVI) was calculated using the formula (TVI = $[V_C-V_E]/V_C$), where V_C and V_E represent the tumor volumes of animals from control and experimental groups respectively. Table 2 depicts TVI (%) for MFD, SFD, MFS and SFS against placebo and Taxol® control groups after 24 and 32 days.

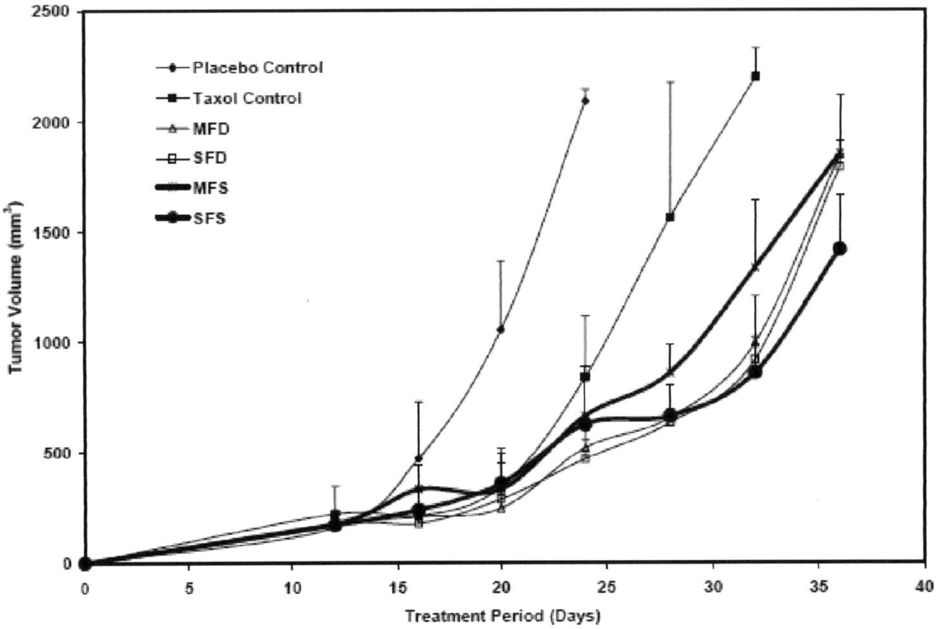

Figure 4: In vivo subcutaneous tumor inhibition profiles for different treatment groups (a*dopted from [3].* (◆) Placebo control; (■) Taxol® control; (△) MFD; (□) SFD; (x) MFS; (●) SFS. Each data point represents five samples (n = 5 animals). Animals were sacrificed after they reached a specified tumor volume (~ 1.7 cm in diameter) irrespective of day of treatment. Error bars represent the standard deviation.

Table 2: In vivo subcutaneous tumor volume inhibition (%) for MFD, SFD, MFS, SFS against placebo and Taxol® control. (TVI: Tumor volume inhibition)

Dosage Forms	TVI (%) against Placebo control (day 24)	TVI (%) against Taxol® control (day 24)	TVI (%) against Taxol® control (day 32)
MFD	75 ± 2	38 ± 6	55 ± 6
SFD	78 ± 6	44 ± 11	59 ± 4
MFS	69 ± 9	21 ± 2	40 ± 13
SFS	71 ± 12	26 ± 2	61 ± 5

In comparison to spray dried microsphere disc having a very low paclitaxel release rate, and EHDA microspheres with a very high initial burst and release rate [1], in the present study the animals having fiber implants had approximately 45% to 60% and 40% to 55% smaller tumors respectively. This shows that a constant paclitaxel release rate within the therapeutic limit could enhance the therapeutic efficacy. The tumor volume inhibition results positively show that the fibrous dosage forms are able to provide sustainable paclitaxel release for at least more than a month *in vivo* which was reflected in the considerable tumor volume inhibition effects when compared to the control groups. Even though Taxol® group could inhibit the tumor significantly initially, it could not extend the effect for long because of its low terminal half life of less than a day *in vivo.* Once the drug cleared off, the tumor growth rate approximated that of placebo group and reached more than 2000 mm^3 after 32 days. But for the fiber treated animals tumor volume was relatively smaller (less than 1000 mm^3)

until day 32 expect for MFS (~ 1300 mm^3). However, fibers discs could be ideal for intracranial chemotherapy because of its low initial burst, sustained drug release and compactness (with higher drug loading).

Comparative study on various PLGA-paclitaxel loaded fiber discs

Fiber diameter, distribution, morphology and paclitaxel encapsulation efficiency

Six fiber disc dosage forms were fabricated using electrospinning process as described earlier and Table 3 shows the process parameters, the co-polymer used, fiber diameter and encapsulation efficiency.

Dosage Forms	Co-polymer used	Polymer concentration (% w/v)	Flow rate (ml/h)	Voltage difference between nozzle & ground (kV)	Fiber mean diameter	Paclitaxel encapsulation efficiency (%)
E1	PLGA 85:15	30	6.0	12 - 16	3.5 ± 0.32 µm	98 ± 4.9
E2	PLGA 85:15	20	0.2	16 - 18	0.92 ± 0.13 µm	93 ± 1.04
E3	PLGA 85:15	8	0.5	18 - 20	92 ±6 nm	90 ± 0.81
F1	PLGA 50:50	30	6.0	12 - 16	3.16 ± 0.3 µm	98 ± 3.0
F2	PLGA 50:50	30	0.3	16 - 18	0.93 ± 35 µm	94 ± 0.57
F3	PLGA 50:50	8	0.2	18 - 20	98 ± 23 nm	90 ± 1.09

Microfibers were fabricated at higher flow rates, polymer concentration and voltage difference compared to sub-microfiber and nanofiber. In nanofiber fabrication, to avoid formation of beads at low polymer concentrations, 20 mM of organic salt (TATPB) was used. The salt increased the conductivity of the solution, thus resulting in continuous nanofibers. Figure 5 shows that the fibers are monodisperse with smooth morphology and have good encapsulation efficiency of 90 to 98 %.

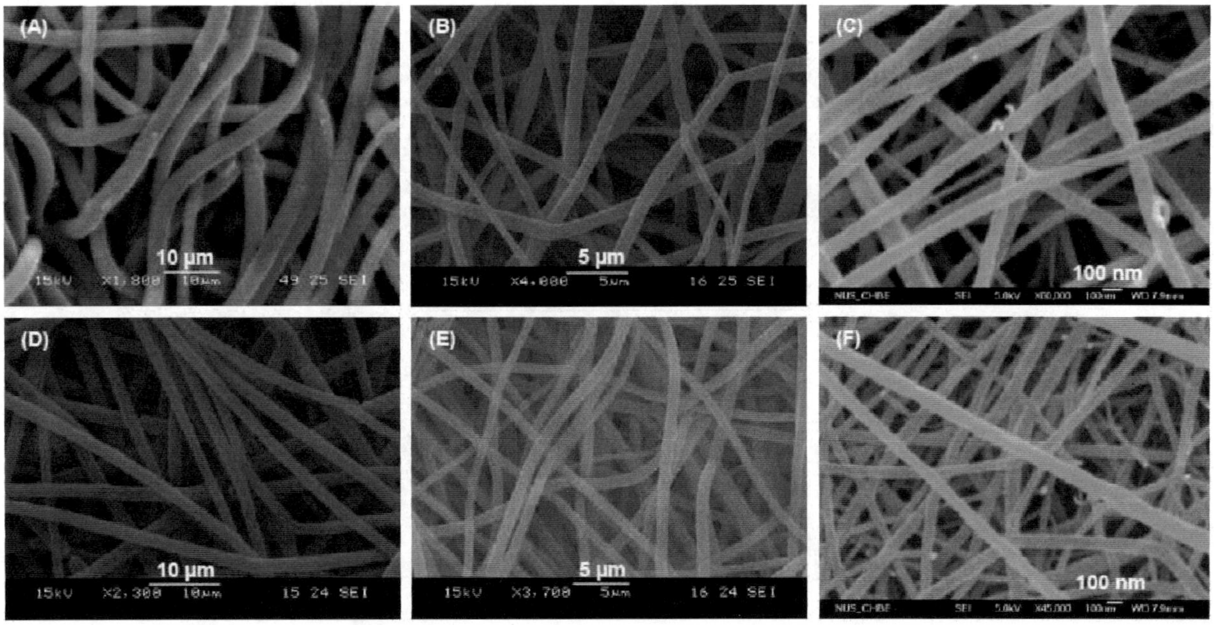

Figure 5: SEM and FESEM images of PLGA-paclitaxel electrospun fibers. (A) PLGA 85:15 microfiber disc (E1); (B) PLGA 85:15 sub-microfiber disc (E2); (C) PLGA 85:15 nanofiber disc (E3); (D) PLGA 50:50 microfiber disc (F1); (E) PLGA 50:50 sub-microfiber disc (F2); (F) PLGA 50:50 nanofiber disc (F3).

In vitro paclitaxel release profile

This study aimed at investigating the in vitro drug release profile from each of the six dosage forms and comparing them to evaluate the effect of co-polymer and fiber diameter on release profiles. Clearly, from Figure 6, we notice that the effect of co-polymer on release kinetics at the same fiber diameter is not that significantas the profiles for PLGA 85:15 and PLGA 50:50 are similar. But it is quite obvious that the fiber diameter has a very significant effect on the release rate especially when nanofibers are compared with sub-micro and microfibers. This highlights the fact that increasing the surface area to volume ratio using nanofiber would augment polymer degradation and drug release. Even PLGA 85:15 degrades at comparable rate to PLGA 50:50 when fabricated as nanofibers. At all the three scales, PLGA 50:50 exhibits slightly higher release rate compared to PLGA 85:15 owing to its greater hydrophilicity.

More importantly, unlike other available dosage forms for delivering paclitaxel [1, 5] these exhibit relatively smaller initial burst and near constant release profile. Owing to slightly higher release rate, for further in vivo studies we screened F3 and F2 as the dosage forms to be applied. F3 has fast release profile while F2 has slow release profile and hence would provide the neurosurgeons to choose between for patient specific glioma chemotherapy.

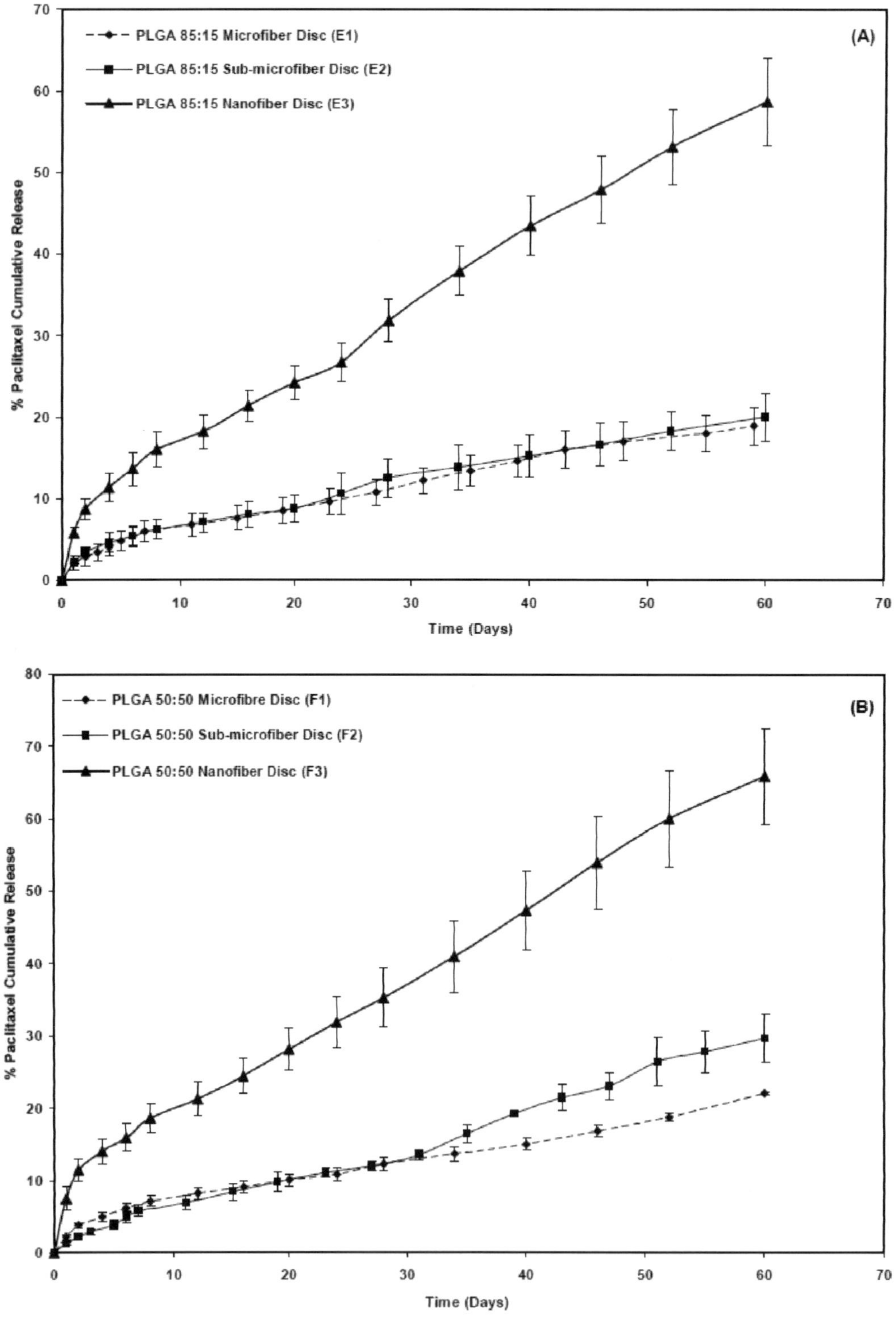

Figure 6: In vitro paclitaxel release from different dosage forms. (A) Paclitaxel loaded PLGA 85:15 fiber discs; (B) Paclitaxel loaded PLGA 50:50 fiber discs. Each data point represents the average of triplicate samples and error bars represent standard deviation.

CONCLUSIONS

The study showed that electrospinning was successfully used to fabricate monodispersed fibers with high drug encapsulation efficiency. As proof of concept, the four dosage forms (MFD, MFS, SFD, SFS) were shown to provide sustained paclitaxel release for more than 80 days with low initial burst which exhibits clear advantage over conventional systemic delivery of Taxol®. Submicrofiber discs/sheets exhibited faster release compared to microfiber discs/sheets. *In vitro* cellular apoptosis study suggested that the fiber discs/sheets were better than acute Taxol® administration against C6 glioma cells in terms of drug sustainability and cytotoxic effect. *In vivo* tumor inhibition study against subcutaneous C6 glioma in BALB/c nude mice showed that the fiber discs/sheets treated animals had much smaller tumors on day 24 and day 32 post tumor inoculation when compared to placebo control and Taxol® control groups confirming sustained release of paclitaxel and improved tumor inhibition. Hence PLGA microfiber discs/sheets could be utilized as drug delivery implants to deliver paclitaxel for post-surgical chemotherapy against malignant brain tumors.

Next, fibers discs were selected for further studies since they could be ideal for intracranial chemotherapy because of low initial burst, sustained drug release and compactness. Six fiber disc dosage forms (E1, E2, E3, F1, F2, F3) were fabricated by varying co-polymer and fiber diameter. It was shown that the co-polymer had very little effect on drug release rate compared to fiber diameter. Nanofiber discs demonstrated a significantly higher release rate in comparison to sub-micro and microfibers of the same co-polymer. But all the six dosage forms exhibited inhibited initial burst and near constant drug release rate. However, due to higher release rates, F3 and F2 were screened for further *in vivo* characterizations involving intracranial experiments such as drug biodistribution study, tumor regression study through non-invasive bio-imaging techniques and are warranted.

REFERENCES

[1] Naraharisetti PK, Ong BYS, Xie JW, Lee TKY, Wang CH, Sahinidis NV. In vivo performance of implantable biodegradable preparations delivering paclitaxel and etanidazole for the treatment of glioma. *Biomaterials* 2007;28(5):886-894.

[2] Wang J, Ng CW, Win KY, Shoemakers P, Lee TKY, Feng SS, et al. Release of paclitaxel from polylactide-co-glycolide (PLGA) microparticles and discs under irradiation. *J Microencapsulat* 2003;20(3):317-327.

[3] Ranganath SH, Wang CH. Biodegradable microfiber implants delivering paclitaxel for post-surgical chemotherapy against malignant glioma. *Biomaterials* 2008;29:2996-3003.

[4] von Eckardstein KL, Patt S, Kratzel C, Kiwit JCW, Reszka R. Local chemotherapy of F98 rat glioblastoma with paclitaxel and carboplatin embedded in liquid crystalline cubic phases. *J Neuro-oncol* 2005;72:209-215.

[5] Li KW, Dang W, Tyler BM, Troiano G, Tihan T, Brem H, et al. Polilactofate microspheres for paclitaxel delivery to central nervous system malignancies. *Clin Cancer Res* 2003;19:3441-3447.

[6] Xie J, Wang CH. Electrospun micro- and nanofibers for sustained delivery of paclitaxel to treat C6 glioma *in vitro*. *Pharma Res* 2006;23(8):1817-1826.

[7] Walter KA, Cahan MA, Gur A, et al. Interstitial Taxol delivered from a biodegradable polymer implant against experimental malignant glioma. *Cancer Res* 1994;54:2207-2212.

[8] Sampath P, Rhines LD, DiMeco F, Tyler BM, Park MC, Brem H. Interstitial docetaxel (Tarotere), carmustine and combined interstitial therapy: a novel treatment for experimental malignant glioma. *J Neurooncol* 2006;80:9-17.

540a Cytocompatibility Evaluation of Novel Thermoresponsive and Chemically Crosslinkable Macromers for Injectable, In Situ Forming Hydrogels

Leda Klouda, Michael C. Hacker, Laura H. Barg-Walkow, James D. Kretlow, and Antonios G. Mikos. Department of Bioengineering, Rice University, 6100 Main Street, MS-142, Houston, TX 77005

Injectable hydrogels provide promising carriers for cell and drug delivery in tissue engineering due to their hydrophilic nature, minimally invasive method of delivery and the ability to fill irregular defects. In this category, thermally responsive hydrogels have the advantage of using temperature as the stimulus for their solidification. They can be liquid at ambient temperature, and gel upon injection in the human body while temperature increases to physiological. However, hydrogels made from exclusively hydrophilic polymers often suffer from limited mechanical stability restricting their use in orthopedic applications. In this study towards the development of injectable hydrogels of controllable and improved mechanical properties, we present the characterization of novel injectable macromers that contain hydrophobic domains for mechanical reinforcement and functionalities for chemical crosslinking in situ. The macromers are designed to exhibit physical gelation at physiological temperature, and the introduction of (meth)acrylate groups allows for subsequent chemical crosslinking. Macromers were synthesized from pentaerythritol diacrylate monostearate, N-isopropylacrylamide, acrylamide and 2-hydroxyethyl acrylate by free radical polymerization. At a next step, the free hydroxyl groups on the macromers were (meth)acrylated. Macromer composition was verified by NMR spectroscopy, and oscillating rheology was employed to characterize the thermally induced phase transition and the resulting gels. It was shown that the degree of modification with (meth) acrylate groups correlated with the transition temperature of the macromers. Cytocompatibility studies on the (meth)acrylated macromers over 24 h revealed that higher degrees of modification increased toxicity; however cell viability was high for up to six hours for most formulations. Moreover, acrylated macromers seemed more cytocompatible than the corresponding methacrylates. These results show potential for the fabrication of hydrogels that solidify almost instantaneously due to physical thermogelation in physiological temperature, and can subsequently slowly crosslink with the addition of thermal initiators in a time interval and manner that does not prove toxic for encapsulated cells.

540b Rapid Hydrolyzing Polyketal Copolymers for Treating Acute Inflammatory Diseases

Chen-Yu Kao, Stephen Yang, Sungmun Lee, and Niren Murthy. Biomedical Engineering Department, Georgia Institute of Technology, 315 Ferst Drive, Atlanta, GA 30332

Acute inflammatory diseases are a major cause of death in the world, and effective treatments are greatly needed. Microparticles formulated from polyketals have great potential to enhance the treatment of inflammatory diseases by targeting therapeutics to macrophages. However, existing polyketals have slow hydrolysis rates, which limit their applications in treating acute inflammatory diseases.

In this study, we demonstrated that the hydrolysis kinetics of aliphatic polyketals can be accelerated by increasing their hydrophilic/hydrophobic balance. Aliphatic polyketals of varying hydrophobicity were synthesized, via the acetal exchange reaction, and their hydrolysis kinetics were investigated at the pH values of 4.5 and 7.4. In particular, one polyketal copolymer, PK3, was developed, which had a hydrolysis half-life of 2 days at pH 5 and 22 days at pH 7.4. Microparticles formulated from PK3 should be suitable for treating acute inflammatory diseases because they will hydrolyze and release therapeutics rapidly in the acidic phagolysosomes of macrophages.

Microparticles were formulated with PK3, which encapsulated the anti-inflammatory drug, imatinib. In vivo experiments demonstrated that PK3 microparticles, which encapsulated imatinib significantly improve the efficacy of imatinib in treating acute liver failure. We anticipate that aliphatic polyketals will have numerous applications for the treatment of acute inflammatory diseases, such as traumatic brain injury (TBI), given their pH sensitivity, tunable hydrolysis kinetics, and biocompatible degradation products.

555a Antimicrobial Performance of Novel Environmentally-Friendly Cationic Polymers

Gang Cheng[1], Shengfu Chen, and Shaoyi Jiang[2]. (1) Department of Chemical Engineering, University of Washington, Box 351750, Seattle, WA 98105, (2) Chemical Engineering, University of Washington, Benson Hall, Box 351750, Seattle, WA 98195

Cationic biocides have been widely used as the antimicrobial agents in paint, water treatment, textile, and food industries, because of their broader antimicrobial spectrum. The major disadvantage of cationic biocides is their potential toxicity for aqueous organisms. To overcome the disadvantage of cationic biocides, we developed a series of novel antimicrobial polymers derived from poly(carboxybetaine acrylic amide) (pCBAA). These pCBAA derivatives are designed to be able to kill bacterial cells efficiently, and then can convert into the non-toxic pCBAA polymers. The efficiency of pCBAA derivatives to kill and inhibit the growth of two bacterial strains (Gram-positive Staphylococcus epidermidis and Gram-negative Escherichia coli) were studied both in bulk solution and on surface. The results showed that the pCBAA derivatives are able to kill both S. epidermidis and E. coli both in bulk solution and on surface as efficient as cationic polymers.

Although cationic biocides have been used as membrane active biocides which attack the integrity or the function of the cytoplasm membrane over 40 years, the precise molecular interactions between bacterial cells and cationic biocides have not been fully discovered. It is generally believed that the electrostatic interaction between positively charged cationic biocides and negatively charged microbial surfaces attract cationic biocides to the surfaces and then the hydrophobic alkyl chains interfere with the cell membrane leading to the death of cells. Numerous researchers have reported that an increase of alkyl chain significantly enhances the antimicrobial activity of cationic biocides. Confusingly it was also reported that the antimicrobial ability was not influenced by the alkyl chain length. In their model, cationic biocides interfere with the other positively charged divalent metallic cations such as $Ca2+$ and $Mg2+$ which are very important to maintain the membrane integrity, and participate in the signal transduction and the enzymatic activity. Therefore the second objective of this study is to further investigate the correlation between the antibacterial efficiency and the structure of pCBAA derivatives and cationic polymers. S. epidermidis and E. coli K12 were used as the model bacteria for gram-positive and gram-negative bacteria. pCBAA derivatives with different structure and cationic polymers with different length of alkyl chain were tested in this study. Cationic polymer with shortest alkyl chain is able to kill both bacterial strains, and the antimicrobial efficiency increases with the increase of hydrophobic moieties of pCBAA derivatives and cationic polymers. In this work, dynamic light scattering were utilized to study the interactions between cationic compounds with different structure and bacterial membranes, and the results show that the cationic compounds carrying more hydrophobic moities showed higher binding affinity with bacterial cells. This result suggested that the long alkyl chain is not necessary for cationic compounds to kill bacterial but can improve the antimicrobial efficiency of cationic polymers.

Preparation of Porous Poly(L-lactic acid) Honeycomb Monolith Structure by Unidirectional Freezing and Freeze-Drying

Jin-woong Kim, Kentaro Taki, Shinsuke Nagamine and Masahiro Ohshima, Kyoto University, Kyoto, Japan

1. Introduction

Recently, the development of biodegradable porous polymeric scaffold attracts much attention. Various techniques have been developed to fabricate polymeric scaffold: progen leaching, thermally induced phase separation, gas foaming. Organs such as nerves, bones, blood vessels, and ligaments require a scaffold with an aligned porous structure for their cell proliferation. Thus, the controlled preparation of aligned porous materials in the form of 2D surface patterns or a 3D monolithic structure is critical for this purpose. A promising and novel technique for creating aligned porous structures is a unidirectional freezing method, which is relatively simple and cost effective. Mukai et al. and Nishihara et al. proposed a unidirectional freezing method to prepare a honeycomb monolith silica gel [1, 2]. Recently, several studies extended the unidirectional freezing method to polymers and prepared porous polymeric materials with aligned microtubes [3-5]. The micro honeycomb structure has great potential as catalyst support, catalyst, substrate for drug delivery, and bio scaffold due to the inherent high contact efficiency and controllability of morphology. Despite of these potentials, the structure, especially prepared by templating a cellular type solvent crystal, has the drawback of poor interconnectivity, which might lead to plugging of microtubes. When micro honeycomb monolith structure is used as a bio scaffold, it can be expected that the plugging of microtubes will easily occur by the proliferation of cell, especially from the top and bottom end of a micro honeycomb monolith structure. Poor interconnectivity can cause slow cell growth during cultivation due to the lack of oxygen and nutrient supply and waste elimination [6, 7]. One of good solutions of preventing the reduction of contact efficiency caused by the plugging is to increase interconnectivity of the structure. It could be achieved by making the wall of microtube to be open porous structure. In the case of bio scaffold made by biodegradable polymers, degradation rate is also affected by the morphology and porosity of 3D structure [8, 9].

In this study, a honeycomb monolith structured porous Poly(L-lactic acid) (PLLA) was prepared by combining unidirectional freezing and freeze-drying methods. The key factors of controlling the morphology of honeycomb monolith structure thoroughly investigated. Furthermore, to prepare a honeycomb monolith structure with micro/nanoscale porous wall, we focused on the possibility of phase separation due to the solubility change induced by decreasing temperature and the crystallization of solvent during unidirectional freezing. Poly(ethylene glycol) (PEG) was selected as a blend polymer to prepare porous PLLA structure because PEG can be easily leached out using water or relatively non-toxic hydrophilic solvent such as ethanol. The addition of PEG into the PLLA/1,4-dioxane solution enabled to prepare honeycomb monolith structures with porous wall. Phase separation of polymers induced by unidirectional freezing of 1,4-dioxane was exploited to create such a honeycomb monolith structure. The porous structure in the wall of honeycomb monolith was prepared by selective leaching of PEG. The effect of PLLA/PEG blend ratio and PEG molecular weight on the morphology was investigated.

2. Experiment

2.1 Materials

PLLA ($\overline{M_W}$ = 2.8x10^5, $\overline{M_w}/\overline{M_n}$ = 2.24) was used as received. PEG600, PEG2000, PEG4000, and PEG6000 was purchased from Wako Chemicals Ind. Ltd. and used as received. Dehydrated 1,4-dioxane (Wako Chemicals Ind. Ltd., Japan) and Ethanol (Wako Chemicals Ind. Ltd., Japan) was used as solvent.

2.2 Preparation of porous structure

PLLA/Dehydrated 1,4-dioxane single phase solution was prepared by dissolving PLLA into dehydrated 1,4-dioxane at 323 K. After it was cooled down to room temperature, the solution was poured into a polypropylene (PP) test tube with a thickness of 1.2 mm, 100 mm in length and 10 mm in diameter. The solution was unidirectionally frozen by soaking the test tube into a liquid nitrogen bath at a constant rate. The tube soaking rate was changed in the range from 3.5 to 17.5 cm hr-1. After freezing the solution completely, the solidified sample was freeze-dried at 268 K for 4 days. To investigate the effect of water on the porous structure, especially the ladder-like tube structure, distilled water was added to the PLLA/1,4-dioxane solution. The water concentration varied in the range from 0.04 wt% to 2.0 wt%. Furthermore, to prepare honeycomb monolith structure with micro/nanoscale porous wall, PLLA and various molecular weight of PEG (PEG600, PEG2000, PEG4000, and PEG6000) were co-dissolved in dehydrated 1,4-dioxane. The solutions were prepared with polymer concentrations of 7wt% and with blend ratios of PLLA to PEG of 90/10, 70/30, and 50/50 so as to see the effect of PLLA/PEG blend ratio on honeycomb monolith structure and morphology of porous wall. Then, the solutions were frozen unidirectionally at a constant rate of 2.0 cm hr^{-1}. Prepared honeycomb monolith structures were then immersed in ethanol for 7 days, and the medium was replaced every 12 hours to assure PEG leaching out. After the completion of leaching, the samples were immersed into *tert*-butyl alcohol and kept there for 1days, and the medium was replaced after 12 hours. Next, the wet samples were freeze-dried at 273K, and finally honeycomb monolith structures with porous wall were obtained.

2.3 Morphology characterization

The microstructures of the obtained sample were observed by scanning electron microscopy (Tiny-SEM 1540, Technex Lab Co. Ltd.). For SEM observation, the freeze-dried samples were further frozen by liquid nitrogen to prevent deformation and cut with a razor blade in both parallel and perpendicular directions to the freezing direction. The prepared sample was coated with gold-palladium *in vacuo*.

2.4 Cloud point measurement

The phase diagram of PLLA/PEG/1,4-dioxane ternary system was determined using a laser scattering scheme. PLLA and PEG were dissolved in 1,4-dioxane as described in Section 2.2. About 0.85g of sample solution was then dropped to glass Petri dish, which was 16.3mm in diameter and 10.8mm in depth, and allowed to evaporate. Petri dish with reflector was mounted on an electronic balance. The laser intensity, which was reflected through polymer solution, was measured vertically using laser censor. The change of laser intensity and the weight change of solution by 1,4-dioxane evaporation were recorded. To avoid humidity effect, all experiments were carried out in a lid (overall dimensions: 18 cm by 18 cm by 10 cm) under 1L/min nitrogen (purity 99.9%) flow at 298±0.5K. Humidity and temperature in the system were measured by a hygro-thermograph (SK-L200T II, SATO KEIRYOKI MFG. Co. Ltd.). The phase separation composition was determined by the polymer concentration at which the light intensity changed drastically

3. Result and discussion

3.1 Formation of honeycomb monolith structure

Honeycomb monolith structure was successfully prepared from a 7 wt% polymer solution without water by unidirectional freezing at a tube soaking rate of 3.5 cm hr^{-1} into a liquid nitrogen bath. After freezing the solution completely, the solidified sample was freeze-dried at 268 K (Fig. 1). Figure 1a and b respectively show a photographic image and scanning electron microscopy (SEM) image of porous PLLA prepared from a 7wt% polymer solution without water by unidirectional freezing at a tube soaking rate of 3.5 cm hr^{-1}. Figure 1b shows that the honeycomb monolith structure was established in the obtained porous PLLA. The microtube structure was created in cross section parallel to the freezing direction, and honeycomb structure was created in cross sectional area perpendicular to the freezing direction. Figure 1c clearly shows the honeycomb structure created in cross section perpendicular to the freeze direction. Figure 1d shows that micro-tubes with smooth walls were aligned in parallel to the freeze direction. The length of the aligned tube was longer than 1 cm.

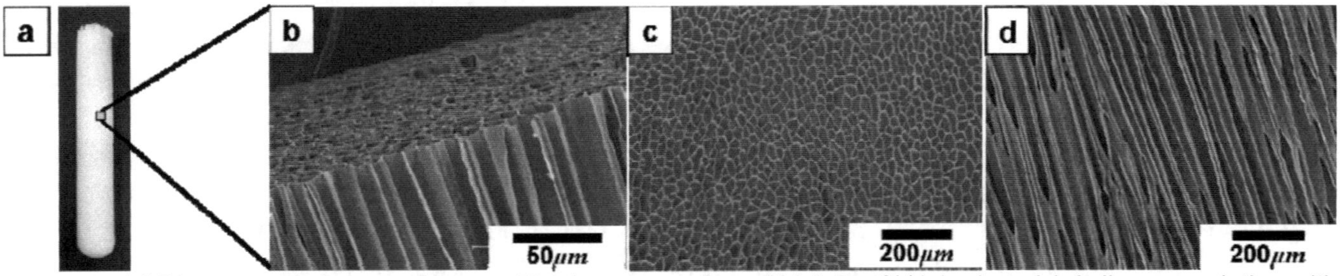

Figure 1. (a) Photographic image of porous PLLA prepared from PLLA 7 wt%/dehydrated 1,4-dioxane solution with a tube soaking rate of 3.5 cm hr-1, (b~d) SEM micrographs of (b) honeycomb monolith structure taken at a tilted angle, (c) Cross sectional area perpendicular to freezing direction, and (d) Cross sectional area parallel to freezing direction.

3.2 Effect of water

To investigate the effect of water as structure modifier, solutions with different water concentrations and different PLLA concentration were frozen by soaking the PP tube into liquid nitrogen at a rate of 3.5 cm hr^{-1}. Figure 2 shows SEM micrographs of the obtained porous PLLA with 7wt% PLLA solutions. As shown in Figure 2, the porous structure was dramatically changed with the addition of water. The porous PLLA prepared from a solution with 0.04 wt% water content showed a transitional surface morphology from a smooth wall microtube structure to a ladder-like microtube with an interconnected structure (Fig. 2a and e). In the case of the solution with 0.12 wt% water content, a highly ordered interconnected structure was observed (Fig. 2b and f). For the solution with 2.00 wt% of water content, the number of interconnected pores between the tubes was drastically increased (Fig. 2c and g). The structure resembles the superdendrite structure observed by Ragnarsson et al [10]. PLLA as well as 1,4-dioxane is likely to contain moisture. The experimental results suggest that small amounts of water has dramatic effects on the porous structure and it is important to control moisture in materials used to prepare honeycomb monolith structures by the unidirectional freeze-drying method. The lack of moisture control may be a reason that honeycomb monolith structures were not created in previous study [11].

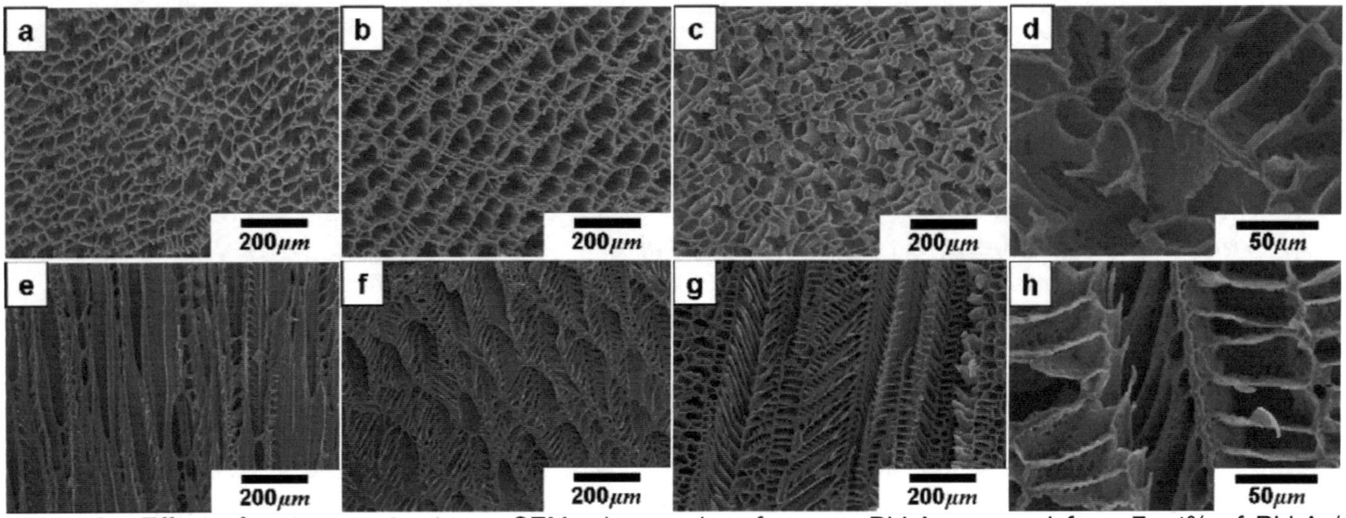

Figure 2. Effect of water on structure - SEM micrographs of porous PLLA prepared from 7 wt% of PLLA / 1,4-dioxane solution: (a~d) Cross sectional area perpendicular to the freezing direction, (e~h) Cross sectional area parallel to the freezing direction, (a and e) 0.04 wt% of water, (b and f) 0.12 wt% of water, and (c~d and g~h) 2.00 wt% of water.

3.3 Tube soaking rate

Figure 3 shows SEM micrographs of a sample cross sectioned perpendicular to the aligned microtubes of the porous PLLA prepared from a 7 wt% solution at two different soaking rates of 7.5, and 17.5cm hr-1. As can be seen, the average diameter of tubes decreased with an increase in the soaking rate, which is equivalent to the effects of the cooling rate. This trend coincides with the result of Zhang et al [4]. They measured average spacing of ice crystal of poly(vinyl alcohol) aqueous solution with temperature gradient plate and microscope. The average diameter and wall thickness decreased with an increase in the soaking rate. This is because rapid cooling increases the degree of super-cooling and leads to the enhancement of crystal nucleation of 1,4-dioxane. This result shows that tube diameter and the size of the honeycomb can be controlled by the cooling rate.

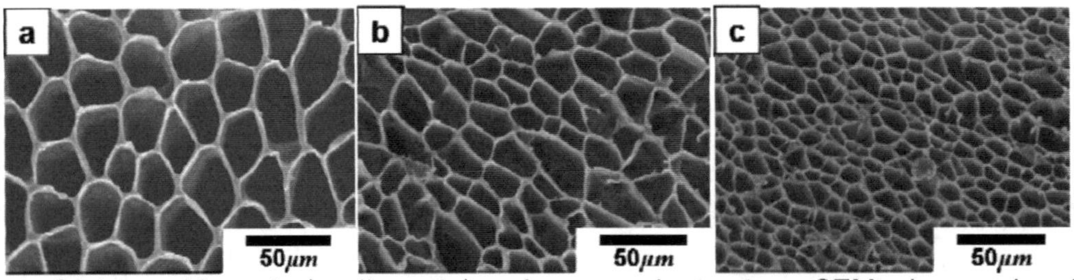

Figure 3. Effect of tube soaking rate (cooling rate) on honeycomb structure- SEM micrographs of porous PLLA prepared from 7 wt% of PLLA/dehydrated 1,4-dioxane solution at the cross sectional area perpendicular to the freezing direction: (a) soaking rate =3.5 cm hr^{-1}, (b) soaking rate =7.5 cm hr^{-1}, and (c) soaking rate =17.5 cm hr^{-1}

3.3 Phase diagram

The phase diagram of PLLA/PEG/1,4-dioxane ternary system was determined using the method described in Section 2.4. The compositions of solutions at which cloud point were measured are displayed in the phase diagram (Fig. 4). We adopted solvent casting method instead of controlled cooling of solution due to the very high viscosity of solution. Even though solvent cast solution shows solvent concentration profile during evaporation of solvent, the result data set had relatively good tendency. The solubility of PLLA/PEG blend in 1,4-dioxane decreased with the increase of PEG molecular weight. This result agreed with the data acquired by PEG/1,4-dioxane binary solution.

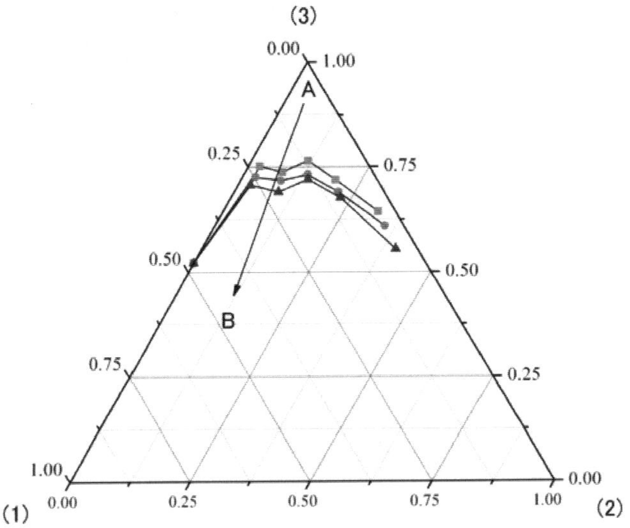

Figure 4. Phase diagram of PLLA (1)/PEG (2) /1,4-dioxane (3) solutions at 25℃.
(▲) PEG600, (●) PEG2000, and (■) PEG6000

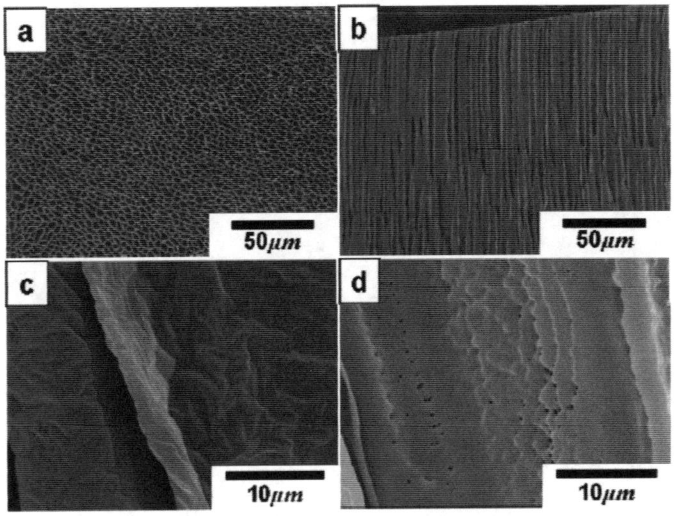

Figure 5. SEM micrographs of honeycomb monolith structure prepared from PLLA/PEG6000 blend (90/10) and dehydrated 1,4-dioxnae solution (a) cross sectional area perpendicular to freezing direction, (b)-(c) cross sectional area parallel to freezing direction, (d) cross sectional area parallel to freezing direction after leaching with ethanol

3.4 Preparation of honeycomb monolith structure with porous wall

A mixture of PLLA/PEG 6000 blend (90/10) and dehydrated 1,4-Dioxane solution (93wt% solvent concentration) was prepared and unidirectionally frozen at a tube soakong rate of 2.0 cm hr^{-1} into a liquid nitrogen bath. After freezing the solution completely, the solidified sample was freeze-dried at 268 K. As shown in Fig. 5, Honeycomb monolith structure was successfully prepared. Figure 5a-b show scanning electron microscopy (SEM) images of PLLA/PEG blend honecomb monolith structure. Figure 5a clearly shows the honeycomb structure created in cross section perpendicular to the freeze direction. Figure 5b shows that micro-tubes with smooth walls were aligned in parallel to the freeze direction. To prepare micro honeycomb monolith structures with porous channel wall, honeycomb monolith structure was leached with ethanol to remove PEG. Figure 5c and d respectively shows the SEM micrographs of cross sectional area parallel to the freeze direction before and after leaching. As can be seen, small pores are successfully prepared by leaching.

3.5 Effect of PEG molecular weight on porous channel wall

The effect of PEG molecular weight in the polymer blend solution on the morphology of channel wall was investigated. Four PLLA/PEG blend (70/30) 1,4-dioxane solutions, of which the PEG molecular weight were changed as PEG600, PEG2000, PEG4000, and PEG 6000, were unidirectionally frozen and freeze-dried. The freeze-dried honeycomb monolith structures were leached with ethanol to remove PEG and freeze dried again using the procedure which was described in experiment section. Figure 6, respectively, shows SEM micrographs of porous channel wall of micro monolith honeycombs prepared from these four different PEG molecular weight solutions. As shown in Figure 6, pore was created on the channel wall of honeycomb monolith structure by leaching out PEG. The pore size increased and the number of pore decreased with increase of the PEG molecular weight.

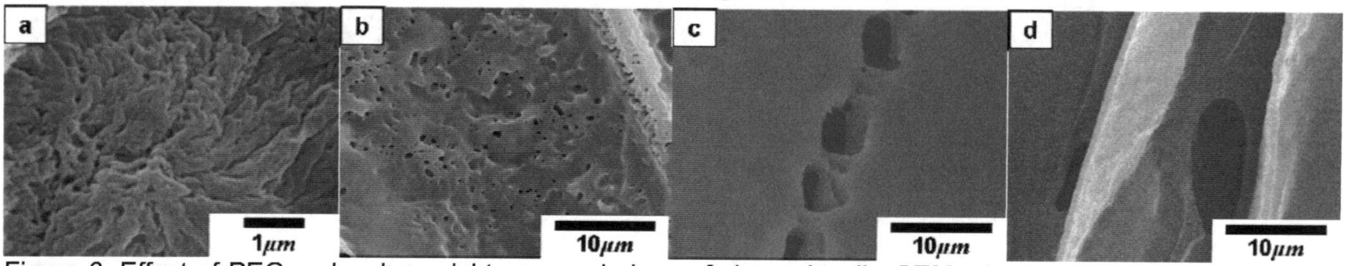

Figure 6. Effect of PEG molecular weight on morphology of channel wall - SEM micrographs of ethanol leached cross sectional area perpendicular to freezing direction prepared from PLLA/PEG blend (70:30) and dehydrated 1,4-dioxnae solution(93wt% solvent concentration). (a) PEG600, (b) PEG2000, (c) PEG4000, and (d) PEG6000

3.6 Effect of PLLA/PEG blend ratio on porous channel wall

Figure 7 respectively shows the SEM micrographs of ethanol leached samples cross sectioned parallel to the aligned microtubes of the porous PLLA materials prepared from PLLA/PEG2000/1,4-dioxane solutions with different PLLA/PEG blend ratio. As can be seen, the pore size on channel wall increased with the increase of PEG blend ratio due to the high content of PEG in honeycomb monolith structure.

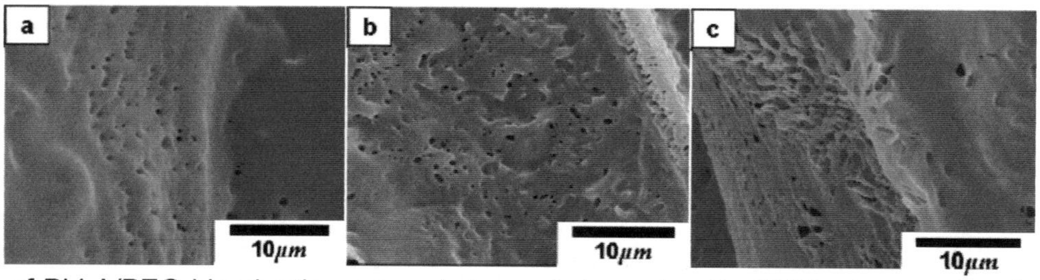

Figure 7. Effect of PLLA/PEG blend ratio on morphology of channel wall - SEM micrographs of ethanol leached cross sectional area perpendicular to freezing direction prepared from different PLLA/PEG6000 blend ratio and dehydrated 1,4-dioxnae solution(93wt% solvent concentration). (a) PLLA:PEG=90:10, (b) PLLA:PEG=70:30, and (c) PLLA:PEG=50: 50.

3.7 Formation mechanism of aligned structures

The porous structure was created by templating from the aligned solvent crystals that were formed during unidirectional freezing. The solvent was crystallized and grown in the freezing direction due to the constitutional supercooling. When the degree of supercooling was large during the crystallization process, Mullins-Sekerka instability occurred and created crystals with cellular (microtube like) or dendrite (ladder like) structures [10, 12]. In the course of directional freezing, polymer was expelled from solvent crystals to the liquid phase and the constitutional supercooling condition was established at the interface. Water and excess polymer chains acted as impurities. Therefore, when water existed, the degree of supercooling increased and the degree of Mullines-Sekerka instability increased due to an impurity-induced reduction of the freezing point of the solution. The instability created dendrite-type branched crystals. As a result, the unidirectional freezing created ladder-like tubes in the presence of water. The interconnectivity of tubes increased as the instability increased. There are some differences in the porous structures of ladder-like tubes prepared in the presence of water than structures produced with a high-concentration of polymer solution. As shown in Figure 2d and h, small pits, 2 μm in diameter, could be observed on the tube wall. The formation mechanism of these small pits can be speculated as follows: water was expelled from 1,4-dioxane crystal and concentrated on the solid-liquid interface due to the lower freezing point of water than that of 1,4-dioxane (m.p. 11.8°C). An increase in water caused a liquid-liquid phase separation of polymer solution on the interface before the solution was completely frozen. Since water acted as an anti-solvent to PLLA [13], the liquid-liquid phase separation created small pits on the tube wall.

3.8 Formation mechanism of pores on the channel wall

The formation mechanism of PEG phase was speculated as follows. As illustrated in Fig. 8a and b, the solvent of polymer blend solution crystallized and grown in the freezing direction. In the course of unidirectional freezing, PLLA and PEG concentration of liquid phase surrounded by 1,4-dioxane crystal increased gradually and 1,4-dioxane content decreased (Fig. 8c and d). This change in solvent concentration caused liquid-liquid de-mixing to PEG rich phase and PLLA rich phase and coalesced until it could not move due to the increase of viscosity by the decrease of 1,4-dioxane in the liquid phase. As explained in Section 3.6, the pore size on the channel wall increased and the number of pore decreased with the increase of PEG molecular weight. This can be explained with the decrease of PLLA/PEG solubility in 1,4-dioxane by increase of PEG molecular weight which was illustrated in Figure 4. In the case of high molecular weight PEG, liquid-liquid phase separation occurs at the relatively high 1,4-dioxane concentration and PEG rich phase has relatively long time to coalesce to large size compared with low molecular weight PEG. The increase of pore size by the increase of PEG blend ratio can be easily expected with increased PEG phase. The decrease of solution viscosity due to the high PEG blend ratio could be another factor of coalescence of PEG domain to large size.

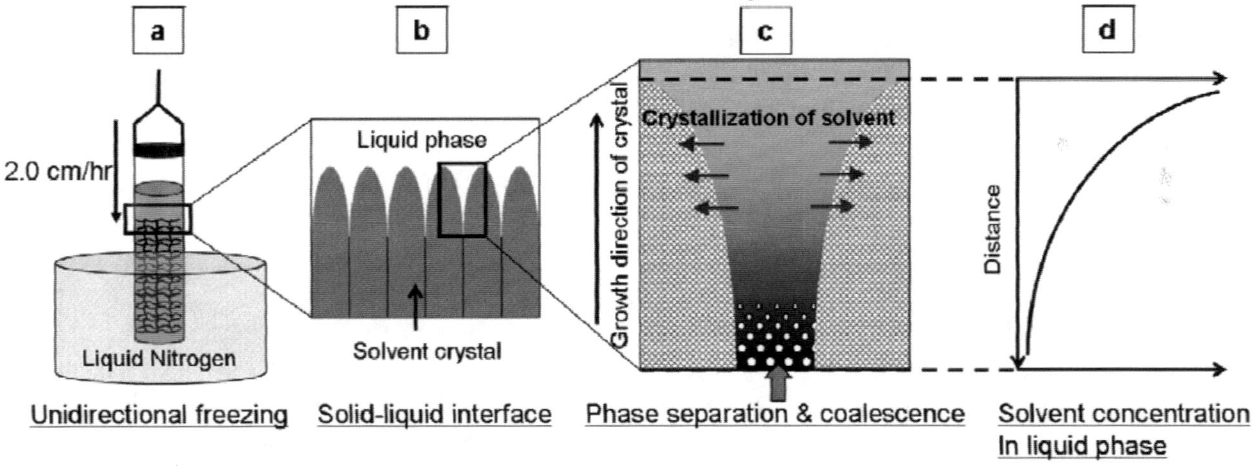

Figure 8. Schematic illustration of the unidirectional freezing and the phase separation

4. Conclusion

In this study, honeycomb monolith structures of PLLA were successfully prepared by thermally induced solid-liquid phase separation and subsequent solvent sublimation. Water content is key factor of creating ladder-like structures in aligned microtubes. Tube diameter and the density of tubes per unit cross section perpendicular to the aligned direction were controlled by tube soaking rate. Furthermore, honeycomb monolith structures of PLLA with micro/nanoscale porous wall were successfully fabricated by combining unidirectional freezing technique and phase separation of polymer blend solution. PEG was used as a blend polymer to prepare phase separated morphology. Porous structure in the channel wall of honeycomb monolith structure was successfully prepared by leaching out PEG domain. Porous structures in channel wall of honeycomb monolith structure were controlled by PEG molecular weight and polymer blend ratio. Based on the PLLA/PEG/1,4-dioxane ternary phase diagram, we speculated the formation mechanism of porous structure on the channel wall as a crystallization induced phase separation. We also suggested that the porous structure changed by PEG molecular weight because PLLA/PEG/1,4-dioxane ternary phase diagram affected by PEG molecular weight. We expect that this method has a potential to improve the contact efficiency of aligned porous materials for tissue, catalyst, organic electronics, micro fluidics, and membrane engineering.

5. Reference

1. Mukai, S.R., Nishihara, H., Tamon, H. (2004), "Formation of monolithic silica gel microhoneycombs (SMHs) using pseudosteady state growth of microstructural ice crystals", Chemical communications, 7, pp. 874-875.
2. Nishihara, H., Mukai, S.R., Yamashita, D., Tamon, H. (2005), "Ordered macroporous silica by ice templating" Chemistry of materials, 17, pp. 683-689.
3. Ma, P.X., Zhang, R., Xiao, G., Franceschi, R. (2001), "Engineering new bone tissue *in vitro* on highly porous poly(α-hydroxyl acids)/hydroxyapatite composite scaffolds" Journal of biomedical materials research, 54, pp. 283-293.
4. Zhang, H., Hussain, I., Brust, M., Butler, M.F., Rannard, S.P., Cooper, A.I. (2005), "Aligned two- and three-dimensional structures by directional freezing of polymers and nanoparticles" Nature materials, 4, pp. 787-793.
5. Zhang, H., Cooper, A.I. (2007), "Aligned porous structures by directional freezing" Advanced materials, 19, pp. 1529-1533.
6. Freed, L.E., Marquis, J.C., Nohria, A., Emmanual, J., Mikos, A.G., Langer, R. (1993), "Neocartilage formation in vitro and in vivo using cells cultured on synthetic biodegradable polymers" Journal of biomedical materials research, 27, pp. 11-23.
7. Freed, L.E., Vunjak-Novakovic, C., Langer, R. (1993), "Cultivation of cell-polymer cartilage implants in Bioreactors" Journal of cellular biochemistry, 51, pp. 257-264.
8. Braunecker, J., Baba, M., Milroy, G.E., Cameron R.E. (2004), "The effects of molecular weight and porosity on the degradation and drug release from polyglycolide" International journal of pharmaceutics, 282, pp. 19-34
9. Wu, L., Ding, J. (2005), "Effects of porosity and pore size on in vitro degradation of three-dimensional porous poly(D,L-lactide-co-glycolide) scaffolds for tissue engineering" Journal of biomedical materials research part B, 75A, pp. 767-777.
10. Ragnarsson, R., Utter, B., Bodenschatz, E. (1998), "Superdendrites in directional solidification of polymer-solvent mixtures" Materials research society, Symposium proceedings, 481, pp. 65-71.
11. Ma, P.X., Zhang, R. (2001), "Microtubular architecture of biodegradable polymer scaffolds" Journal of biomedical materials research, 56, pp. 469-477.
12. Mullins, W.W., Sekerka, R.F. (1964), "Stability of a planar interface during solidification of a dilute binary alloy" Journal of applied physics, 35, pp. 444-451.
13. Hua, F.J., Park, T.G., Lee, D.S. (2003), "A facile preparation of highly interconnected macroporous poly(D,L-lactic acid-*co*-glycolic acid) (PLGA) scaffolds by liquid-liquid phase separation of a PLGA-dioxane-water ternary system" Polymer, 44, pp. 1911-1920.

555c The Structure and Mechanical Relaxation of Foams Made from Wheat Gluten

Thomas O.J. Blomfeldt[1], Mikael S. Hedenqvist[1], and Eva Johansson[2]. (1) Fibre and Polymer Technology, The Royal Institute of Technology, Teknikringen 58, Stockholm, Sweden, (2) Department of Crop Science, The Swedish University of Agricultural Sciences, Alnarp, Sweden

Wheat Gluten protein is a relatively cheap by-product from e.g. biofuel (ethanol) production, a production that may probably increase in the future. Furthermore, gluten is well known to have excellent foaming properties, as observed in e.g. bread making. It has good extensibility, and with the right foaming parameters, it may be a potential insulation material and a competitor to today's oil-based insulating materials, e.g. expanded polystyrene and polyurethane; materials which contribute to green-house effects through their incineration. In addition, as the oil price will probably increase in the future, these materials will become more expensive and less attractive.

This presentation deals with the structure and properties of wheat gluten foams and addresses issues such as pore structure/foam density, solute uptake and compression strength. In addition the effects of reinforcing fillers will be discussed.

555d Amphiphilic Polysaccharides Used for Quantum Dots Encapsulation

Ching-An Peng, Chemical Engineering, National Taiwan University, No. 1, Sec. 4, Roosevelt Road, Taipei, Taiwan and Morris Hsu, Department of Chemical Engineering, National Taiwan University, No. 1, Sec. 4, Roosevelt Road, Taipei, Taiwan.

Colloidal nanocrystal QDs consisting of an inorganic core/shell structure (e.g., CdSe/ZnS) surrounded by a layer of organic ligands (i.e., TOPO) can be converted into hydrophilic nanoparticles by several approaches. The commonly used strategy is based on the exchange of the original organic layer with hydrophilic ligands, however the physical properties (i.e., quantum yield) are usually deteriorated through such surface modification. In this study, we demonstrated the first time that amphiphilic polysaccharides (alginate and chitosan) can encapsulate CdSe/ZnS by intercalating polysaccharide surfactant's hydrophobic pendant moieties (i.e., octyl chains) into the hydrophobic surfactant layer (i.e., TOPO) on the QD surface, thereby resulting in the phase transfer of hydrophobic QDs from organic solvents to aqueous solution via hydrophilic backbone (i.e., polysaccharides). The average size of water-soluble QDs encapsulated by polysaccharide surfactants was determined by DLS and TEM to be the range of 10-30 nm. As long as the number of pendant groups is sufficient high, the linkage of amphiphilic polysaccharide to the QD surface could be very stable and thereby lead to the physical properties of QD intact. The absorption and emission spectra of QD encapsulated with amphiphilic polysaccharide revealed higher absorbance between 525 and 550 nm and strong photoluminescence with the emission wavelength peaked at 534 nm after excited by UV-light. These fabricated nanomaterials were incubated with several cell lines and showed good biocompatibility.

555e Development of Biodegradable Gluten-PLA Laminate Films

Sung-Woo Cho and Mikael S. Hedenqvist. Fiber and Polymer Technology, Royal Institute of Technology, Teknikringen 56-58, Stockholm, 100-44, Sweden

The research interest in the use of natural biomaterials, alternatives to synthetic polymers, has increased with respect to environmental concerns. Several studies, however, have reported that plasticizer migration from biobased films including wheat gluten (WG) causes deterioration of mechanical and barrier properties with time. Also, their high water sensitivity compromises potential applications as packaging materials. For these reasons, laminate films consisting of a WG layer and polylactic acid (PLA) layers were developed by compression molding processes with the aims to limit the effects of aging and to improve the barrier properties, especially against moisture. WG films with different glycerol contents (0, 5, 10, 20 and 30 wt%) were pressed at 110 and 130 °C for 10 min. FT-IR revealed in the intense amide-I region that thermo-forming temperatures led to different degrees of protein aggregation and polymerization. The lamination of WG and PLA layers was performed at 110 °C. Scanning electron microscopy showed that the laminated films were uniform in thickness. The glycerol contents in WG layer had significant effect on tensile properties and bending stiffness of laminates. And the adhesion between WG and PLA layers increased with higher glycerol content. The laminates were very effective to enhance oxygen and moisture barrier. Moreover, the mass loss of these films on a porous paper was not observed during the four-month period and it is assumed that no change in the glycerol content played a key role in maintaining the integrity and properties of the laminates.

555f Poly(L-lactic acid) Toughening

Rahul M. Rasal, Siqiang Richard Zhu, and Douglas E. Hirt. Chemical and Biomolecular Engineering, Clemson University, 127 Earle Hall, Clemson, SC 29634-0909

Poly(L-lactic acid) (PLA) is a biodegradable and bioabsorbable thermoplastic polyester derived from renewable resources like corn, starch, rice, etc. PLA has exhibited excellent biocompatibility and has been FDA approved. However, wide applicability of PLA in consumer and biomedical applications is limited by its poor toughness. This research focused on PLA toughening using biocompatible polymers like poly[(3-hydroxybutyrate)-co-(3-hydroxyhexanoate)] (PHBHHx) and poly(ethylene glycol) (PEG). PLA blends were processed using a twin screw microcompounder operated in co-rotating mode. The resultant blend films were characterized using tensile testing, differential scanning calorimetry (DSC), wide-angle X-ray diffraction (WAXD), and dynamic mechanical analysis (DMA). The resultant blends had very low crystallinity, as characterized using DSC and WAXD analyses. Addition of a small amount (typically less than 20 wt%) of these additives to PLA significantly improved the toughness of the resultant blend.

556b Moisture Sorption In Polylactide with Varying Molecular Weight and Varying Aliphatic Content

Vishesh M. Singh, Giuseppe R. Palmese, and **Richard A. Cairncross**. Chemical & Biological Engineering, Drexel University, 3141 Chestnut St, Philadelphia, PA 19104

Polylactide (PLA) is a biobased–biodegradable polymer which is derived from lactic acid and has numerous potential applications, some of which are limited by its moisture barrier and thermal properties. Prior studies have been inconsistent about how moisture sorption in PLA varies with crystallintity and molecular weight.

This research is a systematic study of how moisture sorption in PLA depends on molecular weight and aliphatic content via end group modification. PLA with varying aliphatic content was achieved by initiating ring opening polymerization of L-lactide with different long chain aliphatic fatty alcohols. Molecular weight was controlled by varying the ratio of L-lactide monomer to alcohol (C16 Palmityl, C10 Decyl, C4 Butyl) initiator and triethylaluminum ethoxide was used as a catalyst because of the better control over molecular weight.

PLA with different molecular weight and with varying end groups have been synthesized and characterized by GPC, DSC, and NMR. Moisture sorption in PLA was measured with Quartz Crystal Microbalance (QCM). Initial experiments of QCM have shown that by increasing the aliphatic content of the end group there was a significant decrease of up to 15% in the moisture sorption of water in PLA.

556c Biodegradability Study of Native Corn Thermoplastic Starch In Laboratory Scale: Influence of Plasticizer Concentration, Ageing, Sample Superficial Area and Soil Source

Diego E. Ballesteros[1], Julie P. Merchan[1], **Isabel C. Jimenez**[1], Jorge A. Medina[2], Felipe Salcedo[1], and Oscar A. Alvarez[1]. (1) Chemical Engineering Department, Universidad de Los Andes, Bogotá, Colombia, (2) Mechanical Engineering Department, Universidad de Los Andes, Bogotá, Colombia

Aerobic biodegradation of different samples of native corn thermoplastic starch (TPS) were accordingly evaluated with standard test method ASTM D 5988 – 03 in order to establish the effect of plasticizer concentration, ageing, sample superficial area and soil source in TPS biodegradation.

Samples were prepared using glycerin as plasticizer, varying its concentration at 30%, 35% and 40%. Dry native corn starch and glycerin were premixed in a blending machine and then extruded in a single screw extruder; pellets and TPS film were obtained after double-step extrusion. Pellet samples were aged during 70 and 90 days in polyethylene bags at room temperature.

Two different soils from a local landfill (Doña Juana, Bogotá/Colombia) and germination peat were used as degradation matrix. Soils were characterized for amylolytic microorganisms, mesophilic aerobic bacteria and fungi, as well as for physical-chemical properties as pH, carbon and moisture content.

Biodegradation rate was assessed by exposing the TPS samples to the soil and measuring the carbon dioxide evolved by microorganisms as a function of time. Carbon dioxide production was measured for TPS samples and was expressed as a percentage of the calculated initial carbon content of samples.

We found that for non-aged samples the increasing plasticizer concentration lead to a higher biodegradation rate. On the other hand, for aged samples, plasticizer concentration negatively affects degradation rate. It was also seen that samples superficial plays a big role in biodegradation rate while no significant difference was observed for different soils.

556d Hydrolytic Degradation Studies of Renewable Copolymers

Mathew D. Rowe and Keisha B. Walters. Swalm School of Chemical Engineering, Mississippi State University, 330 Swalm Chemical Engineering Building, Mississippi State, MS 39762

Hydrolytic degradation behavior as a function of pH, time, and temperature has been studied in novel anhydride and ester renewable copolymers. The degradation rates of these copolymers will influence their potential use as a commodity polymer or biomaterial. The incorporation of both anhydride and ester bonds in the polymer backbone allows for better control of degradation with anhydride bonds hydrolyzing quickly and ester bonds hydrolyzing at a considerably slower rate. From a previous study, copolymers of 1,3-propanediol and malonic acid (PDO-MA) and 1,3-propanediol and itaconic acid (PDO-IA) were synthesized using aluminum chloride at 155 °C. Water absorption and polymer degradation were monitored using gravimetric analysis and contact angle goniometry as a function of pH, aging time, and temperature. Changes in chemical composition of these renewable copolymers were studied under the aging conditions with Fourier transform infrared spectroscopy (FTIR) by monitoring the qualitative and quantitative changes in the presence of anhydride and ester functionality. The changes in molecular weight and polydispersity index were monitored by gel-permeation chromatography (GPC). Surface Contact Angle (SCA) measured changes in surface energy with time and related to changes in molecular weight and chemical composition.

556e Copolymerization of Vegetable-Oil-Based Thermosetting Polymers with Chemically Modified Fatty Acid

Alejandrina Campanella[1], John La Scala[2], and Richard P. Wool[1]. (1) Department of Chemical Engineering and Center for Composite Materials, University of Delaware, 150 Academy St, Newark, DE 19716, (2) U.S. Army Research Laboratory, Attn: AMSRD-ARL--WM-MC, Building 4600, Aberdeen Proving Ground, MD 21005

Polymer and composite materials used in a wide range of fields, such as in aerospace, automotive and industrial applications, are derived from petroleum-based materials. Due to the increased cost of petroleum and its lack of long-term sustainability, there has been increased interest on replacing non-renewable materials with those made from natural resources, such as vegetable oils.

The ACRES group and others have done considerable work in replacing the cross-linking agent in unsaturated polyesters with plant oil-based cross-linkers.[1] However, resins derived from petroleum and modified vegetable oils typically contain high concentrations of reactive diluents (e.g., styrene) to allow resin transfer molding. Styrene, a hazardous pollutant and a volatile organic compound (VOC), poses significant environmental and health risks; therefore, its replacement with alternative monomers is advantageous on many levels. Recently there have been studies employing a non-volatile reactive diluent, chemically modified fatty acid (MFA), as an alternative to styrene in the production of polymers employing vinyl ester (VE) and unsaturated polyesters resins.[2, 3] It has been shown that MFA is an excellent alternative to styrene due to its lower VOC emissions.

The main objective of this work was to formulate a resin containing both plant oil-based crosslinkers and plant oil-based reactive diluents. Different bio-based cross-linkers were produced for this purpose: acrylated epoxidized soybean oil (AESO), maleinated AESO (MAESO), maleinated soybean oil monoglyceride (SOMG/MA) and maleinated castor oil monoglyceride (COMG/MA). Two different reactive diluents with the plant oil cross-linkers: methacrylated lauric acid (MLau) and acrylated epoxidized fatty acid methyl ester (MOct).

The viscosity and mechanical properties of the resulting bio-based polymers were measured and analyzed. Bio-based resins prepared employing MFA had a fairly high viscosity, so blends of MFA and styrene were needed to meet the viscosity requirements established by the composite industry (<1000 cP at RT). The results of these studies showed that the glass transition temperature (Tg) of bio-based thermosets copolymerized with MOct had slightly higher values than those with MLau. The Tg values estimated using the Fox equation were similar to the experimental data obtained in this study. In addition, the Tg of bio-based resin/MFA polymers were significantly lower than the bio based-resin/ styrene polymers. Decreasing the content of MFA improved the Tg and stiffness. Ternary blends of SOMG/MA or COMG/MA with MFA and styrene improved the mechanical properties to acceptable values comparable to commercial resins. Also, in this work chemically modified lignin (butyrated kraft lignin)[4] was added to the bio-based resins, leading to significant improvement of the mechanical properties of the matrix.

1. Khot, S.N.; La Scala, J.J.; Can, E.; Morye, S.S.; Williams, G.I.; Palmese, G.R.; Kusefoglu, S.H.; Wool, R.P J. Applied Polym. Sci. 2001, 82, 703.

2. La Scala, J.J.; Sands, J.M.; Orlicki, J.A.; Robinette, E.J.; Palmese, G.R. Polymer 2004, 45, 7729.

3. Can, E.; La Scala, J.J.; Sands, J.M.; Palmese, G.R. J Appl Polym Sci 2007, 106, 3833-3842.

4. Thielemans W.; Wool R. Biomacromolecules 2005, 6, 1895.

556f Egg Albumin as a Template for Bio-Materials

Naresh K. Budhavaram and Justin & R Barone. Biological Systems Engineering, Virginia Polytechnic Institute and State University, 203 Seitz hall, Biological Sysytems Engineering, Blacksburg, VA 24061

Egg albumin a crystalline protein is currently treated as a waste from food industry. Its current applications are limited because of its crystalline nature. The ratio of ordered to disordered regions in a protein determines its physical properties. It is possible to decrease this ratio in other words increase the disordered regions by using bulky side chain groups (BSG). This can be accomplished by functionalizing the amino and carboxylic groups present in egg albumin with BSG. Here, we have functionalized egg albumin (EA) with BSG (a) ethyl vinyl sulfone (EVS) and (b) Acrylic acid through a simple "green" chemical route known as the Michael addition. The reactions are very benign and are carried out at room temperature, pH 8.0 and in water. Depending on the BSG used and the molar ratios the structure and properties of the end product varied. The resulting product has been studied using FT-IR, X-ray diffraction and Differential Scanning Calorimetry (DSC). FT-IR and X-ray diffraction studies indicated the formation of covalent bonds between BSG and the protein. The DSC results showed the decrease in crystallinity and the emergence of a Tg at a lower temperature indicative of a new amorphous phase.

559a Polymer Dynamics In Single Wall Carbon Nanotube / Polystyrene Nanocomposites

Minfang Mu[1], Nigel Clarks[2], Russell J. Composto, and Karen I. Winey[3]. (1) Department of Materials Science and Engineering, University of Pennsylvania, 3231 Walnut Street, Philadelphia, PA 19104, (2) Department of Chemistry, Durham University, Durham, DH1 3LE, United Kingdom, (3) Materials Science and Engineering, University of Pennsylvania, 3231 Walnut Street, Philadelphia, PA 19104

Polymer nanocomposites provide access to new regimes of polymer dynamics in which the impenetrable filler particles are comparable to and frequently smaller than the polymer chain conformation. In this study, single wall carbon nanotubes (SWCNTs) / polystyrene (PS) nanocomposites were prepared by a coagulation method. Using this method, the nanotubes are evenly distributed in the polymer matrix and the SWCNTs exist as small bundles with a diameter of 9.6 nm and aspect ratio of ~35. An elastic recoil detection (ERD) method was employed to detect the tracer, deuterated polystyrene (dPS), diffusion into the nanocomposites and the tracer diffusion coefficients were determined. Rheological properties were measured in the PS linear viscoelastic regime.

The tracer diffusion coefficient first decreases and then increases with increasing SWCNT loading. The polymer entanglement molecular weight and the relaxation time remain constant according to our linear viscoelastic measurements. Therefore, this does not explain the change in the tracer diffusion coefficient. The transition from decreasing to increasing tracer diffusion corresponds approximately with the onset to rheological percolation and appears to increase to a higher SWCNT loading with decreasing matrix molecular weight. A trap model is proposed to describe the polymer diffusion in these nanocomposites. SWCNT bundles introduce traps around them into the composites system. Within and out of the trap, the polymer diffusion rates are the same, but the diffusion slows down when the tracer moves away from the nanotubes. When the SWCNT loading is below the percolation threshold, the traps are isolated and the diffusion coefficient decreases. With increasing SWCNT loading, the SWCNT bundles percolate and the traps overlap, which allows the tracer polymer to move along the nanotube network and the apparent tracer diffusion coefficient increases. With further increasing SWCNT content, the volume fraction of the trap in the composites becomes larger and larger, the tracer diffusion coefficient recovers back to the diffusion coefficient in the pure polymer. Simulated results from the trap model compare favorably with our experimental results.

559b Demonstration of Continuous Refractive Index Control of a Nanocomposite

Thad Druffel, Matt Lattis, Omar Buazza, and Scott Farmer. R&D, Optical Dynamics, 10106 Bluegrass Pkwy, Louisville, KY 40299

Optically clear, mechanically flexible nanocomposites are created by doping host polymers with nanoparticles having higher or lower refractive indices. The composite materials preserve the viscoelastic properties of the binding polymers and exhibit infinite tunability between the refractive index limits of the system. These properties provide unique benefits for multi-layer thin film optical filters. Additionally, the nanoparticles are dispersed in a fluid or bound in a polymer matrix in use, thus reducing toxic risks that may be associated with raw particles. A simple and safe method of producing highly engineerable optical filters is presented.

By doping a polymer with metal oxide nanoparticles, an engineer can manipulate the refractive index of a thin film while maintaining mechanical flexibility and high transparency to visible light. In this work, we showed that nanocomposites of UV curable monomers and metal oxide nanoparticles and can be used in thin film systems of varying refractive index. On polymer substrates, the nanocomposite filter stacks have outperformed traditional vacuum deposited films at large strains. Several thin film filters for the visible and near-IR regions were constructed using up to 30 layers. The finished stacks exhibited spectral haze of less than 1 percent. These studies indicate that nanocomposites behave as expected optically and maintain the flexibility of the host polymer.

Using a stable dispersion of titanium dioxide nanoparticles and a UV curable monomer, we were able to design and produce several quarter wave filters that demonstrate control of the position and size of the stop band through adjustment of the organic/inorganic ratio. The volume loading of the metal oxides can be adjusted from zero to near the theoretical packing density of spheres, allowing refractive index to be controlled over a large range. Here we demonstrate the abilities of nanocomposites to control filter response through their engineerable refractive indices. With layer counts exceeding 30 (more than twice the layer count of our previous efforts), very little haze is observed. Three formulations of titanium dioxide and UV curable monomers were prepared to produce the following characteristics in the finished films: (1) 1.65 refractive index with a thickness of 73 nm (2) 1.75 refractive index with a thickness of 69 nm (3) 1.85 refractive index with a thickness of 65 nm.

Two pairings of these materials were tested with increasing layer counts. The resulting reflectance spectra are shown in figures 1 and 2. The filters were designed to peak near 480 nm with their widths and heights controlled by the ratio of the refractive indices of the layers and the layer count. The measured results of the final films closely resemble their theoretical models. The goal of this work was to show the capabilities of this simple process in building multilayer thin films that are both optically and mechanically flexible.

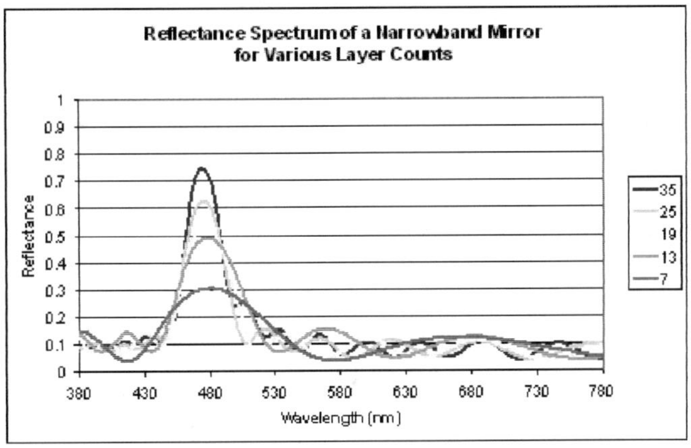

Figure 1: Narrow reflectance filter using refractive indices of 1.75 and 1.85 with layer counts of 7, 13, 19, 25 and 35.

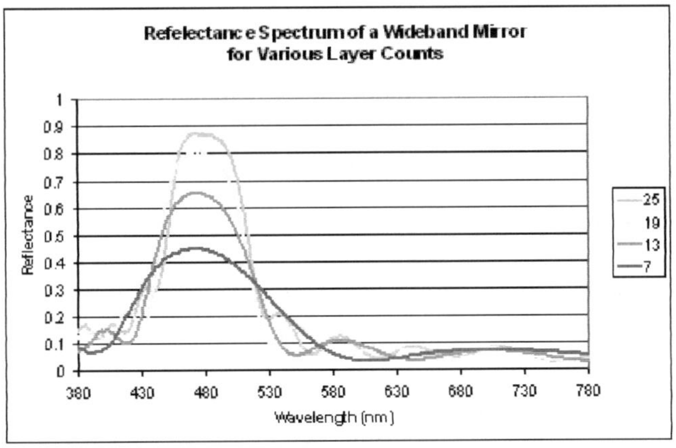

Figure 2: Wide reflectance filter using refractive indices of 1.65 and 1.85 with layer counts of 7, 13, 19, and 25 layers.

559c Electrical Conductivity In Polymer Nanocomposites Containing Metal Nanowires: Simulation and Experiment

Sadie I. White[1], Brian A. DiDonna[2], Lai-Ching Chou[1], Lea V. Nowack[1], Minfang Mu[1], Tom C. Lubensky[3], and Karen I. Winey[1]. (1) Materials Science and Engineering, University of Pennsylvania, 3231 Walnut Street, Philadelphia, PA 19104, (2) Stellar Science, Suite 725, 6565 America's Parkway NE, Albuquerque, NM 87110, (3) Department of Physics and Astronomy, University of Pennsylvania, 209 S. 33rd St., Philadelphia, PA 19104

The study of percolation behavior has resurfaced in recent years, because this phenomenon dominates the electrical conductivity in polymer nanocomposites. Furthermore, carbon nanotubes and metal nanowires access new size and aspect ratio regimes. This paper will explore this topic using both simulations and experiments.

Traditional percolation theories (i.e. the excluded volume model) are only strictly applicable for rods of effectively infinite aspect ratio. Because many processing techniques used for preparing carbon nanotube composites produce nanotubes with aspect ratios of less than 50, a simulation approach to study percolation behaviors of finite-sized rods is needed. In addition, it is useful to combine the simulation of filler morphologies with calculations of a specific composite property, e.g. electrical conductivity. We will present our simulations of conductive rods in an insulating matrix showing the effects of filler loading, aspect ratio, and degree of uniaxial alignment on electrical percolation behavior and conductivity. As the uniaxial alignment increases, there is a dramatic drop in electrical conductivity, because the network structure is destroyed. This result indicates that the traditional polymer processing methods of extrusion and injection molding are likely to produce materials with unintended heterogeneity in the electrical conductivity due to variations in nanofiller orientation.

Our initial experimental work used carbon nanotubes to probe the effect of axial alignment on nanocomposite conductivity. Here, we use metal nanowires, because the size (diameter and length) can be easily controlled by templated synthesis and all the metal nanowires are metallic, unlike carbon nanotubes. We prepare silver nanowires by electroplating into alumina membranes and then free the nanowires by dissolving the membrane. The polystyrene nanocomposites are then made using our group's established coagulation method. Isotropic and uniaxially aligned nanocomposites are prepared by compression molding and melt fiber spinning, respectively. The morphologies of the nanowires and nanocomposites are determined using a combination of optical microscopy, SEM, and X-ray scattering. The electrical conductivity is measured using a two-probe method and compared with our simulation results.

559d Morphology and Properties of Nanocomposites from Poly(ethylene-co-methacrylic acid) Ionomers

D. R. Paul[1], Lli Cui[1], Rhutesh K. Shah[2], and Youngjae Yoo[1]. (1) Department of Chemical Engineering and Texas Materials Institute, The University of Texas at Austin, Austin, TX 78712, (2) Harvard University, School of Engineering and Applied Sciences, Cambridge, MA 02138

Ionomers, where some of the acid groups of ethylene/methacrylic acid copolymers are neutralized to form sodium, zinc or magnesium salts, offer an attractive option in facilitating the exfoliation of the organoclay in polyolefin type matrices. Besides improving the clarity of the polymer, the acid and ionic groups provide the possibility of favorable interactions with the organoclay. Various structural aspects of an ionomer matrix, such as type of neutralizing ion and degree of neutralization of the acid groups, which could have an effect on its ability to exfoliate the organoclays, are discussed in this study. Along with these, some other efforts were made to find the best organoclay structure for nanocomposites prepared from a given ethylene/methacrylic acid sodium ionomer. Nanocomposites studied here were all prepared by melt processing in a twin-screw extruder, followed by a series of characterizations, such as stress-strain analysis, Izod and Dynatup impact tests, wide angle X-ray scattering (WAXS), and transmission electron microscopy (TEM) coupled with particle analysis, to evaluate the morphology and properties of the nanocomposites. The results suggest that nanocomposites prepared from the sodium and the zinc ionomers displayed much better exfoliation of the organoclay and superior levels of reinforcement compared to equivalent nanocomposites prepared from the lithium ionomer. Improved exfoliation of organoclay with elevated degree of neutralization of acid groups in sodium ionomers were observed, indicating that the sodium ionic units in ionomers provide more favorable interaction between the matrix and organoclay as compared to acid units, and lead to better dispersion of the clay particle.

559e A Comparison of Structures and Property Enhancements Associated with Poly(ethylene terephthalate) Nanocomposites Made with Clay or Graphite by Solid-State Shear Pulverization

Cynthia Pierre[1], Wei Qu[1], and John M. Torkelson[2]. (1) Materials Science & Engineering, Northwestern University, 2220 Campus Drive, Cook Hall Room 2036, Evanston, IL 60208-3108, (2) Chemical and Biological Engineering and Materials Science and Engineering, Northwestern University, 2145 Sheridan Rd, TECH E136, Evanston, IL 60208

Poly(ethylene terephthalate) (PET) is widely used for various applications ranging from food packaging to bottles to automotive parts. There have been many studies aimed at optimizing the physical properties of PET through the addition of nanofillers. However, due to the high melt temperature of PET, only a small fraction of studies have led to well-exfoliated fillers and thus, true nanocomposite formation. As a result of the poor nanofiller dispersion, the performance of many PET hybrids falls below expectations. Here we demonstrate the ability to enhance the physical properties of PET via two routes, by simply processing PET using solid-state shear pulverization and through the addition of either clay or graphite to produce well-exfoliated PET nanocomposites.

Solid-state shear pulverization (SSSP) is a continuous, industrially applicable process that employs a cooled, modified twin-screw extruder that maintains the polymer in the solid-state during processing. We demonstrate the ability to very strongly tune the physical and mechanical properties of PET by changing the processing conditions of neat PET during solid-state shear pulverization without addition of any fillers or nucleating agents. Using differential scanning calorimetry, we observe a roughly factor of 3 increase in crystallinity of PET that has been pulverized and subsequently melted relative to the unprocessed PET. We also observe a dramatic increase in the rate of crystallization of the pulverized samples. Rheological characterization has demonstrated an increase in viscosity of the pulverized material, which can be ascribed to chain branching in the pulverized product. We also observe significant reductions in the oxygen permeability of the PET with pulverization as well as enhancements in mechanical properties that are commensurate with the modified crystallization properties of the pulverized PET.

Via a combination of methods, including x-ray diffraction and transmission electron microscopy, we demonstrate that the well-exfoliated state can be achieved via SSSP processing of PET-clay and PET-graphite nanocomposites. Using x-ray diffraction of powder and melt-state samples, we show that SSSP has yielded well-exfoliated states in all of the systems studied. Even more important is the fact that the well-exfoliated state is stable to long-term annealing in the melt state; this means that the exfoliation will be maintained during any subsequent melt processing. Among the properties that have been characterized in these PET nanocomposites include mechanical, barrier and thermal properties. Comparisons will be drawn between the property enhancements achieved through simple processing of the PET and the two fillers used.

THE EXTENT OF FILLER'S ORGANO-MODIFICATION AND THE RESULTING MORPHOLOGY AND THERMO-MECHANICAL PROPERTIES OF POLY(E-CAPROLACTONE)/CLAY NANOCOMPOSITES

S. I. Marras[1], I. Zuburtikudis[1], A. Tsimpliaraki[1], K. Tornikidou[1,2], E. Panayotidou[1], G. Christofidou[1]*

[1]Department of Industrial Design Engineering, TEI of Western Macedonia, 50100 Kozani, Greece
[2]Dentist, Private Dental Practice, I. Christidi 25, 50200, Ptolemaida, Greece

Introduction

The development of polymer/clay nanocomposites has attracted significant attention in recent years due to the exceptional physical property improvements that these materials often exhibit, at relatively low filler loadings [1]. In order to achieve these benefits, efficient dispersion levels of the reinforcing agent is a prerequisite, so that the high aspect ratio and surface area of the clay particles are fully utilized and the resulting nanocomposite properties are isotropic. In general, the inorganic filler neither has good interaction with the organic polymer to achieve good dispersion, nor adequate adhesion, and, as a result, surface organic treatments are common [2]. Organic modifier's packing density within the silicate galleries is a very important factor regarding the dispersibility of the clay into the matrix and the properties of the final hybrid material.

The present study was undertaken to elucidate the effect of the surfactant concentration on the thermo-mechanical behavior of poly(ε-caprolactone) nanocomposites. Hybrids were prepared by the addition of montmorillonite modified by various concentrations of hexadecylammonium cation. Dispersion of the organomodified clay into the polymer matrix was achieved by the melt intercalation technique. The structural characteristics of the as-prepared composites were investigated by X-ray diffraction and their thermal properties were studied by thermogravimetric analysis. The influence of the amphiphilic cation concentration on the material's mechanical properties was explored in tensile loading conditions.

Experimental part

Materials

Poly(ε-caprolactone) (PCL) (Mn=80000) was obtained from Aldrich. Organically modified montmorillonite (C_{16}MMT) was prepared through an ion-exchange reaction between sodium montmorillonite and hexadecylammonium salt (HDA) loadings ranging from 30 to 300% the cation exchange capacity (CEC) of the inorganic material. The structural, thermal and colloidal characteristics of the organophilic clay were studied in a previous work [3].

Preparation of nanocomposites

Nanocomposites were fabricated by the melt intercalation method, at 150°C and 120 rpm, under constant nitrogen flow, using a ThermoHaake Minilab microcompounder. The compounding time was 10 min. The inorganic content in every composite prepared was 3 wt%.

* E-mail address: izub@teikoz.gr

Characterization

The effect of the hexadecylammonium salt loading of C_{16}MMT on the structure of the hybrids was investigated by X-ray diffraction (XRD) using a Rich. Seifert 3003 TT diffractometer with Ni-filtered CuKα radiation (λ=0.154 nm). The 2θ scanning range was varied from 1.5° to 10° with a step of 0.01 and a measuring time of 15 s per step.

Thermogravimetric analyses (TGA) of the hybrids were performed on a Shimadzu TGA-50 Analyzer with a heating ramp of 10 °C/min up to 750 °C both under flowing nitrogen (20 cm^3/min) and static air.

The mechanical properties of the produced composites were studied in tensile loading conditions using an Instron 3345 tensile tester according to ASTM D1708 method. The crosshead speed was 5 mm/min. At least five measurements were taken for each sample and the results were averaged to obtain a mean value.

Results and discussion

Morphology

Figure 1 depicts how the modified clay at various degrees affects the structure of the hybrid material. At low clay modification level (less than 50% of the CEC) mixing of polymer with organo-MMT did not result in the formation of a nanocomposite material. The hybrid displays two 001 reflections indicating an inhomogeneous dispersion of the clay particles into the matrix. Increase of the amphiphilic cation concentration facilitates PCL penetration into the modified mineral interlayer space shifting the clay's characteristic peak towards lower 2θ values.

However, at surfactant loadings higher than twice the clay CEC polymer insertion into the clay basal spacing tends to decrease. Lan et al. [4] reported that large population density of alkylammonium cations restricts intra-gallery penetration of polymer and there is a tendency to form intercalated nanocomposites rather than exfoliated ones. On the contrary, low population density provides more interlayer space to accommodate polymer chains. Mandalia et al. [5] suggested that lower concentration of surfactant aids the penetration of polymer probably due to a higher mobility of alkylammonium chains. Moreover, they concluded that the higher degree of organoclay dispersion can be achieved when the surfactant loading is equivalent to 1.2 to 1.5 times the clay CEC. Vaia et al. [6] reported that hybrid formation via polymer melt intercalation is entropically most favorable when the incorporated organoclay displays intermediate gallery heights and surfactant packing densities.

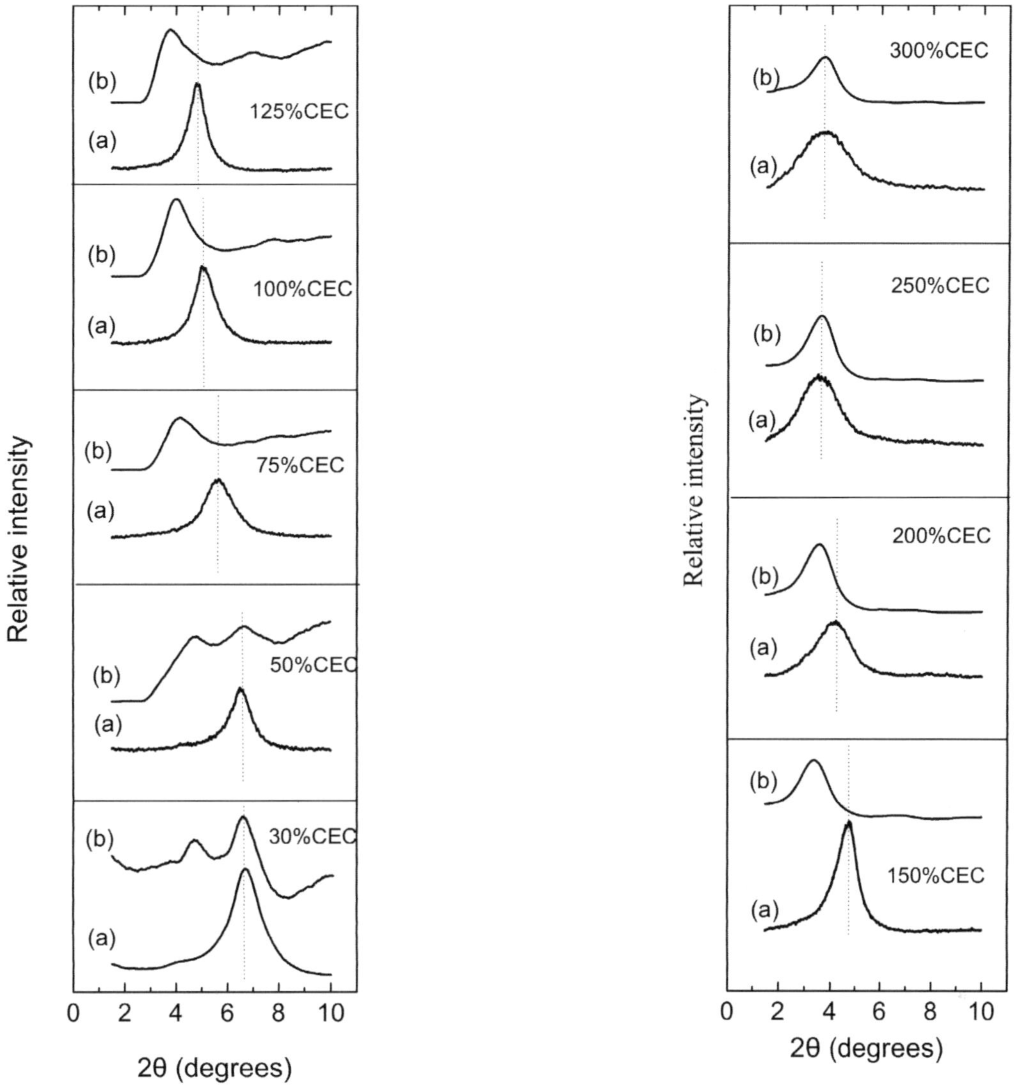

Figure 1. X-ray diffraction patterns of (a) the modified clay and (b) the resultant hybrid.

Thermal characteristics

Figure 2a shows the TGA weight loss curves of the prepared hybrids, which were obtained in a nitrogen atmosphere and for the temperature range between 150°C (where the major part of humidity has been removed) and 750 °C. At low surfactant loadings (30% CEC) thermal decomposition initiation occurs at about 285 °C. Increase of the mineral modification degree reinforces its compatibility with PCL facilitating the dispersion of MMT into the matrix. This results in a significant enhancement of the hybrid thermal stability increasing the onset decomposition temperature. Due to the high aspect ratio, the dispersed clay platelets impede the escape of the volatile degradation products within the nanocomposites. However, at surfactant concentrations higher than 200% the clay CEC, the hybrid displays a reduced thermal stability.

The effect of the amphiphilic cation concentration on the thermal properties of the hybrids in oxidative conditions is shown in Figure 2b. Increase of the surfactant loading up to 200% CEC enhances the thermal stability of the hybrid shifting the onset temperature of the hybrid's degradation toward higher values. The layered silicate incorporation triggers char formation on the surface of the composite which hinders the flow of the volatile decomposition products within the material [7] and limits the access of oxygen to the remaining material. This charring process may be catalyzed by the decomposition of the alkylammonium cation [8].

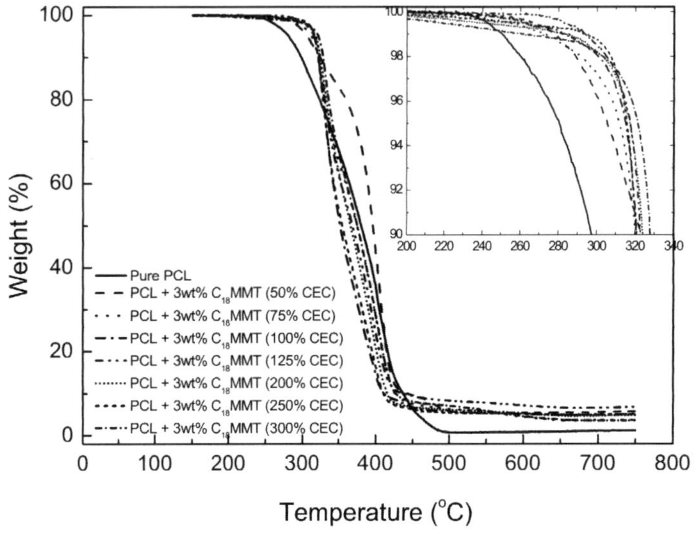

(a)

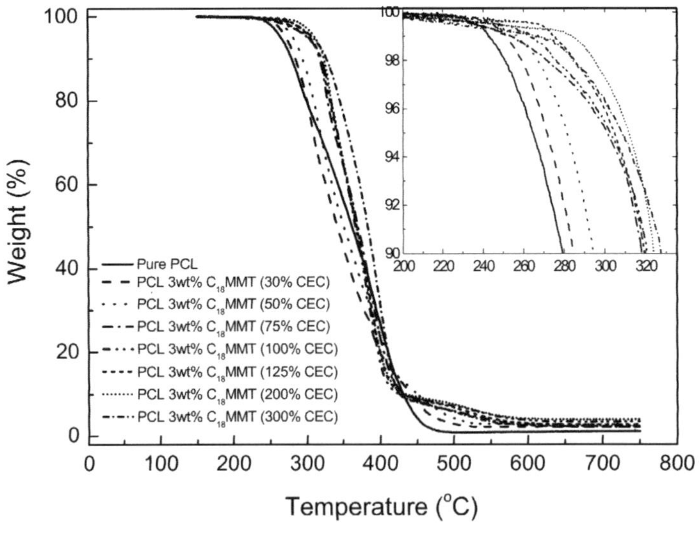

(b)

Figure 2. TGA curves for PCL hybrids prepared by the incorporation of MMT modified to a different extent, in (a) inert and (b) oxidant atmosphere.

The relation between the MMT modification level and the temperature required for 2% of hybrid decomposition in inert and oxidative atmosphere is presented in Figure 3. It is evident that the effect of the surfactant loading is similar in both cases. Moreover, oxygen causes decomposition of PCL/clay hybrids at a lower temperature than nitrogen.

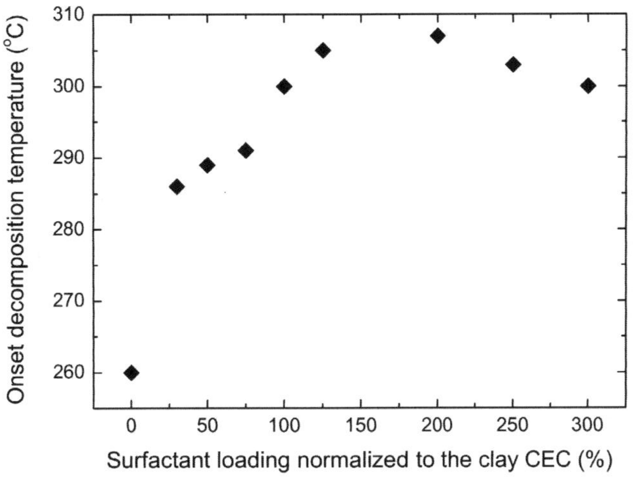

(a)

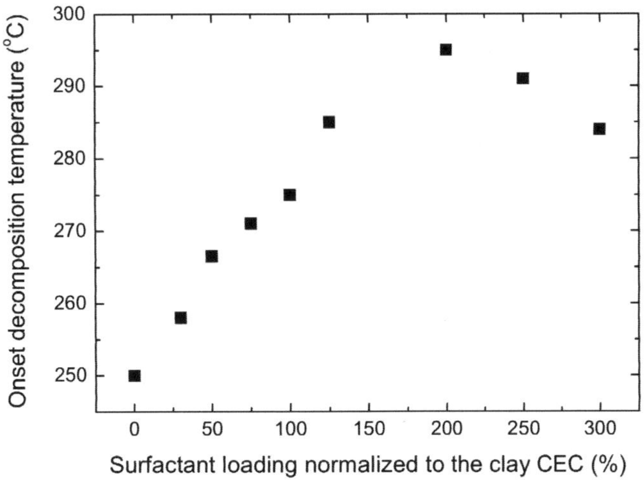

(b)

Figure 3. Effect of clay modification degree on the onset decomposition temperature (estimated at 2% of weight loss) of PCL hybrids, in (a) inert and (b) oxidant atmosphere.

Mechanical properties

Table 1 presents the influence of the clay modification degree on the mechanical properties of the PCL hybrids. It appears that increase of the alkylammonium loading enhances the strength and the elongation at break of the hybrid. Low surfactant loadings cause incomplete dispersion of MMT particles and inadequate polymer chain linkage to the mineral surface. Consequently, large clay agglomerates formed can act as load-concentrated defects which promote crack initiation leading to a decreased strength and material ductility. The nanoscale dispersion achieved at intermediate HDA concentrations provides a substantial rise to the composite's plastic deformation. However, when the alkylammonium coverage exceeds the clay CEC, a drop of the tensile strength along with a material embrittlement is observed.

The trend in Young's modulus values is consistent with the trend of the aforementioned mechanical properties, when the clay modification level is varied. Silicate layers, due to their unique structure which includes a high aspect ratio and a large surface area, significantly improve the modulus of elasticity of the material. This improvement is expected to be larger by decreasing the scale of reinforcement, which is associated with the increase of the ammonium cation concentration. Additionally, an increase of HDA content reinforces polymer affinity and interfacial adhesion to the mineral surfaces and enhances their interaction. Therefore, the stresses are much more effectively transferred to the inorganic filler. Nevertheless, very large excess of the surface modifier triggers a substantial decrease of the material's modulus of elasticity. Ratinac et al. [9] mentioned that at high degrees of clay modification some of the excess modifier may leach or migrate to the matrix and cause plasticization effects reducing the material mechanical properties.

Table 1. Mechanical characteristics of PCL/organomodified MMT hybrids

Modification level of the added MMT (normalized to the % CEC)	Yield Strength (MPa)	Strength at break (MPa)	Elongation at break (%)	Young's Modulus (GPa)
0 (pure PHB)	16.5	33	850	352
30	16	25	603	390
50	16	26	632	417
75	16.2	27	720	438
100	17.2	31	815	498
125	17.4	29	750	494
150	17	27	745	490
200	16.5	26	730	480
250	16	26	712	469

Conclusions

The effect of the modification level of montmorillonite on the structural, thermal and mechanical characteristics of PCL composites prepared by the melt intercalation method is presented in this work. The clay was modified by hexadecylammonium cation with concentrations ranging from 0.3 to 3.0 times the cation exchange capacity of the mineral.

Investigation of the prepared composites' morphology by X-ray diffraction showed that proper swelling of the clay layers can be achieved provided that the alkylammonium concentration is higher than 50% the clay CEC. However, high levels of clay modification restrict intra-gallery penetration.

The influence of the surfactant concentration on the material's mechanical properties was studied in tensile loading conditions. It was found that, initially, an increase of the clay modification degree enhances the compatibility between the mineral and the matrix leading to a significant improvement of the composite's stiffness without sacrificing polymer ductility. However, higher hexadecylammonium concentrations have a negative impact to hybrid's mechanical properties causing their significant deterioration.

Thermogravimetric analysis (TGA) revealed that intermediate clay modification levels enhance the composites thermal stability by increasing the onset decomposition temperature.

Acknowledgements
This work was supported by the Region of Western Macedonia (Greece) through the PEP grant with MIS 105545 to I. Z.

References

1 Ray, S. S., and Okamoto, M. (2003), "Polymer/layered silicate nanocomposites: a review from preparation to processing," *Progress in Polymer Science*, 28(11), pp. 1539-1641.
2 Marras, S. I., Zuburtikudis, I., and Panayiotou, C. (2007), "Nanostructure vs. microstructure: Morphological and thermomechanical characterization of poly(L-lactic acid)/layered silicate hybrids," *European Polymer Journal*, 43(6), pp. 2191-2206.
3 Marras, S. I., Tsimpliaraki, A., Zuburtikudis, I., and Panayiotou, C. (2007), "Thermal and colloidal behavior of amine-treated clays: The role of amphiphilic organic cation concentration," *Journal of Colloid and Interface Science*, 315(2), pp. 520-527.
4 Lan, T., Kaviratna, P. D., and Pinnavaia, T. J. (1995), "Mechanism of Clay Tactoid Exfoliation in Epoxy-Clay Nanocomposites," *Chemistry of Materials*, 7(11), pp. 2144-2150.
5 Mandalia, T., and Bergaya, F. (2006), "Organo clay mineral-melted polyolefin nanocomposites - Effect of surfactant/CEC ratio," *Journal of Physics and Chemistry of Solids*, 67(4), pp. 836-845.
6 Vaia, R. A., and Giannelis, E. P. (1997), "Polymer melt intercalation in organically-modified layered silicates: Model predictions and experiment," *Macromolecules*, 30(25), pp. 8000-8009.
7 Gilman, J. W., Jackson, C. L., Morgan, A. B., Harris, R., Manias, E., Giannelis, E. P., Wuthenow, M., Hilton, D., and Phillips, S. H. (2000), "Flammability properties of polymer - Layered-silicate nanocomposites. Polypropylene and polystyrene nanocomposites," *Chemistry of Materials*, 12(7), pp. 1866-1873.
8 Zanetti, M., Kashiwagi, T., Falqui, L., and Camino, G. (2002), "Cone calorimeter combustion and gasification studies of polymer layered silicate nanocomposites," *Chemistry of Materials*, 14(2), pp. 881-887.
9 Ratinac, K. R., Gilbert, R. G., Ye, L., Jones, A. S., and Ringer, S. P. (2006), "The effects of processing and organoclay properties on the structure of poly(methyl methacrylate)-clay nanocomposites," *Polymer*, 47(18), pp. 6337-6361.

566a Explanation for Benefits of Millisecond Annealing In Ultrashallow Junction Formation

Edmund G. Seebauer and Charlotte T. M. Kwok. Department of Chemical & Biomolecular Engineering, University of Illinois at Urbana-Champaign, 600 South Mathews Avenue, Urbana, IL 61820

Formation of pn junctions in advanced Si-based transistors employs rapid annealing techniques after ion-implantation in order to increase the electrical activation of dopants while minimizing their diffusion. Over the past decade, these techniques have evolved from rapid thermal processing, with time scales of about 1 s, to millisecond methods accomplished by flashlamps or lasers. Although the dopant behavior in terms of diffusion and electrical activation clearly improves as a result of the shortened time scale, the technology transition has taken place on a largely phenomenological basis with little understanding of the physical mechanism for the improvement. The present work provides the key elements of that understanding based on simulations of the reactions between dopant interstitial atoms and interstitial clusters as well as the host semiconductor lattice. The simulations solve the partial differential equations for diffusion and reaction of interstitial atoms, with activation energies for elementary diffusion and reaction steps computed by Maximum a Posteriori paremeter estimation. The fundamental reason for improvements of diffusion and electrical activation in the millisecond regime is that the short time scale promotes exchange of dopant interstitial atoms with the lattice in preference to exchange with interstitial clusters.

566b The Combined Role of Transport Phenomena and Interfacial Attachment Kinetics during Liquid Phase Epitaxy of Mercury Cadmium Telluride

Igal G. Rasin[1], Anne Ben Dov[2], Ilana Grimberg[2], Olga Klin[2], Eliezer Weiss[2], and **Simon Brandon**[1]. (1) Chemical Engineering, Technion, Haifa, 32000, Israel, (2) SCD-Semi-Conductor Devices, P.O. Box 2250/99, Haifa, 31021, Israel

Mercury Cadmium Telluride (HgCdTe) is an important infrared detector material. Relevant devices are often manufactured by depositing thin films of this material, via Liquid Phase Epitaxy (LPE), on lattice-matched CdZnTe substrates. Although generally successful, this process could benefit from significant optimization procedures e.g. aimed at improving uniformity of properties across the as-grown film. As a precursor to the application of such procedures it is important to investigate physics underlying the growth process, thereby uncovering the relevant importance of the different physical phenomena.

In this contribution we will present a combined computational and experimental analysis of basic physical phenomena underlying the HgCdTe LPE growth process. Specifically, we will discuss a dynamic three-dimensional model of mass and momentum transport, combined with interfacial attachment kinetics, which we have developed and applied in the investigation of this system. As will be demonstrated, among the benefits of combining numerical and experimental analyses is our ability to uncover the importance of a specific interfacial attachment kinetics mechanism in this system. Additional results to be presented include a quantitative estimation of kinetic parameters as well as an improved understanding of the role of different transport and kinetic phenomena underlying this growth process.

566c First Principles Determination of Highly Mobile Dopant-Interstitial Complexes and Their Relative Contribution to Dopant Diffusion In Silicon

Kyoung E. Kweon, Electrical Engineering, The University of Texas at Austin, 1 University Station C0400, Austin, TX 78712 and Gyeong S. Hwang, Chemical Engineering, Dept. of Chem. Eng.,The University of Texas at Austin, 1 University Station Stop C0400, Austin, TX 78712.

Scaling complementary metal oxide semiconductor (CMOS) devices beyond the 45-nm node may require the formation of highly doped ultrashallow junctions less than 7 nm deep with high lateral abruptness. To meet these stringent requirements, it is necessary to have a better understanding of the underlying mechanisms of transient enhanced diffusion (TED) of implanted dopants during postimplantation annealing. It is now well accepted that dopant diffusion is mainly mediated by native defects created in the substrate. Some early experimental studies suggested that arsenic (As) TED could be mediated by both interstitials and vacancies while phosphorus (P) and boron (B) TED would exclusively be mediated by interstitials. While various mobile dopant-defect complexes have been identified by extensive first principles quantum mechanical calculations, their relative contribution to evolution of dopant diffusion profiles has not been clarified yet. In this presentation, we will focus on addressing the relative contribution of various dopant-interstitial complexes to diffusion of corresponding dopants, based on extensive first principles calculations. We have found that BI, B2I and B2I2 play an important role in B diffusion, while AsI, AsI2 and PI and PI2 are important diffusing species for As and P diffusion, respectively. The pathways and barriers for their diffusion and dissociation will be presented. Based on the data, we will discuss how these mobile species determine dopant diffusion profiles.

566d Computational Insights into the (Complex) Aggregation Physics of Self-Interstitials

Sumeet Kapur, Univ. of Pennsylvania, 220 S. 33rd St., Philadelphia, PA 19104 and **Talid Sinno**, Chemical and Biomolecular Engineering, University of Pennsylvania, 220 South 33rd St, Philadelphia, PA 19104.

The evolution of self-interstitials and their aggregates during the annealing of ion-implanted silicon has received a tremendous amount of attention because of their strong, non-linear effects on the diffusion of dopants. Ion-implantation is the most common way of introducing dopants into a silicon substrate, as required for microelectronic device fabrication. However, the implantation process leads to extensive lattice damage, and at high enough energies, complete amorphization, which must be recrystallized by thermal annealing following the implantation process. Also generated by the implantation process is a large number of self-interstitials which lead to enhanced dopant diffusion during the annealing process. As the self-interstitial population relaxes towards equilibrium during annealing, the diffusion enhancement varies strongly, a phenomenon that has been described as Transient Enhanced Diffusion, or TED.

A major obstacle to understanding and quantitatively predicting TED is the formation of a variety of self-interstitial aggregates, which range from small amorphous three-dimensional clusters, to planar stacking-faults with various crystallographic orientations. Experimental and theoretical studies have demonstrated that slightly different annealing conditions lead to substantially different TED behavior, which can be linked to differences in the aggregation and dissolution dynamics. While a global picture of TED is now available and the various cluster structures identified, there are still several outstanding issues related to the atomistic mechanisms of self-interstitial clustering that are not understood. In order to develop a fully predictive TED model, a quantitative, temperature dependent understanding of these processes is required.

A major obstacle to understanding and quantitatively predicting self-interstitial aggregation dynamics is their complex morphological transition from small three-dimensional clusters to planar stacking-faults/dislocation loops with various crystallographic orientations. In the present study, we use large-scale constant-stress MD simulations to dynamically simulate the evolution of an ensemble of highly supersaturated self-interstitials at various temperatures and pressures. We show that the simulated interstitial clustering into various types of planar structures exhibits a complex thermodynamic-kinetic phase diagram that is consistent with numerous experimental observations. In particular, we focus on the role of certain "magic" cluster sizes (e.g. four-interstitial and eight-interstitial clusters), which is shown to be extremely rich, exhibiting both temperature and stress dependencies, and leads to a bifurcation in the aggregation pathway of larger cluster sizes. These dependencies are studied with a recently developed approach that maps out the potential energy landscape in the vicinity of the defect cluster and allows for the total (classical) free energy to be analyzed.

566e A Comprehensive Model for Coupled Oxide Precipitation and Point Defect Aggregation In Crystalline Silicon

Rubal Dua and Talid Sinno. Chemical and Biomolecular Engineering, University of Pennsylvania, 220 S 33rd Street, 311A Towne Building, Philadelphia, PA 19104

Continuum modeling of point defect aggregation during silicon crystal growth and wafer annealing has a long-standing history of success. Numerous studies have shown that these models are able to capture, at least qualitatively, the essential features of point defect diffusion and clustering under a variety of crystal growth and wafer annealing conditions, especially in the case of homogeneous vacancy clustering [1,2,3,4].

On the other hand, process modeling of oxygen precipitation in Czochralski-grown (CZ) crystals has been more challenging due to the inherent complexity of oxygen precipitation and its strong coupling to the point defect distributions (e.g. ref. [5]). Oxygen precipitation is a critically important process because bulk oxide precipitates provide essential gettering, or trapping, sites for metallic atoms that are introduced during wafer processing. It is therefore essential to ensure that these precipitates are present in the bulk region of the wafer before metallic elements are introduced, but also that they are not located at the device-active surface region.

We present a comprehensive continuum model, based on a hybrid Master-Fokker Planck equation formulation, of coupled vacancy aggregation and oxide precipitation. This model is an extension of a recently developed model for vacancy aggregation, which has been demonstrated to provide a quantitatively accurate picture of void formation under a wide range of operating conditions during both crystal growth and wafer annealing [1]. An operator splitting scheme is employed to decouple the point defect and cluster equations. It also gives an added advantage in terms of dimensionality reduction of the cluster sub-problem from 2D to 1D for the crystal growth case and from 1D to 0D for the wafer annealing case. The coupling physics between the oxide precipitation and vacancy aggregation behavior is probed in detail and the model results compared to experimental measurements in both crystal growth and wafer thermal annealing environments. Several model descriptions for oxide precipitation are compared and contrasted.

[1] Thomas A. Frewen, Sumeet S. Kapur, Walter Haeckl, Wilfried von Ammon, Talid Sinno, ``A microscopically accurate continuum model for void formation during semiconductor silicon processing'', J. Crystal Growth 279 (2005) 258-271.

[2] Talid Sinno, Robert A. Brown, ``Modeling microdefect formation in Czochralski silicon'', J. Electrochem. Soc. 146 (1999) 2300-2312.

[3] Milind S. Kulkarni, ``A selective review of the quantification of defect dynamics in growing Czochralski silicon crystals'', Ind. Eng. Res. 44 (2005) 6246-6263.

[4] Masanori Akatsuka, Masahiko Okui, Shigeru Umeno, Koji Sueoka, ``Calculation of size distribution of void defects in CZ silicon'', J. Electrochem. Soc. 150 (2003) G587-G590.

[5] Koji Sueoka, Masanori Akatsuka, Masahiko Okui, Hisashi Katahama, ``Computer simulation for morphology, size, and density of oxide precipitates in CZ silicon'', J. Electrochem. Soc. 150 (2003) G469-G475.

566f Optically Stimulated Diffusion In Ultrashallow Junction Formation

Yevgeniy Kondratenko, University of Illinois, Department of Chemical Engineering, 600 S. Mathews, Urbana, IL 61802, Charlotte Kwok, Chemical Engineering, University of Illinois, Department of Chemical Engineering, 600 S. Mathews, Urbana, IL 61802, and Edmund G. Seebauer, Department of Chemical & Biomolecular Engineering, University of Illinois at Urbana-Champaign, 600 South Mathews Avenue, Urbana, IL 61820.

Formation of pn junctions in advanced Si-based transistors employs rapid thermal annealing after ion-implantation in order to increase the activation of dopants. There has long been suspicion that the strong lamp illumination required for this procedure may nonthermally influence the diffusion of dopants. Identification of such effects is difficult in conventional RTA geometries because lamps provide both heating and photostimulation, and because the interpretation of conventional dopant diffusion experiments is impeded by complex dopant-defect interactions. We have circumvented these problems with a new experimental design in which heating and illumination can be decoupled.

Photostimulated effects were studied using the diffusion of isotopically labelled 30Si tracer in an epitaxial 28Si matrix. Results for self-diffusion show that for n-type Si, self-diffusion rates are increased nonthermally by more than an order of magnitude for modest illumination intensities of roughly 1 W/cm2. Results depend on doping type; the rates of both interstitial formation and migration are affected in the case of n-type material. There is no comparable effect for p-type material, however. A physical model based on photostimulated changes in interstitial charge state explains much of the behavior. A non-equilibrium steady-state model was formulated to describe the electronic occupation of defect levels under photo-stimulation. The model is based on Shockley-Read statistics for an arbitrary distribution of energy levels.

Photostimulated diffusion of dopants, however, exhibited more complicated features. Depending on annealing temperature and time, boron diffusion in silicon could be either enhanced or inhibited. Dopant activation was similarly affected. Simulations using continuum equations for the reaction and diffusion of defects were used to determine whether illumination affects cluster dynamics or steady state boron diffusion.

566g Morphological Stability Analysis of Planar Crystalline Solid Surfaces Under the Simultaneous Action of Electric Fields and Mechanical Stresses

Vivek Tomar, M. Rauf Gungor, and Dimitrios Maroudas. Department of Chemical Engineering, University of Massachusetts, 159 Goessmann Laboratory, North Pleasant Street, Amherst, MA 01003

Surface morphological stability underlies various reliability problems of technologically important materials with applications in aerospace, microelectronics, and nanotechnology. Applied mechanical stress has been known to induce surface morphological instabilities in crystalline solid materials. Specifically, linear stability analyses have demonstrated that the competition between elastic strain energy and surface energy can cause the growth of shape perturbations from a planar surface morphology and specified the conditions for the onset of the so-called Asaro-Tiller or Grinfeld instability. In addition, numerical simulations accounting for all the important nonlinearities have addressed the surface dynamics beyond the instability onset and revealed instability mechanisms. In particular, it has been demonstrated that a planar surface of a stressed elastic solid can evolve rapidly into a cusped surface, with smooth tops and deep crack-like grooves by surface diffusion, in agreement with experimental observations. These theoretical predictions are consistent with experimental findings over a broad class of materials. However, the effects of the simultaneous action of an electric field on the surface morphological response of an electrically conducting stressed solid have not been explored systematically.

In this presentation, we report results of linear stability analysis for the morphological response of the planar surface of a stressed elastic solid that is electrically conducting under the simultaneous action of an applied electric field. Our analysis is based on a surface transport model that accounts for curvature-driven surface diffusion, surface electromigration, and stress-driven surface diffusion along with surface diffusional anisotropy. We demonstrate that proper application of an electric field can have an important effect on the dynamical response of a planar stressed solid surface that is otherwise unstable with respect to long-wavelength perturbations. Specifically, we show that application of a sufficiently strong electric field can stabilize the surface of the stressed electrically conducting solid material that would be otherwise vulnerable to surface cracking under certain thermomechanical conditions; therefore, the electric current protects the material against cracking and inhibits its damage. In addition, we report the effects on the surface morphological stability of the misorientation angle between the electric field and fast diffusion directions, as well as the effects of key material properties, such as the strength of the surface diffusional anisotropy that is temperature dependent and the material's texture that is set by the surface crystallographic orientation. We show that (a) there exists an optimum range of misorientation angles for stable morphological response, (b) the strength of the surface diffusional anisotropy can have a stabilizing effect on the surface morphological response, and (c) under given electromechanical conditions, the surface morphological response of <111>-oriented surfaces is superior to that of <100>- and <110>- oriented ones.

In addition to the linear stability analysis, we report computational results for the morphological evolution of a solid surface perturbed from an initially planar morphology under the simultaneous action of an electric field and mechanical stress. The computational predictions for the morphological evolution are based on self-consistent dynamical numerical simulations according to the fully nonlinear surface mass transport model that was employed in the stability analysis; in the simulations, the fully nonlinear model is solved self-consistently with the electric field and stress field distributions on the solid surface, as computed through a Galerkin boundary integral method, and in conjunction with a front tracking method that monitors the surface propagation. The numerical results confirm the main conclusions of the linear stability analysis. In addition, for long-wavelength perturbations, the simulations reveal a tip-splitting instability that causes the appearance of wriggles in the surface morphology. Our findings can be used toward development of systematic surface engineering strategies for improved materials reliability over a broad range of electromechanical conditions.

613a Comparing the Mechanical Properties of Chitosan Films Bound to Titanium Following Deposition, Neutralization, and Sterilization

Holly J. Martin, Dave C. Swalm School of Chemical Engineering, Mississippi State University, 330 Swalm - President's Circle, Mississippi State, MS 39762, Kirk H. Schulz, Vice President for Research and Economic Development, Mississippi State University, 617 Allen Hall, Mississippi State, MS 39762, and Joel D. Bumgardner, Biomedical Engineering Department, Herff College of Engineering, University of Memphis, Memphis, TN 38152.

Mechanical properties, such as strength, weight, and durability, are major factors in determining the metals chosen for use as implants. Titanium is commonly used as an implant material because of its mechanical properties and because it is easily passivated, meaning it does not react with the physiological fluids surrounding the implant. However, implants in general do not possess the ability to promote bone cell attachment and growth, which prevent the integration of the implant into the bone. Bioactive materials attached to the implant surface can improve osseointegration [1]. Several bioactive materials are currently being examined [2-6].

Chitin is the second most abundant form of polymerized carbon in nature and is found in the exoskeletons of shellfish and insects [7,8]. The de-acetylated form of chitin, known as chitosan, is currently being investigated as an implantable material. Chitosan is cationic and has been shown to encourage the attachment and growth of bone cells [9]. Also, chitosan encourages proper bone formation because bone cells retain their desired cell shape, which influences cell-specific functions [10]. Chitosan is also being examined because it is non-toxic and the by-products of degradation are non-toxic [8,11].

At Mississippi State University, four treatment combinations have been designed to attach chitosan to implant quality titanium. These four treatment combinations consist of a surface treatment, either passivation or piranha, and a silane treatment, either aminopropyltriethoxysilane (APTES) or triethoxsilylbutyraldehyde (TESBA) [12,13]. The titanium treated with APTES is then treated with gluteraldehyde before bonding chitosan, resulting in a three step process [12]. A two step process occurs when chitosan is bonded to the titanium treated with TESBA [13]. X-Ray Photoelectron Spectroscopy (XPS) was run during each reaction step, which showed that more silane was deposited on the piranha treated titanium as compared to the passivated titanium [12,13]. XPS was also used on the final films, which demonstrated no significant differences between the films produced using the four treatment combinations [14]. The bulk properties, including hardness and elastic modulus, were unaffected by the treatment combinations [14]. Tensile testing was performed, which demonstrated that there was no statistical difference between the four treatment combinations, but did show that the bond strengths were significantly higher than previous results [7,14].

Before any biological testing can be performed, sterilization of the coating must be performed [15,16]. Sterilization of chitosan has been shown to change some bulk properties, affecting the tensile strength of the chitosan film and contact angle [15,16]. However, these studies were only performed on free chitosan films, not attached to a titanium surface using APTES and TESBA [15,16]. Hardness and elastic modulus, two bulk properties that deal with how the polymer absorbs stress, have not been examined. The research presented will cover the effects of hardness, elastic modulus, and contact angle of chitosan bonded to titanium, following the deposition of the chitosan coating, neutralization of the chitosan coating using sodium hydroxide, and sterilization of the chitosan coating using ethylene oxide. This research will also present bond strength data which will be used to determine if the neutralization and/or the sterilization of the coating affected the silane linker molecules.

[1] B.D. Ratner and A. S. Hoffman. In: *Biomaterials Science: An Introduction to Materials in Medicine*. California: Academic Press, Inc., 1996, Foreword, 1-8.

[2] Y. Yang, C.M. Agrawal, K.H. Kim, H. Martin, K. Schulz, J.D. Bumgardner, J.L. Ong. *Journal of Oral Implantology*, 29, 6, 270-277, 2003.

[3] R.J. Friedman, T.W. Bauer, K. Garg, M. Jaing, Y.H. An, R.A. Draughn. *Journal of Applied Biomaterials*. 6, 231-235, 1995.

[4] A. Nanci, J.D. Wuest, L. Peru, P. Brunet, V. Sharma, S. Zalzal, M.D. McKee. *Journal of Biomedical Materials Research*, 40, 324-335, 1998.

[5] D. A. Puleo. *Journal of Biomedical Materials Research*, 37, 222-228, 1997.

[6] D. A. Puleo. *Journal of Biomedical Materials Research*, 29, 951-957, 1995.

[7] J.D. Bumgardner, R. Wiser, P.D. Gerard, P. Bergin, B. Chestnutt, M. Marini, V. Ramsey, S.H. Elder, J.A. Gilbert. *Journal of Biomaterials Science: Polymer Edition*, 14, 5, 423-438, 2003.

[8] G. Haipeng, Z. Yinghui, L. Jianchun, G. Yandao, Z. Nanming, Z. Xiufang. *Journal of Biomedical Materials Research*, 52, 285-295, 2000.

[9] C. Jarry, C. Chaput, A. Chenite, M.A. Renaud, M. Buschmann, J.C. Leroux. *Journal of Biomedical Materials Research (Applied Biomaterials)*, 58, 127-135, 2001.

[10] A. Lahiji, A. Sohrabi, D. S. Hungerford, C. G. Frondoza. *Journal of Biomedical Materials Research*, 51, 586-595, 2000.

[11] M. Prasitsilp, R. Jenwithisuk, K. Kongsuwan, N. Damrongchai, P. Watts. *Journal of Materials Science: Materials in Medicine,* 11, 773-778, 2000.

[12] H.J. Martin, K.H. Schulz, J.D. Bumgardner, K.B. Walters. *Langmuir*, 23, 6645-6651, 2007.

[13] H.J. Martin, K.H. Schulz, J.D. Bumgardner, K.B. Walters. *Applied Surface Science*, 254, 4599-4605, 2008.

[14] H.J. Martin, K.H. Schulz, J.D. Bumgardner, J.A Schneider. *Thin Solid Films*, 2008, In Press.

[15] S. B. Rao, C. P. Sharma. *Journal of Biomaterials Applications*, 10, 136-143, 1995.

[16] P.R. Marreco, P.d.L. Moreira, S.C. Genari, A.M. Moraes. *Journal of Biomedical Materials Research Part B: Applied Biomaterials*, 71B, 268-277, 2004.

613b Antibiofouling Activity of Natural Phenolic Compounds against Gram Negative Bacteria

Dong-Shik Kim[1], Sumitkumar B. Jagani[1], and Do-Young Yoon[2]. (1) Chemical & Environmental Engineering, University of Toledo, 2805 Bancroft Street, 3048 Nitschke Hall, Toledo, OH 43606, (2) Chemical Engineering, Kwangwoon University, 447-1 Wolgye-Dong, Nowon-Gu, Seoul, South Korea

Biofouling is caused by adhesion of microorganisms to solid surfaces with the production of extracellular polymers (polysaccharides and proteins) that results in namely a biofilm. Biofouling causes serious problems in chemical, medical, pharmaceutical industries, and health care fields. In the medical industry, for example, very high concentrations of antibiotics are used to kill the cells growing in a biofilm, which may lead to antibiotic resistance problems. Because it is difficult to kill or remove biofilm, it is considered better approach to prevent microbial cells from forming. Recent findings indicate that some natural phenolic compounds found in plant extracts have an effect on biofilm formation of Gram-negative bacteria. Natural phenolic compounds are known to possess strong anti-oxidant and anti-microbial activities. In this study, we investigated the effect of anti-oxidant activities of natural phenolic compounds on biofilm formation of Pseudomonas Aeruginosa, Gram-negative bacteria. Pseudomonas Aeruginosa was grown to form biofilm in the presence of several natural phenolic compounds such as phenol, anacardic acid, cardanol, salicylic acid, ethyl linoleate, polyanacardic acid, polycardanol, polysalicylic acid, ascorbic acid, catechin, epigallocatechin, tannic acid and tocopherol. None of the above mentioned phenolic compounds were found affecting the growth of Pseudomonas Aeruginosa below the concentration of 15 ppm. The biofilm formation test was performed using a modified microtiter plate assay. Biofilms were fixed with methanol, stained with crystal violet dye, then the bound dye was released with 33% glacial acetic acid and optical density was measured at 570 nm. The optical density readings give an estimate of amount of biofilm formed on the surfaces of the microtiter plate. Bacteria coordinate their interaction with higher organisms and the surrounding environments by intercellular communication systems, namely quorum sensing. For Gram-negative bacteria, one type of communication system functions via N-acyl homoserine lactone (AHL) signal molecules. In order to investigate the effect of phenolic compounds on AHL, a mutant cell, E. coli JB525, was used. E. coli JB525 is deficient of AHL-producing gene, and harboring green fluorescent protein (GFP) producing gene. Therefore, under normal conditions, E. coli JB525 cannot produce GFP, because AHL is not produced. When AHL is provided from outside, GFP-producing gene sequence is triggered by AHL and GFP is produced. Experiments were performed to compare the different phenolic compounds; in one beaker, E. coli JB525 was cultured with AHL, N-3-oxohexanoyl-L-homoserine lactone at 100 nM (Control). In the other beakers, the cells were cultured with AHL and with various phenolic compounds. Green Fluorescence was measured at excitation wavelength of 475 nm and emission detection at 515 nm. Difference in GFP was observed in the result. This result indicates that different phenols inhibit quorum sensing of the cells. Figure 1 shows the GFP intensity measurement for the phenolic compounds in terms of time. Figure 1. GFP fluorescence measurement for the phenolic compounds in the E. coli JB525 cultures. The control, which had no phenolic compounds (square marks), showed the highest fluorescence while the eppigallocatechin polyphenol (triangle marks) showed the lowest fluorescence. It is thought that the phenolic compounds interfered with AHL molecules, which resulted in low GFP fluorescence. The results show that biofouling is influenced by quorum sensing, and inhibiting or interfering with quorum sensing molecules by natural phenolic compounds may help the system prevent or reduce biofouling.

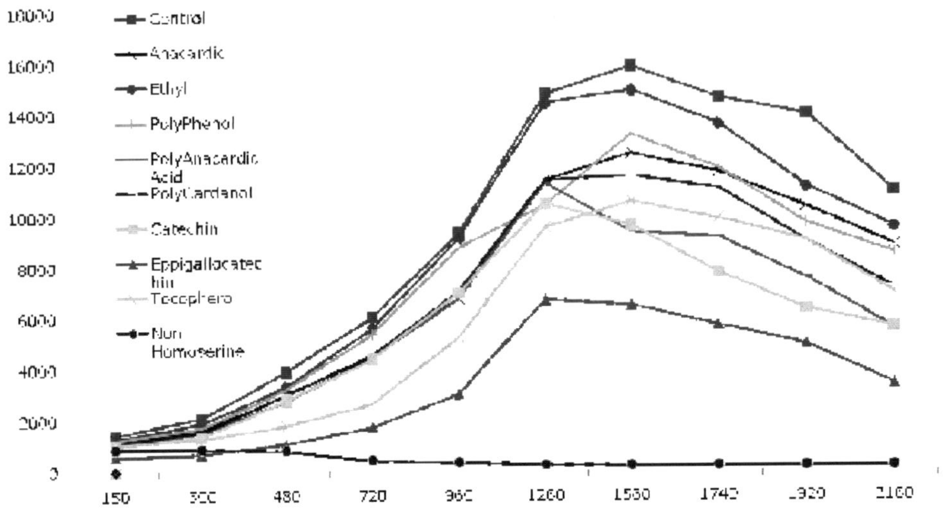

Control
Anacardic
Ethyl
PolyPhenol
PolyAnacardic Acid
PolyCardanol
Catechin
Eppigallocatechin
Non Homoserine
16000
14000
12000
10000
8000
6000
4000
2000
0
150
480
720
1260
2160

Cytocompatibility of Dextran-Based Tissue Sealants for Surgical Wound Closure

Sujata K. Bhatia, **Samuel D. Arthur**, **H. Keith Chenault** and **George K. Kodokian**, Materials Science and Engineering, DuPont Central Research & Development, Experimental Station, E328, P.O. Box 80328, Wilmington, DE 19880-0328

We have developed the ActaMax™ sealant family of hydrogel tissue adhesives, formed by reacting an oxidized polysaccharide with a water-dispersable multi-arm polyether amine. Specifically, we have developed tissue adhesives composed of two components: dextran aldehyde and multi-arm PEG amine, which undergo a Schiff base reaction to form a crosslinked hydrogel (Figure 1). This two-component tissue adhesive system crosslinks in water, cures rapidly (<1 min) at room temperature, adheres to moist tissue, and degrades hydrolytically.

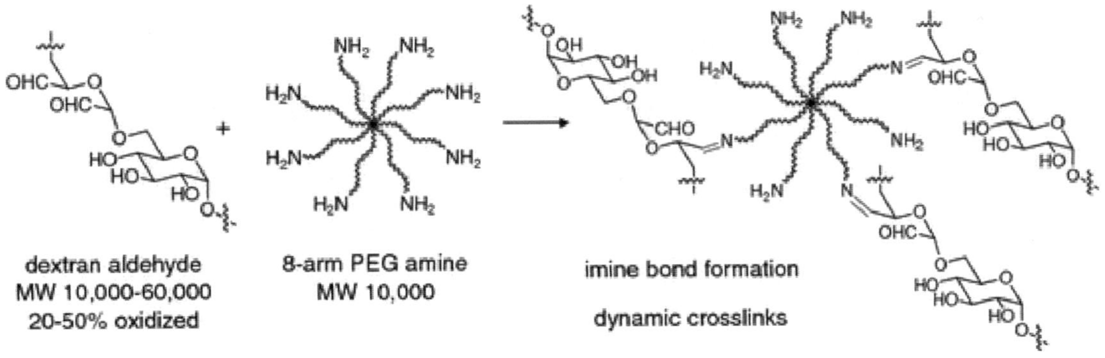

Figure 1. Foundation chemistry for polysaccharide-based tissue adhesives. The oxidized polysaccharide dextran aldehyde reacts with an 8-arm star PEG amine to form a crosslinked hydrogel network.

The effects of dextran-based tissue adhesives on cell survival and inflammatory cell activation were determined using in vitro mouse cell cultures. Cytotoxicity of tissue adhesives was evaluated by placing adhesives in direct contact with 3T3 fibroblast cells. Polysaccharide-based tissue adhesives composed of dextran aldehyde and star PEG amine were non-cytotoxic to fibroblasts; in contrast, a commercial adhesive composed of 2-octyl cyanoacrylate was highly cytotoxic to fibroblasts (Figure 2). The inflammatory potential of tissue adhesives was evaluated by exposing J774 macrophage cells to adhesives, and measuring TNF-alpha release from macrophages. Polysaccharide-based tissue adhesives did not elicit inflammatory TNF-alpha release from macrophages (Figure 3). These results suggest that dextran-based tissue adhesives are non-cytotoxic and non-inflammatory; the results are therefore significant in the design of novel tissue sealants, as well as cell culture systems to study biomaterials.

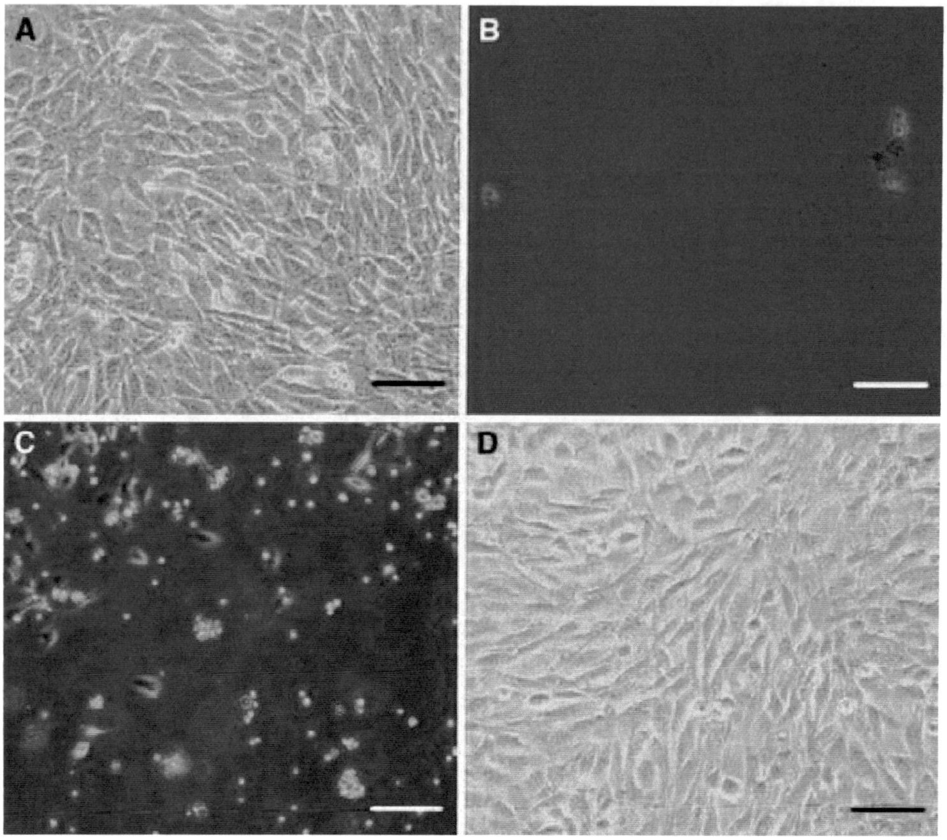

Figure 2. Responses of mouse 3T3 fibroblasts to tissue adhesives. Fibroblast cells were incubated in direct contact with material samples for 24 h at 37°C. Scale bars = 100 µm

(A) Ultra-high molecular weight polyethylene; non-cytotoxic negative control.

(B) Organo-tin PVC; cytotoxic positive control.

(C) Cyanoacrylate-based Dermabond tissue adhesive.

(D) Polysaccharide-based tissue adhesive composed of dextran aldehyde (60,000 MW, 20% aldehyde conversion, 17% solids) + 8-arm PEG amine (10,000 MW, 50% solids).

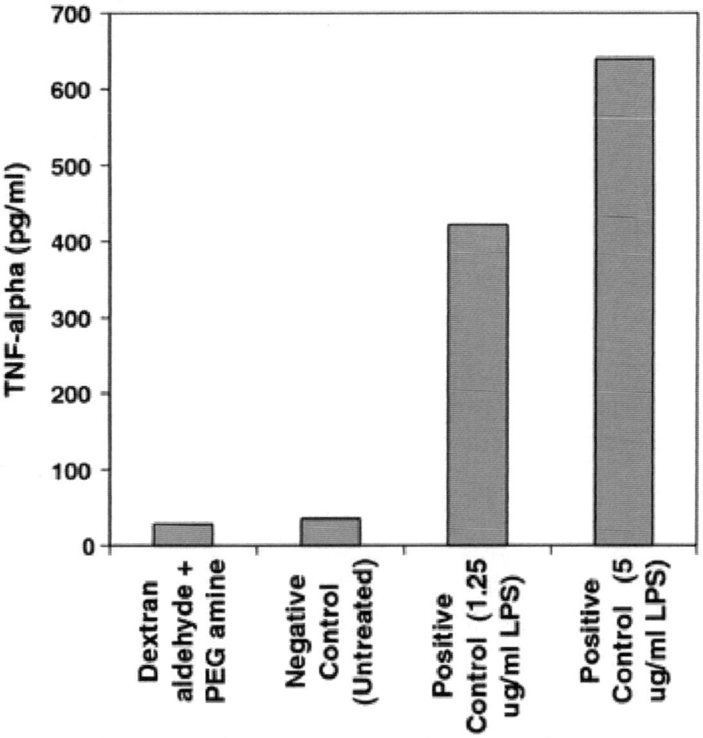

Figure 3. Response of mouse J774 peritoneal macrophages to polysaccharide-based tissue adhesive. Macrophage cells were incubated in direct contact with material samples for 48 h at 37°C. Following incubation, TNF-α release from macrophages was measured by ELISA.

References:

S. K. Bhatia, S. D. Arthur, H. K. Chenault, G. K. Kodokian, "Interactions of Polysaccharide-Based Tissue Adhesives with Clinically Relevant Macrophage and Fibroblast Cell Lines," *Biotechnology Letters* (2007), 29:1645-1649.

S. K. Bhatia, S. D. Arthur, H. K. Chenault, G. D. Figuly, G. K. Kodokian, "Polysaccharide-Based Tissue Adhesives for Sealing Corneal Incisions," *Current Eye Research* (2007), 32:1045-1050.

613d Enzymic Thin Film Coatings for Bioactive Materials

Songtao Wu[1], Xiaodong Tong[1], Archana Trivedi[2], Hongfei Jia[3], Minjuan Zhang[3], Masahiko Ishii[4], and Ping Wang[5]. (1) Bioproducts and Biosystems Engineering, University of Minnesota, 1475 Gortner Ave., 322 Snyder Hall, St. Paul, MN 55108, (2) Chemical and Biomolecular Engineering, University of Akron, 200 E. Buchtel Commons, Whitby 301, Akron, OH 44325, (3) Materials Research, Toyota Motor Engineering and Manufacturing North America, Inc., Ann Arbor, MI 48105, (4) Material Engineering Div. I, Toyota Motor Corporation, 1, Toyota-cho, Toyota, Aichi, 471-8572, Japan, (5) Department of Bioproducts and Biosystems Engineering, University of Minnesota, St Paul, MN 55108

Bioactive materials have been explored for a broad range of applications including biocatalysts, biosensors, antifouling membranes and other functional and smart materials. We report herein a unique method for reparation of bioactive materials through a spin coating process. Specifically,we investigated the preparation of protease Subtilisin Carlsberg-coated plastic films and examined their activities for hydrolysis of chicken egg albumin (CEA). The process generated enzymic coatings with a typical loading of 13 ug/cm2, retaining 46% of the enzyme activity for hydrolysis of CEA in aqueous solutions. Interestingly, the surface-coated protease thin film not only catalyzed the hydrolysis of CEA in aqueous solutions, but also showed good activity for solid-state CEA that was coated on top of the enzyme thin film.

Web Page: www.bbe.umn.edu/faculty/pingwang/

613e Investigating the Biosynthesis of a Tetracycline Analog with Antitumor Properties

Lauren B. Pickens, Chemical and Biomolecular Engineering, University of California Los Angeles, 5667 Boelter Hall, 420 Westwood Plaza, Los Angeles, CA 90066 and Yi Tang, Chemical and Biomolecular Engineering, University of California, Los Angeles, 420 Westwood Plaza, Los Angeles, CA 90095.

Tetracyclines are an important group of natural products that are synthesized by different *Streptomyces* species by way of a type II polyketide biosynthetic pathway. These compounds have played an important role in treating bacterial infections since their discovery in 1948. More recently, several semisynthetic analogs of tetracycline have been shown to have antitumor properties, which has sparked interest in their development as a possible cancer therapy. We have obtained a strain which has been identified as producing a novel tetracycline compound SF2575. This compound has a much more complicated structure than many naturally derived tetracyclines such as chlorotetracycline and oxytetracycline, and has also been shown to have potent antitumor activity. Due to the complexity of the molecule, biosynthetic methods likely provide the most tractable path to producing and manipulating this compound. To this end, we have sequenced the previously unidentified gene cluster responsible for the production of this compound. Through expression of the minimal PKS in a heterologous host, *Streptomyces* CH999, we were able to confirm the production of a decaketide backbone consistant with a tetracycline biosynthesis. We plan to use this new sequence information to systematically investigate the role of individual proteins in the biosynthetic pathway towards the goal of engineering this pathway to produce novel compounds.

613f Elastin-Like Polypeptide-Polyelectrolyte Conjugates for Tissue Engineering

Jonathan Woolfolk, School of Dentistry, University of Mississippi Medical Center, 2500 North State Street, Jackson, MS 39216 and **Amol V. Janorkar**, Biomedical Materials Science, School of Dentistry, University of Mississippi Medical Center, 2500 North State Street, Room D-528, Jackson, MS 39216.

Both tissue engineering and biological science will benefit from a generalized methodology to create multilayered cellular architectures that mimic tissue structures found in vivo. Elastin-like polypeptides (ELPs) are tandemly-repeated polymers derived from a portion of the primary sequence of mammalian elastin. ELPs have previously been used for protein purification, nucleic acid delivery, and thermally-mediated targeting of solid tumors. The ELP was produced in E. coli and purified by the inverse temperature transition purification method. The ELP was conjugated with several polyelectrolytes using carbodiimide activation chemistry. These conjugates were characterized by FTIR spectroscopy, XPS, mass spectroscopy, and the ninhydrin assay. These conjugates were used as scaffold materials for primary rat hepatocytes and H35 rat hepatoma cells. Hepatocyte-specific urea and albumin production were quantified under a variety of culture conditions. These approaches may represent a promising class of biomaterials for a variety of tissue engineering applications.

616a Evidence of Multimolecular Cooperative Charge Regulation In Weak Polyelectrolyte Brushes

Kevin N. Witte[1], Jaehyun Hur[2], Sangtae Kim[2], and You-Yeon Won[1]. (1) School of Chemical Engineering, Purdue University, 480 Stadium Mall Dr, West Lafayette, IN 47907, (2) Chemical Engineering, Purdue University, Forney Hall of Chemical Engineering (FRNY), 480 Stadium Mall Drive, West Lafayette, IN 47907

A novel system for the study of polymer brushes has been developed and utilized to study the thermodynamic stability of weak polyelectrolyte brushes. The brush system was created by at the air-water interface by spreading a unique diblock copolymer composed of poly(n-butyl acrylate) and poly (dimethylaminoethyl methacrylate) (PnBA-b-PDMAEMA). This polymer, synthesized by atom transfer radical polymerization (ATRP), has several characteristics that make it ideal for the study true polymer brushes at the air water interface. Namely, the hydrophobic PnBA anchoring block is non-glassy at ambient conditions, and has a positive spreading coefficient on water. Thus, the diblock copolymer will form a uniform monolayer at the interface that can be readily compressed to vary the available area per chain.

Isotherms for the copolymer were measured and their defining features explored in greater detail. At high area per chain, the behavior is governed by the anchoring block as it was compressed to form a uniform monolayer. The formation of this monolayer is evidenced by a sharp transition in the slope of the isotherm. This effect was further explored and verified by studying PnBA homopolymers of varying molecular length. The spread polymer was compressed to various values of the area per chain and deposited on a graphite substrate via a submersion Langmuir-Blodgett (LB) technique. The formation of the monolayer was then confirmed by fluid atomic force microscopy (AFM). At low area per chain, the behavior is governed by the interactions of the positively charged brush. We here show for the first time experimental evidence for thermodynamic instability in the single component brush system. Using the LB and fluid AFM techniques we show the existence of regions of high and low polymer density consistent with "phase" separation of the single component brush which can be related to the multimolecular cooperative charge regulation inherent in weak polyelectrolyte systems. The system also shows localized "microscopic" separation due to the build up of electrostatic repulsive energy and small ion translational entropy. This effect has been termed "frustrated" phase separation. The electrostatic nature of the behavior was confirmed by addition of 100 mM NaCl to the aqueous subphase. In this case the system remains uniformly dispersed across the interface through out the compression.

616b Ultra Low Fouling Zwitterionic Polymers with Biomimetic Adhesive Groups

Guozhu Li, Department of Chemical Engineering, University of Washington, Box 351750, Seattle, WA 98105 and **Shaoyi Jiang**, Chemical Engineering, University of Washington, Benson Hall, Box 351750, Seattle, WA 98195.

Catechol derivatives inspired from the adhesive proteins found in mussels have been used for the functional modification of material surfaces. This is a very promising method for attaching functional groups to a wide range of surfaces. In recent years, we have developed ultra low fouling polymers based on zwitterionic materials, such as poly(sulfobetaine methacrylate) (polySBMA) and poly (carboxybetaine methacrylate) (polyCBMA). They show strong resistance to non-specific protein adsorption from a single protein solution such as fibrinogen and lysozyme and from complex media such as human serum and plasma. Catechol derivatives are desirable as a versatile grafting way for the efficient, convenient and stable anchoring of ultra low fouling groups onto surfaces.

In this work, we prepared zwitterionic-catechol conjugates. Atom transfer radical polymerization (ATRP) was employed to produce multiple-catechol initiators or catechol-g-zwitterionic polymers with controlled molecular weights. Bindings of these zwitterionic-catechol conjugates to various self-assembly monolayer (SAM) surfaces terminated with CH3, OH, COOH, NH2 have been studied to reveal binding mechanisms. Results are compared with those from bare gold surfaces. Protein adsorption from a single protein solution or blood serum/plasma onto these surfaces grafted with zwitterionic groups was tested by a surface plasmon resonance (SPR) sensor. Results show excellent nonfouling properties of these modified surfaces. Effects of multi-catechol anchor, buffer, ionic strength and temperature on the adhesion were also investigated. This work demonstrates a new, convenient and efficient way for surface modification to achieve ultra low fouling.

616c Nanostructured Polystyrene Films: Graft Polymerization and Organic Sorption Behavior

Gregory T. Lewis, Chemical and Biomolecular Engineering, University of California, Los Angeles, 5531 Boelter Hall, Los Angeles, CA 90095 and Yoram Cohen, Chemical & Biomolecular Engineering, University of California, Los Angeles, 5531 Boelter Hall, 420 Westwood Plaza, Los Angeles, CA 90095-1592.

Recent advances in polymer surface engineering have demonstrated the ability to direct and control nano-scale polymer brush growth from substrate surfaces to create unique chemical functionalities and surface topography. Synthetic routes towards polymer surface structuring are necessary for the development of advanced materials for membrane separations, chemical sensors, and biotechnology. Polymer surface modification has, in the past, relied on traditional adsorption and spin-coating techniques. Such polymer layers, however, can desorb (or delaminate) upon chemical, thermal and shear stresses. In contrast, single molecule polymer chains can be grown directly from the surface to create robust, grafted (i.e., covalently tethered to the surface) chains by graft polymerization. Two main challenges exist, however: 1) creation of a high number density of surface initiation sites, and 2) controlled growth of polymer chains from the surface initiation sites. The classical approach, free-radical graft polymerization, relies on surface immobilized macroinitiators and results in polydisperse polymer growth. In the current work, atmospheric pressure plasma surface activation has been employed to create surface reactive groups, on both organic and inorganic substrates, which effectively function as surface initiators. Polymer growth from plasma-initiated surface sites may still be subject to early termination or chain transfer, leading to a high polydispersity of polymer chain lengths. Therefore, the integration of plasma surface activation/initiation with controlled "living" radical graft polymerization has been investigated, in the present study, to create a high surface coverage of uniform length polymer chains tethered to a inorganic substrate surface. Controlled “living” radical graft polymerization of styrene was achieved by establishing a reversible thermodynamic equilibrium between radical polymer chains and a chemical control agent (CCA). The CCA is a free-radical scavenger that reversibly binds to the growing polymer chains and thus retards the chain termination, thereby enabling control of chain growth. While there are various approaches of controlled "living" polymerization, the current approach has the advantage of not requiring the presence of surface macroinitiators or catalysts and while enabling the synthesis of a polymer brush layer of a relatively low polydispersity. A series of graft polymerization studies were carried out to elucidate the layer growth kinetics and polymer film properties for plasma-induced graft polymerization under both classical and nitroxide-mediated mechanisms. The potential application of the present nanostructured polymer layers toward the synthesis of thin film chemical sensing layers was evaluated via a series of organic sorption studies. The results of this latter part of the study indicated that a higher film diffusivities and organic sorption capacities were feasible with the grafted polymer layers, relative to physically adsorbed or spin-coated sensing layers.

616d Solution Properties, Adsorption and Interfacial Layer Structure of Self-Organized Polystyrene-Polyisoprene Miktoarm Block Copolymers

Juan Pablo Hinestrosa[1], Masashi Osa[2], Kunlun Hong[3], Jimmy W. Mays[4], and S. Michael Kilbey II[4]. (1) Chemical and Biomolecular Engineering, Clemson University, Clemson, SC 29634, (2) Department of Polymer Chemistry, Kyoto University, Kyoto, Japan, (3) Center for Nanophase Materials Science, Oak Ridge National Laboratory, Oak Ridge, TN 37381, (4) Center for Nanophase Materials Sciences and Chemical Science Division, Oak Ridge National Laboratory, Oak Ridge, TN 37831

Solution properties, adsorption and interfacial layer structure of self-organized polystyrene-polyisoprene miktoarm block copolymers

Juan Pablo Hinestrosa, Masashi Osa, Kunlun Hong, Jimmy Mays and S. Micheal Kilbey, II

We have studied the self-assembly at the solid/liquid interface of a set of novel polystyrene (PS) – polyisoprene (PI) block copolymers having a branched architecture in solution and at the solid/liquid interface. Miktoarm (mixed arm) copolymers having a linear PS block connected to a 'star-like' PI-$(PI)_2$ branched block were synthesized by rigorous anionic polymerization (Figure 1). Their solution properties have been studied by static and dynamic light scattering in n-hexane, which is selective for PI. The results suggest that in solution, these novel block copolymers aggregate into spherical 'hairy' micellar structures with the core being composed of the collapsed linear PS blocks and the corona formed by the well-solvated PI blocks. A PS-PI diblock linear copolymer of similar molecular weight and composition showed larger values of apparent hydrodynamic radius, $R_{h,\ app}$, which is consistent with the idea that branched polymers have more compact structure compared to their linear analogs. Atomic Force Microscopy was used to study the morphology of the adsorbed structures. The images show round features that might be attributed to hemispherical micellar aggregates on the surface caused by a micellar relaxation/reorganization process upon adsorption. This conclusion agrees with results from phase modulated ellipsometry studies of adsorption kinetics. By using a model developed for the adsorption of lipid bilayers and vesicle fusion, the self-assembly process of the miktoarm stars onto silicon can be fully characterized proving insights in how these novel block copolymers reorganize at the solid/fluid interface. Technological areas such as deposition of inorganic nanoparticles and control of protein adsorption will be benefited from these fundamental studies.

SHAPE * MERGEFORMAT

PS

PI

PI

PI

Stem

Branches

Anchor

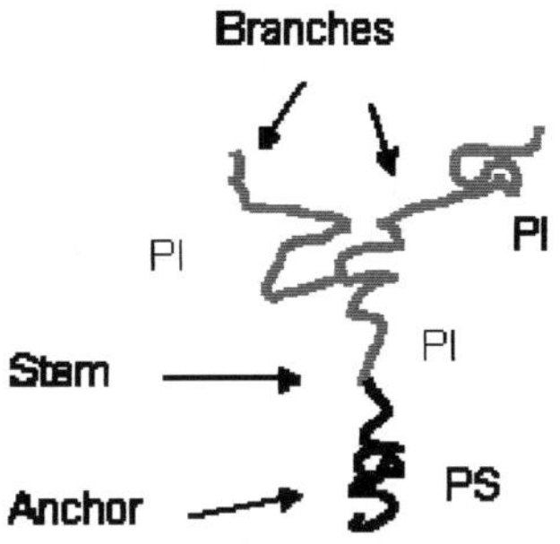

Figure 1. Illustration of a PS-PI-$(PI)_2$ Miktoarm Copolymer Single Chain

616e Generating Surface Energy Gradients for Block Copolymer Thin Film Studies

Julie N. Lawson and Thomas H. Epps, III. Chemical Engineering, University of Delaware, 150 Academy Street, 235 Colburn Lab, Newark, DE 19716

The development of block copolymer materials for future nanotechnologies requires an understanding of how surface energetics affect block copolymer thin film phase behavior. Combinatorial studies of these effects can be accomplished with surface energy gradients. These gradients can be used to identify transitions in thin film phase behavior and microstructure orientation. In this work, gradients are created by vapor deposition of functionalized chlorosilanes on UVO-cleaned silicon substrates under dynamic vacuum. The diffusion profiles of the chlorosilanes are manipulated by the location of the chlorosilane reservoirs in the deposition chamber. X-ray photoelectron spectroscopy (XPS) and diiodomethane contact angle are used to characterize the resulting chemical and surface energy gradients. Then, thin films of cylinder-forming poly(styrene-b-methyl methacrylate) (PS-b-PMMA) are flow coated on the gradient substrates, and the resulting phase behavior is characterized by atomic force microscopy (AFM).

Circularly Polarized Lasers from Solid Films Comprising Chiral Conjugated Oligomers Doped with Functionalized Oligofluorenes

Simon K.-H. Wei[1], Ksenia Dolgaleva[2], Svetlana Lukishova[2], Shaw H. Chen[1] and Robert W. Boyd[2],
(1) Chemical Engineering, University of Rochester, 240 East River Road, Rochester, NY 14623,
(2) Optics, University of Rochester, Rochester, NY 14627,

Since the first successful demonstration of circularly polarized lasers using cholesteric liquid crystal films three decades ago, intensive efforts have been made to overcome hurdles to potential applications as medical diagnostics, quantum cryptography, optical interconnects, optical switches and modulators. The orientational order parameter governing the laser dye's emission dipole is critical to spectral purity. The most commonly used laser dye, **DCM** (4-(dicyanomethylene)-2-methyl-6-(4-dimethyl-aminostryl)-4H-pyran), is capable of a very modest orientational order parameter less than 0.40. In addition to ensuring spectral purity, a high orientation order parameter is expected to increase lasing efficiency and to decrease threshold.

A series of fluorene-based conjugated oligomers have been developed in our laboratory for full-color and white organic light-emitting diodes. These materials are promising for lasing in cholesteric liquid crystal hosts because of their high photoluminescence quantum yields, high orientional order as a dopant or in neat film, and relatively broad emission across the visible spectrum. Here we report the comparison results of **DCM** and one of the oligofluorenes, **OF-r**, or 4,7-bis[5-(9,9-bis(2-ethylhexyl)-9',9',9'',9'', 9''',9'''-hexakis(2-methylbutyl)-7,2';7',2'';7'',2'''-tetrafluoren-2-yl)-thien-2-yl]-2,1,3-benzothiadiazole in fluid choleteric film (the mixture of nematic liquid crystal, ZLI-2244-000 from Merck, with the cholesteric liquid crystal, CB15 from Merck). The molecular structures of light emitters, **OF-r** and **DCM** are depicted in Figure 1. In order to evaluate the emission order parameters of **DCM** and **OF-r** dyes, we measured the emission intensities polarized parallel and perpendicular to the director in the nematic liquid crystals doped with the dyes and then found the emission order parameters to be 0.60 for **OF-r** and 0.36 for **DCM**. As shown in Figure 2, **OF-r**-doped CLCs displayed a much higher temporal and spatial stability in the output radiation, due to the higher value of **OF-r**'s order parameter that prevents strong competition between the high- and low-energy edge frequency modes. In **DCM**-doped CLCs, strong competition between the modes could take place in case of a dye with a low order parameter, which can create temporal and spatial instabilities.

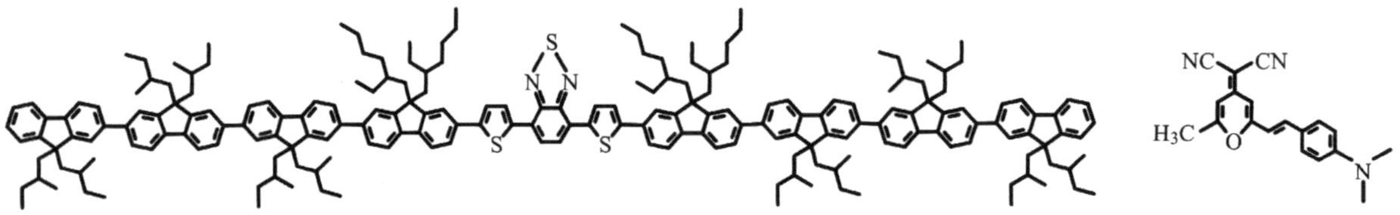

OF-r: *G* 104 °C *N* 304 °C *I* **DCM:** *K* 227 °C *I*

Figure 1: Molecular structures of light emitters, **OF-r** and **DCM**

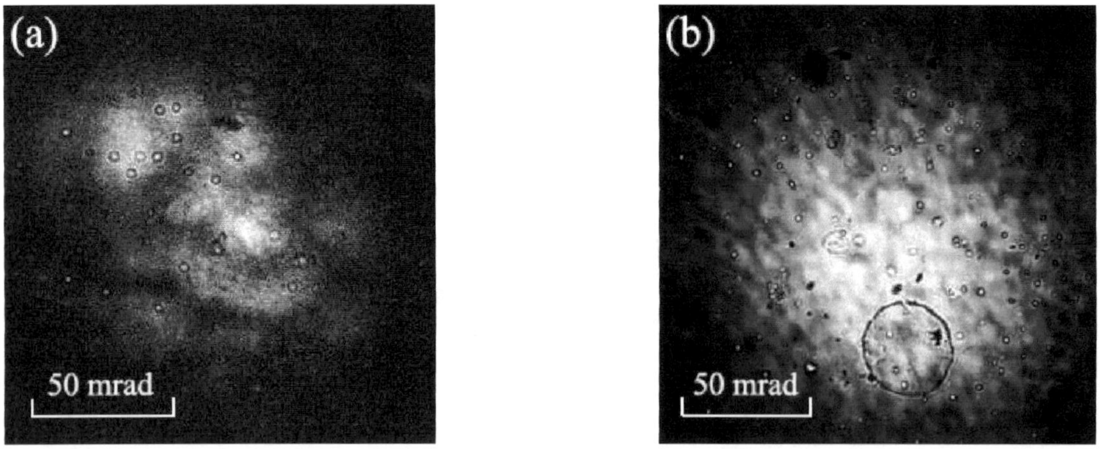

Figure 2: Intensity distribution of the laser output measured using a CCD camera in (a) 1.00 wt. % **DCM**-doped CLC and (b) 2.00 wt. % **OF-r**-doped CLC. The dark ring at the bottom of the picture is a camera artifact.

We also measured the laser output characteristics of **DCM**- and **OF-r**-doped CLCs in two different regimes corresponding to generating a transverse single fundamental spatial mode and a multi-mode ring pattern. As shown in Figure 3, in transverse single-mode regime, the same threshold and slope efficiency were obtained for the two light emitters. However, **OF-r** exhibited superior spatial and temporal stability and a sustained increase in laser output at increasing pump energy, generating output energy five times that of **DCM**. In transverse multi-mode regime, **OF-r** was more than twice as efficient as **DCM**.

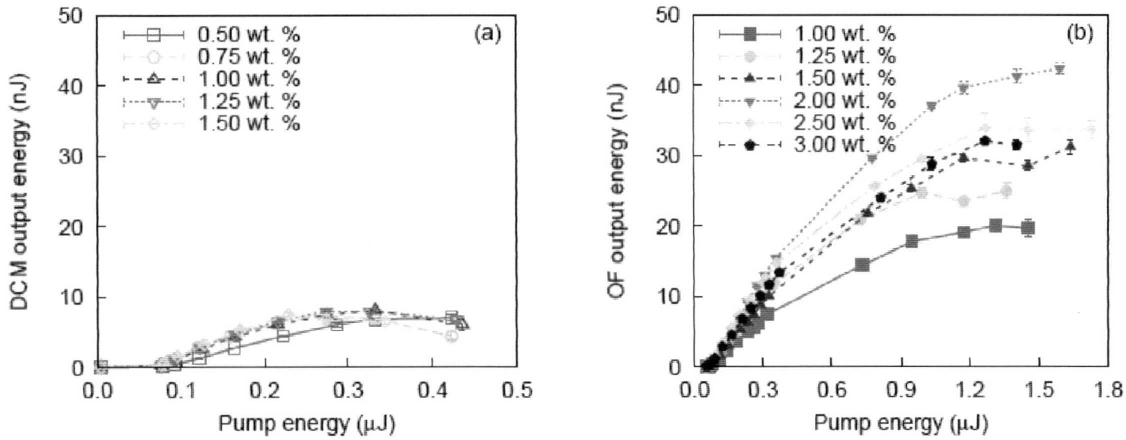

Figure 3: Laser output energy plotted as a function of the incident pump energy of (a) **DCM** and (b) **OF-r**-doped CLC samples in transverse single-mode regime.

Extrinsic factors that may disturb the host film's helical stack or laser dye's orientational order would cause adverse effects on lasing. Optical pumping and lasing could result in heating, causing the stop-band to undergo a red or blue shift depending on the type of cholesteric liquid crystal used as the host. A rise in temperature would also reduce dye molecules' orientational order parameter. Optical torque on the cholesteric liquid crystal structure could lead to a deterioration of laser performance. Yet another potential problem is laser induced fluid flow disturbing the stop-band. In short, cholesteric liquid crystal fluids as the hosts are vulnerable to any or all of these adverse effects, which can be avoided by using solid cholesteric liquid crystals in principle. Prior attempts to use solid cholesteric liquid crystal films, however, suffered from spectral impurity of lasing apparently because of the difficulty preparing highly ordered, disclination-free films.

Our research was motivated to identify potential laser dyes characterized by a high fluorescence quantum yield and a high orientational order parameter governing emission dipoles, and to impart device stability and robustness by using solid cholesteric liquid crystals as the hosts. To ensure device stability and robustness, our promising laser dyes are doped in low-molar-mass cholesteric glassy liquid crystals (GLCs) comprising chiral conjugated oligomers also developed in our laboratory. Here we report the comparison results of **OF-r** doped in fluid CLCs (ZLI-2244-000 and CB15) and glassy CLCs, which is the mixtures of **F(MB)5-Ch,** penta[9,9-bis(2*S*-methylbutyl)fluorene] and **F(MB)5-N**, penta[9,9-bis(2-methylbutyl)-fluorene], as depicted in Figure 4.

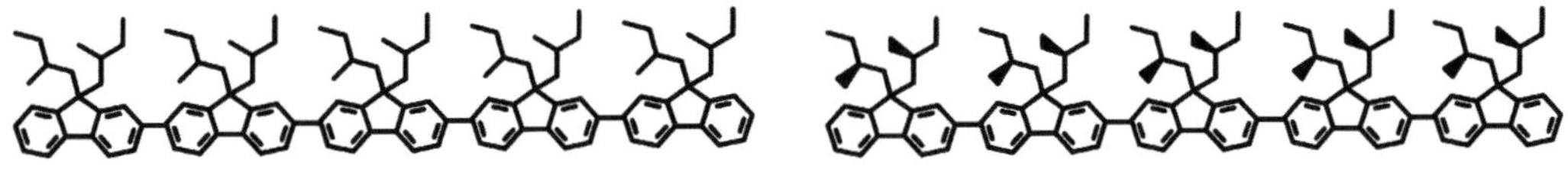

F(MB)5-N: *G* 92 °C *N* 171°C *I* **F(MB)5-Ch:** *G* 91 °C *Ch* 173 °C *I*

Figure 4: Molecular structures of **F(MB)5-Ch** and **F(MB)5-N** used for preparation of glassy CLC host films.

Representative reflection spectrum, fluorescence spectrum, and the lasing peak of **OF-r**-doped glassy CLC and its lasing output energy as a function of pump energy are plotted as Figure 5. At an increasing concentration of **OF-r**, the improvement in laser gain appears to have been overcome by concentration quenching. As shown in Figure 6, the maximum slope efficiency is expected at an optimum concentration between 2.0 and 2.5 wt% of **OF-r** in both glassy and fluid CLC film.

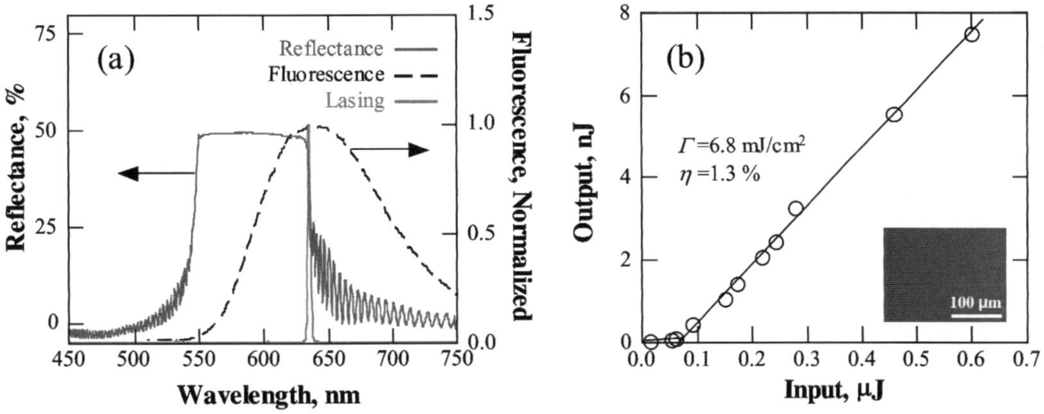

Figure 5: Glassy CLC film consisting of 1.5 wt% **OF-r** in **F(MB)5-Ch:F(MB)5-N** at a 24.0:76.0 mass ratio: (a) Reflection spectrum (solid curve), **OF-r** fluorescence spectrum from a nematic **F(MB)5-N** film (dashed curve), and the lasing peak at 635 nm with a pump fluence of 121 mJ/cm^2 at 10 Hz, and (b) Lasing output energy as a function of pump energy.

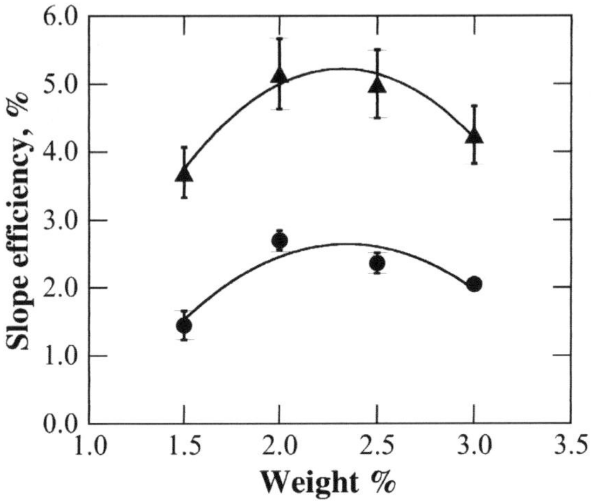

Figure 6: Slope efficiency as a function of wt% **OF-r** in glassy (●) and fluid (▲) CLC lasers. For glassy CLC films with **OF-r** at 1.5, 2.0, 2.5, and 3.0 wt%, the **F(MB)5-Ch:F(MB)5-N** mass ratios are 24.0:76.0, 24.1:75.9, 24.2:75.8, and 24.3:75.7, respectively. For fluid CLC films with **OF-r** at 1.5, 2.0, 2.5, and 3.0 wt%, the **CB-15:ZLI2244-000** mass ratios are 35.5:64.5, 35.6:64.4, 35.7:64.3, and 35.9:64.1, respectively. All the efficiency data were gathered from the initial slopes of output-input relationships.

The expectation that solid CLC films would produce temporally stable lasing output was realized by the data compiled in Figure 7a. A fluid CLC film containing 2.0 wt% **OF-r** in **CB-15:ZLI2244-000** at a mass ratio of 35.6:64.4 (with η=5.2%) was also subjected to stability test. As shown in Figure 7b, the lasing output decays with time, and the decay rate increases with pump fluence most likely due to heating, optical torque, and laser-induced flow that tend to disrupt the CLC structure and the orientational order of **OF-r** molecules.

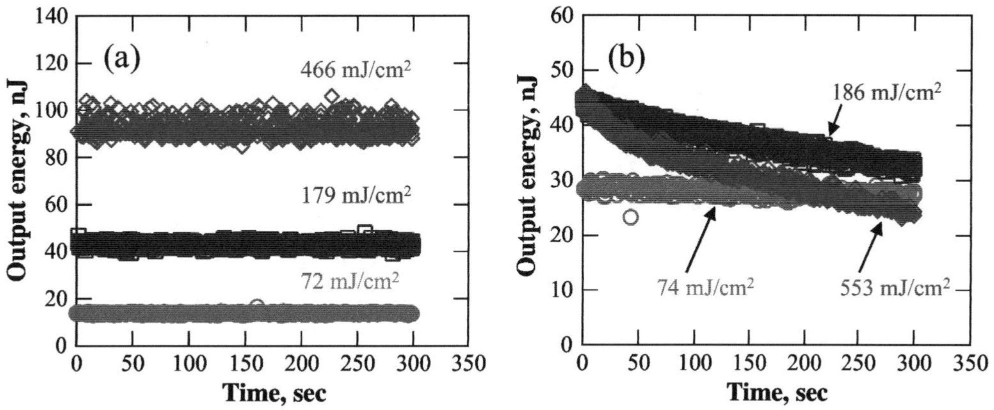

Figure 7: Time dependence of lasing output from (a) a glassy CLC film and (b) a fluid CLC film, both containing **OF-r** at 2.0 wt%, at specified pump fluences and 10 Hz

The fluid CLC film of the same composition as used in Figure 7b was further pumped with a fluence of 550 mJ/cm^2 at 10 Hz. The output energy from the first series of lasing experiments diminished with time, as shown in Figure 8, which is consistent with the observations presented in Figure 7b at a pump fluence of 186 or 553 mJ/cm^2. The second series of experiments was conducted after the CLC film had been left at room temperature for 0.5 h. As presented in Figure 8, the lasing output decreased by 35 to 40 % because fluid CLC film did not have sufficient time to recover from external perturbations incurred during the first series of experiments. The film was then left at room temperature for 20 h before the third series of experiments was performed. The pristine film's output energy as a function of time was restored, indicating a full recovery of order in the fluid CLC film and the absence of photodegradation of materials. The chemical integrity of all material components used in this study is ascertained by the temporal stability of laser output shown in Figures 7a and the full recovery of output energy demonstrated in Figure 8 up to a pump fluence of 466 mJ/cm^2 for the glassy CLC film and 563 mJ/cm^2 for the fluid CLC film, respectively.

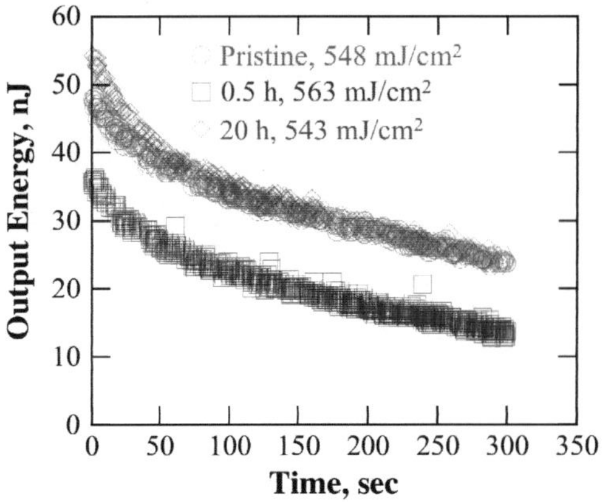

Figure 8: Lasing output from a fluid CLC film containing **OF-r** at 2.0 wt% pumped with 550±12 mJ/cm^2 at 10 Hz rested for 0.5 and 20 h time intervals between three runs.

656a Highly Controllable 3D Scaffolds with Inverted Colloidal Crystal Geometry as Hematapoietic Stem Cell Niches

Meghan J. Cuddihy[1], Jungwoo Lee[2], and Nicholas Kotov[1]. (1) Chemical Engineering, University of Michigan, 3074 HH Dow Building, 2300 Hayward St, Ann Arbor, MI 48104, (2) Biomedical Engineering, University of Michigan, 3074 HH Dow Building, 2300 Hayward St, Ann Arbor, MI 48104

Increasing numbers of studies have demonstrated that the 3D organization of ex vivo bone marrow analogs is essential for hematopoietic stem cell (HSC) expansion, development, and function. These 3D recapitulations of HSC niches could prove useful in the expansion and production of HSCs and their progenitors for therapeutic functions, as well as for drug toxicity testing. The goal of our research is to develop 3D HSC niche analogs for the expansion and differentiation of HSCs for drug toxicity testing. For this purpose, our 3D hydrogel matrix possesses inverted colloidal crystal (ICC) geometry, which consists of highly interconnected empty spherical cavities arranged in an ordered hexagonal crystal lattice. To improve cell adhesion and to deliver cell signals, the surface of hydrogel ICC scaffold is coated with clay nanoplatelets utilizing the layer-by-layer (LBL) assembly technique.

656b Release Characteristics of Rhbmp-2 Conjuagted to Self-Assembled Biodegradable Nanoparticles

Angel Mercado and Esmaiel Jabbari. Chemical Engineering, University of South Carolina, Dept. of Chemical Eng., Rm. 2C11, Swearingen Engineering Center, Columbia, SC 29208

Introduction and Objective: The use of injectable scaffolds along with minimally invasive endoscopic techniques, now in clinical use, requires rhBMP-2 stabilization against denaturation during injection and in situ hardening. Our laboratory has developed novel bioresorbable succinimide-terminated poly (lactide-co-glycolide fumarate) macromers (PLGF-NHS) that self-assemble into NPs. The objective of this work was to determine the release characteristics of rhBMP-2 conjugated to PLGF-NHS NPs.

Methods: Methods: Poly(lactide fumarate) PLAF, PLGF, and poly(lactide-co-ethylene oxide fumarate) (PLEOF) were synthesized by condensation polymerization. Succinimide-terminated macromers (PLAF-NHS or PLGF-NHS) were produced by reacting the hydroxyl end-groups of the PLAF or PLGF macromers with the carbonate group of N,N'-Disuccinimidyl carbonate (DSC). PLAF-NHS and PLGF-NHS NPs were produced by dialysis of the macromers in dimethylsulfoxide (DMSO)/ N, N-dimethylformamide (DMF) against water. The morphology and size distribution of the NPs was examined with ESEM. The size distribution of NPs was measured by dynamic light scattering. Degradation of the NPs was followed by measuring their mass as a function of incubation time. For conjugation, rhBMP-2 (1 μg/ml in PBS) was allowed to react with the succinimide terminated NPs under ambient conditions for 12 h. For determination of release kinetics, 1 mg rhBMP-2 loaded NPs were incubated with 1 ml PBS (pH 7.4) at 37°C with orbital shaking. At each time interval, the suspension was centrifuged at 15,000 rpm, and the supernatant was removed for analysis. The enzymatically active concentration of the BMP released from the microspheres was measured by enzyme-linked immunosorbent assay (ELISA) using the BMP Quantikine kit(R&D Systems).

Results: The average size of PLAF-NHS and PLGF-MHS NPs was 325 and 250 nm, respectively. The conjugation efficiency was determined by measuring the amount of free rhBMP-2 after the conjugation reaction. Conjugation efficiency was >95% for PLAF-NHS and PLGF-NHS NPs. The release of rhBMP-2 from PLAF-NHS NPs was linear with time and >0.4 μg was released after 2 weeks; the release from PLGF-NHS NPs was non-linear and >0.2 μg was released after 2 weeks. PLGF-NHS and PLAF-NHS NPs completely degraded in 25 and 38 days which demonstrated that the release was dominated by erosion of the matrix.

Conclusion: Since proteins can be immobilized on PLGF-NHS NPs by the reaction of succinimide groups with amine groups of the protein at physiological condition, conjugated NPs are potentially useful for immobilization and sustained release of differentiation factors like rhBMP-2 in tissue engineering applications.

656c Mechanical Characterization of Electrospun Laminated Hydrogel/ apatite Nanocomposite

Weijie Xu and **Esmaiel Jabbari**. Chemical Engineering, University of South Carolina, Dept. of Chemical Eng., Rm. 2C11, Swearingen Engineering Center, Columbia, SC 29208

Introduction and Objective: Bone exhibits many orders of structures from macroscopic to microscopic, sub-microscopic, nanoscale, and sub-nanostructure length scales. These levels of organization include the osteons and single trabeculae microstructures (10-500 ìm), the lamellae sub-microstructures (1-10 μm), the fibrillar collagen and embedded mineral nanostructures (100-1000 nm), and the molecular structure of constituent elements like mineral, collagenous, and non-collageneous soluble proteins. The mechanical properties of the bone are determined at the nanoscale with calcium phosphate crystals embedded in collagen fibrils. The objective of this work was to design and measure the mechanical properties of bone-mimetic laminated composites. The composites consists of eletrospun poly(lactide-co-glycolide) (PLGA) nanofibers embedded in a crosslinked hydrogel-apatite nanocomposite.

Methods: PLGA nanofibers were produced by electrospinning. A programmable syringe pump was used to inject the polymer solution (12% PLGA in 1,1,1,3,3,3 hexafluoro-2-propanol) in an electric field (25 kV) at a flow rate of 7.0 ml/h. An aluminum plate connected to the ground electrode was used as the collector. The morphology of bone marrow stromal (BMS) cells on the fibers, stained with Alexa 488 phalloidin and DAPI, was imaged with a confocal scanning laser microscope. The fiber morphology was measured with SEM. The fiber mesh were dipped in a hydrogel precursor mixture containing poly (lactide-co-ethylene oxide fumarate) (PLEOF) macromer, N-Vinyl-2-pyrrolidinone (NVP) crosslinker, water-soluble photoinitiator, and nanoapatite particles (50 nm diameter) in PBS. The dipped layer were stacked and pressed against each other, and the multi-layer structure was crosslinked by UV polymerization to form a degradable bone-mimetic composite network. The tensile strength of the laminated structure was measured by a Dynamic Mechanical Analyzer (DMA; TA Instruments).

Results and Conclusion: The elastic modulus of crosslinked PLEOF hydrogel was 0.47±0.07 MPa. The elastic modulus increased by 65-fold to 31±6 MPa with the addition of 4 layers of electrospum PLGA fibers. Furthermore, the elastic modulus increased to 140 MPa with the addition of 4 PLGA electrospun layers and 10% by weight apatite nanocrystals. As the number of electrospun layers was increased from 3 to 4 and 5, the elastic modulus of the composite structure increased from 145±50 to 195±47 and 560 ±20 MPa, respectively. confocal microscopy results demonstrated that BMS cells elongated along the direction of PLGA fibers. These results demonstrate that laminated apatite nanocomposites are potentially useful as a synthetic bone-mimetic matrix in tissue engineering.

656d Exploring the Size and Composition of Polysaccharide-Based Polyelectrolyte Complex Nanoparticles at Different Charge Molar Ratios

Soheil Boddohi, Chemical and Biological Engineering, Colorado State University, 100 Engineering South/Glover, Fort Collins, CO 80523-1370 and Matt J. Kipper, Chemical and Biological Engineering and School of Biomedical Engineering, Colorado State University, 100 Engineering South/Glover, Fort Collins, CO 80523-1370.

Polyelectrolyte complex nanoparticles (PCN) represent a rapid method for introducing nanoscale features into biomaterials. In this study, PCN are formed based on the interaction of oppositely charged polyelectrolytes. Chitosan and quaternary chitosan (weak and strong polycations) and hyaluronic acid and heparin (weak and strong polyanions) are used to make PCN. Formation of particles was developed by the one-shot addition of oppositely charged polyelectrolytes in 0.1 M, pH 5.0 buffer. PCN size and composition were investigated based on different charge molar ratios (ranging from 15 to 0.06) from different polyanion/polycation pairs. PCN were characterized by scanning electron microscopy (SEM) to look at the morphology and particle size. Sizes of these particles are in the 100 nm range. Dynamic light scattering (DLS) was used to characterize the size distribution of each PCN formulation. PCN composition was assayed by Fourier transform infrared spectroscopy (FT-IR).

656e Nano Bioactive Glass. A High Potential Material for Root Canal Infection Treatment

Dirk Mohn[1], Miguel Gubler[1], Tobias J. Brunner[1], Matthias Zehnder[2], Thomas Imfeld[3], Tuomas Waltimo[4], and Wendelin J. Stark[5]. (1) Department of Chemistry and Applied Biosciences, ETH Zurich, Wolfgang-Pauli-Strasse 10, Zurich, 8093, Switzerland, (2) Department of Preventive Dentistry, Periodontology, and Cariology, University of Zürich Center of Dental Medicine, Zurich, 8093, Switzerland, (3) Departement of Preventive Dentistry, Periodontology, and Cariology, University of Zürich Center of Dental Medicine, Zurich, 8093, Switzerland, (4) Institute of Oral Microbiology and Preventive Dentistry, University of Basel, Hebelstrasse 3, Basel, 4056, Switzerland, (5) Chemistry and Applied Biosciences, ETH Zurich, Wolfgang-Pauli-Str. 10, ETH Hönggerberg, Zurich, 8093, Switzerland

Bioactive glasses are potentially interesting materials in dentistry not only because of their ability to mineralize dentine [1] but also due to their antimicrobial effect in closed systems [2]. We investigated nanometric bioactive glasses because of their higher surface area and their faster mode of action. Furthermore, we tested the hypothesis whether nano-sized bioactive glasses also kill microbiota via mineralization or the release of ions other than sodium. Flame-spray synthesis was applied to produce nanometric glasses (Figure) with different sodium content. Calcium hydroxide was used as a control. Standardized bovine dentine disks with adherent *E. faecalis.* cells were exposed to test and control suspensions.

Sodium containing glasses induced pH levels above 12, compared to less than pH 9 with sodium-free glasses. Calcium hydroxide and the sodium containing glasses killed all bacteria after 1 d. The study revealed that bioactive glasses have both, a not and a directly pH-related antibacterial effect. The former is due to ion release rather than mineralization [4].

Our study shows that nanoparticulate biomaterial can have a strong combined antimicrobial action. This opens wide applications in clinical settings.

Figure: Transmission electron microscopy image of bioactive glass nanoparticles (left). Scannning electron microscopy image of a dentine surface after treatment for 1 week with ultrafine bioactive glass showing no viable bacteria (right).

References

[1] M. Vollenweider, T.J. Brunner, S. Knecht, R.N. Grass, M. Zehnder, T. Imfeld, W.J. Stark, Acta Biomater., 2007, 3 (6), 936-43.

[2] P. Stoor, E. Soderling and J.I. Salonen, Acta Odontol. Scand., 1998, 56(3), 161-5.

[3] T.J. Brunner, R.N. Grass and W.J. Stark, Chem. Commun., 2006, 13, 1384-86.

[4] M. Gubler, T.J. Brunner, M. Zehnder, T. Waltimo, B. Sener and W.J. Stark, Int. Endod. J., 2008, in press.

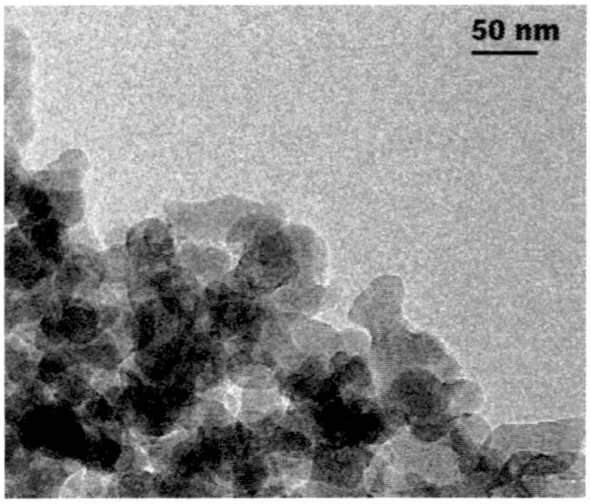
50 nm

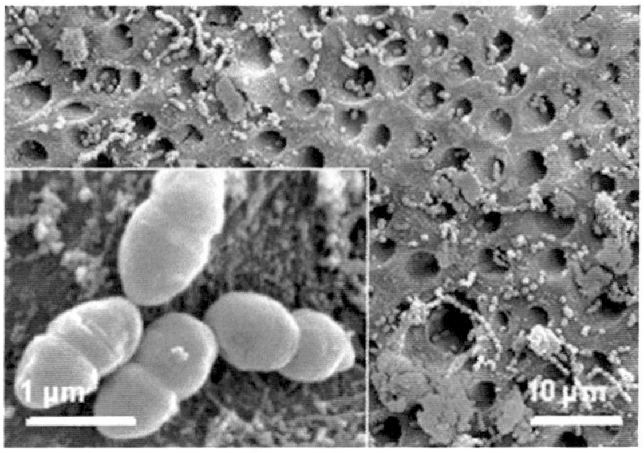
1 µm
10 µm

656f Nanoporous Polyelectrolyte Multilayers: Biomimetic Surfaces for Corneal Epithelial Cells

Christina Hajicharalambous[1], Xiaoxia Sheng[2], William Hix[3], Magdalena Swierczewska[1], Michael F. Rubner[2], and **Padma Rajagopalan**[3]. (1) Bioengineering Program, Lehigh University, Bethlehem, PA 18015, (2) Department of Materials Science and Engineering, Massachusetts Institute of Technology, Cambridge, MA 02139, (3) Chemical Engineering, Virginia Polytechnic Institute and State University, Blacksburg, VA 24061

In vivo, corneal epithelial cells are adhered on basement membranes that exhibit topography on the nanoscale with the diameters of pores and fibers ranging from 20 200nm. Polyelectrolyte multilayers with porosity ranging from the nano to the microscale were assembled to recapitulate the porosity and topography in vivo. The average pore diameter was found to be 100nm and 600nm for the nanoporous and sub-micron porous films respectively. In this study, a purely physical feature, specifically, porosity, provided cues to human corneal epithelial cells. Although both nano- and sub-micron porous surfaces were found to be biocompatible and supported cell adhesion, the nanoscale environment enhanced corneal epithelial cellular response. Corneal epithelial cell proliferation and migration speeds were significantly higher on nanoporous topographies. The actin cytoskeletal organization was well-defined and vinculin focal adhesions were found in cells presented with a nanoscale environment. These trends prevailed for fibronectin-coated surfaces as well suggesting that for corneal epithelial cells, the physical environment plays a defining role in guiding cell behavior. Atomic force microscopy measurements of the substrata did not indicate changes in porosity as a result of cell culture.

656g Surfactant-Assisted Electrospun Metal Nanoparticle-Polysaccharide Based Polymer Blend Nanofibers for Biomedical Applications

Carl D. Saquing[1], Joshua Manasco[1], Jovita M. Saquing[2], Christopher Bonino[1], Francis delos Reyes III[2], and Saad A. Khan[1]. (1) Department of Chemical and Biomolecular Engineering, North Carolina State University, 911 Partners Way, Raleigh, NC 27606, (2) Department of Civil, Construction and Environmental Engineering, North Carolina State University, 2501 Stinson Drive, Raleigh, NC 27695

We examine a one-step process to fabricate metal nanoparticle (NP)-alginate polymer nanofiber (NF) composites via electrospinning for biomedical applications. Alginate, a natural biopolymer extracted from seaweeds used as an additive in pharmaceutical and food formulations, is widely known to promote optimal environment for wound healing. However, we have found that alginate in aqueous solution cannot be electrospun by itself in all concentrations up to saturation. We report the successful generation of electrospun NFs based on alginates with the addition of minute amount of surfactants. The effect of surfactant addition on solution properties as it relates to electrospinning is determined using rheological and microscopy techniques. Furthermore, the novel use of the electrospinning polymer blend acting as both the reducing and protecting agent for the NPs is also investigated. A model system that involves alginate and poly (ethylene oxide) (PEO) blends and a metal salt to generate NF composites is presented. The nanofiber composites have potential as wound dressing with antimicrobial properties. We demonstrate by UV/Vis spectrophotometry that PEO and alginate transform Ag ions to Ag NPs without the addition of any reducing agent and stabilizer. Interesting nanostructures that are well-dispersed and crystalline at or near the NF surface are observed from TEM, XRD and XPS analyses. The antimicrobial property of the NF mats against various model microbes is evaluated.

662a Multifunctional Biomedical Coatings Based on Chemical Vapor Deposition Copolymerization

Yaseen Elkasabi[1], Mutsumi Yoshida, Himabindu Nandivada[2], and Joerg Lahann[2]. (1) Department of Chemical Engineering, University of Michigan, 2300 Hayward, 3434 G.G. Brown, Ann Arbor, MI 48109, (2) Chemical Engineering Department, University of Michigan, 2300 Hayward St, 3434 G.G Brown, Ann Arbor, MI 48109

As artificial materials become more integrated with biological environments, the demand for finely-tuned surface modifications will increase. More specifically, the surface immobilization of multiple functionalities in defined ratios would be of high value for applications such as biological assays, micro-total analytical systems (μ TAS), and biomedical devices. The ability to mix and match surface functionalities which possess contrasting properties would allow one to screen for trends in cellular interactions. This talk will focus on chemical vapor deposition (CVD) copolymerization as a novel method of depositing copolymer coatings with multiple reactivities. Complex substrate geometries do not hinder the CVD process, making it a flexible method for room-temperature surface modification. We utilize a modified Ghoram process to deposit functionalized parylene (poly(*p*-xylylene)) copolymers from their analogous [2.2]paracyclophane derivatives. Two differently-functionalized [2.2] paracyclophanes are copolymerized and deposited. These functionalities are presented at the polymer surface in specific ratios, simply by modifying the CVD process conditions. FTIR spectroscopy and XPS of various poly(*p*-xylylene) copolymers verify that the functional groups are maintained while avoiding *in situ* cross-reaction. Furthermore, these copolymers are reactive and can immobilize two biomolecules in specific ratios. We also demonstrate preliminary biocompatibility of poly(*p*-xylylene) homo- and copolymers. Immobilization of two anti-thrombotic biomolecules onto the same substrate was achieved. Also, we use CVD polymerization to fabricate coatings which possess reactive surface composition gradients. These surfaces are capable of further immobilizing two biomolecules as gradients.

662b Applications of Thin Films of pH-Responsive Hydrogels Synthesized by Initiated Chemical Vapor Deposition

Wyatt E. Tenhaeff and Karen K. Gleason. Department of Chemical Engineering, Massachusetts Institute of Technology, 77 Massachusetts Avenue, Cambridge, MA 02139

Thin functional films of ionic hydrogels have been grafted to silicon and glass substrates by initiated chemical vapor deposition (iCVD). Prior to iCVD, the substrates were primed with 3-aminopropyldimethylethoxysilane, imparting the surface with a free amine group. Three monomers vapors were introduced into the reactor along with the initiator: 0.6 sccm of maleic anhydride (MA), 6 sccm of N,N-dimethylacrylamide (DMAA), and 0.4 sccm of di(ethylene glycol) divinyl ether (DEVE). At a filament temperature of 235 °C and an initiator flow of 0.4 sccm, the polymer film deposited at a rate of 16 nm/min. The DEVE crosslinker created a network by linking all of the polymer chains together, while maleic anhydride groups at the interface between the substrate and hydrogel film formed a covalent bond with the amine to tether the network to the substrate. Covalent linkages prevented the film from delaminating upon exposure to aqueous solutions; the films undergo 11-fold swelling in pH 7 buffer. The carboxylic acids that formed upon hydrolysis of maleic anhydride imparted anionic character to the film. In pH 1 buffer, the swelling ratio was 7. The composition of the film was 76% DMAA, 14% MA and 10% DEVE, as determined by X-ray photoelectron spectroscopy (XPS).

The anhydride groups were reacted with 2-aminoethanethiol prior to placing the films in water to create free sulfhydryl groups at the surface. The functionalized films were soaked in a dispersion of CdSe/ZnS core-shell quantum dots in tetrahydrofuran, and stable linkages formed between sulfhydryls and the quantum dot surfaces, imparting fluorescence to the hydrogel. Three-dimensional implants for biomedical applications have also been coated with the hydrogel in order to improve biocompatibility. Additional examples of using iCVD to create conformal hydrogel coatings on complex structures will be described, along with their applications. In all cases, the chemical functionalities of hydrogel components were maintained and further exploited for applications. Nanoscale thicknesses of the coatings were precisely controlled.

662c A Study of Dewetting during Solvent-Assisted Vapor Deposition of Polymer Films

Xichong Chen, University of Rochester, Department of Chemical Engineering, 206 Gavett Hall, Rochester, NY 14627 and Mitchell Anthamatten, Chemical Engineering, University of Rochester, 206 Gavett Hall, Rochester, NY 14627.

Initiated chemical vapor deposition (iCVD) is a technique to produce polymeric thin films from gaseous reagents.[1-3] At low pressures, monomers and initiator molecules are fed into a vacuum reactor. Once inside, initiator molecules are thermally activated to create free radicals, and they diffuse to a target substrate and initiate a polymerization reaction with adsorbed monomer. We have recently designed and built an axisymmetric iCVD vacuum reactor. Using this reactor, smooth poly (methyl methacrylate) (PMMA) films can be grown from vapor feeds of methyl methacrylate and t-butyl peroxide. Prior studies focused on understanding how polymer film growth rates are related to reactor operating parameters including reagent flowrates, reactor pressure, and hot-zone temperature.[2] The polymer deposition rate was also affected by the presence of solvent vapor (a third gas feed). Under certain operating conditions, solvent vapor leads to dewetting of polymer films from target substrates.

The objective of our current study is to understand and predict the deposition conditions that result in dewetting. We performed several depositions on different substrates including silicon wafers, glass slides, and PMMA-coated surfaces. Solvent vapors with different solubility parameter and vapor pressure including t-butanol, water, and isopropanol were also used. Scanning electron microscopy (SEM) and white light interferomety were used to investigate dewetted structures following deposition. The feature size of the dewetted structures depends on the film thickness, the film growth rate, and the type of solvent. A kinetic model will be discussed that compares rates of film growth, polymerization, and phase separation to explain dewetting behavior.

1. Chan, K. and K.K. Gleason, Initiated CVD of poly(methyl methacrylate) thin films. Chemical Vapor Deposition, 2005. 11(10): p. 437-443.

2. Chen, X.C. and M. Anthamatten, Vapor Deposition Polymerization of Poly(Methyl Methacrylate) in an Axisymmetric Vacuum Reactor. Polymer, 2008. 49: p. 1823-1830.

3. Lau, K.K.S. and K.K. Gleason, Initiated chemical vapor deposition (iCVD) of poly(alkyl acrylates): A kinetic model. Macromolecules, 2006. 39(10): p. 3695-3703.

662d Initiated Chemical Vapor Deposition of Polymer Thin Film Hydrogels

Ranjita K. Bose and Kenneth K.S. Lau. Department of Chemical and Biological Engineering, Drexel University, 3141 Chestnut St., Philadelphia, PA 19104

Initiated Chemical vapor deposition (iCVD) is a one step surface polymerization technique without the use of solvents. The elimination of solvents is especially beneficial in biomedical applications because residual organic solvents in solvent-processed matrices decrease the activity of incorporated factors and promote inflammatory responses in vivo. In iCVD selective thermal decomposition of species is achieved using resistively heated filament wires. Furthermore, iCVD combines the dry environment of chemical vapor deposition (CVD) with a radical polymerization solution chemistry without the liquid phase by thermally creating an initiator radical and combining it with monomer units repeatedly. Additionally, to promote surface adsorption of reactive species, iCVD relies on a cooled particle bed that is at a temperature (usually room temperature) much lower than the initiator activation temperature. This conveniently enables particle materials, for example, pharmaceutical drugs, proteins and polymer systems that may degrade at elevated temperatures to be treated via iCVD without compromising their material integrity. Also, by using controlled polymerization chemistries, iCVD produces exceptionally clean polymers with stoichiometric compositions, high molecular weights, and with no residual solvents, excipients, glidants, or plasticizers.

In this work, we produce thin films of hydrogel-like polymers for potential bio-applications. Poly (ethylene glycol) (PEG) is a technologically important polymer with many biomedical applications including tissue engineering, drug delivery, non-biofouling membranes and spatial patterning of cells. In this work, we created PEG thin films by iCVD without the use of solvents. We also created thin films of poly(2-hydroxyethyl methacrylate) (PHEMA), which is a hydrogel and does not dissolve in water. In addition to synthesis of homopolymers of PEG and PHEMA we also synthesized copolymers of PEG-PHEMA. Depending on the composition, PEG–PHEMA block copolymers may be water soluble or water insoluble. The presence of a second monomer gives us the freedom to change many variables including the percentage of each monomer.

Fourier transform infrared (FTIR) and nuclear magnetic resonance (NMR) spectroscopies showed that the chemistry of PEG obtained by iCVD synthesis matched closely to PEG obtained from solution synthesis. The effect of substrate temperature, system pressure and reactant flowrates were studied to develop a kinetic model for the iCVD of PHEMA and PEG. This work showed that iCVD can be used to create copolymers of PEG-PHEMA. Cationic polymerization of PEG as well as copolymerization in vapor phase was shown to correspond well to those in the liquid phase. This work demonstrates that iCVD can directly apply liquid polymerization chemistries with great versatility.

662e Film Thickness Dependence of Protein Adsorption from Blood Serum and Plasma Onto Poly(sulfobetaine) and Poly(carboxybetaine)-Grafted Surfaces

Wei Yang[1], Gang Cheng[1], and Shaoyi Jiang[2]. (1) Department of Chemical Engineering, University of Washington, Benson Hall 238, Box 351750, Seattle, WA 98195, (2) Chemical Engineering, University of Washington, Benson Hall, Box 351750, Seattle, WA 98195

Protein adsorption is the first event occurring in blood-material interactions. For medical implants and *in vivo* sensors, surface resistance to protein adsorption and cell adhesion from complex media is needed to prevent undesirable responses of the living system to a device or implant. This creates the need for surface modifications that have low reactivity with bloodborne proteins. Recently, zwitterionic materials have been shown to have ultra low protein adsorption. Previous studies have shown that surface packing density and film thickness are important for surface resistance to non-specific protein adsorption.

In this work, poly(sulfobetaine methacrylate) (polySBMA) and poly(carboxybetaine arylic amide) (polyCBAA) were grafted onto surfaces via atom transfer radical polymerization (ATRP) method. Film thickness is controlled between 10-100nm. Protein adsorption from single fibrinogen and lysozyme solutions, serum or plasma was investigated by surface plasmon resonance (SPR) sensors. Results show that protein adsorption strongly depends on grafted film thickness. In addition, the influence of ionic strength on protein adsorption was also studied. Results show that all of these surfaces are highly resistant to non-specific protein adsorption from single fibrinogen and lysozyme solutions and several of them can be considered as "ultra low fouling" surfaces (less than 5 ng/cm^2 protein adsorption) to 100% blood serum or plasma. This work provides a new *in vitro* evaluation standard for the application of biomaterials *in vivo*.

662f Formulation Efforts to Improve the Performance of Military Coatings

Felicia Levine, John J. La Scala, and John A. Escarsega. Army Research Laboratory, Building 4600, Aberdeen Proving Ground, MD 21005

The military has a need for coating systems with improved durability and decreased permeability to chemicals. In an effort to improve these properties, ARL has formulated water-dispersible polyurethane topcoats using small amounts of fluorinated polyol component and through changing the isocyanate to hydroxyl indexing. Increasing the isocyanate to hydroxyl ratio had no effect on Tg of clear coats, but increased the cross-link density as measured by the rubbery modulus through dynamic mechanical analysis. Permeability and solubility measurements correlated well with the cross-link density, as both decreased as isocyanate to hydroxyl indexing was increased. Fluorinated additives produced little effect on thermo-mechanical properties, including Tg, cross-link density, and modulus. X-Ray photoelectron spectroscopy shows that the fluorine additive migrated to the surface and produced fluorine surface concentration 10-90 times that of the bulk. Yet, water contact angle was reduced at low fluorinated polyol content because the ethoxylation of the fluorinated additive to increase its water dispersibility also decreased the contact angle. Phase contrast microscopy showed that the morphology of these films went from being uniform to having distinct aggregates that increased in size and number as fluorinated polyol fraction increased. Permeability and solubility testing is underway to determine whether these changes in morphology affect these properties. Fluorinated additives are supposed to improve weatherability of coatings systems, but color and gloss retention was slightly worse for the fluorinated polyols than the baseline systems.

662g Hydrophobic Coatings Using a Chamberless Atmospheric Pressure Plasma Process

Michael Barankin, Eleazar Gonzalez II, Li Gao, and Robert F. Hicks. Chemical & Biomolecular Engineering, University of California, Los Angeles, Box 951592, Los Angeles, CA 90095-1592

A new process is demonstrated using low-temperature atmospheric pressure plasma to obtain hydrophobic thin films from fluoroalkylsilane precursors on glass and acrylic polymer substrates. An atmospheric pressure plasma is generated at 150 W of RF power using 30.0 L/min of helium and 0.70 L/min nitrogen gases, and is used to polymerize or "cure" the liquid precursor film into a permanent, hydrophobic coating having average water contact angles in excess of 100°. The contact angles of glycerol and diiodomethane are obtained and the Owens, Wendt, Rabel, and Kälble (OWRK) method is used to approximate the total and polar component to the surface energy which were found to be 11 and 1.2 dyn/cm, respectively. These coatings were found to have a relatively low surface roughness and are highly transparent to UV/visible light. The coatings maintained their hydrophobic properties after repeated acetone washing and a year of aging in humid air. The hydrophobicity of these coatings on acrylic were found to depend on pretreatment of the acrylic surface with atmospheric plasma at similar conditions, but with 0.44 L/min of oxygen in place of the nitrogen gas.

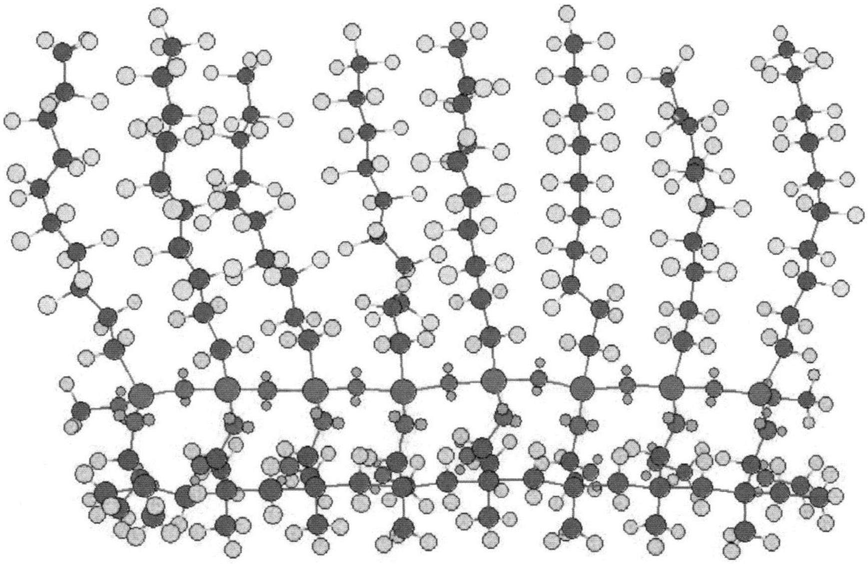

680a Hemocompatibility Enhancement through the Integration of the Antigenic Disguise Protein Tp0483 on a Material Surface

Matthew T. Dickerson, Dept. of Chemical Engineering, University of Kentucky, 177 F. Paul Anderson Tower, Lexington, KY 40506, Dr. Kimberly Anderson, Chemical Engineering, University of Kentucky, 177 F. Paul Anderson Tower, Lexington, KY 40506, and Dr. Leonidas Bachas, Chemistry, University of Kentucky, 207 Chemistry-Physics Building, Lexington, KY 40506.

The first response the body exhibits to a foreign material entering the bloodstream is rapid protein adhesion. Upon binding, adsorbed proteins undergo a conformation change which relays a signal to nearby platelets, resulting in platelet activation. Activated platelets adhere to the adsorbed proteins and at the same time secrete additional clotting factors that induce further platelet activation and result in the eventual formation of a thrombus on the surface. Thrombi pose many dangers including blood vessel occlusion as well as a heart attack or a stroke if the thrombus detaches from the material. To address this issue, it is of interest to develop new methods to improve hemocompatibility, one of these being exploitation of antigenic disguise proteins. Antigenic disguise refers to the property of specific pathogenic organisms to circumvent the body's immune system and avoid a host response. Treponema pallidum, the bacteria responsible for syphilis, is one example of this type of organism. T. pallidum has been known to exhibit latency periods of up to 50 years. It is hypothesized that this ability originates in several proteins presented on the surface of T. pallidum that exhibit the ability to bind host fibronectin (Fn). Fn is a common protein found throughout the body, and it is believed that by binding this protein, T. pallidum forms a protective coating that masks any potential antigenic targets from attack. One such T. pallidum protein is Tp0483, which serves as the focus of this research. In this study, the bond between Tp0483 and Fn as well as potential applications in increasing hemocompatibility are examined. Tp0483/Fn interactions were observed on various surface chemistries and hemocompatibility studies examining plasma protein adhesion and platelet activation were conducted. Binding studies utilized surface plasmon resonance (SPR) to view protein binding on functionalized self-assembled monolayers (SAM's) terminated in COOH, CH3, NH2, or OH. Results indicated that Tp0483 binds equally well to all surface chemistries, but only binds to Fn when immobilized on a COOH surface. This outcome suggests that Tp0483 possesses a wide range of diverse binding domains and that its conformation changes greatly based on the type of surface that it adsorbed to. As a result, the domain responsible for Fn binding is only exposed when Tp0483 is adsorbed on a negatively charged surface (COOH). Likewise, an RGD blocking study resulted in inhibition of Fn binding which supports previous studies indicating that the RGD peptide sequence on Fn mediates Tp0483 binding. Additionally, plasma protein binding studies demonstrated a decrease in binding to Tp0483/FN surfaces and platelet activation experiments showed a decrease in activation on these surfaces, both of which indicate improved hemocompatibility.

680b Examination of the Effects of PEG-Rich Matrices In Multi Drug Resistance-Associated Protein (MRP) and Multi Drug Resistance (MDR) Substrates Transport Utilizing the Caco-2 Cell Model

Nilmarie Santos-Roman, Janet Mendez-Vega, and **Dr. Madeline Torres-Lugo**. Chemical Engineering, University of Puerto Rico - Mayaguez, P.O. Box 5289, Mayaguez, PR 00681

Multidrug resistance (MDR) is responsible for the low absorption of a very wide range of drugs and has been recognized as one of the major hurdles in drug absorption. Crosslinked poly(ethylene glycol) based morphologies are being proposed as multidrug resticance1 (MDR1) and multidrug resistance associated protein (MRP) inhibitors with the added advantage that these can be tailored for controlled drug delivery applications. For this purpose, three design variables were studied; the PEG tethered chain length and crosslinker length, and the particle size. The hydrogels demonstrated that they have MRP inhibitory effects by successfully enhancing the transport of fluorescein sodium salt (FLUO) up to 250%. The transport enhancement appears to be dependent on the hydrogel morphology as well. Temperature effect experiments seem to confirm that this transport enhancement is due to an active interaction with the MRP proteins, since no effect was observed at 4°C. Polymers also demonstrated to have transport enhancement of the MDR1 substrate, Rhodamine 123 (RHO) of up to 350%. This effect was found to be dependent on the length of the tethered chain as well as the concentration of the suspension. These observations suggest a possible competitive action. The potency of the hydrogel inhibition appears to be greater than the known inhibitors verapamil, genistein and probenecid, but similar to the lineal PEG-300 morphology. Therefore, we can conclude that PEG based hydrogels are potential candidates for controlled drug delivery devices aimed at the inhibition of MRP and MDR1 proteins.

Hydrothermal synthesis of biocompatible $CaTiO_3$ particles

SiewShee Lim[1], **Michael Cloke**[1], **Kok Chiang Ng**[2], **Jun Jin**[2] and **George Z. Chen**[2,*],

School of Chemical and Environmental Engineering, University of Nottingham,
[1] Malaysia Campus, Jalan Broga, 43500 Semenyih, Malaysia.
[2] University Park, Nottingham NG7 2RD, UK
*Corresponding author. Phone: +4401159514171, Fax: +4401159514115
Email: George.Chen@nottingham.ac.uk

Extended Abstract

Of late, synthetic grafts have presented themselves as worthy alternative source for continuous bone healing to both the conventional autogenous and allografts. This is due to their reasonable production costs (while harvesting cost is very steep in the conventional methods) and the avoidance of the pathogenic transmission risk as posed in allografts. One of the essential requirements for synthetic grafts is its biocompatibility and this can be achieved through hydrothermal process. Furthermore, this technique of synthesis is also capable of producing a narrow distribution of particle sizes besides its ease in implementation and thus, the reason for its employment in this study.

In the current work, orthorhombic submicro- and nanoparticles of $CaTiO_3$ were synthesised via a simple hydrothermal process using TiO_2 (p25) as the precursor in alkaline media containing excess NaOH and $CaCl_2$. The as-prepared powdery materials were washed with distilled water until the pH value became neutral and dried in oven at 50°C for 24 hours. The dried samples were characterised by X-ray diffraction analysis (XRD) and Scanning Electron Microscopy (SEM).

NaOH in excess provided a basic environment (pH > 12) for the dissolution of TiO_2 and its subsequent conversion to $CaTiO_3$. The formed Ti-OH complexes reacted with the Ca^{2+} ions in the solution to form the perovsite, for example, at 120°C for 24 hours. The chemical reaction involved is shown below.

$2NaOH + CaCl_2 + TiO_2 \rightarrow CaTiO_3$ (or $CaO \cdot TiO_2$) + $2NaCl + H_2O$

The most interesting outcome of this synthesis was that orthorhombic calcium titanate particles with three different mutual edges were generated (shown in Figures 1a and 1b). The crystalline lattice of the produced perovskite was also orthorhombic which was confirmed by the X-ray diffraction angles at 23.1°, 33°, 39°, 41°, 47.3° and 59° (shown in Figure 2).

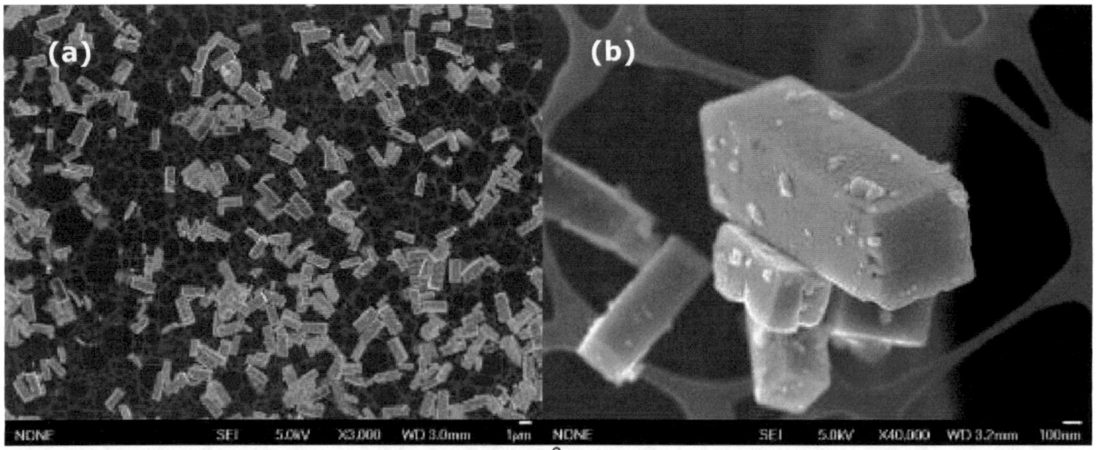

Figure 1 (a) Uniform $CaTiO_3$ generated at 120°C for 24 hours (b) Orthorhombic particles with three different mutual edges

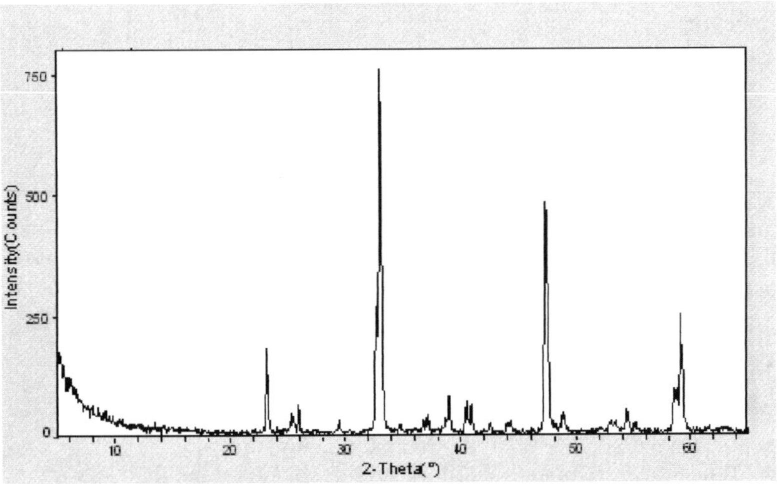

Figure 2. XRD pattern of the perovskite produced at 120°C for 24 hours using P25 TiO_2 and excess NaOH and $CaCl_2$

The size control of the perovskite particles could be achieved by adjusting the amount of NaOH added. For example, when the number of moles of sodium hydroxide was less than 20 mmol than that of calcium chloride in the synthesis, orthorhombic perovskite particles with shorter length were generated as shown in Figure 3. However, the reaction temperature was raised to 200°C for the formation of shorter perovskite. Thus, the formation of shorter orthorhombic particles was more preferable at less basic solution.

Figure 3. The length of the orthorhombic particles reduced at less amount of NaOH.

When NaOH was not added to the mixture of $CaCl_2$ and TiO_2, only anatase was collected after the process. No trace of perovskite peaks is observed in Figure 4.

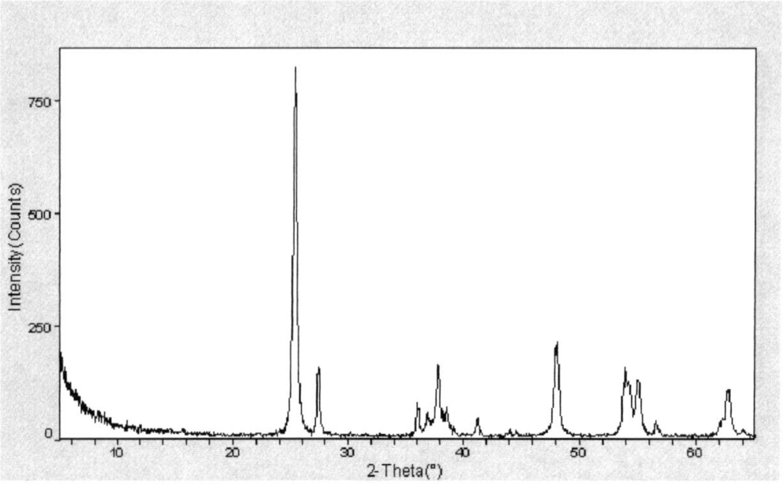

Figure 4. XRD pattern for anatase collected after for 24 hours hydrothermal process at 120°C using only P25 TiO_2 and $CaCl_2$

In conclusion, orthorhombic $CaTiO_3$ particles with narrow size distribution were successfully generated through hydrothermal method. The essential requirement for the formation of $CaTiO_3$ was the presence of NaOH. The amount of this alkaline media also affected the size of the perovskite generated. The formation of longer orthorhombic perovskite was more preferable at pH value of 12. The biocompatibility and osteoconductivity of these novel submicro- and nano-perovskite will be tested in future research[1].

[1] Ohtsu N, Sato K, Saito K, Hanawa T, Asami K, Evaluation of Degradability of $CaTiO_3$ Thin Films in Simulated Body Fluids, Mater Trans 45 (2004), pp. 1778-1781

682a Liquid Crystal Based Protein Assays Developed in Microfluidic Channels

Chang-Ying Xue and Kun-Lin Yang. Department of Chemical and Biomolecular Engineering, National University of Singapore, 4 Engineering Dr 4, Singapore, 117576, Singapore

Protein assays developed in microfluidic channels have many advantages due to its small sample volume, short reaction time, high throughput, and potential for integrating the protein assays into a lab-on-chip device. In this presentation, we report a label-free detection method for protein assays developed in microfluidic channels by using liquid crystal (LC), which can transduce the presence of proteins or antigen-antibody binding events into optical signals clearly visible to the naked eye. Because LC is far more sensitive than fluorescence in distinguishing a very small difference in protein concentration, this new detection method allows the quantification of protein concentration through the length of the bright region in a microfluidic channel, which is similar to a power testing strip on a Energizer® battery. Important issues such as sensitivity and specificity associated with the LC-based protein assay will also be discussed in this presentation.

682b Oxygen-Sensing Microparticles for Probing Cell Microenvironment

Miguel Acosta, Yordan Kostov, and **Jennie B. Leach**. Chemical & Biochemical Engineering, UMBC, 1000 Hilltop Circle, Baltimore, MD 21250

Oxygen concentration is a key parameter in tissue culture. In addition, oxygen diffusion through biomaterial scaffolds plays an important role in maintaining healthy tissues in vitro, as oxygen supply becomes a limiting factor during the growth of highly metabolic tissues and large tissue masses. This is mainly a result of the lack of vascularization in tissues cultured in vitro and the low solubility of oxygen in the culture medium. Here, we present the development and characterization of dual-fluorophore oxygen-sensitive microparticles that can be suspended in any transparent biomaterial scaffold used in cell culture and tissue engineering. By monitoring the sensing microparticles with a fluorescence microscope, ratiometric measurements of spatial and temporal changes in oxygen concentration can be performed in a rapid, automated, and non-invasive manner without consuming the oxygen available to the cells during culture.

The microparticles were prepared in a simple two-step process. The first step consists of electrostatic immobilization of the oxygen-sensitive fluorophore, tris (4,7-diphenyl-1,10-phenanthroline) ruthenium (II) dichloride (Ru), and the reference fluorophore, Nile blue chloride, onto ~10 μm silica gel particles. Ru was chosen because of its high quantum yield, high oxygen quenching, and thermal stability at physiological temperature. The reference fluorophore allows ratiometric (or normalized) measurements that correct for the typical drawbacks of fluorescence intensity measurements such as bleaching, heterogeneous fluorophore concentration, fluctuations in excitation light intensity, and camera sensitivity. The second step consists of the encapsulation of the fluorophore-bound silica gel particles within larger poly(dimethylsiloxane) (PDMS) particles. PDMS is optically transparent, hydrophobic, biocompatible, and highly permeable to gases such as oxygen, nitrogen, and carbon dioxide.

To characterize the microparticles, we measured particle size distribution and compared the performance of the sensing microparticles to that of a conventional dissolved oxygen meter. The 10-40 μm size of the microparticles makes them versatile, allowing them to be suspended in a large range of biomaterial substrates. The microparticle response time was approximately 37% faster than that of a conventional dissolved oxygen meter. Microparticle sensor response was fully reversible and no evidence of hysteresis was observed. Finally, microparticle calibration shows that it is possible to measure oxygen partial pressures in ranges relevant to tissues culture studies and in close proximity to the cells. Sample size is only limited by scaffold transparency and the microscope depth of field; herein, transparent hydrogel samples 1.5 cm wide x 3.8 cm long x 260 μm deep were sampled. A measurement of one ~2 x 2 mm field of view can be performed in about 10 s. The versatility of the described sensing microparticles will allow for future studies where dynamic spatial correlations between cell function and oxygen concentrations can be achieved reliably both in two-dimensional and three-dimensional in vitro environments.

682c Using Functionalized Gold Nanoparticles to Study Uptake and Intracellular Trafficking of Biochemicals Central to Osteonecrosis

Fedena Fanord[1], Korie Fairbairn[1], Harry Kim[2], Venkat Bhethanabotla[3], and Vinay K. Gupta[1]. (1) Chemical & Biomedical Engineering, University of South Florida, 4202 E Fowler Avenue, ENB 118, Tampa, FL 33620, (2) Shriners Hospitals for Children, 12502 Pine Drive, Tampa, FL 33612, (3) Department of Chemical Engineering, University of South Florida, ENB 118, 4202 E. Fowler Ave., Tampa, FL 33620

Osteonecrosis of the femoral head is a common hip disorder in juveniles that causes loss of the structural integrity of bone and deformity, often resulting in premature end-stage osteoarthritis. Current treatments are limited to high risk surgical procedures for young patients. Studies have shown that inhibiting osteoclastic resorption can prevent the development of the femoral head deformity; but much remains to be discovered about the molecular targets, uptake, and intracellular trafficking of the biochemicals that inhibit the osteoclastic resorption. We have focused on using functionalized gold nanoparticles (GNP) and exploiting the optical properties of the GNP for visual imaging of the progression of biochemical action on osteoblast and osteoclast cells. Mineralized matrices treated with functionalized GNP will be used to model cellular uptake from the bone surface. Characterization of the functionalized GNP and cellular response with UV-Vis and infrared spectroscopy, dynamic light scattering (DLS), optical microscopy, and transmission electron microscopy (TEM) will be presented.

682d Two-Step Synthesis of Nanosize Hollow SiO2 Particles for Multiplexing

Gerson R. Aguirre[1], Alexander Couzis[2], Charles Maldarelli[3], and M. Lane Gilchrist[2]. (1) Chemistry Department, City College and Graduate Center of the City University of New York, 140th Street at Convent Avenue, NY, NY 10031, (2) Chemical Engineering, City College and Graduate Center of the City University of New York, 140th Street at Convent Avenue, NY, NY 10031, (3) The Levich Institute and the Chemical Engineering Department, City College and Graduate Center of the City University of New York, 140th Street at Convent Avenue, NY, NY 10031

Entrapment and retention of materials inside synthesized silica particles is an active research field due to its many industrial, biological and chemical activities including as molecular sieves due to their porosity. Many methods include layer-by-layer templating techniques, emulsion droplets, and others use template-assisted synthesis such as latex, colloidal particles to vesicles, even gold nanoparticles. Templating techniques are more effective given the ease with which hollow particles can be obtained. These methods require the removal of the templating agent by acid burning or calcination of the organic material that can be detrimental if a sensitive material needs to be preserved and more importantly to retain its intended function if absorption of such materials is restricted by pore size or chemistry. What follows is a fabrication scheme that will allow for a more benign particle synthesis and allows for retention of introduced material or molecules inside the silica particle that can be beneficial in biological applications.

In this study we report on the formation of silica particles that have entrapped fluorescent quantum dots. We have used an oil-in-water suspension approach by the use of high CMC value of cationic surfactant (CTAB) in a highly alkaline solution. The TBOS (tretrabutoxysilane) oil droplets hydrolyze at the interface, TBOS has a very slow rate of hydrolysis due to its longer alkyl chain, facilitated by the high concentration of charge density provided by the ionized surfactant, according to Huo, et al. At this pH, condensation is enhanced but limited by the rate-limiting hydrolysis step. When the appropriate extent of hydrolysis is achieved at this alkaline regime rapid gelation at the oil/water interface occurs by the driving force of condensation.

In addition to synthesizing these silica micronsize particles we have also encapsulated in the core fluorescing semiconductor nanoparticles, CdSe/ZnS quantum dots (QDs). The distributed oil phase provides the vestibule for the QDs. As the silica particle is formed at the oil-water interface the QDs are trapped in the core. The hydrophobic nature of the QD surface stabilizes their dispersion in the oil phase. The hydrophobic termination of the QDs is provided by a TriOctyl-PhosphineOxide (TOPO) and octadecylamine adsorbed cap. This cap is chemisorbed against the ZnS shell of the QDs and extends its hydrophobic tails into the oil phase, resulting in the stabilization of the dispersion. Confocal microscopy images confirm the quantitative capture of the QDs within the formed silica particle.

682e Preparation of Bioanalytical Sensors by Incorporating Fluorophore In Patternable Poly(ethylene glycol) Diacrylate-Based Membranes

Zhan Gao, Department of Chemical and Biological Engineering, Missouri University Science & Technology, 143 Schrenk Hall, Rolla, MO 65409, Chang-soo Kim, Electrical & Computer Engineering, Missouri University Science & Technology, 141 Emerson Electric Co. Hall, Rolla, MO 65409, and **David B. Henthorn**, Chemical & Biological Engineering, Missouri University Science & Technology, 143 Schrenk Hall, Rolla, MO 65409.

An oxygen sensor used in cell culture was constructed by immobilizing fluorophore in PEGDA-based membrane through *in situ* photografting technology. This technology allows for the grafting sensitive membranes, such as analyte-sensitive hydrogels and enzyme immobilized layers, directly from the negative-tone photoresist SU-8 surface with spatial control, even within completed microfluidic channels. The detection principle of this oxygen sensor was based on the dynamic quenching of the luminescence of a Ruthenium complex to dissolved oxygen. General principles of sensor optimization are presented. These include issues such as fluorophore concentration, thickness control of the membranes and sensitivity enhancement. The characteristic of the oxygen sensor has also been studied here in details, such as production stability, long-term stability and response time. The particular advantages conferred by the use of *in situ* photografting technology are emphasized. Based on the good results achieved, a fluorescent glucose biosensor incorporating glucose oxidase and the Ruthenium complex in PEGDA membrane was also constructed and tested.

Acoustic Wave propagation in a Hexagonal Surface Acoustic Wave biosensor based on $LiTaO_3$: A finite element study

Subramanian K.R.S. Sankaranarayanan and Venkat R. Bhethanabotla
Sensors Research Laboratory, Department of Chemical Engineering,
University of South Florida, Tampa, Florida, 33620, USA.

Abstract — We present a 3-D finite element model of a novel hexagonal SAW biosensor based on $LiTaO_3$ substrate. This SAW biosensor involves the use of one delay path for biological species detection whereas the other delay paths are used to simultaneously remove the non-specifically bound proteins using the acoustic streaming phenomenon, thus eliminating biofouling issues associated with other biosensors. Prior to this biosensor fabrication on any piezoelectric substrate, it is important to establish the type of waves that are generated along the various delay paths. The choice of a delay path for sensing and simultaneous cleaning application depends on the propagation characteristics of the wave generated along the crystal cut and orientation corresponding to that delay path. The frequency response as well as wave propagation characteristics along the delay path corresponding to crystal orientation with on-axis propagation along 36° YX LiTaO3 substrate are analyzed using a coupled field FE model. Similar analysis is extended to the off-axis propagation directions corresponding to Euler rotations by 60° and -60° along the x-z plane. Our findings indicate that the on-axis direction with a significant surface shear horizontal (SH) component should be employed for biological species detection whereas the off-axis directions having mixed modes with a dominant Rayleigh wave component are most suitable for simultaneous cleaning or removal application.

***Keywords* – SAW; Lithium tantalate; Biosensors; Finite element method**

1. INTRODUCTION

Surface acoustic wave (SAW) devices used both individually as well as in arrays find applications in chemical and biological sensing as well as in materials characterization [1]. The conventional SAW devices typically comprise of dual delay-line configurations with one delay-line used as a reference to compensate for environmental variations. Recent experimental efforts have focused on the design and fabrication of acoustic wave devices which can allow for better sensor characteristics as well as materials characterization possibilities [2]. One such novel device with a complicated transducer design is the hexagonal surface acoustic wave (SAW) sensor shown in Fig. 1. It is comprised of three different delay paths aligned to allow for generation of acoustic waves which are different in character in the different directions. It is possible to exploit the generated multiple wave modes to develop SAW devices that can serve as better chemical and biosensor elements.

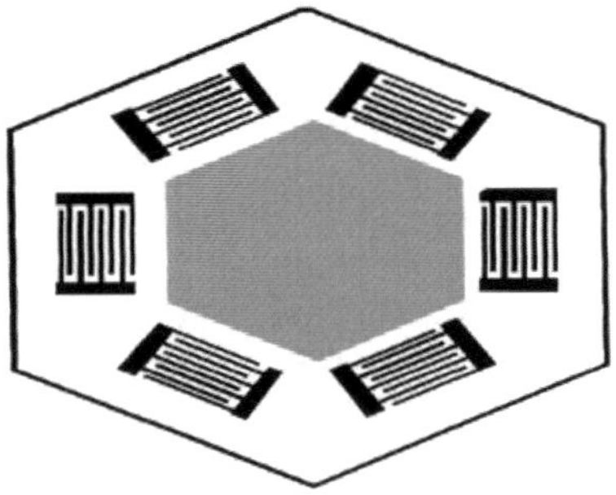

Figure 1. Schematic diagram of the hexagonal SAW device useful for chemical and bio-sensing applications.

In sensor applications, the multiple parameters extracted from the sensing film using the hexagonal SAW device can serve as multiple calibration curves and allow for a more unique and precise characterization of the concentration and type of analyte being sensed. Combined with the array concept, significantly more information can be obtained to better characterize analytes. Additionally, the hexagonal SAW device can also be employed for biological sensing. It is well known that the conventional acoustic wave devices are specific to the phase in which they operate; for example, the Rayleigh wave devices which are suitable for vapor sensing cannot operate in the liquid environment owing to excessive attenuation [3]. If the multiple directions in which the waves are launched in the hexagonal device are different in character, then the hexagonal SAW sensor could allow the same device to be functional in both gas and liquid phases.

In one of our previous investigation, we have shown that the Rayleigh wave devices can be utilized in acoustic cleaning of nonspecifically bound proteins in biosensor applications [4-5]. With the possibility of launching shear- horizontal SAW waves in one direction and Rayleigh waves in another, the hexagonal device should serve as a better biosensor element for liquid phase applications and help to simultaneously eliminate the problems associated with biofouling which render conventional biosensors ineffective. Combined experimental and theoretical investigations to identify the best substrates and propagation directions for potential biosensing applications form the main focus of this research. Our previous theoretical and experimental investigations indicate that such a device fabricated on $LiNbO_3$ was better suited for material characterization than for biosensing applications [4-6].

Identifying the acoustic wave propagation characteristics on hexagonal SAW device based on several different substrates such as lithium tantalate and langasite would help us identify the best possible design for simultaneous removal of nonspecifically bound proteins and sensing utilizing Rayleigh and shear-horizontal waves. In the present work, a hexagonal SAW device based on $LiTaO_3$ substrate is modeled using a FE technique. The wave propagation characteristics along the three different delay paths corresponding to crystal orientation with Euler angles (0, 36, 0), (60, 36, 0) and (-60, 36, 0) respectively, are evaluated. The developed FE model utilizes anisotropic material properties in a three dimensional description of $LiTaO_3$ substrate. In this paper, the particle displacement and voltage profiles obtained at the output Interdigital Transducer (IDT) port as well as along the depth of the piezoelectric substrates along various locations across the three delay paths are analyzed and used to interpret the acoustic wave modes and velocity. Additionally, the response of each of the three delay lines to an applied voltage impulse is utilized to calculate its frequency response. The calculated frequency response and acoustic wave velocities are compared to those obtained experimentally and from a perturbation theory based approach.

2. THEORY

The propagation of acoustic waves in piezoelectric materials is governed by the mechanical equations of motion and Maxwell's equations for electrical behavior [7-9]. A system of four coupled wave equations for the electric potential and the three component of displacement in piezoelectric materials are solved for the piezoelectric substrate:

$$-\rho\frac{\partial^2 u_i}{\partial t^2}+c_{ijkl}^{E}\frac{\partial^2 u_k}{\partial x_j \partial x_l}+e_{kij}\frac{\partial^2 \phi}{\partial x_k \partial x_j}=0 \qquad (2.1)$$

$$e_{ikl}\frac{\partial^2 u_k}{\partial x_i \partial x_l}-\varepsilon_{ik}^{s}\frac{\partial^2 \phi}{\partial x_i \partial x_k}=0 \qquad (2.2)$$

These coupled wave equations can be discretized and solved for generating displacement profiles and voltages at each element/node using the finite element method.

3. COMPUTATIONAL DETAILS

A. Model parameters

The 3-D FE model describes three two-port delay-line structures along each of the Euler direction and consists of three finger pairs in each port. The inter-digital transducer (IDT) fingers are defined on the surface of a lithium tantalate substrate and the fingers are considered as mass-less electrodes to ignore the second-order effects arising from electrode mass, to simplify computations. The periodicity of the finger pairs is 40 microns and the aperture width is 200 microns. The transmitting and receiving IDT's are spaced 130 microns or 3.25λ apart. The substrate for 36° YX $LiTaO_3$ was defined as 1600 microns in propagation length, 300 microns in width and 200 microns in depth. For simulating the other two propagation directions in the hexagonal biosensor, the geometry of the substrate is kept the same, whereas the crystal coordinates are rotated. To achieve this, the material properties, *i.e.*, stiffness, and piezoelectric and permittivity matrices are rotated by 60° and -60° along the x-z plane to model the off-axis directions (60, 36, 0) and (-60, 36, 0), respectively. The simulated models have a total of approx. 250000 nodes and are solved for four degrees of freedom (three displacements and voltage). The model was created to have the highest densities throughout the surface and middle of the substrate.

* Author for correspondence at email venkat@eng.usf.edu; This work was supported in part by NSF grants DGE-0221681, CHE-0722887 and ECCS-0801929, US Army Research Contract No. W81XWH-05-1-0585 and the USF Office of Research by an interdisciplinary research grant.

B. Structure excitation

Two types of analysis were carried out along each of the three delay lines: (1) An impulse input of 10 V over 1 ns was applied to study the frequency response of the device and (2) AC analysis with a 5 V peak-peak input and 100 MHz frequency was utilized to study the wave propagation characteristics.

4. RESULTS AND DISCUSSION

A. Impulse response

By applying an impulse input, the frequency response of the hexagonal SAW biosensor device was also calculated. The calculated frequency responses for an input impulse (1 ns) of 10 V for the on-axis and off-axis directions are shown in Fig. 2. The calculated device frequency along the three directions indicated that least attenuation occurs along the on-axis direction whereas higher propagation losses are observed for the off-axis directions.

The theoretical (Campbell-Jones and finite element method) and experimentally measured velocities corresponding to wave propagation along the three Euler angles are given in Table I. The SAW device frequency is directly proportional to velocity, it is expected that the frequency response would follow the order (0, 36, 0) < (60, 36, 0) < (-60, 36, 0).The simulations of SAW device were performed with three IDT finger pairs. Hence, the passband obtained for the three directions is broad and the calculated device frequencies along the three directions do not show significantly different device center frequencies. The simulated device frequencies along the three directions are close to 100 MHz. Although the center frequencies are not very different, the insertion loss along the three directions varies significantly. These values for the three Euler directions are given in Table II. We find that the least attenuation occurs along the (0, 36, 0) direction whereas the maximum is observed for (-60, 36, 0). This indicates that the (0, 36, 0) direction is most suitable for sensing applications.

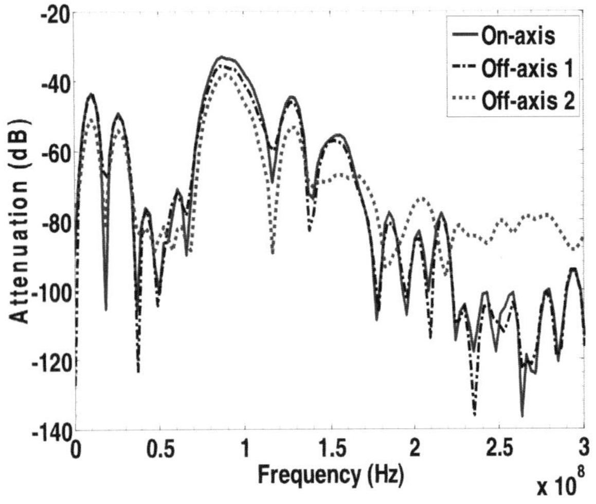

Figure 2. Calculated frequency response along the three Euler directions in the simulated hexagonal SAW biosensor. On axis corresponds to (0, 36, 0) whereas off-axis-1 and off-axis-2 correspond to (60, 36, 0) and (-60, 36, 0) directionson a lithium tantalate substrate.

TABLE I. WAVE VELOCITIES ALONG DIFFERENT SHORTED DELAY PATHS OF LITHIUM TANTALATE

Orientation *Euler angle* *(φ, θ, ψ)*	**Acoustic wave velocities**		
	Numerical (FEM) *(m/s)*	**Theoretical [10]** *(m/s)*	**Experimental** *(m/s)*
(0, 36, 0)	4100	4039	3868 ± 148
(60, 36, 0)	3800	3792	3621 ± 101
(-60, 36, 0)	3600	3662	3593 ± 69

TABLE II. SIMULATED FREQUENCY RESPONSE OF HEXAGONAL SAW DEVICE BASED ON $LiTaO_3$.

Euler direction (φ, θ, ψ)	Simulated insertion loss (dB)
(0, 36, 0)	33.50
(60, 36, 0)	35.00
(-60, 36, 0)	38.00

B. Response to applied AC voltage

The application of AC electrical signal on the input transducer fingers results in generation of mechanical waves which then propagate towards the output transducer fingers. The coupled field finite element model developed in the present study is able to capture the generation and propagation of SAWs on a piezoelectric substrate such as $LiTaO_3$ as shown in Fig. 3.

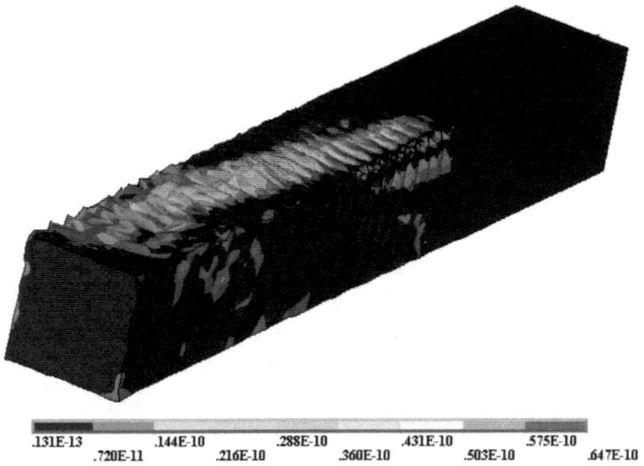

Figure 3. SAW propagation in $LiTaO_3$ along the (0, 36, 0) Euler direction. The displacements shown on the scale bar are in meters. It can be seen that most of the acoustic energy is confined to the surface of the piezoelectric substrate.

The displacement amplitudes of the SAWs are typically in the nanometer to angstrom range. The wave travels outside of the aperture of the IDTs located on the $LiTaO_3$ surface and into the substrate depth, which indicates that part of the energy is being dissipated. The generated wave also propagates in the transverse direction opposite to the receiving transducer fingers and reaches the edge of the substrate at the end of approximately 120 ns. We also observe wave reflections along the edges at longer simulation times. The generated voltage and displacement waveforms for the (0, 36, 0) delay-line configuration are shown in Fig. 4. The response to an applied electrical input was obtained at the output IDT node located 210 microns away from the input IDT fingers. Voltage profiles obtained for the three Euler directions from the AC analysis indicated that the extent of wave attenuation is different for each of the simulated crystal orientations, which is attributed to the crystal anisotropy.

Our analysis of the simulated displacement profiles indicate that the on-axis SH component of displacement at the output IDT is an order of magnitude higher than the surface normal and longitudinal components indicative of SH wave motion (Fig. 4). Thus, a SH SAW propagates along the on-axis direction (36° YX $LiTaO_3$) which is most suitable for biosensing application. For the off-axis direction (60, 36, 0), the surface normal and longitudinal components of displacement at the output IDT are an order of magnitude larger than the SH component indicative of wave motion which is more or less ellipsoidal. This type of wave motion corresponds closely to that of the Rayleigh mode. The displacement profiles of the other off-axis component (-60, 36, 0) showed the smallest amplitude variations amongst the three directions indicative of mixed-wave modes which are a combination of more than one wave type such as pure Rayleigh or SH modes. The wave propagation along the on-axis direction, *i.e.,* 36° YX $LiTaO_3$ substrate is shown in Fig. 3. At approximately 120 ns, the wave has reached the end of the substrate.

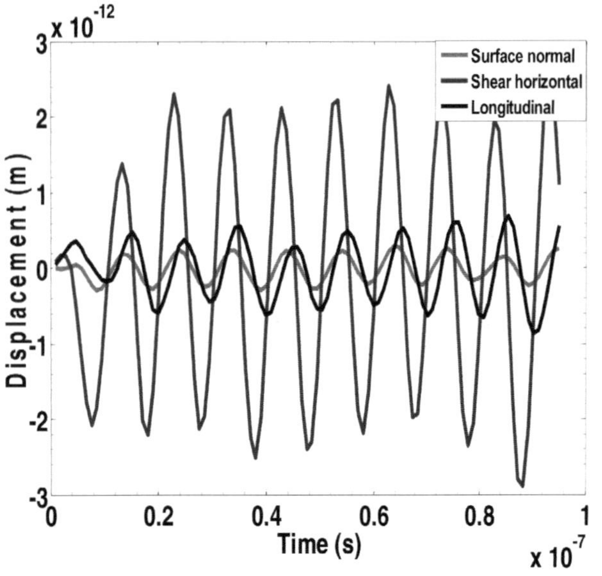

Figure 4. SH displacement component along the on-axis direction in al SAW device based on lithium tantalate.

5. CONCLUSIONS

A 3-D finite element coupled field model of a hexagonal SAW device fabricated in a $LiTaO_3$ substrate is developed in this work. Insights into the acoustic wave generation and propagation along the three Euler directions corresponding to the multiple propagation directions in this transducer configuration are obtained using the developed model. Our simulation results indicate that the waves propagating along the (0, 36, 0) direction have a dominant SH component whereas those propagating along the (60, 36, 0) and (-60, 30, 0) directions involve combination of multiple wave modes with a prominent surface normal displacement component. Such a device can be used for simultaneous biological species detection and non-specifically bound protein removal in SAW biosensing applications. The differences in acoustic wave modes manifest themselves in the form of differing voltage responses. We find a more attenuated voltage response at the output transducer fingers for the (60, 36, 0) and (-60, 36, 0) directions in comparison to the (0, 36, 0) direction. The calculated insertion losses for an applied impulse voltage input along the three propagation directions indicate that the waves along the (0, 36, 0) direction are less attenuated than those along the (60, 36 0) and (-60, 36, 0) directions. The simulation results agree with those obtained using a perturbation technique based method due to Campbell-Jones as well as those observed experimentally. A 3-D FE model can thus be useful in understanding acoustic wave propagation in piezoelectric substrates. Efforts are underway to extend such simulations for hexagonal SAW devices based on other substrates such as langasite which could help identify better biosensor transducers for liquid phase applications.

ACKNOWLEDGEMENT

The authors wish to thank the Engineering Computing Center at USF for providing the computational facilities. The authors thank Samuel Ippolito and Glenn Matthews of RMIT, Melbourne and Reetu Singh of USF for useful discussions.

REFERENCES

[1] S. Cular, V. R. Bhethanabotla, and D. W. Branch, "Hexagonal surface acoustic wave devices for enhanced sensing and material characterization," presented at IEEE Ultrasonics, Rotterdam, Netherlands, 2005.

[2] S. Cular, D. W. Branch, V. R. Bhethanabotla, G. D. Meyer, H. G. Craighead, "Removal of nonspecifically bound proteins on microarrays using surface acoustic waves." *IEEE Sensors Journal* , vol. 8(3), pp. 314-320, 2008.

[3] S. K. R. S. Sankaranarayanan and V. R. Bhethanabotla, "Finite element modeling of hexagonal SAW device based on $LiNbO_3$." *IEEE Sensors Journal*, In Press, 2008.

[4] S. K. R. S. Sankaranarayanan, V. R. Bhethanabotla, "Design of efficient focused surface acoustic wave devices for potential microfluidic applications." *Journal of Applied Physics* vol. 103(6), pp. 064518/1-064518/17, 2008.

[5] S. K. R. S.Sankaranarayanan, S. Cular, V. R. Bhethanabotla, and J. Babu. "Flow induced by acoustic streaming on surface-acoustic-wave devices and its application in biofouling removal: A computational study and comparisons to experiment." *Physical Review E*, vol. 77, pp. 066308/1- 066308/19, 2008.

[6] S. Cular, S. K. R. S. Sankaranarayanan, and V. R. Bhethanabotla, "Enhancing effects of microcavities on shear-horizontal surface acoustic wave sensors: A finite element simulation study." *Applied Physics Letters*, 92(24), 244104/1-244104/3, 2008.

[7] B. A. Auld, *Acoustic Fields and Waves in Solids.*, vol. 1-2. New York, N.Y., USA: John Wiley and Sons, 1973.

[8] S. J. Ippolito, K. Kalantar-Zadeh, D. A. Powell, and W. Wlodarski, "A 3-dimensional finite element approach for simulating acoustic wave propagation in layered SAW devices.," presented at Ultrasonics, IEEE Symposium, 5-8 Oct. 2003.

[9] R. Lerch, "Simulation of piezoelectric devices by two- and three-dimensional finite elements," *Ultrasonics, Ferroelectrics and Frequency Control, IEEE Transactions*, vol. 37, pp. 233 - 247, 1990.

[10] J. J. Campbell and W. R. Jones, "A method for estimating optimal crystal cuts and propagation directions for excitation of piezoelectric surface waves.," *IEEE Transactions on Sonics and Ultrasonics*, vol. 15, pp. 209-217, 1968.

A fractal analysis is presented for (a) the binding of TNF-alpha (tumor necrosis factor) in solution to poly(guanine)-functionalized silica nanoparticles (NOs) (Wang et al., 2006), the binding of different antigens in solution to the anti CD antigen immobilized on a quartz crystal microbalance (QCM) surface (Zeng et al., 2006), (c) the binding and dissociation of cardiomyocytes plus endothelin-1 with and without a DEP (dielectrophoresis) device (Yang et al., 2007), and (d) the binding and dissociation of different concentrations of oxazaborolidine derivatives, BNO1, BNO2, BNO3, and BNO4 + 2 mM sucrose in solution to the enzyme fructosyltransferase (FTF) immobilized on a SPR (surface plasmon resonance) biosensor chip surface (Jabbour et al., 2007). Both, a single- and a dual-fractal analysis are used to model the binding and the dissociation (if applicable) kinetics. The dual-fractal analysis is used only if the single-fractal analysis did not provide an adequate fit. The expressions obtained for the binding and the dissociation rate coefficeints for a single- and a dual-fractal analysis as a function of the fractal dimension indicate a high sensitivity of these rate coefficients on their respective fractal dimensions on the SPR sensor chip surface. Note that the data analysis in itself does not provide any evidence for surface roughness or heterogeneity, and the existence of surface roughness or heterogeneity assumed may not be correct. However, considering the complexity involved on the SPR chip surface, this is not an unreasonable assumption.

Exfoliation of layered platelets and intercalation of polymer chains: effects of molecular weight and quality of polymer solvent

Ras B. Pandey, University of Southern Mississippi, Hattiesburg, MS
Barry L. Farmer, Air Force Research Laboratory, WPAFB, Dayton, OH

Abstract

A Monte Carlo simulation is used to investigate the dispersion of layered platelets in a polymer matrix on a discrete cubic lattice with a coarse-grained model. A polymer chain is a set of L_c nodes tethered together by fluctuating bonds while a platelet (sheet) is described by a set of L_s^2 nodes tethered together by flexible bonds on a square grid. A stack of layered sheets is placed in the center followed by a random distribution of polymer chains on a fraction of lattice sites. Chains and platelets interact and execute their stochastic motion via Metropolis algorithm. Dispersion of sheets and intercalation of chains are studied as a function of molecular weight of polymer chains for attractive and repulsive interaction between polymer and platelets. We find[1] that, with the low molecular weight polymer solvent, the sheets exfoliate in presence of polymer with repulsive interaction while maintain layered structure with attractive interaction via solvent induced interaction. The system becomes much more complex on increasing the molecular weight of polymer as the interplay between the interaction thermodynamics and structural entropy sets in.

1. R.B. Pandey and B.L. Farmer, J. Polym. Sci. B, Polymer Physics (accepted for publication).

702b A Complete Multiscale Modeling Approach for Nanocomposites

Sabrina Pricl, Molecular Simulation Engineering (MOSE) Laboratory - DICAMP, University of Trieste, Piazzale Europa 1, Trieste, 34127, Italy, Paola Posocco, Molecular Simulations Engineering (MOSE) Laboratory - DICAMP, University of Trieste, Piazzale Europa 1, Trieste, 34127, Italy, Giulio Scocchi, Icimsi, SUPSI, Galleria 2, Manno, CH-6928, Switzerland, Maurizio Fermeglia, Molecular Simulation Engineering Laboratory, Department of Chemical Engineering, University of Trieste, Piazzale Europa 1, Trieste, 34127, Italy, Jan-Willem Handgraaf, CULGI B.V., P.O. Box 557, Leiden, Netherlands, and Johannes Fraaije, Leiden Institute of Chemistry, Soft Matter Chemistry, Gorlaeus Laboratories, University of Leiden, Einsteinweg 55, Leiden, Netherlands.

Blending molten polymer and inorganic clays can result in a class of new materials, in which nanoscale clay particles, generally layered silicates, are molecularly dispersed within the polymeric matrix. Such polymer-clay nanocomposites (or PCNs) exhibit dramatic increases in several properties, including mechanical strength and heat resistance, and decrease in gas permeability, when compared to the polymeric matrix alone. Importantly, the improvement in these properties is achieved at very low loadings of the inorganic component, typically, 1-10 weight %, thus rendering PCNs lighter in weight than any other conventionally filled polymer. These unique features make PCNs ideal materials for applications such as high-barrier for food or pharmaceutical packaging, to strong, heat resistant automotive components, just to name a few. Fabricating these materials in an efficient and cost-effective manner, however, poses significant synthetic challenges. To appreciate these challenges, let us discuss briefly the structure of layered silicates by considering montmorillonite (MMT) as a prime example. This inorganic clay consists in stacked silicate sheets, each approximately 200 nm long and 1 nm thick. The spacing between each sheet (or gallery) is also of the order of 1 nm, and this quantity is clearly smaller than the average radius of gyration of any conventional polymer. Therefore, entropy generally constitutes a large barrier that prevents the polymer from penetrating these galleries and become an intercalated material. Accordingly, there is a number of critical issues that need to be addressed in order to optimize the design and production of PCNs. Of foremost importance is the isolation of the conditions that result in a promotion of the polymer penetration into the narrow clay galleries. If, however, the sheets ultimately phase-separate from the polymer matrix, the mixture will not exhibit improved strength, heat resistance, or barrier properties mentioned above. Accordingly, it is also essential to determine the factors that control the macroscopic phase behavior of the mixture. Finally, the properties of the PCN commonly depend on the structure of the material; thus, it is of particular interest to establish the morphology of the final composite.

To date, there are few theories to pinpoint the critical parameters or to predict the thermodynamic stability of a PCN system, forcing synthetic chemists to synthesize all possible mixtures in order to isolate the desired system. Therefore, in order to develop new materials and composites with designed new properties, it is essential for these properties to be predicted before preparation, processing, and experimental characterization. Despite the tremendous advances made in the modeling of structural, thermal, mechanical and transport properties of materials at the macroscopic level (finite element (FE) analysis of complicated structures), there remains a tremendous uncertainty about how to predict many critical properties related to performance. The fundamental problem here is that these properties depend on the atomic level of interactions and chemistry, dealing with the electronic and atomic level of description and at a length/time scale of nanometers and nanoseconds. The material designer, however, needs answers from macroscopical modeling (the finite element paradigm) of components having scales of centimeters and milliseconds, if not larger. To substantially advance the ability to design useful high performance materials, it is then essential that we insert the chemistry into the mesoscopic (MS) and macroscopic (FE) modeling. Currently, atomistic level simulations such as molecular dynamics (MD) or Monte Carlo (MC) techniques allows to predict the structure and properties for systems of considerably large number of atoms and time scales of the order of microseconds. Although this can lead to many relevant results in material design, many critical issues in materials design still require time and length scales far too large for practical MD/MC simulations. Therefore, we need to develop methods treating the mesoscale in between the atomistic length and time scales of MD/MC and the macroscopic length and time scales (microns to mm, and ƒÝs to s) pertaining to FE analysis. This linking through the mesoscale, in which we can describe a system microstructure, is probably the greatest challenge to developing reliable first principles methods for practical and effective material design. Indeed, only by establishing this connection from microscale to mesoscale it is possible to build first principles methods for describing the properties of new materials and composites.

One of our major aims is to reach the domain of materials science and engineering by building from fundamental principles of physics and chemistry. Thus, for fundamental predictions to play a direct role in materials innovation and design, it is essential to fill the micro-meso gap. The problem here is that the current methods of coarsening the description from atomistic to mesoscale (as well as MS to FE) are not as obvious as they are from going to the quantum mechanics (QM) to the atomistic level, being strongly system-dependent and, hence, hardly generalizable. Indeed, it is quite clear that the strategy for polymers should be rather different from that adopted for metals, and again different from that conceivable for ceramic systems.

Given these concepts, it is than necessary to carry out calculations for realistic time scales fast enough to be useful in design. This requires developing techniques useful to design engineers, by incorporating the methods and results of the lower scales (e.g., MD) to mesoscale simulations. In this work, we developed a hierarchical procedure for bridging the gap between atomistic and mesoscopic simulation for PCN design. The Mesoscopic Bead-Field (MBF) hybrid method is adopted as the simulation technique, and all necessary parameters of the mesoscopic model are estimated by a step-by-step procedure involving a) the matching of the atomistic and mesoscopic pair correlation functions to determine the best mesoscopic topology for the organic modifiers, b) the mapping of interaction energies obtained from atomistic simulations onto the mesoscopic bead-bead interaction parameters, and c) the matching of the atomistic and polymer density profiles to calculate the bead-field coupling parameters. Finally, the mesoscopic simulated structures are passed on to the FEM calculations, to estimate the macroscopic properties of the PCN (e.g., Young modulus, diffusivity, etc.).

The global perspective of our work is the complete integration of all available simulation scales, in a hierarchical procedure to provide an efficient and robust simulation protocol for the successful design of PCNs of industrial interest and the prediction of their final performance.

Dynamics of Solvated Chloride Inhibition by Nanoparticle Treated Concrete

Harish Venkateshaiah[1], Jinko Kanno[2], Nicholas Richardson[2], James Phillips[2], Kunal Kupwade-Patil[3], Henry E. Cardenas[3] and Daniela S. Mainardi[1]

[1]*Chemical Engineering Program/ Institute for Micromanufacturing, Louisiana Tech University, Ruston, LA 71272*
[2]*Mathematics Program, Louisiana Tech University*
[3]*Mechanical Engineering Program, Louisiana Tech University*

Abstract

Corrosion of steel reinforcement in concrete is caused largely by ingress of chloride ions from deicing salts and coastal marine environments. Electrokinetic nanoparticle treatment concurrent with electrochemical chloride extraction have been employed to minimize chloride attack and reinstate the passivity of embedded steel. Additionally, pozzolanic nanoparticles such as aluminum- and silicon-containing materials can be injected into concrete pores using externally applied electric fields in order to reduce permeability of chloride ions into concrete. We present results of computational simulations used to investigate dynamics of solvated chloride anion (Cl^-) through a tricalcium silicate nanopore in the presence of alumina (Al_2O_3) and silica (SiO_2) nanoparticles that form the main binder phase of Portland cement. First, equilibrium properties of $Cl^-(H_2O)$ system are explored, using Density Functional Theory with-localized GGA exchange correlation functional of PW91. Microsolvation conformers are thus obtained as a gradual approach to bulk solvation structures. Cl^- atomic size is explored using radial distribution functions (RDF). Molecular Dynamics in the canonical ensemble is used to obtain coefficient of diffusion of solvated Cl^- at 25 °C. Diffusivity calculated from the variation of the mean square displacement of the chloride ions with time show considerable differences depending on their position with respect to nanoparticles. Increase in ion size after solvation and its diffusivity obtained implicate both solvation structure and alumina-silica nanoparticles as controlling factors for Cl^- dynamics in aqueous systems inside concrete pores.

I. Introduction

In the US alone, research programs are established by the Federal Highway Administration for corrosion protection in concrete bridges at high priority to develop economical modes of corrosion containment. Concrete is a composite material that consists essentially of binding medium within which are embedded particles of aggregate that form the porous structure of concrete. Corrosion of reinforcement steel is estimated to have rendered 29.6% of concrete structures deficient or functionally obsolete, as per the US Federal Highway Administration in 1996. Highway bridge repairs alone afflicted an enormous annual direct cost of $8.3 billions [1]. Chloride-induced corrosion is auto-catalytic in nature. The aggressive anion penetrates the porous cementitious material triggering electrochemical reactions corroding the reinforcement [2,3]. Recent application of electrokinetic nanoparticle treatment is known to reduce the permeability, and therefore ionic transport [4]. Adequate porosity reduction was attained to improve the durability of cementitious material considerably. Using electric field, reactive pore blocking chemicals were driven into the saturated concrete pores. The nanoparticles react with cementitious material in presence of moisture to form a main binder phase in Portland cement [5].

Uncertainty is involved with material properties, structure geometry and dimensions, as well as with analytical models of chloride penetration into concrete and onset of reinforcement corrosion though probabilistic models have been developed to understand the extent of damage done [6,7]. The present work investigates diffusion of the chloride anion in the presence of water. Molecular modeling is used to simulate the nanosystem of concrete pore with nanoparticles, analogous to experimental [5] and analytical models [8] developed already. The diffusivity of chloride ions is determined by the MSD method after conducting Molecular Dynamics calculations.

II. Simulation Methods

Solvation of Chloride Anion (Cl^-) Using Density Functional Theory:
The DMOL^3 module of the Material Studio program developed by Accelrys, Inc. has been used to carry out Density Functional Theory (DFT) calculations, to obtain a local minima conformer representative of solvated Cl^-. Vibrational frequencies of the model and the charge population are computed for the geometry optimization step. A customized geometry optimization calculation was employed with a non-localized Generalized Gradient Approach (GGA) exchange correlation functional of Perdew and Wang (PW91). Ground spin state was determined by setting up spin-unrestricted calculation without using symmetry information. Total charge of the water-anion system is set to -1e (charge on anion). With an explicit solvent model, all electrons in the system are included in the calculation. Double Numerical plus polarization (DNP) basis set was used since it includes polarization *p*-function on all hydrogen atoms in water system involving hydrogen bonding. DNP essentially includes one atomic orbital for each occupied atomic orbital, a second set of valence atomic orbitals and a polarization *d*-function on all non-hydrogen atoms. Electrostatic potential-derived Merz-Kollman charges of 0.393 and -0.787 e are applied to the hydrogen atoms and oxygen atom of a single water molecule, respectively.

Tricalcium Silicate (C_3S) Nanopore Using Molecular Mechanics and Dynamics:
Forcite module of the Material Studio program developed by Accelrys, Inc. has been used for Molecular Mechanics (MM) geometry optimization calculations. C_3S unit cell was used to model a concrete nanopore by building a *supercell* of required dimensions. Cylindrical nanopore of about 6 nm diameter and 8 nm height was built. Constraint to atomic motions was applied on parts of the cylinder that do not come in contact with the nanosystem inside the pore for ease of computation. XYZ components of such atomic positions were held constant in the Cartesian space. Geometry minimization calculation was done using the Smart Algorithm, which is a cascade of steepest descent, conjugate gradient, quasi-Newton, and ABNR method. Maximum energy change of ~0.02 kcal/mol was tested as convergence threshold, discounting the force, stress, and displacement. *Universal* forcefield was use to describe interatomic bonded and non-bonded interactions. The non-bond terms in the structure were evaluated by *atom-based* summation method. The electrostatic and van der Waals' forces between all the atoms in the structure participate in the summation. *Normal* convergence was achieved for maximum iterations set to 10,000 optimization cycles. The alumina and silica nanoparticles were modeled

by repeating the above procedure, with no constraints applied to motion of its atoms. Diameters of 0.5 nm for alumina and 2 nm for silica were modeled using respective *supercell*. The models of nanopore and nanoparticles were converted to non-periodic structures. An external electrode of silver (Ag) with adequate positive charge was placed at one end of the cylinder to generate sufficient potential to electrostatically attract negatively charged free Cl^-. Concrete cylinder with alumina-silica nanoparticles, and external electrode was minimized using the Forcite Geometry Optimization calculations setup for the concrete pore previously.

Further Molecular dynamics (MD) calculations were used in this work. MD integrates Newton's law of motion for all atoms in the molecular system with statistical mechanics for the prediction of transport properties [9]. Fundamental transfer property of ions such as diffusivity needs to be evaluated at confined regions to understand its dynamics at the interface. Lower number of hydrogen bonds and reduced density in this region causes interfacial water molecules to translate and rotate at a faster rate than in bulk molecules. Dynamic properties of interfacial water molecules are less affected by ion concentration unlike the bulk water molecules. Hence, interfacial water molecules can be expected to preferentially solvate Cl^- since the solvation free energy depends on the relative location of solute ions at the interface. Hence, MD is effective in providing reasonable free energy data for even systems with many energy minima separated by energy barriers [10].

Mechanism of ionic transfer and free energy of ion across liquid interfaces provides insight on equilibrium properties and dynamics of ions. Understanding the solvation and adsorption of ions in aqueous interfaces is essential in modeling and controlling chemical reactivity [10]. Dynamics of the solvated Cl^- in concrete pore system are studied using MD. The calculations setup for MD has been discussed in detail in further sections.

III. RESULTS AND DISCUSSION

A. Static Properties and Energetics

This section presents the results of simulations carried out to prepare input models for the subsequent dynamics study of Cl^- in a concrete pore system along with alumina-silica nanoparticles.

1. Solvated chloride ion

The chloride anion-water $[Cl^-(H_2O)_n]$ minimum energy structure was obtained from DFT calculations for n=1 to13 and 16 to 25. These geometries correspond to relative minimum on the potential energy surface. The structures are reported as minima for $Cl^-(H_2O)$. Though solvation can occur with 6 water molecules [11], models of n=1 to 6 were found to be surface anions [12].

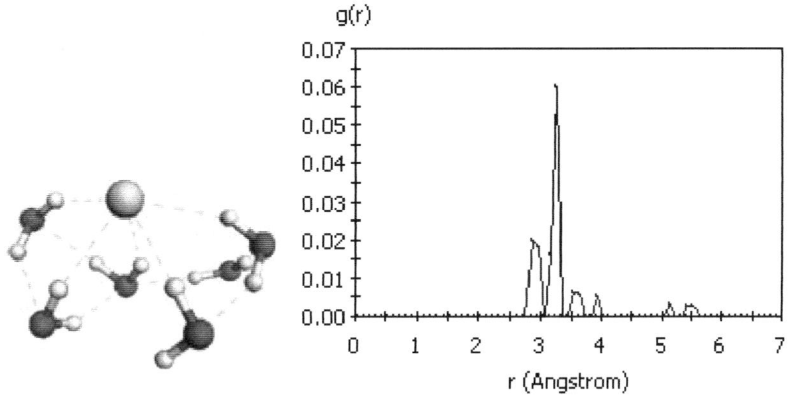

Figure 1: Microsolvation structure $Cl^{-}(H_2O)_{n=6}$ with radial distribution function of O-Cl^{-}

The solvation clusters represent a stationary point on the energy surface where forces on all atoms are zero [13]. The relative energies of such ground states corresponding to stable structure are reported in table 1. Ion position got more interior with higher water molecules. Anion completely solvated is expected to lie within the liquid water network or surrounded by the water molecules [12,14]. Internal solvation was observed for $n>17$ as shown in figure 2, and simulation was continued till $n= 25$ to confirm this transformation from surface to internal solvation [15]. Table 1 also reports the relative ground state energy per water molecule of each model.

Table 1: Electronic energies of ground states for $Cl^-(H_2O)_{n=1\text{ to }13\text{ and }16\text{ to }25}$

No. of Water Molecules per Cl^-	Total Electronic Energy (10^5kcal/mol)	Electronic Energy per water molecule (10^5kcal/mol)
1	-3.37	-3.37
2	-3.85	-1.92
3	-4.33	-1.44
4	-4.81	-1.20
5	-5.29	-1.06
6	-5.77	-0.96
7	-6.25	-0.89
8	-6.73	-0.84
9	-7.21	-0.80
10	-7.69	-0.77
11	-8.17	-0.74
12	-8.65	-0.72
13	-9.13	-0.70
16	-10.57	-0.66
17	-11.04	-0.65
18	-11.52	-0.64
19	-12.00	-0.63
20	-12.48	-0.62
21	-12.96	-0.62
22	-13.44	-0.61
23	-13.92	-0.61
24	-14.40	-0.60
25	-14.88	-0.60

Vibrational frequencies were checked to ensure ground state conformations. At least 3 minima conformers were obtained for each combination of $Cl^-(H_2O)_n$ for n= 1 to13 and 16 to 25, to explore different local minimum configurations. Large reorganization is observed within the hydrogen-bonded (H-bond) water network and the anion seems to travel to the surface of the shell. The smaller clusters continue to show clearly the surface solvation of anions for up to $n<18$. Larger clusters show an indication of liquid water hydration. The anion is completely surrounded by water molecules only after n=18, which gives a good scale [16] for cluster size to be used further in our MD simulations, since $Cl^-(H_2O)_{n=25}$ binds its excess electron more strongly than with less number of water molecules (figure 3). Results show strong H-bonding between Cl^- and hydrogen atoms of water (O-H—Cl^-) and between the individual water molecules (O-H—O). Different minima for same n (number of water molecules) showed the lowest energy configuration to have lesser number of O-H—Cl^- bonds. This indicated more stable structures for clusters with more interwater H-bonding for a particular n [11].

Radial distribution function:

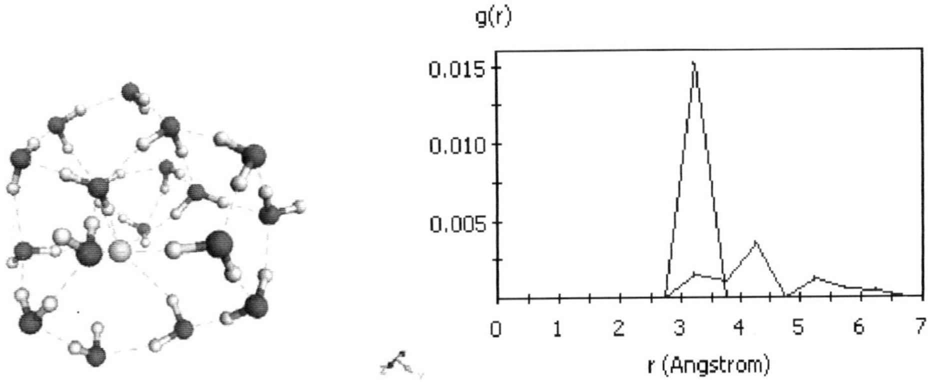

Figure 2: Internal solvation structure $Cl^{-}(H_2O)_{n=18}$ with radial distribution function of O-Cl^{-}

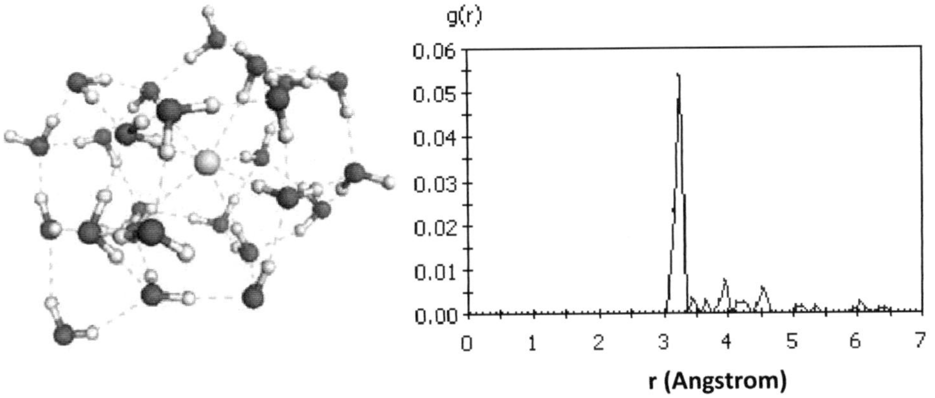

Figure 3: Completely solvation structure $Cl^{-}(H_2O)_{n=25}$ with radial distribution function of O-Cl^{-}

The radial distribution functions (RDFs) of Cl^{-} is shown in figures 1, 2 and 3 for different *n*=6, 18 and 25. The structural influence of Cl^{-} is expected to extend up to a second solvation shell [17]. The plots of distribution function for radial distance for O-Cl distance indicates first hydration shell size of ~ 0.33 to 0.34 nm for system of 6, 18 and 25 water molecules at the highest peaks of respective plots. Figure 1 shows a first peak at ~ 0.28 nm and a second high peak at 0.33 nm due to surface solvated position of Cl^{-} in the shell. There is no peak at 0.28 nm in figures 2 and 3 since the Cl^{-} is more surrounded by water molecules. X-ray and neutron diffraction measurements are known to show hydration of one Cl^{-} with 6 water molecules [11]. Using the atomic radius of oxygen atom of 0.152 nm [8], we observe that the size of the solvated anion is higher than its atomic radius of free Cl^{-}.

2. Alumina-silica Nanoparticles and Concrete Nanopore

Chloride ion penetration in concrete was investigated experimentally [5] for Portland cement concrete cylinder of 3in diameter and 6in length. 2 nm alumina coated over 20 nm silica composites reduced the porosity of concrete by forming a binding phase.

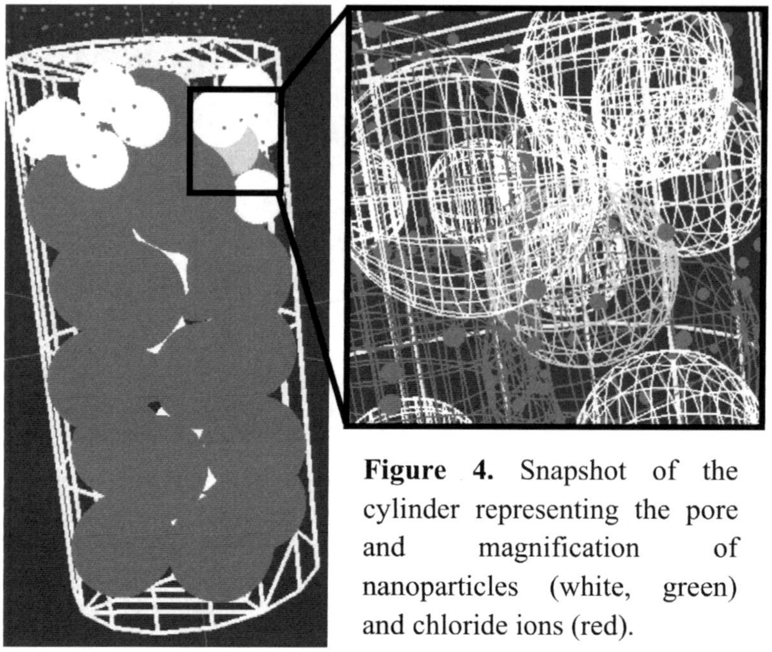

Figure 4. Snapshot of the cylinder representing the pore and magnification of nanoparticles (white, green) and chloride ions (red).

The concrete pore with the nanoparticles and solvated ion were also modeled analytically. The cylinder and spheres representing the system is shown in figure 4. Stacking and packing assembly methods were used such that each particle stays inside the cylinder and also interacts with each other to nearly represent the actual system. Porosity reduction of ~57% was achieved after the model was tested for different combinations of nanoparticles size. Considerable porosity reduction was obtained for a diameter combination of 1:4 and 1:5 [8].

In the present work the sizes of nanoparticles are scaled down for computational ease still maintaining the size ratio. The experimental work showed that particles aggregate with a positive charge making it possible to drive them all the way to the steel reinforcement inside concrete. The alumina on contact with steel adheres to its surface to protect steel from dropping in pH to remain passivated for longer duration [4]. Silica suspension is known to have higher buffer capacity than alumina, and increases the pH of the medium. Though alumina does not aggregate in suspension, on introduction of silica, it forms heteroaggregates owing to opposite charges. Silica suspension is known to have higher buffer capacity than alumina, and increases the pH of the medium. Figure 5 shows the structure optimized to use in MD runs. The nanoparticles were stacked inside the cylinder to cover most of the through space, though not to the extent as in the analytical stacking. It can be observed that the inner surface of the concrete cylinder reacts with few alumina particles adhering to its surface. The silica adsorption is not visually evident due to tight packing. In presence of water the silica is expected to form a binding phase C-S-H.

3. Mean Square Displacement of Chlorides Using Molecular Dynamics (MD)

Forcite module of the Material Studio program developed by Accelrys, Inc. has been used for Molecular Dynamics (MD) calculations to investigate the dynamics of the Cl^- in modeled nanosystem. Four clusters of solvated chloride ion (4 Cl^- ions-100 water molecules) were introduced into the concrete pore system optimized using Molecular Mechanics. The *canonical ensemble* of NVT (number of atoms, system volume and temperature remains constant) for the thermodynamic calculations for atoms in the non-periodic system. *Velocity initialization* for atoms in the system was set using random values with a temperature-dependent Gaussian distribution. A default time step of 1.0 fs is selected with *number of steps* as 5000. These two factors determine the simulation time. A *frame output* of 50 is selected for frequency at which the simulation frame is to be written to the trajectory file. The task parameters controlling accuracy and speed of the MD calculation is set at *coarse* level to better simulation speed, without compromising accuracy of calculation, which would be possible in case of a periodic structure. *Universal* forcefield is chosen to approximate the PES using 'current' system charges. The summation methods for the non-bond interactions- electrostatic and van der Waals, is set to atom based since no groups are defined for this non-periodic system. The dynamics simulation output is then used to analyze the mean square displacement of the chloride ions defined as a distinctive set. From the generated data (a plot of mean square displacement of chloride ions vs. time) the slope of the curve should be equal to 2dD, which gives the final value of chloride diffusivity [9].

Dynamics of solvated chloride- Molecular dynamics

Continuous and discrete models have been adopted for the simulation methods to study ionic mobility. The geometry minimized system after molecular mechanics (figure 5) is used for dynamics run input. The dynamics simulation was conducted at 25 °C and pH 7. The electrostatic forces due to externally applied electric field cause the negative chloride to travel across the nanoparticles. The dynamics of chloride ions with respect to the nanoparticles is studied by fixing the surrounding system of concrete matrix and opposite charge electrode. Only the internal surface of concrete cylinder is found to aggregate with the nanoparticle; which is comparable to the formation of binding phase on cement. The chlorides are found to break out of the solvation shell and travel across the nanoparticles. The majority of system energy is observed to be from bond, angle, torsion, and inversion interactions, as compared to the non-bond energy. System potential energy reduction from initial configuration is lesser when compared to the kinetic energy change. The temperature convergence is observed to be satisfactory with low standard deviation and the average reading being close to required temperature.

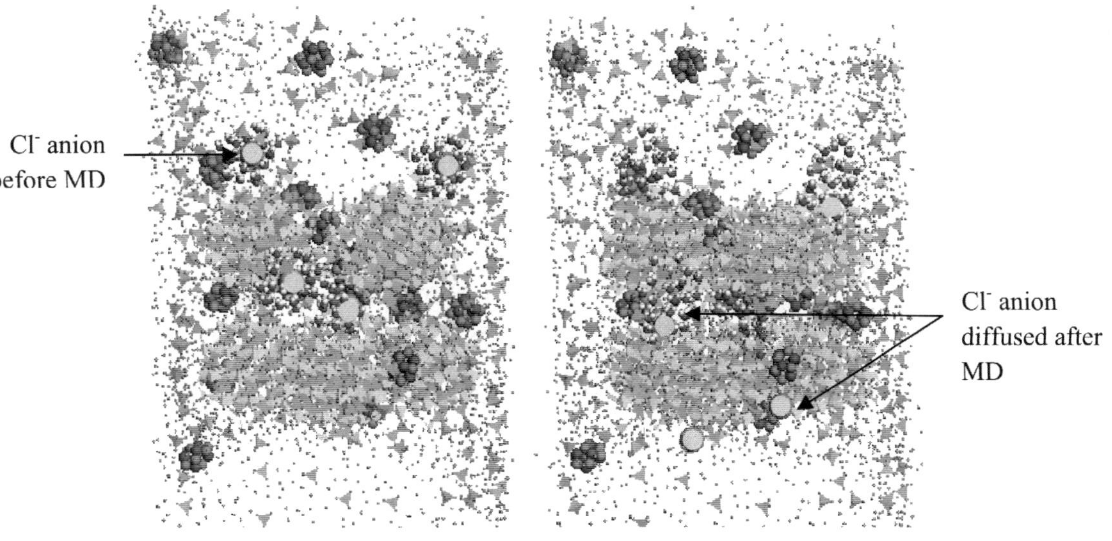

Figure 5: Concrete pore (grey) model with solvated Cl^- (green) and Al_2O_3 (pink)/SiO_2 (yellow) nanoparticles.

The MD simulation output is shown is figure 5. Initially, the 4 solvated chloride ions were seen within the water network. The cylinder model in figure 6 is ~ 8 nm in height. Assuming the bottom of cylinder to be '0 nm' and measuring up to '8 nm' at the top, the diffusion is observed from 8 to 0 nm (top of cylinder to bottom), since the applied external field is a positive charge of +1e or 1.6×10^{-19} coulombs per Ag atom in the electrode, at the 0 nm. The initial position of the 4 anions prior to MD is reported in table 2 with respect position along axial length of the cylinder, starting with anion *A* being the farthest and *D* the closest.

Table 2: Diffusivity of each anion compared to distance from externally applied electric field.

Cl^-	Initial position (Distance measured from 0 to 8 nm)	Individual diffusivity, $D * 10^{-3}$ in cm^2/s
D_A	6.75	1.55
D_B	6.53	5.06
D_C	4.34	2.73
D_D	3.94	3.82

The coefficient of diffusion of each ion calculated from mean square displacement (MSD) of ions is reported in table 2. *A* was placed the farthest across the nanoparticles and showed lowest diffusivity among the 4 ions. On the other hand *B* placed at similar distance showed maximum diffusivity 5.06 cm^2/s. This indicates *B* traversed the cylinder along a path with voids, while path of *A* was blocked by the alumina-silica aggregates.

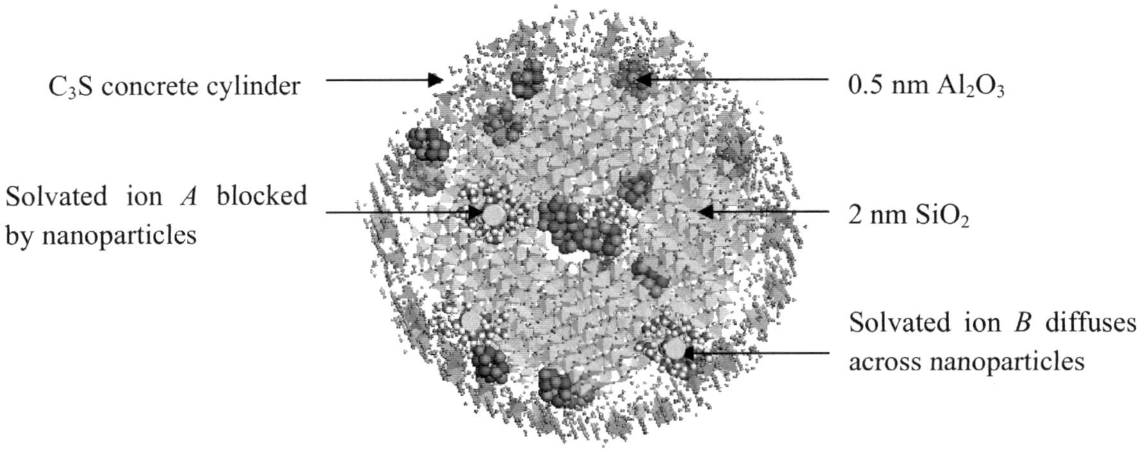

Figure 6: Top view of cylinder highlighting position of Cl^- ions (A & B)

The free chlorides diffuse farther across as compared to the solvated chloride. The solvated anions are defined as unique set in the simulation output trajectory and the dynamics is quantified by mean displacement method using the Einstein's law. The calculation of diffusivity *D* of all 4 chloride ions is shown below along with the MSD profile versus time.

The slope of the linear profile of the curve is 154.9 $Å^2$/ps. Using Einstein's equation:

$r^2(t) = 2dDt$, where D= diffusivity and d=3 for 3-D systems, the diffusion coefficient of chloride ions is $D = 2.58 \times 10^{-3}$ cm^2/s.

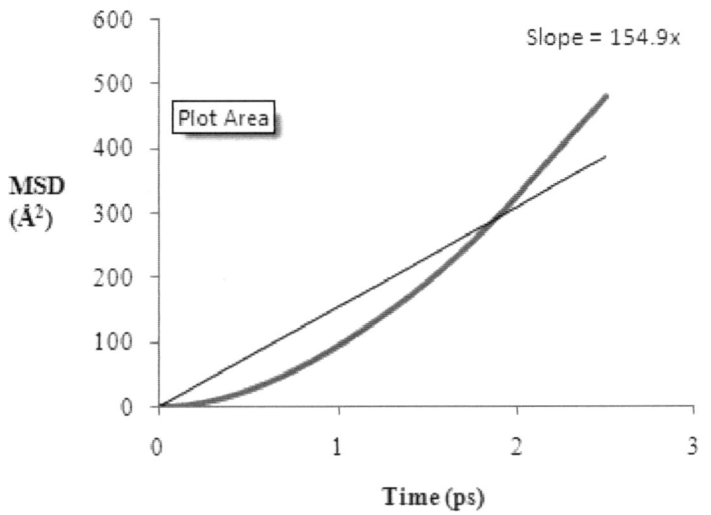

Figure 7: Mean square displacement of solvated Cl^- anions with time

IV. CONCLUSION

Solvation provided Cl^- atomic radius slight increase for n=2 to 25. Larger hydration shells are to be explored to confirm considerable size increase. Size ratio of the nanoparticles modeled (1:5) was in agreement with the analytical model that attained highest porosity reduction. Ionic diffusivity for anions A and B was considerably high; which indicates that the model needs to be investigated with better packing of the nanoparticles inside the pore and different electric field strengths.

References

1. Klieger, P; Lamond, J. F. "Significance of Tests and Properties of Concrete and Concrete-Making Materials," *ASTM International*, 1994.
2. Schweitzer, P. A. "Corrosion Engineering Handbook," *Marcel Dekker*, **1996**.
3. Amleh, L.; Mirza, M. S. "Corrosion Response of Decommisioned Deteriorated Bridge Deck," *Journal of Performance of Constructed Facilities*, **2004**, 18 (4), 185-194.
4. Cardenas, H. E.; Struble, L. J. "Electrokinctic Nanoparticle Treatment of Hardened Cement Paste for Reduction of Permeability," *Journal of Materials in Civil Engineering* **2006**, 554-560.

5. Cardenas, H. E.; Kupwade-Patil, K. V. "Corrosion Mitigation in Concrete Using Electro-Kinetic Injection of Reactive Composite Nanoparticles," *Journal of the American Concrete Institute*, Submitted January **2007**.

6. Mears, R. B.; Brown, R.H. "Corrosion Probability," *Industrial and Engineering Chemistry*, October **1937,** 29 (10); 1087-1091.

7. Lounis, Z. "Probabilistic Modeling of Chloride Contamination and Corrosion of Concrete Bridge Structures," *IEEE Computer Society* **2003,** Proceedings of the Fourth International Symposium on Uncertainty Modeling and Analysis.

8. Kanno, J.; Richardson, N.; Phillips, J.; Kupwade-Patil K.; Mainardi, D.S.; Cardenas H.E. "Modeling and Simulation of Electromutagenic Processes for Multiscale Modification of Concrete" *Proceeding paper of the 12th World Multi-Conference on Systemics, Cybernetics and Informatics: WMSCI 2008*. June 29th - July 2nd, **2008** – Orlando, Florida, USA.

9. Leach, A. R. "Molecular Modeling- Principles and Application," *Prentice Hall*, 2nd Edition **2001**.

10. Chang, T; Dang, L. "Recent Advances in Molecular Simulations of Ion Solvation at Liquid Interfaces," *Chem. Rev.* **2006**, 106, 1305-1322.

11. Pathak, A. K.; Mukherjee, T.; Maity, D. K. "Microhydration Shell Structure in Cl_2^-. nH_2O clusters: A Theoretical Study," *J. Chem. Phys.* **2006**, 125, 074309.

12. Simons, J. "Molecular Anions," *J. Phys. Chem. A.* **2008**, 112, 6401-6511.

13. Reichardt, C. "Solvent and Solvent Effects: An Introduction," *Organic Process Research and Development* **2007**, 11, 105-103.

14. Christopher, H.; Holbrey, J. D.; Nieuwenhuyzen, M.; Youngs, T. G. A. "Structure and Solvation in Ionic Liquids," *Acc. Chem. Res.* **2007**, 40, 1146–1155.

15. Kemp, D. D.; Gordon, M. S. "Theoretical Study of the Solvation of Fluorine and Chlorine Anions by Water", *J. Phys. Chem.* **2005**, 109, 7688-7699.

16. Daguenet, C.; Dyson, P. J. "A Metallacage Encapsulating Chloride as a Probe for a Solvation Scale in Ionic Liquids," *Inorg. Chem.* **2007**, 46, 403-408.

17. Heuft, J. M.; Meijer, E. J. "Density Functional Theory based Molecular-Dynamics Study of Aqueous Chloride Solvation," *J. Chem. Phys.* **2003**, 119(22), 11788-11791.

18. Cerbelaud, M.; Videcoq, A.; Abelard, P.; Pagnoux, C.; Rossignol, F.; Ferrando, R. "Heteroaggregation between Al2O3 Submicrometer Particles and SiO2 Nanoparticles: Experiment and Simulation," *Langmuir*, 2008, 24, 3001-3008.

19. Bizzarri, A. R.; Cannistraro, S. "Molecular Dynamics of Water at the Protein-Solvent Interface," *J. Phys. Chem. B.* **2002**, 106(26), 6617-6633.

20. Lee, S. H.; Rasaiah, J. C. "Molecular Dynamics Simulation of Ion Mobility.2. Alkali Metal and Halide Ions Using the SPC/E Model for Water at 25 deg C," *J. Phys. Chem.* **1996**, 100, 1420-1425.

702d Off-Lattice Monte Carlo Based Nanopaint Design for Coating Scratch Resistance Improvement

Jie Xiao and **Yinlun Huang**. Department of Chemical Engineering and Materials Science, Wayne State University, 5050 Anthony Wayne Drive, Detroit, MI 48202

Nanopaint has drawn great attention in recent years, as it may offer superior mechanical and/chemical properties with the resulting coating (e.g., scratch resistance). The known experimental efforts in this area have improved our understanding on coating microstructure and its correlation with coating properties. However, due to the design and experimental complexity as well as cost concern, how to optimally design nanopaint in a cost effective way is extremely challenging.

In this paper, we will introduce a computational nanopaint design methodology by resorting to off-lattice Monte Carlo modeling and simulation techniques. This methodology contains a unified coarse-grained model that is for characterizing comprehensive interactions among monomers, crosslinkers and spherical nanoparticles. By off-lattice MC, coating microstructures can be generated and computational tensile tests can be then conducted to evaluate the stress-strain behavior of the coating, which is critical to revealing the coating tensile property and scratch resistance performance. In this way, various comprehensive and quantitative correlations among nanopaint material, coating microstructure, coating property, and coating quality can be established. In order to have an in-depth understanding on the structure-property correlations, the microstructures are further characterized by quantifying the spatial distribution of polymer beads surrounding nanoparticles. The introduced characterization method has unique capability for analyzing the nanopaint material, where the distribution of the monomers and the crosslinkers can be differentiated for further investigation. Furthermore, a new stress evaluation method is presented, which can enable a thorough investigation on different stress contributors (i.e., polymer-polymer stress, polymer-nanoparticle stress, nanoparticle-nanoparticle stress) and their evolutions during a tensile test. Such a stress partitioning study is crucial for gaining a better understanding of the deformation behavior of the nanopaint material.

An acrylic-melamine-alumina nanoparticle contained nanopaint design is selected as a case study to demonstrate the attractiveness of the introduced methodology. In this study, the effect of polymer-nanoparticle interaction strength and that of nanoparticle size on coating tensile property and scratch resistance performance are thoroughly investigated. It is found that adding nanoparticles into a convectional paint sample may not be guaranteed for coating performance improvement. A key for achieving the enhancement is to ensure that the polymer-nanoparticle interaction strength is higher than a critical value. It is also identified that a nanopaint material containing smaller nanoparticles can have a better scratch resistance performance, but it requires more energy for curing in coating manufacturing. Thus, nanopaint designers should properly make trade-offs between product performance and process efficiency.

702e Factors Affecting Nanoparticle Dispersion

Deepika R. Gollamandala and Ileana C. Carpen. Department of Chemical Engineering, Tennessee Tech University, Box 5013, Cookeville, TN 38505

Addition of nanosized particles to a polymer solution/matrix can lead to new and improved properties over conventional composites using larger filler particles. In this work, we investigated systems of nanoparticles and polymer using Brownian dynamic simulations. The shape of the nanoparticles was considered to be spherical and the polymer architecture was modeled as bead spring chains. The degree of dispersion and agglomeration in the system were characterized by mean square displacement. Various factors, including size of nanoparticles, nanoparticle-polymer interaction, polymer chain length, and relative volume fractions, affect the dispersion of nanoparticles in the polymer matrix. In this study, the influence of polymer chain length and size of nanoparticles were analyzed and presented.

702f Frontal Flow Effects In 2-D Simulations of Orientation In High Aspect Ratio Particles Composites Made by Injection Molding

Gregorio M. Velez[1], Kevin O. Ortman[2], Aaron P. R. Eberle[3], Peter Wapperom[4], and Dr. Donald G. Baird[2]. (1) Macromolecular Science and Engineering Department, Virginia Polytechnic Institute, 154 Randolf Hall, Blacksburg, VA 24061, (2) Chemical Engineering, Virginia Tech, 154 Randolph Hall, Blacksburg, VA 24061, (3) Chemical Engineering, Virginia Polytechnic Institute, 154 Randolf Hall, Blacksburg, VA 24061, (4) Mathematics Department, Virginia Polytechnic Institute, 154 Randolf Hall, Blacksburg, VA 24061

The prediction of orientation for high aspect ratio particle composites such as nanocomposites or fiber composites is required in order to optimize the mold design of injection molding composites. At present, commercial simulators for fiber composites ignore the viscoelastic behavior of the polymeric matrix and perform the simulations in a decoupled scheme using Hele-Shaw flow approximations. Therefore the predictions of these simulations were highly limited. In this study, the flow through a center gated disk and end-gated plaque geometry is simulated numerically for a highly concentrated short glass fiber in a PBT (Newtonian) or PP (viscoelastic) matrix. The particles are modeled considering Brownian motion for nanoparticles, and the rotary diffusion term and non-affine motion for fibers. The viscoelastic matrix is modeled using a Phan Thien-Tanner model. For this, a 2D Finite Element Method analysis was performed using the traditional Galerkin method for the balance equations and discontinuous Galerkin for the constitutive equations. The impact of the frontal flow region, initial fiber orientation and viscoelasticity matrix was investigated. The predictions were evaluated with experimental data obtained using laser confocal microscopy and optical reflection microscopy.

Molecular Dynamics Studies on the Dispersion of Silica Nanoparticles in Polyethylene Melt Using a Coarse-Grained Model

Yangyang Shen, Aleksey Vishnyakov, and M. Silvina Tomassone
Department of Chemical and Biochemical Engineering
Rutgers, The State University of New Jersey

1. Introduction

Particles are important additives for altering and enhancing the properties of polymers. When the particle size approaching the fundamental length scale of the material, new mechanical, optical, and electrical properties arise, which are not present in the conventional macroscopic counterpart. For example, the dielectric constant of a polymer fluid can be increased by addition of ceramic nanoparticles with high permittivity. Some new polymeric and ceramic materials have been identified as promising candidates for the fabrication of high dielectric strength capacitors. Systems, such as titania, strontium titanate, and barium titanate dispersed in perfluoropoly(ether), poly(dimethyl siloxane), and poly(butadiene), have been investigated in several experimental works. In addition, silica (SiO_2) nanoparticles embedded in a polyethylene (PE) melt has been identified to offer desired dielectric constants. For this system, experimental study by Riman et al. [1] indicates that at nanoparticle filling fractions of 1 vol%, a 50% increase in the electroluminiscence and dielectric constant is observed.

However, at higher concentrations nanoparticles tend to agglomerate in the polymer matrix thus causing a loss of the desired properties of the material. The agglomerates are much larger than the primary particles and they are attributed to the minimum value in both electroluminiscence and dielectric constant. Therefore, the extent of dispersion of the ceramic powder is of crucial importance to avoid dielectric breakdown. It is now highly desirable to develop methodologies that can pinpoint the factors that control the nanoparticle dispersion inside these composites. Again for example, silica (SiO_2) nanoparticles in a polyethylene (PE) matrix have been identified to offer optimum dielectric constants. However, a dielectric breakdown was observed when the filling fraction exceeded 3 vol%. One of the ways to prevent nanoparticle agglomeration is the addition of surfactants to the composites to decrease the interactions between nanoparticles, favoring deagglomeration.

Specifically, when the nanofillers are made out of silica, the dispersion quality is one of the main difficulties, primarily because the non-polar nature of the polymers gives rise to a significant problem in enhancing adhesion between the hydrophilic silica nanofiller. The surface treatment of nanofillers by a coupling agent (i.e. nonionic surfactant) improves the degree of the miscibility of nanoparticles with polymer and its dispersion in the polymer matrix. Adsorption of nonionic surfactants on hydrophilic oxide surfaces has been well studied experimentally and documented in the literatures [2, 3]. Spectroscopic observations have suggested that the nonionic surfactant $C_{14}E_4$ prevents the aggregation of SiO_2 nanoparticles due to a weak hydrophobic interaction between surfactants and nanoparticle surface [4]. Several other experimental studies also indicated that the origin of nanoparticle dispersion might lie in the composition of surfactant molecules [7]. In this work, we study the factors that affect silica nanoparticle dispersion by means of Molecular Dynamics (MD) simulations. We examine both the hydrated surfaces (i.e. surfaces contain a high number of silanols) and the dehydrated surfaces (i.e. surfaces have a low coverage of hydroxyl groups).

During the last few years there have been a significant number of studies using simulation techniques including MD [5, 6, 11-14] and Monte Carlo methods [8, 9] to investigate polymer nanocomposites. Several computational studies have addressed the transport and motion polymer chains in nanocomposites [9-11]. The effect of polymer molecular weight has also been studied by comparing a matrix of polymer chains of length 5 or 20 beads [6]. It was found that the physical dimension of the chain is not important in defining the thermodynamics of the nanoparticle-nanoparticle interaction, which is crucial for nanoparticle dispersion. In the earlier work of Starr et al. [11, 12] the effect of polymer-nanoparticle interactions was explored for a system of one single nanoscopic particle surrounded by a dense polymer melt. Their results show that the elongation and fattening of polymer chains near the nanoparticle surface are independent of the interaction. Later, the performed studies on larger systems where they quantified the degree of agglomeration of nanoparticles in a polymer matrix with the fluctuations of the specific heat. Their results indicate a crossover between dispersed and clustered states. Despite that Elliot and Windle [14] have done Dissipative Particle Dynamics to investigate the how to minimize the void space in polymers containing spherical and non-spherical particles. Their work deals with assemblies of free and fused soft spheres at constant pressure but does not include the effect of surfactants.

2. Coarse-Grained Potential for Polyethylene

Development of reliable and computationally inexpensive models for simulations of polymers is a well-known problem. In this work, PE was described by the model of Guerrault et al. [15]. The polymer is described as a sequence of pseudoatoms or beads; each bead represents eight methylene groups; the diameter of bead is $\sigma = 1nm$. The parameters of the potentials were chosen from the best fit of the radial distribution functions (RDF) obtained with the coarse-grained and molecular models. We simplified the mathematical form of the potentials for simplicity. The bonds between the neighboring beads were described as:

$$w_{\lambda}^{b}(r) = k_{eq}\left(r - r_{eq}\right)^{4} \qquad (1)$$

where r is the distance between the beads, k_{eq} is the bond strength and r_{eq} is the equilibrium distance. The non-bonded interactions were described with a purely repulsive potential:

$$w_{\lambda}^{nb}(r) = a_1 \exp\left(-b_1 r^2\right) + a_2 \exp\left(-b_2 r^2\right) \qquad (2)$$

Both the bonded and non-bonded interaction potentials are demonstrated in Figure 1.

Distinction has to be made between the instantaneous temperature of the molecular

$$k_B T = \frac{\sum_{i=1}^{N} m_i v_i^2}{3N} \qquad (3)$$

and a coarse-grained system. In DPD simulations, the coarse-grained temperature is “mapped” onto that of the real system using the dissipative parameter (which we do not have in the coarse-grained MD) found from the best correspondence between experimental and simulated viscosities and diffusion coefficients. Here we use a different approach obtaining the temperature from the best fit of the velocities of the beads to the averaged velocities of methylene groups that form the bead. First, MD simulations were performed for polyethylene (PE) described by a united atom model [17] with chain length of 80 (denoted by C80) at T = 423K. A system containing 100 chains was compressed to give the density of 0.75 g/cm^3. Then the system was equilibrated for 200ps in an NpT ensemble. Another 200ps simulation was followed for data collection. The real temperature of the system is derived from averaged velocity distribution by fitting with Gaussian function (Figure 2 inset). The mean value for V^2 is 5.0×10^{-5} Å/fs, giving T = 280.77K.

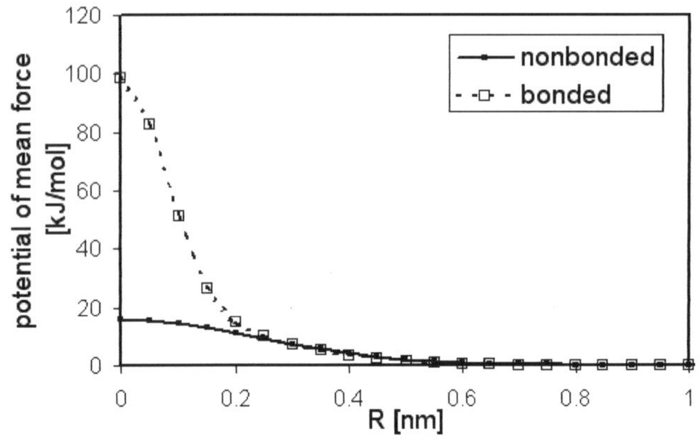

Figure 1. Bonded and non-bonded potentials of mean force.

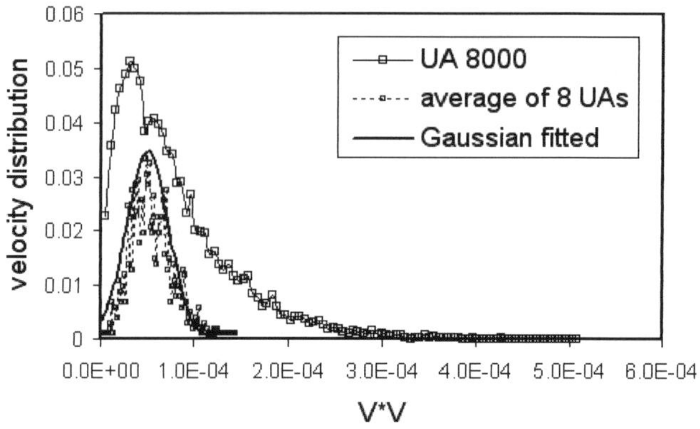

Figure 2. Velocity distribution after equilibration.

3. Nanoparticle Model and Particle-Polymer Interaction Potential

Each nanoparticle was constructed of 108 repulsive beads similar to those of the polymer but rigidly arranged in a body-centered-cubic (BCC) lattice by 251 harmonic bonds (Figure 3). The diameter of each nanoparticle was about 4nm Similarly to the polymers beads, the constituting model silica nanoparticles were soft, but the cumulative repulsion caused by the dense packing of rigidly arranged beads prohibited the deformation of the construction and penetration of other beads into the structure thus making the model nanoparticle behave as solid bodies.

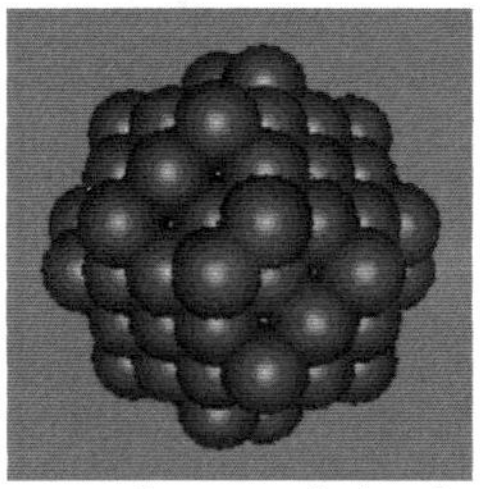

Figure 3. Schematic model of a nanoparticle.

The interaction parameters between the polymer and silica beads were chosen by fitting the surface tension of the polymer-particle interface to the value obtained from atomistic simulations. For particulate matter, the surface tension can be presented as the difference of the free energy of adhesion and the free energy of cohesion divided by the surface area of the interface created. We should note that this quantity differs from the surface tension between amorphous silica. The latter refers to the free energy of creation extra surface of a macroscopic solid body. Our value of the surface tension characterizes the cohesion of two particles. Assuming the canonical conditions, the difference between the cohesion and adhesion free energies can be expressed as:

$$\gamma^*{}_{pe-silica}\Delta A = \Delta F_{pp-pv} + \Delta F_{ss-sv} + \Delta F_{pv-ps} + \Delta F_{sv-sp} = \Delta U_{pp-pv} + \Delta U_{ss-sv} + \Delta U_{pv-ps} + \Delta U_{sv-sp} - T\left(\Delta S_{pp-pv} + \Delta S_{ss-sv} + \Delta S_{pv-ps} + \Delta S_{sv-sp}\right) \quad (4)$$

where ΔF is the total change in Helmholtz free energy, γ_{PS} is the surface tension between the "particulate" silica and PE, $\Delta \boldsymbol{U}_{SS-SV}$ is the change of internal energy related to transfer of the silica ball from silica to vacuum and so on.

Now let us assume that the only entropy change is related to the limitation of polymer conformation due to the creation of the surface in polymer. The silica surfaces don't really "integrate", stay intact and therefore their separation doesn't produce substantial entropic contribution. So we assume $\Delta S_{ss-sv} + \Delta S_{pv-ps} + \Delta S_{sv-sp} << \Delta S_{pp-pv}$ (5)

In this case $\Delta F \approx \Delta U_{pp-pv} - T\Delta S_{pp-pv} + \Delta U_{ss-sv} + \Delta U_{pv-ps} + \Delta U_{sv-sp}$ (6)

Now we take into account that $\Delta U_{pp-pv} - T\Delta S_{pp-pv} = \gamma_p A$ (7)

where γ_P is the surface tension of the polymer-vacuum interface So the equation finally reads:

$$\gamma_{ps}^{*} = \gamma_p + (1/A)(\Delta U_{ss-sv} + \Delta U_{pv-ps} + \Delta U_{sv-ps}) = \gamma_p + (1/A)(\Delta U_{ss-sv} - 2\times\Delta U_{ps}) \quad (8)$$

$\Delta \boldsymbol{U}_{SS-SV}$ is the energy related to the separation of two silica surfaces (positive value), and $\Delta \boldsymbol{U}_{PS}$ is the energy of separation of polymer and silica.

For polymer-vacuum surface tension, we use the experimental value γ_P =0.027J/m^2. The internal energies of separations of two silica surfaces and silica surface from a bulk polymer can be found from a straightforward simulation. The energy for creating a surface from the solid bulk was calculated with Material Studio package from Accelrys. First of all, we created a cell of amorphous silica glass by heating a quartz sample to the melting temperature, and then quenching to 423K and relaxing. Then a surface was cleaved in the glass; the Si-OH groups were created at the places where Si-O bonds were cut off when cleaving. After this, the energy of the sample was minimized and the sample was relaxed in a 300ps MD simulation at 423K. The resulting sample has a hydroxyl density of 4.2/nm^2, typical for a hydrophobic silica surface. On the next step, a layer structure was built by constructing a solid slab shown in Figure 4a. In one case (Figure 4a left one), the solid surfaces are located at very close (or zero) distance. In the other case (Figure 4a right one), they are separated by a large vacuum (20Å wide). Another vacuum layer was added above the second surface so that only one side of the surfaces would interact with each other (otherwise due to the periodic boundary conditions both sides would be considered). Similar procedure was followed for silica and polymer surfaces. Then the surface energy was estimated as the average difference in energy between the (a) and (b) configurations over 300ps MD simulations at our target temperature. The surface energy is obtained using the difference between two close solid surfaces and two separated solid surfaces, giving a value of 0.146 J/m^2. This energy required for creating a surface from the bulk was implemented into the surface tension calculation Eq. 8.

(a)

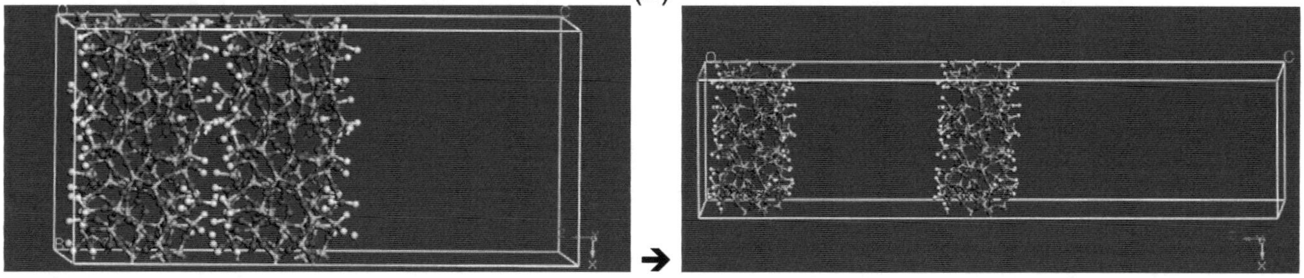

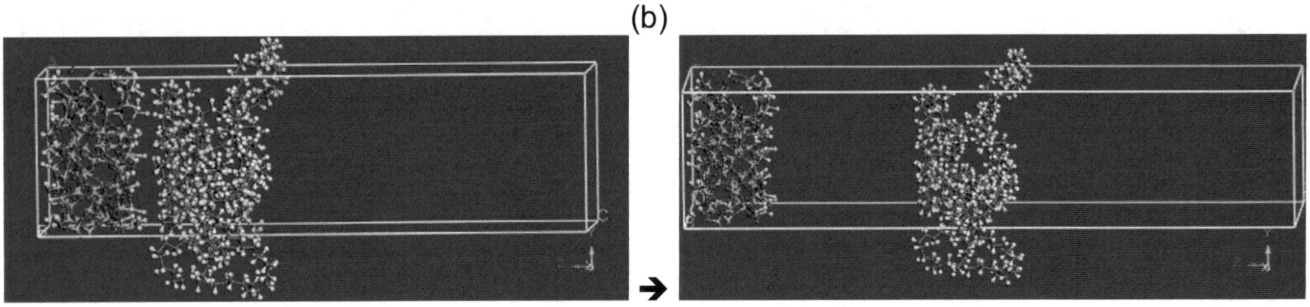

Figure 4. Simulation setup for calculation of the interfacial energy of (a) two silica surfaces and (b) silica-polymer surface.

Now we have to map our simulation onto the experimental value. This was done via a series of simulations of a flat interface between the silica and polymer as described by our coarse-grained models. The simulation box was $10\sigma\times10\sigma\times20\sigma$. The surface tension was obtained from a mechanical formula:

$$\gamma_{\alpha\beta} = -\frac{1}{V}\left\langle \sum_{i}^{N_p} m_i u_{i\alpha} u_{i\beta} + \sum_{i}^{N_p}\sum_{j>i}^{N_p} r_{ij\alpha} F_{ij\beta} \right\rangle \qquad (9)$$

The correlation between the surface tension and the relative strength of repulsive potential between the silica and PE beads is shown in Figure 5. A repulsive coefficient of 2 means that the repulsion between a silica and polymer bead is 2 times stronger then that for a couple of beads of the same component. By interpolating the surface tension obtained from experiments and atomistic simulations onto the correlation, we obtained the solid-fluid interaction parameter for our coarse model.

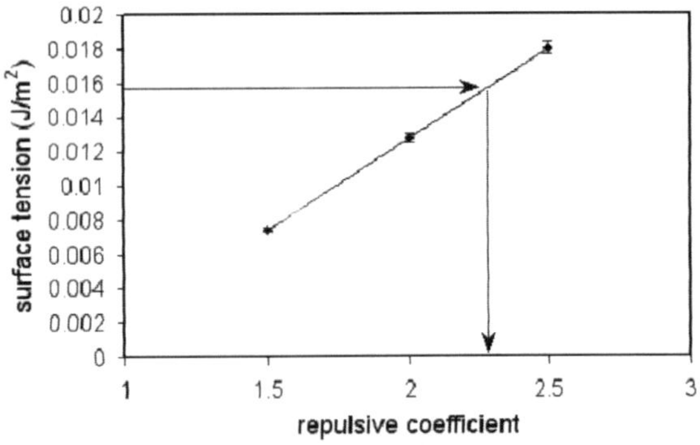

Figure 5. The dependence of silica-polymer surface tension as a function of the relative strength of the repulsive potential.

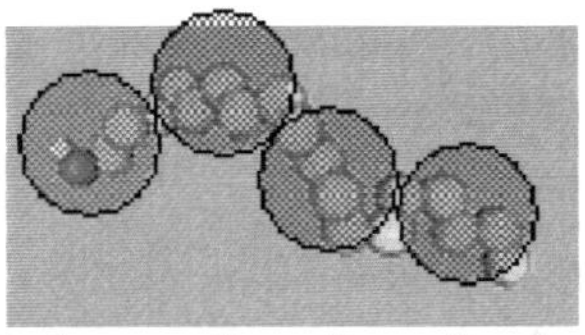

Figure 6. Schematic of surfactant model.

4. Surfactants

As a surfactant, we considered long chain alcohol the structure of CH_3-(CH_2)n-OH, which consists of one hydroxyl group as the hydrophilic unit in the head, and various numbers of units for the hydrophobic tail group. Figure 6 gives a schematic picture of the surfactant model. The CG beads in the surfactant tail should "like" the CG polymer beads, and the head bead "dislike" them. Lengths of the surfactants are expressed in total number of beads connected together in the molecule. Nonbonded interactions between different species are listed in Table 1.

5. Simulation Details

The simulations are based on standard molecular dynamics techniques in an NVT ensemble (constant number of particles, volume, and temperature) [16]. All atoms in the system interact via the derived coarse-grained potential (Equation 1). In our simulations, we have considered three different species: silica nanoparticles, polyethylene (PE) molecules, and nonionic surfactants. The interaction strength parameters between different species in the coarse-grained model are listed in Table 1. Typical simulation included 108,000 coarse-grained beads.

Table 1. Repulsion parameters of different species

	Polymer and surfactant hydrophobic bead (Type 1)	Nanoparticle bead (Type 2)	Surfactant hydrophilic bead (Type 3)
Type 1	1.0	2.3	2.5
Type 2		1.0	1.5
Type 3			1.0

6. Results and Discussion

6.1 Dispersion of Nanoparticles

Simulations of different numbers of nanoparticles in a polymer matrix were conducted systematically. Initially nanoparticles are randomly distributed in the simulation box of $30\sigma\times30\sigma\times30\sigma$. At lower filling fractions (1.6 wt% and 2.4 wt%) smaller clusters of 2 nanoparticles have been found. At higher filling fractions (greater than 3.2 wt%) larger clusters of 3 or 4 nanoparticles have formed. In all the cases, there exist a small number of nanoparticles remaining separated. In order to determine the state of nanoparticle dispersion we can use an explicit measure of structure, such as the radial distribution function $g(r)$ (Figure 7).

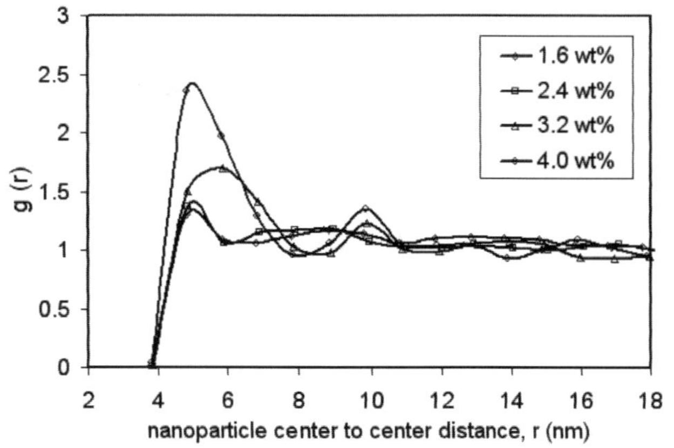

Figure 7. Radial distribution function of nanoparticle center of mass at different nanoparticle filling fractions.

In these simulations, the polymer has a fixed chain length (N = 8), but the number of nanoparticles is varied. $g(r)$ is taken from center to center of nanoparticles. There is a strong tendency for the nanoparticles to aggregate as illustrated by the sharp peak in the nanoparticle-nanoparticle radial distribution function. For a smaller filling fraction, the lower first peak in the radial distribution and the flat distribution at large distance indicate the dispersion of nanoparticles. As increasing the filling fraction, the nanoparticle density is less spatially homogeneous, yielding a second peak at large distance. The coexistence of the first and secondary peaks indicates larger degree of agglomeration.

The state of dispersion of nanoparticles can also be elucidated by the approach proposed by Starr et al. [13]. Potential energy and specific heat as a function of filling fraction are calculated. The potential energy per nanoparticle bead $u_{nn} = U_{nn}/N$ is strongly sensitive to the change of local packing, because it depends on the number of particle-particle contacts (Figure 8a). The potential energy increases with increasing of filling fraction, as contacts between the particles become more frequent. We have also considered the potential energy fluctuations $\delta u_{nn} \equiv u_{nn} - \langle u_{nn} \rangle$, which are thermodynamically quantified by the specific heat per particle of the nanoparticles, $c_V^{nn} = \dfrac{N\langle \delta u_{nn}^2 \rangle}{T^2}$ (Figure 8b).

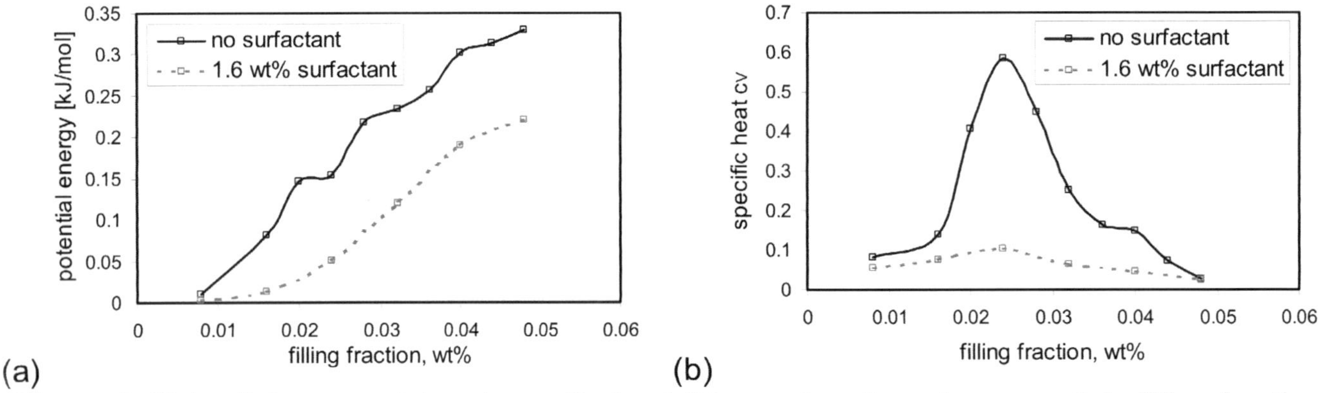

Figure 8. Potential energy (a) and specific heat (b) as a function of nanoparticle filling fraction with and without surfactants.

At the largest and smallest filling fractions, the systems are very stable, thus there are little fluctuations in the potential energy. But for the intermediate state, particles easily aggregate into clusters, and can separate after a short period of time, resulting in large fluctuations in the potential energy, and hence a large value of c_V. As the filling fraction increases, the c_V reached a high peak at 2.7 wt% when there are possibilities for the nanoparticles to form smaller clusters of 2~3 entities. Later as more nanoparticles are included in the system, the formation of larger clusters is taking place, giving a lower peak on the c_V curve. The system is less stable when the smaller clusters are formed (higher peak on the curve), and becomes more stable when the larger clusters are formed (lower peak on the c_V curve). In the presence of surfactants, the transition state is shifted to a larger value of the volume fraction ϕ, showing that it is easier for the system to reach a dispersed state even when there are more nanoparticles embedded in the polymer matrix. The approximate boundary between clustered and dispersed states is shifted accordingly in the presence of surfactants.

6.2 Effect of Polymer Chain Length

The difference in the dispersion efficiency should mainly come from the polymer–nanoparticle interaction, which is a function of polymer chain length and polymer molecular structure. In this work we have also considered the effect of polymer chain length on the dispersion of nanoparticles. The polymer chain length is varied from N = 4 to 16. The choice of the chain length is to avoid the effect of polymer entanglement on the particle dispersion. Figures 9(a) and (b) respectively show the potential energy and specific heat as a function of nanoparticle filling fraction for three different polymer chain lengths. Longer chains tend to hinder the movement of nanoparticles and prevent them from agglomeration. This is also observed from the trajectories.

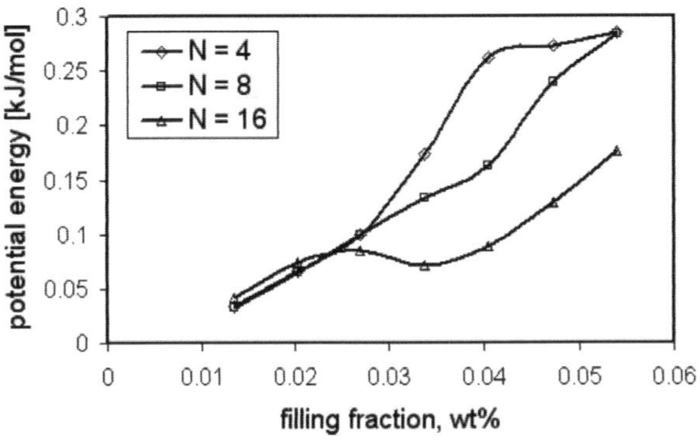

(a)

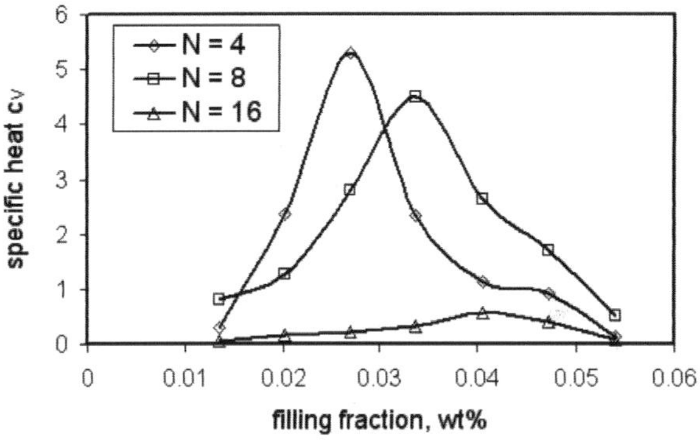

(b)

Figure 9. Potential energy (a) and specific heat (b) as a function of nanoparticle filling fraction for different polymer chain lengths.

In addition, the relative size of the nanoparticle and polymer is the key causing phase separation and nanoparticle agglomeration [18, 19]. The polymer radius of gyration is defined as the root mean square distance between monomers, i.e. $R_g^2 = \frac{1}{2N^2}\sum_{i,j}(r_i - r_j)^2$, which describes the dimension of a polymer chain. First, we simulated the pure system only containing polymer chains, and obtained the initial radius of gyration radius (R_{g0}) for different chain lengths. After adding naoparticles in the system, chain stretching has been observed in some Monte Carlo simulations [20]. In our system the nanoparticle radius was about 2 nm, and the polymer R_g value ranges between 2.3 and 9.4 nm. Figure 10 shows the polymer radius of gyration (R_g) relative to that without nanoparticles (R_{g0}) for three different polymer chain lengths as a function of nanoparticle filling fraction. For nanoparticle size is similar to the polymer radius of gyration, the particles experience fast diffusion, and they move through the polymer melts without necessarily waiting for chains to relax their conformations, so that the chains remain to their initial dimension. However we observed chain expansion in all other cases of longer polymer chains. The nanoparticles stretch the polymer chains, and in turn they show more dispersion with larger R_g. This is due to more adsorbed amount of polymer on to the naoparticles when the chain is longer.

6.3 Force Between Nanoparticles

In this work we have also included two nanoparticles, whose positions were fixed in the simulation box, but the distance between them was varied. The solvation force was calculated simply as $F_s = \left\langle \sum_{i=1}^{N} \sum_{j=1}^{M} l_j f_{ij} \right\rangle$, where *N* is the number of fluid molecules, and *M* is the number of silica pseudoatoms in one nanoparticle. Here, f_{ij} is the force between fluid molecule i and pseudoatom j, l_j is a coefficient that equals 1 if the pseudoatom j belongs to the first nanoparticle and l_j = -1 if j belongs to the second nanoparticle. Each MD simulation lasted for 150ns. The solvation force was averaged over the last 100ns of simulation run. Figure 11 compares the solvation force as a function of nanoparticle center-to-center distance for two different polymer chain lengths. The attractive minimum occurs at approximately 5.6 nm, and then there is a repulsive maximum at about 7 nm for both chain lengths, after which the force seems to decrease slightly and approach to zero at larger separations. When two naoparticles covered with chain molecules approach each other, the chains extend out and overlap with each other. At this point, the polymer chains are being compressed between the surfaces, leading to an unfavorable entropy change. Thus the nanoparticles experience a repulsive force. In the case of longer polymer chains the forces shift up towards the repulsive region, and the repulsive maximum is higher, due to greater amount of polymer adsorption.

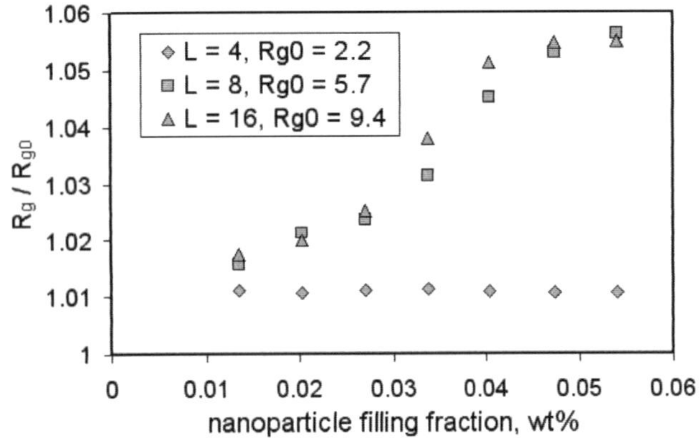

Figure 10. The polymer radius of gyration (R_g) relative to that without nanoparticles (R_{g0}) for three different polymer chain lengths as a function of nanoparticle filling fraction.

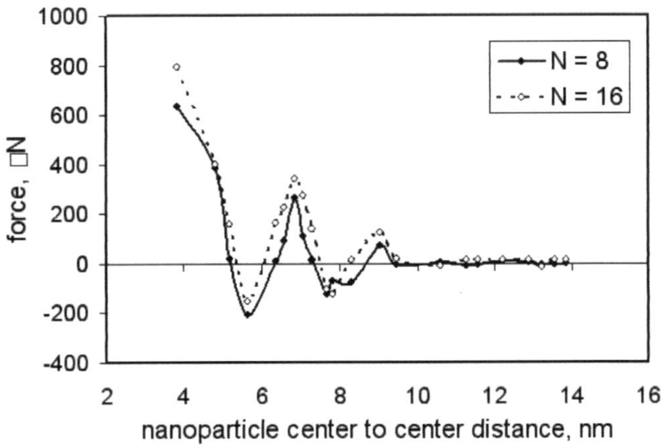

Figure 11. Solvation forces as a function of nanoparticle separation (two polymer chain lengths)

7. Conclusions

We have developed a coarse-grained model of SiO_2 in a polyethylene matrix using a novel approach, which uses experimental data, thermodynamic theory and atomistic simulations. The RDF and specific heat calculations indicate that for filling fractions smaller than 3.6 wt% the system is in a dispersed state and for filling fractions of 3.6 wt% and larger the nanoparticles show agglomeration. We show that thermodynamically stable dispersion of nanoparticles into a polymer melt is enhanced for systems where the radius of gyration of the linear polymer is greater than the radius of the nanoparticle. Dispersed nanoparticles swell the polymer chains, resulting that the polymer radius of gyration grows with the nanoparticle filling fraction. The polymer-mediated forces are also more repulsive in the case of longer chains than that of the shorter ones.

The addition of oleyl alcohol ($C_{18}H_{36}O$) reduces the effective attraction between the nanoparticles, further weakening the agglomeration, which was monitored via particle-particle contribution to the potential energy and the specific heat c_V. Our results show that, for a surfactant concentration of greater or equal than 6.4 wt%, the particles remain in a homogeneously dispersed state, which is consistent with experimental findings.

References

1. Private conversation with A. Pedersson and R. Riman.
2. P. Trens and R. Denoyel, Conformation of poly(ethylene glycol) polymers at the silica/water interface: a microcalorimetric study, *Langmuir* 9 (2): 519-522, 1993.
3. P. Levitz, Aggregative adsorption of nonionic surfactants onto hydrophilic solid/water interface: relation with bulk micellization, *Langmuir* 7 (8): 1595-1608, 1991.
4. W. Wang, B. Gu and L. Liang, Effect of surfactants on the formation, morphology and surface property of synthesized SiO_2 nanoparticles, *Journal of Dispersion Science and Technology* 25 (5): 593-601 2004.
5. D. Bedrov, G. D. Smith, and J. S. Smith, Matrix-induced nanoparticle interactions in a polymer melt: a molecular dynamics simulation study, *Journal of Chemical Physics* 119 (19): 10438-10447 2003.
6. Y. Qin and K. A. Fichthorn, Molecular-dynamics simulation of forces between nanoparticles in a Lennard-Jones liquid, Journal of Chemical Physics 119 (18): 9745-9754 2003.
7. M. S. Lee, S. S. Park, G.-D. Lee, C.-S. Ju and S.-S. Hong, Synthesis of TiO_2 particles by reverse microemulsion method using nonionic surfactants with different hydrophilic and hydrophobic group and their photocatalytic activity, *Catalysis Today* 101 (3-4): 283-290 2005.
8. M. Vacatello, Phantom chain simulations of polymer-nanofiller systems, *Macromolecules* 36 (9): 3411-3416 2003.
9. M. Vacatello, Monte Carlo simulations of polymer melts filled with solid nanoparticles, *Macromolecules* 34 (6): 1946-1952 2001.
10. D. Brown, P. Mele, S. Marceau and N. D. Alberola, A molecular dynamics study of a model nanoparticle embedded in a polymer matrix, *Macromolecules* 36 (4): 1395-1406 2003.
11. F. W. Starr, T. B. Schrøder and S. C. Glotzer, Effects of a nanoscopic filler on the structure and dynamics of a simulated polymer melt and the relationship to ultrathin films, *Physical Review E* 64 (2): 021802-021806 2001.
12. F. W. Starr, T. B. Schrøder and S. C. Glotzer, Molecular dynamics simulation of a polymer melt with a nanoscopic particle, *Macromolecules* 35 (11): 4481-4492 2002.
13. F. W. Starr, J. F. Douglas and S. C. Glotzer, Origin of particle clustering in a simulated polymer nanocomposite and its impact on rheology, *Journal of Chemical Physics* 119 (3): 1777-1788 2003.

14. J. A. Elliott and A. H. Windle, A dissipative particle dynamics method for modeling the geometrical packing of filler particles in polymer composites, *Journal of Chemical Physics* 113 (22): 10367-10376 2000.
15. X. Guerrault, B. Rousseau, and J. Farago, Dissipative particle dynamics simulations of polymer melts. I. Building potential of mean force for polyethylene and cis-polybutadiene. *Journal of Chemical Physics*, 121(13): p. 6538-6546 2004.
16. M. P. Allen and D. J. Tildesley, *Computer Simulation of Liquids*, Clarendon Press, Oxford, 1990.
17. J. D. Moore, S. T. Cui, H. D. Cochran and P. T. Cummings, A molecular dynamics study of a short-chain polyethylene melt: I. Steady-state shear, *Journal of Non-Newtonian Fluid Mechanism* 93 (1): 83-99, 2000.
18. R. B. Thompson, et al., Predicting the mesophases of copolymer-nanoparticle composites. *Science* 292: p. 2469-2472 2001.
19. V. V. Ginzburg, Influence of nanoparticles on miscibility of polymer blends. A simple theory. *Macromolecules* 38(6): p. 2362-2367 2005.
20. Sharaf, M.A. and J.E. Mark, Monte carlo simulations on the effects of nanoparticles on chain deformations and reinforcement in amorphous polyethylene networks. *Polymer* 45(9): p. 3943-3952 2004.

703b Nanoparticle Formation by Amphiphilic Block Copolymer Directed Assembly: A Model Study Using Molecular/brownian Dynamics Simulations

Chungyin Cheng, Chemical and Biological Engineering, Iowa State University, Ames, IA 50011, Monica H. Lamm, Department of Chemical and Biological Engineering, Iowa State University/DOE Ames Laboratory, A510 Zaffarano Hall, Ames, IA 50011, R. O. Fox, Department of Chemical & Biological Engineering, Iowa State University, 2114 Sweeney Hall, Ames, IA 50011-2230, and R. Dennis Vigil, Iowa State University, 2114 Sweeney Hall, Ames, IA 50011.

Nanoprecipitation [1] is a process to produce functional nanoparticles protected by amphiphilic copolymer directed assembly. In this process, organic drug and copolymers are dissolved in solvent and rapidly mixed with nonsolvent in a customized micro-device to create high supersaturation. The growth of nanoparticles is kinetically arrested by block copolymer assembly on the surface of the organic drug, and therefore a tunable and narrow particle size distribution is achievable. The Nanoprecipitation process has been studied in several aspects: CFD simulations were performed to investigate the turbulent flow and mixing evaluation of the micro-reactor, microPIV experiments were conducted to furthermore validate the models, and a population balance model has been developed to account for the particulate flow system. To have a more fundamental understanding of the process, a computational simulation of this system is performed using Brownian dynamics (BD), as well as molecular dynamics (MD), and compared with detailed mechanistic model for micellization kinetics. Each organic particle is represented by a coarse-grained (CG) bead, and the diblock copolymer chain is constructed from a series of CG beads, each of which represents a polymer unit with bead-spring potential. The simulation results are analyzed to obtain information about characteristic precipitation time, aggregation numbers and particle sizes. These data are compared with the experimental data and used to validate a population balance model.

[1] B. K. Johnson and R. K. Prud'homme. Flash nanoprecipitation of organic actives and block copolymers using a confined impinging jets mixer. Australian Journal Of Chemistry, 56(10):1021–1024, 2003.

703c Multi-Scale Modelling Efforts for Diblock Copolymers

Sharon Loverde[1], Vanessa Ortiz, Dennis E. Discher[2], and Michael Klein[3]. (1) Department of Chemical and Biomolecular Engineering, University of Pennsylvania, Philadelphia, PA 19103, (2) Chemical and Biomolecular Engineering, University of Pennsylvania, Room 129 Towne Building, 220 South 33rd Street, Philadelphia, PA 19104-6393, (3) Center for Molecular Modeling, University of Pennsylvania, Philadelphia, PA 19103

Amphiphilic block copolymers self-assemble into vesicles and worm-like micelles in the aqueous phase with several unique characteristics--high stability and morphological tunability, as well efficient integration of hydrophobic drugs. In this simulation study, the thermodynamic stability of worm-like micelles are examined using a complete multiscale modelling effort -- starting from atomistic molecular dynamics that are matched with coarser models, and then progressing to mesoscopic simulations of entire worm-like micelles. Several key issues are addressed: assembly and disassembly, equilibrium scaling of the core size with the hydrophobic length of the copolymer, as well as dynamic limitations on time scales of DPD (dissipative particle dynamics). To begin with, examinations of the phase stability of the micelles probe the worm to spherical micelle crossover region. Fundamental properties of the copolymers in the micelles—including the radius of gyration, end to end distance, as well as the diffusional behavior—are examined. Simple polydisperse mixtures are then used to gain insight into the coupling between local curvature and concentration, as well as the stability of the micelles. Break-up of these mixed assemblies proceeds through dynamic undulation and budding from the end-cap region of the worm-micelles, in agreement with experimental observations.

703d Numerical Study of Polymer-Silica Nanocomposites :Molecular Weight Effect on Their Mechanical Properties

Thomas R. Roussel, Joshua Moore, and Keith Gubbins. North Carolina State University, Engineering Building I, Box 7905, Raleigh, NC 27695-7905

Highly ordered mesoporous polymer-silica nanocomposites have been synthesized by Liu et al (JACS 2006) using a triconstituent evaporation-induced self-assembly method. A Pluronic triblock copolymer (F127) was used as a template. Soluble resol polymer and prehydrolyzed TEOS (tetraethyl orthosilicate) are, respectively, used as organic and inorganic precursors, templated by the hexagonally distributed cylindrical micelles. It has been proposed that ordered mesoporous nanocomposites have "reinforced concrete"-structured frameworks with interpenetrating networks. The presence of silica in nanocomposites drastically inhibits framework shrinkage during the micelle removal by calcination, resulting in highly ordered large-pore mesoporous carbon-silica nanocomposites. The pore size is varied as a function of the surfactant molecular weight, polymerization degree of organic precursor, and the relative precursor ratio (Si/C).

For a better understanding of such materials, we perform Monte Carlo simulations on several bulk atomic structures. First, we built a united atom coarse grained model of resol and silica. The precursors interact weakly with each other and can form chemical bounds independently. We explore, numerically, the structural properties of the nanocomposite by adjusting the initial mass ratios of silica to resol. The degree of the polymerization is adjusted by decreasing the volume of the simulation box at different rates, corresponding to a fictitious evaporation of the solvent. We show how mechanical properties are reinforced as a function of the molecular weight of the polymer (resol). This preliminary study is the first step to investigate the mechanical properties of the nanocomposite templated by micelles.

703e A Multiscale Approach to Predicting Barrier Properties of Polymer Nanocomposites

Youthachack Landry Khounlavong[1], Venkat Ganesan[2], and Victor Pryamitsyn[2]. (1) Chemical Engineering, University of Texas at Austin, 1 University Station, Code C0400, Austin, TX 78712, (2) Department of Chemical Engineering, The University of Texas at Austin, 1 University Station, Code C0400, Austin, TX 78712

Polymer nanocomposites (PNC) have been shown to exhibit extraordinary properties that are counter intuitive. A prime example would be with PNC barrier properties: it has been shown that upon addition of spherical nanofillers to a polymer matrix the membrane's permeability (and diffusivity) is increased. One hypothesis that has been made suggests that the interfacial matrix layer around the nanoparticle has properties that vary significantly from the bulk properties, and that the volume fraction of these layers is high enough to allow its properties to manifest at the macroscopic level. We have developed a multiscale approach that allows us to quantify the influence of the interfacial properties on the macroscopic barrier properties. We present results which illustrate the effects of particle size, interfacial layer thickness, and polymer stiffness on a penetrant's diffusivity.

A 3-D APPROACH TO MODEL DIFFUSION IN RANDOMLY DISTRIBUTED NANOCOMPOSITE

Matteo Minelli, Marco Giacinti Baschetti, Ferruccio Doghieri
DICMA - Department of Chemical Engineering, Mining and Environment Technology
University of Bologna, Via Terracini 28, 40131 Bologna, Italy
Tel. 051/2090428; fax 051/6347788; email. matteo.minelli@mail.ing.unibo.it

Abstract

The mass transport properties of a system with impermeable platelets dispersed in a polymer matrix were investigated, through a numerical algorithm, based on finite volume method. Simulations were carried out on 3-D geometries where different shapes of the lamellae were considered, dispersed in both an ordered and random way; different volume fractions ϕ and aspect ratios α of the platelets were considered, and, in the ordered structures, also different values of the flakes spacing σ were investigated. In the case of random systems, multiple structures were analyzed for a given set of parameters and the results were treated statistically to obtain reliable information.

The results show that the permeability in ordered geometries decreases with the product $\alpha\,\phi$ and that the curves corresponding to different filler shapes have a similar qualitative behavior but quantitative differences in the enhancements of the material barrier properties; circular platelets in particular, showed to be less efficient than other geometries in decreasing the effective diffusivity of the composite. All 3-D structures then showed diminished barrier properties with respect to homologous 2-D structures suggesting that the use of the latter approach would lead to an overestimation of the barrier properties increases in real nanocomposites and thus, despite its simplicity, should not be used for this purpose.

The data relative to 3-D random dispersions, that show barrier properties even worse than those of ordered dispersions of circular lamellae, can be accurately predicted by the Nielsen's model approach which therefore seems more suitable to describe the properties of real materials.

Introduction

Currently nano-technology are widely investigated due to the interesting properties that have been achieved in the last decades by in terms of the mechanical behavior, flame resistance and barrier properties, just with addition of small amounts of a nano-structured inorganic, impermeable phase in a polymeric matrix [1-3]. Such improvements make these materials suitable for several applications and, the effect on the barrier properties in particular has made such nanocomposite materials very attractive in the field of food packaging.

From this point of view, therefore, the study gas molecules diffusion through these heterogeneous membranes is today regarded with renowned interest; this is indeed a classical problem in the field of the transport phenomena that, although widely studied through analytical and computational approaches, has not been completely understood yet [4-7].

In this work a finite volume based algorithm has been applied to the analysis of ordered and random tri-dimensional structures; the enhancement of the barrier properties was studied in relation to the structural parameters such as the platelet aspect ratio α (the ratio between the half-with of the flake to its thickness), the filler volume concentration, the so-called loading ϕ, and, for the case of ordered geometries, the slit shape σ which represents ratio between the flakes spacing and thickness.

Many works have already been carried out on this problem, but most of them have considered just the case of 2 dimensional structures [6-10] that is only a rough approximation of the real structure which one can find in the real nanocomposites. The aim of this work was therefore to analyze the barrier properties in 3-D geometries, definitely more suitable to describe the real case, in order to reach a better understanding of the relationship betweens structure and properties of such nanocomposites materials.

Simulation details

The process, which has been simulated, is a purely diffusive phenomenon considering small penetrant molecules in a two components medium, where the mass transfer is governed by the Fick's law. The heterogeneous lamellar structure as usually considered in 2-D approach has been transposed in a tri-dimensional space by developing 4 different ordered configurations as shown in Figure 1.a,b,c,d. As one can see squares and circles are considered, as well as the octagons, to mimic the shape of the flakes usually dispersed in a polymeric phase. Tapes-shaped geometries were instead obtained by basically extruding the 2-D lamellae in the third direction and they have been considered to have a direct comparison with the results obtained in the bi-dimensional case.

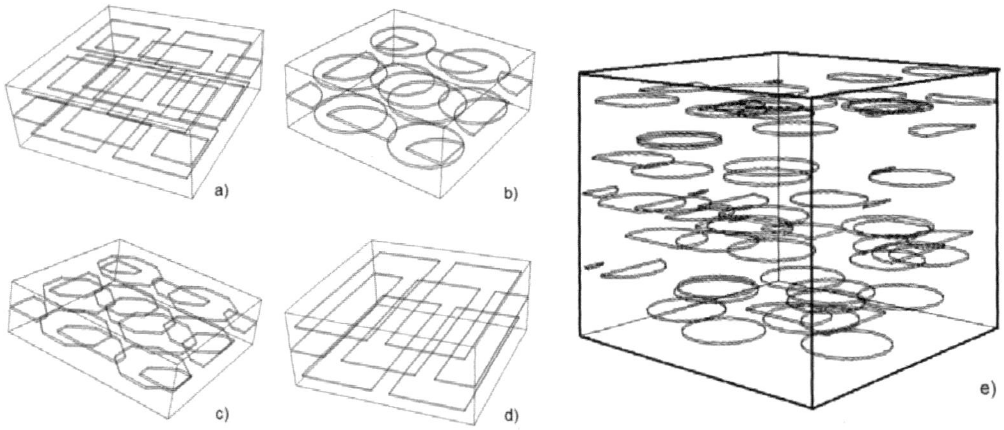

Figure 1, 3-D ordered heterogeneous geometry, squares (a), circles (b), octagons (c) and tapes (d). On the right side the layout of a disordered configuration (e).

These structures are represented by three parameters, as above mentioned, the flakes aspect ratio, the slit shape, and the filler volume fraction whose values have been changed during the simulations in order to investigate all the possible regimes due to the different geometries. In particular the different geometries were built with α values from 20 to 100, slit shapes in the range of 0.01–10, and loading of 0.5 to 20 %. The so-generated structures were then meshed, with a particular care in the critical areas, and the Fick's equation was solved through a finite volume algorithm.

In the case of random systems just circular configuration has been chosen and the geometries were defined by randomly placing a certain number of, not overlapping, platelets in the simulation box. In order to mimic the poly-dispersity of filler dimensions in the real composite the aspect ratio α of the platelets was allowed to vary and a standard deviation of about 10 % of the average value was used to obtain a normal distribution of such parameter in the geometry. In this way, random structures were obtained as the one showed in Figure 1.e. Aspect ratios of 20, 50, 75, 100, as average values, and different filler volume fractions, namely 2.5, 5 and 10 %, were thus considered to investigate the influence of the loading on the composite transport properties.

The same procedure and the same solver were used in case of randomly distributed systems, but, for same set of parameters, simulations were performed on different geometries in order to obtain results independent from the particular structure tested. This procedure was considered to have better statistic on the data and allowed to minimize the standard deviation of the results obtained for a given set of parameters which was indeed kept to a value of about 10 % of the average permeability ratio.

Theoretical background

As previously mentioned the problem of diffusion in a nanocomposite media has been investigated by different authors and different approaches were considered leading to a number of different models proposed both for 2-D and 3-D structure.

In particular, even limiting the analysis to the most interesting models, it should be recalled that one of the first theories in the field was developed by Barrer [11] at first and then modified by Nielsen [12] which proposed the following relationship:

$$\frac{D_0}{D} = 1 + \frac{\alpha\phi}{1-\phi} \qquad (1)$$

where D_0 is the diffusivity through the unfilled material, D is the diffusivity through the nanocomposite membrane and the other symbols have been already introduced.

This model based on ordered 2-D structure was extremely simplified and many authors tried to give more rigorous representation of the barrier effect introduced by filler platelets, this approach lead, among the other, to the following relationship for 2-D ordered structures [9]:

$$\frac{D_0}{D} = 1 + \frac{\alpha^2\phi^2}{1-\phi} + 2\frac{\alpha\phi}{\sigma} \qquad (2)$$

Where the different terms refers to two different types of resistance in the materials: one, linear to the product $\alpha\phi$, was related to the flow constriction in the holes between two adjacent flakes, while the other, of second order with respect to $\alpha\ \phi$, refereed to the tortuous path of the molecule among two different layers of platelets. Such approach was then extended to describe the properties of 2-D and 3-D random systems, in the case of high aspect ratio, obtaining the following equation [13]:

$$\frac{D_0}{D} = 1 + \mu\alpha^2 \frac{\phi^2}{1-\phi} \qquad (3)$$

where a geometrical factor μ, basically dependent on the flake geometrical characteristics, was introduced as a fitting parameter

Other models were then proposed based on numerical analysis (through finite element approach) of the mass transport in nanocomposite materials. In particular in the work due to Gusev, the following formula was presented [14]:

$$\frac{D}{D_0} = \exp\left(-\frac{x}{x_0}\right)^{\beta} \tag{4}$$

where $x = \alpha\,\phi$ and $\beta = 0.71$ and $x_0 = 3.47$ were basically two fitting parameters calculated in the base of the numerical results.

A second work is instead due to Friedrickson and Bicerano, which, based also on physical considerations, obtained a model of the following form [15]:

$$\frac{D}{D_0} = \frac{1}{4}\left(\frac{1}{1+a_1\kappa\alpha\phi} + \frac{1}{1+a_2\kappa\alpha\phi}\right)^2 \tag{5}$$

where: $a_1 = \left(2-\sqrt{2}\right)/4$ and $a_2 = \left(2+\sqrt{2}\right)/4$ and $\kappa = \pi / \ln\alpha$.

Results

Ordered 3D structures

In the present work, the resistance to diffusion in nanocomposite systems was investigated in a wide range of filler concentration and flakes aspect ratios; 4 different configurations of the platelet were also considered to understand the influence of these parameters on the final properties of the materials. The results are presented in Figure 2, that shows the differences between the 4 shapes of platelets in terms of diffusivity ratio as function of α times ϕ; D/D_0, indeed, as most of the authors suggested does not depends on α and ϕ separately but, once the value of σ is defined, is mainly related to their product.

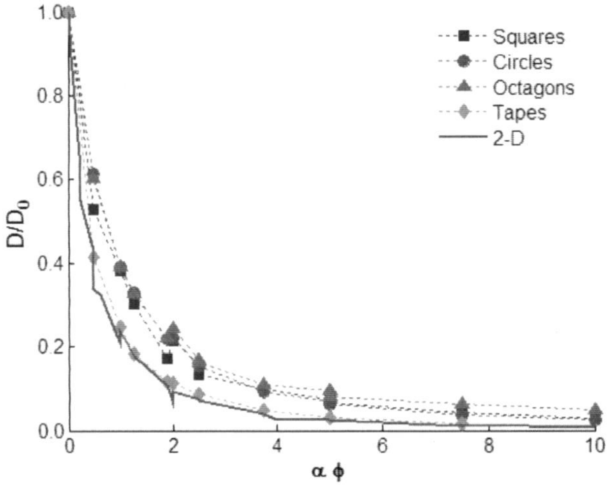

Figure 2, Permeability ratio in ordered systems for the 4 different configurations and the results in 2-D geometries are also reported as comparison. All the geometries are in case of σ=1

In the figure it can be seen that the barrier properties of the material definitely depends on the platelets shape since, for a given set of parameters, the value of D/D_0 ratio results to decrease going from circular flakes, which show lower barrier enhancements, to platelets shaped respectively as octagons, square and tapes. The curve due to the latter geometry then, lies exactly over the results obtained in the 2-D analysis, confirming the internal consistence of the numerical approach, because this configuration has been obtained extending indefinitely the 2-D lamellar geometry in the third direction.

In order to better understand the dependence of the barrier properties as a function of the $\alpha\phi$ product the relative change of diffusivity (D_0/D-1) can be considered. Such parameters which represent a sort of resistance to the diffusion shows a trend that can be approached by a power law with a variable exponent. As shown in Figure 3, indeed, this quantity behaves in two different ways when the entire range of $\alpha\ \phi$ is considered; for diluted systems and small platelets ($\alpha\ \phi$ small) the trend is linear, while increasing $\alpha\ \phi$ the dependence increases and on the right side of the plot becomes approximately quadratic. This behavior closely resemble to what predicted by the simple model for 2-D structured system presented in Eq. 2 so that one can say following the same approach that, also in the case of 3-D ordered system two different behavior are evident, the first valid for low platelet concentration where the resistance is mainly focused in the hole area, while the very few lamellae in the system are not enough to increase significantly the tortuosity of the penetrant path. The second valid for high $\alpha\ \phi$ factor when the "wiggling" of the diffusing molecules between the platelets is predominant and the resistance in the hole is negligible.

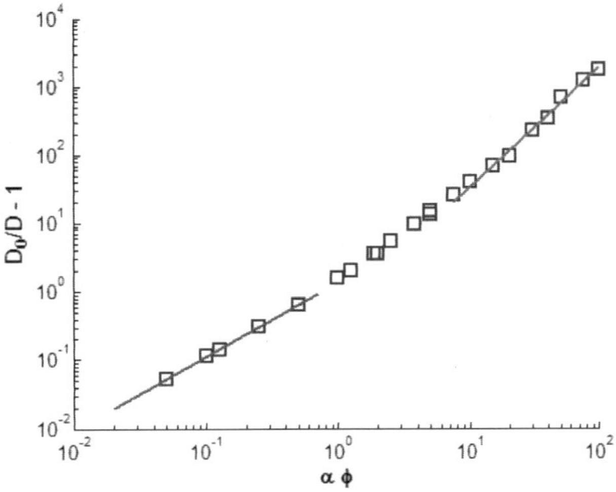

Figure 3, Resistance to the diffusion (D_0/D-1) as function of the factor a f, for the case of circular configuration (σ=10)

Random systems

For random system due to the higher computational cost, only the circular shape of platelets has been investigated; this is indeed the one which has to be considered to mimic as best the clay platelets in the nanocomposite structures. Figure 4 shows the permeability decay in randomly distributed structures in comparison with the one in ordered geometries at different slit shapes; as one can see, there is a large gap between barrier effect observed in ordered structures and the one guaranteed by structures where the platelets are randomly distributed.

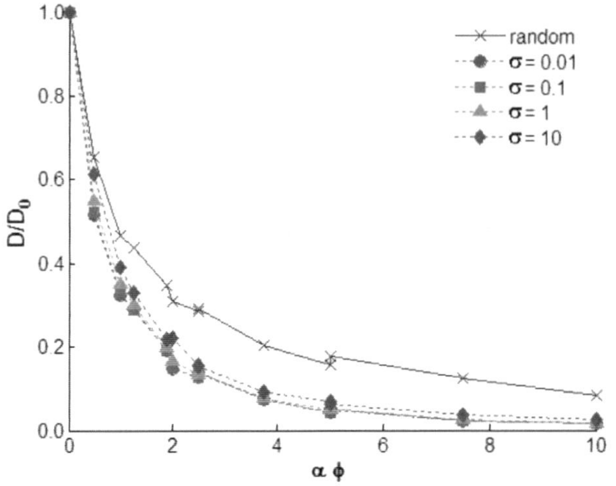

Figure 4, Comparison between the permeability ratio in ordered systems at different values of slit shape and in the randomly distributed systems

Indeed, the random distribution of the platelets is not able to guarantee a remarkable barrier effect because of the formation of clusters of lamella and shortcuts that the penetrant can use to diffuse faster. In this concern, therefore, the simulations have shown that this effect is more pronounced in a three-dimensional system than in 2-D [16]: in fact, contrary to what happens in 2-D systems, even at high values of slit shape (such as of the same order of magnitude of the size of the flake), random structures cannot be modeled by using a ordered approach because there exist in anyway a remarkable difference between the permeability decay in the two cases.

The results in case of random systems are then represented in Figure 5 in comparison with all the models previously mentioned.

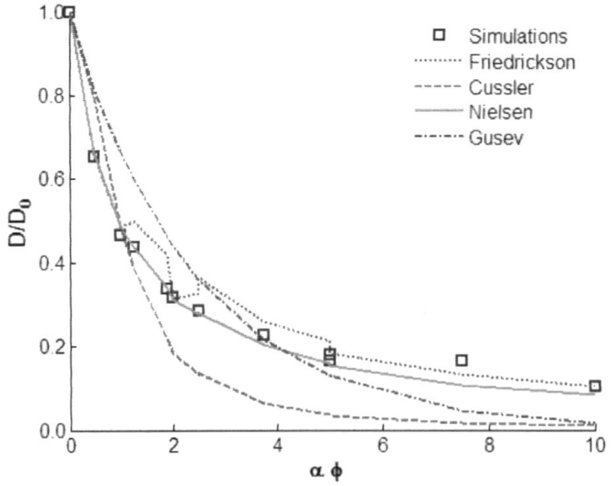

Figure 5, Permeability ratio in randomly distributed system in comparison with the models

As one can see, simulation results show a good agreement with the Friedrickson and an even better match with the prediction of the Nielsen equation; on the other hand the model due to Cussler cannot approach the trend of the simulation even changing the fitting parameter μ and also the equation due to Gusev, substantially exponential refers to a behavior that is not followed to the simulation results obtained in the present work. It has to be considered, indeed, that Cussler developed his approach in the approximation of 2-D structures and the formula subsequently proposed was just an extension for the tri-dimensional case; Gusev on the other side gave was mainly focused on the simulation approach to the problem and the formula he proposed was a sort of fitting equation for his results whose parameter had substantially no physical meaning.

Surprisingly, the model that seems to work better in defining the decrease in permeability due to filler addition is the one due to Nielsen, which in its simplicity is able to catch the essential feature of the problem and results thus able to mimic the resistance to the permeation in nanocomposite material as obtained by the present numerical results.

Conclusions

In this work the barrier properties of a nanocomposite film were modeled and the diffusion of small molecules in a composite media has been simulated from a macroscopic point of view, using a finite volume method. Simulations were carried out on three-dimensional geometries, in which an impermeable phase, consisting in platelets in different configurations, was placed in the simulation box.

At first the effect of the shape of the lamellae have been discussed considering ordered structures; not negligible effects were observed since the barrier properties significantly decreased passing from a tape shape to square or circular platelets. Considering the behavior of the barrier effect in these structure, furthermore, two different trends were observed, the first one linear with the factor $\alpha\ \phi$, and the second one quadratic, related respectively to the hole resistance and to the wiggling of gas molecules around the flakes.

More realistic structures were also developed by placing circular platelets randomly in the simulation box. The behavior of the barrier effect in these structure has been analyzed and compared with the previous one and to the model already describe in the technical literature.

The results showed that the ordered structures hypothesis is too restrictive in modeling diffusion in nanocomposite systems and it overestimates dramatically the barrier effect achieved by the presence of the inorganic filler. On the other hand, Nielsen model, despite its simplicity, proved to be able to represent very accurately the present simulation results and seems to have also high potentialities in the prediction of real materials behavior.

Acknowledgement

This work was performed with financial support of the project "SustainPack: Innovation and Sustainable Development in the Fibre-based Packaging Value Chain" funded by the Nanotechnology RTD programme in the 6th framework research program of the European Union.

References

1. Yano, Kazuhisa et al. (1993), "Synthesis and properties of polyimide–clay hybrid films", *Journal of Polymer Science Part A: Polymer Chemistry*, 31, 2493-2498.
2. Gain, O et al. (2005), "Gas barrier properties of poly(e-caprolattone)/clay nanocomposites: influence of the morphology and polymer/clay interaction' *Journal of Polymer Science, Part B: Polymer Physics*, 43, 205-214.

3. Messersmith, Phillip B. et al. (1995), "Synthesis and barrier properties of poly(ε-caprolactone)-layered silicate nanocomposites" *Journal of Polymer Science Part A: Polymer Chemistry,* 33, 1047-1057.
4. Swannack, Charles et al. (2005), "A three-dimensional simulation of barrier properties of nanocomposite films", *Journal of Membrane Science*, 263, 47–56.
5. Bharadwaj, Rishikesh K. (2001), "Modeling the barrier properties of polymer-layered nanocomposites", *Macromolecules,* 24, 9189-9192.
6. Xiaoming, Chen et al., (2007) "Barrier properties of flake-filled membranes: review and Numerical Evaluation", *Journal of Plastic Film and Sheeting*, 23, 319-328.
7. Aris, Rutherford (1986) "On a problem in hindered diffusion", *Archive for Rational Mechanics and Analysis*, 18, 83-91.
8. Wakeman, William A. et al. (1979) "Diffusion through multiperforate laminae", *Industrial & Engineering Chemistry Fundamental*, 18, 301-305.
9. E.L. Cussler, Edward L. et al. (1988) "Barrier membranes", *Journal of Membrane Science*, 38, 161-174
10. Minelli, Matteo et al. "Analysis of modeling results for barrier properties in ordered nanocomposite systems", Submitted to *Journal of Membrane Science.*
11. Barrer, Richard K. et al. (1961), "Diffusion in Heterogeneous media: lattices of parallelepipeds in a continuous phase", *British Journal of Applied Physics*, 12, 691-697.
12 Neilsen, L.E. (1967), Models for the Permeability of Filled Polymer Systems", *Journal of Macromolecular Science, Part A: Pure and Applied Chemistry*, 1, 929-942.
13. Falla, Wayne R. et al. (1996), "Estimating diffusion through flake-filled membranes", *Journal of Membrane Science*, 119, 129-138.
14. Fredrickson, Glenn H. et al. (1999), "Barrier properties of oriented disk composites", *Journal of Chemical Physics.*, 110 (1999) 2181-2188
15. Gusev, Andrei A. et al. (2001), "Rational design of nanocomposites for barrier applications". *Advanced Materials*, 13, 1641-1643.
16. Minelli, Matteo et al. (2007) "Transport Properties in Nanocomposites: Modeling CFD Approach for Randomly Distributed Systems" AIChE Annual Meeting 2007, Salt Lake City (UT).

MOLECULAR SIMULATION OF NANOSCALE DISTRIBUTION AND MOBILITY OF WATER AND DIMETHYLMETHYLPHOSPHONATE IN NAFION AND SULFONATED POLYSTYRENE

Aleksey Vishnyakov* and Alexander V. Neimark
Department of Chemical & Biochemical engineering, Rutgers, The State University of NJ
* presenting author; email avishnja@rci.rutgers.edu

ABSTRACT

Polyelectrolyte membranes (PEM) such as Nafion (DuPont TM) and sulfonated polystyrene based block copolymers used as compartment separators in fuel cells and other electrochemical applications, are also of interest as permselective diffusion barriers in protective fabrics. By means of molecular dynamics (MD) simulations, we explore the molecular mechanisms of sorption and mobility of water and nerve agent simulant dimethylmethylphosphonate (DMMP) in Nafion and sulfonated PS. It is shown that the composition of the solvent strongly affects the membrane permeability, which cannot be predicted purely from the data for pure compounds. This effect is related to complex microphase segregation in the membranes swollen in aqueous solutions of polar solvents. The results obtained demonstrate possible mechanisms of the diffusion of alcohols and alkylphosphonates in hydrated ionomer membranes and reveal properties controlling the utilization of PEM for protective fabrics.

INTRODUCTION

Two prospective groups of polyelectrolyte membranes (PEM) for protective application include perfluorinated membranes such as Nafion and triblock copolymers, where central polyolefin hydrophobic block is confined between smaller hydrophilic blocks made of sulfonated polystyrene. Permselective properties of both materials are based on their complex nanostructure: upon hydration, they segregate onto hydrophilic and hydrophobic subphases. Water and air supposedly diffuse through the mobile hydrophilic subphase, while organic agents get trapped in a immobile hydrophobic subphase. Because the hydrophilic sidechains of Nafion polymer are relatively small and attached directly to the hydrophobic backbone, the scale of the segregation is very limited: depending on hydration and equivalent weight, the size of hydrophilic aggregates varies between ca 20 to 60Å [1]. Hydrophobic and hydrophilic domains form bi-continuous irregular network. Triblock copolymers form regular hexagonal and lamella structures with segregation scale determined by block length. Conflicting reports were published on the possible nanoscale inhomogeneity of the hydrophilic subphase itself [2,3]. In this work, we will discuss the connection between its nanostructure and solvent mobility.

SOLVATION AND MOBILITY OF WATER AND DMMP IN NAFION

In Nafion simulations, we modeled a thin film of the polymer formed by six four-unit Nafion fragments in a contact with the mixture of solvents composed of 60 DMMP molecules and 420 water molecules, which roughly corresponds to 1:1 vol solvent composition. $K+$ was chasen as counterion. The system was equilibrated in NPT ensemble at 303K and 1atm and proceeded for 1ns for mobility evaluation. Molecular models were developed by us previously [4]. During equilibration period, the hydrophilic sidechains protruded into the solvent. The sulfonate groups were mostly surrounded by water, as radial distribution functions clearly showed. Counterions dissociated from the sidechains and were diffusing in water.

The resulting picture effectively represents an interface between the hydrophobic subphase presented by a film of polymer backbone and the hydrophilic subphase presented by solvents, counterions and sidechains (Figure 1). Surprisingly, we did not find any substantial dissolution of DMMP in the polymer phase. Rather, DMMP molecules were located at the interface between the hydrophilic and hydrophobic subphases, despite water and DMMP being well miscible. In order to demonstrate this observation quantitatively, we calculated a number of simultaneous contacts of DMMP molecules with water molecules and the backbone. It was found that only about 1% (average) of DMMP molecules belonged exclusively to the hydrophobic subphase. About 4% of DMMP molecules belonged neither to the hydrophilic nor to the hydrophobic subphase. That is, they were surrounded by other DMMP molecules, or located in the centers of DMMP mini-clusters, that were observed in that system. Finally, 72% of the DMMP molecules belonged to both hydrophilic and hydrophobic subphases (had at least one hydrogen), and 23% were "dissolved" in water, i. e. had a hydrogen bond to a water molecule, but no neighbors among fluorocarbon groups of the backbone. This means that DMMP tended to be located at the interface between water and Nafion

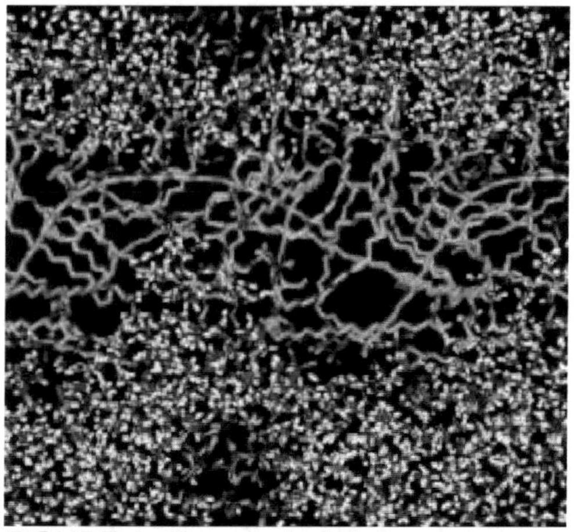

FIGURE 1. Shapshot of Nafion polymer layer solvated in water-DMMP mixture. The presence of Nafion reduces the solubility of DMMP in water. DMMP concentrates on the surface of polymer rather than is dissolved in water. C in green O in red H in white P in blue S in orange K in pink

NANOSTRUCTURE AND SOLVENT DIFFUSION IN SULFONATED POLYSTYRENE

In polystryrene simulations, we considered ca 50 Å fragments of PEM, which were modeled for 4-6 ns after equilibration. We considered 40% and 100% PS sulfonation levels, as well as random 1:1 copolymers of PS and PE at 100% PS sulfonation. The counterion was Ca^{2+} (molecular models are described in ref [5]). The systems are shown in Table 1.

TABLE 1. The list of polymer systems and their properties Solvent content shown in wt % to dry polymer weight. n_{HB} denotes the number of hydrogen bonds per water molecule

Polymer, % sulfonat.	H2O, wt %	DMMP, wt%	ρ, g/cm^3	D_W 10^{-9} m/s^2	D_{DMMP} 10^{-9} m/s^2	n_{HB} ‡ H2O-H2O	n_{HB} ‡ W-DMMP	n_{HB} ‡ W--O3S	n_{HB} ‡ total
sPS, 40	15		1.180	0.030		0.74		0.91	1.65
sPS, 40	30		1.177	0.21		1.17		0.66	1.83
sPS, 40	54	0	1.165	0.91		1.47		0.49	1.96
sPS, 40	54	10	1.168	0.58	0.035	0.96	0.03	0.50	1.49
sPS, 40	54	50	1.178	0.36	0.041	1.40	0.13	0.45	1.98
sPS, 40	54	100	1.188	0.19	0.052	1.30	0.25	0.45	1.95
sPS, 100	15		1.251	0.012		0.28		1.16	1.44
sPS, 100	50		1.242	0.067		0.86		0.94	1.8
sPS-PE, 100	39		1.208	0.109		0.95		0.56	1.51

In all hydrated sPS samples, we observed prominent nanoscale segregation onto hydrophilic and hydrophobic subphases, somewhat similar to that in hydrated Nafion, but at a smaller scale. Similarly to Nafion, the hydrophilic subphase is discontinuous at small level of hydration and becomes continuous at higher hydrations (>15%wt). The segregation is shown in Fig. 2, which shows the areas occupied by water molecules. The lower the sulfonation level, the stronger the segregation: at 100% sulfonation every hydrophobic benzene ring has a sulfonate group attached to it, which reduces the possible spatial scale of segregation to the polymer chain diameter. Because the small segregation length and large interface are, most counterions are still bound to at least one sulfonate group, even at highest water content. Naturally, water diffusion increases radically with the increase of the water content.

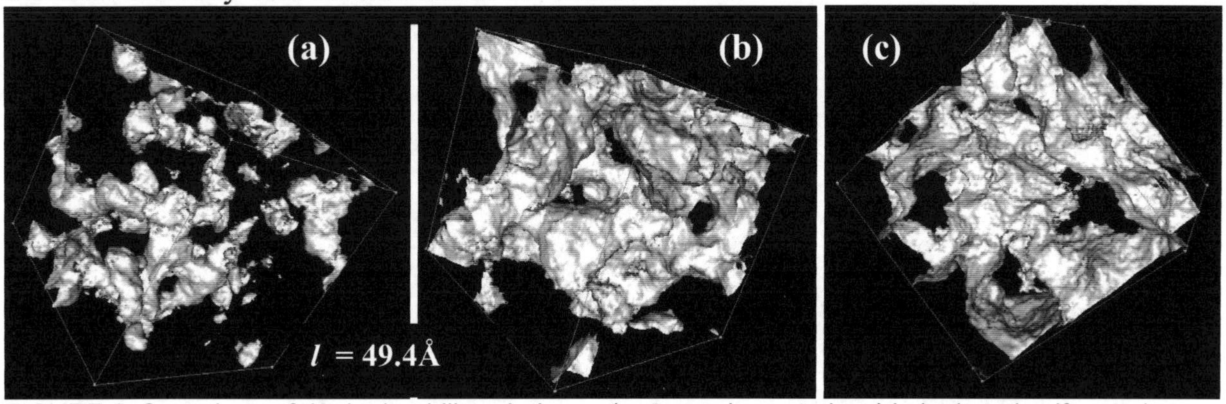

FIGURE 2. Snapshots of the hydrophilic subphase (water and counterions) in hydrated sulfonated polystyrene at 40% sulfonation level. Water content (to dry polymer weight) (a) 15% (b) 30% (c) 54%.

A cluster analysis of the sPS-water-DMMP samples shows that in all systems, the mobile species (water, DMMP, and counterions) form a single continuous subphase. However, at higher DMMP concentrations (50 and 100 wt %), DMMP may be considered as creating a continuous subphase of its own. Therewith, the water diffusion coefficient steadily reduces with the increase of the DMMP content (and, respectively, the increase of the DMMP diffusion coefficient), despite the apparent increase of the cumulative volume occupied by the mobile species, which mix perfectly in the bulk (Table 1). These unforeseen observations indicate that water and DMMP have separate pathways through the system. Within the solvated sPS, DMMP molecules accept only 0.75 hydrogen bonds in average at the DMMP content of 50% and 100% wt compared to 1.3-1.4 bonds accepted by water. This is in contrast to the bulk solution, where DMMP molecules accept 1.5-2 hydrogen bonds from water molecules in average, which is on a par with water molecules

CONCLUSION

The main conclusion of this simulation work is that the nanostructure of PEM in mixture of water and hydrophilic phosphor-organic compounds is very complex and cannot be reasonably predicated from single-component data. Despite DMMP is hydrophilic and water-soluble, its pathways through the membrane are quite different from those of water. Because DMMP does not readily mix with the hydrophobic phase either, the application of PEM for protection will remain problematic as soon as the absorption of DMMP by the polymer remains comparable to the polymer weight [6]. Our current efforts are directed onto the predictions of self organization of triblock copolymers using mesoscale simulation methods

The authors are thankful to Drs. Donald Rivin and Nathan Schneider from Natick RDEC for fruitful; discussions. The support by ARO (grant W911NF-04-0239), DURIP (grant 07030491) and DTRA HDTRA1-08-1-0042 is acknowledged.

REFERENCES

[1] S. J. Paddison, Annual Rev. Materials Res. **33**, 289 (2003). [2] D. Gromadzki, P. Cernoch, M. Janata, V. Kudela, F. Nallet, O. Diat, and P. Stepanek, Europ. Polymer Jl **42**, 2486 (2006). [3] K. Xu, K. Li, P. Khanchaitit, and Q. Wang, Chem. Materials **19**, 5937 (2007). [4] D. Rivin, G. Meermeier, N. S. Schneider, A. Vishnyakov, and A. V. Neimark, J Phys Chem B **108**, 8900 (2004). [5] A. Vishnyakov and A. V. Neimark, J Chem phys. **128**, 164902 (2008). [6] N. S. Schneider and D. Rivin, Polymer **47**, 3119 (2006).

703h Characterization of Entangled Cross-Linked Poly(ethylene glycol) Hydrogels with Dangling Ends

Mahnaz Eskandari[1], Arsun Artel[2], Eric M. Brey[3], and Ali Cinar[1]. (1) Chemical and Biological Engineering, Illinois Institute of Technology, Suite 214A, 10 W. 33rd street, Chicago, IL 60616, (2) Department of Chemical and Biological Engineering, Illinois Institute of Technology, 10 W 33rd Street, Chicago, IL 60616, (3) Biomedical Engineering, Illinois Institute of Technology, Engineering 1 Building, Room 116, 10 W. 32nd Street, Chicago, IL 60616

Characterization of Entangled Cross-linked Poly(ethylene glycol) Hydrogels with Dangling Ends

Poly(ethylene glycol) (PEG) hydrogels have a chemically and physically tied filament self-ensemble structure, high water adsorption, and biodegradability, which makes it a favorable substrate for biomedical application. The addition of cell adhesion sequences as dangling ends allows for engineering substrates for controlled cell growth. While PEG has been used extensively in tissue engineering there have been few attempt to accurately model its properties.

A coarse-grained model is developed for an entangled cross-linked PEG, which includes dangling ends, and is used to characterize PEG's mechanical properties. Dangling ends can be peptide groups intentionally added to make PEG favorable for cell interactions or they can result from the polymerization process. Our model includes four adjustable parameters; percent of a peptide group (Y_{Pep}), percent of dangling ends (Y_{DE}), the average number of persistence lengths between two entanglements (N_E), and the average number of persistence lengths between two cross-links (N_C).

Molecular and chemical process information such as N_E, N_C, Y_{Pep}, and Y_{DE} are used as input parameters to create a numerical ensemble of number of persistence length(s) between entanglements and cross-links, and connecting vectors between entanglements during the preparation step of the model. Different deformation fields at equilibrium are applied to the prepared ensemble to analyze the stress-strain behavior of system in the dynamic step of the modeal. It is assumed that cross-link points and average position of entanglements are deformed affinely. Entanglements are allowed to move along polymer chains; therefore polymer chains are able to create new entanglements or disentangle from each other. The effects of dangling ends and peptide group on mechanical properties are studied under different deformation fields. This model is currently being compared to experimental data. The results indicate that dangling ends and peptide group reduce the mechanical strength that is dominated by cross-links.

722a Development of Double-Layer Hydrogels for Agricultural Applications

Sangjoon Kim, The Department of Chemical & Environmental Engineering, The University of Toledo, Toledo, OH 43606 and **Arunan Nadarajah**, The Department of Bioengineering, The University of Toledo, Toledo, OH 43606.

Hydrogels offer the ability to slowly release moisture and nutrients to plants which can considerably reduce the amount of freshwater and fertilizer needed and the associated labor costs in agriculture. Polyacrylamide (PAAm) hydrogels are being produced commercially for such applications, but they are quite fragile and break apart easily losing their water retention properties. Additionally, while these hydrogels can rapidly absorb large amounts of water, they also dehydrate very rapidly in a matter of hours. In this study, we are developing a two layer PAAm-based hydrogel to overcome these disadvantages. The outer layer is made up of either highly crosslinked PAAm or polyurethane (PU), providing low aqueous permeability and high mechanical strength. Dehydration measurements showed that the double layer hydrogel with a PU outer took longer to dehydrate than the hydrogel with a PAAm outer layer. The hydration/dehydration rates of these PU-PAAm hydrogels are a function of the pore sizes of the PU outer layer hydrogel, which in turn was determined by the molecular weight of the polyethylene glycol used to synthesize PU. The inner layer of these gels consists of soft, low crosslinked PAAm which retains the high water absorbing property, ensuring its ability is to supply sufficient water to plants. The combined system results in a gel with adequate mechanical strength for deployment in soils as well as extending the gel dehydration time to nearly a week.

722b A Modular Approach to Organic Light-Emitting Materials with Tunable Charge Injection and Transport Properties

Andrew C. A. Chen[1], **Jason U. Wallace**[1], Ching W. Tang[2], and Shaw H. Chen[1]. (1) Chemical Engineering, Unicersity of Rochester, 240 East River Road, Rochester, NY 14623, (2) Chemical Engineering, University of Rochester, 101 Gavett Hall, Rochester, NY 14627

Since the discovery of efficient electroluminescence at a relatively low voltage using organic materials, including low-molar-mass fluorescent dyes and p-conjugated polymers, intensive efforts have been devoted to improving the efficiency and lifetime of organic light-emitting diodes (OLEDs). While low-molar-mass materials can be deposited as thin films by sublimation, conjugated polymers can be readily processed into large-area thin films by spin-coating from dilute solutions. In principle, electrons and holes are injected from the cathode and anode, respectively, for the formation of excitons in the emissive layer where radiative decay takes place. To achieve a high quantum yield with long device lifetime, it is crucial that charge injection and transport be balanced and that the recombination zone be spread out in space. Four strategies have been reported in the literature: (i) using a low work function metal as the cathode, such as Ca or Ba capped with Al or Ag, (ii) adding an injection, a buffer, and/or a charge-transport layer, (iii) physical blending of an emissive material with a charge-transporting material, and (iv) chemical modification of an emissive material with charge-transporting moieties. Of the four strategies, chemical modification appears to be the most versatile, and hence has been the most intensively pursued. Low-molar-mass evaporable materials have been constructed by bonding electron- or hole-conducting moieties to light-emitting conjugated molecules through p-conjugation, inevitably affecting individual functionalities. In most conjugated polymers, holes are preferentially transported over electrons. Electron transport has been improved by incorporating p-electron-deficient moieties, such as oxadiazole, triazole, triazine, and quinoxaline, in the polymer backbone, as the pendant, or as the end-cap. In the case of blue OLEDs, hole injection is also a limiting factor because of the high ionization potentials of most blue-emitting materials. This difficulty can be overcome in part by adding a layer of poly(3,4-ethylenedioxythiophene):poly(styrene sulfonate), PEDOT:PSS, between the indium tin oxide, ITO, anode and the emissive polymer layer. Because of its acidic nature, PEDOT:PSS was found to etch ITO, causing device instability. This problem has been addressed using an alternative hole-injection material or a self-assembled monolayer on the ITO anode.

Because of the cost advantage organic materials amenable to solution processing into noncrystalline films are of particular interest. While conjugated polymers hold enormous potential in this regard, it is believed that monodisperse conjugated oligomers with a relatively low molecular weight are advantageous from both the scientific and technological perspectives. Monodisperse conjugated oligomers are characterized by a well-defined and uniform molecular structure as well as superior chemical purity acquired through recrystallization and/or column chromatography. Relatively short and uniform chains are also conducive to the formation of monodomain glassy-nematic films without grain boundaries through thermal annealing under mild conditions. These intrinsic merits are imperative to furnishing fundamental insight into structure-property relationships and to improving OLED device performance, as traces of impurities could result in exciton quenching and device failure. In contrast to polymers, oligomers are less likely to undergo glass transition to form a morphologically stable glassy film. Furthermore, few monodisperse conjugated oligomers are capable of both liquid crystalline mesomorphism and glass transition with an elevated glass transition temperature, Tg, to enable the processing of morphologically stable glassy liquid crystalline films that resist crystallization under ambient conditions. Recently, we have reported the first examples of monodisperse glassy-nematic conjugated oligomers for the demonstration of linearly polarized, full-color and white-light OLEDs. Both the luminance yield and polarization ratio are the best of all polarized OLEDs reported to date. Nevertheless, it was also recognized that charge injection and transport must be varied as desired to further improve device performance. Liquid-crystal conjugated oligomers capable of in-situ polymerization represent a viable alternative to preserving molecular order in solid films; however, heat, UV-irradiation and/or initiators and inhibitors are needed for film processing.

The goal of this study is to create multifunctional molecular materials capable of forming glassy isotropic and liquid crystalline films using light-emitting conjugated oligomers and charge injection and transport moieties as the building blocks. A flexible spacer, such as an alkyl chain, connecting the two building blocks serves to permit light emission and charge injection/transport to be incorporated without mutual interference and to prevent crystallization while encouraging glass formation of the hybrid system. Although the merits of a multilayer device structure are well documented, multifunctional materials will offer devices comprising fewer layers, thereby reducing the fabrication costs and operating voltages while improving device performance. Major findings are as summarized in what follows.

Novel molecular systems are designed to incorporate the following features: (1) an electron- and a hole-conducting core, (2) terfluorene and pentafluorene pendants for blue emission (3) a flexible spacer attaching pendants to cores, thereby enabling independent functions of the two structural elements, (4) the ability to form glassy isotropic and liquid crystalline films by solution processing, and (5) tunable charge injection and transport properties while emitting unpolarized and polarized light. Four representative materials have been synthesized and demonstrated to be capable of forming glassy-isotropic and glassy-nematic films with a glass transition temperature and clearing point close to 140 and 250oC, respectively, an orientational order parameter of 0.75, a photoluminescence quantum yield in neat film up to 51 %, and the HOMO and LUMO levels intermediate between blue-emitting oligofluorenes and the electrodes commonly used in OLEDs (e.g. ITO and Mg:Ag). Work is in progress to exploit these promising materials for the fabrication of highly efficient OLEDs with long-term stability, and the results will be the subject of a future publication.

722c Coarse-Grained Simulation of DNA Dynamics in Sub-Persistence Length Nanochannels

Yeng-Long Chen, Institute of Physics and Research Center for Applied Science, Academia Sinica, 128 Academia Rd Sec. 2, Nangang, Taipei, Taiwan

Recent advances in optical tools and microscopy techniques have spurred investigation of single DNA molecules in micron- and nano-scale fluidic channels. Classical polymer physics has been shown to be extremely successful in predicting the polymer conformation and dynamics of DNA in various experiments. In sub-100nm channels, it has become possible to study the conformation and dynamics of DNA at and below the length scale of the DNA persistence length (~50nm). We employ Brownian dynamics simulations to investigate DNA dynamics in nanochannels. Our study explores how the bending and thermal energy, the conformational entropy, and the DNA-surface interactions contribute to the macromolecular dynamics. Our simulation results are compared to the predictions of the Odijk theory for confined polymers,and we found that the confinement strongly affects the persistence length segment relaxation, which could affect how double stranded DNA interact with enzymes and proteins in biological processes.

722d Flow-Induced Temperature Change and Anisotropic Heat Capacity of a Linear Short-Chain Polyethylene Liquid

Brian Edwards, Chemical and Biomolecular Engineering, The University of Tennessee, Knoxville, TN 37996-2200 and Chunggi Baig, Chemical Engineering, ICEHT-FORTH, Patras, Greece.

The anisotropic configurational changes induced in the linear, short-chain polyethylene liquid C78H158 were simulated using the Siepmann-Karaboni-Smit united atom model under steady-state uniaxial elongation and shear flow. The complementary simulation techniques of Nonequilibrium Monte Carlo (NEMC) and Nonequilibrium Molecular Dynamics (NEMD) were used to determine independently the flow-induced structural changes in these two flow fields, and to calculate physical properties of the system under these conditions. Using the extended Gibbs ensemble of the NEMC method, the configurational temperature of the system was calculated at various values of flow strength, and compared to the set point temperature of the simulation. The same idea was applied to NEMD, where the configurational temperature was compared to the set point kinetic temperature of the system. It was found that the configurational temperature of the system decreased with increasing flow strength, approaching the melting point of the liquid. This seemed to correlate with experimental observations concerning flow-induced crystallization in seemingly isothermal liquids. Flow-induced changes in the heat capacity were also observed, as calculated according to the standard thermodynamic definition of the heat capacity; i.e., the change in internal energy with temperature at constant density and average chain configuration. The heat capacity calculated in this manner was shown to decrease with increasing flow strength. Finally, the simulation results led to the concept of a nonequilibrium temperature in these systems undergoing flow, and a simple expression was derived which quantified this effect. The implications of these findings will be discussed and analyzed in this presentation.

722e Dynamic Relaxation Properties of Polymer Nanocomposites: Effect of Nanoparticles on the Glass-Rubber and Sub-Glass Transitions

Anthony C. Comer[1], Victor A. Kusuma[2], Benny D. Freeman[2], and **Douglass S. Kalika**[1]. (1) Chemical and Materials Engineering, University of Kentucky, 177 Anderson Hall, Lexington, KY 40506-0046, (2) Chemical Engineering, The University of Texas at Austin, 10100 Burnet Rd., Bldg 133, Center for Energy and Environmental Resources, Austin, TX 78758

The glass-rubber and sub-glass relaxation characteristics of model polymer nanocomposites have been investigated via dynamic mechanical and dielectric methods. Two series of nanocomposites were studied; (i) rubbery nanocomposites prepared via UV photopolymerization of poly(ethylene glycol) diacrylate [PEGDA] in the presence of MgO or SiO2 nanoparticles, and (ii) glassy nanocomposites prepared via solution casting of polyetherimide [Ultem PEI] in combination with either native SiO2, or surface-modified (i.e., hydrophobic) silica particles. Both systems are of interest owing to their potential for use as gas separation membranes.

Dynamic mechanical and dielectric measurements revealed offsets in the position of the glass-rubber relaxation in the nanocomposites as compared to the unfilled polymer that reflect the character of the particle-polymer interaction as well as possible physical confinement. Depending upon the specific nanocomposite system, both an upward shift in the bulk Tg, as well as the emergence of a second segmental relaxation event at higher temperature were detected. These phenomena, as well as potential sub-glass perturbations, are examined as a function of particle loading, particle surface chemistry and nanocomposite morphology.

722f Unraveling Entanglements and Polymer Rheology: New Rules

Richard P. Wool, Department of Chemical Engineering and Center for Composite Materials, University of Delaware, 150 Academy St, Newark, DE 19716

In this paper we propose a solution to an unsolved problem in solid state physics, namely, the nature and structure of the glass transition in amorphous materials. The development of dynamic percolating fractal structures near Tg is the main element of the Twinkling Fractal Theory (TFT) presented herein and the percolating fractal twinkles with a frequency spectrum F(w) ~ wdf-1 exp Delta E|/kT as solid and liquid clusters interchange with frequency ù. The Orbach vibrational density of states for a fractal is g(w) ~ wdf-1, where df = 4/3 and the temperature dependent activation energy behaves as Delta E ~ (T2 – Tg2). The key concept of the TFT derives from the Boltzmann population of excited states in the anharmonic intermolecular potential between atoms, coupled with percolating solid fractal structures near Tg. The twinkling fractal spectrum F(w) at Tg predicts that the spatio-temporal thermal fluctuation autocorrelation relaxation function C(t) behave as C(t) ~ t-1/3 (short times), C(t) ~ t-4/3 (long times) and C(t) ~ t-2 (w < wc), which were found to be in excellent agreement with published nanoscale dielectric AFM experiments on a glassy polymer near Tg. Using the Morse potential, the TFT predicts that Tg = 2Do/9k, where Do is the interatomic bonding energy ~ 2-5 kcal/mol and is comparable to the heat of fusion DeltaHf. Since anharmonicity controls both the thermal expansion coefficient CTEL, and Tg, the TFT uniquely predicts that CTEL*Tg = 0.03, which is found to be universal for a broad range of glassy materials from pyrex to polymers to glycerol. Below Tg, the glassy structure attains a frustrated non-equilibrium state by getting constrained on the fractal structure and the thermal expansion in the glass is reduced by the percolation threshold pc as CTEg = pcCTEL. The change in heat capacity DeltaCp = CpL - Cpg at Tg was found to be related to the change in dimensionality from Df to 3 in the Debye approximation as the ratio CpL/Cpg = 3/Df, where Df is the fractal dimension of the glass. The TFT describes the molecular weight dependence of Tg, the role of crosslinks on Tg, the Flory-Fox rule of mixtures and the WLF relation for the time-temperature shift factor aT., which are traditionally viewed in terms of Free-Volume theory. The TFT offers new insight into the behavior of nano-confined glassy materials and the dynamics of physical aging. It also predicts the relation between the melting point Tm and Tg as Tm/Tg = 1/[1-pc] = 2.

722g Kinetic Modeling of Atom Transfer Radical Polymerization: Linking Control to Reaction Rates

Dagmar R. D'hooge, Marie-Françoise Reyniers, and Guy B. Marin. Chemical Engineering, Ghent University, Laboratory for Chemical Technology (LCT), Krijgslaan 281 (S5), Ghent, 9000, Belgium

1. Introduction

Atom transfer radical polymerization (ATRP) is a recently developed and industrially attractive controlled radical polymerization technique (Matyjaszewski and Xia, 2001). It results in well-defined macromolecules, i.e. macromolecules with a controlled molecular mass, a controlled chain architecture and chain composition, a low polydispersity index and possible end group functionality in their structure. The latter allows further chemical modification of the polymer, such as the synthesis of block copolymers.

Figure 1 presents the principle of ATRP. Macroradicals (*R*) can be deactivated by a transition metal complex ($M_t^{n+1}L_yX_2$, deactivator) into a dormant form (*RX*) by transfer of the end group functionality (*X*) of the polymer from the deactivator to the macroradical. The end group functionality *X* of the polymer is generally a halogen atom. The transfer of *X* between the deactivator and the macroradical results in the formation of a transition metal complex ($M_t^nL_yX$, activator) with a lower oxidation number for the transition metal. Reaction of *RX* and $M_t^nL_yX$ regenerates the active form *R* and allows an increase of the polymer chain length by propagation of *R* with monomer (*M*). Termination of radicals results in the formation of dead polymer molecules with loss of the end group functionality *X*. However, for high deactivation reaction rates a low radical concentration compared to the concentration of the dormant species, and hence a low termination reaction rate, can be obtained. Therefore ATRP allows to synthesize polymers with a limited loss of end group functionality *X* during the polymerization.

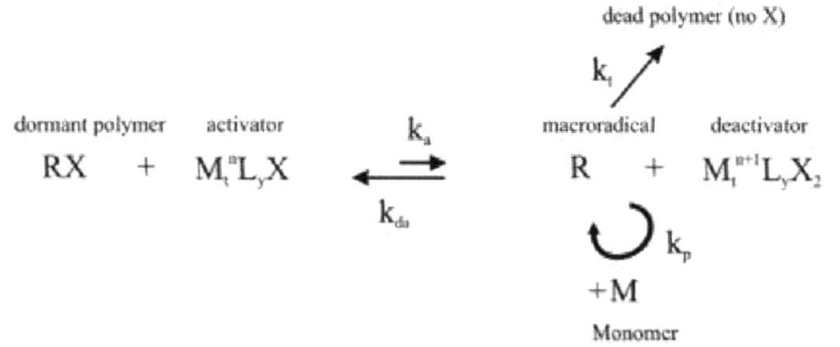

Figure 1: Principle of ATRP; k_a, k_{da}, k_p and k_t = rate coefficient of activation, deactivation, propagation and termination ($m^3mol^{-1}s^{-1}$); M_t = transition metal; n(+1)= oxidation number; L = Ligand; y = number; X = end group functionality of the polymer (halogen atom)

2. Methodology

In this contribution a methodology is presented for the kinetic modeling of ATRP that allows to simulate the monomer conversion and important structural characteristics of the polymer as a function of polymerization time and process conditions.

A first important structural characteristic of a polymer is its molecular mass distribution (MMD). The MMD of a polymer determines important physical properties of the polymer, such as the polymer strength and the polymer plasticity. The developed methodology allows to simulate the full MMD of the dormant polymer molecules and averages of the MMD of the polymer (i.e. dead and dormant polymer molecules) as a function of polymerization time and process conditions.

Secondly, for controlled radical polymerizations the living character of the polymer (i.e. the number of polymer molecules with end group functionality *X*) is important. The fraction of the polymer molecules having end group functionality *X*, f_p, is defined as and calculated from:

$$f_p = \frac{[RX_{tot}]}{[RX_{tot}]+[P_{tot}]} \qquad (1)$$

In equation (1) $[RX_{tot}]$ and $[P_{tot}]$ are the total concentration of respectively the dormant polymer molecules and the dead polymer molecules.

In the kinetic model considered in this study, the most important reaction steps to describe ATRP (i.e. propagation, termination, activation, deactivation and chain transfer to monomer) are included. For all considered reaction steps diffusional limitations are systematically accounted for by introducing an apparent rate coefficient for each reaction step, consisting of a chemical and a diffusional contribution (de Kock et al., 2001). The required diffusion coefficients for the latter contribution are calculated using the free volume theory (Vrentas and Vrentas, 1998).

To solve the rate equations, a methodology based on the method of moments (Achilias, 1988) and on the quasi-steady state approximation for intermediate reactive species (Monteiro et al, 2001; Russell, 2002) has been developed. An extension of the method of moments is used allowing a better description of the contribution of the polymer molecules with a high molecular mass to the MMD of the polymer (i.e. the right side of the MMD of the polymer). Using this approach, moment dependent average apparent rate coefficients for all reaction steps involving macromolecules are introduced. For example the zeroth order average apparent rate coefficient for termination by recombination ($<k_{tc},0>$) is defined by:

$$<k_{tc},0> = \frac{\sum_{i=1}^{\infty}\sum_{j=1}^{\infty} k_{tc}^{ij}[R_i][R_j]}{\left(\sum_{i=1}^{\infty}[R_i]\right)^2} \qquad (2)$$

In equation (2) k_{tc}^{ij} is the apparent rate coefficient for termination by recombination between a macroradical with chain length i and a macroradical with a chain length j and $[R_i]$ the concentration of the macroradicals with chain length i. The $[R_i]$ required for the calculation of the average apparent rate coefficients of the reaction steps can be obtained by application of the quasi-steady state approximation (QSSA) to all the macroradicals (i.e. by solving a set of algebraic equations).

However, the calculation of $[R_i]$ requires also the knowledge of the concentration of the dormant polymer molecules with chain length i, $[R_iX]$. The latter cannot be obtained by application of the QSSA to the dormant polymer molecules, as they are not intermediate reactive species. Instead, $[R_iX]$ is calculated by directly solving the corresponding rate equations. Hence, both $[R_i]$ and $[R_iX]$ are obtained by solving a set of differential algebraic equations.

3. Results

The developed methodology is applied to the batch isothermal ATRP of methylmethacrylate (MMA). The extent to which control of the ATRP is obtained is analyzed in terms of the rates of the occurring reaction steps. The ATRP of MMA was studied within a representative range of process conditions (polymerization temperature (T_{pol}) range: 313-363 K; volume fraction of solvent: 0-0.7; initial M/R_0X molar ratio: 50-200).

Figure 2 (a) and Figure 2 (b) present respectively the first three averages of the MMD of the polymer (i. e. M_n^{pol}, M_m^{pol} and M_z^{pol}) and the f_p of the polymer as a function of monomer conversion (x_m) and this for the following reaction conditions: a T_{pol} of 363 K, a volume fraction of solvent of 0.5, an initial $M/R_0X/M_t^nX/L$ molar ratio equal to 100/1/1/1 and no deactivator present at the beginning of the polymerization.

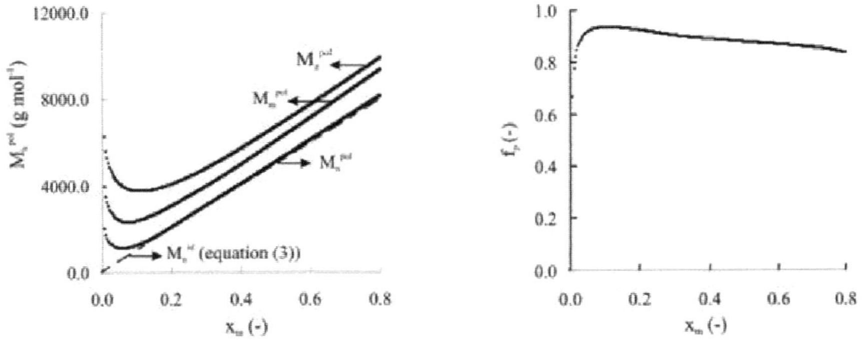

Figure 2: (a) First three averages of the MMD of the polymer (i.e. M_n^{pol}, M_m^{pol}*and* M_z^{pol}; full lines) and ideal average number molecular mass (M_n^{id}; dashed line) as a function of monomer conversion (x_m) (b) Fraction of polymer molecules with end group functionality X (f_p; equation (1)) as a function of monomer conversion (x_m);

It can be seen from Figure 2 (a) that all averages of the MMD of the polymer have a similar profile: a decrease at low x_m, followed by a linear increase at high x_m. The linear increase of the first average of the MMD of the polymer (M_n^{pol}) at high x_m approaches the ideal number average molecular mass of the polymer (M_n^{id}) as a function of x_m (equation (3)).

$$M_n^{id} = MM_{R_0X} + MM_M \frac{[M]_{t=0}}{[R_0X]_{t=0}} x_m = MM_{R_0X} + MM_M PD^{id} x_m \qquad (3)$$

In equation (3) MM_A ($A=M$, R_0X) is the molar mass of the reaction component A and PD^{id} the desired degree of polymerization at final monomer conversion.

From Figure 2(b) it follows that the f_p of the polymer (equation (1)) decreases after an initial increase as a function of x_m. The f_p profile is strongly related to the termination reaction rate, as every termination reaction results in loss of the end group functionality X of the polymer. Figure 3 presents the termination reaction rate (R_{term}) as a function of x_m. It can be seen from this figure that R_{term} is significant at low x_m and then decreases rapidly as a function of x_m. However, at high x_m R_{term} is still high enough to result in further loss of end group functionality of the polymer.

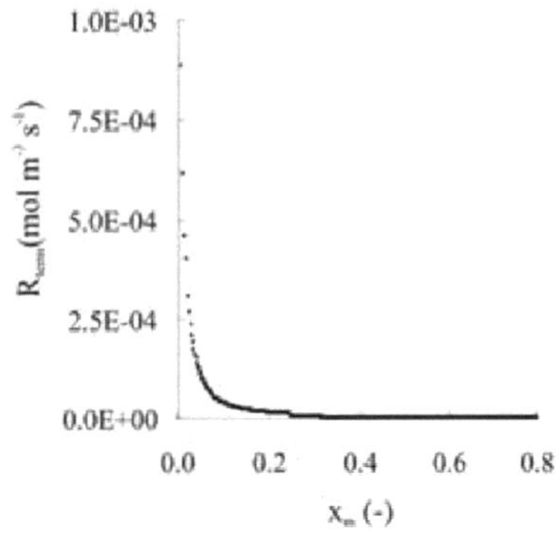

Figure 3: Termination reaction rate (R_{term}) as a function of monomer conversion (x_m)

Acknowledgements

The Institute for the Promotion of Innovation through Science and Technology in Flanders (IWT Vlaanderen) and the Belgian Government (IAP/IUAP/PAI P6/27: "Functional Supramolecular Systems") are acknowledged for financial support.

References

Achilias D., Kiparissides C. (1988) Modeling of diffusion-controlled free-radical polymerization reactions. *J. Appl. Polym. Sci.*, 35: 1303

de Kock J.B.L., van Herk A.M., German A.L. (2001). Bimolecular free-radical termination at low conversion. *J. Macromol. Sci.-Pol. R.*, C41(3):199

Matyjaszewski K. and Xia J.H. (2001). Atom Transfer Radical Polymerization. *Chem. Rev.*, 101: 2921

Monteiro M.J., Subramaniam N., Taylor J.R., Pham B.T.T., Tonge M.P. and Gilbert R.G. (2001). Retardative chain transfer in free radical polymerizations of vinyl *neo*-decanoate in low molecular weight polyisoprene and toluene. *Polymer*, 42:2403

Russell G.T. (2002) The kinetics of free radical polymerizations: Fundamental aspects. *Aust. J. Chem.*, 55: 399

Vrentas J.S. and Vrentas C.M. (1998). Predictive methods for self-diffusion and mutual diffusion coefficients in polymer-solvent. *Eur. Polym. J.*, 34 (5/6):797

722h Open-Pore Polyurethane for Vacuum Insulated Panels

Jun Ji, Marine Engineering Department, Shanghai Maritime University, Pudong Ave. 1550, Shanghai, China, Houde Han, Marine Enginering Department, Shanghai Maritime University, Pudong Ave. 1550, Shanghai, China, and Ankang Kan, Marine Enginering Depatment, Shanghai Maritime University, Pudong Ave. 1550, Shanghai, China.

Vacuum insulated panels (VIPs) are distinguished by their relatively thin thermal insulation layers, about 20 mm, and their outstandingly low thermal conductivity, which is only 33% to 10% that of the traditional heat preservation materials. The core material is an important element of VIP construction, which is mostly made of a special kind of open-pore polyurethane. The manufacturing process of the polyurethane is put forward, and the influence of the open porous ratio, pore-diameter, density, and intensity of the polyurethane on the characteristics of VIPs is discussed. With the pre-heat treatment of the core material, the thermal conductivity of VIPs can reach less than 0.01 W/(m•K), energy can be saved by more than 26.82%.

722i Fibrillar Morphology of Syndiotactic Polystyrene Synthesized Over Heterogeneous Metallocene Catalyst

Joong Jin Han[1], Sang Bok Lee[2], and Kyu Yong Choi[1]. (1) Department of Chemical and Biomolecular Engineering, University of Maryland, College Park, College Park, MD 20742, (2) Department of Chemistry and Biochemistry, University of Maryland, 0107 Chemistry Building, University of Maryland, College Park, MD 20742

The morphologies of semi-crystalline syndiotactic polystyrene (sPS) polymerized over heterogenized metallocene catalysts have been investigated. It has been observed that sPS grows in fibrillar morphology with its diameter of about 30-50 nm. When silica-supported catalyst is used, the fragmentation of original silica support material takes place as confirmed by energy dispersive X-ray (EDX) spectroscopy. The polymer fibrils grow at the solid catalyst surface and intertwine quickly to larger nanofibrils. When the polymerization occurs in a confined geometry such as silica nanotube reactor (SNTR), these nanofibrils intertwine further to larger size and extrude out from the pores. The crystalline structure of the sPS analyzed by X-ray diffraction was of the d-form. Our experimental results show that the intrinsic morphology of sPS synthesized over heterogeneous metallocene catalysts is a fibrillar morphology because of rapid crystallization and the intercalation of solvent or monomer molecules between the polymer chain segments.

722j Modeling Viscoelasticity and Stress Generation In Solidifying Coatings

Daniel J. O'Neal and Alon V. McCormick. Department of Chemical Engineering & Materials Science, University of Minnesota, 421 Washington Ave SE, Minneapolis, MN 55414

As a polymer solution coating solidifies by drying, or a coating of monomer or oligomer solidifies by polymerization and crosslinking reactions, it acquires an elastic modulus. Subsequent departure of solvent or advance of curing reactions causes the coating's elastic stress-free state to shrink isotropically but possibly nonuniformly. Adhesion to a solid substrate frustrates the solidification-induced shrinkage in the coating plane and produces elastic strain and, consequently, elastic stress in the coating. The magnitude and directions of the state of stress depend on the type of material, its physical property and shrinkage history, and the topography of the coated surface. Predicting stress generation can show how undesired defects are formed in the coating and can suggest avenues for optimizing the solidification process to minimize the residual internal stress and reduce defects.

Calculating the stress generation mathematically shoes the entire state of internal stress in the coating domain. To calculate how a drying or curing material behaves, equations of momentum conservation and a constitutive equation must be combined and solved simultaneously. This research presents a novel implementation of the mathematics behind the viscoelastic phenomena seen in a solidifying coating material by manipulating the constitutive equation before inserting it into the equation of momentum conservation. The solution method is computationally less expensive than with standard constitutive and momentum equations. The partial-differential equations are solved using Galerkin's method with finite element basis functions. The final internal state of stress depends on the physical property change with solidification, shrinkage, heterogeneity, and substrate topography. In general, residual stress is caused by competing factors; modulus increase and shrinkage tend to create stress while relaxation destroys it. By varying the rates of shrinkage, modulus development, relaxation time development, and the substrate topography, two-dimensional stress profiles are generated that allow us to see spatially and time-dependent states of stress in a coating. We present the 2-dimensional stress profiles for an ultraviolet light cured system and show how the curing time and substrate topographic height and length change the stress maximum magnitude and location. Such profiles can be used to optimize curing schedules and substrate geometry.

722k Synthesis and Characterization of pH-Responsive P(MAA-co-EGMA) Hydrogel Microparticles for Intelligent Drug Delivery System

Bumsang Kim[1], Youngsik Lee[1], Juseung Yang[1], Eunmi Lee[1], Won Sun Ryoo[1], Sang Min Lee[2], and Kyu Sik Kim[2]. (1) Chemical Engineering, Hongik University, Mapo-gu, Sangsu-dong, Seoul, 121-791, South Korea, (2) Nabion Co. Ltd, Hyundai I valley 605, Sangdaewon-dong, Sungnam-si, 462-714, South Korea

Environmentally responsive materials show drastic changes in their swelling ratio in response to changes in their external pH, temperature, nature and composition of the swelling agent, and electrical or magnetic stimulus. The drastic swelling change can be used in the design of an intelligent drug delivery system. In this study, we developed the novel method to synthesize the pH-responsive P(MAA-co-EGMA) hydrogel particles with micro- or nanoscales using dispersion photopolymerization or supercritical fluid system. We also evaluated the feasibility of the P(MAA-co-EGMA) hydrogel particles as the biological on-off switch for an intelligent drug delivery triggered by an external pH change. The particles having a spherical shape were obtained with an average size in the range between 500 nm and 3 μ m depending on the synthesis method. In addition, the P(MAA-co-EGMA) hydrogel particles showed a typical pH-responsive swelling and release behaviors of the anionic hydrogels.

Preparation of a Fine Porous Thin Film by Phase Separation induced in the Course of Drying an Immiscible Polymer Blend Solution

Jae-Kyung Kim, Kentaro Taki, Shinsuke Nagamine and Masahiro Ohshima, Kyoto University, Kyoto, Japan

Introduction

Porous polymeric films with fine pores in the range from micrometer to sub-micrometer or nano scale order have attracted much attention for various applications, such as membrane separation and purification,[1] solid supports for sensors and catalysts,[2] scaffolds for biological cells,[3] low-dielectric constant materials for microelectronic devices,[4] and battery separators in fuel cells and lithium ion batteries.[5] Recently, phase separation[6] in immiscible polymer blend solutions to prepare porous material has been of great interest due to the simplicity and cost-efficiency of the process. Phase separation in immiscible polymer blends is induced by evaporating the solvent (solvent quenching).[7] It has been known that the dispersed domain (droplet) size and its density can be changed by the coalescence of domains during the phase separation,[8] and the process often results in low domain density and large domain size.[9] Various factors for determining the dispersed domain size and the density, such as solvent type,[10] concentration,[11] molecular weight of the polymer,[12] and weight ratios of polymers[13] have been identified. However, there still remains a great challenge in controlling the pore size and pore density and clarifying the formation mechanisms of porous structures due to the versatility of polymer blends' morphologies.

In this paper, we investigate the PS/PEG/solvent systems in terms of controllability of pore size and pore density in the film. The effects of the polymer solubility in the solvents and drying rate on the porous structure were investigated by using six different solvents in the PS/PEG polymer blend as well as different casting methods were applied to the PS/PEG blend film to control the pore size. Furthermore, the formation mechanism of the structure is also discussed.

Experimental

Materials

Polystyrene (PS) and Polyethylene glycol (PEG) were purchased from Aldrich Chemical Company and Wako Pure Chemicals Ind. Ltd., Japan, respectively. The weight average molecular weight, M_w, the number average molecular weight, M_n, and polydispersity, M_w/M_n, of these polymers were measured by gel permeation chromatography (Shimadzu, model DGU-20A_3, Column; Shodex GPC KF-806L, Eluent; Chloroform). The retention time and molecular weight were calibrated using PS standards. Six different organic solvents, including toluene-dehydrate, tetrahydrofuran-dehydrated (THF), methyl ethyl ketone -dehydrated (MEK), chlorobenzene-dehydrated, 1,4-dioxane-dehydrated, and chloroform -dehydrated (Wako Pure Chemicals Ind. Ltd. Japan, Purity 99.5%), were used without further purification. The characteristics of the materials used in this work are listed in Table 1.

Table 1 Characteristics of PEG, PS, and various solvents used in this work

Materials	Mw	Mn	Mw/Mn	Vapor pressure [mmHg] at 20°C	δ [MPa]$^{0.5}$
Polystyrene (PS)	222,000	99,000	2.24		18.6[14]
Polyethylene glycol (PEG)	280	260	1.07		26.1[15]
Chlorobenzene				12	19.4[14]
Methyl ethyl ketone (MEK)				71	19[14]
Tetrahydrofuran (THF)				129	18.6[14]
1,4-dioxane				27	20.5[14]
Toluene				22	18.2[14]
Chloroform				159	19[14]

Film preparation

PS and PEG 200 were blended at different weight ratios and dissolved into a solvent with different polymer concentrations. The different casting methods are applied to this study. At first, the solution was cast by a micro-injector onto an aluminum Petri-dish, which was 2.1 cm in radius and 0.5 cm in depth. The film thickness was controlled by changing the amount of injected solution. The volume of solution to be injected was 100 µl in the nominal case. The initial thickness of solution was about 300 µm. Secondary, the solution drops onto a substrate and sequentially performing the spin casting. Glass slides (2.5 × 2.5 cm) are used to the substrate in the spin-casting process. The spin-casting is performed at room temperature, at a casting speed of 3000 rpm, and with 10 s of processing time. Then, the cast samples were dried under a nitrogen purge in a temperature-controlled incubator.

Observation of film morphology

The porous structure of the obtained film was observed by scanning electron microscope (Tiny-SEM 1540, Technex Lab Co. Ltd.) with the treatment of gold-palladium sputter-coating under a vacuum atmosphere. AFM (SPM-9500J3, Shimadzu, Japan) is also used to observe the morphology in the nano-scale size. AFM images are taken in tapping mode at room temperature. For the pore size and the density calculation, we counted the number of pores observed in a micrograph of the cross sectional area of the film and divided the number by the area of the cross section.

Results and discussion

Resulting porous structure

A mixture of polystyrene/polyethylene glycol 200 (70/30) blends and toluene solution (90 wt % solvent concentration) was prepared and cast on an aluminum Petri-dish. Then, it was dried under nitrogen purging in a temperature-controlled incubator. For a reference, a homogeneous solution of polystyrene alone with toluene solvent was prepared and dried with the same procedure. Figure 1 shows SEM micrographs of surface and cross-sectional areas of the resulting PS/PEG films. As shown in Figure 1(a), the ordered porous structure about 12 µm in pore size was formed on the surface. The porous structure was exclusively formed on both the surface and interface between the substrate and solution. No porous structure was observed inside the film as illustrated in Figure 1(b). Figure 2 shows a SEM micrograph of the cross-sectional area of the resulting PS film. As shown in Figures 1 and 2, porous structure was well-formed in the polymer blend films but not in the film of PS alone. Comparison of the micrographs in Figure 1 with that in Figure 2 clearly shows that the presence of PEG composition affects the morphology in the polymer blend film. In other words, the film fabricated without PEG has no porous structure, while the films made from PS/PEG solutions showed porous structures.

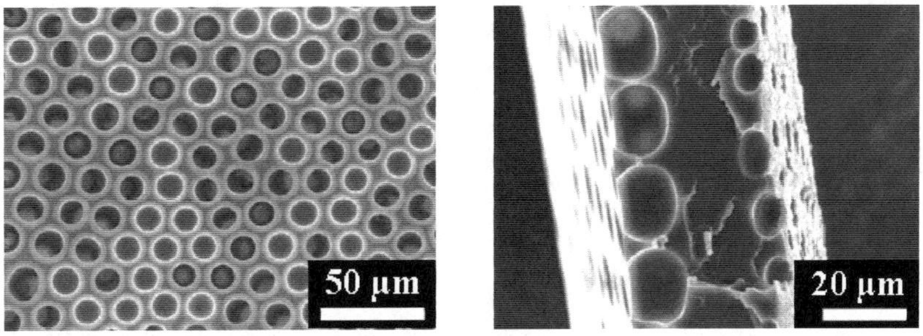

(a) (b)

Figure 1. SEM micrographs of PS/PEG 200 (70/30) (w/w) cast from 90 wt % toluene solution under 3 L/min N_2 flows: (a) Surface micrograph; (b) Cross-sectional micrograph.

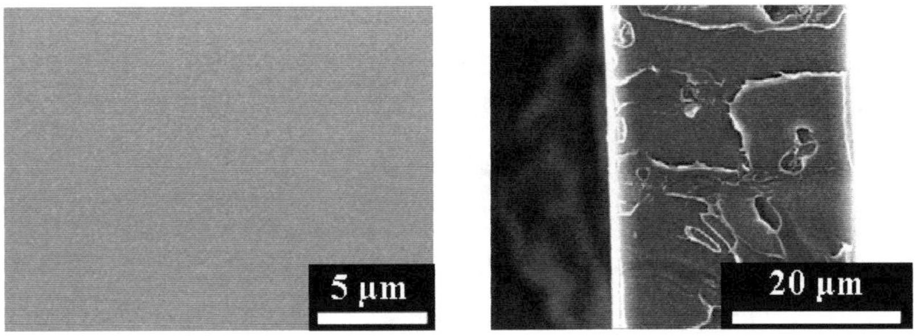

(a) (b)

Figure 2. SEM surface and cross-sectional micrograph of PS cast from 90 wt% toluene solution under 3 L/min N_2 purging.

In our previous work,[12] we showed a formation mechanism of porous structures via two step phase separation in the course of drying a ternary polymer solution. As illustrated in Table 1, the solubility parameters of PS and PEG are different, and the solubility of PS and PEG in the toluene is different as well. The solubility parameters of polymers and solvents indicate that toluene was poor for PEG. Thus, in the course of drying the polymer solutions, the phase separation into PEG-rich and PEG-poor (i.e., PS-rich) phases occurred. The PEG-rich phase dispersed domain formed droplets and the PEG-poor phase became a matrix where PS and solvent coexisted as a single phase solution. Then, secondary phase separation followed in the PEG droplets, which was induced by further solvent evaporation, and formed into solvent-rich and PEG-rich domains within the droplets. The solvent evaporation and secondary phase separation created the pores in the film.

Effect of solvent on the formation of porous structure in PS/PEG blend film

Six homogeneous solutions of PS/PEG blends with different solvents were prepared. All solutions were made at the 70/30 PS/PEG weight ratio and 10 wt% of polymer concentration. Figure 3 shows SEM micrographs of the surface and the cross-sectional area of the resulting films made from the six different solutions. As shown in Figure 3, the pore size and the density of pores in the film were different with the various solvents. The pore size became largest and density became smallest in the film made from the toluene solution, and the size became smallest and density became largest in the film made from the chloroform solution.

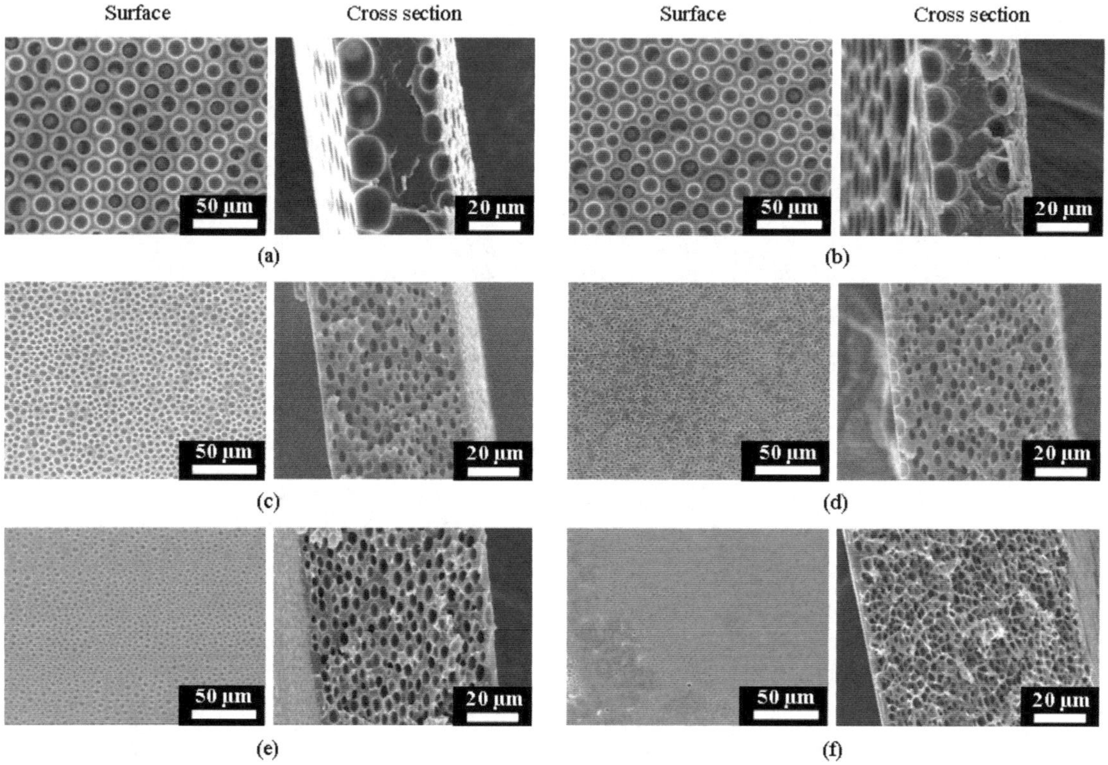

Figure 3. SEM micrographs of surface and cross sectional area in PS/PEG 200 (70/30) (w/w) cast from 90 wt% solvent under 3 L/min N_2 flows ((a) toluene; (b) chlorobenzene; (c) THF; (d) MEK; (e) 1,4-dioxane; (f) chloroform).

The resulting porous structures indicated that the occurrence of phase separation alone could not explain the formation mechanism of the pore size and the density of pores. The factors controlling the process, such as drying rate and change in viscosity of the polymer solution, also affect the porous structure.

The drying rates in each solution were measured by weighing the solution over the course of drying, and the solubilities of PEG to each solvent were estimated by the Flory-Huggins interaction parameter (χ). The interaction parameters were calculated from the solubility parameters of PEG and solvents by:

$$\chi = \frac{V_i}{RT}\left(\delta_i - \delta_j\right)^2 + \beta \qquad (1)$$

where V is the molar volume of the solvent, δ is the Hildebrand solubility parameter for the solvent and polymer, R is the gas constant, and T is the absolute temperature.[16] β corresponds to the entropic component, and a value of 0.34 is used for non-polar systems. The interaction parameters in this study were calculated without an entropic component (β=0) because of the polarity of PEG.[16]

Figure 4 shows some experimental results of the drying rate. The value of drying rate was calculated from the initial slope of the weighting curve against drying time.

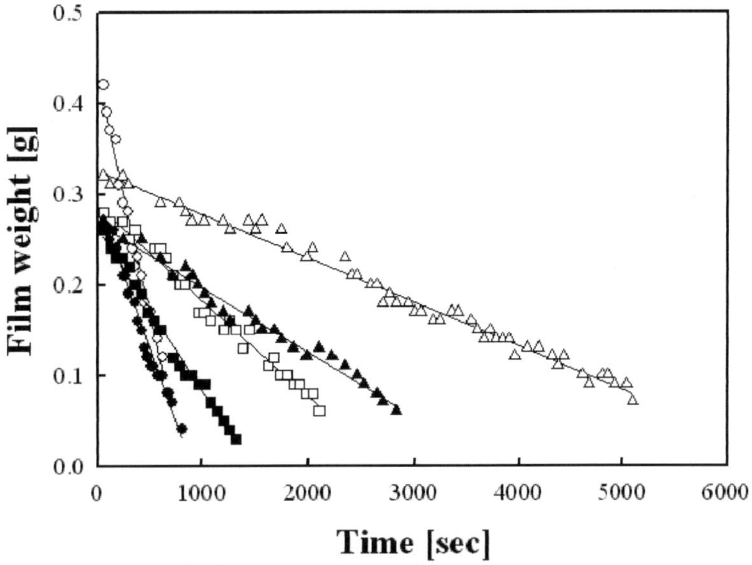

Figure 4. Drying rate for several polymer blend and solvents systems ((○) Chloroform; (●) THF; (□) 1,4-dioxane; (■) MEK; (Δ) Chlorobenzene; (▲) Toluene).

The calculated Flory-Huggins interaction parameter is listed in Tables 2. Figure 5 plots average pore size and density of pores in the cross-sectional area of polymer blend/solvent systems in the coordinate system of drying rate and Flory-Huggins interaction parameter (χ) between its solvent and PEG. Figure 5 clearly shows that the higher drying rate and lower Flory-Huggins interaction parameter made porous structure in the film finer, i.e., pore size smaller and density larger.

Table 2 Flory-Huggins interaction parameter (χ) of solvents and polymers

Materials	Molar volume [cm^3/mol]	$\chi_{PEG\text{-}solvent}$ at 30 °C	$\chi_{PS\text{-}solvent}$ at 30°C
Toluene	106.8	2.64	0.35
Chlorobenzene	102.1	1.81	0.37
Chloroform	80.7	1.61	0.35
THF	81.7	1.82	0.34
MEK	90.1	1.80	0.35
1,4-dioxane	85.7	1.07	0.46

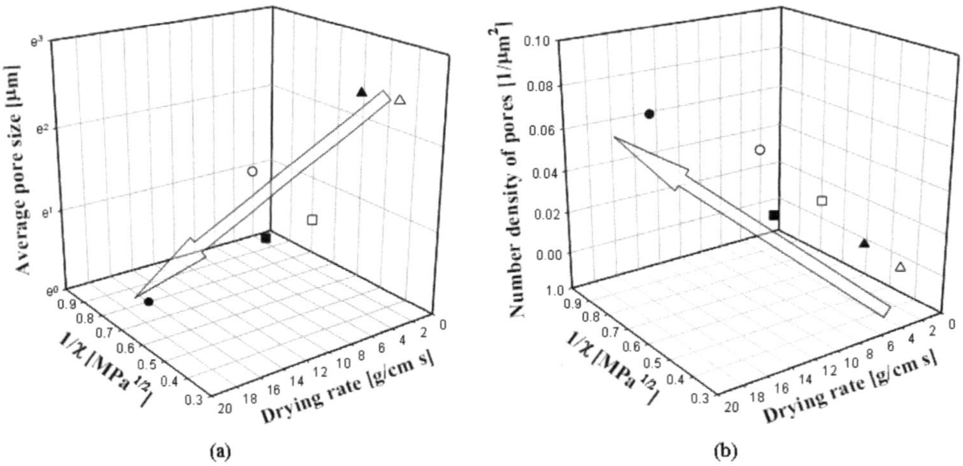

Figure 5. Average pore size (a) and number density of pores (b) in polymer blend/solvent systems on a diagram of drying rate and Flory-Huggins interaction parameter (χ) between its solvent to PEG ((Δ) Toluene, (▲) Chlorobenzene; (□) MEK; (○) 1,4-dioxane; (■) THF; (●) Chloroform).

Formation mechanism of a fine porous structure in polymer blend film

In the course of drying a ternary composition polymer solution, solvent evaporation induces phase separation. The solvent concentration decreases from the surface and phase separation occurs from the surface. As drying proceeded, the solvent concentration profile propagated inside the solution and PEG-rich domains were created inside of the solution and coalesced. The growth rate of the PEG-rich domains was affected by drying rate and the solubility of PEG to solvent. Thus, when fast drying proceeds, the increase of drying rate increases polymer concentration as well as the viscosity of the PS-rich phase (matrix phase) quickly. High polymer concentration and high viscosity of the solution could suppress the coalescence of PEG-rich domains and made the domain size smaller and increased the density of domains in the film.

The effect of solubility of PEG to solvent on the controllability of pore size and density of pores was analyzed as follows, by means of the Flory-Huggins interaction parameter (χ). The Flory-Huggins interaction parameter corresponds to the solubility of the polymer to solvent. Grulke stated that a polymer was considered to be miscible to the solvent when χ became lower than 0.5. It is well known that the smaller the value of χ, the higher the solubility of polymer to solvent. As listed in Table 3, the solubility of PEG200 to solvent increased from 1,4-dioxane, chloroform, MEK, THF, chlorobenzene, and toluene. When the solubility of PEG to solvent increases, the polymer concentration at the onset of PEG phase separation increases. The increase in the solubility of PEG to solvent increases the polymer concentration, and the viscosity of the solutions at the onset of phase separation increases. The higher viscosity of the solution could suppress coalescence of the PEG-rich domain after the phase separation. Therefore, the pore size decreased with the increase in the solubility of PEG to solvent. Thus, the frequency of the coalescence of the PEG-rich domain was controlled by the viscosity of the matrix (PS-rich phase) at the onset of the PEG phase separation i.e., the solubility of PEG to solvent.

When the higher drying rate with the higher solubility of PEG to solvent system was conducted, the PEG-rich domains created inside of the film could not easily coalesce due to the higher viscosity of the matrix at the onset of phase separation. Therefore, the PEG-rich domains remained in the interior of the film, and a fine porous structure could be created in the PEG droplets by the secondary phase separation in the film.

Effect of solvent on the formation of porous structure in PS/PEG blend film

A single phase solution of PS and PEG200 at ratios of 90/10, 80/20, 70/30 in a 1 wt% benzene solvent is cast on a glass slide. Figure 6 shows AFM images of the surface of the resulting PS/PEG films.

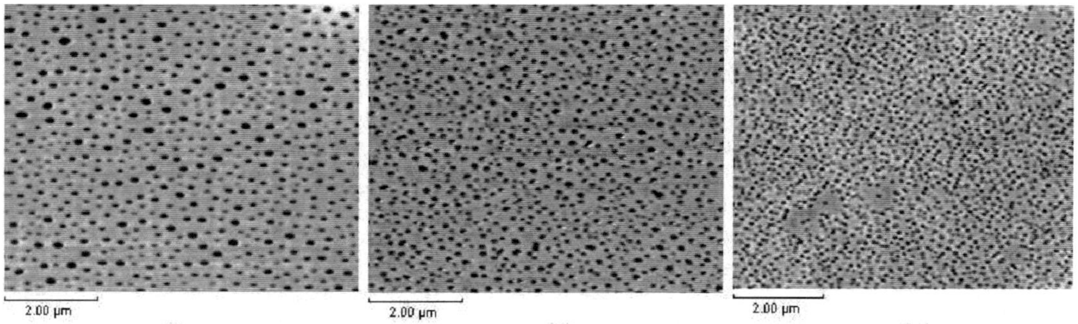

(a) (b) (c)

Figure 6. AFM height images of PS/PEG200 film from benzene solution: (a) 70/30, (b) 80/20, (c) 90/10 of PS/PEG

As shown in Figure 6, pores appear uniformly on the entire surface of the films and when compared to drop casting, pore size drastically decreases from micron-size to sub-micron or nano-scale order. During spin casting, at the initial stage, the solution flows on the substrate in radial direction, and actually a few amounts of the solution remains on the substrate. It caused short drying time in the solution. The growth of PEG domains is induced by coalescence, but due to short drying time, the concentration of PS solution which forms the matrix phase increases fast and suppresses the coalescence of PEG domains. Therefore, there is not enough time for phase-separated PEG domains to grow and then very small pores are created on the film surface.

Conclusions

A polymeric film with a fine porous structure was prepared by dry-casting polystyrene (PS) and polyethylene glycol (PEG) and solvent solutions. By changing the solvent to PS/PEG blend, the pore size and density as well as the porous structure were drastically changed with the different solvent types and the casting methods. The increase of drying rate and high solubility of PEG to solvent increased the polymer concentration of the PS-rich phase at the onset of phase separation of PEG. The higher concentration and higher viscosity of PS at the onset of the PEG phase separation may have suppressed the growth of the PEG-rich domain, and phase-separated PEG domains were left in the inside of the film. Moreover, the short drying time due to spin casting increased fast the concentration of the polymer solution and suppressed the coalescence of phase-separated PEG domains. It caused the very small sized domains. Then pores were created in the PEG-rich domain by secondary phase separation. Finally, a fine porous structure could be formed in the film.

Reference

1. Akolekar, D. B.; Hind, A. R.; Bhargava, S. K. *J Colloid Interface Sci* 1998, 199, 92.
2. Lin, V. S. Y.; Motesharei, K.; Dancil, K. P. S.; Sailor, M. J.; Ghadiri, M. R. *Science* 1997, 278, 840.
3. Nishikawa, T.; Nonomura, M.; Arai, K.; Hayashi, J.; Sawadaishi, T.; Nishiura, Y.; Hara, M.; Shimomura, M. *Langmuir* 2003, 19, 6193.
4. Hedrick, J. L.; Miller, R. D.; Hawker, C. J.; Carter, K. R.; Volksen, W.; Yoon, D. Y.; Trollsas, M. *Adv Mater* 1998, 10, 1049.
5. Gineste, J. L.; Pourcelly, G. *J Membr Sci* 1995, 107, 155.
6. Yamamura, M.; Nishio, T.; Kajiwara, T.; Adachi, K. *Drying Technol* 2001, 19, 1397.
7. Bergues, B.; Lekki, J.; Budkowski, A.; Cyganik, P.; Lekka, M.; Bernasik, A., Rysz, J.; Postawa, Z. *Vacuum* 2001, 63, 297.
8. Kim, W. K.; Char, K.; Kim, C. K. J *Polym Sci Part B: Polym Phys* 2000, 38, 3042.
9. Song, S. W.; Torkelson, J. M. *J Membr Sci* 1995, 98, 209.
10. Dekeyser, C. M.; Biltresse, S.; Marchand-Brynaert, J.; Rouxhet, P. G.; Dupont-Gillain, Ch. C. *Polymer* 2004, 45, 2211.
11. Bormashenko, E.; Pogreb, R.; Musin, A.; Stanevsky, O.; Bormashenko, Y.; Whyman, G.; Gendelman, O.; Barkay, Z. *J Colloid Interface Sci* 2006, 297, 534.
12. Kim, J. -K.; Taki, K.; Ohshima, M. *Langmuir* 2007, 23, 12397.
13. Cha, B. J.; Char, K.; Kim, J. J.; Kim, S. S.; Kim, C. K. *J Membr Sci* 1995, 108, 219.
14. Brandrup, J.; Immergut, E. H. Polymer handbook, 3rd ed.; Wiley-Interscience: New York, 1989.
15. Kia, S.; Samuel, H. Y. AAPS PharmSciTech 2006, 7, E26.
16. Mark, J. E. Physical properties of polymers handbook; Springer: New York, 1996.

Preparation and Evaluation of Thermosensitive Microcapsules Using Double Tube Nozzle

Junichi Ida*, Takuro Takahagi, Tatsushi Matsuyama and Hideo Yamamoto,
Department of Environmental Engineering for Symbiosis, Soka University, 1-236 Tangi-cho, Hachioji, Tokyo, Japan

1. Introduction

Recently, stimuli-responsive polymers have been developed and application of these polymers is also being tried. Among them, N-isopropyl acrylamide (NIPAAm) is well known to have thermo-responsive properties, and Kuckling et al reported to prepare thermo-sensitive gel beads which consisted of NIPAAm and alginate as an application of thermo-responsive polymers [1).

On the other hand, microcapsule has attracted interests for applications of controlled release delivery or micro-reactor. Application of the stimuli responsive polymers for the microcapsule is an interesting topic, and for the application, it is important to establish simple preparation method of stimuli responsive microcapsule, which can easily produce capsules with controlled capsule size and shell thickness. Microcapsule preparation using double tube nozzle is one of the method which is possible to achieve above conditions. Therefore, in this study, preparation of thermo-responsive microcapsule which consisted of NIPAAm and alginate was tried using double tube nozzles.

2. Experimental

Fig. 1 shows the experimental set up to prepare microcapsules using double tube nozzles. At first, microcapsule formation using the double tube nozzles was tested using only sodium alginate as shell material. As a core material, cocoa suspension was used. In the experiment, 2 wt% sodium alginate solution was pumped out from the outer tube, and cocoa suspension was pumped out from inner tube. The formed double layered droplets were then dropped into consolidation solution (0.4 M $CaCl_2$ solution). In the solution, gellation and consolidation of the alginate in the outer layer occurred due to the calcium cation crosslinkage of the alginate molecules. After 90 min., the capsules were recovered from the solution. The effect of the flow rate of the solution from the outer tube (F_{out}) and that from the inner tube (F_{in}) on the size and shell thickness of the capsules was examined.

Then, the preparation of thermo-responsive microcapsules was tried. In the experiment, NIPAAm was also used to form capsule shell in addition to sodium alginate. The solution which contained NIPAAm monomer (9.6 wt%), and sodium alginate (1 wt%) was pumped out from outer tube, and cocoa suspension was pumped out from inner tube. Consolidation solution contained ammonium peroxide sulfate (APS) as polymerization initiator (0.5 wt%) and N,N,N',N'-Tetramethyl ethylenediamine (TEMED) as promoter (0.25 vol%) and 0.1 M $CaCl_2$. In the solution, at first gellation and consolidation of the alginate in the outer layer occurred and it formed support structure of the capsule. Since polymerization reaction of NIPAAm monomer is not so fast, NIPAAm monomers were then polymerized within the alginate gell layer, and the capsules with alginate-NIPAAm complex shell were obtained. Crosslinkage of the NIPAAm polymer in the capsule shell was also tried by using N, N'-Methylenebisacrylamide (MBAAm) as cross linking agent. Furthermore, alginate-NIPAAm beads were prepared by using single tube nozzle without changing other conditions for comparison. Crosslinked alginate-NIPAAm beads were also prepared. Thermo-responsive volume change of the capsules and beads were measured with changing temperature from 27 to 40°C using microscope.

3. Results and Discussion

The effect of the flow rate of the solution from the outer tube (F_{out}) and that from the inner tube (F_{in}) on the size and shell thickness of the alginate capsules were shown in Fig. 2 and 3. The results showed that the size of the capsules increased and the shell thickness decreased with the increase of the flow rate ratio (F_{in} / F_{out}). This suggests that capsule size and shell thickness could be controlled by changing F_{in} / F_{out}. Figure 3 also showed that although F_{out} was different, if F_{in} / F_{out} was the same, shell thickness was essentially the same for three different F_{out}. This means that shell thickness is not dependent on the absolute value of the flow rate in the range used in this study. However, when the flow rate ratio (F_{in} / F_{out}) exceeded 0.45, formation of the capsule became difficult.

Then, temperature-responsive volume change of the alginate-NIPAAm capsules and beads was measured. The result showed that the capsule started to shrink when the temperature increased, and it swollen when the temperature decreased. The temperature responsive volume change was reversible. The swelling ratio was calculated based on the measured volume change. To calculate the swelling ratio, the volume at each temperature was standardized by the volume at the lowest temperature for each capsules and beads. The results were shown in Fig. 4. From the results, it was found that the tendency of the volume change in the case of capsule and beads was almost the same. For both cases, the volume changed greatly with temperature in the case with crosslinkage.

Reference

1. Kuckling D, Schmidt T, Filipcsei, Adler HJ P, Arndt KF, Preparation of filled temperature -sensitive poly(N-isopropylacrylamide) gel beads, Macromol. Symp. 2004; 210, 369-376

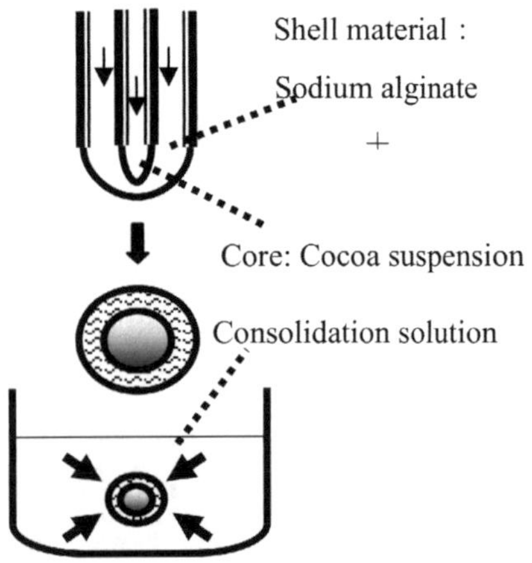

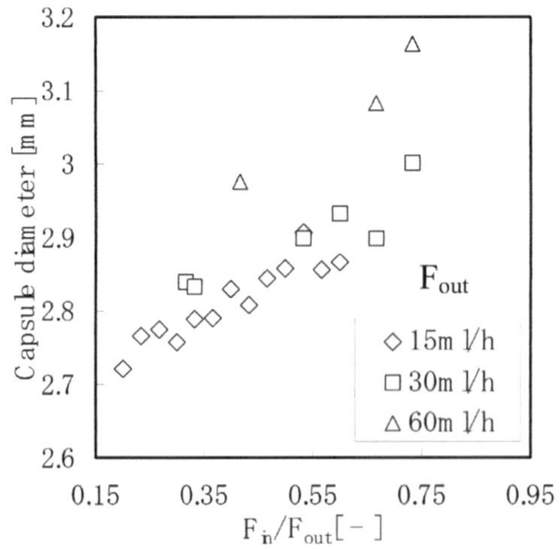

Fig.1 Experimental set up to prepare micro-capsules using double tube nozzle.

Fig.2 Effect of F_{in}/F_{out} on capsule diameter

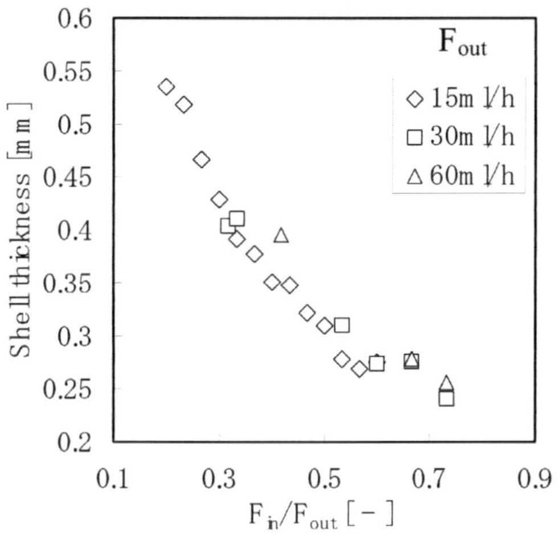

Fig.3 Effect of F_{in}/F_{out} on shell thickness

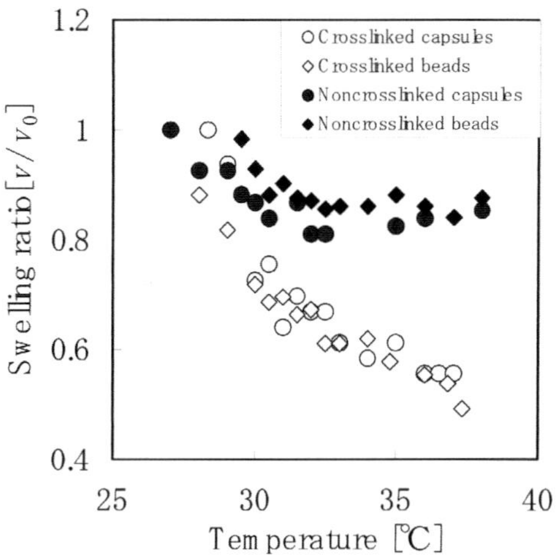

Fig.4 Swelling ratio of the capsule and beads vs temperature

722n Interactions of Chloride and 3-Mercapto-1-Propanesulfonic Acid In Acidic Copper Sulfate Electrolyte

Hung-Ming Chen, **Satish J. Parulekar**, and Alan Zdunek. Chemical and Biological Engineering Department, Illinois Institute of Technology, 127 Perlstein Hall, 10 W. 33rd Street, Chicago, IL 60616

Along with the decrease in the critical dimension of semiconductor chips, there are strong demands for higher electron mobility metal interconnects to permit faster signal transmission and reduce resistor-capacitor (RC) delay of the integrated circuit (IC). In addition, the metal deposition process must be able to create void-free deposits in the chip interconnects, which are as narrow as 65 nm. Replacing the traditional aluminum interconnect with copper interconnect can reduce the wiring resistance by up to 45%. The void-free electrochemical copper deposition (ECD) deposition is often referred to as superfilling or superconformal deposition and is enabled by several additives at optimal concentration ranges and proper processing conditions. In a typical copper electroplating bath, chloride, poly(ethylene) glycol, and mercapto acids, such as MPS and SPS, are the necessary additives to achieve superconformal metal interconnects in sub-micron vias or trenches.

At negative plating potentials, Cu2+ ions in the vicinity of the cathode surface adsorb onto the surface, which is followed by two single electron transfer steps. Chloride ions are the most common additive in copper plating chemistries. Affinity for adsorption of chloride onto copper surfaces has been studied with a variety of techniques. Chloride ions, at trace (parts-per-million, ppm) level in the electrolyte, are able to catalyze the Cu2+/Cu+ reaction by changing the reaction mechanism from an outer-sphere reaction (water-water bridge) to an inner-sphere reaction (chloride bridge), and therefore the copper deposition potential is depolarized. Two related chemicals, 3-mercapto-1-propanesulfonic acid (MPS) and disodium bis(3-sulfopropyl)disulfide (SPS), are well-known accelerators or brighteners in the electrochemical deposition bath chemistry. The two brightening agents are unstable at open circuit potential and during electrolysis. Formation of cuprous thiolate intermediates is the key step which leads to the brightening of electrodeposited copper.

The individual roles of and the interactions between two additives, chloride ions (Cl-) and 3-mercapto-1-propanesulfonic acid (MPS), in copper electrochemical deposition (ECD) from an acidic copper sulfate solution are studied here with electrochemical methodologies such as linear sweeping voltammetry (LSV), chronoamperometry, and alternating-current electrochemical impedance spectroscopy (EIS). The LSV results reveal that the current density is not increased by MPS without the presence of chloride in the electrolyte. The EIS results reveal a low frequency capacitive loop associated with chloride interaction with cupric ions (Cu2+) and MPS and a high frequency capacitive loop related to double layer charging. A detailed reaction mechanism is proposed based on the experimental observations and an EIS model based on the reaction mechanism is developed. The EIS results are used to estimate parameters in a simple equivalent circuit model and the EIS model. The validity of the mechanism is established by comparing predictions of the EIS model with experimental EIS data. The EIS model predictions reflect, in a qualitative manner, the features observed in the EIS experimental data.

722o Synthesis and Characterization of Interfacially Modified Block Copolymers

Nripen Singh and Thomas H. Epps III. Chemical Engineering, University of Delaware, 150 Academy Street, Newark, DE 19711

This work presents a methodology to develop innovative and durable materials for their potential membrane and templating applications. Towards this goal, we are strategically manipulating the interface between blocks in poly(styrene-b-isoprene) [PS-PI] block copolymers using anionic polymerization techniques. By following this approach, we are able to control the interfacial interactions between blocks independent of molecular weight and block constituents. A library of copolymers is prepared with various interfacial modifications to enable the examination of interfacial effects on copolymer self-assembly. The molecular characteristics of the block copolymers are characterized by gel permeation chromatography (GPC), NMR, and ozonolysis. The morphology of the microphase-separated domain structures is investigated by small-angle X-ray scattering (SAXS), transmission electron microscopy (TEM), and dynamic mechanical analysis (DMA).

722p Development of Ion Gels with Addition of Ionic Liquid

Jiao Guo and Kyonsuku Min. Polymer Engineering, The University of Akron, 250 S Forge st, Akron, OH 44325

In-situ polymerization of ion gel system was conducted in this study. The ion gel systems include methylmethacrylate (MMA) monomer, crosslinking agent, and ionic liquid (IL) 1-butyl-3-methylimidazolium hexafluorophosphate (BMIPF6). The transparent, self-standing and flexible t ion gel products were obtained. The ion gel systems with different weight ratio of MMA to IL exhibit high room temperature conductivity and good thermal stability. The ion transfer behavior and its temperature dependency were also investigated. The structure of crosslinked network of ion gel was studied by swelling ratio.

722q Development of a Multi-Variate Statistical Non-Destructive Test for Measuring Aging In Materials

Derrick Rollins, Iowa State University, Department of Chemical Engineering, 2114 Sweeney Hall, Ames, IA 50011-2230, Ai-Ling Teh, Statistics, Iowa State University, Snedecor Hall, Ames, IA 50011, and Krishna Rajan, Material Science and Engineering, Iowa State University, 2220H Hoover Hall, Ames, IA 50011.

This work discusses the application of statistical techniques based on principal component analysis (PCA) in determining fatigue and other effects of aging in materials in a non-destructive manner.

Web Page: www.public.iastate.edu/~drollins/Publications

722r A Unifying Approach for Melt Rheology of Linear Polystryene

Sung Hun Kim, Chemical R&D Center, Cheil Industries Inc., Samsung Corporation, 332-2, Gocheon-Dong, Uiwang-Si, Gyeonggi-Do, 437-711, South Korea

Several studies of melt rheological properties of polystyrene have been conducted over the past 50 years. Several approaches, including empirical models, have been developed to understand the behavior of materials using simple equations.

The existing melt rheology models are best suited for high-molecular-weight polymers whose Tg does not vary. In this work, a semiempirical viscosity equation has been derived, including the effect of Tg dependence on molecular weight, to describe the melt rheology of low-molecular-weight polymers. The equation is derived based on a combination of well-known concepts, such as the effects of free volume and molecular dynamics on polymer rheology. This provides a better understanding of the rheological behavior in the low-molecular-weight regime with respect to temperature and molecular weight.

Because of the industrial trend towards lower molecular weight materials for applications such as high solids coatings, this unifying approach, based on the free volume theory with a simple expression, is of extreme practical significance. This equation can predict the zero shear viscosity behavior for different molecular weights, including low-molecular-weight regions, and temperatures. Viscosity calculations using the empirical equation agree with published experimental data.

722s Macroscopic Inspection for Nanofiller Dispersion and Polymer Degradation: Fundamental Characterization in Polymer Nanocomposites

Paul J. Hubert, Joseph N. Kuterbach, Christopher E. Roman, and **Katsuyuki Wakabayashi**. Department of Chemical Engineering, Bucknell University, Lewisburg, PA 17837

Polymer-based nanocomposites involving various organic and inorganic nanofillers continue to be of great scientific and commercial interest. The common challenge is the fabrication of effective nanocomposites with well-dispersed and well-separated (exfoliated) filler morphology, and there have been different fabrication methods reported in the literature, ranging from solution to melt and solid-state processing.

One fundamental characterization that is often overlooked in polymer nanocomposites is simply of the physical appearance of the resulting specimens; macroscopic inspection of the fabricated nanocomposites is often ignored or left undiscussed in the literature. Considering that many of these materials are designed for real-world applications in consumer products and packaging, this is rather surprising.

In this paper, post-fabrication, physical appearance characterization of polymer-based nanocomposites is conducted. A series of specimens of prototypical nanocomposites are considered, including base polymers like polypropylene, polystyrene and poly(lactic acid), and nanofillers such as organically-modified clay, graphite and carbon nanotubes. The effects of different fabrication methods and post-production processing methods are evaluated.

In melt-mixed nanocomposites, organically-modified nanofillers have the highest tendency to degrade the base polymer, resulting in non-uniform, dark brown color unsuitable for packaging applications. Low-temperature, solid-state processing can suppress the degradation and discoloration of the polymer nanocomposites to some extent. Carbon-based nanofillers such as carbon black, graphite and carbon nanotubes, when properly processed, lead to uniformly black specimens, which can be suitable for certain commercial applications.

722t Fabrication of Self-Supporting, Hierarchical 3D Nanocomposites with Multi-Scale Controllable Features

Christine M. Andres, Chemical Engineering, University of Michigan, 2300 Hayward, Ann Arbor, MI 48109 and Nicholas A. Kotov, Chemical Engineering Department, Materials Science Department, Biomedical Engineering Department, University of Michigan, 2300 Hayward, Ann Arbor, MI 48109.

A practical and multipurpose technique has been developed to create 3D free-standing nanocomposites with well controlled nano- and micro-scale features. The internal organization, controlled molecular structure and versatility of conventional layer-by-layer (LBL) assembly techniques have been combined with the nearly perfect order and control over interconnecting channels of inverted colloidal crystals to create a hierarchically controlled 3D nano-assembly. LBL assembly of polyelectrolytes and nanoparticles is performed on polystyrene beads arranged in a hexagonal crystal lattice. Upon dissolution of the colloidal crystal a structure with multi-scale and multidimensional response to external stimuli is produced. This simple, low cost, highly repeatable fabrication technique allows for optimized surface area, mass transport, and mechanical properties. The resulting composite displays many properties of natural materials which often exhibit hierarchical combination of nano- and microscale structural elements. Additionally, fairly large 3D samples of hierarchical nanocomposites can be produced and tested in this way. As 2D LBL assembled films have been employed over many disciplines, the self-supporting 3D structure produced has properties that are extremely attractive to multiple disciplines including tissue engineering, catalysis, porous, high strength materials, and aviation.

Stress-Induced Crystallization of Poly(Trimethylene Terephthalate) Fibers by Molecular Dynamic Simulations

Min-Kang Hsieh and Shiang-Tai Lin,
Department of Chemical Engineering, National Taiwan University, Taipei, Taiwan

Introduction

The phenomena of stress-induced (or flow-induced, strain-induced, force-induced) crystallization is characterized by the preferential of orientation of polymer segments along the direction of applied force. Many researches report that the drawing and spinning process would increase the crystallinity of polymeric materials, resulting in oriented segments that would explosively improve the crystallization by oriented nuclei.

There are many works showing the evidences of the structural changes during drawing and spinning processes. From the behavior of stress-draw ratio curve of the melt-quenched amorphous PTT film during uniaxial drawing, a significant strain hardening could be observed simultaneously with crystallinity (measured by density) abruptly rising after draw ratio of 2.5.[1,2] Besides, the development of oriented structure has been quantified by orientation factor calculated from IR dichrocic ratio of characteristic bands[3] and intensity of WXRD[4] of the melt-quenched amorphous PTT. Moreover, the birefringence of PTT increases with spinning speeds.[5] On the other hand, many theoretical works on stress-induced crystallization using molecular simulations, most of which focus on linear polyethylene (PE). It is found that the torsional ordering[6] and is followed by increase in orientation factor[7] in the simulation with the pre-oriented models. The shish-kebab crystallization was observed by extensional flow process,[8] and the secondary nucleation around a knot in the polymer system has been studied.[9]

The change in the torsional conformation of a propylene glycol segment ($O-CH_2-CH_2-CH_2-O$) during the drawing or spinning process is of great interest. The torsion angle distribution of the segment has been reported as *tans-gauche-gauche-trans* (t-g-g-t) in the crystalline state.[10,11] However, what the distribution in amorphous state is and how they transfer form amorphous to crystalline state remain unclear. The conformation of amorphous state may either be dominated by *g-t-t-g*, state[3] or in random[12] states; some even estimated the content of the gauche state of ϕ_2 to be 33.4%.[2] Furthermore, the mechanism of the conformation transition from amorphous to crystalline phase has been studied using IR spectroscopy[2,3,12-14], X-ray scatting[15], and ^{13}C solid-state NMR.[16] Form those resuts, in crystalline phase, the ϕ_2 exists largely in the *gauche* state, and ϕ_1 conformation exists largely in the *trans* state. In amorphous phase, ϕ_2 conformation exists mostly in the *trans* state, but ϕ_1 conformation is unclear. Moreover, during the annealing process, ϕ_2 conformation changes form *trans* state to *gauche* state but ϕ_1 conformation is unclear. During drawing process, ϕ_2 conformation changes form *trans* state to *gauche* state, if polymer chains undergo sufficient thermal relaxation; however, ϕ_1 conformation is unclear.

In this work we focus on the development of ordering of structures in the stress-induced crystallization process. (The thermal effect of the development ordering structure has been studied in our previous works.) The temperature dependence of the amount of ordered clusters, the behavior of ordered structures and torsion angle transition during drawing and annealing processes are analyzed. The results presented here help us better understand the mechanism of the nucleation at the early stage of crystallization.

Computational methods

Atomistic molecular dynamic simulations are used to study the dynamic behavior of the nucleation process. The molecular models of PTT are prepared using commercial package Cerius2.[17] Each model contains 4 chains of PTT molecules, each having a degree of polymerization of 27. Thus each unit cell contains 108 repeating units of PTT, or equivalently 2708 atoms. In all the simulations, the generic force field Dreiding[18] is used to describe the interactions among atoms in a system. The potential energy includes the bond stretching energy (U_b), angle bending energy (U_θ), torsional angle rotation energy (U_ϕ), electrostatic interaction (U_{coul}), and van der Waals interactions (U_{vdw}). Atomic (Mulliken) charges are determined from density functional theory (DFT) calculations for the monomer of PTT using the B3LYP functional and 6-31G** basis set in Gaussian 98[19]. Moreover, the torsion terms of the propylene glycol segment (O-CH_2-CH_2-CH_2-O) being modified to better reproduce quantum mechanics (QM) results, according the energy barrier results of rotating torsion angles.[20] This force field could reproduce reasonably the thermal properties (Tm and Tg), mechanical property (Young's modulus), and crystal structural properties.[20]

The computer code LAMMPS [21] is used for all subsequent molecular dynamic simulations. A series of expansion and compression steps were applied to the initial structures in order to obtain equilibrium samples at 600K (which is above the melting temperature T_m). The 100 ps equilibrated sample is then cooled to 50 K at a rate of 1 K/ps. At each 50 K interval during the quenching process, the samples are then drawn with different drawing speeds ($1x10^8 s^{-1}$(only at 400K), $1x10^9 s^{-1}$, $1x10^{10} s^{-1}$, and $1x10^{11} s^{-1}$), and isothermally relaxed at several draw ratios (DR=1, 2, 3, and 4) up to 12 ns with pressure at 0 bar and $NTL_x\sigma_{yy}\sigma_{zz}$ ensemble with zero normal stress in y and z directions.[22] Nose-Hoover thermostat is used for temperature control. The integration time step used is 0.5 to 1 fs.

Results and discussions

Bulk Properties

The changes in bulk properties are observed (Fig. 1) during stress-induced crystallization in our simulation. The increase of the amount of precursors in the draw process is much more remarkable than that in isothermal crystallization. Furthermore, the subsequent process of relaxation (NVT simulation after the drawing process) would lead to further increase in the fraction of oriented precursors. The rapid density decrease in each simulation indicates that the defects (voids) in the bulk phase are created during the drawing processes. At high draw speeds ($1x10^{10}$ s^{-1}, and $1x10^{11}$ s^{-1}) and the temperatures below T_g the formation of defects is prominent; on the other hand, at the slower draw speeds ($1x10^8$ s^{-1}, and $1x10^9$ s^{-1}) and temperatures above T_g the system relaxation is fast enough to reduce the amount of defects.

The changes in van der Waals interaction and the torsional energy reflect the transition of the structure under various drawing conditions. The van der Waals interaction increases with DR in all cases. These interaction energies also have a significant effect in the system density. For example, the relaxation of van der Waals interaction leads to a constant density at lower draw speed ($1x10^9$ s^{-1}). A rapid increase in van der Waals interaction leads to a rapid drop in density at higher draw speed ($1x10^{11}$ s^{-1}). Besides, the van der Waals interaction increases with temperature at constant DR. The enhancement of van der Waals interactions slows down at temperatures higher than T_g at lower draw speed. These results indicate that the mechanism of the chain packing is different in the various draw speeds and temperatures. This could be understood as a competition between thermal effect, which is prominent at slower draw speed and higher temperatures, and stress effect, which is prominent at faster draw speed and lower temperatures. On the other hand, the torsional energy decreases with DR at lower draw speeds ($1x10^9$ s^{-1}, and $1x10^{10}$ s^{-1}), but fluctuates at the highest draw speed ($1x10^{11}$ s^{-1}). It is almost a linear function of temperature at constant DR. These results support that the stress effect influents the backbone chain rotation much more than thermal effect.

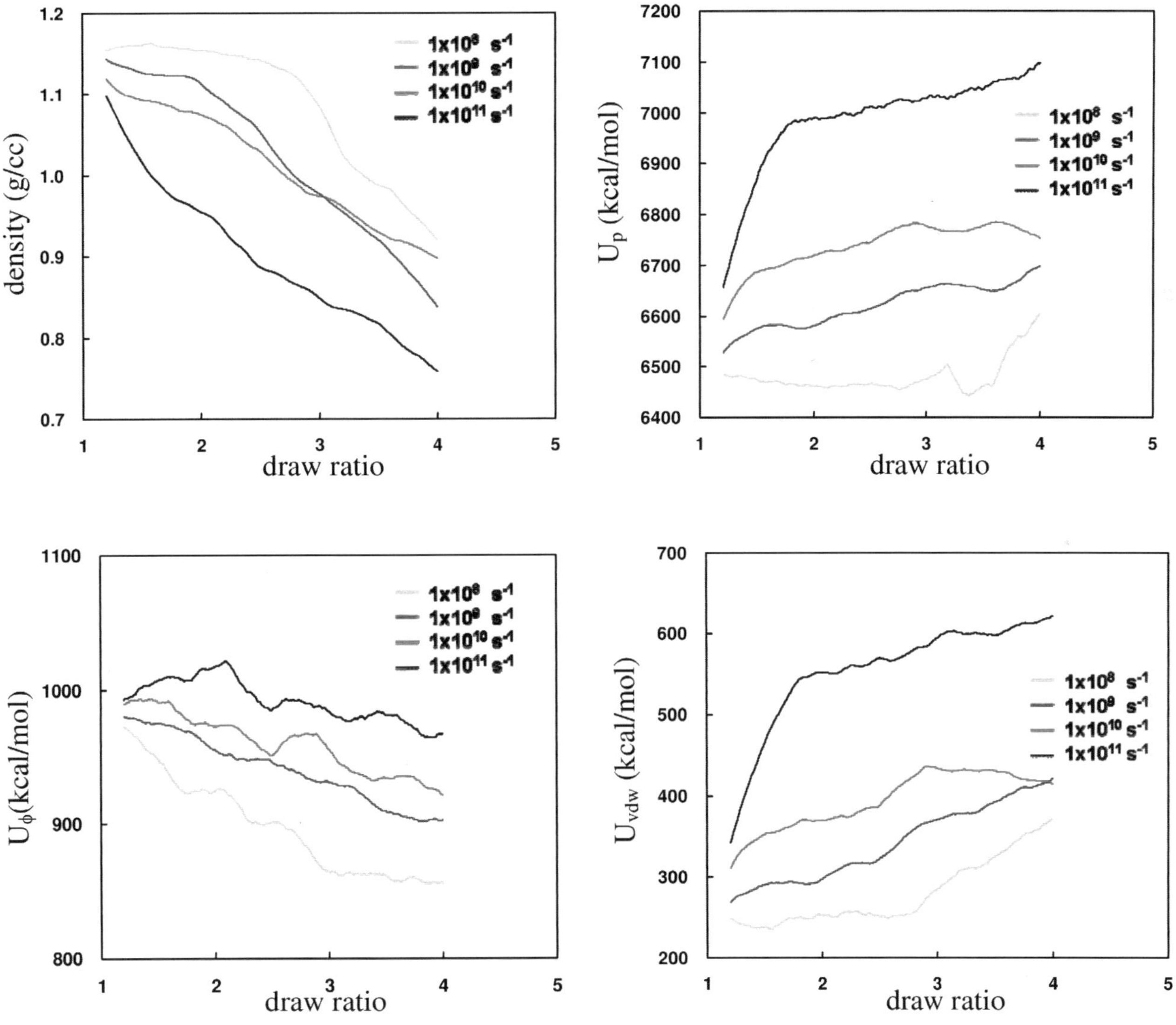

Figure1 The energy and density of the overall system.

The Fraction of Ordering Structure

The fraction of ordering structure, *X*, has been defined in previous work.[20] As shown in Fig. 2, at small DR (DR=2), the fraction of precursors increases in all temperatures with slower draw speeds ($1x10^9 s^{-1}$ and $1x10^{10} s^{-1}$), but decreases at temperatures 350K and 400K with higher draw speed ($1x10^{11} s^{-1}$). At larger DR (DR>3), the saturation of induced precursor is observed for all temperatures with the slowest draw speed ($1x10^9 s^{-1}$). On the other hand, the development of precursors continues at the higher draw speed ($1x10^{10} s^{-1}$ and $1x10^{10} s^{-1}$). Besides, the orientation factor increases with DR, and the curves are similar in all situations, indicating that the orientation factor may be a function of draw ratio only. This implies that the stress-induced precursor is similar at the same DR with each draw speed.

The subsequent thermal relaxation (following the drawing process) could stabilize and facilitate in the growth of oriented precursors. The results of the fraction of precursors shows a temperature depence similar to that of the isothermal crystallization, but the position of highest amount of precursors shifts to a lower temperature (300K with drawing vs 400 K without drawing[20]). The local motion of polymer segments would be induced by uni-axis stress. (Note the maximum capacity of precursor formation should be found between the temperature allowing for short rage motion and that for entire chain motion.)

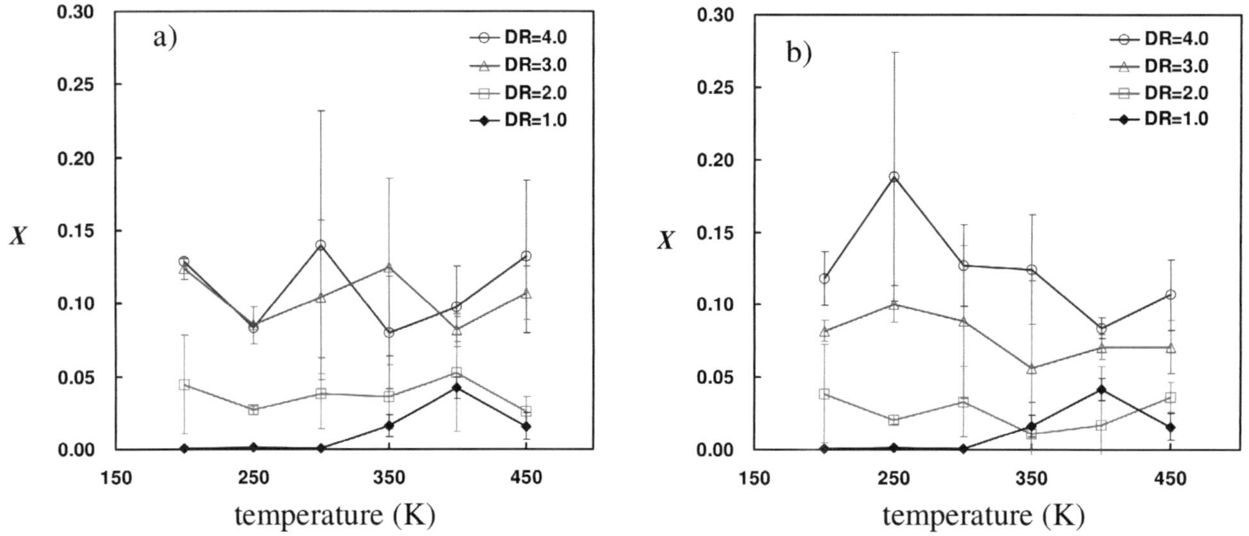

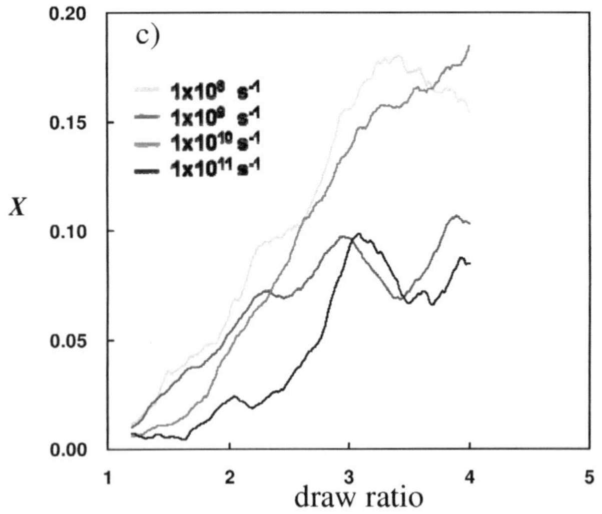

Figure2 The fraction of ordering structure was shown as the function of temperature at a) $1x10^9 s^{-1}$ and b) $1x10^{11} s^{-1}$, and c) the function of draw ratio at 400K with all draw speeds.

The trosional angle distribution

To understand the distribution of torsional angles in the PTT backbone (O-CH_2-CH_2-CH_2-O), we analyze the percentage of the trans and gauche states during the stress-induced crystallization. Different behaviors observed at different draw speeds. The percentage of ϕ_1 and ϕ_2 in trans state fluctuates at small DR, and increases at larger DR with slower draw speed. At higher draw speeds, this percentage decreases at small DR and increases at larger DR. The percentage value rebounds to the original value with draw speed $1 \times 10^{11} s^{-1}$. On the other hand, the percentage of gauche state decreases with DR at all draw speeds. During thermal relaxation, the percentage of the trans rebounds in all draw speeds, but that of gauche rebounds only in faster draw speed. These results suggest that torsional angles change to transition state in the beginning drawing process, and then become trans state at higher DR or under thermal relaxation.

Detailed analysis also is performed on the two feature torsions (ϕ_1 and ϕ_2) in the PTT backbone (Fig. 3). It is found that the trans-to-gauche ratio in ϕ_1 increases rapidly (from 3 to 6) at lower drawing speed and higher temperatures; however, the increase is less significant at high drawing speed and low temperatures. In contrast, the gauche-to-trans ratio for ϕ_2 seems to be insensitive to processing conditions (varies from 1.6 to 0.8 regardless of drawing speeds). After thermal relaxation, the value of trans-to-gauche ratio in ϕ_1 improves significantly in the slower draw speeds, but remains almost constant in the fastest draw speed. On the other hand, the gauche-to-trans ratio for ϕ_2 decreases at small DR with slower draw speeds ($1 \times 10^{9} s^{-1}$ and $1 \times 10^{10} s^{-1}$), but increases at larger DR with faster draw speed.

Before drawing, the fraction of trans state is ~34%, and that of gauche is ~25% in the bulk phase; the most populated conformations in ϕ_1 is trans, and that in ϕ_2 is gauche; i.e., the most populated conformations in the backbone torsions are t-g-g-t (ϕ_1- ϕ_2- ϕ_2- ϕ_1). During draw process, the t-t-t-t conformation increases. Our results are in agreement with the result of Chuch's experimental observation.[14] The thermal relaxation would change the conformations in some of our simulations, but the changes are not obvious.

The states of backbone torsional angles in precursors evolve with DR differently in different processing conditions. The fraction of ϕ_1 in trans increases with DR, but the fraction of ϕ_2 in gauche decreases with DR in most precursors. The stress and thermal effects are different in these two torsional angles. Slower draw speed and lower temperature would improve the change of ϕ_1 to the trans state; on the other hand, the faster draw speed and medium temperatures would improve the change of ϕ_2 to the gauche state. Furthermore, the fraction of ϕ_1 in trans is very high (~70%) in each precursor, but the fraction of ϕ_2 in gauche only 20~30%. That indicates that the stress-induced transition in ϕ_2 is faster than that in ϕ_1. In addition, the faction of ϕ_2 in gauche fluctuates in some cases, indicating the torsional angle ϕ_2 would favorably move to the gauche after forcing to the trans state by the external stress.

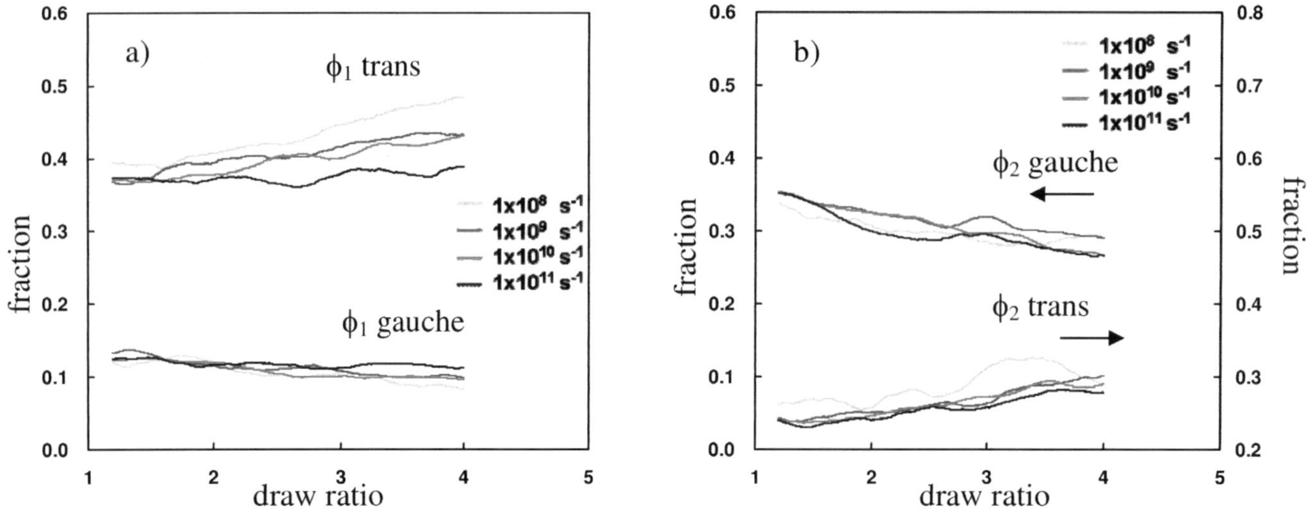

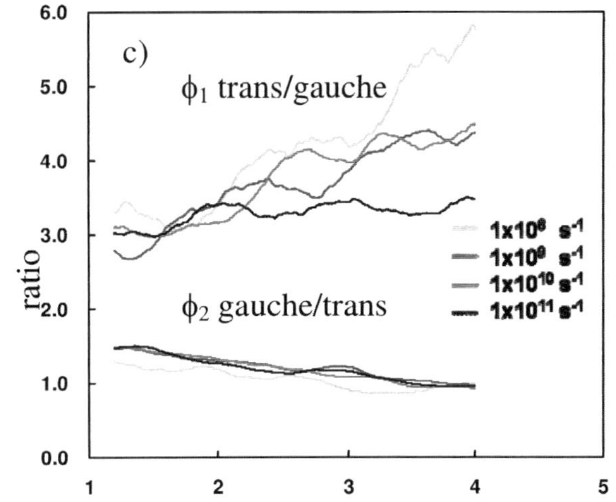

Figure3 The fraction of torsional angle in a) ϕ_1 and b) ϕ_2 was shown as the function of draw ratio, and c) the ratio of each torsional type faction at 400K.

Conclusion

The formation of oriented precursor in the stress-induced crystallization is observed in our simulation. The torsional and van der Waals forces are the dominant interactions, similar to the case observed in isothermal crystallization. The amount of stress-induced precursor increases in all regions of temperature. The maximum size of oriented precursor is larger than that created only by thermal stimulation. During stress-induced crystallization, the torsional distribution of the polymer backbone for segments rapidly rearrange to the t-t-t-t conformation in bulk phase. Within oriented precursors, the response of the torsional angle induced by stress is faster than that only induced by thermal stimulation, especially trans in ϕ_1(the transition rate of trans state in ϕ_1 is faster than that of gauche state in ϕ_2).

Acknowledgment

We would like to thank the financial support from Grant NSC 96-2221-E-002-106 by the National Science Council of Taiwan, and computation resources from the National Center for

High-Performance Computing of Taiwan.

References

1. Lee, H. S., S. C. Park and Y. H. Kim, "Structural changes of poly(trimethylene terephthalate) film upon uniaxial and biaxial drawing," *Macromolecules*, **33**, 7994 (2000).
2. Jeong, Y. G., W. J. Bae and W. H. Jo, "Effect of uniaxial drawing on surface chain structure and surface tension of poly(trimethylene terephthalate) film," *Polymer*, **46**, 8297 (2005).
3. Kim, K. J., J. H. Bae and Y. H. Kim, "Infrared spectroscopic analysis of poly(trimethylene terephthalate)," *Polymer*, **42**, 1023 (2001).
4. Wu, J., J. M. Schultz, J. M. Samon, et al., "In situ study of structure development in poly(trimethylene terephthalate) fibers during stretching by simultaneous synchrotron small- and wide-angle X-ray scattering," *Polymer*, **42**, 7141 (2001).
5.
6. Koyama, A., T. Yamamoto, K. Fukao, et al., "Molecular dynamics simulation of polymer crystallization from an oriented amorphous state," *Physical Review E*, **65**, (2002).
7. Lavine, M. S., N. Waheed and G. C. Rutledge, "Molecular dynamics simulation of orientation and crystallization of polyethylene during uniaxial extension," *Polymer*, **44**, 1771 (2003).
8. Dukovski, I. and M. Muthukumar, "Langevin dynamics simulations of early stage shish-kebab crystallization of polymers in extensional flow," *Journal of Chemical Physics*, **118**, 6648 (2003).
9. Saitta, A. M. and M. L. Klein, "Influence of a knot on the stretching-induced crystallization of a polymer," *Journal of Chemical Physics*, **116**, 5333 (2002).
10. Poulindandurand, S., S. Perez, J. F. Revol, et al., "Crystal-Structure of Poly(Trimethylene Terephthalate) by X-Ray and Electron-Diffraction," *Polymer*, **20**, 419 (1979).
11. Desborough, I. J., I. H. Hall and J. Z. Neisser, "Structure of Poly(Trimethylene Terephthalate)," *Polymer*, **20**, 545 (1979).
12. Ohtaki, M., T. Kameda, T. Asakura, et al., "Structural characterization of drawn and annealed poly(trimethylene terephthalate) fibers," *Polymer Journal*, **37**, 214 (2005).
13. Park, S. C., Y. R. Liang, H. S. Lee, et al., "Three-dimensional orientation change during thermally induced structural change of oriented poly(trimethylene terephthalate) films using polarized FTIR-ATR spectroscopy," *Polymer*, **45**, 8981 (2004).
14. Chuah, H. H., "Orientation and structure development in poly(trimethylene terephthalate) tensile drawing," *Macromolecules*, **34**, 6985 (2001).
15. Wu, J., J. M. Schultz, J. M. Samon, et al., "In situ study of structure development during continuous hot-drawing of poly(trimethylene terephthalate) fibers by simultaneous synchrotron small- and wide-angle X-ray scattering," *Polymer*, **42**, 7161 (2001).
16. Kameda, T., M. Miyazawa and S. Murase, "Conformation of drawn poly(trimethylene terephthalate) studied by solid-state C-13 NMR," *Magnetic Resonance in Chemistry*, **43**, 21 (2005).
17. Cerius2 (2003), Accelrys Inc., San Diego.
18. Mayo, S. L., Olafson, B. D. and Goddard, W. A. (1990), "*Dreiding - a Generic Force-Field for Molecular Simulations*," Journal of Physical Chemistry, pp. 8897-8909.
19. Gaussian 98 (2001), Revision A.11, Gaussian, Inc., Pittsburgh, PA.
20. Hsieh, M-K, and ST Lin, "Molecular Dynamic Simulations of Polymer Crystallization at the Early Stage", AIChE annual meeting, 2007.

21. Plimpton, S. (1995), "*Fast Parallel Algorithms for Short-Range Molecular-Dynamics*," Journal of Computational Physics, pp. 1-19.
22. Capaldi, F. M., M. C. Boyce and G. C. Rutledge, "Molecular response of a glassy polymer to active deformation," Polymer, 45, 1391 (2004).

722v Physical Chemical Properties of Phase Separated Polymersomes

Aiwei Tian[1], David A. Christian[1], Dennis E. Discher[2], and Tobias Baumgart[3]. (1) Chemical & Biomolecular Engineering, University of Pennsylvania, Philadelphia, PA 19104-6391, (2) Chemical and Biomolecular Engineering, University of Pennsylvania, Room 129 Towne Building, 220 South 33rd Street, Philadelphia, PA 19104-6393, (3) Department of Chemistry, University of Pennsylvania, Philadelphia, PA 19104

Polymersomes made of neutral poly(ethylene oxide)-polybutadiene (PEO-PBD) and negatively charged poly(acrylic acid)-polybutadiene (PAA-PBD) diblock copolymers can laterally segregate into coexisting domains in a certain pH range by calcium cross-bridging between PAA-PBD chains. The phase diagram for phase separation reveals that domain formation occurs in a surprisingly small region of pH and calcium concentration. The sensitivity of these domains to solution conditions after their formation was investigated by varying pH and calcium concentration. 3D images of polymersomes were recorded via confocal fluorescence microcopy imaging, which allowed to obtain domain area fractions using a 2D mapping algorithm. We furthermore studied the effect of temperature on polymersome phase separation as well as the mechanical properties of pure PAA-PBD polymersomes.

722w Kinetics of Thermal Decomposition of Expanded Plystyrene In Different Gaseous Environments

Pravin Kannan[1], Joseph J. Biernacki[1], Donald P. Visco[1], and William Lambert[2]. (1) Chemical Engineering, Tennessee Technological University, Department of Chemical Engineering, P.O. Box 5013, Cookeville, TN 38505, (2) Chemical Engineering, University of Nebraska at Lincoln, Lincoln, NE

The characteristics of thermal decomposition of expanded polystyrene (EPS) foams were studied in various gaseous environments, both oxidizing and non-oxidizing, over a narrow range of heating rates between 1 and 30 K/min using non-isothermal thermo gravimetric analysis (TGA). A modified integral-optimization technique was developed for multiple rate-controlling decomposition reactions. Results from kinetic data analysis shows that the activation energy of EPS foam decomposition in air is much less than in non-oxidizing environments, including nitrogen, helium and argon. Qualitative mass spectrometric studies of EPS pyrolysis gases revealed the differences in EPS foam decomposition mechanism between various gaseous environments. Furthermore, it was found that the non-oxidizing gases, especially helium, have a significant effect on the decomposition efficiency of the EPS foam.

722x Particle Dynamics and Adhesion In a Biomimetic System: Skipping, Rolling, and Arrest Governed by Spatially Varying Colloidal Interactions

Jeffrey M. Davis[1], Ranojoy D. Duffadar[1], and Maria M. Santore[2]. (1) Chemical Engineering, University of Massachusetts, 159 Goesmann Lab, 686 North Pleasant St., Amherst, MA 01003, (2) Polymer Science and Engineering, University of Massachusetts, Conte Polymer Laboratory, 120 Governor's Drive, Amherst, MA 01003

Biological phenomena related to cell rolling and arrest, such as the motion signatures of neutrophils over the endothelial lining, have been studied extensively. Transient bond formation and breakage between sialyl-lewisx groups (receptors) on neutrophils and selectins (ligands) on the endothelium cause the neutrophils to move slowly over the interior of the blood vessel. This results in the engagement of integrins that arrest neutrophil motion. This behavior has been reproduced in vitro by employing selectin-functionalized microparticles, a model that relies on specific interactions between biomolecules. Our work is focused on understanding and mimicking such phenomena by reproducing the motion signatures of skipping, rolling, and arrest in an entirely non-biological system based on colloidal interactions with a surface bearing randomly distributed polymeric nanoconstructs on a repulsive background field.

A novel and predictive computational model is validated through extensive comparisons to experiment for the non-biological system. This model is then used to interpret experimental results by providing detailed insight into the comparative influence of each of various functional parameters: ionic strength, particle size, nanoconstruct length scales, shear rate, etc. The model combines hydrodynamic forces and torques exerted by the fluid flow to the heterogeneous colloidal landscape presented by the nanopatterned surface to simulate particle motion. Frictional forces are incorporated upon contact with the surface, giving rise to the motion signatures of rolling, skipping and firm adhesion.

Simulation of particle dynamics using the model yields predictions of adhesion thresholds that are in excellent quantitative agreement with experimental data. Multi-variable phase-space maps, or "adhesion state diagrams," demarcate the regimes of different motion signatures (no contact, rolling, skipping, arrest) over a range of parameter space. Comparison of these phase-space maps with the state diagrams for neutrophil rolling reveals qualitative agreement in the structure of the diagrams along with similar quantitative trends, thereby further establishing the relevance of this work to understanding biomimetic behavior. Comparisons of characteristic nanoconstruct density, the distribution of particle rolling velocities, and the mean force and energy per nanoconstruct are made with analogous parameters in biological systems for different conditions of particle adhesion. This work explores how systems can be designed so that biomimetic particles exhibit similar dynamic motion signatures to neutrophils. Extensions are made to particle-surface interactions governed by both colloidal forces from a heterogeneous surface and physical bonds between ligands and receptors.

722y Structural and Thermo-Mechanical Characteristics of High-Amylose Starch/bionolle Nanocomposite Blends

Sotirios, I. Marras[1], **Ioannis Zuburtikudis**[1], Kyriaki Tornikidou[2], Athanasia Tsimpliaraki[1], Elpiniki Panayiotidou[1], and Georgia Christofidou[1]. (1) Industrial Design Engineering, TEI of Western Macedonia, Kozani, Greece, (2) Industrial Design Engineering & Private Dental Practice, TEI of Western Macedonia, Kozani, Greece

In the past decade, significant attention has been focused on biodegradable and biocompatible polymers, both from biomedical and ecological perspectives. In this direction, starch has been considered as a low-cost alternative. However, this material displays some drawbacks compared to conventional polymers, such as limited processability and high stiffness and brittleness. Blending of starch with other synthetic, biodegradable, thermoplastic polymers could lead to the development of a new range of environmentally friendly polymeric materials with improved properties. A further enhancement of the material mechanical properties can be achieved by the addition of small amounts (typically less than 5%) of nanometer-thick clay platelets.

In the present work, starch-polybutylene succinate adipate (Bionolle) nanocomposites blends were fabricated by the addition of various loadings of organomodified montmorillonite using a twin screw extruder. The structure of the prepared hybrids was studied by X-ray diffraction (XRD) and their thermal characteristics were investigated by thermogravimetry (TGA) and differential scanning calorimetry (DSC). The thermomechanical response of the nanocomposites was explored by dynamic mechanical and thermal analysis (DMTA), while the material's mechanical properties were studied in tensile loading conditions. The results revealed that the nanoclays presence affect significantly the properties of the blends.

Acknowledgements

This work was supported by the Region of Western Macedonia (Greece) through the PEP grant with MIS 105545 to I. Z.

OXYGEN PERMEABILITY OF NOVEL ORGANIC-INORGANIC HYBRID MATERIALS COATED ON POLYETHYLENE

M.G. De Angelis, [a] *M. Minelli,* [a] *F. Doghieri,* [a] *M. Marini,* [b] *M. Toselli,* [c] *F. Pilati* [b]

[a] *Department of Chemical, Mining and Environmental Technologies Engineering (DICMA) University of Bologna, Via Terracini 28, 40131 Bologna, Italy, email. grazia.deangelis@unibo.it*
[b] *Department of Materials and Environmental Engineering, University of Modena and Reggio Emilia*
[c] *Department of Applied Chemistry and Material Science, University of Bologna*

Abstract

The oxygen barrier properties of composite materials, consisting of a 45 μm thick LDPE substrate coated by a thin layer (0.7-1 μm) of nanostructured hybrid organic/inorganic materials based on SiO_2 and polyethylene-polyethylene glycol (PE-PEG) block copolymers, obtained via a sol-gel technique have been characterized at 35°C and 50°C. A significant decrease (25-40%) of the oxygen transfer rate has been observed for coated samples with respect to bare substrate both at 35°C and 50°C.
The barrier effect was further improved by adding of a second organic component capable of forming hydrogen bonds, namely poly(4-hydroxy styrene) (PHS): use of PHS leads to a remarkable decrease (50 - 70 %) of the oxygen transfer rate both at 35°C and 50°C. It has been determined that the optimal formulation is the one containing 22% of PHS and 33% of silica; the thickness required for the multilayer sample to have the same barrier properties as a 45 μm thick, homogenous film composed of HDPE, oPP and PLA is equal to 2.5, 0.2 and 0.1 μm respectively.

Introduction

In the last few years a new class of materials with nanophasic morphology, the ceramers, has been proposed and developed by exploiting the sol–gel chemistry. [1-3] Ceramers are hybrid materials composed of organic and inorganic moieties, the inorganic fraction being generally based on silica. The peculiar properties of this kind of organic/inorganic hybrids result from the synergism between the properties of the polymer composing the organic phase (tenacity, flexibility, transparency, processability, low cost and good adhesion to substrates) and those of the inorganic component (toughness, flame retardancy, chemical and thermal stability, gas barrier effect). Ceramers are characterized by a stronger interaction between the organic and inorganic phase with respect to conventional nanocomposite materials, due to the covalent bonds that form between them. Many works have been carried out in order to investigate the properties of these materials: in particular, the oxygen barrier properties are crucial features for the use in food packaging industry because they affect the shelf life of the packaged food [4-6].

It is indeed the main objective of this work to focus on the oxygen barrier properties of a common polymeric material, such as LDPE, coated by a thin layer of organic-inorganic hybrid, where a copolymer PE-PEG has been chosen as organic phase and TEOS is the silica precursor. Different organic/inorganic ratios and different PE-PEG molecular weights have been investigated and the effect of the addition of a hydrogen-bonding-capable group (PHS) in the organic phase was tested [7-8].

Materials and methods

The organic inorganic hybrids were prepared following a procedure described elsewhere [7]. Briefly, most of the hybrid coatings were prepared by dissolving TEOS and triethoxysilane terminated copolymers (PE-PEG(n)-Si) in warm THF under magnetic stirring at the concentration of 30% wt./v. Water, EtOH and catalyst (HCl or NaOH) were added in the following molar ratios with respect to the overall ethoxide groups (deriving both from TEOS and functionalized copolymers): EtO-:H_2O:EtOH:catalyst = 1:1:1:0.05, and finally partially cured in a closed vial at 60°C for 2 h before deposition on the LDPE substrate by spin-coating. Table 1 summarizes the main composition data for all the hybrid coatings tested and reports data about the film thickness, as measured by SEM from the cross-sectional view of the coated films (see Figure 1 as an example).

Table 1, List of uncoated and coated LDPE films analyzed

Code	Sample	Film thickness (lm)	M_n (PE-PEG) (g/mol)	Coating composition		
				PE-PEG	SiO_2	Additive (type [(d)])
LDPE	Uncoated	45	-	-	-	-
H67	coated PE-PEG-Si/SiO_2	45 + 0.72	2250	0.33	0.67	-
H50	coated PE-PEG-Si/SiO_2	45 + 0.82	2250	0.50	0.50	-
H33	coated PE-PEG-Si/SiO_2	45 + 1.00	2250	0.67	0.33	-
L67	coated PE-PEG-Si/SiO_2	45 + 0.72	920	0.33	0.67	-
L50	coated PE-PEG-Si/SiO_2	45 + 0.82	920	0.50	0.50	-
L33	coated PE-PEG-Si/SiO_2	45 + 1.00	920	0.67	0.33	-
H00	coated PE-PEG	45 + 5.0	2250	1.00	0.00	-
H33-S22	coated PE-PEG-Si/SiO_2/PHS	45 + 1.00	2250	0.45	0.33	0.22 (PHS)
H67-S11	coated PE-PEG-Si/SiO_2/PHS	45 + 0.75	2250	0.22	0.67	0.11 (PHS)
H33-S44	coated PE-PEG-Si/SiO_2/PHS	45 + 1.00	2250	0.22	0.33	0.44 (PHS)
H00-S20	coated PE-PEG/PHS	45 + 5.0	2250	0.80	0.00	0.20 (PHS)
H00-S40	coated PE-PEG/PHS	45 + 5.0	2250	0.60	0.00	0.40 (PHS)

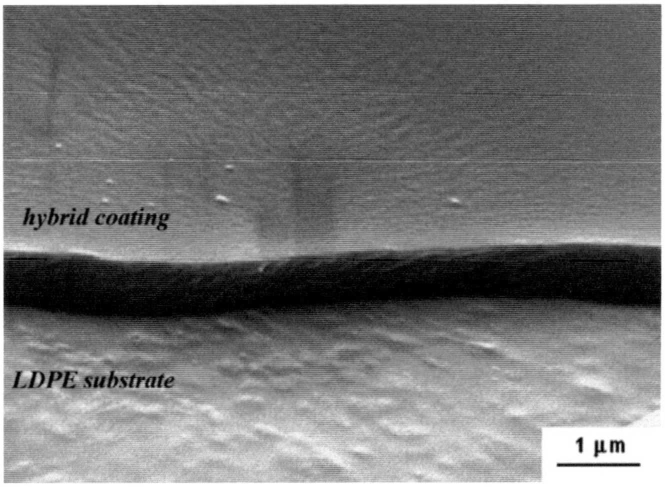

Figure 1, SEM micrograph of the edge view of LDPE film coated with H50 hybrid coating.

Pure oxygen permeability (0% R.H.) in the films was investigated by means of a closed-volume manometric apparatus, already described elsewhere [7-9] and especially designed to characterize ultra barrier films. The value of the Oxygen Transfer Rate, *O.T.R.*, which is defined in terms of the steady state gas flux, J_{SS}, and the pressure difference, Δp, is then calculated from the experimental data as:

$$O.T.R. \overset{DEF}{\equiv} \frac{J_{SS}}{\Delta p} = \left(\frac{dp_d}{dt}\right)_{t\to\infty} \cdot \frac{V}{R \cdot T \cdot A} \cdot \frac{1}{\Delta p} \qquad (1)$$

where A is the area of the sample, V is the downstream volume, R is the universal gas constant, and p_d is the downstream pressure. The *O.T.R.* value in this work is expressed in cm^3(STP)/(cm^2 d atm). The characteristic time of the diffusion process can be evaluated using the time lag value (θ_L), that is the intercept on the t axis of the pd vs. t curve after steady state, represented by a constant value of the slope, has been reached (as shown in Figure 2). If p_d is negligible compared to the upstream pressure, as it is the case of our experiments, and the film is initially gas-free, the time lag value (θ_L) is related to the diffusivity, D, of the gas in the material, through the following equation [10]:

$$\vartheta_L = \frac{l^2}{6 \cdot D} \qquad (2)$$

where l is the sample thickness.

The permeability of each layer, P_i, can be evaluated combining the *O.T.R.* of the multilayer sample with the one of the substrate through the series-resistance formula:

$$\frac{1}{O.T.R.} = \sum_i \frac{1}{(O.T.R.)_i} = \sum_i \frac{l_i}{P_i} \qquad (3)$$

where l_i is the thickness of each layer i. The permeability, which is a material property not depending on the geometry of the sample, was used to compare the transport properties of different coatings. In this work, the permeability is numerically expressed in Barrer (1 Barrer = 10^{10} [(cm3(STP)cm)/(cm^2 s cmHg)]).

Results and discussion

Experimental values of the oxygen transport properties (*O.T.R.* and the related time-lag) of the various samples measured at 35 and 50°C are listed in Table 2.

Table 2. Oxygen transfer rate and time-lag values in the various samples at 35°C and 50°C.

	35°C		**50°C**	
Code	**O.T.R.** $cm^3(STP)/(cm^2 \cdot d \cdot atm)$	**time-lag** s	**O.T.R.** $cm^3(STP)/(cm^2 \cdot d \cdot atm)$	**time-lag** s
LDPE	0.675	4.7	1.413	1.9
H67	0.390	7.3	0.916	3.3
H50	0.436	7.0	1.001	3.1
H33	0.410	7.1	1.064	2.9
L67	0.523	6.5	1.135	2.8
L50	0.538	6.7	1.201	2.9
L33	0.501	6.8	1.122	3.0
H00	0.316	6.2	0.826	1.3
H33-S22	0.194	6.2	0.615	2.6
H67-S11	0.358	6.0	0.737	4.2
H33-S44	0.296	9.4	0.572	1.4
H00-S20	0.253	5.3	0.843	1.0
H00-S40	0.291	10.3	0.836	2.5

(Maximum uncertainty: ±2.5 % for *O.T.R.*; ±7.5 % for time lag values)

Although the overall thickness of the hybrid coating is typically lower than 1 μm, in most cases the permeance of LDPE is remarkably reduced after addition of the coating, and the improvement is comparable with the one observed in case of plastic filled with Montmorillonite [11-12]. The results in terms of time-lag values show that the hybrid coating also allows to slow down remarkably the permeation process, due to a slower kinetics of diffusion.

While there is only a limited effect of the organic–inorganic ratio on both *O.T.R.* and time-lag, significant increase of *O.T.R.* (10–20%) and a slight decrease of time-lag (about 5%) are due to the decrease of the molecular weight of the PE-PEG organic component from 2250 to 920 g/mol; the higher free volume in organic domains associated with the lower molecular weight may be responsible of this behavior.

In order to have a more meaningful comparison between the properties of the different hybrids, the permeabilities of the pure coatings have been evaluated according to Eq. (3). The calculated values at 35 and 50°C are reported in Barrer and listed in Table 3.

Table 3, Oxygen permeability in LDPE and pure coatings at 35 and 50°C

Code	Permeability (Barrer)	
	35°C	50°C
LDPE	4.625	9.683
H67	0.101	0.285
H50	0.154	0.428
H33	0.159	0.657
L67	0.247	0.615
L50	0.332	1.003
L33	0.296	0.831
H00	0.451	1.512
H33-S22	0.041	0.166
H67-S11	0.087	0.176
H33-S44	0.080	0.146
H00-S20	0.307	1.590
H00-S40	0.389	1.561

The so-obtained results were analyzed on the basis of the Maxwell model which predicts a permeability reduction as follows [13]:

$$\frac{P}{P_0}=\frac{1-\phi}{1+\phi/2} \qquad (4)$$

where P_0 is the permeability in the pure material and ϕ is the volumetric fraction of the inorganic phase, (estimated by assuming the density of the organic and inorganic phases equal to 1 and 2 g/cm^3, respectively). The data calculated from eq. 4 for the case of PE-PEG/Si-SiO_2 hybrids are plotted as dotted lines in Figure 2, together with the experimental data.

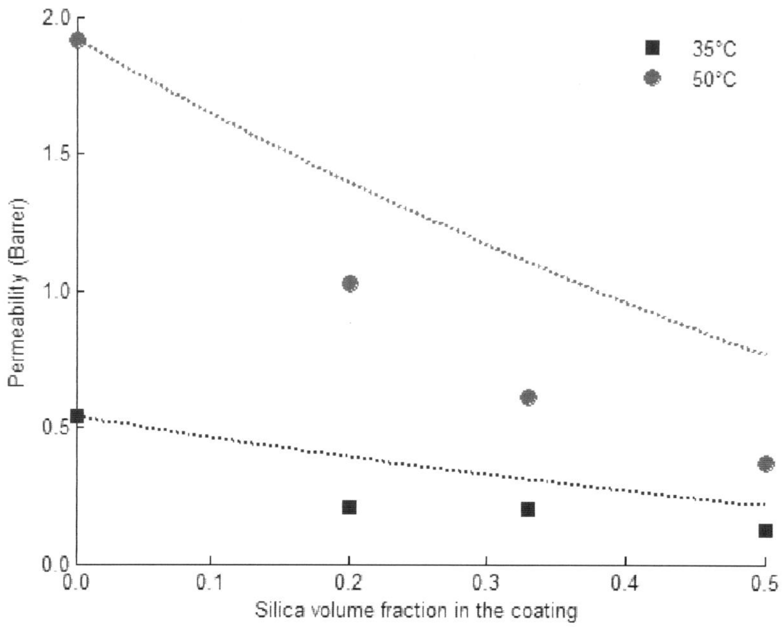

Figure 2, Oxygen permeability of coatings containing PE-PEG with $M_n = 2250$ g/mol at 35 and 50°C. Experimental results (squares, 35°C, and circles, 50°C) are reported along with the curves (dotted lines) calculated from the Maxwell model

As one can see, the experimental points, both at 35°C or 50°C, lie below the curves calculated by the Maxwell equation. These results suggest that, for the organic-inorganic hybrid material prepared under acidic conditions, the improvement of the barrier properties is more significant then predicted by Eq. 4 for conventional composite materials in which the two phases do not interact, due to a synergetic effect existing between the two phases.

On the other hand, all coatings containing PHS showed permeabilities quite lower then that in the PE-PEG-Si/SiO_2 hybrids, both at 35°C and at 50°C and stronger barrier effects were obtained for the coating with intermediate amount of silica and PHS (H33-S22). However, the effect of the addition of PHS on permeability is not monotonous with PHS content, but there exists a minimum, that has to be located at around 20 wt. % of PHS. The existence of a minimum is not surprising and may be due to different issues that occur at high concentrations of PHS such as the saturation of the hydrogen bonds that silanol groups can form with PHS, and, to a lower extent, with ether groups of PE-PEG or poor miscibility between the two organic components, that can induce phase separation of PHS, limiting the formation of hydrogen bonds with silanols.

To describe the permeation behavior of the polymer blend we used a simple mixing rule that neglects the mutual interactions between the two polymers:

$$P_{org} = x_{PHS} P_{PHS} + x_{PE-PEG} P_{PE-PEG} \quad (5)$$

where x_{PHS} and $x_{PE\text{-}PEG}$ are the molar fractions (calculated considering the average molecular weights of the organic species) of the two components in the organic phase and P_{PHS} and $P_{PE\text{-}PEG}$ are the pure polymers permeabilities; once the theoretical permeability of the organic phase is calculated, Eq. 4 can be then applied to predict the barrier properties for a heterogeneous material.

In Figure 3, the data estimated by Eqs. (4) and (5) are reported as continuous lines and are compared to the experimental ones and to the value of pure PHS permeability taken from the literature at 35°C [14].

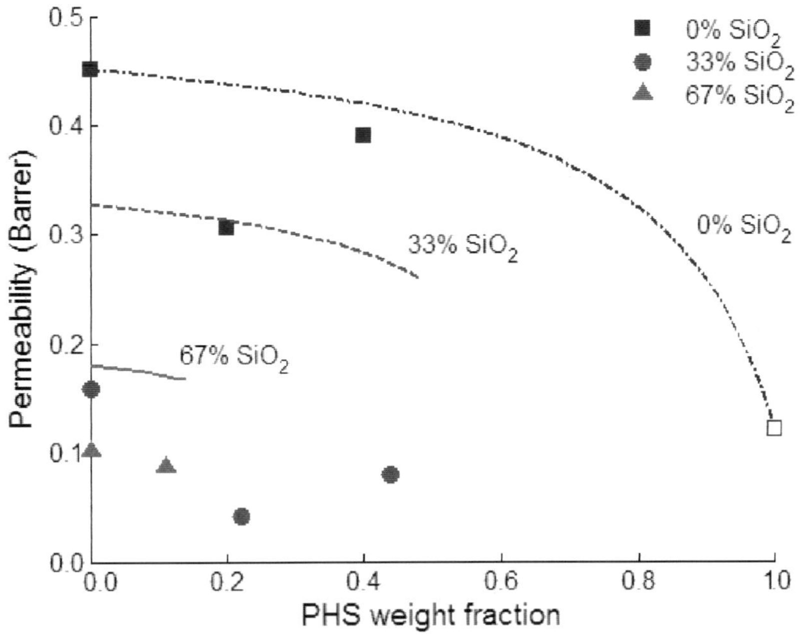

Figure 3, Permeability in organic/inorganic hybrid at 35°C, curves are due to the model in Eqs. (4,5). The open symbol is taken from the literature [14]

Considering at first just the polymer blend by itself, and comparing the experimental data with those calculated by the additive rule expressed by Eq. (4), we notice that at low PHS content the experimental permeability of the system is lower than that predicted by the additive rule, meaning that a synergic effect takes place between the two polymers, that may be due to the formation of hydrogen bonds between PHS and PE-PEG segments. For higher PHS content, the system seems to follow more closely the additive behavior represented by Eq. (4).

When considering the ternary system, containing also the inorganic component, it can be seen that the hybrid permeability is, as previously observed for the case with no PHS, much lower than that calculated on the basis of Eq. (4), representative of the permeability of a filled system obtained by physically dispersing the same volume fraction of spherical silica. This result can indeed be attributed to the fact that the coating molecular architecture consists of a highly interpenetrated network of organic and inorganic phases rather than of a physical dispersion of spherical silica particles in the organic phase.

Conclusions

It has been shown how a thin layer of organic-inorganic hybrid coating can be used to improve oxygen barrier properties of plastic films. The use of a suitable polymer (PE-PEG) within the coating allows to obtain good adhesion to the LDPE substrate without any previous surface treatment. The oxygen permeability decreases with increasing molecular weight of the organic phase and with increasing silica weight content. Furthermore, the inclusion of a second polymeric phase, PHS, able to form hydrogen bonds has been explored and it has been shown that it allows to increase further the barrier effect. The experimental results have been compared with those calculated by using simple models describing the permeability of polymer blends and conventionally filled particulate composites, allowing to evidence significant contributions of morphology and hydrogen bonds to the reduction of the permeability.

The easy processability of these coatings suggests that they are a convenient alternative approach to the improvement of oxygen barrier properties with respect to the conventional composites obtained by clay dispersion in polymeric matrices.

Acknowledgement

This work has been funded by the Project ''Organic–inorganic hybrid Coatings for Packaging innovative Films" supported by MIUR (PRIN 2004, Prot. 2004030304). We also acknowledge Alessandro Andreoni and Greg Simmonds for the experimental support.

References

1. Schottner, Gerhard (2001), "Hybrid Sol-Gel-Derived Polymers: Applications of Multifunctional Materials", *Chemistry of Materials,* 13(10), 3422-3435.
2. Wen, Jianye et al. (1996), "Organic/Inorganic Hybrid Network Materials by the Sol-Gel Approach", *Chemistry of Materials*, 8(8), 1667-1681
3. Sanchez, Clement et al. (2005), "Applications of hybrid organic–inorganic nanocomposites", *Journal of Materials Chemistry*, 15, 3559–92.
4. Malucelli, Giulio et al. (2007), "Surface and barrier properties of hybrid nanocomposites containing silica and PEO segments". *Journal of Applied Polymer Science*, 103(6) 4107–15.
5. Amberg-Schwab, Sabine et al. (1998), "Inorganic–organic polymers with barrier properties for water vapor, oxygen and flavors". *Journal of Sol-Gel Science and Technology*, 13(1–3), 141–146.
6. Montenero, Angelo et al. (2007), "A laminated material having a high oxygen-barrier effect", PCT Int Appl 2007; [WO 2007042993]
7. Minelli, Matteo et al. (2008), "Oxygen permeability of novel organic–inorganic coatings: I. Effects of organic–inorganic ratio and molecular weight of the organic component", *European Polymer Journal,* 44, 2581–2588
8. Toselli, Maurizio et al. (2008) "Oxygen permeability of novel organic–inorganic coatings: II. Modification of the organic component with hydrogen-bond forming polymer", *European Polymer Journal*, in Press.
9. Pizzi, Diego et al. (2005), "Moisture sorption and oxygen transport in a nylon-6/montmorillonite composite", *Pierucci S. Chemical Engineering Transactions*, 6, AIDIC, 515–520.
10. Crank J. (1956), "The mathematics of diffusion", *Oxford* at the *Clarendon press*.
11. Choia, Woo J. et al. (2004), "Synthesis of chain-extended organifier and properties of polyurethane/clay nanocomposites", *Polymer*, 45(17) 6045–6057.
12. Picard, Emile et al. (2007), "Barrier properties of nylon 6-montmorillonite nanocomposite membranes prepared by melt blending: influence of the clay content and dispersion state". *Journal of Membrane Science*, 292(1–2), 133–144
13. Maxwell, James C. (1904), "Treatise on Electricity and Magnetism", Oxford, *Clarendon*.

14. Puleo A.C. et al. (1989), "Gas sorption and transport in substituted polystyrenes", *Journal of Polymer Science Part B: Polymer Physics*, 27(11) 2385-2406.

Rigorous Dynamic Model of the High-Pressure Polymerization of Ethylene in Tubular Reactors Able to Predict the Full Molecular Weight Distribution

Mariano Asteasuain, Adriana Brandolin

Planta Piloto de Ingeniería Química (CONICET – UNS), Bahía Blanca, Argentina.

Introduction

High-pressure polymerization of ethylene in tubular reactors is a widely employed industrial process. It allows obtaining branched low density polyethylene (LDPE) with characteristics that have not been reproduced by the more modern low or medium pressure polymerizations. The process is carried out under rigorous conditions. For instance, the reactor is operated at very high pressure, between 1300 to 3000 bar at the inlet. The temperature rises from about 50 °C at the reactor inlet to 330 °C at the hottest points due to the exothermic polymerization. Axial velocities are also high, usually around 10 m/s. Besides, the reactor configuration is complex. The main feed usually consists of ethylene, inert components, transfer agents to control the molecular weight, and eventually oxygen as initiator. Most of the polymerization reaction, however, is initiated by peroxides that are fed to the reactor through lateral injections. Monomer and/or transfer agents can also be fed through side injections. The polymerization takes place in short reaction zones following the peroxides injections, exhibiting high heat generation and steep temperature profiles. The rest of the reactor is mainly used as a heat exchanger, heating or cooling the reaction mixture in order to reach appropriate temperatures for peroxides addition or for downstream units. The control of the reaction mixture temperature is achieved by circulating vapor or liquid water through different jacket zones, which may be interconnected or not. Moreover, some of these reactors are provided with a let-down valve located at the reactor exit that is periodically opened to produce a pressure pulse which sweeps out the polymer from the walls. Besides, the interaction between the different reacting species is very complex, and there is an intricate connection between polymer quality and process conditions.

In this context, a mathematical model is fundamental tool to study safely and economically the influence of the different design and operative variables on production performance and product quality, as well as for process optimization. In order to achieve high-quality results that could be useful for actual plants, comprehensive models with accurate predictive capabilities are required. In spite of mathematical modeling of industrial LDPE tubular reactors being a complex task, a considerable amount of research work has been devoted to this task. Rigorous steady state models of this process have been reported in the literature.[1-3] In general, these models consider realistic reactor configurations and include detailed predictions of physical and transport properties along the axial distance. Most of these models have focused on the polymer average molecular properties, like the average molecular weights or average degrees of branching. However, the prediction of their distributions, such as the full molecular weight distribution (MWD) or two-dimensional distributions in molecular weight and branching frequency have also been reported.[4-6]

On the other hand, less attention has been devoted to the dynamic operation of these reactors. Dynamics is an important component of the performance of this process. As is usual in continuous polymerization plants, different polymer grades are produced in the same equipment. Therefore, normal operation requires changes from one steady state to another in order to switch among different final products. In the transient states during these changes, off specification, low-value product is generally produced. Dynamic operation is also important in start-up or shut-down procedures. Although it has been stated that quasi-steady state can be assumed in this process,[7] recent works have shown that dynamics cannot be ignored under certain conditions, specially when material recycle is involved, because in this case time constants may change from minutes to hours.[8] There are few works dealing with the dynamic operation of this process available in the literature. In these, usually very simplified and small models have been used.[9,10] A few dynamic models involving a more detailed description of the process have been reported, [8,11] but they have not included prediction of distributions of molecular properties.

In previous works we developed a rigorous steady state model of a LDPE tubular reactor.[4,12] This model was validated against experimental data from an actual industrial reactor. In this work, we present an extension of that model consisting in the incorporation of the process dynamics. Comprehensiveness of the former model was kept. This involves employing rigorous correlations for the calculation of physical and transport properties, such as density, viscosity and heat capacity of the reaction mixture and coolant, and heat-transfer coefficient along the axial distance and time. Realistic reactor configuration is also considered. The resulting model is capable of predicting the full MWD, as well as average branching indexes, monomer conversion and average molecular weights along time and reactor length. This model is used for analyzing the influence of different dynamic scenarios on process and product quality variables. In particular, grade transition policies are studied.

Process Description and Mathematical Model

A simplified flow sheet of the LDPE process analyzed in this work is displayed in Figure 1. The polymer is produced by high-pressure polymerization of ethylene in a tubular reactor, using oxygen and peroxide mixtures as initiators. The make-up stream, consisting of ethylene, inert components, oxygen and transfer agents, is mixed with the low-pressure recycle stream, and then compressed to about 250 bars in a multi-stage primary compressor. This stream is mixed with the high-pressure recycle stream, and then compressed in a hyper-compressor up to the reaction pressure, which is usually around 2000-2800 bars. The compressor outcome is fed to the jacketed tubular reactor, where the ethylene is polymerized with a conversion per pass of approximately 25 – 30 %. Although the reaction is partially initiated by the oxygen, most of the polymerization is initiated by organic peroxide mixtures. These peroxides are added at additional feeds located at different axial positions, producing short reaction zones with sharp temperature and conversion increases. The reactor sections apart from these reaction zones are mainly used as heat exchangers in order to reach appropriate temperatures for peroxide addition or for downstream units. Monomer and/or transfer agents can also be fed through side feedings. The control of the reactor temperature is achieved by circulating vapor or liquid water through independent tube jackets. The reactor

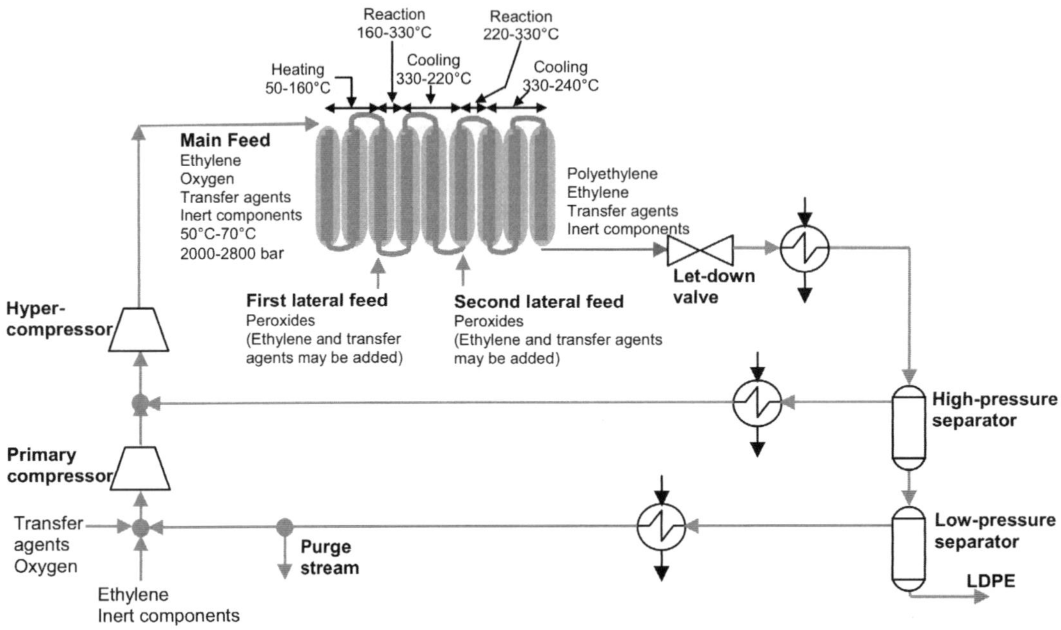

Figure 1. Simplified flow sheet of the LDPE production plant

output is expanded in a let-down valve, which produces a pressure pulse that travels back along the reactor removing polymer build ups from the reactor wall. The expanded reactor output is separated first in a high pressure and then in a low pressure separator. Unreacted ethylene and transfer agents, and inert components, are cooled and then recycled. The polymer is obtained from the low pressure separator. The particular configuration of the reactor analyzed in this work considers eight jacket zones and two lateral peroxide feeds.

This work is focused on the analysis of the tubular reactor unit. The mathematical model of the polymerization reactor is based on a previous, comprehensive steady state model developed by the authors, [4,12] which is extended by incorporating the process dynamics. The model assumes plug flow and supercritical reaction mixture. Besides, it considers variation of physical and transport properties (i.e. axial velocity, heat capacity, thermal conductivity, viscosity and density) along the axial distance, calculated with rigorous correlations. Detailed calculation of the heat-transfer coefficient along the axial distance is also included.[3] The kinetic mechanism considered in the reactor model is shown in Table 1. It includes all the kinetic steps commonly proposed in the literature for this process. In order to avoid iterative calculations that increase the computational burden, the jacket temperature at each one of the eight reaction zones is assumed uniform, and the pressure pulse is neglected. The peroxide mixtures are treated as single fictitious species. The same is done for the transfer agent mixtures. These simplifications were validated in a previous work by the authors[12] against several data sets from an industrial tubular reactor. The same transfer agent mixture, and therefore the same single fictitious transfer agent (S), that is employed for the main reactor feed is considered for possible lateral feeds. However, peroxide mixtures for the first and second lateral feeds are different in composition,[12] and hence they are represented in this model by two different fictitious peroxides (I_1 for the first lateral feed and I_2 for the second one).

The set of kinetic constants reported by Asteasuain et al.[12] is used here. The main model equations are outlined in Table 2. Reaction mixture physical and transport properties, such as density ($\rho(z,t)$), heat-capacity ($\mathrm{Cp}(z,t)$), heat-transfer coefficient ($U(z,t)$), and others like thermal conductivity and viscosity, are calculated along axial distance and time using rigorous correlations that are function of the reaction mixture composition, reactor temperature and pressure, and dimensionless numbers such as the Nusselt, Reynolds, Prandtl and Pecklet numbers. The well-known method of moments and the probability generating function (pgf) transformation technique[13,14] are employed for calculating the average molecular weights and the full MWD, respectively. Briefly, the pgf method is based on a transformation of the mass balances of macroradicals and dead polymer ($R(m)$ and $P(m)$, respectively, in Table 1) into the pgf domain. In this way, balances for the pgf transforms of the macroradicals and dead polymer number, weight and differential log distributions are obtained (see Eq. (20), with $\phi(a)$ and $\varphi(a)$ representing macroradicals and dead polymer transforms, respectively, and a = 0, 1, 2 standing for number, weight and differential log distributions, respectively). Pgf transforms obtained by solution of the pgf balances are inverted by an inversion algorithm (function f in Eq. (19)), to obtain the corresponding distributions. This technique allowed modeling the MWD easily and efficiently, in spite of the reactor model complexity. Expressions corresponding to the reaction rates $r_j(z,t)$ in Eq. (20) and the pgf inversion algorithm can be found elsewhere.[1,4]

Table 1. Kinetic mechanism

Peroxide initiation		*Oxygen initiation*	
$I_k \xrightarrow{k_{ik}} 2R(0) \quad k=1,2$	(1)	$O_2 + M \xrightarrow{k_o} 2R(0)$	(8)
Monomer thermal initiation		*Generation of inert*	
$3M \xrightarrow{k_{mi}} R(1) + R(2)$	(2)	$O_2 + R(m) \xrightarrow{f_o k_o} X$	(9)
Propagation		*Termination by combination*	
$R(m) + M \xrightarrow{k_p} R(m+1)$	(3)	$R(n) + R(m) \xrightarrow{k_{tc}} R(n+m)$	(10)
Thermal degradation		*Chain transfer to monomer*	
$R(m) \xrightarrow{k_{tdt}} P(m) + R(0)$	(4)	$R(m) + M \xrightarrow{k_{trm}} P(m) + R(1)$	(11)
Chain transfer to polymer		*Chain transfer to transfer agent*	
$R(n) + P(m) \xrightarrow{m k_{trp}} P(n) + R(m)$	(5)	$R(m) + S \xrightarrow{k_{trs}} R(m) + R(0)$	(12)
Backbiting		*β-Scission of secondary radical*	
$R(m) \xrightarrow{k_{bb}} R(m)$	(6)	$R(m) \xrightarrow{k_{\beta 1}} P(m) + R(0)$	(13)
β-Scission of tertiary radical			
$R(m) \xrightarrow{k_{\beta}} P(m) + R(0)$	(7)		

Peripheral units, such as compressors, separators and heat exchangers in the recycle section are not modeled, because this work focuses only on the qualitative effects of the material recycle on the reactor performance. Instead, a perfect separation of the polymer from the remaining components of the reactor output is assumed. The resulting recycle stream and the make-up stream are considered to be both at the reactor feed temperature when they mix with each other. However, an important aspect of this process is the presence of several time delays.[15] In this work, time delays are incorporated into the plant model as a pure delay in the recycle stream. Model implementation was carried out in the commercial software gPROMS (Process Systems Enterprise, Ltd.).

Table 2. Outline of the main model equations

Global mass balance $\frac{\partial(\rho(z,t)v(z,t))}{\partial z}+\frac{\partial \rho(z,t)}{\partial t}=0 \quad (14)$ *Pressure drop* $\frac{\partial P(z,t)}{\partial z}=c \quad c: \text{constant parameter} \quad (15)$ *Number average molecular weight* $\text{Mn}(z,t)=\text{Mw}_{\text{M}}\frac{\lambda_1(z,t)+\mu_1(z,t)}{\lambda_0(z,t)+\mu_0(z,t)} \quad (16)$	*Mass balances of components, MWD moments and pgfs* $\frac{\partial C_j(z,t)}{\partial t}+\frac{\partial v(z,t)C_j(z,t)}{\partial z}=r_j(z,t)$ $j=O_2(\text{oxigen}), M(\text{ethylene}), I_k(k\text{th initiator}, k=1,2),$ S(transfer agent), Me (Methyl groups), Vi (vinyl groups), Vd (vinylidene groups), LCB (long chain branches), (20) λ_a(*ath* order moment of the macroradical MWD, a=0,1,2), μ_a(*ath* order moment of the dead polymer MWD, a=0,1,2), ϕ_a(*ath* pgf transform of the macroradical MWD, a=0,1,2), φ_a(*ath* pgf transform of the dead polymer MWD, a=0,1,2)
Weight average molecular weight $\text{Mw}(z,t)=\text{Mw}_{\text{M}}\frac{\lambda_2(z,t)+\mu_2(z,t)}{\lambda_1(z,t)+\mu_1(z,t)} \quad (17)$ *Short chain branches / 1000 C* $\text{SCB/1000C}\ (z,t)=500\frac{\text{Me}(z,t)}{\lambda_1(z,t)+\mu_1(z,t)} \quad (18)$	*Energy balance* $\rho(z,t)\text{Cp}(z,t)\frac{\partial T(z,t)}{\partial t}+\rho(z,t)v(z,t)\text{Cp}(z,t)\frac{\partial T(z,t)}{\partial z}=$ $-\frac{4U(z,t)(T(z,t)-T_j)}{D}+r_{\text{pm}}(z,t)(-\Delta H) \quad (21)$
Full MWDs $n(z,t)=f\left(\phi_0(z,t)+\varphi_0(z,t))/(\lambda_0(z,t)+\mu_0(z,t)\right)$ $w(z,t)=f\left(\phi_1(z,t)+\varphi_1(z,t))/(\lambda_1(z,t)+\mu_1(z,t)\right)$ $\frac{dW}{d(\log_{10}Mw)}(z,t)=f\left(\phi_2(z,t)+\varphi_2(z,t))/(\lambda_1(z,t)+\mu_1(z,t)\right)$ $n(z,t)$: number fraction distribution (19) $w(z,t)$: weight fraction distribution $dW/d\log_{10}Mw\ (z,t)$: differential log distribution $f(.)$: algebraic equations of the pgf inversion algorithm	*Vinyl groups / 1000 C* $\text{Vi/1000C}\ (z,t)=500\frac{\text{Vi}(z,t)}{\lambda_1(z,t)+\mu_1(z,t)} \quad (22)$ *Vinylidene groups / 1000 C* $\text{Vd/1000C}\ (z,t)=500\frac{\text{Vd}(z,t)}{\lambda_1(z,t)+\mu_1(z,t)} \quad (23)$ *Long chain branches / 1000 C* $\text{LCB/1000C}\ (z,t)=500\frac{\text{LCB}(z,t)}{\lambda_1(z,t)+\mu_1(z,t)} \quad (24)$

Results and Discussion

In order to check the model validity, simulations were performed for different initial conditions. It was observed that model outputs evolved towards the expected steady state values. An example of this analysis is presented in Figure 2, which shows the evolution of the temperature profile during a reactor start-up. The reactor is considered to be initially filled with monomer at 76 ºC. The recycle stream is disconnected, and operating variables are set at their steady state values at time 0. It can be observed that the steady state temperature profile matches reasonable well the plant data. It is also interesting to note the extremely rapid increase of the reactor temperature. For instance, the steady state value of the first temperature peak is reached in less than 20 s, involving an increment of 255 ºC with respect to the initial value. At increasing times, steady state temperature values along the reactor are reached after an overshoot. However, it can be seen that the temperature value of 345 ºC, above which reactor runaway can occur, is never surpassed.

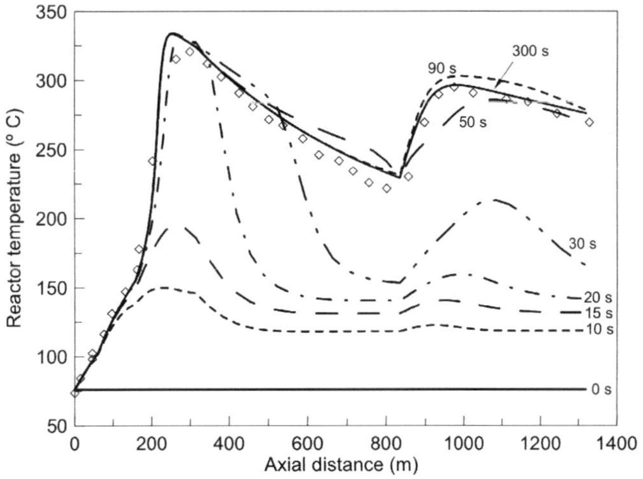

Figure 2. Axial temperature profiles calculated during a reactor start-up. Symbol: plant steady state values; lines: model

The model was used to asses the influences of different process perturbations, with and without recycle. Changes in the flow rates of monomer, initiators and transfer agents and in the feed temperature were analyzed. As an example, results for the flow rate of monomer are presented, which was the variable that most significantly affected the process for the same relative change in its original value. Figure 3 shows the time profiles of the position and temperature values of the temperature peaks after a step reduction of 30 % in the monomer feed rate, for the reactor without recycle. Temperature peaks are very important parameters for the operation of these reactors. It can be seen that the temperature of the first peak surpasses a dangerous value of 345 °C before reaching its steady state value, and that its position moves towards the reactor entrance. The second peak shows larger differences between its transient and steady state values. This illustrates the importance of the capability of predicting the dynamic behavior of these reactors.

Figure 4 shows the time profiles of the temperature peaks for the same perturbation, but for the reactor with recycle. It can be seen, noting the different time scales of the graphics, that the behaviors of the second temperature peak with and without recycle are similar. However, dynamics for the first peak are slower for the reactor with recycle. The time profile of the position of this peak is similar to the one without recycle, but slightly delayed in time. On the contrary, the temperature value of this peak reaches a much higher value than in the previous case, showing a delay in reaching the new steady state more than ten times greater than before. This slower dynamics is an expected result of the influence of the recycle of material, since the recycle loop needs a time to settle down after the feed perturbation. Consistent results about the influence of the recycle loop on the reactor dynamics have been reported in the literature.[8]

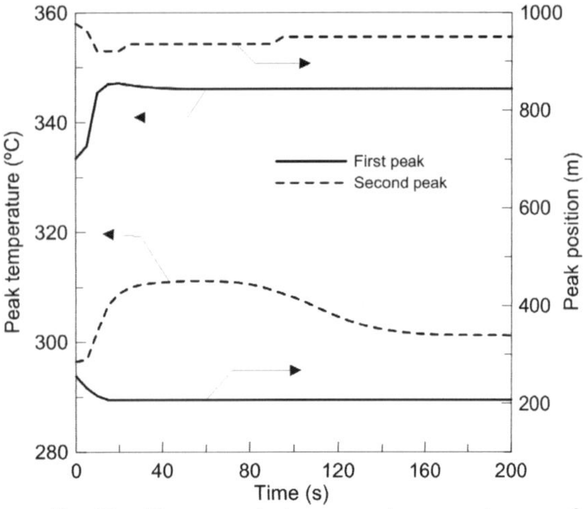

Figure 3. Position and temperature values of the temperature peaks after a step reduction of 30 % in the monomer feed rate without recycle

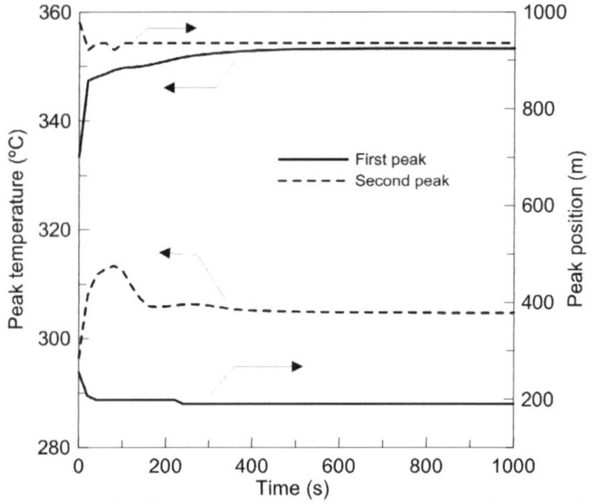

Figure 4. Position and temperature values of the temperature peaks after a step reduction of 30 % in the monomer feed rate with recycle

Transitions between the productions of different polymer grades are frequent and are an important component of the operation of this process. A usual method of achieving these transitions is to modify the transfer agent flow rate,[15] which directly affects the polymer molecular weights. Deviations of the monomer feed to the side feeds may also be applied.[4] Figure 5 shows the evolution of the MWD during a transition from a polyethylene grade of Mw = 140000 g/mol to a grade of Mw = 220000 g/mol, considering the reactor with recycle. The new operating scenario involves a reduction of the transfer agent flow rate at the reactor inlet from 0.27 kg/h to 0.23 kg/h, and a deviation of 7 % of the monomer feed to the second lateral feed. Also, a different jacket temperature profile is applied. The transition policy consists of a step change in each of these variables to the operating point value for the new grade. The shift of the MWD to higher molecular weights can be observed in Figure 5. It is interesting to note that the intermediate portion of the MWD partially shifts back to lower molecular weights after a certain time, while the high molecular weight tail of the distribution becomes more important towards the end of the transition with a marked formation of a shoulder. The grade transition finishes approximately 2.2 h after the change in operating conditions. The detailed tracking of the evolution of the MWD provided by the model has a great potential for a careful analysis of transition policies.

The transition time is markedly different for the same transition policy but without material recycle. In this case, the time needed to achieve the transition is only 0.08 h. This difference may be attributed to the fact that in the no-recycle case the transmission of the change in the make-up stream to the reactor inlet is instantaneous. On the contrary, when the recycle stream is mixed with the make-up stream, the changes in the latter are masked by the return of unreacted monomer and transfer agent. This is particularly important for the transfer agent, because the consumption per pass of this reactant is very small. Hence, a long time is needed before the excess of it is consumed through several passes through the reactor. This is illustrated in Figure 6, which shows the transfer agent flow rate at the reactor inlet and in the make-up stream for the grade transitions with and without recycle. A similar behavior is observed for the monomer flow rate, although the time delay is shorter.

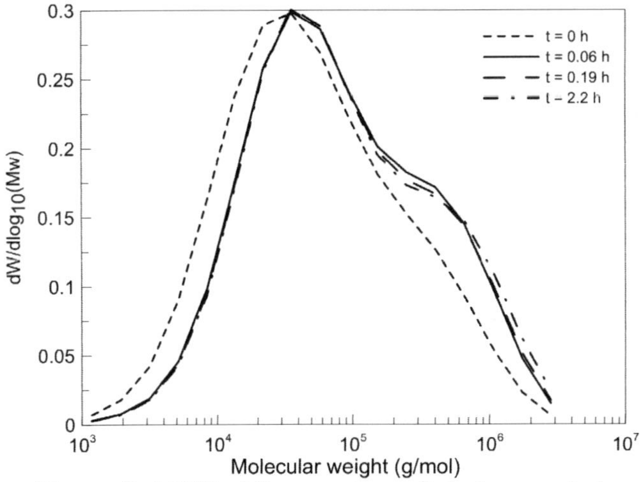

Figure 5. MWD at the reactor exit during grade transition

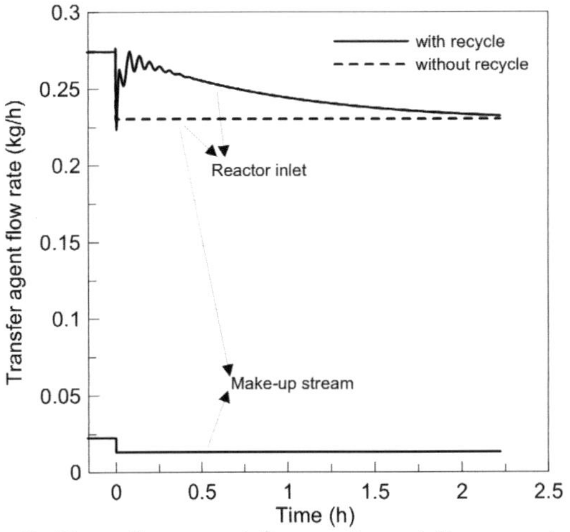

Figure 6. Transfer agent flow rates at the reactor inlet during grade transition. Make-up flow rates are switched to final values at t = 0

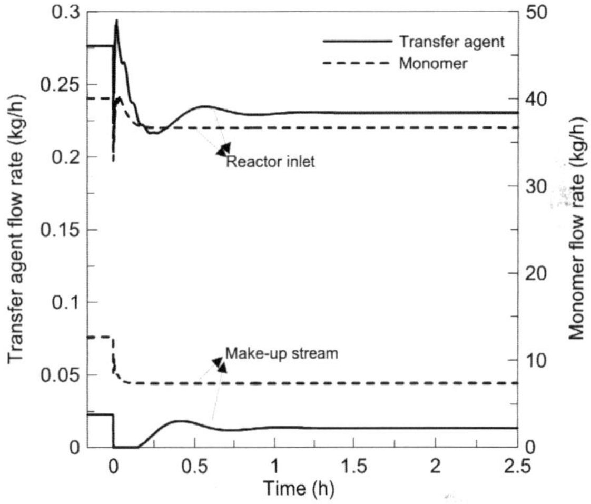

Figure 7. Monomer and transfer agent flow rates during grade transition. Make-up flow rates are manipulated by PI controllers

A corollary of these results is that an important benefit can be obtained by optimizing the transition policy. Figure 7 shows the flow rates of transfer agent and monomer at the reactor inlet and in the make-up stream, when the make-up flow rates are manipulated by PI controllers with the desired flow rates at the reactor inlet as set points. Although the PI controllers had only been roughly tuned by hand, the improvement in transition time is significant.

Conclusions

A comprehensive dynamic model of the high-pressure polymerization of ethylene in tubular reactors was developed. The model, which considers variable physical and transport properties calculated by means of rigorous correlations, can predict the full MWD, monomer conversion, reactants concentrations, average molecular weights and average degree of branching along the axial distance and time. In particular, the pgf technique allowed modeling the full MWD easily and efficiently in spite of the reactor model complexity. The results obtained show that the model has great potential for analyzing dynamic responses of MWD and other process variables under different transition policies, as well as to predict the influence of process perturbations and for studying start-up and shut-down operations. The influence of the material recycle on the process dynamics was assessed. In particular, it was shown that the recycle stream must be taken into account when analyzing transition policies. Besides, it was also shown that a significant improvement can be obtained by optimizing the transition procedure.

References

1. Brandolin A, Lacunza MH, Ugrin PE, Capiati NJ. High pressure polymerization of ethylene. An improved mathematical model for industrial tubular reactors. Polym React Eng. 1996; 4: 193-241.
2. Kiparissides C, Verros G, Kalfas G, Koutoudi M, Kantzia C. A comprehensive mathematical model for a multizone tubular high-pressure ldpe reactor. Chem Eng Commun. 1993; 121: 193–217.
3. Lacunza MH, Ugrin PE, Brandolin A, Capiati NJ. Heat transfer coefficient in a high pressure tubular reactor for ethylene polymerization. Polym Eng Sci. 1998; 38: 992-1013.
4. Asteasuain M, Brandolin A. Modeling and optimization of a high-pressure ethylene polymerization reactor using gproms. Comput Chem Eng. 2008; 32: 396-408.
5. Kim D, Iedema PD. Molecular weight distribution in low-density polyethylene polymerization; impact of scission mechanisms in the case of a tubular reactor. Chem Eng Sci. 2004; 59: 2039-2052.
6. Schmidt C, Busch M, Lilge D, Wulkow M. Detailed molecular structure modeling - a path forward to designing application properties of ldpe. Macromol Mater Eng. 2005; 290: 404-414.
7. Kiparissides C, Verros G, MacGregor JF. Mathematical modelling, optimization and quality control of high-pressure ethylene polymerization reactors. J Macromol Sci Rev Macromol Chem Phys. 1993; C33: 437–527.
8. Häfele M, Kienle A, Boll M, Schmidt CU. Modeling and analysis of a plant for the production of low density polyethylene. Comput Chem Eng. 2006; 31: 51-65.
9. Asteasuain M, Tonelli SM, Brandolin A, Bandoni A. Dynamic simulation and optimization of tubular polymerization reactors in gproms. Comput Chem Eng. 2001; 25: 509-515.
10. Kiparissides C, Verros G, Pertsinidis A. On-line optimization of a high-pressure low-density polyethylene tubular reactor. Chem Eng Sci. 1994; 49: 5011-5024.
11. Mummudi M, Fox RO. A comprehensive dynamic model for high-pressure tubular ldpe reactors. J Chin Inst Chem Engrs. 2006; 37: 1-18.
12. Asteasuain M, Pereda S, Lacunza MH, Ugrin PE, Brandolin A. Industrial high pressure ethylene polymerization initiated by peroxide mixtures: A reduced mathematical model for parameter adjustment. Polym Eng Sci. 2001; 41: 711-726.
13. Asteasuain M, Sarmoria C, Brandolin A. Recovery of molecular weight distributions from transformed domains. Part i: Application of pgf to mass balances describing reactions involving free radicals. Polymer. 2002; 43: 2513-2527.
14. Asteasuain M, Sarmoria C, Brandolin A. Peroxide modification of polyethylene. Prediction of molecular weight distributions by probability generating functions. Polymer. 2002; 43: 2363-2373.
15. Cervantes A, Tonelli SM, Brandolin A, Bandoni A, Biegler L. Large-scale dynamic optimization of a low density polyethylene plant. Comput Chem Eng. 2000; 24: 983-989.

722ab Probing Mechanical Function of Vertical Multi-Layered Systems by Enhancing the Adhesion at the Interface

Tiffani B. Abernathy, Ravi Ayyer, Eric Baer, Anne Hiltner, and LaShanda Korley. Department of Macromolecular Science and Engineering, Case Western Reserve University, NSF Center for Layered Polymeric Systems, Cleveland, OH 44106

Forced-assembly is a process by which immiscible polymers are co-extruded through a series of die elements termed as 'multipliers' to form multi-layered films. This layering process, which offers advantages of continuous film formation and selective tuning of interphase thickness and layer spacing, provides a unique method to probe the nature of the interphase and to determine its relationship to mechanical function of the polymeric system. Recent advances in microlayering have focused on the development of vertical multi-layers. It is anticipated that the interphase between layers A and B is inherently weak along the length of film when subjected to applied tension along the transverse direction of the film. Through the use of dynamic mechanical analysis and tensile testing, it has been shown that these materials are mechanically anisotropic by nature, and thus, are interesting systems for probing mechanical function. The modulus of these films drops significantly when the force is applied perpendicular to the direction of flow in the system. This can be attributed to the weak adhesion properties of the interphase in vertical multi-layered films. Probing chemical interactions between the polymers in the vertical layered systems would lead to control of interfacial adhesive properties, hence increasing the mechanical properties of the system.

Multi-layer co-extrusion will be utilized to produce novel materials with enhanced mechanical properties through selective manipulation of the interphase. It is expected that control of interfacial adhesion through ultraviolet (UV) crosslinking or ionic interactions, enhanced mechanical properties can be achieved, particularly along the length of the film.

"This material is based upon work supported by the National Science

Foundation under Grant No. DMR 0423914."

722ac Synthesis of Structured Polymer Particles by Micro Dispersive Suspension Polymerization

Yunju Jung[1], Carla Luciani[1], Joong Jin Han[2], and Kyu Yong Choi[2]. (1) Chemical and Biomolecular Engineering, University of Maryland, Stadium Drive, Bldg 090, College Park, MD 20742, (2) Department of Chemical and Biomolecular Engineering, University of Maryland, College Park, College Park, MD 20742

A novel suspension polymerization technique is presented to synthesize micron-size acrylate polymers with porous or hollow interior structure. The precipitation of polymer inside a liquid droplet is induced by controlling the solubility parameter of the liquid phase. It has been found that the morphology of the polymer particles is surprisingly complex as determined by thermodynamic phase separation, particle precipitation, phase inversion, and interfacial mass transfer. For example, multi-hollow particles, porous particles, single hollow particles, severely deformed particles, and particles with external holes can be produced by controlling the reaction conditions. Under certain reaction conditions, thick cellular membrane or shell is formed with a core that consists of micron-sized and uniform particles. The observed morphological complexity of the polymer particles suggests that a simple suspension polymerization process can be modified to synthesize a new class of polymer particles that can have a broad spectrum of applications. In this paper, we also present a mathematical model to calculate the local distribution of monomer, solvent, and polymer inside a polymerizing droplet to understand the morphological development.

Synthesis, characterization and surface properties of fluorinated methacrylic polymers for the protection and conservation of stone

Stella K. Papadopoulou, Aristotle University of Thessaloniki, Thessaloniki, Greece
Chrysa Michailof, Aristotle University of Thessaloniki, Thessaloniki, Greece
Ioannis Karapanagiotis, “Ormylia” Art Diagnosis Centre, Ormylia, Greece
Ioannis Zuburtikudis, TEI of Western Macedonia, Kozani, Greece
Costas Panayiotou, Aristotle University of Thessaloniki, Thessaloniki Greece

Abstract

Physicochemical properties of two fluorinated methacrylic homopolymers (PTFEMA and PHFIMA) were studied by Inverse Gas Chromatography (IGC) and the Contact Angle method. The dispersive component of the polymers' surface energy was calculated by both methods, while the total surface energy by the contact angle method. Moreover the polymer-solvents interactions were investigated by IGC. The results revealed that both polymers are hydrophobic and insoluble in many organic solvents.

Introduction

Perfluoroalkyl methacrylic polymers exhibit low surface energy and good stability towards degradation compared to their unfluorinated analogs. Moreover, the introduction of fluorine atoms into methacrylic polymers' structure results in good adhesion and film-forming properties[1,2]. The above mentioned merits qualify these polymers as promising materials for stone protection. The goal of the present work was to study the effects of the fluorine group contained in the polymer chain on the surface properties of the material. The following homopolymers were included in the study: poly(2,2,2-trifluoroethyl methacrylate) (PTFEMA) and poly(1,1,1,3,3,3 hexafluoroisopropyl methacrylate) (PHFIMA). The polymers were synthesized by free radical polymerization using AIBN as initiator in tetrahydrofuran. Supercritical carbon dioxide was also used as an alternative, “green” polymerization medium for the synthesis of all polymers. Polymer surface properties were studied with Inverse Gas Chromatography (IGC) and with the Contact Angle Method. In most polymers the dispersive component is the main contributor to the surface energy. The dispersive component is a very valuable indication about the polymers' hydrophobicity. It is well known that polymers with low surface energy are hydrophobic.

Theory

Contact angle method

The Young contact angle, θ, between a liquid drop and a solid is provided by the following equation[3]:

$$\gamma_S = \gamma_L \cos\theta + \gamma_{SL} \quad (1)$$

Where γ_S, γ_L and γ_{SL} are the surface energy of the solid, the surface energy of the liquid and the interfacial energy between the solid and liquid, respectively.
Fowkes suggested that the surface energy of a solid or a liquid consists of two components, the dispersive and polar.

$$\gamma^{T} = \gamma^{d} + \gamma^{p} \quad (2)$$

Where γ^T is the total surface energy, γ^d is the dispersive component and γ^p the polar component of surface energy. The polar component of surface energy γ_p can be expressed as:

$$\gamma^{p} = 2(\gamma^{-}\gamma^{+})^{\frac{1}{2}} \quad (3)$$

Where γ^+ and γ^- are representing the electron-acceptor and electron-donor parameters of γ, respectively.

According to Van Oss[3] $\gamma^{SL} = \gamma_S + \gamma_L - 2[(\gamma_L{}^{d}\gamma_S{}^{d})^{\frac{1}{2}} + (\gamma_S{}^{-}\gamma_L{}^{+})^{\frac{1}{2}} + (\gamma_S{}^{+}\gamma_L{}^{-})^{\frac{1}{2}}]$ (4)

Equation (1) in conjunction with equation (4) gives

$$\gamma_L(1+\cos\theta) = 2[(\gamma_S{}^{d}\gamma_L{}^{d})^{\frac{1}{2}} + (\gamma_S{}^{-}\gamma_L{}^{+})^{\frac{1}{2}} + (\gamma_S{}^{+}\gamma_L{}^{-})^{\frac{1}{2}}] \quad (5)$$

The calculation of all components of the surface energy of a polymer ($\gamma_S{}^{d}$, $\gamma_S{}^{+}$, $\gamma_S{}^{-}$) is possible when equation 5 is applied for three liquids with known values of $\gamma_L{}^{d}$, $\gamma_L{}^{+}$, $\gamma_L{}^{-}$. The liquids are usually two polar and an apolar ($\gamma_L = \gamma^{d}_L$). In this study, water, diiodomethane and formamide were employed.

Inverse gas chromatography (IGC)

IGC is a useful and reliable technique to characterize the surface and thermodynamic properties of polymers. The dispersive component of polymers' surface energy and their solubility parameters can be evaluated by the use of IGC.

The term inverse implies that the material under study is the stationary phase and probes of known properties are injected at infinite dilution. When the measurements are carried out at infinite dilution the probe–probe interactions are negligible and stationary phase–probe interactions are the governing ones.

The keynote measurement in IGC experiments is the net retention volume V_N. The retention volume of a probe is the volume of the carrier gas required to sweep out an injected probe from the column. The specific retention volume (V_g) can be used for the calculation of thermodynamic properties[4,5] is: $V_g = \frac{273 V_N}{W_S T} = \frac{273}{W_S T_F} \cdot j \cdot F_M \cdot \left(1 - \frac{p_W}{P'}\right) \cdot (t_R - t_M)$ (6)

Where Ws is the polymer's mass, F_M is the uncorrected flow rate of the carrier gas measured by bubble flow meter, t_R is the retention time of the probe, t_M is the retention time of non adsorbing marker (methane). T and T_F are the column and flow meter temperatures, respectively. p_W is the vapor pressure of water at T_F and P' is the pressure at the flow meter.

The free energy of adsorption is expressed as the standard free energy of transferring 1 mole of molecules from the standard gaseous state to a standard adsorption state and it is related to the retention volume Vn according to: $\Delta G^{ads} = - \cdot R \cdot T \cdot \ln(V_N) + C$ (7)

Where R is the gas constant, T is the column temperature and C is a constant depending on the reference state and on the total surface area of the solid contained in the column. From the slope of the plot of $R \cdot T \cdot \ln V_N$ versus $\alpha\left(\gamma_L^d\right)^{1/2}$ the dispersive component of the surface free energy can be calculated.

Another way to calculate the dispersive component of the surface free energy of a polymer is through the use of Dorris and Gray equation:

$$\gamma_s^d = \left(\frac{1}{4 \cdot \gamma_{CH_2}}\right) \cdot \left(\frac{-\Delta G^{CH_2}}{N \cdot \alpha_{CH_2}}\right) n \quad (8)$$

Where α_{CH_2} is the area occupied by a -CH_2- group (0.06 nm^2), and γ_{CH_2} is the surface tension of a surface consisting of CH_2 groups. $\gamma_{CH_2} = 36.8 - 0.058 \cdot T$ (9)

ΔG^{CH_2} can be calculated by the slope of a straight line obtained from the plot of ΔG^{ads} of n-alkanes versus their number of carbon atoms.

The weight fraction activity coefficient of the solvent at infinite dilution, Ω_1^∞, is given by the following equation;

$$\Omega^\infty = \frac{273{,}15 \cdot R}{V_g P_1^0 M_1 \exp\left(\frac{-P_1^0\left(B_{11} - V_1\right)}{RT}\right)} \quad (10)$$

Where R is the universal gas constant, T is the column temperature, M_1 is the molecular weight of the solute, B_{11} is the second virial coefficient of the solute, V_1 molar volume and P_1 is the vapor pressure of the solute. All constants are known[6].

The weight fraction activity coefficient of the solvent at infinite dilution, Ω_1^∞, gives an idea of the polymer–solvent compatibility indicating that; $\Omega_1^\infty < 5$ for good solvents, $5 < \Omega_1^\infty < 10$ for moderate solvents and $\Omega_1^\infty > 10$ for bad solvents.

Experimental Section

Materials

Solvents: Tetrahydrofuran (THF) (Aldrich) was dried over molecular sieves before polymerization. Carbon dioxide (purity >99.98%) and nitrogen (purity >99.999%) were purchased from Air Liquide Mediterranee. All solvents were of the highest purity available (Aldrich).

Initiator: 2, 2'-azobisisobutyrronitrile (AIBN) was purchased from Acros Organics and was recrystallized twice from ethanol.

Monomers: 2,2,2-trifluoroethyl methacrylate (TFEMA), 1,1,1,3,3,3 hexafluoroisopropyl methacrylate (HFIMA) were purchased from Acros Organics. All monomers were used as received without further purification.

Polymerization using THF as the polymerization medium

The polymerization reactions were carried out in glass test tubes with a narrow neck. The first step was to insert the monomer in the tube and then the polymerization solvent. AIBN was then added to the solution at 0.3 wt % ratio with respect to the monomers mass. The mixture was emerged in an ice-bath and purged with nitrogen gas for 10 min in order to remove oxygen and create inert atmosphere. Afterwards, the glass tube was carefully sealed and placed in a water bath at 65°C. The polymerization time was 24 hours. The resulting polymers were recovered as precipitate after the addition of methanol. The polymer was washed with excess volume of methanol in order to remove any unreacted monomer, filtered and dried under vacuum overnight[7]. The yield was 73% and 57% for PTFEMA and PHFIMA, respectively.

Polymerization using carbon dioxide as the polymerization medium

Polymerizations were conducted in a 15 ml high pressure cell. The mixture (monomer and AIBN 0.3 wt% of monomer) was added in the reactor and purged with nitrogen. Carbon dioxide was charged in the reactor at room temperature. The initial pressure was 73 bar. Then it was placed in a furnace and heated at 65°C. The pressure after heating was 78 bar. 24 hours after the cell left to cool at room temperature and then carbon dioxide was depressurized slowly through a glass trap filled with methanol. At the end of reaction pressure became 91 bar. Pressure variations can be attributed to the changes of the monomer volume while being converted to polymer. Both polymers showed the same trend as far as pressure changes are concerned. The resulting polymer was recovered by the procedure previously described. The apparatus consists of a high pressure cell, an autoclave valve, a pressure gauge, a thermocouple, a temperature indicator, a back-pressure regulator and a glass trap[8,9]. The yield was 85% and 79% for PTFEMA and PHFIMA, respectively.

Polymer characterization

Fourier transform infrared measurements were performed on a Bio-Rad FTS 175 FTIR spectrophotometer. The dominant carbonyl group absorption appears at circa 1760 cm^{-1}. SEC Analysis was carried out using a PLGPC210 + Viscometer system equipped with a refractive index detector, model PL-210, a viscometry detector, Viscotek Model 210R and 2x PLgel 10um mixed B columns. Chloroform was used as the eluent. All polymers had high molecular weight and low polydispersity index. Differential scanning calorimetry measurements were performed using a Shimadzu DSC-50Q analyzer. Glass transition temperatures were 74°C and 69°C for PTFEMA and PHFIMA, respectively. TGA measurements were performed with a Shimadzu TGA-50 thermogravimetric analyzer. The onset of the first decomposition step (10% weight loss) occurs at around 250°C.

Contact angle measurements

Polymers' solutions of 5% wt in THF were spin coated on silicon wafers for 30 seconds at a spin rate of 6000 rpm. Tetrahydrofuran was left to be evaporated overnight. Measurements preformed in a Kruss DSA 100 goniometer. An average volume of 5 μl drop of each test liquid was deposited on the surface under study. Up to ten measurements of each liquid were made for the calculation of the surface energy. The liquids used were water, formamide and diiodomethane.

Inverse gas chromatography (IGC)

Polymer solutions 2%wt were coated on Chromosorb W HP (80/100 mesh) from Supelco. Polymers' loading on Chromosorb was 14.5%. Chromosorb was carefully packed with mechanical vibration in stainless steel columns with O.D 1/8 inch and 65 cm length. The ends of the columns were closed with glass wool. Inverse gas chromatography measurements were conducted using a Shimadzu GC14-A gas chromatograph ecquiped with a flame ionization detector (FID). The carrier gas was helium at a flow rate of 20 ml/min. Flow rates were measured with a calibrated soap bubble flowmeter. The injector and detector temperatures were set at 150 and 200°C, respectively. All columns were conditioned overnight at the working temperature. The temperature under study was 50°C. Methane was used as the non-interacting reference marker to determine the dead volume of the column. Four injections for each probe were made manually with a Hamilton 1μl syringe.

Results and Discussion

Contact angle measurements

The measured contact angles and the calculated components of the surface energy of each polymer are shown in tables 1 and 2, respectively.

Table 1. Contact angles of various liquids on polymer's surface.

Polymer	Contact angles θ (^{0})		
	Water	Diiodomethane	Formamide
PTFEMA	97	77	80
PHFIMA	103	88	95

Table 2. Components of surface energy

Polymer	Components of surface energy (mJ/m^2)				
	γ^d	γ^+	γ^-	γ^P	γ^T
PTFEMA	19,1800	0,4119	2,7471	2,1	21,3
PHFIMA	13,4039	0,0046	4,5008	0,3	13,7

Water contact angles are larger than 90° for both polymers, indicating thus the hydrophobic nature of the tested materials. The latter may be therefore further studied for stone protection. The surface energies of the tested materials at 25°C follow the same trend with their contact angles. PHFIMA exhibits very low dispersive surface energy with respect to PTFEMA. That can be attributed to the fact that PHFIMA has a higher fluorine content than PTFEMA.
The contribution of the polar part to the total surface energy appears to be small in both cases.

Inverse gas chromatography (IGC)

All measurements were conducted at 50°C. The apolar probes used for the determination of the dispersive component of surface energy were n-alkanes. Figures and show the plot of $RTlnV_N$ versus the number of carbon atoms of alkanes, from n-hexane to n-decane. For both polymers the above function was linear and the regression coefficient was 0.999.

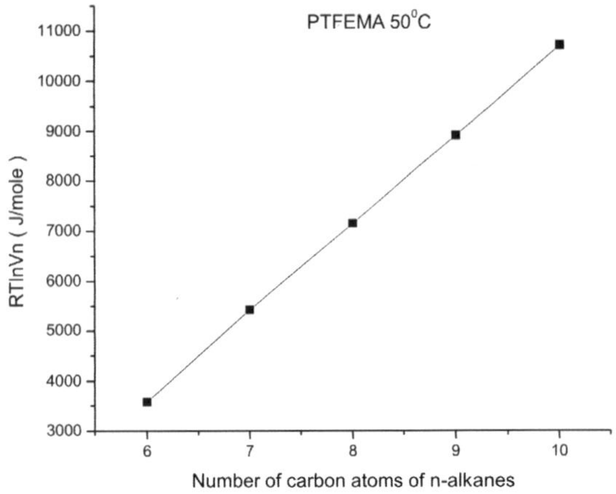

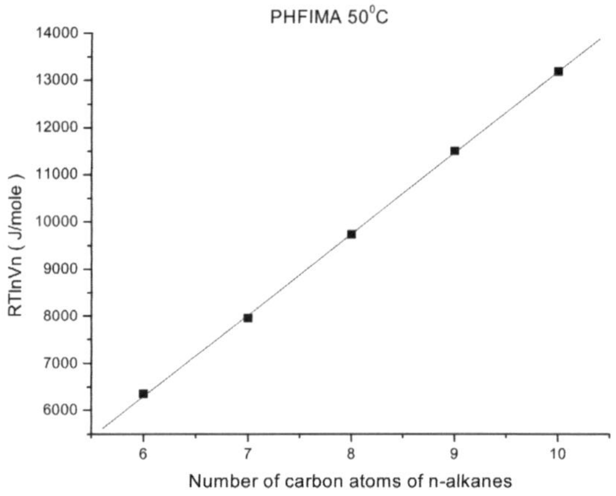

Figure 1. Adsorption free energy versus the number of carbon atoms of n-alkanes

Table 3. Dispersive component of surface energy

Polymer	γ^d (mJ/m^2)
PTFEMA	17,78
PHFIMA	16,74

Comparing the values of the dispersive component of surface energy γ^d obtained by the contact angle method (Table 5) with that obtained by IGC (Table 6), we see a difference between results of the two methods. The value of γ^d of PTFEMA seems to decrease from 19,18 mJ/m^2 at 25°C (contact angle method) to 17,78 mJ/m^2 at 50°C (IGC). That seems to be normal according to the fact that the surface free energy decreases linearly with an increase of temperature. In the case of PHFIMA the value of γ^d decreases with the increase of temperature. The results derived from the two methods are not uniform due to the fact that IGC measures the surface energy by means of zero coverage adsorption, whereas a multilayer adsorption is utilized in contact angle measurement[10]. The weight fraction activity coefficient of a solvent at infinite dilution, Ω_1^∞ was calculated for both polymers at 50°C. The solvents under study, apart from n-alkanes, were: Acetonitrile, ethanol, 1- propanol, n-butanol, terrahydrofuran, nitropropane, chloroform, 2-butanone and pyridine.

Solvents	V_g (cm^3/g)		Ω_1^∞	
	PTFEMA	PHFIMA	PTFEMA	PHFIMA
n-Hexane	13,63	37,85	36,86	13,27
n-Heptane	27,07	68,69	45,27	17,84
n-Octane	51,49	133,07	58,13	22,49
n-Nonane	99,37	257,16	73,95	28,57
n-Decane	194,11	477,99	94,32	38,30
Acetonitrile	50,67	36,70	33,61	46,40
Chloroform	37,45	30,33	7,51	9,28
2-Butanone	67,45	71,01	13,31	12,64
THF	49,94	94,67	11,01	5,81
Ethanol	19,75	18,90	85,94	89,80
2-Propanol	37,08	40,55	43,15	39,46
n-Butanol	82,04	77,18	81,29	86,41
Pyridine	230,14	206,59	13,11	14,61
Nitropropane	261,51	181,30	19,46	21,35

Table 4 Retention volumes and activity coefficients of various solvents.

The weight fraction activity coefficient of the examined range of solvents for PTFEMA and PHFIMA reveals the fact that they are insoluble in many organic solvents[10]. The results in table 4 s how that only chloroform is a good solvent for both polymers[7]. Moreover tetrahydrofuran[1,8] is nearly moderate solvent for PTFEMA and good solvent for PTFEMA. The second solvent that could be considered as a moderate solvent for PTFEMA and PHFIMA, is 2-butanone[11].

Conclusions

The fluorinated methacrylic polymers PTFEMA and PHFIMA were synthesized by two free radical polymerization processes. Apart from using an organic solvent as the polymerization medium, carbon dioxide substituted THF successfully.

The surface properties of two polymers with different fluorine content were investigated by means of the contact angle method and inverse gas chromatography. The study indicates that both polymers are hydrophobic but the polymer with the highest fluorine content, PHFIMA, is more hydrophobic than PTFEMA. The results of the dispersive component of surface energy calculated with both methods are inconsistent due to the fact that IGC mainly evaluates high-energy sites, whereas contact angle measurement represents mainly low-energy regions of a surface. In terms of polymer-solvent compatibility, IGC provided valuable information. The solubility of the polymers in various solvents was established experimentally for the first time through the use of IGC.

Acknowledgements

This work was supported by the Greek Ministry of Development – GSRT (25%) and the European Union – European Social Fund (75%) through the research grant PENED 2003 with MIS 03EΔ 91

References

1. Signori, F., Lazzari, M., Castelvetro, V., and Chiantore, O. (2006), "Copolymers of isopropenyl alkyl ethers with fluorinated acrylates and fluoroacrylates: Influence of fluorine on their thermal, photochemical, and hydrolytic stability." *Macromolecules* 39, 1749-1758.
2. Chiantore, O., Lazzari, M., Aglietto, M., Castelvetro, V., and Ciardelli, F. (2000), "Photochemical stability of partially fluorinated acrylic protective coatings I. Poly(2,2,2-trifluoroethyl methacrylate) and poly(1H,1H,2H,2H-perfluorodecyl methacrylate-co-2-ethylhexyl methacrylate)s." *Polymer Degradation and Stability* 67, 461-467.
3. Vanoss, C. J., Good, R. J., and Chaudhury, M. K. (1988), "Additive and Nonadditive Surface-Tension Components and the Interpretation of Contact Angles." *Langmuir* 4, 884-891.
4. Conder, J. R.;Young, C. L. (1980), "Physicochemical measurement by gas chromatography." Wiley (Chichester).
5. Panzer, U., and Schreiber, H. P. (1992): "On the Evaluation of Surface Interactions by Inverse Gas-Chromatography." *Macromolecules* 25, 3633-3637.
6. Daubert, T. E. Danner., R. P. "Data Compilation Tables of Properties of Pure Compounds; AIChE Symp Ser. No. 203" *American Institute of Chemical Engineers*: New York 1985.
7. Liu, W. H., Tang, K., Guo, Y. Z., Koike, Y., and Okamoto, Y. (2003): "Tacticity control in the radical polymerization of 2,2,2-trifluoroethyl methacrylate with fluoroalcohol. " *Journal of Fluorine Chemistry* 123, 147-151.
8. Kwon, S., Bae, W., Lee, K., Byun, H. S., and Kim, H. (2007): "High pressure phase behavior of carbon dioxide plus 2,2,2-trifluoroethyl methacrylate and plus poly(2,2,2-trifluoroethyl methacrylate) systems. "*Journal of Chemical and Engineering Data* 52, 89-92.
9. Kwon, S., Bae, W., and Kim, H. (2004), "The effect of CO2 in free-radical polymerization of 2,2,2-trifluoroethyl methacrylate. " *Korean Journal of Chemical Engineering* 21, 910-914.
10. Ticehurst, M. D., Rowe, R. C., and York, P. (1994): "Determination of the Surface-Properties of 2 Batches of Salbutamol Sulfate by Inverse Gas-Chromatography. " *International Journal of Pharmaceutics* 111, 241-249.
11. Castelvetro, V., di Mirabello, L. M., Aglietto, M., and Passaglia, E. (2001), "Homo- and copolymers of hexafluoroisopropyl methacrylate and alpha-fluoroacrylate with alkyl vinyl ethers: Microstructure and thermal properties." *Journal of Polymer Science Part a-Polymer Chemistry* 39, 32-45
12. Huang, H. L., Goh, S. H., and Wee, A. T. S. (2002): Miscibility and interactions in poly (2,2,3,3,3-pentafluoropropyl methacrylate-co-4-vinylpyridine)lpoly(p-vinylphenol) blends. " *Polymer* 43, 2861-2867.

722ae Evolution of Interfacial Water Layers on a Highly Hydroxylated Silica Film: Implications for Cellular Interactions

A. Anderson and W. Robert Ashurst. Chemical Engineering, Auburn University, 212 Ross Hall, Auburn, AL 36849

Attenuated total reflection Fourier transform infrared spectroscopy (ATR-FTIR) is a powerful surface analysis tool. As an example, valuable details on the structure of water layers that form on oxidized silicon surfaces in humid environments have been extracted from this technique. Other work suggests that the structure of this interfacial water layer depends largely on the abundance of Lewis sites on the surface that can compete with interfacial water for hydrogen bonding partners. In this work, ATR-FTIR spectroscopy is employed to investigate the water structure that exists on a highly hydroxylated silica surface that is formed from the vapor-phase hydrolysis of silicon tetrachloride. Comparisons to native silicon oxide surfaces and water plasma treated silicon surfaces will be made, and implications for biomaterials science are discussed.

722af Maximizing or Eliminating the Effects of Nanoscale Confinement on the Glass Transition In Freely Standing, Supported, and Multilayer Polymer Films

Soyoung Kim[1], Connie B. Roth[1], and John M. Torkelson[2]. (1) Chemical and Biological Engineering, Northwestern University, 2145 Sheridan Rd., Evanston, IL 60208-3120, (2) Chemical and Biological Engineering and Materials Science and Engineering, Northwestern University, 2145 Sheridan Rd, TECH E136, Evanston, IL 60208

Using fluorescence spectroscopy and ellipsometry, we are able to characterize the effects of nanoscale confinement on the glass transition temperature (Tg) in supported single-layer and multilayer films and in freely standing polymer films. We have recently demonstrated several ways in which we can maximize or minimize the perturbations from bulk Tg response caused by confinement and in which we can address fundamental issues regarding the fundamental reasons behind the effects of nanoconfinement on polymer properties.

First, we have shown that we can eliminate the effect of nanoconfinement on polymer Tg via the addition of several weight percent of low molecular weight anti-plasticizer molecules to the polymer. In the case of poly(vinyl acetate) (PVAc), water, which is naturally sorbed at levels of several weight percent in PVAc, is an anti-plasticizer. Bone-dry PVAc films exhibit a significant Tg reduction with decreasing film thickness. However, PVAc films containing several weight percent water exhibit a film-thickness-independent Tg down to thicknesses of ~10 nm. Also, we determined via ellipsometry that water molecules increase the density of PVAc films, and this effect indicates that water acts as an anti-plasticizer in PVAc. These results are consistent with a recent theoretical analysis by Riggelman et al. (PRL 97, 045502 (2006)) and related experimental studies done previously in our group (Ellison et al., PRL 02, 095702 (2004) and Mundra et al., Eur Phys J Spec Topics 141, 143 (2007)) indicating that the presence of anti-plasticizers leads to a suppression of confinement effects on the polymer dynamics. Our results also indicate that researchers must be aware of these effects when developing polymeric systems used under conditions of nanoconfinement (e.g., as ultrathin films or in nanocomposites) that may sorb significant amount of water from the atmosphere.

Second, we have developed a new fluorescence method to measure the effect of confinement on the properties of freely standing polymer films, i.e., films that are not supported by a substrate. Related studies have been reported by only a small number of other research groups around the world because of experimental difficulties associated with freely standing films, especially when the films are less than 100 nm thick. In our studies, we measure Tg via the temperature dependence of the shape of the fluorescence spectrum of a dye covalently attached to the polymer at trace levels. We have found that a 40-nm-thick polystyrene (PS) freely standing film exhibits a Tg value that 36 K below that of the bulk Tg. We are exploiting our novel characterization method to obtain a better fundamental understanding of the large effect of confinement on Tg in freely standing films and in particular to test the theory proposed in 2000 by de Gennes (Eur Phys J E, 2, 201 (2000)).

Finally, we have discovered that the Tg of a 14-nm-thick PS film at the free surface of a bilayer film can be tuned over the range of 318 K to 418 K by simply varying the polymer species of an underlayer film supporting the ultrathin PS layer. This shows that the cooperative segmental dynamics of two immiscible polymers are strongly coupled over length scales of several tens to hundreds of nanometers and that sufficiently thick underlayers of selected polymers can essentially cause the Tg dynamics of the ultrathin PS layer to be "slaved" to the dynamics of the underlayer. This discovery suggests that nanostructured multilayer films may yield properties that cannot be achieved by conventional polymer blends, providing a new route to the design of novel materials.

Surface Polymerization of Iron Particles for Magnetorheological Elastomers (MREs) and Their Potential Application as Sensors

A. Fuchs[1], J. Sutrisno[1], F. Gordaninejad[2], M. B. Caglar[2], W. Xiaojie[2], and L. Yanming[2]

1) *Chemical Engineering Department, University of Nevada, Reno – Nevada, USA*
2) *Mechanical Engineering Department, University of Nevada, Reno – Nevada, USA*

Magnetorheological Elastomers (MREs) are intelligent materials which are fabricated from magnetic particles and a crosslinked polymer matrix. MRE have viscoelastic, magnetorestrictive, and magnetoresistive properties which can be used for vibration damper and sensing devices. This is possible because of the alignment of the magnetic particles in the polymer matrix. Chainlike magnetic particles are formed by the induced magnetic field. These chains are locked in place during chemical cross linking of the elastomer. In addition, changes in the magnetostrictive and magnetoresistive properties are controlled by varying the magnetic field. Chain-like magnetic particles provide rapid time response of milliseconds.

The magnetostrictive effect of MRE is caused by changing the distance between magnetic particles in the elastomer matrix under the influence of an external magnetic field. The magnetostrictive effect includes stiffness change and dimensional change in MRE. Under the influence of a magnetic field the MRE becomes stiffer due to decreased distance between the magnetic particles. This phenomenon is suitable for the vibration damper application. The MRE magnetostrictive effect is the most important feature for the vibration control application. Another phenomenon, the magnetoresistive effect, is important for sensor applications. The magnetoresistive effect is due to the change in electric impedance of the material. The impedance consists of a real and imaginary part. The real impedance represents the material resistance and the imaginary part represents the capacitance. The applied external magnetic field onto MRE only causes the magnetostrictive effect. The addition of an external force onto magnetized MRE will cause the deformation of the chain-like structure of magnetic particles. The electric properties of MRE will be changed by deformation of the chain-like structure.

Oxidation takes place when the MRE is in contact with air and a magnetic particle. This effect can be reduced by surface treatment of the carbonyl iron surface. “Bridging” between iron particles due to surface coating is an additional concern. Two approaches to deal with this “bridging” issue include: 1) using a hard polymer coating in which the particles do not agglomerate or 2) using a soft polymer coating with diluents in which the coated particles are maintained in liquid until they are added to the elastomer. The polymer matrix provides thermal oxidative and chemical resistance. Silicone and polyurethane are good candidates as polymer matrices which have good oxidation resistance and mechanical properties.

Many of living polymerizations have been studied including ionic polymerization, emulsion polymerization, nitroxide radical polymerization (NMP), ring opening polymerization (ROP), atom transfer radical polymerization (ATRP), and reversible addition fragmentation chain transfer (RAFT). The application of ATRP for surface polymerization of iron particles has been investigated by Fuchs et. al. The iron particles has been coated using poly(butyl acrylate). Surface modified iron particles were used in magnetorheological fluid (MRF) application. As a result, MRF with surface coated particles has shown better performance in compared with regular MRF. The grafting technique of thermo responsive poly(*N*-isopropylacrylamide) poly(NIPAAm) onto silica nanoparticles using ATRP has been investigated by Wu et. al.

In this present work, MREs were synthesized from polyurethane (PU) elastomer and silicone RTV (Room Vulcanizing Temperature) eslatomer with iron particles as magnetic particles. The iron particles concentration was varied from 10 wt.%, 30 wt.%, 50 wt.%, and 70 wt.%. The ATRP technique was used for surface polymerization of iron particles with fluorinated styrene as monomers. As a comparison, the radical polymerization was also used for surface polymerization of iron particles. The mechanical properties and electrical properties of MREs, thermal properties of surface coating polymer, and the alignment of iron particles within MRE have been characterized.

The mechanical properties and electrical properties of MREs have been characterized using force – displacement test device (instron) and electrochemical impedance spectroscopy (EIS) respectively, thermal properties of surface coating polymer were characterized using differential scanning calorimetry (DSC), and the alignment of iron particles within MRE have been characterized using scanning electron microscopy (SEM). MREs (70 wt.%) with surface coated iron particles have higher oxidation stability based on the force – displacement test results. The EIS and force – displacement test results have showed that the impedance and real resistance value at low frequency (1 Hz) of MRE was inversely proportional to the displacement or strain, and supplied current (magnetic fields). As results, MREs have potential to be used as a smart material for vibration isolator, force – displacement sensor and magnetic field sensor, and combination of damping and sensor material.

722ah Synthesis of Polypyrrole/azocalix[4]Arene Salts: Electrical Properties and Thermal Stability

Ayse Gul Yavuz[1], Aysegul Uygun[1], and Songul Sen[2]. (1) Chemistry, Suleyman Demirel University, Isparta, Turkey, (2) Chemistry, Mehmet Akif Ersoy University, Burdur, Turkey

The inclusion of cage molecules into electroconducting polymers, ECPs, has opened new perspectives of application for the construction of ionic sensors. It is noted that sulphonated calixarenes can be immobilized in conducting polymers like PPy as dopants during electropolymerization. The grafting of calixarene to polythiophene has already been reported. Host-guest chemistry has also been used to facilitate the preparation of conducting polymers in aqueous solution using nonpolar monomers.

In present work we reported the synthesis of polypyrrole in the presence of water-soluble azocalix[4] arene which obtained from diazo coupling reaction between calix[4]arene and diazonium salts. Azo groups bring to calixarenes a chromogenic activity.The (5,11,17,23-tetrakis[(p-carboxyphenyl)azo] 25,26,27,28 tetra hydroxy calix[4]arene) was used as dopant anion. The effect of the content of azocalix [4]arene onto properties of polypyrrole was investigated. Polypyrrole/azocalix[4]arene salts were characterized by FTIR, TGA, SEM and conductivity measurements. The conductivity of polypyrrole increased in the presence of azocalix[4]arene. TGA results indicated that the polypyrrole/azocalix[4] arene salts have higher thermal stability than PPy. It was observed from SEM analysis that the particle diameter of PPy decreased with increasing content of azocalix[4]arene.

722ai Imprint Lithography and Characterization of Photosensitive Polymers for Advanced Microelectronics Packaging

Sue Ann Bidstrup Allen, Chemical Engineering, Georgia Tech, Atlanta, GA 30332-0100, Venmathy Rajarathinam, Chemical and Biomolecular Engineering, Georgia Tech, Atlanta, GA 30332-0100, and Paul Kohl, Chemical and Biomolecular Engineering, Georgia Institute of Technology, 311 Ferst Dr., Atlanta, GA 30332.

Imprint lithography can potentially produce complex, high aspect ratio and three-dimensional patterns at low cost through mold replication. The process requires a durable stamp with sufficient spatial resolution, and a polymeric material to make a conformal replica of stamp topography. In this project, imprint technology is applied to three microelectronics packaging applications: high aspect ratio chip to substrate interconnects, ultra-low loss coaxial interconnects on organic substrates, and high density lines and vias. Polycarbonates and polynorbornenes have been selected as the imprint materials. Relationships between imprint material properties and imprint fidelity are assessed with particular attention to thick film photo-imprint applications. Intermediate soluble layers are investigated to facilitate separation between high pattern density, high aspect ratio stamps and imprinted materials. The development of imprint lithography processes for these packaging applications have the potential to advanced integrated circuit performance capabilities with high throughput and yield while maintaining low cost.

Effect of Asphaltenes on Water Wetting & Internal Corrosion in Oil-Water Pipe Flow

Pankaj Ajmera
Institute for Corrosion and Multiphase Technology, Ohio University

Introduction

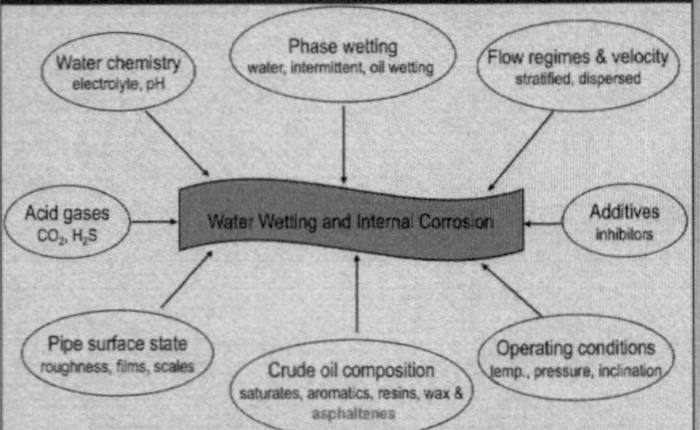

Literature review - asphaltene

Asphaltenes are nonvolatile constituents which are found in crude oil, along with resins, aromatics and saturated hydrocarbons.

Characteristics of asphaltenes:

- Dark brown to black friable solids that have no definite melting point and usually foam and swell on heating to leave a carbonaceous residue. [1]
- Defined as the portion of crude oil insoluble in *n*-alkanes such as *n*-pentane or *n*-heptane yet soluble in aromatics such as benzene or toluene. [2]
- The highest molecular weight and most polar constituents in crude oil. The molecular weight is in the range of 400 to 1500 amu with a maximum around 750 amu. [1]
- Mainly composed of polyaromatic units with oxygen, nitrogen, and sulfur, (NSO-compounds) combined with minor amounts of heavy metals, particularly vanadium (V) and nickel (Ni) which occur in porphyrin structures. [3]

Figure 1: Asphaltene Structure (MW 708 amu) [4]

- Since asphaltenes are a portion of crude oil, their identity changes as the source of crude oil changes.
- So complex that there is no model compound available which resemble asphaltenes, to carry out tests asphaltenes needs to be extracted from crude.

References:

[1] J.G. Speight, "Petroleum asphaltene part 1: asphaltene, resins and the structure of petroleum", Oil & Gas Science and Technology-Rev. IFP, Vol. 59, No. 5, pp 467-477, 2004.

[2] P.K. Kilpatrick et al., "Aggregation and solubility behavior of asphaltenes and their subfractions", Journal of Colloid and Interface Science, Vol. 267, pp 178-190, 2003.

[3] J.R. Becker, "Crude oil waxes, emulsions and asphaltenes", Tulsa: Penn Well Books,1997.

[4] Oliver C. Mullins, "Asphaltenes and polycyclic aromatic hydrocarbons", Schlumberger-doll research presented at SSRL 30th conference.

Research objectives

The following objectives are important to study effects of asphaltenes on water wetting and internal corrosion in oil-water pipe flow.

- Study the corrosion inhibitive benefits of asphaltenes in oil-water pipe flow.
- Study the effect of asphaltenes on the flow pattern of two phase oil-water flow.
- Study the impact of asphaltenes on pipe wall wetting.

Asphaltene precipitation procedure

- Arab heavy crude oil stored under argon blanket
- Mix well to ensure homogeneity
- Diluted with n-heptane
- Remove precipitated asphaltene by vacuum filtration
- n-heptane wash to remove maltenes
- Redissolve asphaltene in methylene chloride to remove inorganic compounds
- Rotary drying and Argon stream drying
- Oven drying to get pure asphaltene

Figure 2: Asphaltene Crystals

Experimental set-up and test matrix

Glass cell (corrosion inhibition tests)

Table 1: Glass cell test matrix

Asphaltene source	Arab heavy crude oil
Oil phase	Toluene* (solvent)
Asphaltene concentration	0, 0.1, 1 and 5 wt% in toluene
Material	Carbon steel (C-1018)
Water phase	1.0 wt% NaCl
Partial pressure of CO_2	1 bar
System pressure	1 bar
pH	5.0
Temperature	25°C
Rotation speed	1000 rpm

*Toluene is neutral solvent and representative of crude oil.

Figure 3: Schematic of glass cell

1. Reference electrode 2. Gas outlet 3. Temperature probe 4. Pt counter electrode 5. Rotator 6. Gas inlet 7. pH-electrode 8. Luggin capillary 9. Working electrode 10. Hot Plate

Experimental procedure

1. Glass cell (corrosion inhibition tests)

Step 1: Partitioning

This is the initial condition. Coupon is in water phase.

Solvent + Asphaltene added at the top. Keeping coupon in water phase, corrosion rate measurement done every 20 minutes.

Figure 4: Partitioning step

Step 2: Corrosion inhibition

Initial condition coupon wetted by water — Coupon wetted by asphaltene for 15 min. — Coupon put back in water phase measure corrosion rate

Figure 5: Corrosion inhibition step

Step 3: Persistency

After repeating corrosion inhibition step several time, coupon (working electrode) is permanently placed back in the water phase and corrosion rate measurements are done.

2. Contact angle & Interfacial tension tests (surface active properties)

Table 2: Surface active properties test matrix

Oil Phase	Toluene + Asphaltene
Asphaltene concentration	0, 0.1, 1, 5% wt in toluene
Water Phase	1.0 wt% NaCl, pH 5.0, CO_2 saturated, 25°C
Material (wettability test)	C-1018

Effect on Wettability : Steel-oil contact angle tests
Effect on flow pattern : Oil-water interfacial tension tests

Wettability test design

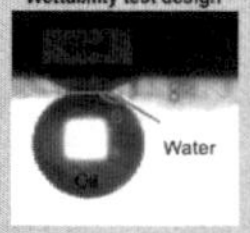

Figure 6: Oil droplet in water phase

Test results

Oil-water interfacial tension tests

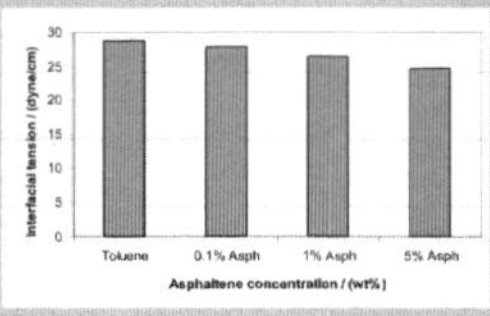

Oil-water interfacial tension is an important parameter to understand the effect of addition of asphaltene on oil water flow pattern.

After addition of asphaltene, oil water interfacial tension does not decrease by significant amount. Hence addition of asphaltene does not affect oil-water flow pattern.

Test results

Glass cell (corrosion inhibition tests)

Step 1: Partitioning

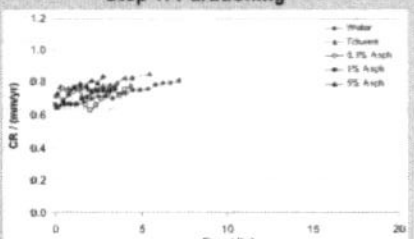

Corrosion rates for asphaltene test matrix is nearly same as pure water inhibition case. Hence it can be concluded that asphaltene does not partition into water phase.

Step 2: Corrosion inhibition

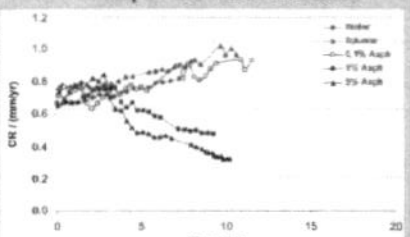

Pure toluene and toluene + 0.1% asphaltene solution has no effect on corrosion rate. Higher concentrations of asphaltene(1% & 5%) causes significant reduction in corrosion rate.

Step 3: Persistency

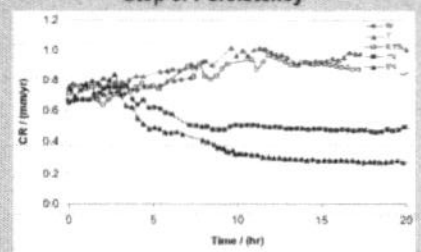

1% and 5% asphaltene has reduction in corrosion rate which means a protective layer of asohaltene formed on coupon. From persistency test it was observed that corrosion rate stabilizes hence the formed protective layer is non reversible and permanent.

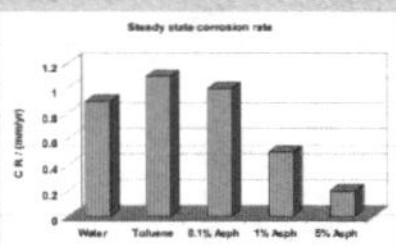

For 1% asphaltene the steady state corrosion rate falls by 50% and by 75% for 5% asphaltene, as compared to pure water corrosion rate.

Wettability tests (Steel-oil contact angle measurement)

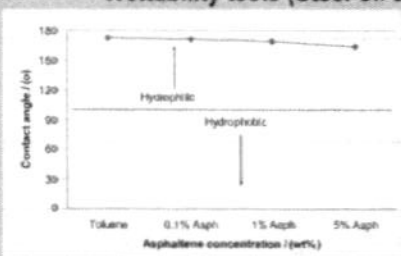

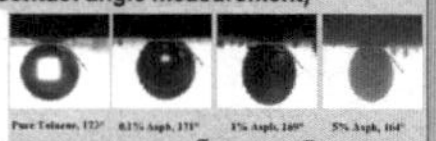

Addition of asphaltene to oil do not affect the wettability of steel surface, and it continues to be hydrophilic

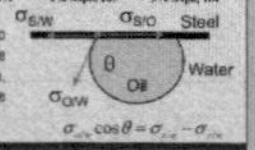

Conclusions

- Asphaltene causes significant reduction in corrosion rate.
- As concentration of asphaltene increases corrosion rate decreases.
- Asphaltene forms strong protective layer on steel surface which acts as a barrier for corrosive agent and hence reduces the corrosion rate.
- Asphaltene (dissolved) does not affect oil water flow pattern in pipe flow.
- Asphaltene does not affect the wettability of the steel surface, it continued to be hydrophilic.

Sponsors: ConocoPhillips

Acknowledgements: Advisor: Professor Srdjan Nesic, Project leader: Dr. Sonja Richter

POLYURETHANES BASED ON POLYOLS FROM CASTOR OIL, STARCH GRANULES AND STARCH-DERIVED GLYCOL AND GLYCEROL GLYCOSIDES: MORPHOLOGY, SYNTHESIS, CHEMICAL, MECHANICAL, AND THERMAL PROPERTIES

Alvaro Ramírez[a*], Manuel F. Valero[a], Jorge E. Pulido[a], Zhengdong Cheng[b]

[a]*Polymers Research Group, School of Chemical Engineering, Universidad Industrial de Santander A.A. 678, Bucaramanga, Colombia.*

[b]*Chemical Engineering Department, Texas A&M University, College Station, 77840, TX, USA.*

*Corresponding author. Telephone: 57-7-6344000 ext. 2434, Fax: 57-7-6344684. E-mail: argarcia@uis.edu.co. Address: Cra 27 calle 9 Campus Universitario, Escuela de Ingeniería Química, U.I.S, Bucaramanga, Colombia.

Abstract

A series of polyols were prepared from castor oil (CO) and cassava starch (S) by two routes in order to obtain products with high hydroxyl content: a) Incorporation of starch granules into the original castor oil and pentaerythritol- modified castor oil and b) Modification of starch by glycosylation and transesterification with castor oil and castor oil modified. The resultant polyols were characterized by hydroxyl value, *Fourier* transform infrared spectroscopy (FTIR) and Matrix Assisted Laser Desorption/Ionization Time of Flight Mass Spectroscopy (MALDI TOF MS). MALDI TOF MS was used to detect and characterize the components of castor oil and to determine the products of the transesterification reaction. Several polyurethanes (PU) were synthesized by reacting the different polyols derived from castor oil and cassava starch with 4,4-diphenyl methane diisocyanate (MDI) and isophorone diisocyanate (IPDI). The crosslink density of polyurethanes was determined by swelling tests using the Flory-Rehner's equation. The variation of the crosslink density of polyurethanes was studied based on the hydroxyl value of the polyol utilized in the synthesis. The effect of polyol modification on the mechanical, chemical and thermal properties of the resulting polyurethanes was evaluated as a function of the percent of modifier agent (pentaerythritol, starch or starch-derived glycosides). The properties of the elastomeric polyurethanes were studied related to their chemical structure and morphology.

Introduction

Castor oil is a renewable recourse and a natural polyol and has been a successful candidate in the formation of industrially useful materials like cost-effective polyurethane elastomers, it is yet to be exploited for the modification of polymers [1, 2, 3]. The polyurethanes obtained from castor oil are typically water resistant and flexible due the presence of a long fatty acid chain and lend themselves to thermosetting-type structures due to their tri-functional nature. The long pendant chains of the fatty acids impart flexibility and hydrolysis resistance to the network and the double bond present in the castor oil serves as a grafting center. The properties of material obtained from castor oil include low hydroxyl number leading to inherently low modulus materials, a sluggish rate of curing of secondary hydroxyl groups and structural irregularity due to steric hindrances caused by the long pendant fatty acid chains during polyurethane formation, resulting in low tear strength [4, 5, 13].

To modify these properties, oil-based polyurethanes are mixed in the form of interpenetrating networks with others monomers like styrene, methacrylate, formaldehyde, etc. They are also transesterified or alcoholized with alcohols thereby introducing reactive hydroxyl groups. Most of the literature published has illustrated the use of glycerol for transesterification [4, 12]. However, glycerol (a component of the triglyceride molecule) is readily destructible at high temperatures. This disadvantageous property stems from the presence of hydrogen atoms in position β relative to the hydroxyl group in the glycerol molecule. This structural feature is conductive to the partial defragmentation of the molecule and the formation of unsaturated compounds. The problem can be solved by replacing glycerol with another polyhydric alcohol that does not contain β-hydrogen atoms, i.e pentaerythritol.

Other route evaluated to increase the hydroxyl functionality of the castor oil was glycosylation of starch, followed by transesterification with the castor oil [6, 7]). Starch was reacted with ethylene glycol to form glycol glycoside. Polyols were synthesized by reacting castor oil with glycol glycoside.

The goal of this work was to study the effect of polyol modification on the mechanical, chemical and thermal properties of the resulting polyurethanes was evaluated as a function of the percent of modifier agent (pentaerythritol, starch or starch-derived glycosides). The properties of the elastomeric polyurethanes were studied related to their chemical structure and morphology. There were also evaluated the effects due to the structure of the soft segments, consisting on the polyol, and the hard segments, consisting on the polyfunctional isocyanate and crosslinker. The morphology of polyurethane was investigated as related to compatibility and microphase segregation between the two segments. The change in the crosslink density and physical interactions in the polyurethane were studied in function of the chemical modification of castor oil and starch.

Experimental

Materials

Castor oil and yucca starch were obtained from the local market. Pentaerythritol (Fisher, USA) and catalyst litharge (Merck, USA) were used for transesterification. Glycerol and ethylene glycol along with sulfuric acid were reactants for glycosylation of the starch. Isophorone diisocyanate (IPDI) (Aldrich, USA) was employed to obtain polyurethanes.

Transesterificacion castor oil using Pentaerythritol

The castor oil transesterification was carried out in a flask with four openings equipped with agitator, reflux condenser, nitrogen gas entry and a thermo regulated warming blanket. 400 g of castor oil and the transesterification modifier were charged in a previously established proportion, along with a 0.05% of PbO catalyst, based on the oil weight. The reaction took place at 200°C for 2.5 hours

Preparation of Glycosides

Starch was reacted with ethylene glycol and catalyst sulfuric acid (0.5%), in a 1 L reactor connected to a constant temperature system, under nitrogen atmosphere. The glycosylation reaction was carried out for 2 h at 1108C. Glycerol was also used in place of glycol for some experiments that look for variation in the glycoside structure. After reaction, the residual sulfuric acid was precipitated with barium hydroxide. The residual ethylene glycol or glycerol after reaction was also separated by vacuum distillation [10, 11].

Preparation of Polyols and Polyol-glycosides

Polyols were synthesized from castor oil by transesterification reactions with pentaerythritol at different proportions, under nitrogen atmosphere with catalyst litharge (0.05%) [12]. The time of reaction was 2.5 h at 2008C in a 1 L reactor equipped with agitator, reflux condenser, and a thermo regulated warming blanket. Two types of these polyols were obtained: P0 and P1 with increasing hydroxyl values of 186 and 233mg KOH/g, respectively. Original castor oil had hydroxyl value of 160mg KOH/g. Either glycol glycosides or glycerol glycosides were mixed with the castor oil polyols in amounts of 5 and 10% along with catalyst litharge (0.05%) and reacted at 2108C for 2 h under nitrogen atmosphere, in a reactor of similar characteristics as the one used for transesterification. Thus, two kinds of products were obtained: polyol-ethylene glycol glycosides (PEG) and polyol-glycerol glycosides (PGG). As a point of reference, original castor oil was also reacted with glycol glycoside and glycerol glycoside to obtain, respectively, a castor oil-glycol glycoside (CO-EG) and a castor oil-glycerol glycoside (CO-GG).

Characterization

Castor oil polyols, glycosides, and polyol-glycosides were characterized by measuring the hydroxyl value (ASTM D1957-86), specific gravity (ASTM D792-86), viscosity on a Brookfield DV-III rheometer at 200 rpm and 268C and Fourier Transform Infrared spectroscopy (FTIR) on a Nicolet Avatar 360 spectrometer. FTIR spectra for polyurethanes were also obtained. Glycosides and polyol-glycosides were analyzed using MALDI TOF mass spectroscopy; iodometric tests were also used to check the presence of any traces of starch not converted to glycoside.

Chemical and solvent resistances were measured according to ASTM D543-67 method. Each polymer specimen of known dimensions was immersed in solvents with increasing order of solubility parameter δ from 8.2 to 14.5 $(cal/cm^3)^{1/2}$. This procedure was continued till equilibrium swelling was achieved. The swelling behavior of the polyurethanes in various solvents was investigated and the solubility parameter was determined based on the average structure of each polyurethane repeating unit, as determined by the method of matrix-assisted laser desorption/ionization time-of-flight mass spectroscopy (MALDI TOF MS). The crosslink density of polyurethanes was determined by swelling tests using the Flory-Rehner's equation. The variation of the crosslink density of polyurethanes was studied based on the hydroxyl value of the polyol utilized in the synthesis. The stress-strain properties were determined using an Instron universal testing machine according to ASTM D-527 procedure and hardness measurements were made on a Shore-A hardness test apparatus using ASTM-785 method. Dynamic mechanical thermal analyses (DMTA) of the selected polyurethanes were carried out on a Rheometrics DMTA V. The storage modulus (E'), the loss modulus (E'') and loss factor (tanδ) were measured. The temperature corresponding of the maximum of the tanδ peak position was taken as glass transition temperature (Tg). Scanning electron micrographs were obtained on a JEOL JSM-6400 electron microscope. The samples were previously coated with a gold layer and the morphology of the materials was studied through their fractures after immersion into liquid nitrogen.

Results and discussion

Figure 1 shows the relation between the hydroxyl value of the polyol and the concentration of modifier agent for the polyols obtained from castor oil since a. starch incorporating without modification, b. oil transesterification with ethylene glycol and c. oil transesterification with glycerol glycoside. It can be seen that the hydroxyl value of polyols obtained from castor oil modified with glycosides is higher than the polyols obtained from incorporating starch without prior modification.

Figure 1. Relation between the hydroxyl value of the polyol and the concentration of modified agent.

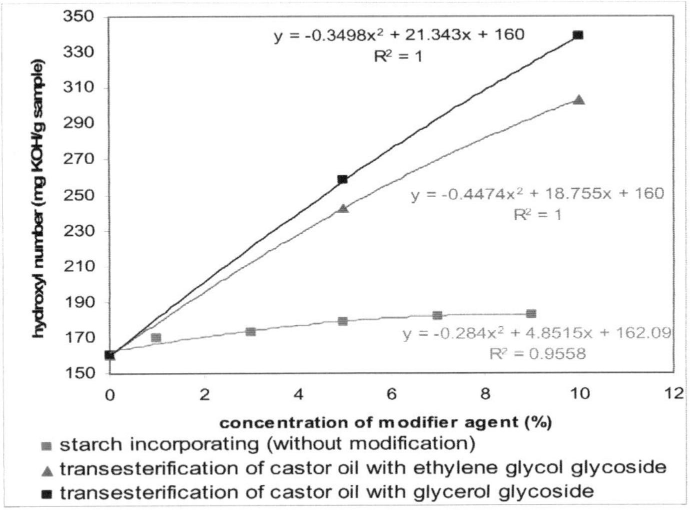

Figure 2 shows the relation between crosslinking density and the concentration of modifier agent. It is observed that crosslinking density of polyurethanes obtained from starch incorporating is almost constant. Without prior modification, starch cannot be reacted with the diisocyanate to obtain polyurethane. In contrast, polyurethanes obtained from transesterification of castor oil with glycosides exhibited an increment of crosslinking density in function of the modifier agent due to the hydroxyl groups of the glycosides participating in the formation of urethane bonding. The data suggests that The modification of starch also affects the physical crosslinking by hydrogen bonding between polar groups.

Figure 2. Relation between crosslinking density of the polyurethane and the concentration of the modifier agent.

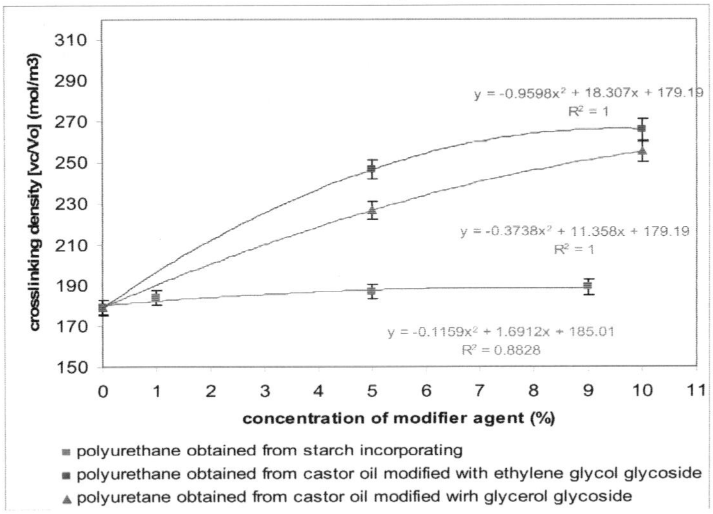

The data of tensile strength and shore A hardness of polyurethanes obtained from castor oil original and modified are shown in Figure 3. It is observed that polyurethane obtained from castor oil modified with glycosides showed higher tensile strength and shore A hardness compared to polyurethane obtained from original castor oil and starch incorporating. Polyurethane prepared by castor oil modified with glycosides improved properties over unmodified castor oil due to more crosslinking density due to an increase in hydroxyl content. The incorporation of starch without modification results in a increase in stiffness due to starch acts as a rigid filler.

Figure 3. Tensile strength and shore A hardness results of polyurethanes obtained from castor oil (original and modified)

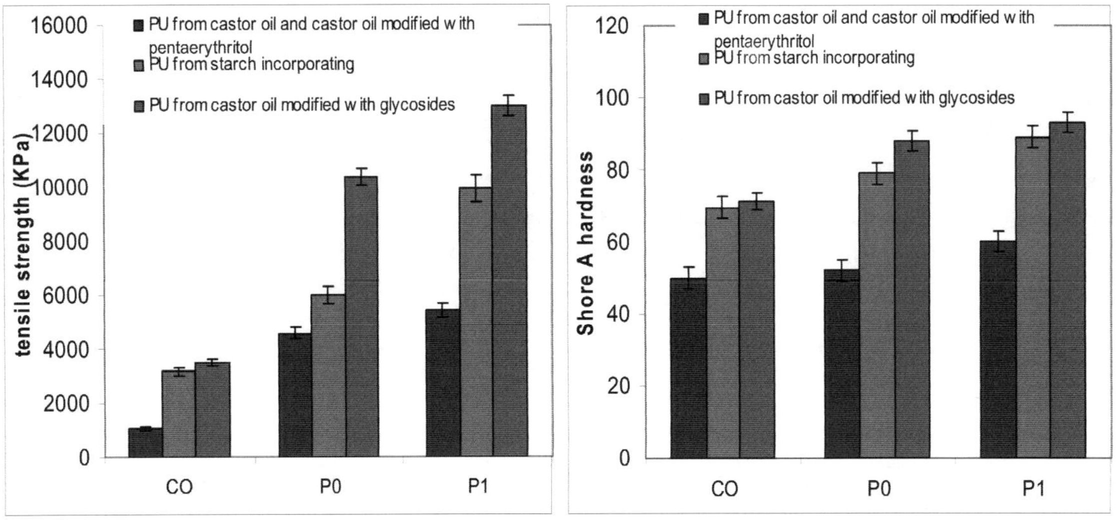

Figure 4.a shows the storage modulus, *E'*, versus the temperature for the studied polyurethanes. It can be seen that the storage modulus of polyurethane obtained from castor oil modified with glycosides is higher than the polyurethane from starch

incorporating and original castor oil, indicating increased rigidity in these materials. In the Figure 4.b, it is observed that the glass transition temperature (Tg) is greater for the PU obtained from castor oil modified with glycosides; it is also observed that the intensity of the peak of tanδ for this polyurethane decreases (the restriction in the mobility of the chains increases as a consequence of the higher density of crosslinking). These results demonstrate the greater degree of crosslinking of polyurethane obtained from castor oil modified with glycosides due to the high hydroxyl functionality of this polyol.

Figure 4. a. Storage modulus and b. tanδ versus temperature for the polyurethanes obtained from castor oil (original and modified)

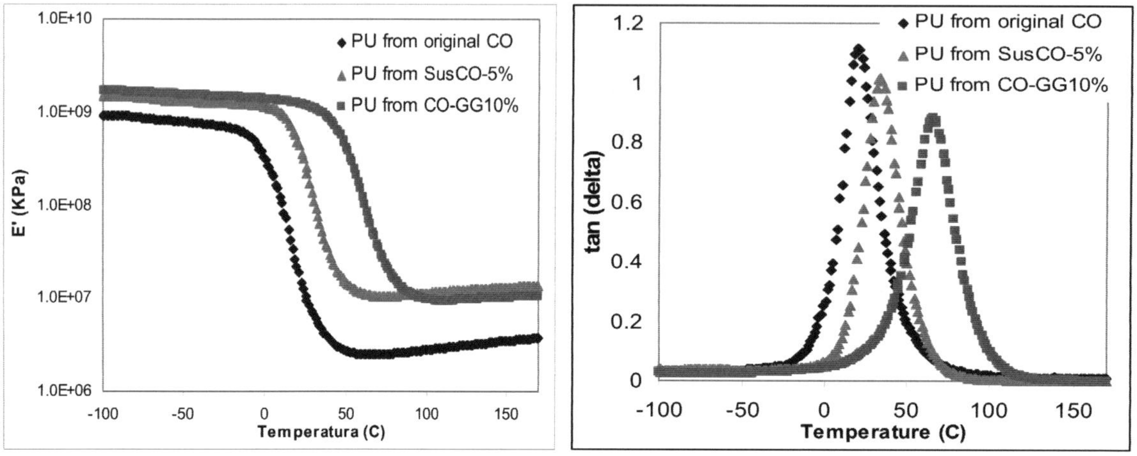

The Figure 5 shows the dependence between the glass transition temperature determined by DMTA and the number of hydroxyl groups of polyol, for polyurethanes obtained from the original castor oil and castor oil modified by the routes raised. We found a linear relationship between the hydroxyl value of the polyols and the glass transition temperature, with a correlation parameter of 0,946. By increasing the crosslinking density increases Tg due to increasing restrictions on the movements of the chains. Petrovic and colleagues reported a similar result for different polyols derived from vegetable oils [5]. Polyurethanes with high crosslinking density are obtained from polyols with high functionality. For the other hand, soft and elastic polyurethanes are obtained from polyols with low hydroxyl value [6]. Using the results of Figure 5, its possible predicts the values of the crosslinking density of the PU with the hydroxyl value of the polyol (for a given relationship NCO / OH).

Conclusions

Polyurethane networks prepared from IPDI and castor oil modified by transesterification with pentaerythritol and by reaction with starch glycosides showed excellent chemical resistance, hardness and tensile strength properties compared to the ones synthesized from unmodified castor oil. All materials prepared showed a uniform polyurethane phase without any presence of starch particles which indicates that polyol-glycosides of high hydroxyl content have been effectively reached and reacted to obtain a network formation in which starch glycosides are chemically incorporated. The degree of crosslinking in these networks can be controlled according to the type of polyol-glycoside used, which gives higher levels when glycerol is employed instead of ethylene glycol.

Figure 5. Dependence between the glass transition temperature determined by DMTA and the number of hydroxyl groups of polyol

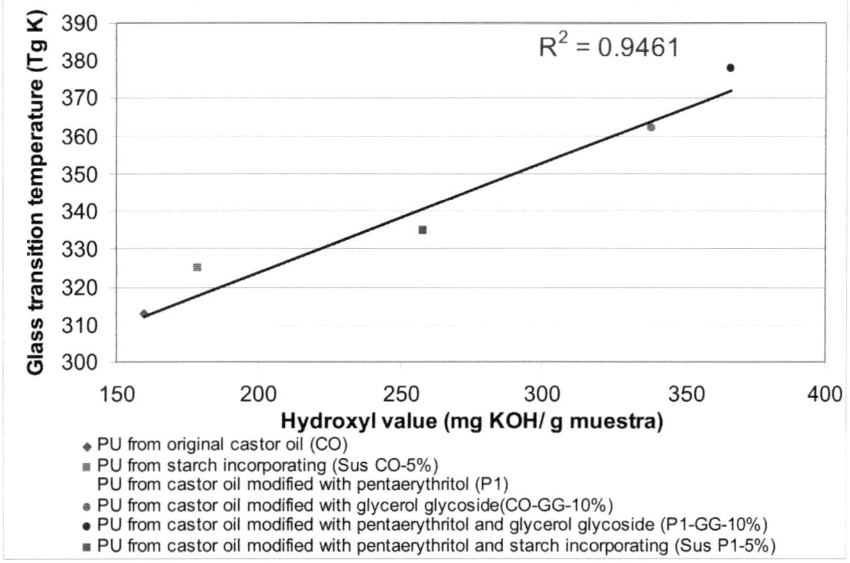

References

1. Ho¨ffer, P., Daute, P., Gru¨tzmacher, R. and Westfechtel, A. **(1997).** *Journal Coatings and Technology*, 69(869): 65–72.

2. Carraher Jr, C.E. and Sperling, L.H. **(1981).** p. 304, Plenum Press, New York.

3. Yenwo, G.M., Manson, J.A., Pulido, J., Sperling, L.H., Conde, A. and Devia, N. **(1977).** *Journal of Applied Polymer Science*, 21(6): 1531–1541.

4. Athawale, V. and Kolekar, S. **(1998).** *European Polymer Journal,* 34: 1447–1451.

5. Petrovic′, Z.S. and Fajnik, D.J. **(1984).** *Journal of Applied Polymer Science,* 29: 1031–1040.

6. Sonal, D., Thakore, I.M., Sarawade, B.D. and Devi, S. **(2000).** *Polymer Engineering and Science,* 40(5): 1200.

7. Seung Kyu, H. and Broecker, H. **(2003).** *Macromolecular Materials and Engineering,* 288(7): 569–577.

8. Siddaramaiah, J. and Kendaganna, B. **(2003).** *Journal Applied Polymer Science*, 90: 2945–2954.

9. Cao, X. and Zhang, L. **(2005).** *Biomacromolecules,* 6: 671–677.

10. Otey, F.H., Bennet, F.L., Zagoren, B.L. and Mehltretter, C.L. **(1965).** *Product Research and Development*, 4: 224–227.

11. Carr, M.E. **(1991).** *Journal of Applied Polymer Science,* 42: 45–53.

12. Prashantha, P., Vasanth Kumar Pai, K., Sherigara, B.S. and Prasannakumar, S. **(2001).** *Bulletin of Materials Science,* 24(5): 535–538.

13. Kaushik, A. and Paramjit, S. **(2005).** *International Journal of Polymer Analysis and Characterization,* 10: 373.

14. Alvaro Ramírez, Manuel F. Valero, Jorge E. Pulido, Zhengdong Cheng **(2008).** *Journal of Elastomers and Plastics.* DOI: 10.1177/0095244308091785.

One-part geopolymer mixes from geothermal silica and sodium aluminate

Ailar Hajimohammadi*, John L. Provis, and Jannie S.J. van Deventer

Department of Chemical and Biomolecular Engineering

University of Melbourne, Victoria 3010, Australia

* *a.hajimohammadi@pgrad.unimelb.edu.au*

Abstract

At present, the most commonly used building material is ordinary Portland cement (OPC). However, OPC has a negative environmental effect during synthesis, with the release of significant amount of CO_2 greenhouse gas. The cement industry is responsible for 5-8% of total global anthropogenic CO_2 emissions. Geopolymerization is a technology capable of turning industrial wastes into strong and chemically durable cement-like binders, with significant Greenhouse emission savings. The synthesis of geopolymers can start from a variety of aluminosilicate sources such as fly ash or metakaolin. In geopolymer technology, alkali silicate solutions are frequently used to dissolve the solid aluminosilicate precursor to produce the binder. Sodium aluminate solutions have also been used. These corrosive and often viscous solutions are not user friendly, and would be difficult to use for bulk production. Developing geopolymers as a 1-part mixture ("just add water") similar to OPC increases their commercial viability. Here, the geopolymer system consisting of geothermal silica and solid sodium aluminate (providing the solid silica, alkali and alumina sources), activated by addition of water, is studied. The effects of water content, high early silica and high early alumina in the formation of one part mix geopolymers are also investigated. XRD shows that the formulation with less water has an unexpected greater extent of crystallinity. It is also observed that a high early Al concentration inhibits geopolymerization, while a high early Si concentration enhances the reaction

1. INTRODUCTION

The geopolymer structure is mainly an aluminosilicate gel network, where the silica gel structure is partially substituted with tetrahedral Al^{3+} sites charge-balanced by alkali cations. The first stage of geopolymerization is the release of aluminate and silicate species from a solid source, which is usually achieved by alkali attack on an aluminosilicate material. First, the surface of the solid contacts the activating solution, and hydrolysis reactions begin to occur, later accompanied with the formation of oligomers and finally polycondensation of the oligomers to form a three-dimensional aluminosilicate network.[1] Soluble silicates are frequently used in geopolymer production for helping dissolution of the aluminosilicate starting material and governing the mechanical properties of the binder such as its compressive strength [2, 3]. In the present work, a solid silicate material and sodium aluminate are used as potential silica and alumina replacements in the production of aluminosilicate cementitious binders.

The solid silicate material used in the current work is geothermal silica. The production of geothermal power results in large quantities of residual silica, frequently removed as scale build up in pipes [4]. At the Cerro Prieto geothermal plant in Baja California, Mexico, approximately 5 tonnes of silica per month is removed from process pipes and discarded to an evaporation lake [5]. This waste residue is a source of solid amorphous silica, which can potentially be used as a replacement for soluble silicate solution in geopolymerization.

2. EXPERIMENTAL PROCEDURE

Raw silica waste from the Cerro Prieto plant was washed with distilled water to remove salts and dried before use. The purified geothermal silica is 96% SiO_2 and contains small amounts of various salts. The silica particles should have a high enough surface area for reaction, but not so high as to compromise workability of the geopolymer paste. Geothermal silica has a very small fundamental particle size; however the particles are aggregated. [6]

Geothermal silica was mixed with solid sodium aluminate to attain Si/ Al molar ratios of 1.5:1, 2:1 and 2.5:1, then water was added to these solid mixtures to attain samples with effective H_2O/Na_2O molar ratios of 7:1 and 12:1. The samples are named by these two ratios. The number outside parentheses refers to the Si/Al molar ratio and the number in the parentheses is the H_2O/Na_2O molar ratio. For instance, sample 1.5(12) is the geopolymer sample with Si/Al ratio of 1.5 and H_2O/Na_2O molar ratio of 12.

ATR-FTIR spectra were collected using a Varian FTS 7000 FT-IR spectrometer, with a Specac MKII Golden Gate single reflectance diamond ATR attachment with KRS-5 lenses and heater top plate. Absorbance spectra were collected from 4000-400 cm^{-1} at a resolution of 2 cm^{-1} and a scanning speed of 5 kHz with 32 scans. Scanning electron microscopy was also performed, using a Philips XL30 Field Emission Gun Scanning Electron Microscope (FEG-SEM). Fractured specimens were mounted on stubs and gold coated before analysis.

All geopolymer pastes were cured at 40°C in an oven. X-ray diffraction (XRD) (Phillips PW-1800) was conducted using CuKα X-rays, 30mA and 40 kV with 0.02° 2θ steps, 2s $step^{-1}$. Attenuated total reflectance Fourier-transform infrared (ATR-FTIR) spectroscopy was also used.

3. RESULTS AND DISCUSSION

The results presented in Figure 1 show the XRD diffractograms obtained for cured pastes after two weeks. The main crystalline phases identified in high water sample are quartz (SiO_2, minor contaminant in amorphous geothermal silica) and faujasite (approximately $Na_2Al_2Si_{3.3}O_{10.6}\cdot 7H_2O$), and in the sample with lower water content new peaks can be observed which are attributed to Zeolite A (approximately $Na_2Al_2Si_{1.85}O_{7.7}\cdot 5.1H_2O$). The hump from 20 to 40° 2θ in both samples demonstrates that a significant amount of the synthesized materials is X-ray amorphous, consistent with the majority of the geopolymers literature. The sharpness of the peaks in the lower-water sample may also be an indication of larger crystals.

It is of interest to note that the sample with less water added contained a greater extent of crystallite formation – this is contrary to many reports of reduced crystallinity with water addition to 'traditional' geopolymer systems, and suggests that there are additional effects occurring in the 'one-part mix' system that are not observed during silicate or hydroxide activation of aluminosilicate materials.

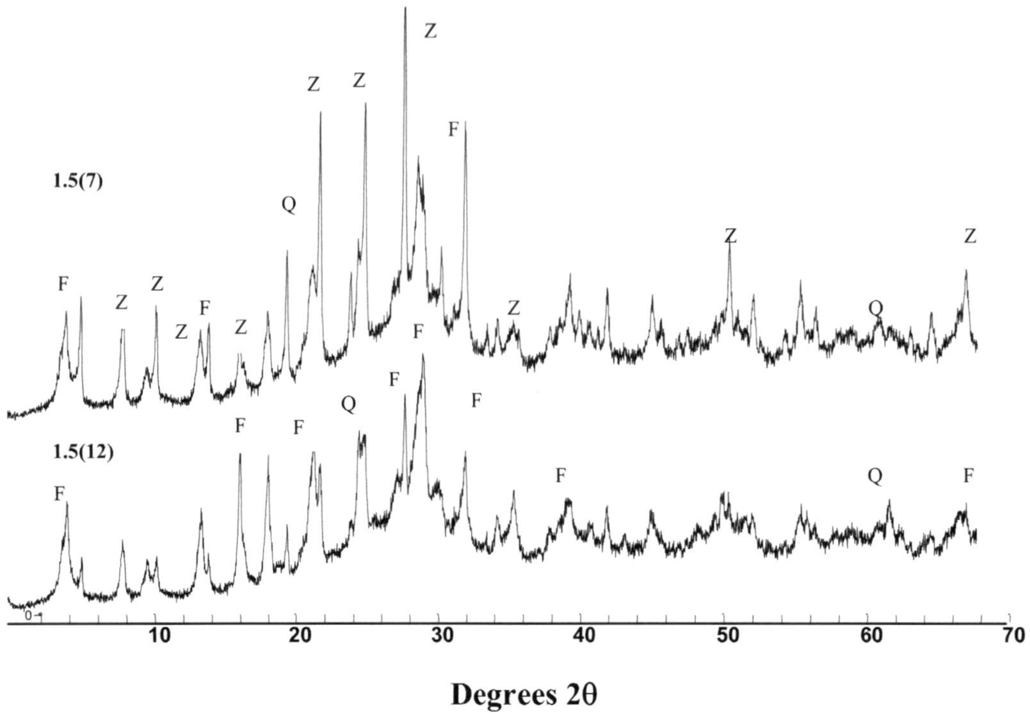

Figure 1. XRD diffractograms of geopolymers derived from geothermal silica and sodium aluminate cured for two weeks. Prefixes refer to H_2O/Na_2O molar ratio. Q = quartz, F = faujasite, Z= zeolite A. Data from reference [7].

Although XRD is an analytical technique commonly used for investigation of the geopolymer systems, it is important to note that this technique has significant limitations due to the apparent amorphicity of these materials[8]. So, other analytical techniques are required to provide insight about the structural characteristics of these materials, as well as the kinetics of structural evolution following mixing of the reactants.

The structural evolution of the geopolymer samples was therefore studied using ATR-FTIR (Figure 2). The spectra of geopolymer pastes for two samples are shown, as a function of reaction time over a 15-day period. Bands at 1060 cm^{-1} are assigned to stretching of Si-O-Si bands at the surface of the unreacted silica particles [9], and bands at 800 and 475 cm^{-1} relate to bending and rocking of the Si-O-Si bonds in the network of the unreacted geothermal silica [10].Bands at 625 cm^{-1} are related to the Al-O vibrations in the unreacted solid aluminate and bands at approximately 700 cm^{-1} are assigned to AlO_3 vibrations and a band at about 545 cm^{-1} is related to a single Al-O-Al mode in the aluminate [11].

The position of the main Si-O-T stretching band gives an indication of the length and angle of the Si-O bands in the silicate network, and for amorphous silica this peak occurs at approximately 1100cm^{-1}[12].This region is attributed to stretching vibration of inner SiO_2 groups [9].

In the present work, main Si-O-T stretching band occurs at 1060 cm^{-1} indicating the stretching vibrations of surface SiO_2 groups[9] . Over time, main band shifts to lower wavenumbers in both samples that can be the indication of some probable changes in silicate network including an increase of non bridging oxygen in silicate sites, charge balancing by sodium cations in the system [13]or increasing the substitution of tetrahedral Al in the silicate network [14].

Over time, there is a reduction in intensity of the main Si-O-Si band indicating that the solid silica is dissolving and changing in molecular structure. At the same time, a new band starts forming at about 950cm^{-1} and the intensity of this band increases over time. This particular band is associated with the stretching vibrations of the Si-O-T (T: Si or Al) bond of the geopolymer network. The intensity of this band continues increasing gradually over time until there is only one clear band at 940 cm^{-1} for both samples. The shoulder in 1060 cm^{-1} is attributed to unreacted geothermal silica remaining in the samples.

Water plays a critical role during the dissolution, polycondensation and hardening stages of geopolymerization. Figure 2 shows that the dissolution of geothermal silica and the formation of geopolymer network bonds are much faster in the sample with lower water content.

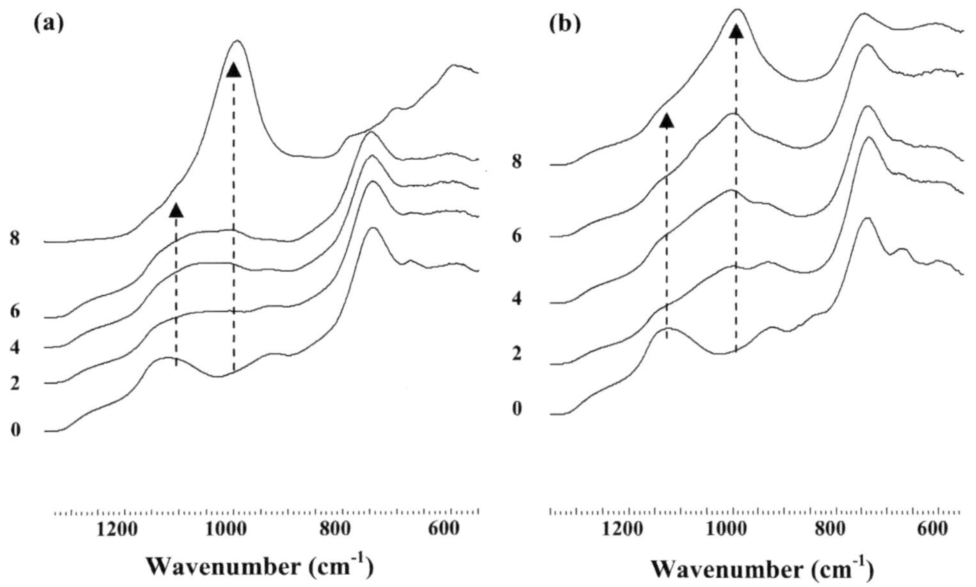

Figure 2. ATR-FTIR spectra for geopolymer samples 1.5(12) (plot A) and 1.5(7) (plot B), showing the process of synthesis from geothermal silica and sodium aluminate. Numbers refer to the geopolymer age in days. Data from reference [7].

Figure 3 shows the structural evolution of one-part mix geopolymer samples with different Si/Al ratios and the same amount of water. It is clear from these spectra that dissolution of geothermal silica and formation of geopolymer are faster in the sample with the lower Si/Al ratio

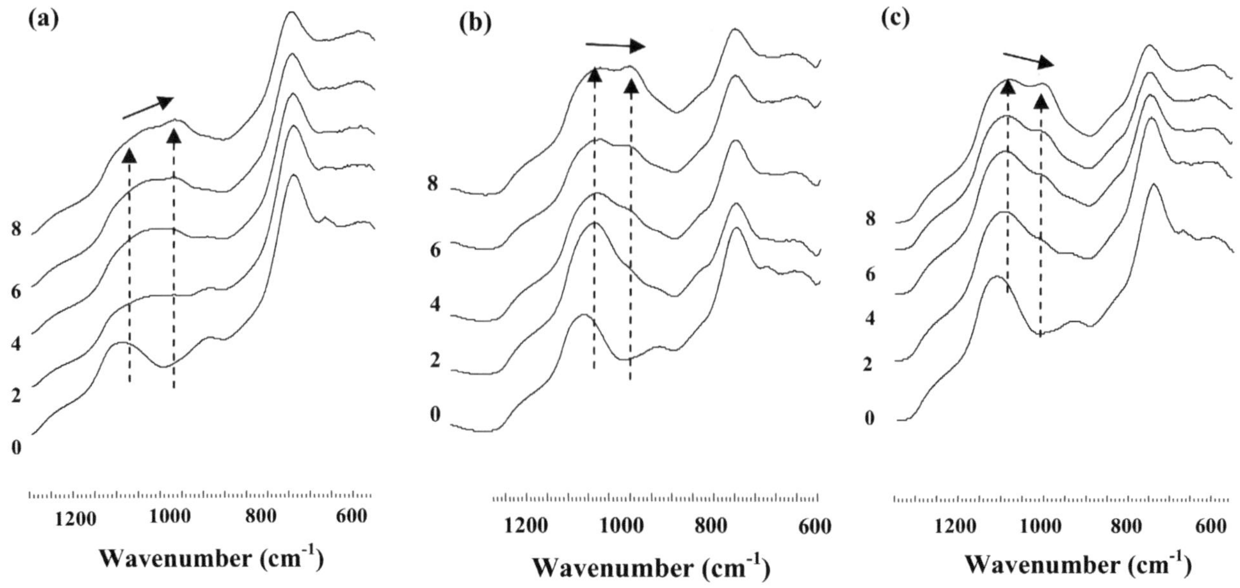

Fig 3. ATR-FTIR spectra of (a) 1.5(12), (b) 2(12) and (c) 2.5(12) geopolymer samples. Numbers refer to the geopolymer age in days. Data from reference [7].

These results show that the process of development of the geopolymer gel in the 'just add water' system studied here is relatively similar to the gel formation process in fly ash geopolymerization as was previously studied by this technique.[15-18] Both systems, although chemically very different, show the gradual destruction of the solid aluminate and silicate sources (which is very rapid in the case of solid sodium aluminate), and the formation of a new aluminosilicate gel phase characterized by a strong Si-O-T asymmetric stretching peak.

The durability of a cement structure strongly depends on its physical properties, such as permeability and ionic diffusivity, which are controlled by the microstructural characteristics of cement paste.[19] The porosity and pore size distribution are vital components of the microstructure of cement paste. The pore structure determines the permeability of cement, and thus the extent of penetration by aggressive agents.[20]

In order to study the microstructure of 'just add water'-synthesized geopolymers, scanning electron microscopy (SEM) is used. Examples of the microstructures of sodium aluminate-geothermal silica geopolymers after two weeks are shown in Figure 4. An obvious difference can be observed between the two samples with different water contents. These micrographs show the greater extent of large pores, as well as larger size and higher density of gel particulates, in the sample with higher water content. Fine pore structure enables a material to better resist chemical attack as it leads to lower permeability.[21] Resistance to chloride penetration of mortar and concrete is one of the most important issues regarding the durability of concrete structures.[21-23]

(a) (b)

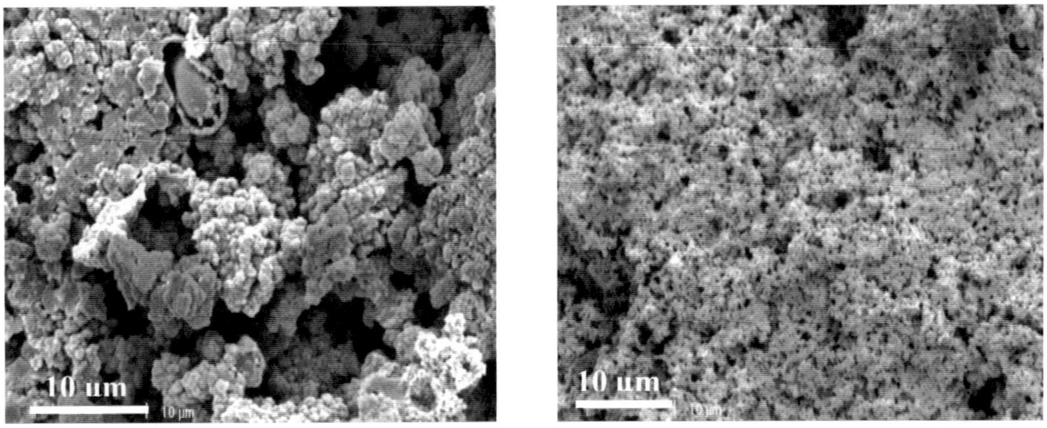

Fig 4. SEM micrographs of (a) 1.5(12) and (b) 1.5(7) geopolymer samples after curing for two weeks at 40°C. Data from reference [7].

In order to study the effect of Si/Al ratio on the microstructure of one part mix geopolymers, 2(12) and 2.5(12) samples were made using geothermal silica and sodium alumina, and their microstructures were also studied after two weeks' curing at 40°C (Figure 10). In traditional two-part geopolymer systems, the microstructure changes from containing large pores to being more homogenous with small pores as the Si/Al ratio increases.[24] Comparison of Figure 4 with Figure 5 shows the same behavior in the one-part system.

(a) (b)

Fig 5. SEM micrographs of (a) 2(12) and (b) 2,5(12) geopolymer samples after curing for two weeks at 40°C. Data from reference [7].

4. CONCLUSIONS

The development of a solid replacement for the silicate activator in geopolymer synthesis will increase the commercial viability of this material by removing the difficulties associated with the use of viscous alkaline solutions. Also, the use of siliceous industrial wastes as silicate replacements in geopolymer cements for the construction industry can create an economical and environmentally friendly replacement for Portland cement, and finally a one-part mix or 'bag' of geopolymer cement can be made.

In this work, synthesis of a one part mix geopolymer from solid sodium aluminate and geothermal silica was investigated. This system demonstrates that making geopolymers from solid sources by "just adding water" is possible. Water plays a critical role during geopolymerization. Samples with less water have a greater extent of crystallite formation, which was unexpected because this is contrary to many reports of reduced crystallinity with water addition to 'traditional' geopolymer system. This difference may be due to the rapid dissolution of the aluminate source and relatively slower dissolution of silica source in the early stages of geopolymerisation. SEM micrographs clearly show the difference between samples with different water content and different Si/Al ratio. ATR-FTIR enables structural analysis of geopolymer samples during the geopolymerisation reaction.

5. ACKNOWLEDGMENTS

The assistance and guidance of Dr. Catherine Rees in this project is gratefully acknowledged. This work was funded in part through the Particulate Fluids Processing Centre (PFPC) (a Special Research Centre of the Australian Research Council), and by the Geopolymer Alliance.

6. References

1. J.L. Provis and J.S.J. van Deventer, *Geopolymerisation kinetics. 2. Reaction kinetic modelling.* Chemical Engineering Science, 2007, 62: p. 2318 - 2329.
2. W.K.W. Lee and J.S.J. van Deventer, *Use of infrared spectroscopy to study geopolymerization of heterogeneous amorphous aluminosilicates.* Langmuir, 2003, 19(21): p. 8726-8734.
3. M. Rowles and B. O'Connor, *Chemical optimization of the compressive strength of aluminosilicate geopolymers synthesized by sodium silicate activation of metakaolinite.* Journal of Materials Chemistry, 2003(13): p. 1161-1165.
4. J.M. Rincon, *Thermal behaviour of silica waste from a geothermal power station and derived silica ceramics.* Journal of Thermal Analysis and Calorimetry, 1999, 56(1): p. 1261-1269.
5. C. Diaz and F.J. Valle-Fuentes, *Use of silicon residue in ceramics and glass.* American Ceramic Society Bulletin, 1999, 78(8): p. 112-115.
6. C. Rees, *Mechanisms and Kinetics of Gel Formation in Geopolymers*, Ph.D. Thesis, University of Melbourne, 2007
7. A. Hajimohammadi, J.L. Provis, and J.S.J. van Deventer, *One-part geopolymer mixes from geothermal silica and sodium aluminate.* Industrial & Engineering Chemistry Research, 2008: p. In press.

8. P.A. Jacobs, E.G. Derouane, and J. Weitkamp, *Evidence for X-ray-amorphous zeolites.* Journal of the Chemical Society, Chemical Communications, 1981: p. 591 - 593.
9. J. Osswald and K.T. Fehr, *FTIR spectroscopic study on liquid silica solutions and nanoscale particle size determination.* Journal of Material Science 2006, 41: p. 1335-1339.
10. R.J. Bell and P. Dean, *Atomic vibrations in vitreous silica.* Discussions of the Faraday Society, 1970, 50: p. 55-61.
11. R.J. Moolenaar, J.C. Evans, and L.D. McKeever, *The structure of the aluminate ion in solutions at high pH.* Journal of Physical Chemistry, 1970, 74(20): p. 3629-3638.
12. M. Handke and W. Mozgawa, *Vibrational spectroscopy of the amorphous silicates.* Vibrational Spectroscopy, 1993, 5(1): p. 75-84.
13. J.R. Sweet and W.B. White, *Study of sodium silicate glasses and liquids by infrared reflectance spectroscopy.* Physics and Chemistry of Glasses, 1969, 10(6): p. 246-251.
14. B.N. Roy, *Infrared spectroscopy of lead and alkaline-earth aluminosilicate glasses.* Journal of the American Ceramic Society, 1990, 73(4): p. 846-855.
15. C.A. Rees, J.L. Provis, G.C. Lukey, and J.S.J. van Deventer, *Attenuated total reflectance fourier transform infrared analysis of fly ash geopolymer gel aging.* Langmuir, 2007, 23: p. 8170-8179.
16. C.A. Rees, J.L. Provis, G.C. Lukey, and J.S.J. van Deventer, *In situ ATR-FTIR study of the early stages of fly ash geopolymer gel formation.* Langmuir, 2007, 23: p. 9076-9082.
17. A. Fernández-Jiménez, A.G. de la Torre, A. Palomo, G. Lopez-Olmo, M.M. Alonso, and M.A.G. Aranda, *Quantitative determination of phases in the alkali activation of fly ash. Part I. Potential ash reactivity.* Fuel, 2006, 85(5-6): p. 625-634.
18. A. Fernández-Jiménez, A.G. de la Torre, A. Palomo, G. López-Olmo, M.M. Alonso, and M.A.G. Aranda, *Quantitative determination of phases in the alkaline activation of fly ash. Part II: Degree of reaction.* Fuel, 2006, 85(14-15): p. 1960-1969.
19. P. Chindaprasirt, C. Jaturapitakkul, and T. Sinsiri, *Effect of fly ash fineness on compressive strength and pore size of blended cement paste.* Cement & Concrete Composites, 2005, 27: p. 425-428.
20. P. Chindaprasirt, S. Homwuttiwong, and V. Sirivivatnanon, *Influence of fly ash fineness on strength, drying shrinkage and sulfate resistance of blended cement mortar.* Cement and Concrete Research, 2004 34: p. 1087-1092.
21. P. Chindaprasirt , S. Rukzon , and V. Sirivivatnanon *Resistance to chloride penetration of blended Portland cement mortar containing palm oil fuel ash, rice husk ash and fly ash.* Construction and Building Materials, 2008 22: p. 932-938.
22. M.D.A. Thomas, P.B. Bamforth, and F.T. Bingham, *Modelling chloride diffusion in concrete. Effect of fly ash and slag.* Cement and Concrete Research, 1999, 29: p. 487-495.

23. C.S. Poon, Y.L. Wong, and L. Lam, *The influence of different curing conditions on the pore structure and related properties of fly-ash cement pastes and mortars.* Constntction and Building Materials, 1997, 11: p. 383-393.
24. P. Duxson, J.L. Provis, G.C. Lukey, S.W. Mallicoat, W.M. Kriven, and J.S.J. van Deventer, *Understanding the relationship between geopolymer composition, microstructure and mechanical properties.* Colloids and Surfaces A: Physicochemical and Engineering Aspects, 2005, 269(1-3): p. 47-58.

722am Development and Characterization of Copolymers from Renewable Resource Monomers

Mathew D. Rowe and Keisha B. Walters. Swalm School of Chemical Engineering, Mississippi State University, 330 Swalm Chemical Engineering Building, Mississippi State, MS 39762

Renewable polymers have been developed from a selection of poly-alcohol and poly-acid monomers available from renewable resources. A simple polycondensation reaction was selected to produce anhydride and ester copolymers. The renewable resource monomers glycerol, 1,3-propane diol, malonic acid, fumaric acid, and itaconic acid were selected based on biomass availability and functional groups. Reaction conditions and catalysts were selected to maximize yield and molecular weight and minimize impacts on the environment though a reduction of energy and solvent required during polymerization and processing. Effects of catalyst ratio, monomer ratio, and temperature on copolymer yield and bulk properties were studied. Polymer synthesis was verified and the chemical composition was determined using Fourier transform infrared spectroscopy (FTIR), X-ray photoelectron spectroscopy and nuclear magnetic resonance (NMR). Gel permeation chromatography (GPC) and differential scanning calorimetry (DSC) were used to determine average molecular weights, molecular weight distributions, and phase transitions in these new copolymers. A series of anhydride and ester copolymers have been developed and chemically and thermally characterized as possible renewable replacements for petroleum based polymers.

722an The Rheology, Degradation, and Processing of Renewable Resource Polymers

Graham H. Harrison, Chemical and Biomolecular Engineering and Center for Advanced Engineering Fibers and Films, Clemson University, 127 Earle Hall, Clemson, SC 29634-0909 and Jason D. Conrad, Chemical & Biomolecular Engineering, Clemson University, 127 Earle Hall, Clemson, SC 29634.

Renewable resource polymers have shown potential as a replacement to conventional fossil fuel-based polymers in many applications including single use products. This study presents recent results on poly-lactic acid (PLA) and a series of polyhydroxyalkanoate (PHA) copolymers. These polymers are derived from renewable resources and degrade under the appropriate conditions. The PHA copolymers are poly-3-hydroxybutyrate-co-3-hydroxyvalerate (poly-3HB-3HV) with 5, 8, and 12 mol % HV comonomer. A rheological characterization in shear, extensional, and transient modes examines their flow behavior under varying flow fields and identifies the degradation of the materials occurring at typical processing conditions. Suitable conditions are determined for processing the polymers using a DSM Xplore microcompounder with the cast film attachment. Blends of PLA and PHA are produced and subsequent testing characterizes the films as a function of PHA composition using DSC, WAXD, tensile testing, and SEM to examine the thermal, mechanical, and structural properties.

722ao One-Step Functionalization of Cellulosic Nanowhiskers Using Renewable Reagents

Birgit Braun and John R Dorgan. Chemical Engineering, Colorado School of Mines, 1500 Illinois Street, Golden, CO 80401

Escalating fuel prices and increasing environmental concerns have sparked interest in materials based on renewable resources. Biopolymers have become attractive alternatives to oil-derived plastics, and research efforts are being made to further improve their properties thus making them suitable to a wider range of applications.

Composite formation is an effective approach to enhance mechanical and thermo-physical characteristics of polymers. Ideally, the rigid filler introduced into a biopolymer is also derived from renewables. Cellulosic nanowhiskers (CNW) can be isolated from various cellulose sources via controlled acid hydrolysis, and have been shown to be very efficient in improving mechanical properties. However, when they are to be incorporated into hydrophobic polymers their hydrophilic nature is problematic as it limits dispersion, ultimately leading to inferior reinforcement. The addition of surfactants or chemical alterations of the CNW surface are commonly employed strategies to address this issue; these approaches typically require multiple reaction steps.

Accessible hydroxyl groups can be functionalized in a new one-step chemistry in the presence of an acid catalyst. This one step method can be practiced simultaneously with the hydrolysis of amorphous cellulose regions to substitute a fraction of the surface hydroxyl groups on the resulting CNW. This renders the nanowhiskers more hydrophobic. Fourier-transform infrared spectroscopy in conjunction with multi-angle laser-light scattering for dimensional analysis allows determination of the average surface coverage. Consequently, the potential of these functionalized hydrophobic CNW as reinforcement for biopolymers to form nanocomposites is demonstrated.

722ap Palm Oil Based Epoxides as Pvc Plasticizers: Synthesis, Characterization and Incorporation and Their Thermomechanical and Chemical Properties

José Rodrigo Gómez-Díaz[1], Alejandro Boyaca[2], and Luis Francisco Boada Eslava[2]. (1) Department of Chemical Engineering,, Universidad de los Andes,, Calle 19 1A-40, Mario Laserna Engineering Building, Bogotá, Colombia, (2) Chemical Engineering Department, Universidad Nacional de Colombia, Sede Bogotá, Carrera 30 No. 45-03, Edificio Aulas de Ingeniería, oficina 301, Bogotá, Colombia

This century opens new opportunities for the engineering of chemical reactions involving renewable materials, coupled with product design requirements as specified by the polymer processor. Environmentally conscious design of transformation process is a necessity for the plastics industry. Colombia as the fourth palm oil producer in the world has an increasing crop of palm per hectare. Palm hybrid olein, a product of palm oil refining, was epoxidized by the in-situ production of peracetic acid from hydrogen peroxide and glacial acetic acid using acid catalysis. The effect of several process parameters were studied to improve the epoxide yield thus favoring the plasticizing effect of the functionalized triglyceride; the process simulation was held using some adjusted kinetic parameters from Gan et Al allowing reproducible production of palm olein epoxides.

After stabilization and purification, the epoxide was incorporated on PVC from local industry (in the Andean region supplied by Mexichem-Petco) using di-octyl phthalate (DOP) as primary plasticizer. By varying the amount of DOP/epoxide ratio and through a statically design of the experiment, the insertion of the epoxide during PVC extrusion, allowed the replacement of 30-50% of DOP content as plasticizer. Selected thermomechanical properties were measured including the glass transition temperature, Tg, which for the PVC-Plasticizer mix was adjusted by the use of the Cox-equation, allowing the thermodynamic validation of the mixture. Several solvents were employed to measure the solubility parameter (8.06 cal/cm3) matching several models for the Gibbs Excess energy with adjustment for the prediction of compatibility via activity coefficients. Long term stability was measured through the detection of permeation rates from films subjected to external pressure; also oxygen and water permeability was measured. Finally a product film was manufactured in an ISO recognized industrial colombian laboratory and key properties measured and adjusted to match industry specifications. Due to high variation in crop composition the same study was performed with oil produced from different regions of Colombia. The determination of the large scale viability for the incorporation of palm oil epoxides on PVC is ongoin on several industrial extrusion trials to be performed with Colombian PVC processors. The production cycle is robust enough in order to respond not only to price variations in crude oil but also to the struggle between vegetable oils as their prices rise, a direct consequence of the biofuels boom, this was examined using a process model to analyse multifeedstock variation (palm and soybean oil).

ENVIRONMENTALLY FRIENDLY CHEMICAL SYNTHESIS AND CATALYSIS FOR IMPROVED POLYETHYLENE PRODUCTS

Zoe Ziaka* and Savvas Vasileiadis
Chemical & Materials Engineering, ZIVATECH, 535 Plum str., Youngstown, Ohio, 44503, USA

Summary

Environmentally–friendly polymer processes have been developed for polyethylene product synthesis and catalysis. The developed processes seek ethylene and energy conservation coupled with elimination of toxic and hazardous solvents, catalysts and other byproduct reaction media. Increased polyethylene yields, ethylene recycling within the process, reduction in hazardous waste generation, reduction of process equipment and of capital and operational costs (including monomer and energy costs) are some of the improvements of the new processes. Ethylene can come from a variety of sources including a dehydrogenation membrane reactor which produces increased yields of the valuable monomer. Hydrogen or hydrogen and ethylene separation can occur via the walls of the catalytic dehydrogenation membrane reactor. Related modeling and optimization of the multiphase polyethylene reactor in the downstream is also underway to describe its functionality, operational range, capacity and polymer selectivity. The models used can simulate relevant acquired laboratory and industrial data. They describe numerically reaction, transport, recycling, polymerization, and polymer separation operations in the polyethylene reactor. Relative compositions of feedstock ethylene diluted with other components can be used in this reactor to produce polyethylene of variable molecular weight, structure, and tensile properties. Moreover, some forms of copolymers can be also produced directly during ethylene polymerization by adjusting the operation conditions and temperature of the upstream membrane dehydrogenator.

Process and Materials Description

This new communication reports current research analysis on reaction-separation systems for polymer-grade ethylene production via initial ethane based dehydrogenation reactions [1-4]. We present new process designs based on ethane dehydrogenation processes including work in design of experiments, operation and best parameter selection and optimization of such systems. These processes are of current significance in ethylene processing and conversion to valuable chemicals such as polyethylene and its copolymers [1-4,6]. This is because of the unique design characteristics of the examined reactor-separator systems to perform multiple operations. Among others, permeable catalytic carriers (membrane reactors) can be utilized by applying the equilibrium shift principles in ethane conversion and ethylene yield through the rapid removal of permselective species (mainly hydrogen) out of the catalyst and reaction zones.

*corresponding author, bookeng@hotmail.com, tel. +30-2310-275473

In the downstream, the ethylene products are used for direct polyethylene production through successive polymerization reactions [1-4,6].

Further, the work targets to develop efficient reactors and systems with increased processing capacity, conversion and selectivity, separation, and additional utilization (e.g., via recycling and side-feed flows) of primary and secondary ethane based streams. Ethane can come from a variety of sources including fossil and renewable resources after passing the proper purification steps and is directed in the entrance of the catalytic dehydrogenator. There is a substantial effort today to increase conversion of renewable ethane feedstocks into valuable end products such as polymers, from an engineering, economic, and environmental point of view [1,6].

Various combinations of permselective materials, including metals, inorganics, and composites can be used if a membrane reactor or a membrane permeator is selected for the operation. In these cases the hydrogen, or hydrogen and ethylene separation from the ethylene-hydrogen-unreacted ethane mixture is the operation that needs to be optimized in terms of the separative products yield and selectivity [1-4,8-11].

The use of such integrated systems of membrane reactors and permeators in comparison with previous impermeable-tubular reactors and separators, are well suited for low molecular weight alkane dehydrogenation reactions for production and separation of the corresponding olefins. These can be also considered sample reactions for stoichiometric hydrogen production from low-carbon alkane molecules. According to the proposed integration, product H_2 can be recovered/separated via the membranes from ethylene and ethane molecules and used in successive chemical synthesis (e.g., various hydrogenations) or as clean additive fuel in power generation equipment, such as turbines and fuel cells [1,6,10,11]. PEM and Alkaline fuel cells as example, are well suited to operate with the separated hydrogen from the membranes, and even SOFCs can operate within an acceptable thermal framework [6]. Further, the separated ethylene can be fed in polymerization reactors (such as coordination-catalysis type vessels) for direct polyethylene production, [7].

Related modeling of such multifunctional membrane dehydrogenation and polymerization reactors for the described applications is underway to describe their function, operational range, capacity, and product yield and selectivity. The models used can simulate relevant acquired laboratory and industrial data and predict the conditions for best dehydrogenator, separator, and polymerization reactor operation by variation of the process key-intrinsic parameters [1,6,11].

Kinetic and transport parameters used in the reaction/separation modeling can be obtained experimentally by separate kinetic and transport experiments or calculated by empirical correlations [6-11]. Moreover, the modeling of permeable and tubular reactors shows how the ethane conversion and ethylene yield is affected by varying key-design parameters such as the extent of the catalytic ethane dehydrogenation reaction, the permeability and selectivity factors of the membrane reactor wall material, the magnitude of the equilibrium constant for the dehydrogenation reaction, the ratio of pressures and flowrates of the reactive and inert gases within the catalytic bed and in the external side of the membrane reactor. Variations in the composition of the exit product stream from the dehydrogenator-separator assembly affects the polymerization in the consecutive reactor and the range of molecular weight, structure, and properties of the produced polyethylene. Moreover, copolymers can be also produced directly in polymerization by adjusting the operational conditions of the catalytic membrane dehydrogenator.

As a conclusion, the proposed membrane based dehydrogenation-polymerization configurations for ethylene and polyethylene production from ethane based streams seek to perform multiple and combined unit operations by using single type or integrated membrane reaction-separation modules. These effects/operations make them advanced in comparison with up to now utilized conventional dehydrogenation reactors. Thus, the introduced membrane catalytic systems project a promising replacement technology for currently utilized ethane to ethylene production reactors (such as crackers or catalytic dehydrogenators) and in conjunction with downstream polyethylene yield reactors. Economic and environmental advantages are also applicable and promising.

References:

1) Z. Ziaka and S. Vasileiadis, "New integrated processes for olefin and polyolefin production", paper#14, in Novel Reactors and Processes, *ISCRE 17*, Hong Kong, China, Aug. (2002).

2) Z.Ziaka and S.Vasileiadis, "Polymer membrane reactors for enhanced hydrocarbon conversion and upgrading", *AIChE Annual meeting*, paper #275c, Nov. (1997).

3) Z. Ziaka and S.Vasileiadis, "Polymer grade olefin production with multifunctional permreactor-permeator systems; *NASCRE 1st* Meeting, Houston, TX Jan. (2001).

4) Z. Ziaka and S. Vasileiadis, "Multifunctional hydrocarbon processing systems for production of fuels and valuable chemicals" paper #361i, in Kinetics, Catalysis and Reaction Engineering, *AIChE annual meeting*, Nov. (2000).

5) Z. Ziaka and S. Vasileiadis, *Chem. Eng. Comm.*, 156, 161-200 (1996).

6) S. Vasileiadis and Z. Ziaka, "Environmentally benign hydrocarbon processing applications of single and integrated permreactors, in "Reaction Engineering for Pollution Prevention", *Elsevier Science*, pp. 247-304, (2000).

7) F.W. Billmeyer, "Textbook of Polymer Science", *John Wiley & Sons*, 3rd Ed., (1984).

8) Z. Ziaka, R.G. Minet, and T.T. Tsotsis, *AIChE J.*, 39, 526-529, (1993).

9) Z. Ziaka, R.G. Minet, and T.T. Tsotsis, *J. Membrane Sci.*, 77, 221-232, (1993).

10) T. Tsotsis, A. Champagnie, S. Vasileiadis, Z.Ziaka and R.Minet, *Sep. Sci. & Technol.*, Vol.28, No1-3, pp.397-422, (1993).

11) T. Tsotsis, A. Champagnie, S. Vasileiadis, Z.Ziaka and R.Minet, *Chem. Eng. Sci.*, Vol.47, No.9-11, pp.2903-2908, (1992).

722ar Nafion-Based Anion Exchange Membranes for the Alkaline Fuel Cell

Holly Schaeffer and Yossef A. Elabd. Chemical and Biological Engineering, Drexel University, 3141 Chestnut St., Philadelphia, PA 19104

Polymer electrolyte membrane fuel cells (PEMFCs) are actively being explored to replace conventional power sources in automobiles and portable electronic devices. However, alkaline fuel cells (AFCs) have several key advantages over traditional PEMFCs, such as higher efficiencies at low temperatures and the ability to use non-precious metal catalysts (e.g., nickel) instead of platinum. Problems that have deterred the advancement of the AFC are electrolyte leakage (KOH(aq)) and electrode degradation due to carbonate crystal formation (due to carbon dioxide in the fuel) and adsorption onto the catalyst. Electrode degradation can be prevented with the use of pure fuel streams, however this is cost prohibitive. Replacement of the liquid electrolyte with an anion exchange membrane (AEM) (a solid polymer film with immobile cations) can solve the current AFC problems. This would provide a fuel cell that is significantly more cost-effective than current PEMFCs. Recently, several research groups have demonstrated a working AEM-based AFC. However, there are still several scientific challenges that must be overcome to produce a high power density device with a long operational lifetime, such as poor chemical stability of the AEM under AFC conditions (high pH and >60°C) and low hydroxyl ion conductivity of the AEM.

In this study, the synthesis of a new anion-exchange form of the Nafion membrane was demonstrated. In this work, the amination of the precursor form (perfluoroether branches terminated with (–SO2F) of Nafion was performed in contrast to the typical hydrolysis reaction, which produces the common cation-exchange form of Nafion terminated with sulfonic acid (-SO3H). Synthesis was confirmed with infrared spectroscopy and elemental anaysis. Reactions were carried out over a wide range of times and temperatures and quantified with infrared spectroscopy. Furthermore, ionic conductivities were measured using electrochemical impedance spectroscopy as a function of reaction conditions.

722as Wetting and Absorption of Water Drops on Nafion Films

Sharonmoyee Goswami, Chemical Engineering, Princeton University, Princeton, NJ 08544, Shannon Klaus, Chemical Engineering, University of Missouri, Columbia, MO 65201, and Jay B. Benziger, Department of Chemical Engineering, Princeton University, Princeton, NJ 08544.

Water drops on Nafion films caused the surface to switch from being hydrophobic to being hydrophilic. Contact angle hysteresis of greater than 70° between advancing and receding values were obtained by the Wilhelmy plate technique. Sessile drop measurements were consistent with the advancing contact angle; the sessile drop contact angle was 108°. Water drop adhesion, as measured by the detachment angle on an inclined plane, showed much stronger water adhesion on Nafion than Teflon. Sessile water and methanol drops caused dry Nafion films to deflect. The flexure went through a maximum with time. Flexure increased with contact area of the drop, but was insensitive to the film thickness. Methanol drops spread more on Nafion and caused larger film flexure than water. The results suggest that the Nafion surface was initially hydrophobic but water and methanol drops caused hydrophilic sulfonic acid domains to be drawn to the Nafion surface. Local swelling of the film beneath the water drop caused the film to buckle. The maximum flexure is suggested to result from motion of a water swelling front through the Nafion film.

Web Page: pemfc.princeton.edu

722at Microstructure Transitions of Lithium Salt Doped PS-B-PEO Copolymer

Wen-Shiue Young, Paul Brigandi, and Thomas H. Epps, III. Chemical Engineering, University of Delaware, 150 Academy St., 235 Colburn Lab., Newark, DE 19716

Salt-doped poly(ethylene oxide)-based block copolymers have been considered to be promising candidates for polymer electrolytes which require strong mechanical property as well as high ionic conductivity. We studied the phase behavior of poly(styrene-b-ethylene oxide) block copolymers doped with various salts over a range of EO:ion ratios. Small-angle X-ray scattering (SAXS), transmission electron microscopy (TEM), and differential scanning calorimetry (DSC) experiments were used to characterize the phase behavior of our neat and salt-doped copolymers. Samples were prepared in an argon atmosphere and rigorously dried to reduce the effects of moisture uptake on phase behavior. We found that the copolymer microstructure can be tuned by variations in the salt and counterion, as well as the salt doping ratio. Additionally, we observed thermal order-order transitions that were caused by the crystallization of confined PEO-lithium complexes.

722au Diffusion of Water and Methanol In Nafion: In Situ Infrared Experiments and Multicomponent Models

James A. Throckmorton, Daniel T. Hallinan Jr., and Yossef A. Elabd. Chemical and Biological Engineering, Drexel University, 3141 Chestnut St., Philadelphia, PA 19104

Understanding multicomponent diffusion in polymers is important for many applications, including membrane-based separations and fuel cells. Although there are a number of publications documenting multicomponent diffusion models in polymers, there is limited in situ experimental evidence of multicomponent diffusion in polymers. In this work, a classic problem in fuel cells has been investigated. Specifically, for the direct methanol fuel cell (DMFC), methanol crossover (high methanol flux) is a critical concern as it deters both fuel efficiency and power output by interfering with the cathode reduction reaction. Therefore, a fundamental understanding of multicomponent diffusion of methanol and water in Nafion is desired. In this study, the diffusion of methanol/water mixtures in Nafion was measured using time-resolved Fourier-transform infrared, attenuated total reflectance (FTIR-ATR) spectroscopy. This technique not only provides in situ molecular-level contrast between diffusants and polymers in real time, but also can measure chemical interactions between diffusants and polymers through shifts in the infrared spectra. The multicomponent diffusion data from this work has been compared to classic multicomponent diffusion models using the generalized Stefan-Maxwell and Onsager formalisms derived from irreversible thermodynamics. More specifically, "main-term" and "cross-term diffusion coefficients have been regressed from the data. These results will be discussed in relation to model constraints presented in the literature, additional models that account for diffusant-diffusant and diffusant-polymer interactions, multicomponent diffusion effects (the effect of one diffusant's concentration gradient on the flux of another diffusant), and the overall impact of these results on methanol fuel cells.

722av Novel Organometallic Low Energy Gap Materials and Synthetic Models of Carbon Monoxide Hydrogenase (CODH) In Biological System

Jibin Sun, Chemistry and Chemical Engineering Department, University of California, Berkeley, Tan Hall B81, Chemistry and Chemical Engineering Department, UC, Berkeley, Berkeley, CA 94720-1460

Organometallic polymers combine the good physical, mechanical, and electrical properties of organic polymers with the optical, electrical, magnetic, and catalytic properties inherent to transition metals. In order to clearly understand and compare the functionality of transition metal containing materials with their organic analogues, we developed a set of conjugated transition metal containing building blocks which can be assembled into 1-D linear polymers, zig-zag polymers, and two dimensional sheet materials. Our methodology's key step is based on the alkyn-metathesis mechanism developed by Richard R. Schrock in 1980s. The introduction of metals into the conjugated organic backbones is high yield, fast, and scaleable. The energy gap of HOMO-LUMO bands can be adjusted by choosing different ligands and oxidation states of the metals. The first conjugated organometallic polymer was prepared by the polymerization of these building blocks with various organic linkers. These materials maintain the backbone's conjugation after the introduction of metal ions, unlike previous organometallic polymers. A sticking property of this type of materials is the HOMO-LUMO energy gap can be readily tuned to around or below 2.0 eV by using different ligands. Another interesting property of these types of materials is that they can be processed into mixed oxidation state materials by applying appropriate oxidants.

Furthermore, we studied the length effects of organic linkers; with shorten organic linkers, the metal centers become strong coupled. A new type of conjugated metalloyne material has been prepared by using -C≡C-C≡C- as the linker. These types of materials are exact analogues of polyynes, which are highly unstable, and insoluble materials at room temperature. Other applications of this metalloyne material, such as a potential proton transfer wire, will be also presented.

In addition, in earth carbon cycle, $CO + H_2O \rightarrow CO_2 + 2H^+ + 2e^-$ plays a very important role. In biological system, there are Fe and Ni containing clusters, where this type of transformation happens. Analyzing this mechanism via chemical methods is a longstanding challenge. We used E. J. Corey's retrosynthetic analysis to separate this complex into pieces, independently synthesize each part, and assemble the complex at the last step. Dramatic progress has been achieved toward the synthesis of the complete synthetic model complex.

722aw Inverse Opal Topology of TiO2 Electrodes for Photoelectrochemical Cells

Sonia S. Mathew and Ilona Kretzschmar. Dept. of Chemical Engineering, City College of New York, New York, NY 10030

Harnessing solar energy with inexpensive materials and manufacturing methods is an important challenge in today's energy market. A dye-sensitized solar cell (DSC) is a photo-electrochemical device that uses cheap electrodes like TiO2. It has been demonstrated that adding order to the TiO2 layer of a DSC improves the efficiency of the cell, by reducing electron recombination events [1]. Using this idea, Mor et. al. synthesized electrodes using TiO2 nanotube arrays. In our research, we are investigating TiO2 electrodes that have an inverse opal topology. Such structures are ordered and provide a (comparatively) larger surface area, which allows higher concentrations of adsorbed dye per unit area. It also provides close contact between the dye-TiO2 anode and the electrolyte. Using the method of colloidal self-assembly, we have prepared TiO2 electrodes that have pores in a hexagonal pattern. These electrodes are prepared by assembling polystyrene particles into molds, and subsequently infiltrating them with a polymer-TiO2 mixture. The cured polymer-TiO2-polystyrene film is washed with solvent to remove the polystyrene templates, resulting in a porous film. In this poster, we will report the mechanical, optical, and electrical properties of both porous and non-porous TiO2-polymer films.

[1] G. K. Mor, K. Shankar, M. Paulose, O. K. Varghese and C. A. Grimes, Nano Lett., 2006, 6, 215

722ax Synthesis and Reversible Wettability of Conducting Polyaniline Films Consisting of Helical Microfibers

Lianbin Xu, Key Lab for Nanomaterials, Ministry of Education, Beijing University of Chemical Technology, Beijing, 100029, China, Yushan Yan, Department of Chemical and Environmental Engineering, University of California, Riverside, CA 92521, and Jian-Feng Chen, Research Center of the Ministry of Education for High Gravity Engineering and Technology, Beijing University of Chemical Technology, Beijing, 100029, China.

Conducting polyaniline (PAni) films consisting of helical microfibers were electrochemically deposited in acetonitrile-water electrolyte containing aniline monomer and perfluorooctanesulfonic (PFOS) acid. The films have extended network structure composed of interconnected helical PAni fibers. Typical length of the fibers is about 10-20 microns and the diameter ranges from 200-800 nm but most around 400 nm. The helical fibrous structure is supposed to form through a supermolecular templating process. The surface of the PFOS-doped PAni films exhibited superhydrophobicity with a water contact angle of 153°. Reducing the PFOS-doped PAni films in tetraethylammonium perfluorooctanesulfonate (TEAPFOS) solution by negative potential led to dedoped PAni (in leucoemeraldine base form) films, which were superhydrophilic (water contact angle close to 0°). When oxidizing the superhydrophilic dedoped PAni film by positive potential, the superhydrophobic PAni (in PFOS-doped emeraldine salt form) surface was regenerated. By controlling the electrical potential, PAni films were changed between doped state and dedoped state, resulting in reversibly switchable superhydrophobic and superhydrophilic surfaces. Details on the preparation and characterization of the conducting PAni films are presented including scanning electron microscopy (SEM), transmission electron microscopy (TEM), energy dispersive X-ray spectroscopy (EDS), UV-Vis spectroscopy, cyclic voltammetry (CV), contact angle, and conductivity measurements.

722ay Electrochemical Supercapacitors Based on Polymerizable Ionic Liquids

Surya Sekhar Moganty[1], Joshua Close[1], Pubudu Goonetilleke[2], Sitaraman Krishnan[1], Ruth Baltus[1], and Dipankar Roy[2]. (1) Department of Chemical Engineering, Clarkson University, 8 Clarkson Ave, Potsdam, NY 13699, (2) Department of Physics, Clarkson University, 8 Clarkson Ave, Potsdam, NY 13699

Electrical energy storage and utilization has become a primary focus of the world energy community. There has been a great interest in developing and modifying energy storage devices. Supercapacitor is one such device that has emerged with the potential to meet the current energy demands. Supercapacitors are electrochemical double layer capacitors, intermediate energy systems between batteries and conventional dielectric capacitors. Despite greater capacitances than conventional capacitors, supercapacitors have yet to match the energy densities of batteries and fuel cells. This presentation reports the applicability of polymerizable ionic liquids as the promising advanced materials towards optimizing the performance of supercapacitors. Carbon nanotube based polymer films are prepared by insitu-polymerization of ionic liquid monomers and these plastics are used to build the solid state electrochemical supercapacitor. Supercapacitor performance was characterized by charge-discharge cycles and impedance spectroscopy.

722az Cyclic Acetal Hydroxyapatite Nanocomposites for Bone Tissue Engineering Applications

Minal Patel and John P. Fisher. Fischell Department of Bioengineering, University of Maryland, 3740 Jeong.H.Kim Engineering Bldg, College Park, MD 20742

Hydroxyapatite (HA) is a promising material for bone tissue engineering as its chemistry and structure resembles the mineral component of bone tissue. HA particle size and shape influences the biological outcome of tissue repair. Increasing applications of HA nanoparticles in the field of orthopedics and dentistry have been noted due its biocompatibility, bioactivity, and osteoconductive properties. Addition of HA nanoparticles to biomaterials can enhance the surface properties of the nanocomposite leading to cellular adhesion, migration, proliferation, as well as endogenous cell signaling. We have investigated a biological role of HA nanoparticles, particularly whether a nanocomposite can augment endogenous osteogenic signaling among an embedded osteoprogenitor (OP) cell population within a cyclic acetal biomaterial. Cyclic acetal biodegradable materials are a novel class of materials which degrade into nonacidic and cytocompatible products, critical for maintaining endogenous signaling. We have synthesized nanocomposites consisting of HA nanoparticles (20-70nm, 100nm, 500nm) incorporated within cyclic acetal hydrogels for bone tissue engineering applications. We hypothesize that creation of such nanocomposites can provide a cell scaffold environment to enhance OP cell proliferation and differentiation by facilitating molecular signaling. The objectives of this study are to (1) investigate material properties and cell viability of nanocomposites fabricated by HA nanoparticles of varying sizes within cyclic acetal hydrogels and (2) to determine whether encapsulated OP cells can promote expression of bone morphogenetic protein-2 (BMP-2) within these three-dimensional nanocomposites. Cyclic acetal hydroxyapatite nanocomposites (6x2mm discs, 20-70nm, 100nm, 500nm) were fabricated by preparing six solutions of 5-ethyl-5-(hydroxymethyl)-β,β-dimethyl-1,3-dioxane-2-ethanol diacrylate : poly(ethylene glycol) diacrylate (EHD:PEGDA) and incorporating into each 50 and 0ng/ml HA nanoparticles. Distribution of HA nanoparticles within the nanocomposites was assayed using Field Emission Environmental Scanning Electron Microscope (FE-ESEM). Rat bone marrow stromal cell viability was assayed after 1 week culture in HA/cyclic acetal nanocomposites (50 and 0ng/ml HA in cyclic acetal, 6x2mm). Endogenous expression of BMP-2 was measured at the mRNA level. This study describes a novel nanocomposite and demonstrates that HA nanoparticles may promote osteogenic signaling within the biomaterial.

722ba Synthesis and Characterization of Alpha-Helical Peptide-Based Anchors for Tether Supported Membranes

Lina Zhong, Department of Chemical Engineering, City College of New York (Graduate Center, CUNY), T308 Steinman Hall, 140th St at Convent Ave, New York, NY 10031, Raymond Tu, Chemical Engineering, City College of New York (of CUNY), T313 Steinman Hall, 140th St at Convent Ave, New York, NY 10031, and M. Lane Gilchrist, Chemical Engineering, City College and the Graduate Center of the City University of New York, Steinman Hall, 140th St @ Convent Ave, New York, NY 10031.

We have applied rationally designed peptide anchors to a lipid bilayer interface in order to enhance the viability of a supported lipid membrane anchored to a substrate via polymer tethers. Our aim is to design these supramolecular complexes to enhance their stability and biological feasibility for the study of membrane proteins and their interactions. In current work, we employ alpha-helical peptides (K3A4L2A7L2A3K3) as anchoring molecules, where conjugation of the peptide with N-hydroxysuccinimide (NHS)-esters through amine-based coupling allows one to access a variety of chemistries for orthogonal modification.

These peptides partition within surfactant micelles, which function as in vitro models for interactions with the membrane. Moreover, the secondary structure of peptide within these micelles is characterized with circular dichroism. Lateral fluidity of the fluorescently tagged peptide is analyzed via fluorescence imaging microscopy (Confocal Microscopy) and quantified using fluorescence recovery after photobleaching (FRAP) techniques. Lastly, we inspect the peptide-based anchor in detail via Nuclear Magnetic Resonance (NMR), which will supply data to evaluate the structural properties of the anchor within the bilayer. Variations in the peptide sequence allow us to rationally investigate the influence of sequence on peptide anchor stability.

722bb Replication of Intestinal Basement Membrane Via Chemical Vapor Deposited Silica Thin Films

Brian J. McMahon, Rebecca Carrier, and Daniel Burkey. Chemical Engineering, Northeastern University, 360 Huntington Ave, 342 Snell Engineering, Boston, MA 02114

Cell culture models are frequently used to determine if drug compound candidates will be absorbed across the intestinal barrier, however, these in vitro studies are not always predictive. It has been shown that culture substrate chemistry influences cell phenotype and it is hypothesized that topography may also influence cellular function. The topography of the porcine intestinal basement membrane, which is similar to that of human, consists of finger-like projections called villi and well-like invaginations called crypts. In the crypts, undifferentiated cells begin to migrate toward the top of the villus, changing their cellular function as they move along the topography; indicating that cell culture substrate topography may be an important factor.

Chemical vapor deposition (CVD) provides an opportunity to recreate the intestinal basement membrane topography at the nano- and micro-scale for use in such drug transport studies as well as intestinal tissue engineering. CVD of silica films has been shown to replicate biological structures as small as 150 nm[ ]. These silica films are produced in a surface phase reaction between organic silanes and hydrogen peroxide at ambient temperatures, with no external excitation required. The gross dimensions of intestinal villi range in the order of 100-200*f*Ým, well within the range of CVD's ability to coat.

Removal of cells from the porcine intestinal basement membrane and preparation for CVD was accomplished through a multi-step fixation, maceration and dehydration procedure. Replication of the basement membrane was accomplished through chemical vapor deposition of a thin silica film. We have shown through SEM analysis that the topography of the porcine intestinal basement membrane was maintained, exhibiting visible crypt and villus structures. Ongoing work is being done to increase the mechanical properties of the replica by optimizing the formation of the silica matrix, verified through Fourier transform infrared (FTIR) spectroscopy.

1 Cook, G et al, Angewandte Chemie International Edition, 2003. 42(5): p.557-559.

722bc Biomimetic Polymer Membranes Exhibiting Curvature-Dependent Adhesion

Maria M. Santore and Jin Nam. University of Massachusetts Amherst, Conte Polymer Laboratory, 120 Governor s Drive, Amherst, MA 01003

Cell adhesion initiates with ligand receptor binding, but the immediate consequences are typically also of interest. This can mean signaling, cell rolling and arrest, engulfment or phagocytosis, or endocytosis (for instance relevant to the uptake of drug delivery particles and viruses.) This paper concerns situations where an adherent object or cells is (partially) engulfed by the cell of interest. Such a process involves substantial rearrangement of the cell membrane and underlying cytoskeletal structure. With a series of bilayer polymer membranes, we demonstrate a coupling between adhesion and membrane mechanics which produces qualitative differences in the resulting ability of a "cell" (polymer vesicle) to engulf a target, including metastable binding states and arrested spreading.

722bd Localized Drug Delivery System for the Treatment of Cancer

Marlyn Colon, Eva Christabel Williams, Ryan Toomey, and Norma Alcantar. Chemical and Biomedical Engineering, University of South Florida, 4202 East Fowler Avenue-ENB 118, Tampa, FL 33620

Localized delivery of drugs can provide better treatment for residual cancer tumors. Our research proposes a "package within a package" drug delivery system which results in an effective controlled release of the drug to the targeted site and also reduces its toxic effects to other parts of the body. The first part of the system is a non-ionic surfactant vesicle or niosome and the second a biodegradable and pH sensitive cross-linked hydrogel. The vesicle is prepared by hydration of amphiphillic films with a PBS buffer solution and a fluorescent marker, 5,6-carboxyfluorescein. We chose to encapsulate this particular marker to imitate and test the release rate of the chemotherapy drug from the vesicle interior. After hydration, the vesicles were extruded to constrict their size distribution. Un-entrapped dye was removed using ultracentrifugation and gel exclusion chromatography. The dye encapsulated vesicles were placed in a semi-permeable cellulose membrane and immersed in a solution of either PBS (a high ion solution) or milli-Q water (a low ion solution). The environment in the PBS solution would closely mimic the vesicle's behavior in the hydrogel and the water environment would closely follow that of in the body. Diffusion of the dye through the membrane was determined at different time intervals by using fluorescence spectroscopy and its release rate was determined. Using this methodology, we hope to provide a better treatment technique for drug delivery in cancer patients.

722be A Two-Step Process for the Synthesis of Poly(L-lactic acid)

In Hak Baick[1], Yuesheng Ye[1], and Kyu Yong Choi[2]. (1) Chemical and Biomolecular Engineering Department, University of Maryland, Stadium Drive, Bldg 090, College Park, MD 20742, (2) Department of Chemical and Biomolecular Engineering, University of Maryland, College Park, College Park, MD 20742

Poly(L-lactic acid) (PLLA) is a biodegradable polymer that can be synthesized by ring-opening polymerization of lactide of by direct polycondensation of L-lactic acid. High molecular weight PLLA can be obtained by ring-opening polymerization; however, the process requires a production of high purity lactide and the overall process is complex and costly. Direct polycondensation converts L-lactic acid to PLLA via step-growth polymerization but high molecular weight PLLA is difficult to obtain. In this work, we investigate a two-step direct polycondensation of lactic acid. The first step consists of dehydration of aqueous lactic acid and the prepolymerization at low pressure or by gas purging. The prepolymer is then solid-state polymerized to increase it molecular weight. The polymerization conditions that affect the performance of the polymerization include reaction temperature, catalyst concentration, and water removal efficiency. Discoloration of the reaction mixture at high catalyst concentration is also a problem to be resolved. The polymerization kinetics and the effect of condensate removal by diffusional mass transfer are investigated to develop a quantitative understanding of the polymerization of lactic acid.

722bf Structure and Stability of Electrospun Pvdf/Chitosan Nanofiber

France Noelle B. Villaruel and Hern Kim. Department of Environmental Engineering and Biotechnology, Myongji University, San 38-2 Nam-dong Cheoin-gu, Yongin, Kyonggi-do, 449-728, South Korea

Nanofibers composed of polyvinylidene fluoride(PVDF) with the incorporation of different amounts of chitosan were prepared by electrospinning. The structure, morphology and stability in water of the electrospun nanofibers were investigated. The addition of chitosan (0, 2, 5, 8, 10, 15 w%) in the polymer solution modified the structure and morphology of the nanofiber such that increasing the amount of chitosan increased the nanofiber diameter. Also, the stability of each nanofiber composite under neutral, acidic and alkaline environment was also studied. Results showed that the electrospun PDVF/chitosan, could be a potential biomaterial for tissue engineering or for drug delivery.

722bg Adhesive Properties and Thermodynamic Stability of Hpc:PEG Films

Marlena Brown, Chemical and Biochemical Engineering/Food Science, Rutgers University, 65 Dudley Road, Lab 401, New Brunswick, NJ 08901 and Paul Takhistov, Food Science, Rutgers University, 65 Dudley Rd., New Brunswick, NJ 08901.

Polymer blends are created by mixing polymers with different chemical structures together in order to create new properties intermediate between the two individual polymer components. Film structure, surface properties, and stability of immiscible hydroxypropyl cellulose (HPC) polyethylene glycol (PEG) polymeric blend films were investigated. The effect of substrate properties on the phase separation behavior of a polymer films has been investigated. We found that the amplitude of this process may be modified by a substrate material. This is due to the reduced surface attraction of HPC for the increased surface coverage by PEG. It was observed that regime of spinodal decomposition of films could be controlled through manipulation of key parameters of film formation process. Control of these parameters result in guided surface morphology where surface roughness, pore size, and segregation can be predicted and directed

722bh Development of Sustained Fluoride Releasing Elastomeric Rings for Orthodontic Applications

Shadeed M. Khan[1], Ozge Guney-Altay[1], and Eser Tufekci[2]. (1) Chemical and Life Science Engineering Department, Virginia Commonwealth University, 601 W. Main St., Richmond, VA 23284, (2) Department of Orthodontics, Virginia Commonwealth University, 520 North 12th Street, Richmond, VA 23298-0566

Acidogenic bacteria in the dental plaque produce acids while metabolizing carbohydrates. As these acids diffuse through the plaque into the enamel, hydrogen ions are released. During this process, the acidogenic oral pH favors the dissolution of the minerals on the enamel surface. The diffusion of calcium and phosphate ions out of enamel produces a chalky white spot lesion. During orthodontic treatment, the formation of white spot lesions around the brackets due to poor oral hygiene has long been a recognized problem. Approximately, 50% of orthodontic patients will develop these lesions during treatment of approximately 2 years. These incipient caries can appear in as little as 2-3 weeks after plaque accumulation in gingival areas of teeth.

In this study, a novel approach will be used to develop a fluoride-delivery system to provide fluoride ion release in a controlled and continuous manner in an effort to eliminate the formation of white spot lesions. Our attention is focused on orthodontic ligature ties (o-rings), which are renewed during the routine adjustment appointments of an orthodontic patient during treatment. In this study, we present our initial results on the performance of a delivery matrix that is capable of releasing fluoride in a pre-designed manner at a constant rate during the 30 to 45 day period between these appointments. This method of fluoride delivery presents the potential to eliminate any need for patient compliance and would ensure delivery of fluoride at each orthodontic visit by simply replacing the elastomeric rings.

722bi Photo-Induced Locomotion of Chemo-Responsive Polymer Gels

Pratyush Dayal[1], Olga Kuksenok[2], and Anna C. Balazs[1]. (1) Department of Chemical and Petroleum Engineering, University of Pittsburgh, 1249 Benedum Hall, 3700 Ohara St., Pittsburgh, PA 15261, (2) Chemical Engineering Department, University of Pittsburgh, 1248 Benedum Hall, Pittsburgh, PA 15261

The need to translate chemical energy into a mechanical response, a characteristic of many biological processes, has motivated the study of stimuli-responsive polymer gels. Recently, it has been shown experimentally that by coupling the mechanical properties of the gel with the Belousov-Zhabotinsky (BZ) reaction it is possible to induce self-sustained oscillations in the gel. One of the means for controlling these chemical oscillations is using light as an external stimulus. To study the effect of light on the mechanical behavior of the gel, we use our recently developed a 3D gel lattice spring model (gLSM) which couples the BZ reaction kinetics to the gel dynamics. In this model, the polymer-solvent interactions were taken into account by adding a coupling term to the Flory-Huggins free energy. By virtue of this coupling term, the swelling—de-swelling behavior of the gel was captured in 3D. In order to include the effect of the polymer on the reaction kinetics, the Oregonator model for the photo-sensitive BZ reaction was also modified. Using gLSM model, we probed the effect of non-uniform light irradiation on the gel dynamics. We were able to manipulate the direction and velocity of locomotion of the gel using light as a control parameter. This ability to control the movement of the gel can be utilized in a variety of applications, ranging from bio-actuators to controlled drug release systems.

739a First-Principles Analysis and Monte Carlo Simulations of Surface Segregation In $ZnSe_{1-x}S_x$ Nanostructures

Sumeet C. Pandey[1], Tejinder Singh[1], T. J. Mountziaris[1], and Dimitrios Maroudas[2]. (1) Department of Chemical Engineering, University of Massachusetts - Amherst, 213 Goessmann Laboratory, 686 North Pleasant Street, Amherst, MA 01003, (2) Department of Chemical Engineering, University of Massachusetts, 686 N Pleasant Street, Amherst, MA 01003

Nanocrystals of compound semiconductors, such as the II-VI compounds ZnS, CdSe, and ZnSe, exhibit size-dependent optoelectronic properties and form the basis for a new generation of highly integrated nanoelectronic and photovoltaic devices, as well as biological labels. Nanocrystalline structures over the 2-10 nm size range (quantum dots) exhibit size-dependent luminescence due to quantum confinement of optically excited electron-hole pairs (excitons), which leads to an unprecedented tunability in band gap that can be controlled by varying the size of the nanocrystals. Synthesis of core-shell quantum-dot structures, such as ternary (e.g., $ZnSe_{1-x}S_x$) nanocrystals, produces materials with even wider band gap.

Various synthesis routes to core/shell structures exist, including the coating of a wider-band-gap semiconductor core with a shell of narrower-band-gap material. However, the underlying mechanisms for formation of such core/shell structures remain elusive.

In this presentation, we report theoretical results toward understanding some of the underlying physics that governs such core/shell structure formation. Specifically, we address the problem of equilibrium surface segregation in ternary II-VI semiconductor structures using Monte Carlo (MC) and conjugate gradient (CG) relaxation methods based on classical force fields in conjunction with first-principles density functional theory (DFT) calculations. Atomic-scale (MC & CG) simulations of combined structural and compositional relaxation have been conducted for $ZnSe_{1-x}S_x$ slabs and nanocrystal particles with well-defined facets in order to attain the concentration profile of group VI atoms (S and Se). The slab supercell models expose two (001), (111), or (110) free surfaces. Our MC method employs a multi-step sequence including one MC sweep for compositional relaxation (exchange between Se and S atoms) followed by many continuous-space MC sweeps over all atoms for structural relaxation and an MC step for strain relaxation after each such sweep; in all the steps, trials are accepted or rejected according to the Metropolis criterion. The MC simulation is preceded and followed by energy minimization with a CG scheme to account for local structural relaxation. Our DFT calculations are carried out within the generalized gradient approximation (GGA) and employ plane-wave basis sets, ultra-soft pseudopotentials, and supercell models. We use the DFT calculations to test thoroughly our MC/CG simulation scheme for bulk $ZnSe_{1-x}S_x$ and $ZnSe_{1-x}S_x$ (001)-(2x1) slabs.

We present results for the compositional distribution in ternary $ZnSe_{1-x}S_x$(001), (111) and (110) slabs, as well as in ternary $ZnSe_{1-x}S_x$ faceted nanocrystals. In both cases, we analyze the underlying surface segregation phenomena and the resulting equilibrium state. For the ternary slabs, surface segregation is studied as a function of the composition, x, for various slab thicknesses. For the ternary nanocrystal particles, surface segregation is studied as a function of x and of the size or diameter of the nanocrystal. The underlying surface segregation phenomena are analyzed and plausible core/shell structure formation mechanisms are discussed.

739b Hybrid Organic/inorganic Solar Cells Featuring Microcontact-Printed Quantum Dot Films

Matthew J. Panzer[1], Alexi C. Arango[2], Tim P. Osedach[2], Scott M. Geyer[3], Moungi G. Bawendi[3], and Vladimir Bulovic[2]. (1) Research Laboratory of Electronics, Massachusetts Institute of Technology, Room 13-3082, 77 Massachusetts Avenue, Cambridge, MA 02139, (2) Electrical Engineering & Computer Science, Massachusetts Institute of Technology, (3) Chemistry, Massachusetts Institute of Technology

Colloidal quantum dots (QDs) are of great interest for next generation photovoltaic (PV) device architectures because of their amenity to large scale solution-based synthesis and processing techniques, as well as tunability of the QD bandgap with its size. As an early step towards realizing a low-cost, entirely printed PV platform, we investigated a hybrid device structure in which thin films of CdSe QDs are deposited onto an organic hole transport layer via microcontact printing using a polydimethylsiloxane (PDMS) stamp. Control of QD film morphology by adjusting the surface energy of the PDMS stamp showed a significant improvement in the open circuit voltage of the device. In addition, chemical treatment of the QD films in situ on stamps was used in an effort to improve carrier transport through the QD film in the completed PV device by exchanging the bulky native trioctylphosphine (TOP) ligands for several amines with shorter molecular lengths. Solar cells with optimized QD layers achieved open circuit voltages of 1.3 V and short circuit currents of nearly 0.1 mA/cm^2 under green light illumination (78 mW/cm^2; peaked at 532 nm). An internal quantum efficiency of ~30% was also obtained (external quantum efficiency ~0.5%). Further improvements to our hybrid device architecture are currently under investigation, including the possibility of using a multilayer tandem structure.

739d Electrochemical Impedance Spectroscopy of ZnO Nanowire and Nanoparticle Dye Sensitized Solar Cells

H. Majidi, Chemical and Biological Engineering, Drexel University, 3141 Chestnut Street, Philadelphia, PA 19104 and J.B. Baxter, Chemical & Biological Engineering, Drexel University, 3141 Chestnut St, Philadelphia, PA 19104.

Dye sensitized solar cells (DSSCs) are one of the most promising designs for low-cost solar cells. In DSSCs, a mesoporous, wide band gap semiconductor is coated with a monolayer of sensitizing dye and infiltrated with a liquid electrolyte containing I-/I3- redox couple. Photons are absorbed by the dye molecules, which inject photoexcited electrons into the semiconductor conduction band where they are then transported to the conducting substrate. The dye is regenerated by reduction by I-, and the resulting I3- diffuses to the counterelectrode where it is reduced to complete the circuit. Optimization of DSSCs for commercial use requires fundamental understanding of electronic and ionic transport within the individual materials and at the interfaces within the cell. Electrons are transported to the substrate by diffusion, and achieving high charge collection efficiencies requires that diffusion times be much less than recombination times. Electron transport in the ZnO electrode can be improved by using nanowires instead of nanoparticles because nanowires provide a direct path for electron diffusion, as opposed to nanoparticles where many interparticle hopping steps are required.

In this paper, we will compare electron transport and recombination in DSSCs built with ZnO electrodes employing nanoparticle and nanowire morphologies. Transport, accumulation, and recombination of charge were investigated using electrochemical impedance spectroscopy (EIS). In EIS, the frequency-dependent complex impedance is determined by measuring the current response to an AC voltage perturbation. EIS spectra are plotted in Bode or Nyquist form, and are used to determine an equivalent circuit model. The model describes electronic and ionic processes in different parts of the cell using appropriate circuit elements. For example, ionic diffusion in the electrolyte is modeled using a bounded Warburg impedance, charge transfer at the platinum counterelectrode by an RC circuit, and electron diffusion and recombination in the ZnO electrode by a Warburg impedance in parallel with a capacitance. We will discuss the differences in electron transport observed between nanoparticle and nanowire ZnO DSSCs determined by EIS; as well as how EIS data can be used in conjunction with I-V curve characteristics to determine which fundamental processes limit efficiencies in both cells.

739e Structure and Low Temperature Spectral Emissivity of Erbium Doped Metal Oxide Nanofibers for Application as Selective Emitters In Thermophotovoltaic Devices

Ruofeng Wang and E. A. Evans. Department of Chemical and Biomolecular Engineering, The University of Akron, Whitby Hall 401, Akron, OH 44325-3906

A series of novel, thermally excited selective emitter material (e.g., $Er_2Ti_2O_7$) for thermophotovoltaic (TPV) energy conversion is described in this work. The photovoltaic material's properties are very sensitive to the crystal structure and crystal field environment. The X-ray diffraction (XRD), transmission electron microscopy (TEM), X-ray photoelectron spectroscopy (XPS), and Fourier transformation IR (FTIR) have been applied to investigate the crystal structure and optical properties of the selective emitter materials. It has been shown that different crystal structures were formed by controlling the annealing conditions. Both the crystal structure and host materials will affect emission properties of selective emitter materials. Lattice parameter, grain size and particle size distribution profiles have been investigated. Gaussian multipeak fitting module has been applied to analyze center peak bandwidth of Er^{3+} emission bands at 1.53μ m. The experimental measurements show that power densities can be obtained up to 125 W/mole of Er, in the wavelength range between 1.3 and 1.7 μ m, at temperature of 500°C. Excitation mechanisms of thermal induced Er^{3+} emission in polycrystalline metal oxide matrices have been proposed. Optimize surrounding environment of Er^{3+} ion will inhibit the non-radiative de-excitation of Er^{3+} and enhance energy transfer efficiency from matrices to 4f electrons of Er^{3+} ion. The emissive power spectrums of selective emitter materials are matched very efficiently to the response of GaSb and InGaAs photovoltaic cell in TPV devices.

745a Controlling Microbial Adhesion and Biofilm Formation by Self-Assembled Monolayers of Alkanethiols Presenting Mannitol Group

Shuyu Hou[1], Erik A. Burton[2], Yan-Yeung Luk[2], and Dacheng Ren[1]. (1) Department of Biomedical and Chemical Engineering, Syracuse University, 121 Link Hall, Syracuse, NY 13244, (2) Department of Chemistry, Syracuse University, 1-014 Center Sci & Tech, Syracuse, NY 13244

Biofilms are sessile colonies of microbes attached to a variety of surfaces. Such ubiquitous structures cause serious problems such as persistent infections and biocorrosion. According to NIH, biofilms are involved in approximately 80% of human bacterial infections. We recently reported that self-assembled monolayers presenting tri(ethylene) glycol (TEG) are highly resistant to adhesion and biofilm formation of *Escherichia coli*. To further understand microbe-surface interactions and to develop effective biofilm control strategies, novel self-assembled monolayers (SAMs) of alkanethiols presenting mannitol group on gold films were analyzed for their resistance to biofilm formation in this study. The cell adhesion and biofilm formation were monitored with confocal laser scanning microscopy. The three dimensional structure of biofilms was analyzed with the COMSTAT software to quantify the biomass, surface coverage, and thickness. The mannitol-terminated SAM exhibited strong resistance not only to the biofilm formation of *E. coli* RP437 (84.1% ± 4.5% reduction), but also to that of the opportunistic human pathogens *Pseudomonas aeruginosa* PA01 (88.8% ± 9.0% reduction) and *Candida albicans* SC5314 (86.6% ± 9.0% reduction) compared to bare gold surface. This suggests that the bioinertness of mannitol-terminated SAMs is not species-specific because they are resistant to both bacterial and fungal biofilm formation. In addition, the long-term biofilm assay in a flow cell showed that mannitol-terminated SAMs were resistant to *E. coli* biofilm formation for at least 4 weeks, which is four times longer than TEG-terminated SAMs. To our best knowledge, mannitol-terminated SAM has the longest resistance to microbial biofilm formation compared to other SAMs reported to date. Such well-defined surfaces are an ideal platform for studying microbes-surface interactions and developing effective control strategies for antifouling and anticorrosion applications.

Keyword: self-assembled monolayers, biofilm, mannitol, tri-(ethylene glycol)

745c Self-Assembly of Novel Hybrid Vesicles from Phospholipids/ dendrimers/block Copolymers

Anthony J. Kim[1], Mark S. Kaucher[2], Dalia H. Levine[1], Virgil Percec[3], and Daniel A. Hammer[4]. (1) Chemical and Biomolecular Engineering, University of Pennsylvania, 210 South 33rd St, Philadelphia, PA 19104, (2) Chemistry, University of Pennsylvania, 210 South 33rd St, Philadelphia, PA 19104, (3) Chemistry, U. Pennsylvania, 231 South 34th Street, Philadelphia, PA 19104-6323, (4) Bioengineering, University of Pennsylvania, 240 Skirkanich Hall, 210 South 33rd Street, Philadelphia, PA 19104

Phospholipid vesicles and polymersomes (polymer vesicles) are biocompatible nanocarriers that have an enormous potential for fundamental membrane studies and in vivo imaging, diagnostics, and drug delivery applications. However, phospholipid vesicles and polymersomes each have their own drawbacks that limit their use in a number of biomedical applications. In this presentation, I will first discuss a new class of vesicles, “dendrosomes”, assembled from a library of amphiphilic dendrimers. Second, I will describe novel hybrid vesicular structures assembled from phopholipid/dendrimer/block copolymer blends. These hybrid vesicles have a number of very unique structural properties and their various applications will be discussed. Finally, in vitro study of Doxorubicin encapsulated hybrid vesicles that target different cell lines will be discussed.

745d Tailoring of Self-Assembling Peptide Scaffolds with Biofunctional Epitopes

Ying Chau[1], Ying Luo[2], Qianqian Li[1], and Alex C.Y Cheung[1]. (1) Chemical Engineering, Hong Kong University of Science and Technology, Clear Water Bay, Kowloon, Hong Kong, Hong Kong, (2) Biomedical Engineering, Peking University, 60 YanNan Yuan, HaiDian Dist, Beijing, China

We have discovered a template for inserting biofunctional epitopes into self-assembling peptides. Flanking the centrically positioned cue sequence by three repeats of RADA on each side retains the gelation properties of the self-assembling peptides. Using this bottom-up approach, new nanofibrous hydrogel scaffolds responsive and instructive to cell behavior can be readily prepared for 3-D cell culture and tissue engineering. To illustrate its application, we have synthesized building blocks containing short peptides to provide the following functions: 1) degradability sensitive to matrix metalloproteinases (MMPs), enzymes involved in extracellular matrix turnover during cell expansion; and 2) stimulation of neural differentiation. These examples demonstrate that two-way communication between cells and materials can be achieved with the designer peptide scaffolds.

745e Self-Assembly and DNA Binding Properties of Bzip Peptide Amphiphiles

Rachel S. Marullo[1], Raymond Tu[2], and Matthew Tirrell[1]. (1) University of California, Santa Barbara, Materials Research Laboratory, Santa Barbara, CA 93106-5121, (2) Chemical Engineering, City College of New York (of CUNY), T313 Steinman Hall, 140th St at Convent Ave, New York, NY 10031

GCN4 is a well-characterized protein in yeast that is a member of the AP-1 family of transcription factors. This work involves a thirty eight amino acid "bZip" peptide derived from GCN4 that contains a leucine zipper based dimerization segment and a basic DNA binding region. Conjugating bZip to a hydrophobic carbon chain tail promotes the formation of cylindrical micelles in aqueous solution. Circular dichroism has shown that the peptides adopt a higher degree of alpha-helical secondary structure in the micelle corona compared to the unaggregated monomers. Enhanced secondary structure facilitates the DNA binding of bZip peptide amphiphiles in a fashion similar to the native protein, in contrast to the unmodified peptide. The specificity of the DNA binding to the AP-1 target site versus other sequences is currently under examination. Modification of the peptide amphiphiles to drive assembly into different shaped aggregates is also being studied.

Biofunctional bZip constructs may have applications as artificial transcription factors or gene delivery vehicles. Future work will include in vitro cell assays of bZip peptide amphiphile micelles to determine their biological potential. The micelles could be engineered to incorporate peptides with other functionalities such as cellular uptake and nuclear targeting.

745f Self-Assembly of Comb-Rod Dendritic Block Copolymers

Shujun Chen and Paula T. Hammond. Chemical Engineering, Massachusetts Institute of Technology, 77 Massachusetts Ave, Cambridge, MA 02139

Linear-dendritic block copolymers combine the multi-functionality of dendrimers and the phase segregated morphological behavior of block copolymers. In this work, the aqueous self-assembly of a family of comb-rod dendritic block copolymers, which consists of a monodisperse alkyl-modified hydrophobic peptide rod block, poly(n-alkyl-L-glutamate) or PALG, and a biodegradable hydrophilic polyester dendron (PED) block, was investigated using dynamic light scattering (DLS), transmission electron microscopy (TEM), and cryogenic transmission electron microscopy (Cryo-TEM). Above the critical micelle concentration (CMC), as the concentration increases, the morphology of these PALG-PED copolymers changes from micelles to vesicles. At intermediate concentrations, non-spherical nanostructures such as nanoprisms and nanotubes are observed. Reducing the length of the linear PALG block shifts the micellar phase to higher concentrations. These amphiphilic biodegradable PALG-PED copolymers offer great promise as functional tunable drug delivery carriers.

745g Structure and Nanomechanics of Collagen Complexes for Novel Biomaterials

David V. Svintradze[1], **Vamsi K. Yadavalli[2]**, and Ramana Pidaparti[1]. (1) Mechanical Engineering, Virginia Commonwealth University, 401 W Main St., Room E2234, Richmond, VA 23284, (2) Chemical and Life Science Engineering, Virginia Commonwealth University, 601 W Main Street, Richmond, VA 23284

Collagen is a fibrous protein that undergoes a complex self-organizing process to form hierarchical fibers. The self-assembly of the triple-helical collagen into higher order structure governs the mechanism of macro fibril formation. The mechanical properties and anisotropy of the biomaterial are affected by the orientation and composition of individual fibers. The process of fiber nano and microstructure formation is extremely sensitive to the biophysical and biochemical environment. However, the general principles that allow collagen molecules to form diverse supramolecular structures are not well understood. Here we describe the formation of single collagen fibers under altering environmental conditions including pH, ionic strength and an external voltage. High resolution imaging by an atomic force microscope (AFM) allowed us to probe the self-assembly of collagen molecules into fibers and the structural changes induced by these environmental factors. The nanomechanical properties of the collagen fibers thus formed were determined via force spectroscopy. Based on our results we identify a set of rules critical to fibrillogenesis and describe a design strategy to synthesize composite fibrous materials from a mixture of collagen, DNA, and synthetic polymers. Experiments outlining the formation of self-organizing collagen/DNA-based materials in vitro will be described. Our results investigate the self-organizing process and allow the development of novel hierarchical biomaterials with unique mechanical and structural properties. Potential applications include self-organizing nano-membranes, and self-healing artificial tissues to facilitate wound healing.

746a Synthesis and Swelling Response of Block and Random Copolymer Brushes Incorporating Thermo- and pH-Responsive Constituents

S. Michael Kilbey II, Center for Nanophase Materials Sciences, Oak Ridge National Laboratory, Oak Ridge, TN 37831, J. Alaina Floyd, Chemical and Biomolecular Engineering, Clemson University, Clemson, SC 29634, and Santosh B. Rahane, Department of Chemical and Biomolecular Engineering, Clemson University, 127 Earle Hall, Clemson, SC 29634-0909.

Layers of end-tethered polymer chains – so-called "polymer brushes" – are of interest for their ability to straddle the fluid-solid interface and impart new and useful properties to the underlying surface. Because brushes adjust their structure when environmental conditions change, a strategy to enrich the properties of the interfacial layer is to integrate multiple components into the brush and use constituent materials that exhibit, for example, pH-, temperature-, or photo-responsive behavior. In this work, polymer brushes consisting of poly(N-isopropylacrylamide) and poly(methacrylic acid) were grown by surface initiated, photoiniferter-mediated photopolymerization. To create bi-level, block copolymer brushes, tetraethylthiuram disulfide (TED), a source of deactivating radicals, must be added to establish and maintain the equilibrium between dormant and active (growing) chains during photopolymerization. Layers consisting of either block or random copolymers were synthesized and characterized in the dry and solvated states using multi-angle ellipsometry. The swelling response of the solvated homopolymer and copolymer brushes as a function of pH, temperature and ionic strength will be discussed, highlighting how the chemical information encoded into the brush chains determines swelling behavior.

746b Towards the Development of Biodegradable Sensors from Smart Hydrogels

Diana K. Snelling, Chemical Engineering, The University of Texas at Austin, 1 University Station C0400, Austin, TX 78712 and Nicholas A. Peppas, Chemical Engineering, Biomedical Engineering, Division of Pharmaceutics, The University of Texas at Austin, 1 University Station C0400, Austin, TX 78712.

The performance of environmentally responsive hydrogels can be greatly improved by fabricating them on the micro and nanoscale. A smaller spatial scale offers enhanced mass transfer and chemomechanical response. As part of a microscale sensing platform these networks can offer unriveled reliability and reproducibility. We have previously shown the effectiveness of this approach in the design of micropatterned ultrasensitive microcantilever pH-sensors. The work here describes the design of pH-responsive hydrogels capable of tunable hydrolytic degradation for use in microscale sensing applications. A novel hydrogel consisting of poly(methacrylic acid), (PMAA), crosslinked with poly (caprolactone) was synthesized via UV-initiated free-radical polymerization. The swelling kinetics closely followed those of nondegradable PMAA networks. The degradation kinetics were studied by swelling hydrogel disks in phosphate buffered saline (PBS) solution at 37°C for 60 days to mimic biological conditions. The molecular weight between crosslinks was found to increase linearly with time. Overall, this hydrogel demonstrates favorable swelling and degradation kinetics to function in degradable microsensing applications.

746c Prediction of Phase Behavior In Surface-Tethered Poly(N-isopropylacrylamide) Networks from Demixing Behavior of Linear Poly (NIPAAm) Solutions

Ajay Vidyasagar and Ryan Toomey. Chemical Engineering, University of South Florida, 4202 E. Fowler Ave. ENB 118, Tampa, FL 33620

Lower critical solution temperature (LCST) polymers experience a volume-phase transition when subjected to small perturbations in external stimuli. For example, poly(N-isopropylacrylamide) (poly (NIPAAm)) undergoes a hydrophilic/hydrophobic transition at roughly 32°C. The critical point of poly (NIPAAm) occurs at a temperature of 29.5 °C and a volume fraction, χ, of 0.43. As the polymer concentration is reduced towards zero, the demixing temperature approaches 34 °C. To explain this phenomenon, the effective Flory interaction parameter χ must be expanded in powers of ϕ, with a minimum of three terms, wherein each term is a function of temperature.

A key question concerns the use of a concentration dependent χ parameter derived from the solution phase diagram to describe the behavior of constrained systems, including end-tethered or cross-linked polymers. We report on the swelling of surface-tethered poly(NIPAAm) networks as characterized by neutron reflectivity. In this study, surface-tethered networks were prepared from photo-cross-linkable poly(NIPAAm) copolymers with benzophenone-pendant monomers. Ultraviolet radiation (≤f = 350 nm) triggers the n,π* transition in the benzophenone moieties leading to a biradicaloid triplet state that abstracts a hydrogen from a neighboring aliphatic C-H group, forming a stable C-C bond. Neutron reflection reveals that the discontinuity in the volume transition of the surface-tethered networks coincides with the miscibility gap of non-cross-linked linear poly(NIPAAm). This result signifies that the concentration dependent χ interaction parameter is unaffected by cross-linking and can be used to model volume phase transitions in constrained systems.

746d "Smart" Surfaces of Polymer Brushes

Dong Meng, Chemical & Biological Engineering, Colorado State University, Fort Collins, CO 80521 and Qiang Wang, Chemical and Biological Engineering, Colorado State University, Fort Collins, CO 80521.

Using a continuum self-consistent field theory in real space, we have investigated the structure and response of two types of "smart" polymer brushes: (1) Poly(N-isopropyl acrylamide) (PNIPAM) brushes in water. Here with a concentration-dependent Flory-Huggins parameter obtained from experiments, we have studied the structure and thermo-response of PNIPAM brushes as a function of chain length and grafting density. (2) Charged two-component polymer brushes. Here we have studied diblock copolymer (DBC) brushes with one block charged. When charge effects dominate, it is ineffective to switch the brush surface by different solvents. Rather, applied electric field and solution pH are more effective, with sharp transitions for the surface-switching. We have also compared the charged DBC brushes with corresponding binary brushes. Our result can be used to guide the experimental design of "smart" surfaces from these polymer brushes.

746e Influence of Binding Strength on the Structure of Supramolecular Polymer-Surfactant Complexes

Chinedum Osuji[1], Manesh Gopinadhan[1], Evan Beach[2], and Paul Anastas[2]. (1) Chemical Engineering, Yale University, 302 Mason Laboratory, 9 Hillhouse Avenue, New Haven, CT 06511, (2) Chemistry, Yale University, 302 Mason Laboratory, 9 Hillhouse Avenue, New Haven, CT 06511

Specific interactions between small molecule species and host binding sites on a polymer chain can be used to engineer supramolecular complexes which display liquid crytalline order. In particular, hydrogen bonding interactions represent a flexible platform for the creation of graft-copolymer like structures by the reversible, specific association of small molecules with complementary sites on the polymer backbone. We use hydrogen bonding between the imidazole termination of mesogenic species and the carboxylic acid groups of poly(acrylic acid) to form side-chain liquid crystalline polymers in solution and in the melt state. We investigate their phase behavior and binding using a combination of calorimetry (DSC), x-ray scattering (SAXS), infrared spectroscopy (FTIR) and optical microscopy. We find that there is a critical interaction strength and critical stoichiometry required for the formation of a liquid crystalline mesophase in these systems. The manipulation of the binding equilibria via temperature makes for an interesting class of stimuli responsive materials in solution and in the melt state.

746f Studying the Influence of Structure on Hydrogel Behaviour: A Molecular Simulation Study of Poly(N-isopropyl acrylamide)

Sanket Deshmukh, Mateusz Landowski, Damian A. Mooney, and J. M. D. Macelroy. School of Chemical and Bioprocess Engineering, University College Dublin, UCD Engineering and Materials Science Centre, Belfield Campus, Dublin, Ireland

Poly (N- isopropyl acrylamide) (PNIPAM) is a temperature sensitive hydrogel, with a lower critical solution temperature (LCST) between 305 – 307K, across which it exhibits a volume change from a swollen state (< 305K) to a collapsed state (>307K). The temperature at which this volume transition occurs, as well as the degree and dynamics of swelling, can be altered by changing the method of synthesis, the functionality of the crosslinker used, the temperature of synthesis as well as by the co-monomer used. However, the link between specific molecular (microscopic) and macroscopic properties is still not clearly understood. As an attempt to provide this connection, we have utilised molecular simulation techniques, including specially developed Monte Carlo (MC) and molecular dynamics (MD), to study the effect of using three different crosslinkers, namely N, N-methylene bisacrylamide (BIS) (tetra-functional), ethylene glycol dimethacrylate (EGDMA) (tetra functional) and glyoxal bis (diallylacetal) (GLY) (octa-functional) on hydrogel behaviour. The effect of these crosslinkers on hydrogel properties were monitored using local structural analysis, as well as dynamical properties, including water diffusion coefficients, hydrogen bonding lifetimes and energies, as well as the transport behaviour of selected probe molecules, including 5-fluoro-1H-pyrimidine-2,4-dione (used in the treatment of certain types of cancer). Insights are provided regarding the a-priori design of these materials for specific functions, including drug delivery applications.

746g Development of Environmentally Responsive Hydrogel Systems Over Micro- and Nanodevices Via ATRP

Hariharasudhan D. Chirra, Department of Chemical and Materials Engineering, University of Kentucky, 177 F. Paul Anderson Tower, Lexington, KY 40506 and J. Zach Hilt, Chemical & Materials Engineering, University of Kentucky, 177 F. Paul Anderson Tower, Lexington, KY 40506.

The incorporation of intelligent responsive hydrogel systems at the micro- and nanoscale has resulted in a wide variety of applications in the diagnostic and therapeutic field. Though conventional methods of the semiconductor industry have been successfully employed for the patterning of hydrogels in devices, methods for fabricating precise 3-D patterns of hydrogels at the micro- and nanoscale over material surfaces are limited. Herein, microcontact printing (μCP) for XY control, followed by atom transfer radical polymerization (ATRP) for Z/thickness control was applied to synthesize a controlled array of environmentally responsive hydrogels (e.g., N-isopropylacrylamide) over material surfaces. The controlled growth and the responsive behavior of the hydrogels to external stimuli were characterized using optical microscopy, fourier transform infrared (FTIR) imaging, atomic force microscopy (AFM), quartz crystal microbalance with dissipation (QCM-D) and scanning electron microscopy (SEM). Using this method, it is possible to synthesize hydrogels over various surfaces for potential applications in the biomedical field.

754a Fabrication of Adhesive Structures Based on the Gecko

Noshir S. Pesika, Chemical Engineering, University of California, Interfacial Science Lab, Engineering II, Room 3357, Santa Barbara, CA 93106 and Jacob N. Israelachvili, Chemical Engineering, University of California at Santa Barbara, 3357 Engineering II, Santa Barbara, CA 93106-5080.

The gecko adhesive system is composed of complex hierarchical structures, which play specific roles to achieve high adhesion and friction forces. A "crowding model" is derived based on geometrical (pillar diameter, spacing, and tilt angle) consideration of a square lattice of pillars analogous to setal arrays (the adhesive structures on the gecko feet) of geckos. The model is used to understand the crowding events that occur upon normal compression of a setal array which limit the achievable tilt angles of the pillars. Good agreement is found between experimental data and prediction based on the "crowding model". The model is also used to predict the optimum packing density of setae (the individual pillars that make up the setal array) of approximately 14,200 setae/mm^2 in agreement with the measured value of 14,400 setae/mm^2. Predictions from the model can be used to guide the fabrication of synthetic dry adhesives inspired by the gecko. In addition, preliminary synthetic structures and novel fabrication techniques will be presented for the fabrication of gecko-like adhesive structures.

754b Biomimetic Membrane Properties and the Morphology of Giant Hybrid Vesicles with Phospholipid/block Copolymer Coexistence

Jin Nam[1], Paul A. Beales[1], and T. Kyle Vanderlick[2]. (1) Department of Chemical Engineering, Princeton University, Engineering Quadrangle, Princeton, NJ 08544, (2) Department of Chemical Engineering, Yale University, 10 Hillhouse Ave, Dunham Laboratory 236, New Haven, CT 08544

A number of elaborate liposome systems and theories based on fluid phospholipid membranes have been investigated and their synthetic analogue, amphiphilic diblock copolymers have provided an alternative to phospholipids as mimics of the cell membrane with the unique morphology and a variety of tunable properties. Though phospholipids can afford lots biological functionalities in the flexible membrane the application is limited due to the low stability in circulation. The biomimetic membrane system made of diblock copolymer can impart structural stability into lipid based vesicles bilayer architectures. Therefore, with each of the advantages of lipid membrane biofunctionalities and amphiphilic copolymers structural stability we create self-assembled giant hybrid lipid/diblock copolymer vesicles and study changes in the morphology and membrane mechanical properties of the hybrid vesicle membrane. The effect of lipid to copolymer ratios on the vesicle membrane is studied with a micropipette aspiration technique and the morphology is characterized by confocal microscope. Phospholipids formed micron sized multi-spot domains within a diblock copolymer matrix and the number of domains increased at higher lipid concentrations. We also study the possibility of pore formations from the spot domains generated by lipid extraction with surfactant seeking formation of multi-functional constructs whose bilayers are perforated or decorated with biofunctional molecules such as functional liposome carriers with biological windows. The unique vesicular morphology having pores is highly beneficial in biological applications not only in such fields as delivery-release but also micro-machinery vehicles with incorporation of biological materials.

754c Lipid Raft Formation In Bilayer Membranes with Biomimetic Cytoskeletons

Noah Malmstadt, Mork Family Department of Chemical Engineering and Materials Science, University of Southern California, 925 Bloom Walk, HED 216, Los Angeles, CA 90089

Nanostructured phase-separated domains—so-called lipid rafts—are widely thought to act as control elements in the plasma membrane. They have been implicated in a wide variety of biological processes, including signal transduction, viral infection, and antibiotic activity. The ability to control and manipulate lipid nanostructures in cell membranes is a potential novel therapeutic approach for a wide variety of diseases. Insulin resistance underlying type-2 diabetes, for instance, has been linked to a pathological association between the insulin receptor protein and membrane domains enriched in the lipid ganglioside GM3. Understanding the biophysical bases of such interactions and developing pharmaceutical approaches to altering them would be greatly aided by the development of simple in vitro systems for studying lipid nanostructure formation and raft/protein interactions.

Simplified model systems for studying lipid biophysics in vitro are typically formed from synthetic or purified natural lipids that spontaneously self-assemble into bilayer structures. Given the proper lipid compositions, such systems display the phase separation behavior thought to underlie the lipid raft phenomenon. This phase separation, however, takes place on the wrong scale: existing biomimetic membrane systems produce separated domains tens of microns wide rather than the physiologically relevant nanoscale lipid domains. Recent work has suggested that this scale discrepancy originates with the mechanical connectivity between the cell membrane and the cytoskeleton. Membrane anchoring to the cytoskeleton results in the formation of membrane compartments between which lipids cannot freely diffuse, limiting the long-range diffusivity of lipid molecules and restricting the growth of phase-separated domains to nanoscale rafts.

We are building lipid-based systems that properly recapitulate biological raft formation by mimicking the mechanical attachments between the lipid bilayer membrane and the cytoskeleton. These systems are based on giant unilammelar vesicles (GUVs) that contain hydrophilic monomer solutions in their interior and lipids with vinyl-modified headgroups in the membrane. Upon initiation of free-radical polymerization, a cross-linked polymer hydrogel is formed and covalently attached to the modified lipids. The hydrogel then serves as a biomimetic cytoskeleton, mechanically restricting diffusion in the bilayer.

Nanoscale lipid behavior is observed using a combination of fluorescence resonance energy transfer techniques and total internal reflection fluorescence microscopy. These techniques allow us to probe supermolecular complexes containing only a few molecules, and to observe early events in raft formation. These observations promise insights on the thermodynamics and kinetics of raft formation.

We are further developing a microfluidic system that facilitates the automated production of GUVs with highly tailored structures, including asymmetric bilayer compositions and precise control over vesicle contents. This system is based on a multiphase droplet flow configuration that allows for the sequential assembly of the bilayer from two independently constituted lipid monolayers. Together with the technology for engineering biomimetic cytoskeletons, this system promises to serve as a flexible platform for studying raft/protein interactions and developing pharmaceuticals that target lipid nanostructures.

Web Page: chems.usc.edu/malmstadt/research/nanoscale-lipid-structures.htm

754d Development of An Enzymatically Crosslinkable Biomimetic Collagen That Exhibits Collagen-Like Molecular Architecture with Improved Cellular Recognition

Yen Wah Tong and Shih Tak Khew. Department of Chemical and Biomolecular Engineering, National University of Singapore, 4 Engineering Drive 4, Singapore, 117576, Singapore

This work focuses on establishing a molecular strategy to engineer a functional biomimetic collagen that exhibits stable collagen-like triple-helical conformation with cell-binding activity, in addition to an enzyme-mediated crosslinking by tissue transglutaminase (tTGase). A novel sequence derived from human fibrillin-1 was identified as an amine donor substrate for tTGase, using a previously characterized APQQEA derived from human osteonectin as an amine acceptor probe. Subsequently, collagen-mimetic peptides (CMPs) supplemented with a cell-binding sequence (GFOGER) and the identified tTGase substrate sequences were conjugated onto a generation 2 poly(amidoamine) dendrimer, resulting in a crosslinkable collagen-mimetic dendrimer, denoted as CMD-K and CMD-Q, respectively. Both CMD-K and CMD-Q exhibited enhanced triple-helical stability and supported cell adhesion in an integrin-dependent manner. Finally, tTGase-mediated crosslinking between CMD-K and CMD-Q resulted in a supramolecular structure that exhibited stable collagen-like triple-helical conformation and improved cellular recognition. The results show that the triple-helical structure is important in preserving the GFOGER cell-binding site while the tTGase-mediated protein crosslinking may also be crucial for the recognition by cell surface integrin receptors.

754e Photopatternable Synthetic Materials with Biorecognition Abilities for Use In Analytical Microdevices

Youyou Zheng, Department of Chemical and Biological Engineering, Missouri University of Science and Technology, 143 Schrenk Hall, Rolla, MO 65409-1230 and David B. Henthorn, Department of Chemical & Biological Engineering, Missouri University of Science and Technology, 143 Schrenk Hall, Rolla, MO 65409-1230.

Molecular recognition is crucial to the function of many biological systems including substrate binding by enzymes, molecular signaling, and immunological response. Synthetic materials produced with these molecular recognition traits would find use in a number of different applications, especially in areas where harsh processing conditions would denature or destroy naturally produced biomolecules. In this work, we investigate the fabrication of synthetic polymeric films with surface exposed binding sites specific to biomolecules of interest (immunoglobulins, etc.) as a means to create sensitive devices. We have developed a novel surface imprinting technique where the target protein is used as a template in the creation of binding sites. The use of a templating mask lends well to traditional photolithographic processes and enables the construction of surfaces with spatial control on molecular recognition. Recognitive sites show a 5-fold preference for the target IgG over control materials with minimum affinity for competing IgG molecules. Reuse of the template mask is shown to be possible, allowing for a number of synthetic recognitive surfaces to be created from a small amount of the template molecule. The work is being extended to include other species of interest including cell adhesion molecules.

754f Incorporating Stimulus-Responsive Character into Filamentous Virus Assemblies

Harry Bermudez and Adam P. Hathorne. Polymer Science and Engineering, University of Massachusetts, 120 Governors Drive, Amherst, MA 01003

Controlling interactions between building blocks, in either guided- or self-assemblies, is becoming increasingly important for the creation of functional materials. We have focused our attention on the well-known model assembly, the filamentous bacteriophage, where our strategy is to selectively alter surface features by focusing on spatially distinct capsid proteins. Towards introducing stimulus-responsive behavior in these flexible, rod-like particles, we have introduced elastin-like polypeptide (ELP) motifs of isoleucine and tyrosine guest residues by recombinant DNA methods. Our hypothesis is that modification of the major coat capsid protein would be greatly amplified by the 2700 copies per particle. Characterization of ELP-phage particles was carried out by microbiological assays, zeta potential, dynamic light scattering, and calorimetry. Bacteria producing ELP-phage particles grow more slowly and surprisingly, ELP-modified phages display a significant reduction in viral infectivity. For the lengths of ELP inserts studied, modified phages do not aggregate from solution as monitored by DLS. However, the hydrodynamic size of the phages depends on the details of the ELP motif. Zeta potential measurements reveal the particles are electrostatically stabilized, and this contributes in part to the energetic barrier against aggregation. Preliminary calorimetric data indicate subtle thermal transitions in the range 35–45°C, suggesting that the ELP motif may collapse without triggering macroscopic aggregation. The results are consistent with the classical picture of critical solution phenomena at low concentrations, where to drive phase separation, solvent quality must be increasingly poor. Apart from being model systems to study basic questions of self-assembly, extending these modular systems is likely to result in improved understanding and control over self-assembly in various applications.

754g Fabrication of Tubular Scaffolds from Silk Fibroin Using Biologically Inspired Gel Spinning Technique

Chris Cannizzaro[1], Michael Lovett[1], Gordana Vunjak-Novakovic[2], and David Kaplan[1]. (1) Biomedical Engineering, Tufts University, Science and Technology Center, 4 Colby Street, Medford, MA 02155, (2) Biomedical Engineering, Columbia University, New York, NY

The applications for tubular scaffolds within the field of tissue engineering are numerous, provided they could be engineered with the proper mechanical, biological, and degradation properties. Previously, we demonstrated the utility of tubes fabricated from silk fibroin to fulfill this need (Lovett et al., 2007). Tubes covering a wide range of diameters (0.2 to 6.0 mm) were produced quickly and simply using a dipping method, whereby a rod was successively dipped in a silk fibroin solution and a methanol solution to induce beta-sheet formation. We have since developed a gel spinning technique to fabricate tubes with improved biological and mechanical properties. In this process, the silk is driven through a small gauge needle, which induces a shear stress upon the amorphous concentrated fibroin (silk I) helping to exclude water from the protein solution, align the silk fibrils, and induce silk II (antiparallel β-sheet, aqueous insoluble) structure. This process mimics the process of protein spinning in the native silkworm, where fibroin concentration and physical shear play critical roles in the spinneret. The gel spinning process allows properties such as winding pattern, tube porosity, and tube composition to be controlled in the new process described here, with further options during post-winding processing via treatment with methanol, air-drying, and/or lyophilization. Silk fibroin tubes generated using this new process have applications within tissue engineering, from blood vessel grafts and nerve guidance channels to in vitro migration assays, permeability studies, and novel composite scaffolds in general.

Lovett, M., Cannizzaro, C., Daheron, L., Messmer, B., Vunjak-Novakovic, G., and Kaplan, D. L. (2007): Silk fibroin microtubes for blood vessel engineering. Biomaterials 28, 5271-5279.

Web Page: ase.tufts.edu/terc/

770a Acid Sensitive Polymers for Drug Delivery

Eric Bachelder, S. E. Paramonov, J Dashe, T. T. Beaudette, and J. M. J. Frechet. Chemistry, UC Berkeley, UC Berkeley, 719 Latimer Hall, Berkeley, CA 94720

Polymers that are acid sensitive are very important for immunotherapy and cancer treatment. The lower pH present in the lysosome of a macrophage, or of a tumor cell can be targeted by the selective degradation of the pH sensitive polymer leading to the release of an encapsulated drug inside the targeted cell. Towards this goal, a library of polyurethanes with different hydrophobicities containing the same acid-degradable dimethyl ketal moiety embedded in the polymer main chain have been prepared. The polymers were designed to hydrolyze at different rates in mildly acidic conditions as a function of their hydrophobicity to afford small molecules only with no polymeric byproduct. The library of polymers was screened for the formation of microparticles using a double emulsion technique. The microparticles that were obtained degraded significantly faster at acidic pH (5.0) than at physiological pH (7.4) with degradation kinetics related to the hydrophobicity of the starting polymer. In vitro studies demonstrated the ability of the FITC-BSA loaded microparticles to be phagocytosed by macrophages resulting in a 10-fold increase in the protein uptake compared to a free protein control; in addition, the microparticles were found to be nontoxic at the concentrations tested of up to 1 mg/mL. The particles generated drastically increased the ability of macrophages to present antigens to T-cells compared to nonencapsulated protein. Overall, these new class of polymers are easily tunable to deliver the protein in a pH sensitive manner and have the potential to increase the efficacy of protein vaccines.

770b Reductively Degradeable Polymeric Assemblies Constructed from Polyethyleneoxide-S-S-Polycaprolactone Diblock Copolymer for Applications In Drug Delivery

Karthikan Rajagopal and Dennis E. Discher. Chemical and Biomolecular Engineering, University of Pennsylvania, Philadelphia, PA 19104

Filamentous worm-like micelles constructed from amphiphilic diblcock copolymers are attractive nano-scale structures that have shown potential for application in drug delivery. The utility of these structures stems from their favorable physical attributes that resemble natural viruses. As a result, the worm-like micelles have shown significantly longer in-vivo circulation times compared to spherical assemblies such as micelles and vesicles. The functionality of these structures can be further enhanced if their degradation rates can be controlled in response to changes in physiologically relevant stimuli such as pH or temperature. Alternatively, we envisioned that the use of reduction potential as a trigger for worm-like micelle degradation would be an attractive strategy, particularly for the intra-cellular delivery of drugs as one can take advantage of the differences in reduction potential between the inside and outside of the cells. With that objective herein, we present the design, synthesis and characterization of a diblock copolymer, polyethyleneoxide-S-S-polycaprolactone (PEO-S-S-PCL), which self-assembles to form worm-like micelles and specifically undergoes degradation within a reducing environment. The stability of the PEO-S-S-PCL worm micelles investigated at different concentrations of reducing agents show that at millimolar range significant degradation is observed. Time lapse fluorescence microscopy studies show that the instantaneous rupture of the supra-molecular worm micelles occurs as a result of the molecular level reduction of the disulfide bond. Further, using a similar synthetic approach and by changing the block sizes a vesicular morphology can be obtained. Encapsulation and release of hydrophilic molecules from vesicles and hydrophobic molecules from worm micelles are currently being investigated.

770c Synthesis of Thermo-Responsive Polymeric Micelles of Pnipaam-B-Omma as a Drug Carrier for Loading and Controlled Release of Prednisolone

Dapeng Cao, Wei Li, and Weixia Tu. College of Chemical Engineering, Beijing University of Chemical Technology, Bei-San-Huan East Road 15, Beijing, 100029, China

We synthesize a drug delivery system of poly(N-isopropylacrylamide)-b- oligo(methyl methacrylate) (PNIPAAm-b-OMMA) via polycondensation of two homopolymers in 1,4-dioxane. The products are characterized by FT-IR and 1H NMR spectra and TEM. The PNIPAAm-b-OMMA copolymer micelles in aqueous solution present the same lower critical solution temperature (LCST) as the unmodified PNIPAAm, owing to the formation of a core-shell micellar structure shielding the hydrophobic inner OMMA core from interacting with water. The micelle carriers exhibit two heterogeneous microdomains: a hydrophobic inner core capable of highly solubilizing hydrophobic prednisolone molecules, plus a hydrated outer shell that stabilizes this micellar structure below its LCST. Moreover, the micelle carriers show reversible thermo-responsive aggregation/dispersion in responsive to temperature cycles through the LCST. By using the anti-inflammation drug prednisolone as model drug, it is found that the PNIPAAm-b-OMMA drug carrier could prolong the release time, and control the release amount by changing the temperature. Accordingly, this copolymer micelle may provide as an effective drug carrier for drug control and release.

770d Surface Modification of Monodisperse Magnetic Nanoparticles with Fluorescent-Thermo-Responsive Polymeric Shells for Biomedical Applications

Adriana P. Herrera, Yashira Zallas, and Carlos Rinaldi. Chemical Engineering, University of Puerto Rico Mayguez Campus, Mayaguez, PR 00680

Magnetic nanoparticles such as magnetite (Fe_3O_4) or cobalt ferrite ($CoFe_2O_4$) were synthesized in presence of oleic acid by the thermodecomposition method. These nanoparticles were modified with silane-functional molecules to graft acrylate ($CH_2=C(CH_3)-C(=O)-O$) end groups onto the nanoparticles surface. Functionalization of these modified nanoparticles with N-isopropyl acrylamide monomers such as NIPAM and NIPMAM were carried out by free radical polymerization using an azo-initiator such as AIBN and a methyl bis-acrylamide (MBA) crosslinking agent. The hydrodynamic size of these functionalized nanoparticles was studied as function of temperature. Measurements of the specific absorption rate (SAR) of magnetic nanoparticles coated with thermo-responsive polymeric shells were performed in presence of a magnetic field to determine their potential therapeutic applications. Fluorescent magnetic nanoparticles were also synthesized by the incorporation of a fluorescent modified acrylamide monomer (FMA). The temperature of the medium surrounding a collection of nanoparticles was determined by monitoring fluorescence changes in the FMA monomer resulting from the decrease in polarity of the NIPAM/NIPMAM copolymer chain with increasing temperature.

770e Plasma Polymerization of Thermoresponsive Poly(N-isopropyl acrylamide) for the Release of Multiple Cell Types

Heather E. Canavan[1], Jamie A. Reed[2], Adrianne E. Lucero[2], and Steven Candelaria[2]. (1) University of New Mexico, 1 University of New Mexico, MSC 01 1120, Albuquerque, NM 87131, (2) Chemical and Nuclear Engineering, University of New Mexico, 1 University of New Mexico, MSC 01 1120, Albuquerque, NM 87131

Poly(N-isopropyl acrylamide) (pNIPAM) undergoes a sharp property change in response to a moderate temperature drop around physiological temperatures. This behavior has generated great interest in the biomaterials community, and pNIPAM has already been used to harvest intact cell sheets for cell sheet engineering applications. Currently, many techniques are used to deposit pNIPAM, including electron beam irradiation and solution deposition of silanes and self-assembled monomers. Our research uses plasma polymerization of NIPAM from the vapor phase (ppNIPAM), a sterile, solvent-free method that is compatible with surfaces of any geometry or chemistry, making it extremely useful for cell and tissue culture. Following their characterization (e.g., using X-ray photoelectron spectroscopy for surface chemistry, interferometry for film thickness, contact angles for thermoresponse), ppNIPAM-derivatized substrates were used to culture multiple cell types, including bacterial (Halomonas marina), yeast (Saccharomyces cerevisiae), and mammalian (bovine aortic endothelial) cells. We find that each of these cell types adhere to pNIPAM above its LCST, including yeast, which are not normally adherent to culture substrates. Furthermore, we find that each of the cell types is released from pNIPAM, although the method of release may differ (i.e., detachment of sheets vs. release of isolated cells). These findings indicate that the ability of pNIPAM to reversibly adhere cells has many other applications for the study of cell/surface interactions, in addition to its well-established use for cell sheet engineering.

770f Creasing Instability of Stimuli-Responsive Polymer Gels

Jungwook Kim and **Ryan C. Hayward**. Department of Polymer Science and Engineering, University of Massachusetts Amherst, 120 Governors Drive, Amherst, MA 01003

Coating a rigid surface with a thin layer of a stimulus-responsive polymer gel provides a convenient means to create surfaces with responsive properties such as biocompatibility, adhesion, and tribology. However, as the gel increases its level of swelling, the constraint imposed by the substrate can result in large in-plane compressive stresses, in some cases leading to a creasing instability of the gel surface. While this instability has been known in practice for well over a century, it remains poorly understood. Using model polyacrylamide and poly(N-isopropyl acrylamide) copolymer gels, we have studied the onset of creasing as a function of material properties and gel thickness, and addressed basic questions regarding crease morphologies and growth mechanisms. Using the understanding gained from these studies, we are developing this instability as a route to create active surfaces, where we can exercise control over surface topography and chemical patterns in both space and time.

772a Sorption Equilibria of CO2 In Atactic Polystyrene by Molecular Simulation

Theodora Spyriouni[1], Georgios C. Boulougouris[2], and **Doros N. Theodorou**[2]. (1) Institute of Physical Chemistry, NCSR "Demokritos", 15310 Aghia Paraskevi Attikis, Athens, Greece, (2) School of Chemical Engineering, National Technical University of Athens, 9 Heroon Polytechniou Street, Zografou Campus, Athens, 157 80, Greece

A molecular simulation strategy allowing the computation of sorption isotherms, up to high penetrant activities, in glassy polymer matrices will be presented. The new scheme permits an equilibrium repartition of the penetrant molecules in the polymer matrix. Glassy matrices of polystyrene (PS) were obtained from reverse mapping of equilibrated coarse-grained configurations (Spyriouni et al., *Macromolecules* **2007**, *40*, 3876). The sorption isotherms of CO_2 in atactic PS and the induced polymer swelling have been calculated in the temperature range from 308 K to 405 K (below and above the glass transition temperature of the pure polymer) and for pressures up to 300 bar. The results compare favorably with available experimental data. Derivative properties, such as the partial molar volume and the partial molar enthalpy of CO_2, have been calculated in good agreement with experimental values. The segmental dynamics of PS have been analyzed and the pressure induced glass transition has been calculated for various temperatures below the T_g of pure PS.

The glassy matrices are loaded with CO_2 at various concentrations by insertions of CO_2 molecules, after mapping the accessible free volume. The loaded matrices are subjected to molecular dynamics (MD) simulations at constant temperature, pressure and number of molecules (N_1N_2PT ensemble). The penetrant is redistributed among the accessible cavities in order to achieve an equilibrium repartitioning in the thermally fluctuating polymer matrix. This is accomplished by selecting periodically one configuration and attempting a certain number of CO_2 displacements between the accessible cavities. The latter are determined by Delaunay tessellation followed by volume and connectivity analysis, as implemented by Greenfield and Theodorou (*Macromolecules* **1993**, *26*, 5461). Following acceptance or rejection of the attempted repartitioning moves, we proceed with the MD simulation of the polymer-penetrant system.

The excess chemical potential of CO_2 is calculated using the Direct Particle Deletion (DPD) method, a generalization of Staged Particle Deletion (Boulougouris et al. *Mol. Phys.* **1999**, *96*, 905; *J. Chem. Phys.*, **2001**, *115*, 8231). This method is very powerful for dense systems and/or large solute molecules in comparison to traditional Widom insertion. Moreover, the DPD version of the method permits the calculation of the excess chemical potential from a single simulation run.

The fugacity of CO_2 in the polymer matrix is calculated from the excess chemical potential via the relationship: $f_{CO2} = N_2 k_B T \exp(\beta\mu^{ex})/\langle V\rangle$, where N_2 is the number of CO_2 molecules in the polymer matrix and $\langle V\rangle$ is the average volume under *NPT* conditions. A new pressure is calculated via an equation of state for pure gaseous CO_2, such as SAFT, by setting the gas-phase fugacity equal to f_{CO2}. The calculations are repeated at the new pressure, until the CO_2 fugacity in the polymer matrix and in the gas are equal. The scheme converges rapidly, since the CO_2 fugacity in the polymer is not very sensitive to the pressure.

Web Page: comse.chemeng.ntua.gr

772b Atomistic Simulation of Poly(dimethylsiloxane): Structure, Thermodynamic and Diffusion Properties to Gases and N-Alkanes

Ioannis G. Economou and Zoi A. Makrodimitri. Institute of Physical Chemistry, National Center for Scientific Research "Demokritos", GR-15310, Aghia Paraskevi, Greece

Poly(dimethyl siloxane) (PDMS) is a widely used polymer for a number of industrial applications. In order that PDMS is selected for a specific application, accurate knowledge of its physical properties is necessary. Physical properties can be either measured or calculated based on reliable suitable methods. Molecular simulation using realistic models is a powerful tool for the elucidation of microscopic structure of polymers and the subsequent estimation of macroscopic physical properties.

In this work, a force field is developed for the prediction of thermodynamic and structure properties of PDMS melts. Force field development was based on existing force fields for PDMS together with fitting to experimental thermodynamic data at ambient conditions. Extensive NPT Molecular Dynamics (MD) simulations were performed at different temperature and pressure values. In all cases, good agreement was obtained between literature experimental data and model predictions for the melt density. Calculations are reported also for the solubility parameter of the polymer melt at different temperatures. The new force field is used subsequently for the calculation of solubility of seventeen (17) different compounds in PDMS using the Widom test particle insertion method. The solubility of n-alkanes from methane to n-hexane at 300 and 450 K and different pressures was calculated. In addition, solubility calculations for n-perfluoroalkanes at 300 K and 450 K and for noble and light gases at 300 K, 375 K and 450 K and ambient pressure were performed. Model predictions are in very good agreement with experimental data, in all cases. The infinite dilution solubility coefficient is shown to increase with temperature for very light gases and decrease for the heavier ones.

Finally, the atomistic force field is used to calculate the permeability properties of the polymer to light gases and to n-alkanes. The torsional potential barrier is re-estimated compared to the original force field. Chain dynamics and chain sizes are affected by this change while thermodynamic properties of the melt and of PDMS mixtures remain unaffected. The diffusion coefficients of penetrants to PDMS are calculated at different temperatures based on long molecular dynamics simulations. Subsequently, the permeability coefficients are estimated and ideal mixture selectivities are evaluated for binary hydrocarbon mixtures. In all cases, agreement with literature experimental data is ranged from good to excellent. Calculations for mixed penentrant permeation reveal that, in the presence of a second penetrant species, solubility and diffusion coefficients increase, in agreement with recent experimental evidence.

772c Heteropolymers with Adjustable Monomer Sequences (HAMS) as Compatibilizers for Homopolymer Blends

Ravish Malik, Carol K. Hall, and Jan Genzer. Chemical and Biomolecular Engineering, North Carolina State University, 911 Partners Way, Raleigh, NC 27695-7905

Heteropolymers with adjustable monomer sequences (HAMS), which are also called protein-like copolymers represent a new type of functional copolymer, exhibiting large-scale compositional heterogeneities and long-range correlations along the sequence. HAMS were first introduced by Khokhlov and coworkers who used computer simulations to demonstrate that random copolymers with tunable monomer sequences could, in principle, be generated by adjusting the compactness of a parent homopolymer (H), and then converting selected segments on the polymer surface into P segments by reacting them with species in the surrounding solution. In this talk we describe the results of a computer simulations study aimed at supporting the development of HAMS as compatibilizing agents for a polymer blend containing two homopolymers. Large-scale discontinuous molecular dynamics simulations (DMD) are used to explore the effect of compatibilizer sequence on the miscibility and interfacial characteritics of the polymer blend. The results are compared to those for systems in which the compatibilizers are block, regular alternating and random copolymers.

Web Page: turbo.che.ncsu.edu

772d Athermal Contribution to the Excess Entropy in the Long Chain Limit

J. Richard Elliott[1], Neil H. Gray[2], and Amir Vahid[1]. (1) Chemical and Biomolecular Engineering, University of Akron, Akron, OH 44325-3906, (2) Chemstations Inc., 2901 Wilcrest Drive, Houston, TX 77042

Mixtures of hydrocarbons have been simulated with discontinuous potential models to characterize the Helmholtz energy of the repulsive reference fluids (A0) as functions of density and composition. A previous study focused on methane, ethane, propane, n-butane, n-hexane, n-heptane, n-decane, and benzene.[1] Unfortunately, a slight inconsistency was encountered when the trend observed for these small molecules was extrapolated to the long chain limit. In the present work, we extend the analysis to methane with C10, C20, and C40 molecules of: nalkanes, ethyl-styrenes, and ethyl-propylenes. Compositions varied from 0.1 to 0.9 mole fraction for each binary mixture. Packing fractions varied from 0.3 to 0.5.

The goal of the present study was to develop a general strategy for interpolating the perturbation contributions between simulations of individual state points in order to facilitate usage of molecular simulation results in engineering applications. We find that the athermal entropy of mixing deviates significantly from ideality, but still follows the van der Waals mixing formula. This leads to an accurate characterization of the entropic contribution for mixtures of all sizes and shapes. A general rule is developed for predicting the athermal entropy of mixing based on knowledge of the volume ratios of the constituent molecules. The simulations are compared to several theories, including the MCSL theory for hard spheres,[2] the SAFT model,[3] and the Guggenheim-Staverman theory.[4]

Keywords: Thermodynamic properties, molecular simulation, polymers, entropy, mixtures, phase equilibria.

Reference:

1. Gray, N. H.; Elliott, J. R. In Quadratic Mixing in Perturbation Theory, AIChE Fall National Meeting, Austin, TX, 2004; Austin, TX, 2004; p 160e.

2. Mansoori, G. A.; Carnahan, N. F.; Starling, K. E.; Leland, T. W., Equilibrium Thermodynamic Properties of the Mixture of Hard Spheres. J. Chem. Phys 1971, 54, 1523-1525.

3. Chapman, W. G.; Gubbins, K. E.; Jackson, G.; Radosz, M., SAFT: Equation of State Solution Model for Associating Fluids. Fluid Phase Equilib. 1989, 52, 31.

4. Abrams, D. S.; Prausnitz, J. M., Statistical Thermodynamics of Liquid Mixtures: A New Expression for the Excess Gibbs Energy of Partly or Completely Miscible Systems. AIChE J. 1975, 21, 116.

772e Polymer-CO2 Interfaces In Equilibrium. A Density Functional Study

Manish Talreja[1], Isamu Kusaka[2], and David L. Tomasko[2]. (1) The Ohio State University, Room 125A, Koffolt Labs., 140W. 19th Ave., Columbus, OH 43210, (2) Chemical and Biomolecular Engineering, The Ohio State University, Room 125A, Koffolt Labs., 140W. 19th Ave., Columbus, OH 43210

Supercritical CO2 has been established as a very promising polymer processing solvent in many experimental studies. However, theoretically these systems haven't been studied in great detail. This work is aimed at bridging the gap between theoretical and experimental knowledge through molecular level analysis of the system using polymer density functional theory (PDFT). This theory offers semi-quantitative information at lower computational costs as compared to computer simulation. The basic idea of PDFT is to express the free energy as a functional of spatially varying density distribution, from which equilibrium density distribution and other thermodynamic information can be derived. Using a PDFT model parameterized to represent the octacosane(C28)-CO2 mixture, we can make predictions for the properties of a polymer-CO2 binary system. Of particular interest to polymer processing applications are the properties like surface tension, surface adsorption and width of the interface. In this talk we will present a detailed account of the equilibrium behavior of octacosane near the critical region of CO2. We will show how density profiles of the different segments and other thermodynamic properties of the system are affected by the introduction of CO2, and how these change in the vicinity of its critical point. We will use these results to show how the behavior of real polymers can be predicted in the supercritical CO2 environment.

772f Modeling the Phase Behavior of Polydisperse Rodlike Molecules with Attractive Interactions

Micah J. Green[1], Nicholas G. Parra-Vasquez[2], Natnael Behabtu[2], Virginia A. Davis[3], and Matteo Pasquali[4]. (1) Department of Chemical and Biomolecular Engineering, Rice University, 6100 Main Street, Houston, TX 77005, (2) Chemical and Biomolecular Engineering, Rice University, 6100 Main Street, Houston, TX 77005, (3) Department of Chemical Engineering, Auburn University, Auburn, AL 36849, (4) Chemical and Biomolecular Engineering, Chemistry, Rice University, 6100 Main Street, MS-362, Houston, TX 77005

The Onsager theory for the phase behavior of rodlike molecules is extended to rods with polydisperse length and solvent-dependent attractive intermolecular forces for the first time. This combination of polydispersity effects and solvent effects is critical for developing a quantitatively accurate model for phase separation in solutions of rodlike molecules. We utilize a phenomenological square-well potential to capture these attractive interactions and combine these interactions with the Onsager excluded volume potential. For polydisperse systems, the isotropic-liquid crystalline phase separation is a function of the initial rod concentration, unlike monodisperse systems. We use Newton's method and arclength continuation to compute isotropic/liquid-crystalline phase separation and length fractionation as a function of solvent quality.

We show that the resulting phase diagram can capture the phase behavior of experimental systems featuring rods with attractive interactions simply by fitting the adjustable parameters of the square-well potential. Specifically, we show that the model closely replicates experimental data points for isotropic/liquid crystalline phase separation of single-walled carbon nanotubes (SWNTs) in various super-acids. The model predicts that the isotropic cloud point goes to zero at a certain solvent quality, and this prediction was subsequently confirmed by experiments. This result means that for acids below a threshhold acid strength, the liquid-crystalline phase is present even at dilute concentrations. The model yields critical insights into both the fundamental thermodynamics as well as practical applications of SWNT/super-acid systems.

772g Compressible Lattice Model for CO2+ Cosolvent + Polymer Systems

Yanhui Yuan, Anupama Kasturirangan, and **Amyn S. Teja**. Georgia Institute of Technology, 311 Ferst Drive NW, Atlanta, GA 30332-0100

We have recently developed a compressible lattice model for representation of both low-pressure and high-pressure phase equilibria in CO2 + polymer systems. The model explicitly accounts for weak complex formation in these systems, using two parameters that are obtained by fitting experimental cloud point pressures or sorption equilibria. Furthermore, we have shown that these parameters can be obtained from spectroscopic measurements, or molecular dynamic simulations. In the present work, we extend the model to CO2 + cosolvent + polymer systems, using parameters obtained from cloud point data in the binary systems. Applications of the model to the study cosolvent effects of acetone, dimethyl ether, and chlorodifluoromethane in CO2 + poly (caprolactone) and poly (lactic acid) systems are described, combined with experiments using FTIR spectroscopy. Finally, we demonstrate that the model can also predict solubilities of multicomponent gases in polymers membranes.

772h Extension of the Friction Theory to the Description of the Rheological Behavior of Polymer Systems

Sergio E. Quiñones-Cisneros[1], Juan P. Aguayo[1], Octavio Manero[1], Torben Laursen[2], and Ulrich K. Deiters[3]. (1) Instituto de Investigaciones en Materiales, Universidad Nacional Autónoma de México, Apdo. Postal 70-360 Cto. Exterior, Ciudad Universitaria., Mexico D.F., 04510, Mexico, (2) VLXE ApS, Scion-DTU, Diplomvej, Building 376, Lyngby, 2800, Denmark, (3) Institute of Physical Chemistry, University of Cologne, Luxemburger Str. 116, Cologne, 50939, Germany

The rheological behavior under simple shear for a wide number of Newtonian fluids has been accurately described by the friction theory, from pure compounds to complex mixtures such as petroleum fluids. Recently, this approach has been further extended to the modeling of the non-Newtonian behavior found in waxy-heavy oils. Thus, the f-theory is now capable of reproducing the viscosity reduction with the shear-rate and additionally, the presence of the two Newtonian regions, located at low and high values of the strain-rate. This has been achieved for reservoir fluids using cubic equations of state, widely used in the oil industry, with the friction theory.

The extension to polymeric systems requires the use of a more advanced fundamental model intended for chain molecules, such as the PC-SAFT equation of state. A shear-thinning viscosity is typically observed in concentrated polymer solutions. Initial results using the PC-SAFT friction theory successfully described the rheological behavior of polystyrene/polystyrene systems and notoriously predicted the rheological behavior of polystyrene/carbon dioxide mixtures. On the basis of these encouraging results, the generalization of the f-theory approach to additional polymer systems is further examined.

AUTHOR INDEX

AUTHOR INDEX

AUTHOR INDEX

AUTHOR INDEX

AUTHOR INDEX

AUTHOR INDEX

AUTHOR INDEX

AUTHOR INDEX

AUTHOR INDEX

AUTHOR INDEX

AUTHOR INDEX

AUTHOR INDEX

AUTHOR INDEX

AUTHOR INDEX

AUTHOR INDEX

AUTHOR INDEX

AUTHOR INDEX

AUTHOR INDEX